AIRFIELD AND HIGHWAY PAVEMENTS

EFFICIENT PAVEMENTS SUPPORTING TRANSPORTATION'S FUTURE

PROCEEDINGS OF THE 2008 AIRFIELD AND HIGHWAY
PAVEMENTS CONFERENCE

October 15–18, 2008
Bellevue, Washington

SPONSORED BY
The Transportation & Development Institute (T&DI)
of the American Society of Civil Engineers

EDITED BY
Jeffery R. Roesler, Ph.D., P.E.
Hussain U. Bahia, Ph.D.
Imad L. Al-Qadi, Ph.D., P.E.
Scott D. Murrell, P.E.

Published by the American Society of Civil Engineers

Library of Congress Cataloging-in-Publication Data

Airfield and Highway Pavements Conference (2008 : Bellevue, Wash.)
 Airfield and highway pavements : efficient pavements supporting transportation's future : proceedings of the 2008 Airfield and Highway Pavements Conference, October 15-18, 2008, Bellevue, Washington / sponsored by the Transportation & Development Institute (T&DI) of the American Society of Civil Engineers ; edited by Jeffery R. Roesler ... [et al.].
   p. cm.
 Includes bibliographical references and index.
 ISBN-13: 978-0-7844-1005-9
 1. Runways (Aeronautics)--Materials--Congresses. 2. Express highways--Materials--Congresses. 3. Pavements--Congresses. I. Roesler, Jeffery R. II. Transportation & Development Institute (American Society of Civil Engineers). III. Title.

TL725.3.R8A8255 2008
629.136'34--dc22   2008039154

American Society of Civil Engineers
1801 Alexander Bell Drive
Reston, Virginia, 20191-4400

www.pubs.asce.org

Any statements expressed in these materials are those of the individual authors and do not necessarily represent the views of ASCE, which takes no responsibility for any statement made herein. No reference made in this publication to any specific method, product, process, or service constitutes or implies an endorsement, recommendation, or warranty thereof by ASCE. The materials are for general information only and do not represent a standard of ASCE, nor are they intended as a reference in purchase specifications, contracts, regulations, statutes, or any other legal document. ASCE makes no representation or warranty of any kind, whether express or implied, concerning the accuracy, completeness, suitability, or utility of any information, apparatus, product, or process discussed in this publication, and assumes no liability therefore. This information should not be used without first securing competent advice with respect to its suitability for any general or specific application. Anyone utilizing this information assumes all liability arising from such use, including but not limited to infringement of any patent or patents.

ASCE and American Society of Civil Engineers—Registered in U.S. Patent and Trademark Office.

Photocopies: Authorization to photocopy material for internal or personal use under circumstances not falling within the fair use provisions of the Copyright Act is granted by ASCE to libraries and other users registered with the Copyright Clearance Center (CCC) Transactional Reporting Service, provided that the base fee of $35.00 per article is paid directly to CCC, 222 Rosewood Drive, Danvers, MA 01923. The identifier for this book is 0-7844-0761-4/06/ $35.00. Requests for special permission or bulk copying should be addressed to Permissions Publications Division, ASCE; email: permissions@asce.org

Copyright © 2009 by the American Society of Civil Engineers. All Rights Reserved.
ISBN 978-0-7844-1005-9
Manufactured in the United States of America.

Foreword

The success of the 2006 ASCE International Airfield and Highway Specialty Conference in Atlanta, Georgia, which for the first time combined highway and airfield pavements, was the motivation behind continuing this biennial gathering of practitioners and academicians interested in exchanging ideas, introducing innovations, and defining the current state-of-the-practice in the areas of airfield and highway pavement materials, design, analysis, and construction. Pavement engineers are being challenged today more than ever to design and build long lasting cost effective pavement systems, use marginal materials in the man-made layers, reduce the delay time for the traveling public, and consider environmental sustainability. The theme of this conference, *Efficient Pavements Supporting Transportation's Future*, supports the complex interaction between the various objectives pavement engineers must consider in meeting future pavement infrastructure demands.

This conference proceeding contains recent advances in Flexible Pavement Mechanistic-Empirical Modeling; Airfield Pavement Rehabilitation; Airfield Pavement Analysis and Behavior; Flexible Pavement Field Performance; Airfield Pavement Evaluation and Performance; Performance of Stabilized Pavement Layers; Warm and Foamed HMA Characterization; Airfield Pavement Materials and Design Optimization; Factors Affecting Airfield Pavement Design; HMA Laboratory Characterization; Airfield Pavement Management; Pavement Roughness: Evaluation, Analysis, and Prevention; Advanced HMA Modeling; Unbound Layer Evaluation and Behavior; Recycled Materials in Pavements; Effect of Modified Binder on HMA; Highway Pavement Design and Performance; and Effect of Loading on HMA Pavements. The papers published in this proceeding are ***fully refereed.*** Each paper submitted for this conference was peer-reviewed by two to three professionals in their respective technical fields.

There continues to be a significant participation by the international pavement community in ASCE specialty conferences. The conference was organized around three concurrent tracks of presentations related to pavement modeling, performance, and testing of construction materials. There were also three workshops held on the first day of the conference: Back-To-Basics: Airport Concrete Pavement Mixture Design and Construction; FAARFIELD: New FAA Design Software for Airport Pavement Thickness; and Sustainability of Pavements. A field trip to the Boeing Corporation facilities and the Seattle-Tacoma International Airport (SEATAC) was also part of the program.

We would like to express our gratitude to the many anonymous reviewers, who volunteered their time to thoroughly examine the papers and offer constructive technical comments to the authors. We acknowledge the following steering committee members for their contributions in planning the various technical activities and field trips: Gary Mitchell, Dr. David Lee, Ray Rawe, Dr. Halil Ceylan, Jim Scherocman, Ken DeBord, Mike Roginski, and Rich Thuma. We would like to also gratefully acknowledge the efforts of Mr. Cristian Gaedicke, who served as the proceeding secretary. He was invaluable in organizing the abstracts, submitted papers, technical reviews, and correspondence with the authors. Finally, we would like to thank the following ASCE and T&DI staff for their efforts and guidance: Jon Esslinger, Elaine Watson, and Donna Dickert. The Conference was chaired by Imad L. Al-Qadi.

Jeffery R. Roesler, Ph.D., P.E.
Department of Civil and Environmental Engineering
University of Illinois

Hussain U. Bahia, Ph.D.
Department of Civil and Environmental Engineering
University of Wisconsin

Imad L. Al-Qadi, Ph.D., P.E.
Department of Civil and Environmental Engineering
University of Illinois

Scott D. Murrell, P.E.
Port Authority of New York and New Jersey

Contents

Pavement Modeling

Comparison between Mechanistic Analysis and In-Situ Response of Full-Depth Flexible Pavements........................1
  Hao Wang and Imad L. Al-Qadi

Mechanism of Mitigating Shear-Induced Rutting of Asphalt Pavement Using Geotextile........................16
  Yinghao Miao and Jinxi Zhang

Highway Pavement Damage and Cost Due to Routine Permitted Axles........................28
  D. H. Timm, K. D. Peters, and R. E. Turochy

Use of Artificial Neural Networks to Detect Aggregates in Poor-Quality X-Ray CT Images of Asphalt Concrete........................40
  M. Emin Kutay, Edith Arambula, Nelson Gibson, Jack Youtcheff, and Katherine Petros

A Probabilistic Approach to Account for Temperature Impact on Flexible Pavement Stiffness........................52
  Sameh Zaghloul, Nicholas Vitillo, and T. Joseph Holland

Speed Up Discrete Element Simulation of Asphalt Mixtures with User-Written C++ Codes........................65
  Yu Liu and Zhanping You

Finite Element Modeling of Reflective Cracking under Moving Vehicular Loading: Investigation of the Mechanism of Reflective Cracking in Hot-Mix Asphalt Overlays Reinforced with Interlayer Systems........................74
  Jongeun Baek and Imad L. Al-Qadi

Distribution of Permanent Deformations within HMA Layers........................86
  Charles W. Schwartz and Regis L. Carvalho

Implications of Complex Axle Loading and Multiple Wheel Load Interaction in Low Volume Roads........................99
  Minkwan Kim and Erol Tutumluer

Characterization of Pavement Materials

Measuring the Specific Gravity and Absorption of Steel Slag and Crushed Concrete Coarse Aggregates: A Preliminary Study........................111
  Julian Mills-Beale and Zhanping You

Optimizing Low Density Concrete Behavior for Soft Ground Arrestor Systems........................122
  E. Heymsfield, W. M. Hale, and T. L. Halsey

Mechanistic Characteristics of Moisture Damaged Asphalt Matrix and Hot Mix Asphalt Mixtures........................134
  Mohammad J. Khattak and Vikram Kyatham

A Comparative Study of Laboratory Measured and Predicted Dynamic
Modulus for Characterizing Florida Asphalt Mixtures 147
    W. Virgil Ping and Yuan Xiao

Determination of the Elastic Modulus and Poisson's Ratio of Asphaltic
Mixtures Using Uniaxial Creep Recovery Tests 159
    H. Taherkhani and A. C. Collop

Properties of Asphalt Mixtures with RAP in the Mechanistic-Empirical
Pavement Design of Flexible Pavements: A Preliminary Investigation 171
    Shu Wei Goh and Zhanping You

Laboratory Evaluation of Warm Asphalt Properties and Performance 182
    Amy Hearon and Stacey Diefenderfer

Laboratory Simulation of Warm Mix Asphalt (WMA) Binder Aging
Characteristics 195
    Tejash Gandhi and Serji Amirkhanian

Volumetric Properties of Warm Rubberized Mixes Depending on Compaction
Temperature 205
    Chandra K. Akisetty, Soon-Jae Lee, and Serji N. Amirkhanian

Impacts of Laboratory Curing Condition on Indirect Tensile Strength of Cold
In-Place Recycling Mixtures Using Foamed Asphalt 213
    Hosin "David" Lee, Soohyok Im, and Yongjoo Kim

Experimental Study on Gilsonite-Modified Asphalt 222
    Juanyu Liu and Peng Li

Experimental Research on Preparing Conductive SMA Doped Graphite 229
    Qingjun Ding, Xuewei Wu, Xinquan Liu, and Shuguang Hu

Research and Optimization on Flame-Retarding Asphalt System Based on ATH 241
    Qingjun Ding, Fan Shen, Xinquan Liu, and Shuguang Hu

On the Mechanical Modeling of Asphalt Matrix and Hot Mix Asphalt Mixtures 253
    Mohammad J. Khattak, Zhanping You, and Vikram Kyatham

Effect of Loading and Temperature on Dynamic Modulus of Hot Mix Asphalt
Tested under MMLS3 267
    Sudip Bhattacharjee, Rajib B. Mallick, and Jo Sias Daniel

Pavement Management, Evaluation, and Rehabilitation

Using PCI Data to Define Major Rehabilitation Projects at Washington Dulles
International Airport 279
    Richard G. Thuma, Gary K. Fuselier, and Peter K. Yip

Fast-Track Construction of Runway 14-32 Pavement Rehabilitation
at the Sarasota-Bradenton International Airport 289
    E. M. Vélez-Vega and D. R. Bardt

Materials and Pavement Evaluation for the New Doha International Airport
Using Mechanistic-Empirical Technology 301
    R. B. Leahy, L. Popescu, C. Dedmon, and C. L. Monismith

25 Years of NDT Analysis: Runway 1L-19R at Washington Dulles
International Airport 322
    Stanley M. Herrin, Roy D. McQueen, Michael J. Darter, Gary Fuselier,
    and Joseph S. Grubbs

The Life Cycle of a Runway Pavement: A Case Study of Runway 1L-19R
at the Washington Dulles International Airport 334
    Gary K. Fuselier, Joseph S. Grubbs, and Roy D. McQueen

Accuracy of Pavement Management Predictions: A Case Study at Washington
Dulles International Airport 346
    Stanley M. Herrin and Gary Fuselier

Remaining Service Life Analysis of Concrete Airfield Pavements at Denver
International Airport Using the FACS Method 358
    Michael T. McNerney

Pavement Surface Mixture, Texture, and Skid Resistance: A Factorial Analysis 370
    M. Alauddin Ahammed and Susan L. Tighe

Evaluating International Roughness Index Data Quality at Project
and Network Levels 385
    G. P. Ong and K. C. Sinha

Managing Utility Installation/Maintenance Activities in Advance to Reduce
Pavement Utility Cuts Using Spatiotemporal Objects Database 397
    Chien-Cheng Chou, Yi-Ping Chen, and Chien-Ming Chiu

Performance of Stabilized and Unbound Pavement Layers

Dynamic Modulus and Fatigue Testing of Lightly Cementitiously Stabilized
Granular Pavement Materials 410
    Piratheepan Jegatheesan and C. T. Gnanendran

Performance Evaluation of Asphalt Pavement with Fly Ash Stabilized FDR
Base: A Case Study 423
    Haifang Wen and Bruce Ramme

Investigation into the Use of Cement Stabilized Sand in Road Pavement
Construction in Bangladesh 434
    Waliur Rahman, Richard Freer-Hewish, and Gurmel S. Ghataora

Sustainable Base Course 442
    Casimir J. Bognacki and Marco Pirozzi

Minimum Standards for Using Recycled Materials in Unbound Highway
Pavement Layers 453
    Athar Saeed and Michael I. Hammons

Effect of Aircraft Load Wander on Unbound Aggregate Pavement Layer
Stiffness and Deformation Behavior 465
    Phillip Donovan and Erol Tutumluer

Pavement Subgrade Evaluation and Value Engineering Solution for H-JAIA
End-Around Taxiway (Taxiway Victor) 477
    Raghuram N. Tadimalla, Richard L. Boudreau, Subash Kuchikulla,
    Viswanath Dokka, and Robert C. Briggs

Performance of Flexible Pavements over Two Subgrades with Similar CBR
but Different Soil Types (Silty Clay and Clay) at the FAA's National Airport
Pavement Test Facility 487
    Navneet Garg and Gordon F. Hayhoe

Use of Recycled Concrete as Unbound Base Aggregate in Airfield and Highway
Pavements to Enhance Sustainability..497
    Athar Saeed and Michael I. Hammons

Pavement Design and Analysis

Location and Timing of Fatigue Cracks on Jointed Plain Concrete Pavements............509
    Jacob E. Hiller and Jeffery R. Roesler

Maximizing Pavement Design for Highway Design Build Projects: Lessons
Learned from I-5 Everett HOV 521
    Kurt Pedersen and Dan Peterson

Synthesis on Composite Pavement Systems: Benefits, Performance, Design,
and Mechanistic Analysis 535
    Orlando Núñez, Gerardo W. Flintsch, and Brian K. Diefenderfer

Recalibration of Airport Pavement Structural Design System 547
    Greg White

Pavement Design Issues and Embankment Construction for the Second
Runway at Cancún International Airport, Mexico 560
    George Nowak, Hector Saldivar Moguel, and David Martinez Salazar

Impact of New-Generation Aircraft at Hartsfield-Jackson Atlanta
International Airport 573
    Quintin Watkins, Robert Rau, and Richard L. Boudreau

Landing Gear Factors for Pavements with FAA Ratings 585
    Kenneth J. DeBord

Pavement Performance Studies

George Washington Bridge Asphalt-Wearing Course and Bond Coat Analysis 596
    Jami M. Bjornstad, Casimir J. Bognacki, and Joseph Marsano

Comparison of Butt and Notched Wedge Longitudinal Joints Constructed
in Connecticut..608
    A. Zofka, J. Mahoney, S. Zinke, and G. Shaffer

A Quest for Successful Pavements in Texas 620
    Carlos M. Chang Albitres, Paul E. Krugler, and Ahmed Eltahan

Tales from the Pacific: Airport Activities and Experiences in Micronesia 632
    Frank V. Hermann

Indexes

Subject Index 645

Author Index 647

Comparison between Mechanistic Analysis and In-Situ Response of Full-depth Flexible Pavements

Hao Wang[1], Imad L. Al-Qadi[2]

Accurate prediction of tensile strain at the bottom of hot-mix asphalt (HMA) is crucial for mechanistic pavement analysis. In this study, transverse tensile strain at the bottom of HMA was calculated using mechanistic analysis and compared to measured values from test sections exposed to accelerated pavement testing (APT).

Three full-depth flexible pavement sections having 152, 254, and 420mm of HMA placed on 300mm of lime-stabilized subgrade were constructed and exposed to APT using the Advanced Transportation Loading ASsembly (ATLAS), housed at the University of Illinois at Urbana-Champaign. Transverse tensile strain histories at the bottom of HMA under various loading conditions (load, tire pressure, and speed) were measured through embedded strain gauges at the HMA-stabilized subgrade interface. Indirect tensile complex modulus tests for three HMA materials were conducted using an Instron Universal Test Machine (UTM). The elastic modulus of subgrade was backcalculated from FWD testing conducted on the test sections. The pavement responses of test sections were calculated using the NCHRP 1-37A Mechanistic-Empirical Pavement Design Guide (MEPDG) procedure as well as a finite element model.

In general, the MEPDG approach underestimates the transverse and longitudinal strains when the pavement thickness is equal to or smaller than 254mm; the difference is more manifested at shallow depths below 152mm. Hence, the pavement thickness design based on MEPDG procedure could underestimate the pavement fatigue damage for thin pavements. Inaccurate calculation of the loading pulse period and the empirical conversion between loading period and frequency are among the assumptions that contribute significantly to the ill-prediction of pavement responses when current NCHRP 1-37A is used.

A 3D finite element model, on the other hand, was developed in order to predict pavement response to tire loading. The model considers measured tire-pavement contact stresses, continuous moving-wheel loading, implicit dynamic analysis, and HMA viscoelastic characteristics. Difference between measured and predicted strains was found to be within 5%.

Keywords: Accelerated Pavement Testing, Mechanistic Analysis, Dynamic Modulus, Loading Frequency

[1] Graduate Research Assistant, Department of Civil and Environmental Engineering, University of Illinois at Urbana-Champaign, 205 N Mathews MC-250, Urbana, IL 61801, E-mail: haowang4@uiuc.edu

[2] **Corresponding Author**, Founder Professor of Engineering, Illinois Center for Transportation, Director, University of Illinois at Urbana-Champaign, 205 N Mathews MC-250, Urbana, IL 61801, E-mail: alqadi@uiuc.edu

1. Introduction

Due to the current trend toward mechanistic flexible pavement design and the need for more reliable design procedures, the accurate prediction of pavement response under vehicle loading is critical. To achieve an accurate pavement response prediction, the hot-mix asphalt (HMA) must be considered as a viscoelastic material.

Generally, two methods (direct and indirect) can be used to consider the viscoelasticity of HMA in pavement mechanistic analysis. The direct method considers the HMA modulus as a function of time (loading frequency) and temperature. The stress-strain relationship is formulated as hereditary integral and can be solved either analytically or numerically (Ferry 1980). The indirect method maintains the stress-strain relationship as an elastic relationship by replacing the elastic modulus with the applicable complex modulus at the pertinent temperature and loading frequency. The direct method uses fundamental properties of HMA, while the indirect method requires less computation cost and can be easily incorporated into the current multi-layer elastic analysis method. However, the current approach to calculate the loading frequency from vehicular speed is questionable at best (Al-Qadi et al. 2008a,b). Using an inaccurate frequency would result in an unreliable selection of complex modulus and, hence, wrong prediction of pavement response.

A major milestone of the proposed NCHRP 1-37A Mechanistic-Empirical Pavement Design Guide (MEPDG) is that HMA is no longer described as a purely linear elastic material. The indirect method is used to characterize the HMA viscoelasticity. The enhanced integrated climatic model (EICM) is adopted to predict the temperature distribution in the HMA layers. However, a simplified procedure is used to calculate the loading frequency as a function of the vehicle speed and the cross-section of pavement structure (NCHRP 1-37A 2004).

There are two major concerns about the simplified procedure for loading frequency calculation: 1) it is based on elastic theory and doesn't consider the HMA viscoelasticity; especially the adoption of the Odemark's method and the erroneous 45° load distribution angle; and 2) the conversion of loading period to frequency ($t = 1/f$) is not consistent with the assumption usually made in the rheology field, $t = 1/\omega = 1/(2\pi f)$ (Dongre et al. 2006; Loulizi et al. 2006). On the other hand, accurate pavement response prediction may be established using an analytical approach that accurately simulates tire-pavement stresses and incorporates the fundamental HMA viscoelastic properties.

In this study, the tensile strains at the bottom of HMA in three full-depth flexible pavement sections were calculated using the elastic theory and a developed finite element model. The calculated tensile response strains were compared to in-situ responses due to APT loading. The elastic theory uses circular static compression load and the applicable complex modulus is obtained at the loading frequency calculated in accordance with the MEPDG procedure. On the other hand, the developed finite element model uses measured 3D tire-pavement contact stresses, continuous moving-wheel loading, implicit dynamic analysis, and HMA viscoelastic characteristics.

2. Experimental Program

2.1 Accelerated Pavement Testing

The accelerated pavement testing program made use of three existing HMA test sections built as part of a project focused on the extended-life pavement (Garcia and Thompson 2004). Table 1 presents the cross sections of the three full-depth flexible pavement sections. The HMA was prepared in accordance with the SuperPave™ volumetric design procedure. The laboratory mix-design criterion is based on 90 gyrations to achieve 4% air void. Two asphalt binders were used in the HMA layers, a PG 64-22 for the standard binder course, and a SBS PG 70-22 for the polymer modified binder courses and dense graded surface. The asphalt content of the standard and polymer modified binder courses is 4.5%; while the dense-graded asphalt content is 5.4%. No liquid anti-stripping was used in any mixture. The aggregate used in all mixes is limestone. The subgrade is 305mm lime-stabilized to address the high water content of the nature soil.

Two pavement instrumentations were embedded during construction: H-type HMA strain gauges and thermocouples. The H-type HMA strain gauges were embedded at the stabilized subgrade-HMA interface. T-type copper-constantan thermocouples were placed on top of the subgrade as well as at various locations within the HMA layers. The strain gauges and thermocouples were connected to a data acquisition system uses Labview software. To compare the pavement response at various loading conditions, strain measurements were shifted to a reference temperature (25°C) using a correction factor (Wang and Al-Qadi 2008).

Table 1 Full-depth Flexible Pavement Sections

Sections	HMA Layers
F	51mm dense graded surface + 101mm standard binder course
D	51mm dense graded surface + 114mm polymer modified course + 89mm standard binder course
B	51mm dense graded surface + 114mm polymer modified course + 254mm standard binder course

The APT was conducted using the Advanced Transportation Loading ASsembly (ATLAS) housed at the Advanced Transportation Research and Engineering Laboratory (ATREL) at the University of Illinois at Urbana-Champaign. The test program included three tire types (dual-tire assembly, wide-base 425 tire, and wide-base 455 tire); five loading levels (26, 35, 44, 53, and 62kN); three tire inflation pressures (550, 690, and 760kPa); and two speeds (8 and 16km/h). Only the test data under dual-tire assembly loading and its analysis results are presented in this paper.

2.2 Laboratory Material Characterization

Several tests can be used to characterize HMA viscoelasticity, including the time-dependent creep test, the time-dependent relaxation test, or the frequency-sweep complex modulus test. In this paper, complex modulus, in an indirect tensile setup (IDT), was conducted to characterize the three HMA materials used in the full-depth flexible pavement sections. The indirect tensile setup was used to allow testing thin HMA layer cores taken from the field. The dimensions of indirect tensile specimens were 150mm diameter by 51mm high. These specimens were cut from the field cores taken from the full-depth pavement sections. It was reported that the tensile stress state in the perpendicular diametrical direction of cylindrical IDT specimen is close to field stress condition at the bottom of HMA layer (Buttlar and Roque 1994). Although the authors are aware of the shortcomings of the IDT test, including the concentrated loading under the strip and shear development at very small deformation, this test has an advantage over the compression cylindrical test.

The IDT complex modulus test was performed by applying sinusoidal vertical loads, and the corresponding horizontal and vertical deformations were measured. Unlike the uniaxial test, the stress-strain distribution in the IDT setup is biaxial tension instead of uniaxial compression (Figure 1(a) and (b)). Based on the linear viscoelastic solution derived by Kim (2004), the applied load and measured deformation history were analyzed and fitted using Equation (1). The IDT complex modulus was calculated using Equation (2):

$$f(t) = a + bt + c \cdot \cos(wt + \varphi) \tag{1}$$

where,
$f(t)$ = load or deformation time history;
$a, b,$ and c = regression coefficients;
φ = phase angle; and
w = angular frequency.

$$|E^*| = 2\frac{P_0}{\pi a d} \frac{\beta_1\gamma_2 - \beta_2\gamma_1}{\gamma_2 V_0 - \beta_2 U_0} \tag{2}$$

where,
E^* = complex modulus;
P_0 = amplitude of applied load;
a = loading strip width;
d = thickness of specimen;
U_0 and V_0 = amplitude of horizontal and vertical displacement history; and
$\beta_1, \beta_2, \gamma_1,$ and γ_2 = coefficients for Poisson's ratio depending on specimen diameter and gauge length (38mm in this paper).

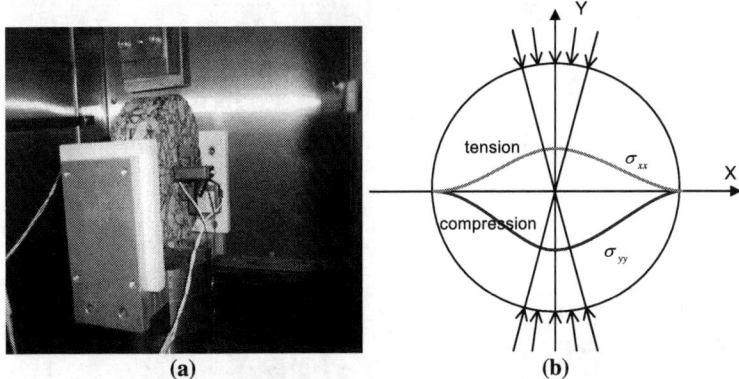

Figure 1 Indirect Tensile Test: (a) Specimen Set-up for Testing and (b) Schematic Stress State under Strip Load (Kim et al. 2004)

The complex modulus tests were conducted at three temperatures (-10, 5, and 25°C) and seven loading frequencies (0.01, 0.1, 0.5, 1, 5, 10, and 25Hz). The reduced testing at high temperature can be compensated by the test at low frequency (0.01Hz) based on the concept of time-temperature superposition. The sigmoidal function in the MEPDG was used to describe the complex modulus master curve, Equation (3). The maximum limiting modulus was estimated from the HMA volumetric properties (void in mineral aggregate (VMA) and voids filled with asphalt (VFA)) using the Hirsch model and a limiting binder modulus of 1GPa (Bonaquist and Christensen 2005). The least square error technique available in the solver module in Excel was used to obtain the fitting parameters by minimizing the residual errors that were generated by fitting the complex modulus data to the predicted values. The time-temperature shift factors were also calculated automatically in the fitting process.

$$\log|E^*| = \delta + \frac{(Max - \delta)}{1 + e^{\beta + \gamma \left\{ \log(t) - \frac{\Delta E_a}{19.14714} \left[\left(\frac{1}{T}\right) - \left(\frac{1}{295.25}\right) \right] \right\}}} \qquad (3)$$

where,
$|E^*|$ = dynamic modulus;
t = loading time;
T = temperature in ° Rankine;
δ, β and γ = fitting parameters; and
Max = limiting maximum modulus.

The fitted master curves for the dense grades course (DG) and polymer modified course (PB) at a reference temperature of 25°C are shown in Figures 2 (a) and (b). As expected, under a constant loading frequency, the magnitude of the complex modulus decreases with an increase in temperature; and under a constant testing temperature, the magnitude of the complex modulus increases as frequency increases.

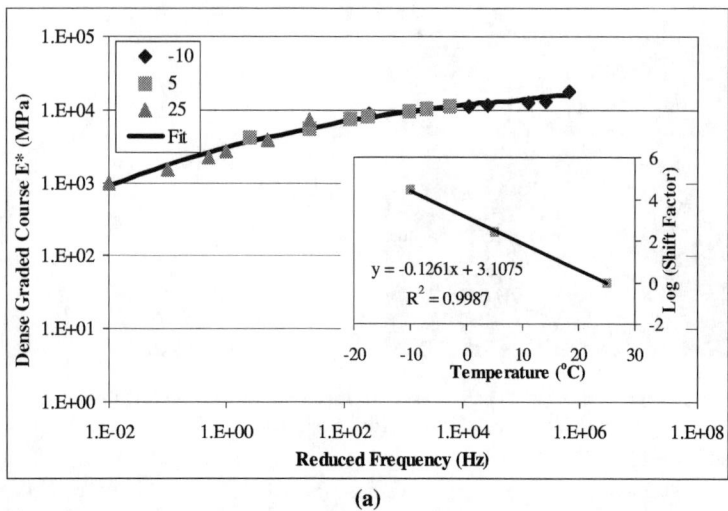

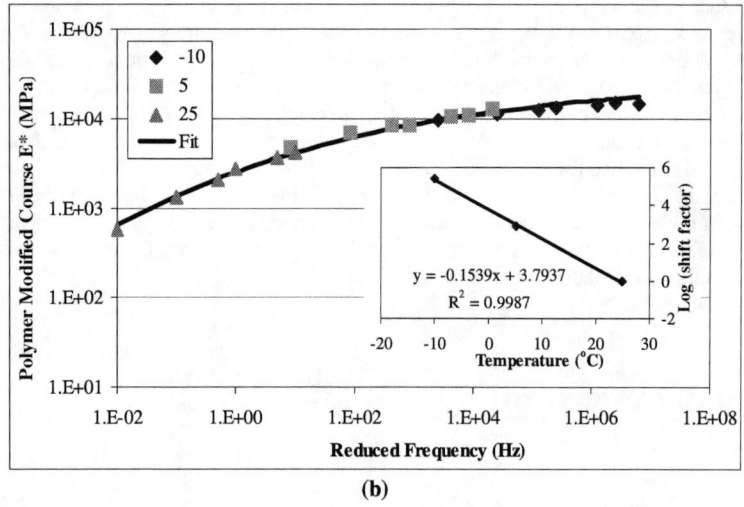

Figure 2 Measured and Fitted Complex Modulus Curves: (a) Dense-Graded Course, and (b) Polymer-Modified Course

2.3 FWD Backcalculation of Subgrade Modulus

Falling weight deflectometer (FWD) tests were conducted on the full-depth pavement sections to backcalculate the elastic modulus of subgrade. The temperature profile along pavement depth during the FWD test was recorded using the embedded thermocouples. Because the load pulse of the FWD test was around 0.03s, the loading frequency of 5.3Hz ($\omega = 1/t = 33.3$ Hz and $f = \omega/2\pi$) was used to choose the applicable HMA complex modulus in the backcalculation

(Louizi et al. 2006). The approach of using $f = \omega/2\pi$ is considered acceptable when one loading pulse is applied; as in the case of FWD. However, $f = \omega/2\pi$ may not be applicable for a moving load. In the backcalculation, the subgrade was divided into two layers: the first 305mm lime-stabilized layer and the second infinite layer.

The backcalculated subgrade moduli are presented in Table 2. As expected, there is a significant difference in modulus between the lime-stabilized soil and the deeper natural soil. It was found that the lime-stabilized and natural subgrade moduli at section F were lower than the corresponding values for other sections due to the high moisture content of section F's subgrade. The backcalculated moduli were found to be in good agreement with the California Bearing Ratio (CBR) test results collected at the time of subgrade construction (Garcia and Thompson 2004).

Table 2 Backcalculated Soil Elastic Modulus

Sections	Lime-stabilized soil (MPa)	Natural soil (MPa)
A,B	442	152
D	393	173
F	138	97

3. Comparison between Mechanistic Analysis Using MEPDG Procedure and In-Situ Response

In accordance with MEPDG (NCHRP 1-37A), the flexible pavements were modeled as multilayer linear elastic systems. The elastic modulus and Poisson's ratio of individual layers were required as input. The total HMA layer was divided into several sublayers. The elastic modulus of each sublayer was determined from the complex modulus master curve at the pertinent loading frequency and field temperature (measured from thermocouple). BISAR, a multilayer linear elastic software, was used to compute the pavement elastic response. The responses from the JULEA program integrated in MEPDG could not be extracted. The output responses of BISAR and JULEA, both use Burmister's theory (Huang 1993), are similar for identical pavement system under the same loading conditions.

The loading pulse duration within the pavement structure was calculated, in accordance with MEPDG approach, as a function of vehicle speed and effective length, Equation (4). The loading frequency in Hz was calculated using Equation (5).

$$t = \frac{L_{eff}}{17.6 v_s} \quad (4)$$

$$f = \frac{1}{t} \quad (5)$$

where,
t = time of loading (sec);
L_{eff} = effective length (in);
v_s = vehicle speed (mph); and
f = loading frequency (Hz)

In accordance with the MEPDG approach, to calculate the effective length at a specified depth, the Odemark's method of thickness equivalency is used, and the stress distribution in a single layer system is assumed to be at 45°. The effective depth (Z_{eff}), at which the effective length of load pulse is computed, is calculated by Equation (6). No overlap occurred between adjacent axles at shallow depth in the case of a single axle. The effective length is calculated using Equation (7).

$$Z_{eff} = \sum_{i=1}^{n-1}\left(h_i \sqrt[3]{\frac{E_i}{E_{SG}}}\right) + h_n \sqrt[3]{\frac{E_n}{E_{SG}}} \qquad (6)$$

$$L_{eff} = 2(a_c + Z_{eff}) \qquad (7)$$

where,
Z_{eff} = effective depth of interest;
h_n = thickness of the layer of interest (layer n);
E_{SG} = subgrade modulus of elasticity; and
E_n = modulus of elasticity of the layer of interest.
a_c = radius of the tire contact area; and
L_{eff} = effective length at the effective depth of interest.

3.1 Loading Pulse Duration

The calculated loading pulse durations at the bottom of HMA layer at three test sections (F, D, and B) are shown in Table 3. As expected, the loading pulse durations are dependent on the depth and vehicle speed. The time of the loading pulse is increased as the depth increases within the same pavement structure; and as the vehicle speed decreases. The calculated loading pulse is also dependent on the material modulus. For example, different loading pulse durations resulted at the same depth of three sections due to variation in material moduli.

Table 3 Calculated Loading Pulse Durations at Three Test Sections

Sections	F		D			B		
Depth (mm)	50	152	50	165	254	50	165	420
Loading pulse at 8km/h (sec)	0.28	0.55	0.24	0.45	0.62	0.23	0.44	0.92
Loading pulse at 16km/h (sec)	0.14	0.28	0.12	0.23	0.32	0.12	0.22	0.46

A limited number of researchers have attempted to study the stress pulse duration at field loading conditions. Barksdale (1971) reported that the pulse shape varied from a sinusoidal at the surface to a more nearly triangular at greater depth. Brown (1974) formulated an equation relating the loading time to vehicle speed and depth. Recently, Al-Qadi and his co-workers (2002) measured the compressive stress pulse in flexible pavement at the Virginia Smart Road for various speeds and pavement thicknesses. One of the flexible pavement structures at the Smart Road, data from which is presented in this paper, includes 188mm HMA layer, 75mm open-graded drainage layer (OGDL), 150mm cement-treated aggregate (CTA), and 175mm aggregate layer. As shown in Table 5, the calculated loading pulse durations in this study were found to be greater than the calculated values using Brown's equation and smaller than the Virginia Smart Road measurements at the same depth. *It should be noted that the effects of different pavement structures were not accounted for here.* However, it was presented to show the trend.

Table 4 Comparison between Calculated Pulse Durations with Other Measurements

Speed (km/h)	8			16		
Sections	F	D	B	F	D	B
Depth (mm)	152	254	420	152	254	420
Calculated loading pulse duration (sec)	0.55	0.62	0.92	0.28	0.32	0.46
Brown's equation (sec)	0.11	0.12	0.14	0.06	0.06	0.08
Smart Road measurements (sec)*	0.79	1.02	1.21	0.41	0.53	0.67

* Different pavement structure than the tested sections at ATREL, UIUC

3.2 Pavement Response

The calculated tensile strains were compared to the measured strains at the bottom of the HMA, as shown in Figures 3 and 4, respectively for the longitudinal and transverse strains under 35kN dual-tire loading. The calculated longitudinal tensile strains were less than the measured values at sections F and D, while greater than the measured strains at section B. The calculated transverse tensile strains were less than the measured values at sections D and B (transverse strain gauges at section F were not functional at testing). If the longitudinal strain is considered as the critical strain for bottom-up fatigue cracking, it is evident that the simplified procedure underestimates the critical strain response for relatively thin pavements. Actually, at a thickness lower than 152mm, the difference is expected to significantly increase according to the trend in Figure 3.

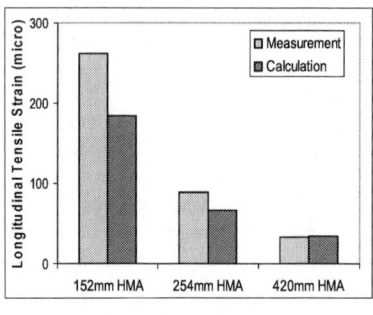
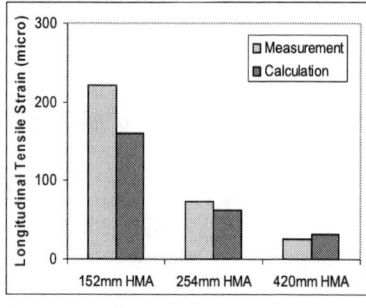

(a)   (b)

Figure 3 Measured and Calculated Longitudinal Strains Using MEPDG Procedure at (a) 8km/h, and (b) 16km/h

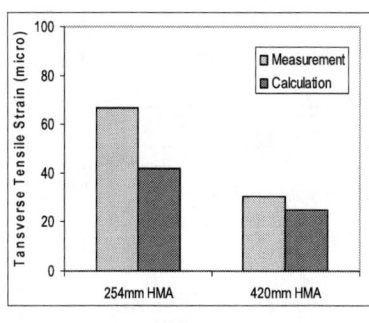
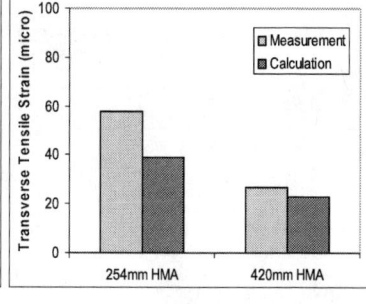

(a)   (b)

Figure 4 Measured and Calculated Transverse Strains Using MEPDG Procedure at (a) 8km/h, and (b) 16km/h

The average difference ratios between the MEPDG calculated to measured strains, under various loading conditions (26, 35, 44, 53, and 62kN), are shown in Table 5. The difference depends on pavement thickness and vehicle speed. It appears that the accuracy of the MEPDG procedure improves as depth increases, which is consistent with the recent finding by Al-Qadi and his co-workers: the effect of the loading frequency calculation on pavement response is more pronounced at depths below 152mm (Al-Qadi et al. 2008b).

Table 5 Difference between Measured Strains and MEPDG Calculated Strains.

Speed (km/h)	8			16		
HMA thickness (mm)	152	254	420	152	254	420
Longitudinal strain difference (%)	-30%	-21%	+8%	-30%	-11%	+8%
Transverse strain difference (%)	/	-32%	-23%	/	-30%	-20%

The underestimation of tensile strains could be related to the erroneous calculation of loading pulse period and the empirical conversion between loading period and frequency. The real loading pulse durations within the HMA under vehicle loading could be greater because the HMA viscoelasticity was not considered in the calculation. Actually, there is an intrinsic difference between laboratory loading function and in-situ stress pulse under vehicular loading. The loading function in the laboratory complex modulus test is pure sinusoid/haversine function, while the stress pulse within the pavement structure under vehicle loading results from many "pulses". Thus the simple conversion of loading period to frequency through inverse relationship is inaccurate. A Fourier Transform (FT) method has been proposed to decompose the measured loading pulse into many sinusoid pulse component and obtain effective loading frequency (Al-Qadi et al. 2008a). In addition, the delayed-response of HMA as well as the stress residual is usually encountered under real vehicle loading. Observation of in-situ stress pulse clearly showed that the rise and decay time when the vehicle is approaching and leaving could not be precisely captured by sinusoid/haversine function (Louizi and Al-Qadi 2002).

4. Comparison between Mechanistic Analysis Using 3D Finite Element Model and In-Situ Response

In addition to the aforementioned concerns about the loading pulse period and loading frequency, other factors such as non-uniform contact stress distribution and dynamic loading effect may affect pavement response prediction. The MEPDG method assumes uniform contact stresses distribution within circular contact area and stationary vehicular loading. Unfortunately, these assumptions are inconsistent with realistic loading conditions (Al-Qadi and Yoo 2007). Considering that the aforementioned data is necessary to accurately predict pavement response, a 3D finite element model was developed using ABAQUS version 6.5. The 3D FE model is more appropriate, compared to axisymmetric or 2D plane model, because it allows considering measured tire-pavement contact stress distribution under each rib, mass inertia, and damping ratio. In this model, the HMA layer is characterized as a viscoelastic material and its relaxation modulus at field temperature (measured from thermocouple) is used as an input. The transient dynamic tire loading is simulated using a continuous moving load in the implicit dynamic analysis. More details about the developed finite element model and dynamic analysis can be found elsewhere (Al-Qadi et al. 2008c).

Figures 5 and 6 present the measured and FE model calculated strains for the three pavement sections under 44kN dual-tire loading. Good agreement is noticed between the measured and calculated strains. In general, the difference between measured and calculated strains is within 5% under various loading conditions for the various pavement structures. This difference is considered acceptable.

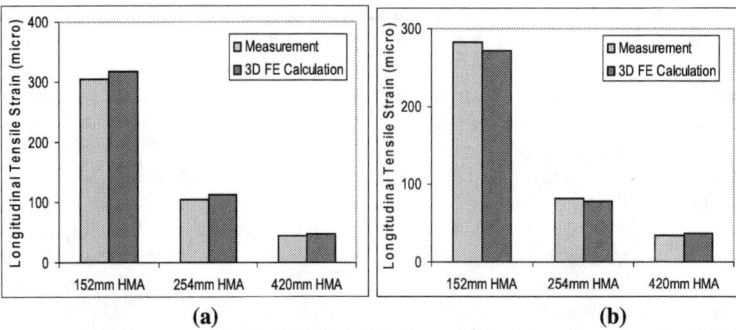

Figure 5 Measured and 3D FE Model Calculated Longitudinal Strains at (a) 8km/h, and (b) 16km/h

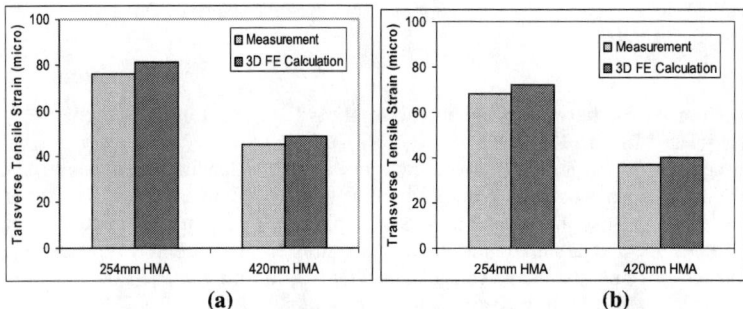

Figure 6 Measured and 3D FE Model Calculated Transverse Strains at (a) 8km/h, and (b) 16km/h

It has to be noted that a recent study using the developed 3D FE model reported that alternative parameters were potentially responsible for flexible pavement damage. "Near surface" vertical shear strain in HMA was reported as a major contributor to fatigue cracking in perpetual pavements (Al-Qadi et al., 2008c). At intermediate to high temperatures, the vertical shear strains at the shallow depth (up to 100mm below the surface) are significantly greater than the critical transverse and longitudinal tensile strains at the surface and bottom of HMA layers. This introduces a new phenomenon of crack development referred to by the second author as "near-surface" cracking (Al-Qadi and Yoo, 2007; Yoo and Al-Qadi 2008). Therefore, although this study evaluates the potential fatigue damage as suggested by the NCHRP 1-37A approach, the authors believe that "near-surface" cracking is more critical than bottom-up cracking for perpetual pavements.

5. Conclusion

The applicable complex modulus at pertinent loading frequency and temperature is critical for the proposed mechanistic pavement design by MEPDG. However, the MEPDG approach underestimates the transverse tensile strains by 11% to 32% when the pavement thickness is equal or smaller than 254mm. This underestimation is more manifested at shallow depths and thin pavements. It suggests that the pavement thickness design based on MEPDG procedure could underestimate the pavement fatigue damage for thin pavement. This could be caused by the erroneous calculation of loading pulse period and the empirical conversion between loading period and frequency, among other assumptions, such as static uniform tire contact stress within circular area, that contribute significantly to the ill-prediction of pavement responses when the current NCHRP 1-37A is used.

A 3D finite element model was developed in order to predict pavement response to tire loading. The model considers measured 3D tire-pavement contact stresses, continuous moving-wheel loading, implicit dynamic analysis, and HMA viscoelastic characteristics. Difference between measured and predicted strains was found to be within 5%.

Acknowledgement

This publication is based on the results of ICT-R59, Evaluation of Pavement Damage Due to New Tire Designs. ICT-R59 is conducted in cooperation with the Illinois Center for Transportation; the Illinois Department of Transportation, Division of Highways; and the U.S. Department of Transportation, Federal Highway Administration. The authors would like to acknowledge David Lippert's support of the wide-base tire research and the assistance of the following members of the Technical Review Panel for ICT-R59: Mark Gawedzinski (Chair), Amy Schutzbach, Bruce Peebles, Charles Wienrank, and Rich Telford. The authors would like also to acknowledge their colleagues Samer Dessouky and Jim Meister for their help in accelerated loading testing and data management. The earlier work by Jun Yoo is greatly appreciated. The contents of this paper reflect the view of the authors, who are responsible for the facts and the accuracy of the data presented herein. The contents do not necessarily reflect the official views or policies of the Illinois Center for Transportation, the Illinois Department of Transportation, or the Federal Highway Administration. This paper does not constitute a standard, specification, or regulation.

References

ARA, Inc., ERES Division (2004), "Development of the 2002 Guide for the Design of New and Rehabilitated Pavements" *NCHRP 1-37A*, Transportation Research Board, Washington, D.C.

Al-Qadi, I. L. and P. J. Yoo (2007), "Effect of Surface Tangential Contact Stress on Flexible Pavement Response," *The Journal of the Association of Asphalt Paving Technologists*, Vol. 76, 2007, pp. 663-692.

Al-Qadi, I. L., Wei X., and M. A. Elseifi (2008a), "Frequency Determination from Vehicular Loading Time Pulse to Predict Appropriate Complex Modulus in MEPDG" Accepted for publication at *Journal of the Association of Asphalt Paving Technologists*, Vol. 77

Al-Qadi, I. L., M. A. Elseifi, P. J. Yoo, S. H. Dessouky, N. H. Gibson, T. P. Harman, J. A. D'Angelo, and K. A. Petros (2008b), "Accuracy of Current Complex Modulus Selection Procedure from Vehicular Load Pulse in NCHRP 1-37A Mechanistic-Empirical Pavement Design Guide." Accepted for publication in *Transportation Research Record*, Paper No. 08-0528, TRB, National Research Council, Washington, D.C.

Al-Qadi, I. L., H. Wang, P. J. Yoo, and S. H. Dessouky (2008c), "Dynamic Analysis and In-Situ Validation of Perpetual Pavement Response to Vehicular Loading" Accepted for publication in *Transportation Research Record*, Paper No. 08-0612, TRB, National Research Council, Washington, D.C.

Barksdale, R. G. (1971), "Compressive Stress Pulse Times in Flexible Pavements for Use in Dynamic Testing." Highway Research Record 345, Highway Research Board, pp. 32-44

Bonaquist, R. and D. W. Christensen (2005), "A Practical Procedure for Developing Dynamic Modulus Master Curves for Pavement Structural Design." *Transportation Research Record,* No. 1929, Washington, D.C., pp. 208-217.

Brown, S. F. (1974), "Determination of Young's Modulus for Bituminous Materials and Pavement Design." *Highway Research Record* 431, Highway Research Board, pp. 38-49.

Buttlar, W. and R. Roque (1994), "Development and Evaluation of the Strategic Highway Research Program Measurement and Analysis System for Indirect Tensile Testing at Low Temperatures" *Transportation Research Record*, No. 1454, pp. 163-171

Dongre, R., L. Myers, and J. D'Angelo (2006), "Conversion of Testing Frequency to Loading Time: Impact on Performance Predictions Obtained from the M-E Pavement Design Guide." Paper No. 06-2394, Presented at the 85th Transportation Research Board Annual Meeting

Ferry, J. D. (1980), "Viscoelastic Properties of Polymers" 3rd edition, Wiley, New York

Garcia, G. and M. Thompson (2004), "Validation of Design Concepts for Extended Life Hot Mix Asphalt Pavements" IHR-R39 Project Review

Huang, Y. H. (1993), "Pavement Analysis and Design" 1st ed., Prentice Hall, NJ

Kim, Y. R., Youngguk S., Mark K., and Mostafa M. (2004), "Dynamic Modulus Testing of Asphalt Concrete in Indirect Tension Mode" *Transportation Research Record,* No. 1891, pp. 163-173

Loulizi, A., I. L. Al-Qadi, S. Lahouar, and T. E. Freeman(2002), "Measurement of Vertical Compressive Stress Pulse in Flexible Pavements and Its Representation for Dynamic Loading Tests." *Transportation Research Record*, No. 1816, pp. 125-136.

Loulizi, A., G. W. Flintsch, I. L. Al-Qadi, and D. W. Mokarem (2006), "Comparing Resilient Modulus and Dynamic Modulus of Hot-Mix Asphalt as Material Properties for Flexible Pavement Design" *Transportation Research Record*, No.1970, pp. 161-170.

Wang, H. and Al-Qadi, I. L. (2008), "Full-depth Flexible Pavement Fatigue Response under Various Tire and Axle Load Configurations" submitted for 3rd International Conference on Accelerated Pavement Testing, Madrid, Spain

Yoo, P. J. and I. L. Al-Qadi (2007), "Effect of Transient Dynamic Loading on Flexible Pavements." In *Transportation Research Record*, No.1990, TRB, National Research Council, Washington, D.C., pp. 129-140

Yoo, P. J. and I. L. Al-Qadi (2008), "The Truth and Myth of Fatigue Cracking Potential in Hot-Mix Asphalt: Numerical Analysis and Validation." Accepted for publication in *Journal of Association of Asphalt Paving Technologists*, Vol. 77

Mechanism of Mitigating Shear-induced Rutting of Asphalt Pavement Using Geotextile

Yinghao Miao[1], and Jinxi Zhang[2]

[1]Transportation Research Center, Beijing University of Technology, 100 Pingleyuan, Chaoyang District, Beijing 100124 P. R. China; PH (8610) 6739-6181 ext. 204; FAX (8610) 6739 1509; e-mail: miaoyinghao@bjut.edu.cn

[2]Transportation Research Center, Beijing University of Technology, 100 Pingleyuan, Chaoyang District, Beijing 100124 P. R. China; PH (8610) 6739-1583; FAX (8610) 6739 1509; e-mail: zhangjinxi@bjut.edu.cn

Abstract: To study the mechanism of mitigating shear-induced rutting in semi-rigid based asphalt pavement using geotextile, a series of finite element simulations are carried out to evaluate the benefits of applying geotextile reinforcement. Finite element analysis is performed by ABAQUS with consideration of plastic behavior of asphalt concrete. Parametric study is also conducted to investigate the influence of geotextile properties and layout. Outer-directed transverse shoving in middle of AC layer, the major contributor for shear-induced rutting, can be effectively reduced by integrated geotextile there and such effect can be enhanced through using high elastic modulus geotextile or more layers of geotextile. However, geotextile laid at AC bottom makes no contribution to rutting mitigation. In addition, almost all common geotextile can fulfill the requirement on tensile strength.

Introduction

Permanent deformation, rutting, is a major distress of asphalt pavement, which brings severity impacts on pavement durability and traffic safety. Rutting is commonly considered comprising two components (White et al. 2002):
1. Vertical compression of asphalt concrete (AC) surface, base, subbase, even subgrade, which derived from densification;
2. Transverse shoving deformation of AC surface, which derived from shearing yield.

To asphalt pavement with rigid or semi-rigid base, AC surface deformation is the main contributor of rutting for deformation of rigid or semi-rigid base and below are negligible.

Some measures were taken to mitigate rutting in asphalt pavement by participators and practitioners, such as using modified asphalt binder, innovating aggregate gradation, and adding various fibers. Applying geosynthetics reinforcement was also thought as effective measure. Laboratory and in-situ experiments were conducted by researchers (Appea 1997; Wasage et al. 2004; Perkins et al. 2005) to evaluate rutting mitigation of asphalt pavement with geosynthetics reinforcement. Mechanical analysis was also performed by researchers (Saad et al. 2006) using finite element method to account rutting resistance from geosynthetics application in asphalt pavement. However, only unbound granular base was considered in most conducted researches. Benefits and mechanism of geosynthetics reinforced asphalt pavement with rigid or semi-rigid base was rarely analyzed. Rutting in asphalt pavement with rigid or semi-rigid base is usually shear-induced which is dissimilar with unbound granular based asphalt pavements. It is necessary to perform mechanism research on rutting resistance of geosynthetic reinforced asphalt pavement with rigid or semi-rigid base.

The aim of this work was to study the mechanism of mitigating shear-induce rutting in semi-rigid based asphalt pavement using geotextile. Finite element analysis was conducted by ABAQUS with consideration of plastic deformation of AC surface. Factors of geotextile properties and layout were also taken into account to evaluate their influence. Mechanical explanation for mitigating shear-induced rutting in asphalt pavement using geotextile was brought. The results also provided some references for design of geotextile reinforced asphalt pavement with semi-rigid base.

Finite Element Model

To pavement system, three-dimension (3D) analysis was confirmed more effective than axisymmetric or plane strain analysis by many researchers (Saad et al. 2006; Chen et al. 2007; Su et al. 2007). This work also adopted 3D analysis to simulate pavement system.

Loading System
Standard axle load of 100KN distributed on a set of axle with dual wheels each side was considered in *"Specifications for Design of Highway Asphalt Pavement (JTG D50-2006)"* (MOC of China 2006). Tire-pavement contract press of 0.7MPa was also taken into account in such specification. This work adopted above-mentioned load parameters and assumed wheel footprints were squares. Figure 1 gives the

details of tire-pavement contact areas. Friction between tire and pavement was also brought into this work and frictional coefficient of 0.5 was applied in calculation, which is identical to 0.35MPa tangential stress along traffic direction. Transverse stress was not considered in this study.

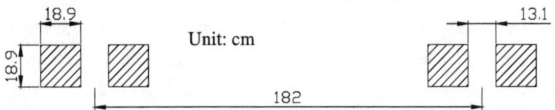

Figure 1. Details of tire-pavement contact areas

Model Size and pavement structure
A finite element model should be large enough to avoid excessive edge error. Length, width and depth of the model were determined according to trial calculations and symmetric property of the problem. Figure 2 gives final size of the finite element model. A typical structure of semi-rigid based asphalt pavement, which is shown in figure 2, was considered in this work.

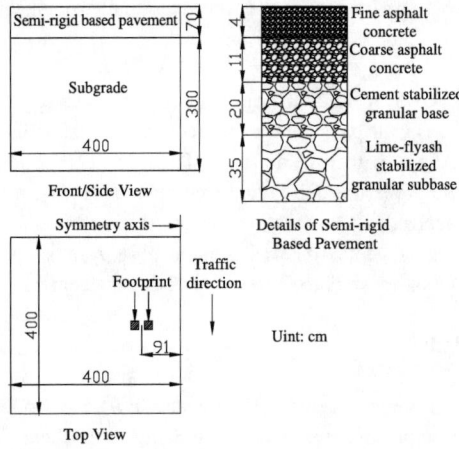

Figure 2. Size and Structure details of FE model

Element Type
Geotextile was modeled as membrane layer with M3D8R element of ABAQUS. The other parts were all modeled as solid with C3D20R element.

Material Constitutive and Typical Material Parameters
Permanent deformation of asphalt surface is the main contributor of rutting in semi-rigid based asphalt pavement. Mohr Coulomb constitutive model (illustrated in

figure 3) is considered to represent the elastoplastic behavior of AC and many researches (He 2000; Xie 2006) were performed to obtain friction angle and cohesion of AC by triaxial test. So, Mohr Coulomb constitutive model was applied to simulate behavior of AC in this study. While liner elastic constitutive was used for layers except AC to effectively simplify the problem with negligible errors. Geotextile was also assumed as elastic material in this work. Typical parameters of road layers are listed in table 1. Because AC is a temperature sensitive material, this work adopted parameters corresponding with 60°C which is identical with requirement of *"Technical Specifications for construction of Highway Asphalt Pavements (JTG F40-2004)"* (MOC of China 2004). Geotextile parameters, as variables in parametric study, will be discussed in following text.

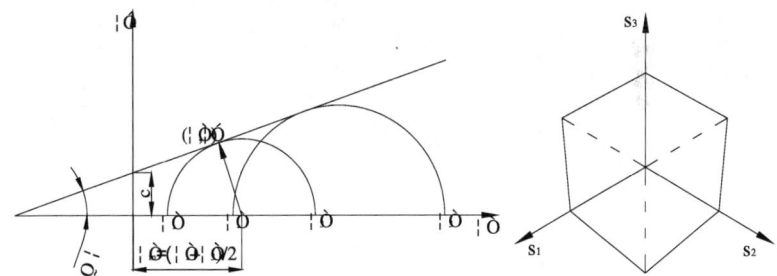

(a) Failure criterion (b) Yield surface
Figure 3. Mohr Coulomb constitutive used for AC

Table 1. Typical parameters of road layers

Layer	Elastic modulus [MPa]	Poisson ratio	Frictional angle [°]	Cohesion [KPa]
Fine AC	150	0.3	28	180
Coarse AC	180	0.3	30	160
Base	1500	0.25	-	-
Subbase	1000	0.25	-	-
Subgrade	40	0.35	-	-

Boundary Conditions
Normal displacement restriction was applied at all four vertical boundaries and bottom. Such boundary conditions have been successfully used by Miao (2005).

Meshing Consideration
In general, smaller mesh size should be adopted to ensure calculation precision. However, extremely small mesh size will result in unacceptable computation time and storage requirements. A compromise solution was used which is achieved through biased meshing method. Mesh size was smaller under and around tire

footprints, while larger mesh size was used far from tire footprints. Elements just under tire footprints are 3.15cm in length and width and 1.33cm in depth, which are the minimum sizes. Figure 4 shows a selected 3D finite element mesh.

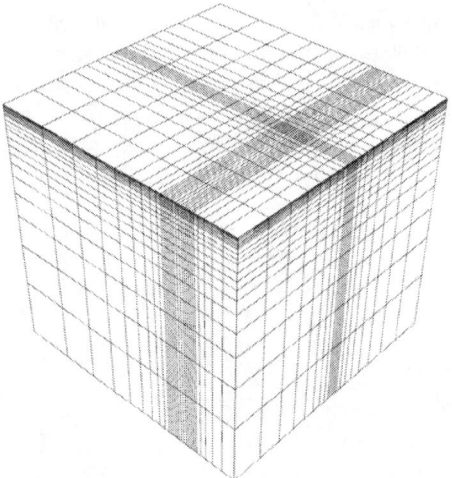

Figure 4. Selected 3D finite element mesh

Parametric Study

Behavior of geotextile reinforced asphalt pavement on semi-rigid base and its variation with layout and elastic modulus of geotextile were analyzed through parametric study. Table 2 gives geotextile layouts considered in this work. Geotextile parameters are listed in table 3. The elastic modulus was adopted same value for each geotextile layer when more than one layers of geotextile was applied. For the convenience of depiction, FEM Models were named by geotextile layout ID in table 2 and elastic modulus of geotextile in MPa, such as TB200, M800 and so on.

Table 2. Geotextile layouts

ID	Geotextile location in pavement
N	No geotextile
T	4cm below pavement surface
M	9cm below pavement surface
B	15cm below pavement surface
TM	4cm, 9cm below pavement surface
TB	4cm, 15cm below pavement surface
MB	9cm, 15cm below pavement surface
TMB	4cm, 9cm, 15cm below pavement surface

Table 3. Geotextile parameters

Elastic modulus [MPa]	Poisson ratio	Thickness [mm]
200, 400, 600, 800, 1000, 1200, 1400	0.35	1

Results and Discussions

Transverse total displacement (hereinafter called transverse displacement) and transverse plastic strain distribution of AC layer along "Path A" and "Path B" was selected to evaluate mitigation of shear-induced rutting in asphalt pavement. Details of "Path A" and "Path B" are illustrated in figure 5. Stress distributions of geotextile are also important which reveals some requirements for choosing geotextile. So, distributions of tension stress in geotextile along centerline of wheel footprints which is normal to traffic direction were considered.

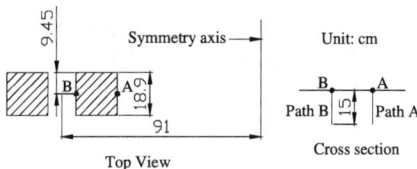

Figure 5. Details of "Path A" and "Path B"

Transverse Displacement

Typical distributions of transverse displacement along "Path A" and "Path B" are illustrated in figure 6 and figure 7. For "Path A", negative displacement is outer-directed based on wheel footprint and positive displacement is inner-directed. For "Path B", it is just opposite to "Path A". Figure 6 and figure 7 show that shoving transverse displacement is prominent around the wheel footprint, especially between 0.04m and 0.1m in depth, though inner-directed displacement occurs in top (and bottom) of AC layer. Distribution form of transverse displacement is not changed but amount is decreased significantly for applying geotextile.

Figure 8 and figure 9 give the distributions of transverse displacement decrease percentage derived from reinforcement along "Path A" and "Path B" with various getextile layouts. Near the points at which sign of displacement switches, decrease percentage extremely fluctuate because displacement there in Model N is nearly zero which is denominator in calculating decrease percentage. It is negligible in analysis of rutting mitigation. Figure 8 and figure 9 illustrate that decrease percentage of B1400 is very little and curves of TB1400, MB1400, and TMB1400 almost coincide with T1400, M1400, and TM1400 respectively. Therefore, it is concluded that geotextile laid at AC bottom (15cm below pavement surface) almost make no

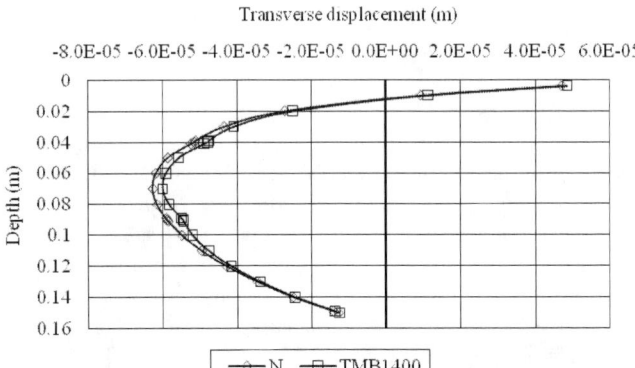

Figure 6. Typical distribution of transverse displacement along "Path A"

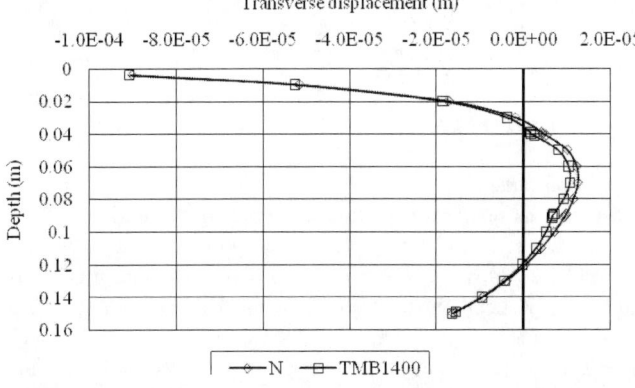

Figure 7. Typical distribution of transverse displacement along "Path B"

contributions to rutting mitigation. Near the geotextile (except geotextile laid at AC bottom), transverse shoving is largely reduced, but the influence is decreasing rapidly with distance from geotextile. Curves of T1400, M1400 and TM1400 show that geotextile layout in TM1400 is more effective for rutting mitigation than in T1400 and M1400 because influence area of one layer geotextile is not enough.

Figure 10 and figure 11 depict distributions of transverse displacement decrease percentage derived from reinforcement along "Path A" and "Path B" with TM geotextile layout and different elastic modulus of geotextile. Though distribution curves are similar, the amount of transverse displacement decrease percentage is sharply down with reducing of geotextile elastic modulus. When geotextile elastic modulus is down to 200MPa, influence of reinforcement on rutting mitigation becomes very small. So, if possible, geotextile with high elastic modulus should be applied.

Transverse plastic strain
At the given loading condition, some of AC under or around tire footprints achieved plastic state. For Path A and Path B, only top 4cm have transverse plastic strain which is illustrated in figure 12 and figure 13. It is showed that curves of N, M1400,

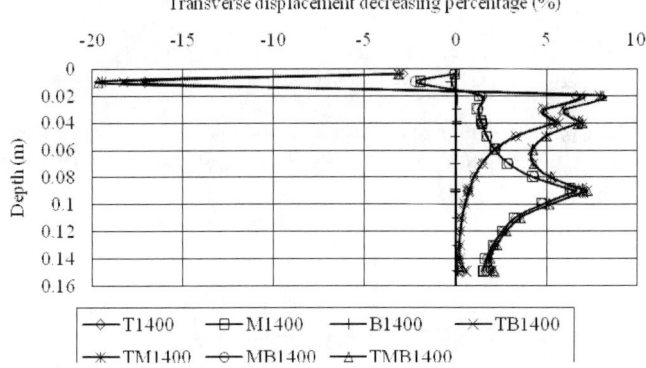

Figure 8. Distribution of transverse displacement decrease percentage along "Path A" with various geotextile layouts

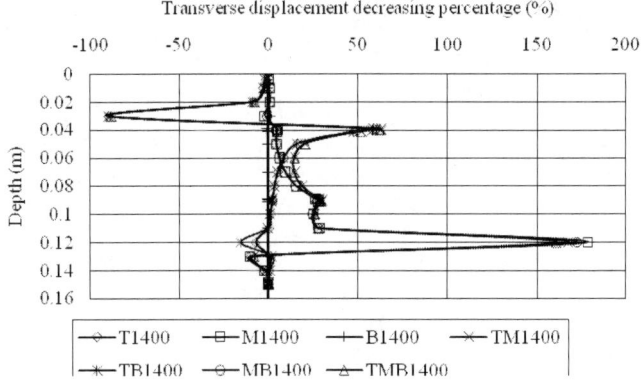

Figure 9. Distribution of transverse displacement decrease percentage along "Path B" with various geotextile layouts

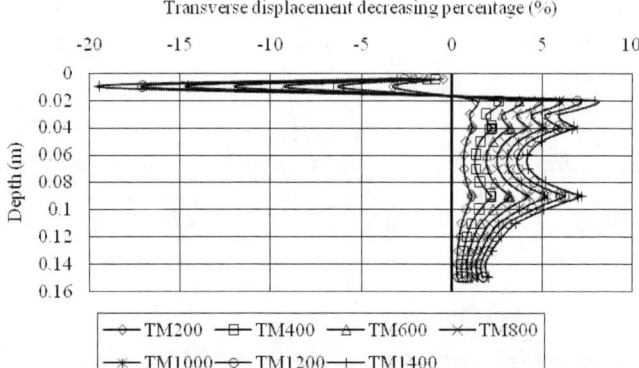

Figure 10. Distribution of transverse displacement decrease percentage along "Path A" with various geotextile elastic modulus

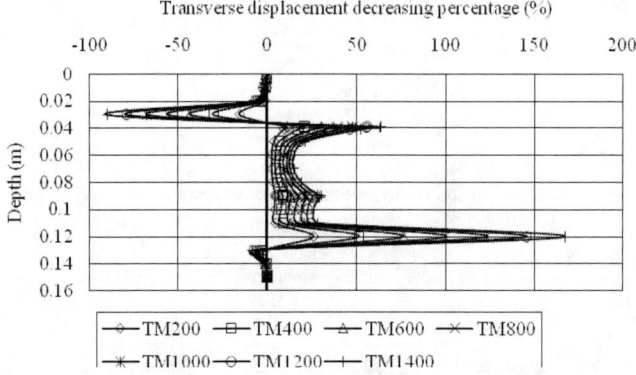

Figure 11. Distribution of transverse displacement decrease percentage along "Path B" with various geotextile elastic modulus

B1400, and MB1400 almost coincide with each other and so do curves of T1400, TM1400, TB1400, and TMB1400. Transverse plastic strains of models whose names have "T" are less than the others. So, geotextile laid at 4cm below pavement surface is efficient for restraining transverse plastic strain. And it is also concluded that influence area of one layer of geotextile is rather small.

Tension Stress of Geotextile
In order to obtain tensile strength and other requirements of mitigating shear-induced rutting in asphalt pavement on geotextile, typical distribution and magnitude of

tension stress in geotextile should be considered as reference. Figure 14, in which TM geotextile layout is considered, gives typical distributions of tension stress in geotextile along centerline of wheel footprints which is normal to traffic direction. It shows only narrow range of geotextile near wheel footprint is distributed tension stress. Stress in geotextile at 9cm below pavement surface is more than in geotextile at 4cm below pavement surface. Elastic modulus is also an active variable. With its reducing, stress in geotextile is down. All tension stress is extremely less than tensile strength of common geotextile though distribution curves are not similar.

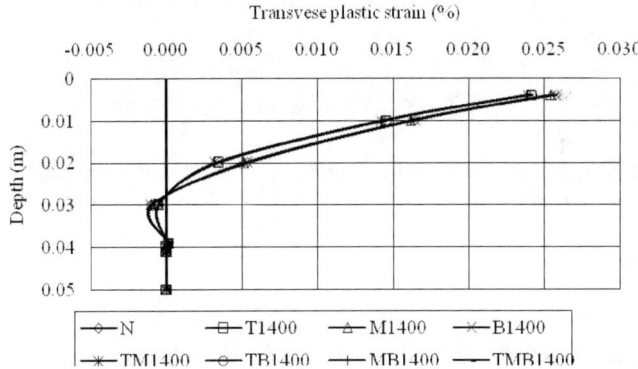

Figure 12. Distribution of transverse plastic strain along "Path A"

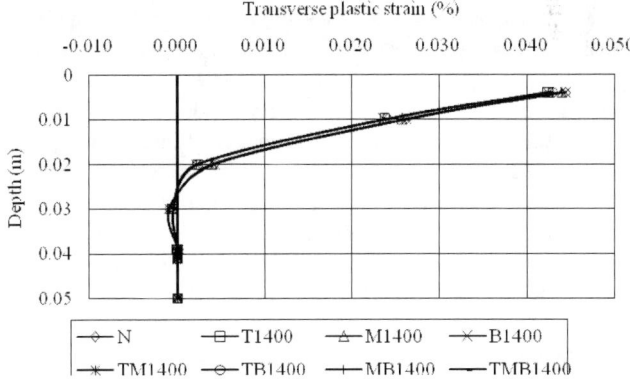

Figure 13. Distribution of transverse plastic strain along "Path B"

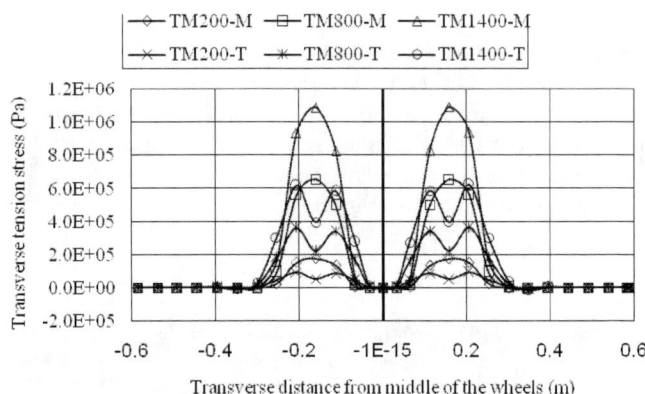

Figure 14. Typical distribution of tension stress in geotextile along centerline of wheel footprints which is normal to traffic direction

Conclusions

A series of finite element simulations were performed to evaluate the benefits of applying geotextile reinforcement to mitigating shear-induced rutting in semi-rigid based asphalt pavement. Parametric study was carried out to investigate how such effects are influenced by geotextile layout and elastic modulus. Following conclusions were drawn from this work:
1. Outer-directed transverse shoving is severe in middle of AC layer of semi-rigid based asphalt pavement. And in the given loading, top of AC layer achieves plastic state.
2. Outer-directed transverse shoving in AC layer can be effectively reduced by integrated geotextile and such effect can be enhanced through using high elastic modulus geotextile. Transverse plastic strain is also restrained by geotextile.
3. More than one layers of geotextile should be laid in middle of AC part for mitigating shear-induced rutting because influence area of one layer of geotextile is rather small. However, geotextile should not be laid at AC bottom for its effect is negligible for rutting mitigation.
4. No special tensile strength requirement should be fulfilled for geotextile stress is far lower than tensile strength of almost all common geotextile.

References

Appea, A. K. (1997). "In-situ behavior of geosynthetically stabilized flexible pavement." *Master of Science Thesis*, Virginia Polytechnic Institute and State University, Blacksburg, Virginia, USA.

Chen, S., Peng, C., Wang, S., and Ma, Q. (2007). "Finite element analysis of stress caused by load on lean concrete-asphalt overlays composite pavement." *Journal of Chang'an University (Natural Science Edition)*, 27(6), 1-5.

He, X. (2000). "Mechanical characteristics of bitumen concrete under condition of triaxial stress." *Journal of Yangtze River Scientific Research Institute*, (17)2, 37-40.

Miao, Y., Hu, C., and Wang, B. (2005). "3-D FEM analysis of pavement crack at cut to fill location of subgrade." *Journal of Traffic and Transportation Engineering*, (5)4, 43-47.

Ministry of Communications of the People's Republic of China (2004). "Technical Specifications for construction of Highway Asphalt Pavements (JTG F40-2004)." China Communications Press, Beijing, China.

Ministry of Communications of the People's Republic of China (2006). "Specifications for Design of Highway Asphalt Pavement (JTG D50-2006)." China Communications Press, Beijing, China.

Perkins, S. W., and Cortez, E. R. (2005). "Evaluation of base-reinforced pavements using a heavy vehicle simulator." *Geosynthetics International*, 12(2), 86-98.

Saad, B., Mitri, H., and Poorooshasb, H. (2006). "3D FE analysis of flexible pavement with geosynthetic reinforcement." *Journal of Transportation Engineering*, 132(5), 402-415.

Su, K., Sun, L., Wang, Y., and Ye H. (2007). "Mechanics analysis of loads and pavement structure type to asphalt pavement rutting by 3D finite element method." *Journal of Tongji University (Natural Science)*, 35(2), 187-192.

Wasage, T. L. J., Ong, G. P., Fwa, T. F., and Tan, S. A. (2004). "Laboratory evaluation of rutting resistance of geosynthetics reinforced asphalt pavement." *Journal of the Institution of Engineers, Singapore*, 44(2), 29-44.

White, T. D., Haddock, J. E., Hand, A. J. T., and Fang, H. (2002). "Contributions of pavement structural layers to rutting of hot mix asphalt pavements." *NCHRP Report 468*, Transportation Research Board, National Research Council, Washington D. C., USA.

Xie, Z. (2006). "The research on triaxial test of asphalt mixture high-temperature stability." *Master of Engineering Thesis*, Changsha University of Science & Technology, Changsha, China.

Highway Pavement Damage and Cost Due to Routine Permitted Axles

D.H. Timm[1], P.E., K.D. Peters[2] and R.E. Turochy[3], P.E.

[1]Associate Member, Gottlieb Associate Professor of Civil Engineering;
email: timmdav@eng.auburn.edu
[2]Graduate Research Assistant; email: peterkd@auburn.edu
[3]Member, Associate Professor of Civil Engineering; email: rodturochy@auburn.edu
Auburn University Highway Research Center, Auburn, AL 36849
Tel: 334-844-4320

Abstract
Routine permitting of overloaded trucks and axles is a common practice within the U.S. These permits only require the filing of paperwork and the payment of a fee without any further infrastructure damage evaluation. As cited in the literature, the fees are based primarily on administrative fees rather than costs associated with more rapid deterioration of the highway infrastructure. This investigation evaluated the impact of routine permitting on flexible and rigid pavement deterioration using the Mechanistic-Empirical Pavement Design Guide (MEPDG). Life cycle cost analyses were also conducted to estimate increased pavement costs due to permitting of heavier loads. Similar trends were noted between both flexible and rigid pavement analyses. Namely, the predicted reduction in pavement life as a function of the volume of permitted axles can be modeled well as a negative exponential relationship. There was a corresponding exponential increase in life cycle cost from increases in permitted axle volume.

Introduction
It is generally well known that gross vehicle weight (GVW) and axle weight limits on non-interstate highways tend to follow federal regulations for the U.S. interstate system (FHWA, 2000). In all but 11 states, the GVW is equivalent between the interstate and non-interstate systems. Only six states exceed the 20,000 lb limit for single axles and ten states exceed the 34,000 lb limit for tandem axles (FHWA, 2000). The uniformity of these limits across the country has helped improve trucking efficiency and road operations.

While the weight limits are relatively consistent among states, so-called "routine permitting" programs are highly state-specific. A routine permit can be defined as a permit that is applied for and obtained by a trucking company without any further investigation by the issuing agency. Most states have routine permits that can be issued based upon GVW or axle type and weight. For example, 32 states have single axle routine permits that exceed 20,000 lb (FHWA, 2000).

The need for routine permitting is due, in part, to the large volume of permit requests that could not be handled if more detailed analysis were completed on a case-by-case basis. Straus and Semmens (2006) reported that over 2.9 million permits were issued by state agencies in 2003 with Indiana alone issuing over 200,000 permits.

Many studies have estimated the amount of infrastructure damage caused by overweight vehicles. The following are just a small sample of estimated costs:
- In 1987, Terrell and Bell reported annual costs due to overloaded trucks on the federal-aid highway system approaching $1 billion/year.
- In 1995, FHWA estimated overloaded trucks in the U.S. cost taxpayers $160-$670 million/year in pavement damage (Chen et al., 2005).

Despite the general understanding that overloaded trucks can cause significant pavement damage, the fee structure has historically not been damage-based (FHWA, 2000), but rather set by state legislatures. According to the Truck Size and Weight Study (FHWA, 2000), the associated permitting fees are usually established to recover the administration costs of the permitting program itself, including enforcement activities. The fees themselves have undergone little change from 1989 to 2000. In 1989, state permit fees for an 84,000 lb GVW vehicle ranged from $6-$61 (FHWA, 2000).

Tying permitting fees to pavement damage resulting from heavier loadings requires analysis of pavement-vehicle interaction. In recent years, there has been a national movement in the U.S. toward mechanistic-empirical (M-E) pavement design and analysis. This approach utilizes mechanistic pavement models to simulate pavement responses under applied traffic loads. The responses can then be empirically correlated to expected pavement performance. While there are existing M-E approaches, such as those developed by the Asphalt Institute (1982) and the Portland Cement Association (1984), recent efforts under NCHRP Project 1-37A (Eres, 2004) to develop the M-E Pavement Design Guide (MEPDG) have helped to unify design standards for both flexible and rigid pavement structures. In coming years, it is expected that AASHTO will adopt the MEPDG as its design standard.

Objective and Scope of Work

Given the demand for routine permits and the need for agencies to account for accelerated pavement damage under heavier axle loads, the main objective of this study was to develop a damage and cost analysis framework for evaluating routine permitted axles. The primary tool in this investigation was the MEPDG (version 0.94). Hypothetical pavement scenarios were considered and evaluated under a variety of permitting conditions. Cost analyses were conducted to estimate increased agency costs due to heavier permitted axles.

M-E and Cost Analysis Framework Overview

The concepts described above are combined in Figure 1 illustrating the entire M-E and life cycle cost analysis framework for evaluating permitted overloads (Scenario B) against a baseline load distribution (Scenario A). This framework can be expanded to accommodate numerous alternative scenarios, but only two are shown here for demonstration purposes. Regardless of the number of scenarios, they are all applied to the same pavement section that is entered into the mechanistic model for the computation of pavement response. This is to ensure that various loading scenarios are evaluated equally.

After establishing the pavement cross section, the traffic must be characterized. In Figure 1, Scenario A can be thought of as the "baseline" load distribution. This is the traffic that is currently using the facility under normal or expected conditions. Scenario B represents some other overloaded condition. According to the M-E approach, the load spectra for each scenario must be defined by the constituent axle types, loadings and frequencies. These are then simulated on the pavement section in the mechanistic model.

The M-E model in Figure 1 predicts the pavement distress over time for the various loading scenarios. Presumably, Scenario B, having heavier loads, will result in more frequent maintenance and rehabilitation efforts to maintain the structure at the same level of performance.

Life cycle cost analysis (LCCA) is used to determine the increase in cost resulting from more frequent maintenance and rehabilitation efforts. To maximize accuracy in determining actual costs, a detailed study of interest rates, material and labor cost rates would be required. In some cases, this level of detail may be warranted and agency records can be consulted to determine these costs. In other cases, however, simply knowing the percent increase in cost would be very useful from a planning perspective. In this situation, actual costs would not be needed as long as the cost per work activity was held constant between the scenarios. The end result is a direct net present value (NPV) comparison between scenarios.

Traffic Characterization
In the context of permitting heavier axles, two general scenarios were considered:
1. Constant Volume – Increased Weight
2. Decreased Volume – Constant Weight

Constant Volume – Increased Weight (CV-IW)
As the name implies, this scenario fixes the total axle volume while reassigning a certain percentage of the axle repetitions to the permitted (or overloaded) axle weight. By way of further explanation, consider the hypothetical single axle load distributions shown in Figure 2. The baseline distribution was obtained from the default traffic data in the MEPDG software. Adding up the number of axles at each load level results in 1,535 single axles/day. Multiplying each load magnitude by the corresponding volume results in 7 Mtons/day. The permitted distribution represents the condition where 5% of the axle volume below the permitted weight (24,000 lb in this case) has been permitted to operate at 24,000 lb. 5% of the volume at each load level below 24,000 lb has been reassigned to the 24,000 lb category; thus the dramatic change from less than 5 axles/day at 24,000 lb to nearly 80 axles/day. Slightly less dramatic is the 5% change in the volume for each load level less than 24,000 lb. In this scenario, it was assumed that loads heavier than the permitted weight were left unaffected. The net result for the permitted scenario is the same number of axles per day (1,535), but an increase of 570,000 tons/day due to the increased number of 24,000 lb axles.

Decreased Volume – Constant Weight (DV-CW)
The second scenario fixes the total weight which results in fewer total trips due to more efficient use of the trucking fleet. Figure 2 also illustrates this situation with the same

hypothetical baseline single axle spectrum and targeted single axle load of 24,000 lb. However, rather than reassigning 5% of the *volume* up to the permitted axle, 5% of the *weight* up to the permitted axle was calculated. The number of additional 24,000 lb axles needed to carry this load was then determined. The net result of these two computations is the same total load level (7 Mtons/day) but with 48 fewer axle passes than the CV-IW scenario.

It should be noted that many other scenarios where the traffic volume remains constant are possible. For example, it is possible that only the heavier axles, already close to the permitted weight, would begin operating at the permitted weight, leaving the more lightly loaded axles unaffected. A more detailed investigation of how permitting activities affect load spectra would allow for a more informed and realistic representation of changes in load spectra but is beyond the scope of this investigation.

Traffic in the MEPDG
Like most other M-E methods of pavement analysis and design, the MEPDG uses load spectra to characterize the traffic. Within the software, there are a number of modules that facilitate data entry and modification of default load spectra. Though these modules could be used to enter various load spectra, it was found that interacting with the "temporary" load spectra files created by the MEPDG during program execution was much more efficient. A simple spreadsheet template was developed that enabled a target axle weight and percent of weight or volume to be selected from which modified load spectra were automatically generated and used in analysis. Full details of the template have been documented by Timm et al. (2008).

Cost Analysis
Translating the reduction in pavement life resulting from increased loads into cost is an important consideration for agencies responsible for load regulation, permitting and enforcement. To achieve that end, life cycle cost analysis (LCCA) is a powerful and much needed tool. FHWA (1998) has published a technical guide on LCCA best practices that contains both simple analysis tools and complex theory on accounting for variability of the input data and work zone user costs. For pavement type selection, these considerations are important since they may vary between design alternatives and may be the driving factor between selecting one pavement over another. However, in the context of overload analysis, where a single pavement structure is to be evaluated with various loading scenarios, factors such as input variability and work zone activities would not vary and therefore could be ignored for simplicity.

As presented by Barros (1985), and discussed in the FHWA LCCA guide (1998), the straightforward NPV calculation can be used as a fair comparison between scenarios:

$$NPV = I.C. + \sum_{k=1}^{N} R.C._k \left[\frac{1}{(1+i)^{n_k}} \right] \quad (1)$$

where: I.C. = initial construction cost
R.C. = recurring rehabilitation/maintenance costs
k = rehabilitation/maintenance activity
i = interest rate
n = year in which rehabilitation/maintenance occurs

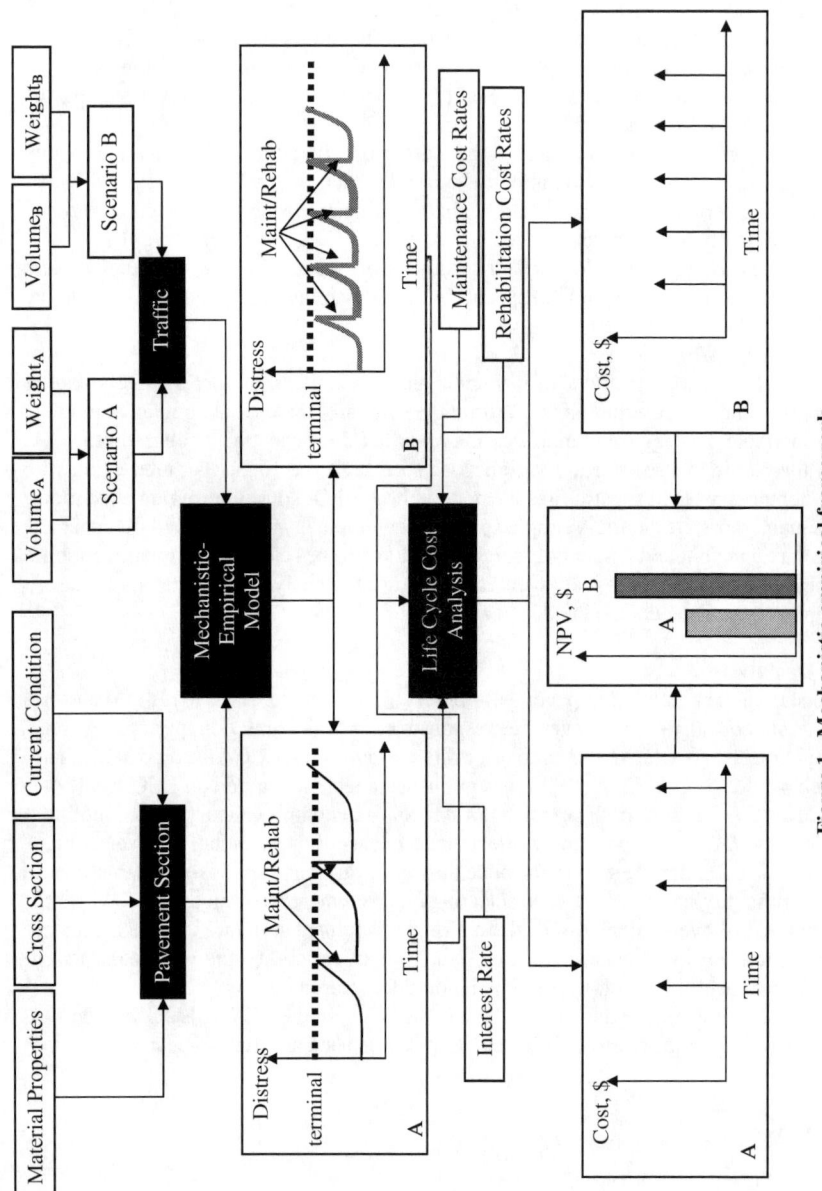

Figure 1. Mechanistic-empirical framework.

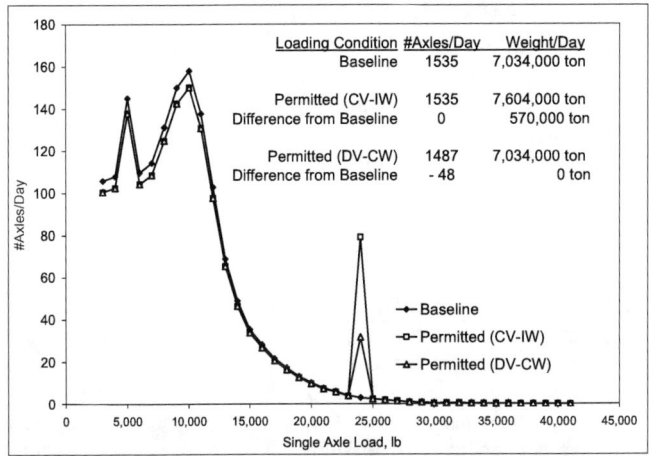

Figure 2. Permitting examples.

Because the overload analysis considers the same pavement structure under various loading conditions, the initial construction costs are equal. This leaves only the recurring rehabilitation/maintenance costs to vary as a function of loading scenario. Assuming for the purposes of overload analysis that the same activity would be required regardless of the loading scenario, just at different frequencies, then a ratio of the NPV of the costs associated with the overload scenario to the NPV of the costs associated with the baseline case can be developed. This ratio, or "cost factor" can be defined by:

$$CF = \frac{NPV_{overload}}{NPV_{baseline}} \qquad (2)$$

Where:
CF = cost factor
$NPV_{overload}$ = net present value in permitted or overloaded condition
$NPV_{baseline}$ = net present value in baseline condition

Representative Cases - Inputs

To demonstrate the framework detailed above, a number of hypothetical scenarios were analyzed with the MEPDG. The following sub-sections detail the performance parameters, climate, pavement cross sections, material property and traffic inputs used in the simulations. It should be noted that an infinite number of analyses could be conducted using this framework. This investigation was not meant to be exhaustive, but rather to exercise the framework and gain some understanding regarding the interaction of several input variables on pavement damage and cost.

Performance Parameters and Climate

The MEPDG predicts and designs for specific distresses. Three criteria (rutting, fatigue cracking and pavement roughness) were used for flexible pavement design. Slab cracking, joint faulting and roughness were used for rigid pavement design. All the parameters were set to the default levels within the MEPDG and analyzed at 90%

reliability. A pavement was considered failed when any of predicted distresses exceeded the pre-set limit. A 20-year design period was targeted for the baseline case. Once the pavement cross section was established to meet this time frame, the various loading scenarios were applied to measure the reduction in pavement life. It must be noted that the Alabama climate was simulated within this investigation. Future investigations could look at varying climate conditions.

Pavement Cross Sections
The scenarios all considered a pavement built upon 15 inches of good quality crushed stone base (M_r = 30,000 psi) over an A-6 soil (M_r = 15,000 psi). The moduli were automatically estimated by the MEPDG based upon the material selection.

Both asphalt and concrete pavement surfaces were considered in this analysis. The asphalt was a PG 76-22, selected to meet the requirements of the warm Alabama climate. The other properties were selected to be representative of typical Superpave mixtures. For the rigid pavement designs, the concrete properties were selected to be representative of typical paving-grade mixtures. A modulus of rupture equaling 690 psi was entered from which the MEPDG calculated a Young's modulus of 4,403,280 psi. Joint spacings of 15 ft with dowels providing load transfer were also selected. The full complement of material properties for each sub-structure pavement was documented elsewhere (Timm et al., 2008).

Traffic
The overload analysis, as described previously, examined changes in load spectra due to permitting specific axles. Another factor was the volume of traffic. Four levels were selected so that a range of pavement thicknesses would be generated.

The MEPDG requires traffic volume to be entered as the average annual daily truck traffic (AADTT). Four levels (250, 1000, 4500 and 8000 AADTT) of two-way truck volumes were selected to represent typical pavements in Alabama ranging from lower-volume routes to higher-volume interstates. A 50% directional split and 95% lane distribution factor were assumed, in addition to a 4% compound annual growth factor. The baseline axle load spectra for the single, tandem and tridem axles were generated automatically by the MEPDG based upon the default vehicle distribution built into the software.

Permitting of specific axles considered both "Constant Volume – Increased Weight" and "Decreased Volume – Constant Weight" scenarios as stipulated previously. For each of these cases, average (Case D) and maximum (Case E) routine permitted axles were simulated. The average permitted weights (24,000 lb single and 46,000 lb tandem) were computed from the routine permitting weights across the U.S. published by the Comprehensive Truck Size and Weight Study (FHWA, 2000). The maximum permitted weights (31,000 lb single and 62,000 lb tandem) were taken from the same data set (FHWA, 2000) and represent the routine permitting in Washington, D.C. For this study, each case considered both the single and tandem axles simultaneously. Also, four percentages of permitted axles (0%, 1%, 10% and 20%) were used to cover a wide range of conditions.

Representative Cases Results and Discussion

The design thicknesses determined through the MEPDG are summarized in Table 1. As expected, there was a general trend of increasing the required thickness with increasing truck traffic volume. It is important to note that the minimum allowable concrete thickness in the MEPDG is 6 in. Therefore, it can be surmised that the pavement designed for 250 AADTT was overly conservative since the 1,000 AADTT case also only required 6 in. It is also important to note that the HMA at 8,000 AADTT is extremely thick and may not be a practical pavement design. However, as it was a result of using the MEPDG, it was included in this analysis.

Table 1. Pavement thickness summary – baseline cases

Traffic (AADT)	Pavement Depth, in.	
	HMA	PCC
250	6	6
1000	10.25	6
4500	16	11
8000	27	12.5

Damage Analysis

After the thicknesses were determined for the baseline traffic scenarios, the pavement cross-sections were re-evaluated under each of the permitted load spectra scenarios. In total, two pavement types, four truck volumes, two sets of permitted weights, four levels of percent permitted axles and two loading scenarios (CV-IW vs. DV-CW) were considered. This resulted in 128 MEPDG simulations. For conciseness, only a limited number of results are presented here to illustrate important trends. Full results can be found in (Timm et al., 2008).

Figure 3 illustrates the change in flexible pavement life under the CV-IW loading condition. As expected, a decrease in pavement life was observed as the proportion of permitted axles increased from 0 – 20%. The effect of permitted weight (Case D vs. E) also was as expected; the heavier weight (Case E) resulted in shorter pavement lives compared to Case D. These trends were consistent for all the cases considered.

An interesting situation was noted in Figure 3 (4,500 AADTT with Case E). A sharp decrease in pavement life was observed between the baseline and 1% case which was not observed for the other scenarios in Figure 3. These simulations were rechecked and found to be correct. Therefore, it evidently shows that small numbers of permitted axles in certain cases can have a dramatic impact on pavement life. This stresses the need to consider pavements on a case-by-case basis.

A comparison of the two loading scenarios (CV-IW vs. DV-CW) indicated an average life increase of approximately 3.5 years when a reduction in volume is included in the analysis. This demonstrates that a more efficient trucking fleet, carrying the same total freight with fewer trips, is less detrimental to the pavement infrastructure.

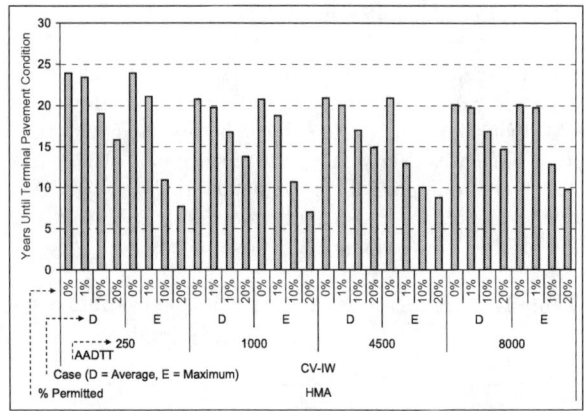

Figure 3. Pavement life – flexible – CV-IW.

The relationship between pavement life and percent of permitted axles, as illustrated in Figure 3, was found to be modeled well by a negative exponential relationship:

$$Life = \lambda_1 e^{\lambda_2 \cdot \% Permitted} \tag{3}$$

Where:
Life = years until terminal pavement condition
% Permitted = percent of axles that are at the permitted weight
λ_1, λ_2 = regression coefficients

Regression analyses were performed for all the cases investigated and the results are summarized in Table 2. The high R^2 in almost all the cases indicates a strong relationship between pavement life and the percentage of permitted axles. The one exception was observed in the flexible scenarios (Case E, CV-IW, 4500 AADTT). As noted above, there was a sharp decrease in life between baseline and 1% permitted axles, resulting in a relatively poor fit of the exponential model for this case. This fact carried through to the cost factor analysis as well.

The λ_2 columns in Table 2 can be interpreted as the impact that increasing the percentage of permitted axles has on the pavement life. Since these values were not strongly influenced by the truck traffic volume, they were averaged by case and scenario to bring more meaning to the data. It is interesting to note that λ_2 is consistently two to four times greater for the CV-IW cases than the comparable DV-CW cases. Similar ratios were observed when comparing the averages between the Case D and E scenarios. It must be finally noted that these relationships are specific to the cases investigated in this study and though they may represent general trends, would not necessarily hold true for all cases. For example, if the Case E loadings were much greater, the resulting pavement life (in terms of λ_2) would be much shorter.

Table 2. Pavement life regression summary.

Case	Scenario	AADTT	Flexible Pavement λ_1	λ_2	R^2	Rigid Pavement λ_1	λ_2	R^2
D	CV-IW	250	23.8	-2.09	0.997	27.9	-1.39	0.930
		1000	20.5	-1.99	0.996	21.7	-3.49	0.998
		4500	20.5	-1.67	0.987	21.3	-1.38	0.993
		8000	20.0	-1.59	0.996	20.4	-1.53	0.997
		Average	-1.83			Average	-1.95	
	DV-CW	250	24.0	-0.80	0.998	28.4	-0.92	0.984
		1000	20.7	-0.79	0.991	21.8	-0.98	0.918
		4500	21.0	-0.51	0.997	21.5	-0.37	0.997
		8000	20.1	-0.35	0.948	20.5	-0.39	0.997
		Average	-0.61			Average	-0.66	
E	CV-IW	250	22.3	-5.69	0.965	27.3	-5.26	0.994
		1000	19.9	-5.40	0.987	19.6	-10.73	0.951
		4500	16.3	-3.48	0.715	21.4	-2.37	0.999
		8000	19.9	-3.70	0.981	20.4	-2.46	0.996
		Average	-4.57			Average	-5.21	
	DV-CW	250	23.6	-2.83	0.988	28.0	-1.73	0.903
		1000	20.3	-2.40	0.990	21.4	-5.83	0.993
		4500	20.9	-2.04	0.997	21.5	-0.54	0.999
		8000	20.1	-1.28	1.000	20.6	-0.64	0.998
		Average	-2.14			Average	-2.19	

Cost Analysis

Following the procedures described above, a life cycle cost analysis was conducted for the permitted axle scenarios. A 60-year time frame, to cover at least two rehabilitation cycles, with an interest rate of 4% were used. Future analyses could certainly examine different time periods and interest rates. Figure 4, which was representative of all the data (Timm et al., 2008), summarizes the results for the CV-IW cases and corresponds to the life predictions in Figure 3.

Generally speaking, the infrastructure costs became quite high at or above 10% permitted. Even for the average case (Case D), at 10% permitted axles, the infrastructure costs were generally 1.5 times the baseline cost. In other words, a $50 million project becomes a $75 million project due to changes in the load spectra.

As expected, the cost factors were lower for the DV-CW cases compared to the CV-IW cases. As explained above, the reduction in costs is derived from carrying the same total weight more efficiently on fewer total axle passes. Similar to the life reductions, the cost factor was found to be well modeled with an exponential relationship:

$$CF = \phi_1 e^{\phi_2 \cdot \% Permitted} \quad (4)$$

Where:
CF = Cost Factor
% Permitted = percent of axles that are at the permitted weight
ϕ_1, ϕ_2 = regression coefficients

The R^2 values resulting from the regression analysis exceeded 0.89 in all cases. Though not shown here due to space limitations, a few observations from the data are important to note. As expected, the costs tended to increase more rapidly (increased ϕ_2) for the CV-IW cases compared to the DV-CW cases. Furthermore, the heavier loaded axle cases (E) were greater than their counterparts (D). Finally, similar ϕ_2 values were noted between the flexible and rigid pavement counterparts indicating a similar impact on costs for the two pavement types. The full data set has been documented by Timm et al. (2008).

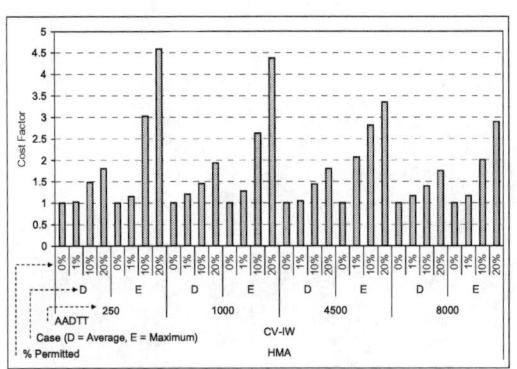

Figure 4. Cost factors – flexible – CV-IW.

Conclusions and Recommendations

This investigation developed and demonstrated a framework for evaluating the effects of overloaded axles on pavement life and associated cost factors. The MEPDG, LCCA concepts and accompanying spreadsheet templates were used to execute the analyses. Based upon the data presented herein, the following conclusions and recommendations can be made:

1. The MEPDG can be used to study effects of overloads on pavement damage, but it requires the analyst to have detailed knowledge of the MEPDG-generated temporary files and develop a separate spreadsheet template to modify the load spectra files.
2. Cost analysis must be handled separately from the MEPDG.
3. Cost factors are an effective means of comparing baseline to alternative loading scenarios. These factors can be used as multipliers to scale costs to different sized projects.
4. With respect to reduced pavement life and cost increases, the overall truck volume is less important than the load magnitudes and numbers of overloaded axles.
5. For the cases considered here, reduced pavement life was well characterized by a negative exponential relationship with the percentage of permitted axles operating at a specific weight.
6. For the cases considered here, cost factor was well characterized by a positive exponential relationship with the percentage of permitted axles operating at a specific weight.
7. Cost factors became significant (>1.4) when the permitted axles met or exceeded 10%, but were certainly dependent upon the cases considered.

8. There did not appear to be large differences between flexible and rigid pavements in terms of reductions in pavement life or increasing cost factors due to increasing the number of permitted axles.
9. Other loading scenario conditions to be simulated with the MEPDG should be identified and characterized.
10. The effect of changing analysis period and interest rate on cost factors should be explored.
11. A methodology to more accurately predict changes in load spectra resulting from changes in legal limits, or from permitting heavier axles, should be developed.
12. Though trends were observed in the data, it is important to evaluate pavements on a case-by-case basis for reduction in pavement life and associated cost factors.

Acknowledgements
The authors wish to thank the FHWA for their support of this research.

References
1. FHWA (2000). *Comprehensive Truck Size and Weight Study Volume II – Issues and Background*. Publication FHWA-PL-00-029 (Volume II). U.S. Department of Transportation.
2. Straus, S.H. and Semmens, J. (2006). "Estimating the cost of overweight vehicle travel on Arizona highways." *Proc. of the 85th Annual Meeting of the Transportation Research Board*, Washington, D.C.
3. Terrell, R.L. and Bell, C.A. (1987). *Effects of Permit and Illegal Overloads on Pavements*. NCHRP 131, National Cooperative Highway Research Program.
4. Chen, D-H, Bilyeu, J. and Chang, J-R. (2005). "A Review of the Superheavy Load Permitting Programme in Texas." *International J. of Pavement Engineering*, Vol. 6, No. 1, 47-55.
5. Asphalt Institute (1982). *Thickness Design, Asphalt Pavements for Highways and Streets*. Report MS-1.
6. Portland Cement Association (1984). *Thickness Design for Concrete Highway and Street Pavements*. Skokie, IL.
7. Eres Consultants Division (2004). *Guide For Mechanistic-Empirical Pavement Design of New and Rehabilitated Pavement Structures*. Final Report, NCHRP 1-37A.
8. Kilareski, W.P. (1989). "Heavy Vehicle Evaluation for Overload Permits." *Transp. Research Record No. 1227*, TRB, National Research Council, Washington, D.C., 194-204.
9. Barros, R.T. (1985). "Analysis of Pavement Damage Attributable to Overweight Trucks in New Jersey." *Transp. Research Record No. 1038*, TRB, National Research Council, Washington, D.C., 1-9.
10. Timm, D.H., Turochy, R.E. and Peters, K.D. (2008) *Correlation Between Truck Weight, Highway Infrastructure Damage and Cost*. Final Report, Federal Highway Administration, Washington, D.C.
11. Walls, J. and Smith, M.R. (1998). *Life Cycle Cost Analysis in Pavement Design – Interim Technical Bulletin*. Publication FHWA-SA-98-079, FHWA, U.S. Department of Transportation.

Use of Artificial Neural Networks to Detect Aggregates in Poor-Quality X-ray CT Images of Asphalt Concrete

M. Emin Kutay[1], Edith Arambula[2], Nelson Gibson[3], Jack Youtcheff[4], Katherine Petros[5]

[1]Asphalt Mixture Scientist and Laboratory Manager; mkutay@fhwa.dot.gov
[2]Postdoctoral Research Associate; edith.arambula@fhwa.dot.gov
[3]Materials Research Engineer; nelson.gibson@fhwa.dot.gov
[4]Pavement Materials and Construction Team Leader; jack.youtcheff@fhwa.dot.gov
[5]Pavement Design and Performance Modeling Team Leader; katherine.petros@fhwa.dot.gov
Federal Highway Administration, Turner-Fairbank Research Center, 6300 Georgetown Pike, McLean, VA 22101

ABSTRACT: Different laboratory compactors and protocols are employed to simulate field compaction using a reduced representative asphalt mixture specimen. Studies show that the mixture density, air voids, and mechanical properties vary within the results of each compaction method and between different compaction protocols, which may yield to estimates that mislead the design and performance prediction of the asphalt pavement. X-ray Computed Tomography (X-ray CT), a non-destructive technique for generating three-dimensional (3D) imaging of the internal structure of opaque materials, has commonly been used to quantify the air void distribution of asphalt mixtures. However, aggregate location, orientation and aggregate-to-aggregate contact points have rarely been successfully quantified using the same technique mainly because of image noise and poor contrast between the coarser aggregates and the other phases of the asphalt mixture (i.e. the air voids and the blend of the finer fraction aggregates and the asphalt binder). To overcome these shortcomings, an advanced tool has been developed utilizing 3D X-Ray CT image processing and artificial neural networks (ANN) to perform image segmentation and identify the coarser aggregates even in poor contrast X-ray CT images. This paper presents the details of the ANN tool and its application in determining the approximate size and location of the coarse aggregates in asphalt specimens.

INTRODUCTION

Numerous compactors and protocols such as the Superpave Gyratory, Marshall,

Hveem, French Roller, and German Sector are being used all over the world for laboratory compaction of asphalt mixtures. The common goal of these protocols is to simulate field compaction so that a representative sample can be produced in the laboratory using a reduced amount of material. It has been reported in the literature that the variability within the results of the same compaction protocol performed in different laboratories and between different compaction methods may yield estimates that mislead the design and performance prediction of the asphalt pavement (Khan et al. 1998). Studies comparing existing compaction methods focus mainly on the mixture density, air voids, and mechanical properties (Buchanan and Brown 2001; Masad et al. 2002; Peterson et al. 2004; Saadeh et al. 2002; Tashman et al. 2002). However, limited information exists about the microstructure achieved by different compactors primarily due to the limited number of resources and techniques available to measure the three-dimensional (3D) microstructure characteristics of the asphalt mixture specimens.

X-ray Computed Tomography (X-ray CT) is a state-of-the-art non-destructive technique for generating 3D images of the internal structure of opaque materials. Even though X-ray CT has commonly been used to quantify the air void distribution of asphalt mixtures, the aggregate location, orientation, and aggregate-to-aggregate contact points (in 3D) have rarely been successfully quantified due to the resulting image noise and poor contrast between the coarser aggregates and the other phases of the asphalt mixture (i.e. the air voids and the blend of the asphalt binder with the finer fraction aggregates also called fine portion of the mixture). This challenge significantly limited the use of invaluable X-ray CT technology in asphalt pavement applications.

Advanced mathematical pattern recognition (i.e. artificial neural networks or AAN) is a popular and common used technique for many purposes. In the field of road materials characterization, AAN have been employed to classify the size of aggregates (Kim et al. 2004), predict pavement layer moduli (Ceylan et al. 2007; Kim and Kim 1998), simulate rutting and fatigue performance of asphalt mixtures (Huang et al., 2007; Tarefder et al. 2005), estimate the thickness of the pavement layers (Gucunski and Krstic 1996), approximate the resilient modulus of base materials (Tutumluer and Seyhan 1998), relate mixture variables to permeability and roughness (Choi et al. 2004; Tarefder et al. 2005b), etc. In this study, an advanced tool has been developed utilizing 3D X-Ray CT image processing and ANN to perform image segmentation and distinguish the coarser aggregates from the air voids and the fine portion of the mixture. The tool was able to perform segmentation even when poor contrast X-ray CT images were input. This paper describes the development of the tool and presents examples of its application to asphalt mixtures specimens.

The results of this research are part of an ongoing study, where image analysis methods are used to directly measure the 3D microstructure characteristics of asphalt mixture specimens made with a wide variety of material compositions and prepared using multiple compactors. These directly-measured comprehensive parameters that represent the 3D internal structure of the asphalt mixtures will be utilized to compare different laboratory compaction methods to field compaction.

X-RAY COMPUTED TOMOGRAPHY IMAGING

X-ray CT is a non-destructive image acquisition technique for generating 3D images of the internal structure of materials. As shown in Figure 1a, the X-ray CT equipment consists of an X-ray source and a detector. The source emits X-rays of known intensities and the detector records the intensities after the X-rays pass through a rotating specimen, which is placed in between the source and the detector (Figure 1a). Once a full rotation is completed, the specimen is shifted vertically by a fixed amount (i.e. slice thickness) until the end of the specimen is reached. The intensity values are used to calculate the distribution of the linear attenuation coefficients within the specimen, which are then mapped into grayscale CT scan images. The grayscale intensity of each pixel ranging from 0 (black) to 255 (white) in the image is directly related to the density of the specimen at that point. Each image represents a slice of the specimen of a known thickness. Its order of presence within the stack corresponds to its vertical location, and by simply stacking the slices, a 3D image can be obtained as shown in Figure 1b. The X-ray CT device used in this study included a 420 keV continuous X-ray source and a linear array detector of 512 channels.

The minimum aggregate size that can be measured using the X-ray CT technique is highly dependent on the spatial resolution of the images. In this study, the resolution of the resulting images, corresponding to 150 mm diameter specimens, were approximately 0.3 mm/pixel in the horizontal direction and 0.8 mm/pixel in the vertical direction. Therefore, it is assumed that approximately 5-10 voxels are needed to be able to segment an aggregate. As a result, the minimum size of the aggregate that can be measured was of about 3 mm. This size can be reduced if smaller size specimens are scanned.

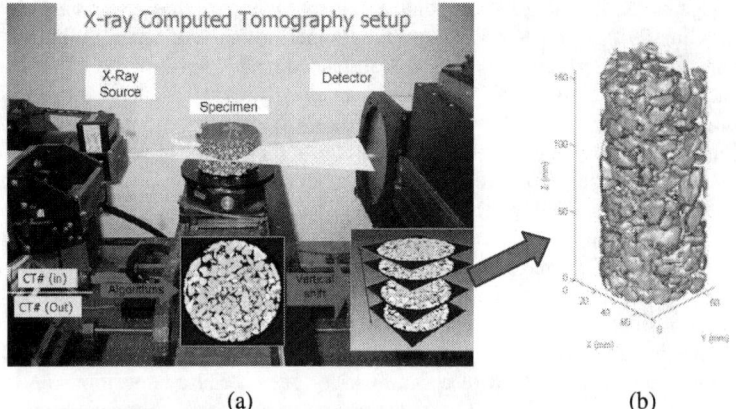

Figure 1. Illustration of an X-ray CT system; (a) equipment components, (b) reconstructed 3D image of an asphalt specimen.

USE OF TRADITIONAL IMAGE PROCESSING ALGORITHMS TO EXTRACT AGGREGATE DATA FROM X-RAY CT IMAGES

The quality of the X-ray CT image and the resulting intensity contrast between different phases of the asphalt mixture (which depends on the differences in density between the coarser aggregates, the fine portion, and the air voids) determines whether traditional image processing methods can be utilized to obtain good quantitative information. Figure 2 shows examples of good (image A) and poor (image B) quality/contrast X-ray CT images. Through rigorous image processing, image A can be converted to a binary image such that the different aggregates are represented by distinctly separate white regions. These white regions can then be labeled and the properties such as volume, surface area, and angularity of each aggregate can be computed from the voxel coordinates in each labeled region. The scope of this paper does not include the image processing and analysis methods that are utilized to process relatively good quality/contrast images. Therefore, details of these methods are not presented here and will be published elsewhere. The processing of poor quality images such as image B in Figure 2 using the same traditional image processing methods is time consuming and yield poor results. Therefore, non-traditional computing tools such as the ANN need to be utilized.

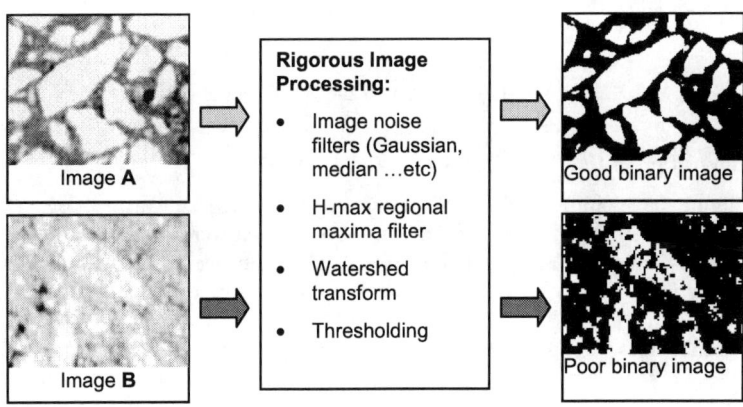

Figure 2. Illustration of the application of traditional image processing to good and poor quality images to obtain binary images.

USE OF ARTIFICIAL NEURAL NETWORKS TO EXTRACT AGGREGATE DATA FROM X-RAY CT IMAGES

In the field of Computer Science, ANNs have been extensively utilized for pattern recognition in images, with special emphasis to the application of face detection (Propp and Samal 1992, Rowley et al. 1998, Sung and Poggio 1998). In this study, an

ANN was developed and trained to recognize coarser aggregates in X-ray CT images of asphalt mixtures. The architecture of the ANN used in this study was similar to those used in the field of face detection.

Figure 3 illustrates the input and expected output of the ANN model. The ANN is trained such that when a box of voxels cropped from the X-ray CT image is input, it recognizes if an entire aggregate is present in the box (i.e., if the box tightly fits one entire aggregate). The ANN should be able to successfully recognize aggregates even if when poor contrast and noisy images are used (e.g., Image B in Figure 2). Furthermore, it should still provide a correct answer even if there are pieces of other aggregates as well as smaller size aggregates are present in the box. By varying the size of the box, aggregates with different sizes should be detected.

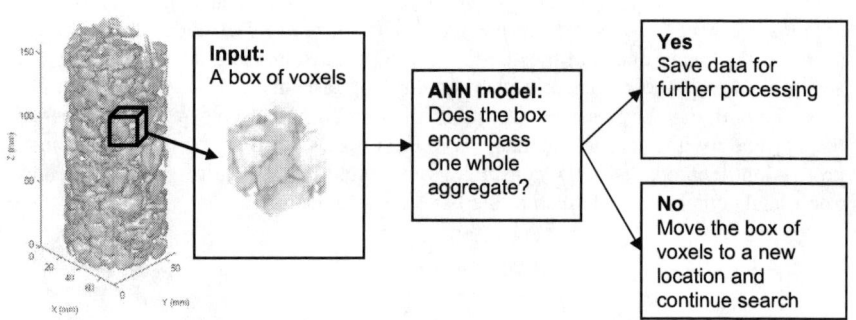

Figure 3. Overview of the ANN model process

Preparation of the Input Dataset for the ANN

Before developing of the ANN structure, the first step was to develop input datasets (for both training and testing the ANN) which were composed of images of *aggregates* and *non-aggregates*. For this, an algorithm was developed in Matlab to manually crop and save boxes (see input 3D image in Figure 3) from the X-ray CT images. Two types of input datasets (sets of boxes) were prepared: (1) *Aggregates:* boxes encompassing only one whole coarse aggregate (the box may also include other smaller size material) and (2) *non-aggregates*: random voxels which does not include one whole coarse aggregate (e.g., composed of sections of coarser aggregates, fine portion of the mixture, and air voids).

X-ray CT images from 11 different specimens were used to obtain 20 *aggregate* and 20 *non-aggregate* input boxes yielding a full dataset of 440 boxes. Out of the 11 specimens, 7 of them were utilized in training the ANN (i.e., total 7 × 40 = 280 input boxes) and 4 of them were used for testing the ANN (i.e., total 4 × 40 = 160 input boxes).

Due to the different size and shape of the aggregates in different specimens, the dimensions of the input boxes were naturally quite different. Additional processing was necessary for preparation of the training and testing datasets (i.e., input image boxes) in order to be able to develop a comprehensive ANN that could recognize aggregates with different sizes. Therefore, all the input boxes were resized using 3D interpolation to give a uniform box size of 20×20×20 voxels (Figure 4).Then, the resized image is reshaped to give a column vector of 8000×1 voxels. This column vector is utilized to train the ANN.

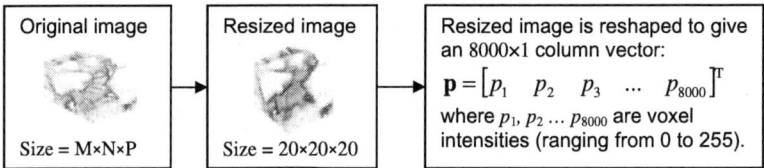

Figure 4. Preparation of the input for the ANN

Structure of the ANN

A feed-forward (backpropagation) network of one hidden layer of 160 neurons and a single neuron output layer was utilized in this study. Figure 5 illustrates the structure of the ANN, where the abbreviated notation suggested by Demuth and Beale (2004) was used for clarity. In Figure 5, $\mathbf{W}^H$ and $\mathbf{W}^o$ are the weight matrices (factors) of hidden and output layers, respectively. Similarly, $\mathbf{b}^H$ and $\mathbf{b}^o$ are the bias vectors of hidden and output layers, respectively. During training of the ANN, weight matrices and bias vectors are adjusted iteratively to produce the desired output.

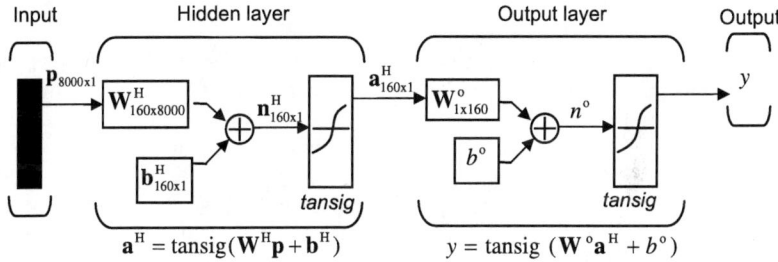

Figure 5. The feed-forward ANN structure

In the ANN shown in Figure5, a scalar output is desired for given input vector $\mathbf{p}$. The ANN is trained such that an arbitrary positive number is generated when an *aggregate* image (i.e., vector $\mathbf{p}$) is input and a negative number is generated when a *non-aggregate* or random image is input. In this study, the values for the output y were selected as $y = 0.9$ for *aggregates* and $y = -0.9$ for *non-aggregate* input images.

Training the ANN

As discussed previously, the input training dataset was prepared such that only two different outputs were obtained from a given input vector of $\mathbf{p}$. If the input $\mathbf{p}$ was an *aggregate* (i.e., a box of voxels that encompass a whole coarse aggregate), y equals 0.9 and if the input is a *non-aggregate* (i.e., a box of voxels that does not encompass a whole coarse aggregate), y equals -0.9. The following steps describe the calculation of y from the ANN and the training of the ANN such that desired output is obtained.

1. The output of the Hidden Layer ($\mathbf{a}^H$) is computed using Eq. 1. The variables in bold letters in Eq. 1a indicate that they are matrices (or vectors) and the multiplication and summation in the equation are matrix operations. The *tansig* function in Eq. 1b, however, is applied to the each element of the vector.

$$\mathbf{n}^H = \mathbf{W}^H \mathbf{p} + \mathbf{b}^H \quad (1a)$$

$$\mathbf{a}^H = \text{tansig}(\mathbf{n}^H) \quad (1b)$$

where $\mathbf{p}$ is the input vector (8000×1), $\mathbf{W}^H$ is the weight matrix (160×8000) and $\mathbf{b}^H$ is the bias vector (160×1) of the Hidden Layer, and the '*tansig*' is the transfer function that is given as:

$$\text{tansig}(x) = \frac{2}{1+\exp(-2x)} - 1 \quad (2)$$

2. Similarly, the output of the Output Layer was computed as follows:

$$n^o = \mathbf{W}^o \mathbf{a}^H + b^o \quad (3a)$$

$$y = \text{tansig}(n^o) \quad (3b)$$

where y is the scalar output of the entire network, $\mathbf{a}^H$ is the output of the Hidden Layer (160×1), $\mathbf{W}^o$ is the weight matrix (1×160) and b^o is the bias constant of the Output Layer.

3. Steps 1 and 2 are repeated for all input (training) boxes to compute a set of y values.
4. The mean square error between the computed y and the target y_t were computed for all images. It is noted that, target y_t values were previously assigned to be either 0.9 or -0.9 for each of *aggregate* and *non-aggregate* input boxes, respectively.
5. Steps 1 through 4 were repeated by adjusting the weight and bias values in the Hidden and Output layers until the overall error computed in step 4 is less than a threshold value. This was accomplished using the scaled conjugate gradient method (*trainscg*) provided in Matlab's Neural Network toolbox (Demuth and Beale, 2004).

Figure 6 shows the change in the mean square error (MSE) as the *epochs* or number of cycles increase during the training. The graph shows both training and validation datasets. The ANN only uses the information from the training dataset and adjusts the weights and biases accordingly. The MSE of the validation images are also computed during training at each *epoch* to make sure that the ANN does not loose its general nature, i.e., it is not over-trained for the given number of training input boxes. Ideally, the training and validation curves in Figure6 should be relatively close to each other until the threshold or performance goal is met. If the training and the validation curves diverge before the training goal is met, it may indicate that the ANN is being over-trained for the given training dataset.

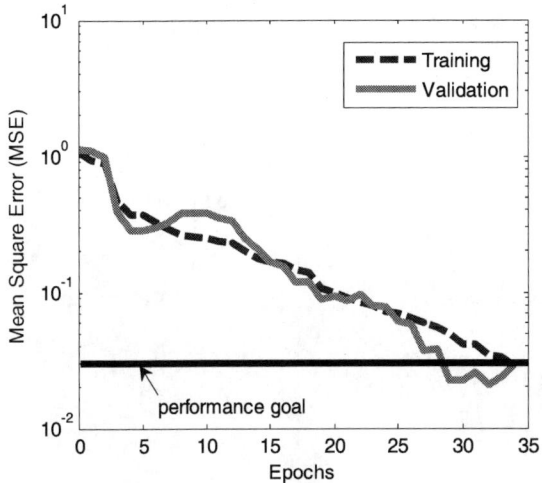

Figure 6. Change in the mean square error (MSE) as the *epochs* or number of cycles increase during the training

Performance of the ANN

The performance of the trained ANN was tested using the input boxes of the specimens not utilized in the training. As mentioned previously, out of the 11 different specimens considered (total 440 input boxes), 7 specimens (280 input boxes) were used in training, and 4 specimens (160 input boxes) were used for testing the performance of the ANN. After the ANN was trained, the weights and bias vectors were used to calculate the output scalars (y) of the testing images, then, the output was compared with the known target values (y_t).

Figure 7 shows the performance of the trained ANN in predicting whether the given testing input box is an *aggregate* or a *non-aggregate* image. For each specimen in Figure7, the top graph shows ANN simulated y versus the known y_t values which were either 0.9 for *aggregate* or -0.9 for *non-aggregate* images.

The success of the ANN was determined through the analysis of simulated y values; if the simulated y is greater than 0, algorithm classifies the image as *aggregate*, and if y is less than 0, the image is classified as *non-aggregate*. Figures 7a and 7b indicate a 100% success for those particular specimens, whereas Figures 7c and 7d indicate 95% success in the predictions. The 5% error in each of Figures 7c and 7d indicate that 2 out of 40 input boxes for each represented specimen were classified as *aggregate* while they were in-fact *non-aggregate* images. A possible explanation for this behavior is that smaller aggregates could have been present in the *non-aggregate* image boxes in those particular instances. In general, however, the performance of the

ANN can be classified as exceptional.

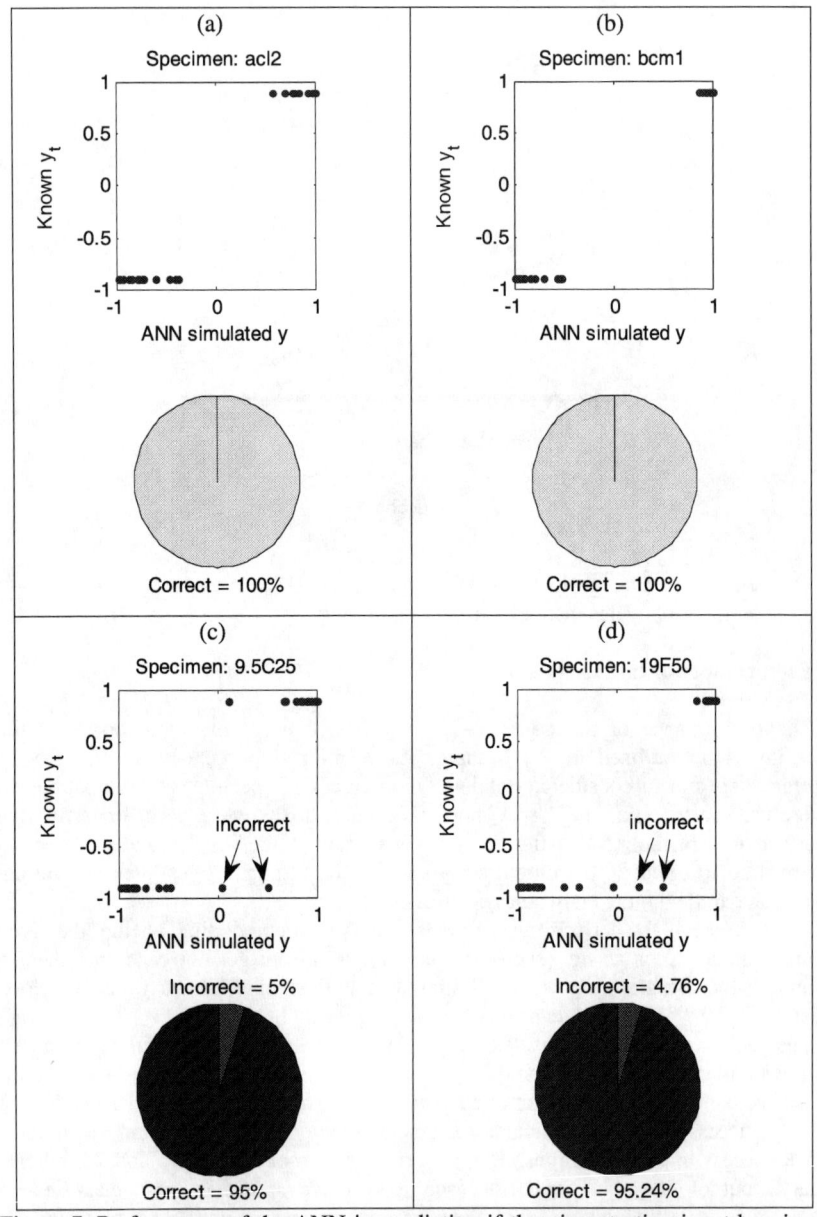

Figure 7. Performance of the ANN in predicting if the given testing input box is an *aggregate* or *non-aggregate* image for four different specimens.

Detecting the Location and Sizes of Coarse Aggregates using ANN

The ultimate goal of developing the described ANN is to be able to utilize it for detecting the location and size of aggregates in a 3D X-ray CT image of an entire asphalt mixture specimen. This will be accomplished by moving a box (with a certain size) through the entire image (see Figure 3) and using the ANN determine if there is an aggregate tightly fitting in the given box. By varying the size of the box, aggregates of different sizes will be detected. In addition, the method will be able to provide the location and size of the aggregate bounding-box, which can be used to approximately determine the sieve size in which the aggregate in retained and develop the grain size distribution of the coarse aggregates. Also, the locations of the aggregates can be used to determine the contact points between different aggregates.

CONCLUSIONS

A feed-forward artificial neural network (ANN) was developed as a part of an effort towards the goal of extracting coarse aggregate properties from the 3D X-Ray CT images of asphalt specimens. The developed ANN is able to determine if there is a tightly fitting aggregate in a box of voxels extracted from an X-ray CT image. The ANN was able to perform successfully even when poor contrast/poor quality X-ray CT images were input. This paper describes the development and training procedure of the ANN and presents examples of its application to asphalt mixtures specimens.

The results of this research are part of an ongoing study, where image analysis and ANN methods are being developed to directly measure the 3D microstructure characteristics (such as aggregate contact points, orientation, size distribution, segregation, etc.) of asphalt mixture specimens. The goal is to study the microstructure characteristics, including the aggregate angularity, sphericity, segregation, orientation and aggregate-to-aggregate contact points in asphalt specimens made with a wide variety of material compositions and prepared using different types of laboratory and field compactors.

ACKNOWLEDGMENTS

This research was performed while Edith Arambula held a National Research Council Research Associateship Award at the Federal Highway Administration Turner-Fairbank Highway Research Center.

REFERENCES

Buchanan, M.S. and Brown, E.R. (2001). "Effect of Superpave gyratory compactor type on compacted hot-mix asphalt density." *Transportation Research Record: Journal of the Transportation Research Board*, No. 1671: 50-60.

Ceylan, H., Gopalakrishnan, K. and Guclu A. (2007). "Advanced Approaches to Characterizing Nonlinear Pavement System Responses." *Transportation Research*

Record: *Journal of the Transportation Research Board*, No. 2005: 86-94.
Choi, J.-H., Adams, T.M. and Bahia, H.U. (2004). "Pavement roughness modeling using back-propagation neural networks." *Computer-Aided Civil and Infrastructure Engineering*, Vol. 19 (4): 295-303.
Demuth, H. and Beale, M. (2004). *Neural Network Toolbox for use with Matlab (User's guide)*. Version 4. The Mathworks, Inc., Natick, MA.
Gucunski, N. and Krstic, V. (1996). "Backcalculation of pavement profiles from spectral-analysis-of-surface-waves test by neural networks using individual receiver spacing approach." *Transportation Research Record: Journal of the Transportation Research Board*, No. 1526: 6-13.
Huang, C., Najjar, Y.M. and Romanoschi, S.A. (2007). "Predicting asphalt concrete fatigue life using artificial neural network approach." Transportation Research Board Annual Meeting, Washington, DC.
Khan, Z.A., Wahab, H.I.A.-A., Asi, I. and Ramadhan, R. (1998). "Comparative study of asphalt concrete laboratory compaction methods to simulate field compaction." *Construction and Building Materials*, Vol. 12: 373-384.
Kim, H., Rauch, A.F. and Haas, C.T. (2004). "Automated quality assessment of stone aggregates based on laser imaging and a neural network." *Journal of Computing In Civil Engineering*, Vol. 18 (1): 58-64.
Kim, Y. and Kim, R. (1998). "Prediction of layer moduli from falling weight deflectometer and surface wave measurements using artificial neural network." *Transportation Research Record: Journal of the Transportation Research Board*, No. 1639: 53-61.
Masad, E., Jandhyala, V.K., Dasgupta, N., Somadevan, N. and Shashidhar, N. (2002). "Characterization of air void distribution in asphalt mixes using x-ray computed tomography." Journal of Materials in Civil Engineering, Vol. 4 (2): 122-129.
Peterson, R.L., Mahboub, K.C., Anderson R.M., Masad, E. and Tashman, L. (2004). "Comparing Superpave gyratory compactor data to field cores." *Journal of Materials in Civil Engineering*, Vol. 16 (1): 78-83.
Propp, M. and Samal, A. (1992) "Artificial neural network architecture for human face detection". *Intell. Eng. Systems Artificial Neural Networks* 2, pages 535–540.
Rowley, H., Baluja, S. and Kanade, T.(1998). "Neural network-based face detection" In *IEEE Patt. Anal. Mach. Intell.*, volume 20, pages 22–38.
Saadeh, S., Tashman, L., Masad, E. and Mogawer, W. (2002). "Spatial and directional distribution of aggregates in asphalt mixes." *Journal of Testing and Evaluation*, Vol. 30 (6): 1-9.
Sung, K. and Poggio, T. (1998) "Example-based learning for view-based face detection", In *IEEE Patt. Anal. Mach. Intell.*, volume 20, pages 39–51.
Tarefder, R.A., White, L. and Zaman, M. (2005). "Development and application of a rut prediction model for flexible pavement." *Transportation Research Record: Journal of the Transportation Research Board*, No. 1936: 201-209.
Tarefder, R.A, White, L. and Zaman, M. (2005b). "Neural network model for asphalt concrete permeability." *Journal of Materials in Civil Engineering*, Vol. 17 (1) 19-27.
Tashman, L., Masad, E., D'angelo, J., Bukowski, J. and Harman, T. (2002). "X-ray tomography to characterize air void distribution in Superpave gyratory compacted

specimens." *The International Journal of Pavement Engineering*, Vol. 3 (1), pp. 19-28.

Tutumluer, E. and Seyhan, U. (1998). "Neural network modeling of anisotropic aggregate behavior from repeated load triaxial tests." *Transportation Research Record: Journal of the Transportation Research Board*, No. 1615: 86-93.

A Probabilistic Approach to Account for Temperature Impact on Flexible Pavement Stiffness

Sameh Zaghloul, Ph.D., P.E., P.Eng.,[1] Nicholas Vitillo, Ph.D.[2], and T. Joseph Holland, Ph.D., P.E.[3]

[1]H.W. Lochner Inc, 310 Fullerton Ave, Suite 200, Newburgh, NY 12550; PH (845) 863-0960; FAX (845) 863-0961; e-mail: szaghloul@hwlochner.com
[2]Research Consulting Services, LLC, 61 Springwood Ct, Princeton, NJ 08540-9404; PH (609) 915-4556; FAX (732) 329-8328; e-mail: nvrcs@comcast.net
[3]California Department of Transportation, Division of Research and Technology, MS # 5, 5900 Folsom Blvd, Sacramento, CA 95819; PH (916) 227-5835; FAX (916) 227-7075; e-mail: t_joe_holland@dot.ca.gov

ABSTRACT

In recent years, recognition of the influence of environmental factors on pavement performance has greatly increased and many research studies have focused on this area. Although there is agreement on the significance of temperature and moisture impact on pavement performance, only limited field measured data is available to quantify its magnitude. Most available data is based on laboratory studies, such as the impact of temperature on asphalt mix stiffness. However, in reality pavements perform as one structure, and overall pavement performance is of more interest than the performance of individual layers.

In this paper, field data collected from two New Jersey and California pavement sections is used to develop a probabilistic approach to address the daily changes in pavement stiffness and how these changes can be considered at the design stage. This data was collected as a part of field studies performed to quantify the impact of temperature and moisture on pavement performance. This probabilistic approach yielded distributions that were extremely similar for the New Jersey and California sections, despite the major differences in environmental zone, temperature range, and pavement structure between the sections.

BACKGROUND

Pavement stiffness is a dynamic parameter that changes on an hourly, monthly, and yearly basis, as a result of environmental changes and deterioration due to traffic. The concept of assessing pavement damage based on the pavement stiffness when loads are applied is very valid and has a sound background. This approach has been adopted by the Mechanistic-Empirical Pavement Design Guide (MEPDG) in which the Enhanced Integrated Climatic Model (EICM) is used to predict the hourly changes in environmental parameters. These predictions are then used to adjust the pavement stiffness. As a result, the pavement stiffness used to calculate the expected damage due to traffic loads also varies hourly. Previous studies [Ahmed et.al. 2005;

Zaghloul et.al. 2006] have indicated that more validation is required for EICM predictions, especially with the significant impact it has on the MEPDG outcomes.

Although there is global agreement on the significance of temperature and moisture impact on pavement stiffness, there is only limited field measured data available to quantify the magnitude of this impact on the overall pavement performance. Most of the available data is based on laboratory studies, such as the impact of temperature on the asphalt mix stiffness. However, in reality pavements perform as one structure and the overall pavement performance and service life is of more interest than the performance of individual layers.

In a study funded by the New Jersey Department of Transportation (NJDOT) and the Federal Highway Administration (FHWA) to develop temperature and seasonal adjustment models that suit New Jersey conditions, twenty-four test sections were instrumented to continuously measure environmental and climatic parameters. Deflection testing (using the Falling Weight Deflectometer (FWD)) and seismic testing (using the Seismic Pavement Analyzer (SPA)) were performed on the test sections on a monthly basis for two years. In addition, 24-hour testing cycles (testing every hour for a 24-hour period) were performed on selected sections to capture the short-term changes in pavement stiffness. The data collected in this study was analyzed to determine the daily and seasonal changes in pavement stiffness.

In another study performed for the California Department of Transportation (Caltrans) to develop seasonal adjustment and temperature correction models based on California conditions, 18 pavement sections located in California's 6 main environmental zones were selected for frequent FWD testing. The FWD testing program consisted of monthly testing to monitor the seasonal changes (month-to-month) and 24-hour testing (testing every 2 hours for a 24-hour period) on selected sections to monitor the short-term variability, mainly due to temperature. Three pavement classes were considered in the study: thin flexible, thick flexible, and rigid pavements. As a minimum, each test section included 11 test locations in order to have a large enough dataset. The models developed in this study will be used to account for the temperature and seasonal variations among FWD tests performed under different environmental conditions.

Daily Changes in Pavement Stiffness

The daily changes in pavement stiffness were investigated using data collected during 24-hour FWD testing cycles at one flexible pavement section from each of the above two studies. In these testing cycles, FWD testing was repeated every hour (for the NJDOT section) or every 2 hours (for the Caltrans section) on the same locations (11 locations) for a 24-hour period (24 measurements for each location for the NJDOT section and 12 measurements for the Caltrans section). Backcalculation analysis was performed to estimate the in-situ pavement structural capacity in terms of the maximum number of load repetitions the pavement can carry before it reaches a terminal condition (N).

Figures 1 and 2 show the temperature variation within the 24-hour period and the corresponding N values, respectively, for one test location from the NJDOT section. As can be seen, during the 24-hour period the pavement temperature ranged from 8°C to 20°C, while N ranged from 13.2 million to 23.2 million. Although the change in temperature was not extreme (only 12°C and not in the very hot or very cold regions), the impact on N was significant.

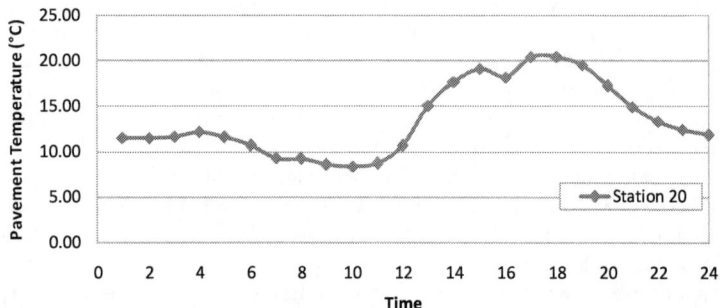

Figure 1: Pavement Temperature during 24-Hour Period (NJDOT Section)

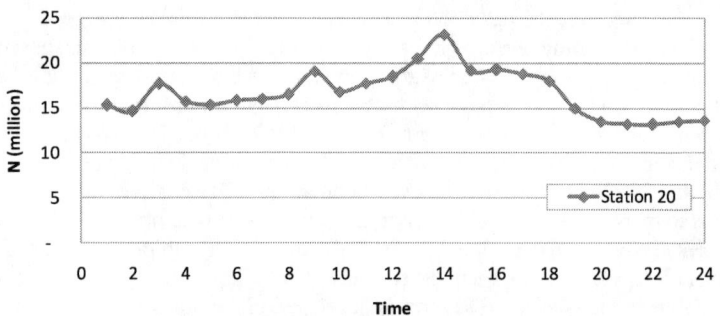

Figure 2: Maximum Number of Load Repetitions (NJDOT Section)

Similar analysis was performed on one of the Caltrans test sections that was also tested for 24 hours. Figures 3 to 4 show the pavement temperature variation within the 24-hour period and N, respectively, for one test location from the Caltrans section. As can be seen, during the 24-hour period the temperature ranged from 15°C to 30°C, while N changed from 36.6 million to 79.8 million, i.e. it more than doubled.

It should be noted that for both NJDOT and Caltrans studies, all repeated tests were performed using the same equipment and testing protocols, i.e. the equipment contribution to the difference among repeated tests is very small.

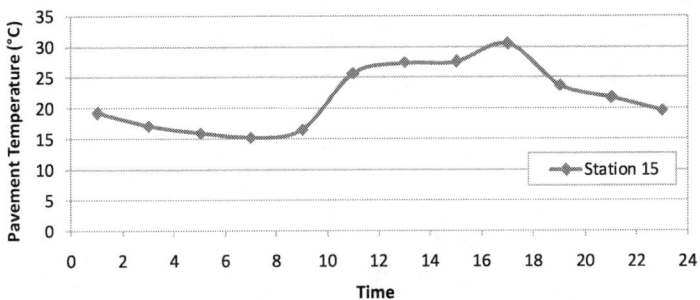

Figure 3: Pavement Temperature during 24-Hour Period (Caltrans Section)

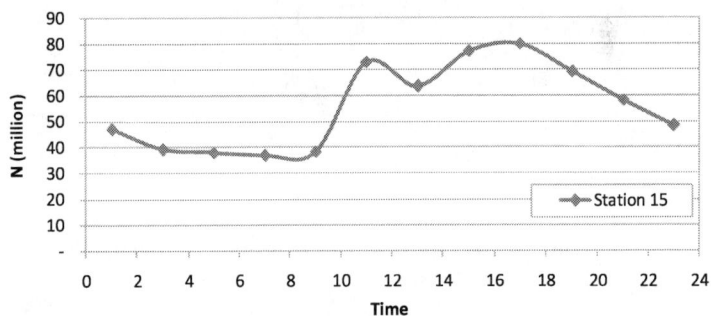

Figure 4: Maximum Number of Load Repetitions (Caltrans Section)

Analysis Approach

The results of the 24-hour testing cycles of NJDOT and Caltrans sections indicate that the changes in N within a 24-hour period can be significant – in the above cases it almost, or more than, doubled. These changes are related mainly to changes in temperature because tests were performed using the same FWD equipment and testing protocol. As can be seen from Figures 2 and 4, the in-situ pavement structural capacity for a single point, in terms of N, changed significantly within a 24-hour period. Therefore, it would be more appropriate to present N as a probability distribution, instead of a single value.

As such, a probabilistic approach was used to address the impact of temperature changes on pavement stiffness, as outlined in Figures 5 and 6. In this approach, backcalculation analysis was performed to calculate N for all stations within the section and all testing rounds (hours) within the 24-hour period. The result of this analysis was a dataset that includes, in the case of the New Jersey section, 264 data points (11 locations and 24 rounds of testing). Since the main variable among all these data points is the time of testing, and hence the temperature during the testing, it is assumed that the variation in N is due to changes in temperature. A normal distribution was then used to present this dataset. An example of this distribution is shown in Figure 6. In this distribution, the probability that the pavement will carry N equal to or less than 3.8 million is represented by 1 minus the hatched area (1 - 16% = 84%).

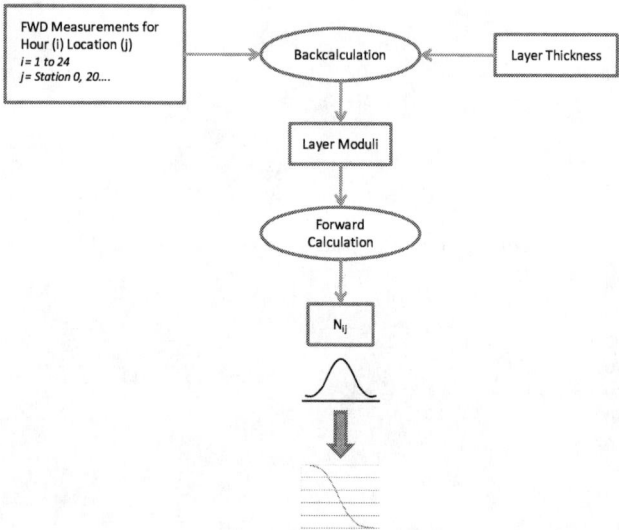

Figure 5: Procedure for Probabilistic Approach

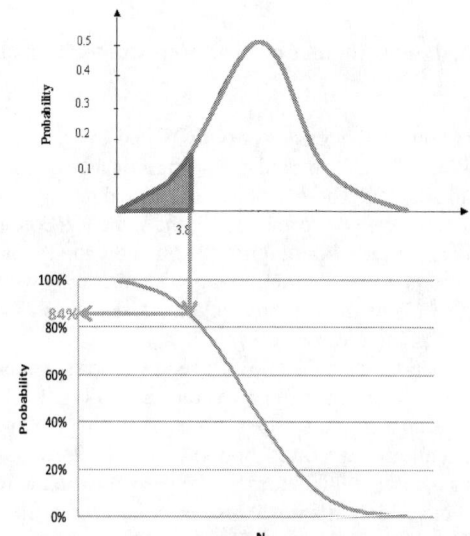

Figure 6: Conversion to Accumulated Probability Distribution

Figure 7 shows the probability distribution of N for Station 20, the same location shown in Figure 1, from the NJDOT study. Figure 8 shows the individual probability distribution of N for all locations tested within the NJDOT section. As can be seen, there is significant difference between the locations along the test section. This variation is mainly attributed to construction related issues, such as layer thickness and subgrade condition. The impact of construction variability on pavement performance is not the main focus of this paper and has been discussed in other references [Zaghloul et. al. 2005].

Figure 9 shows the overall probability distribution of N for all locations within the NJDOT test section. The number of data points used to construct this distribution is 264 data points (11 test locations x 24 measurements per location). As can be seen, the probabilities that N is equal to or less than 8, 15 and 20 million are about 74%, 10% and 0.5%, respectively.

Since the main focus of this paper is the effect of temperature changes on the pavement structural capacity, rather than the effect of construction variability on the pavement structural capacity, the probability distribution of the overall section will be used in the subsequent analysis. If the pavement shown in Figure 9 was designed and built for N equal to 10 million, then the probability that the pavement will carry this traffic is only 53.7%. It should be noted that this is as a result only of the daily temperature changes. This is because the data used to construct this distribution is collected from repeated tests performed using the same equipment and testing protocols and within 24-hour period, i.e. no expected pavement deterioration due to traffic.

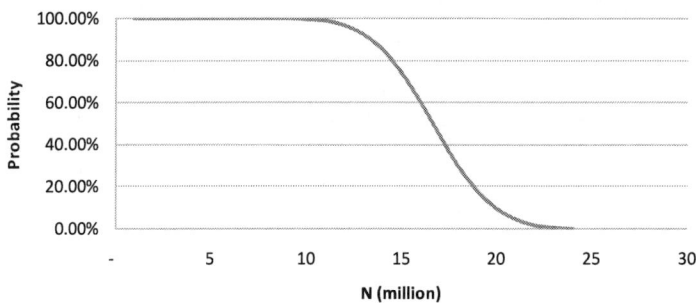

Figure 7: Probability Distribution of N (NJDOT section) - Station 20

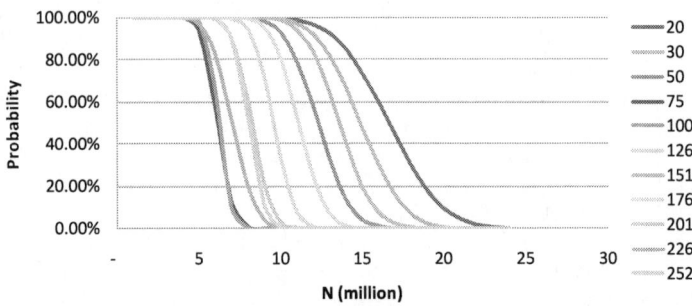

Figure 8: Probability Distribution of N (NJDOT section) - All Stations

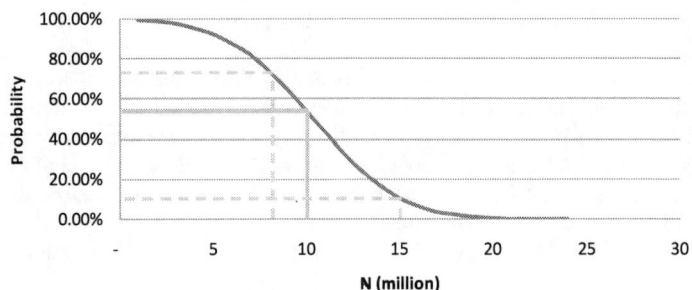

Figure 9: Probability Distribution of N (NJDOT section) - All Locations

If the impact of temperature is ignored in the design, then the probability that the pavement will survive the design traffic will be low. The impact on the probability distribution of N of adding an additional one inch of asphalt to the pavement structure is shown in Figure 10a. As can be seen from this figure, the probability of N being equal to or less than 10 million for the pavement structure with the additional one inch of asphalt, Figure 10a, is 89.4%. If the pavement structure is made thicker by only half an inch of asphalt, then the probability of N equal to or less than 10 million is 77.2%, as shown in Figure 10b.

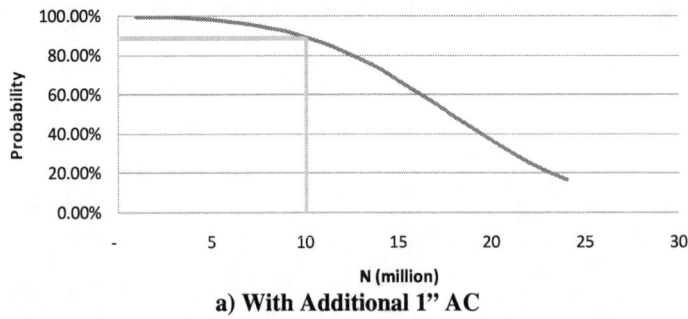

a) With Additional 1" AC

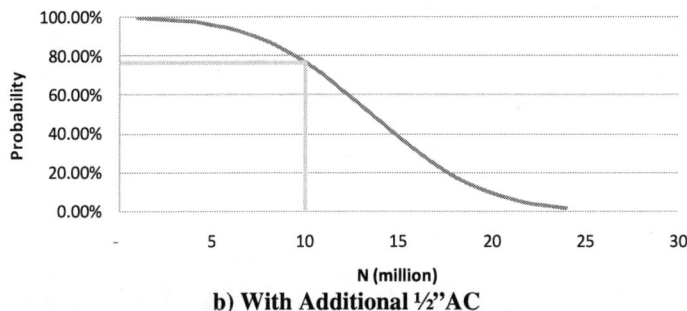

b) **With Additional ½"AC**

Figure 10: Probability Distribution of N for All Locations (NJDOT section)

The same analysis was repeated on the Caltrans section. Figure 11a shows the overall probability distribution of N for all the tested locations within the Caltrans section. Based on this figure, if this pavement was designed and built to carry 30 million, then the probability that the pavement will carry this traffic (i.e. that N will be equal to or less than 30 million is about 62%). The probabilities of N being equal to or less than 20 million or 45 million are about 89% and about 14%, respectively.

Figures 11b and c show the probability distributions of N if the pavement structure of the Caltrans section is made thicker by one inch and one half-inch of asphalt, respectively. The probability that N is equal to or less than 30 million for the pavement structure with the additional one inch of asphalt, is about 93%, while this number for the pavement with an additional one half-inch of asphalt is about 83%.

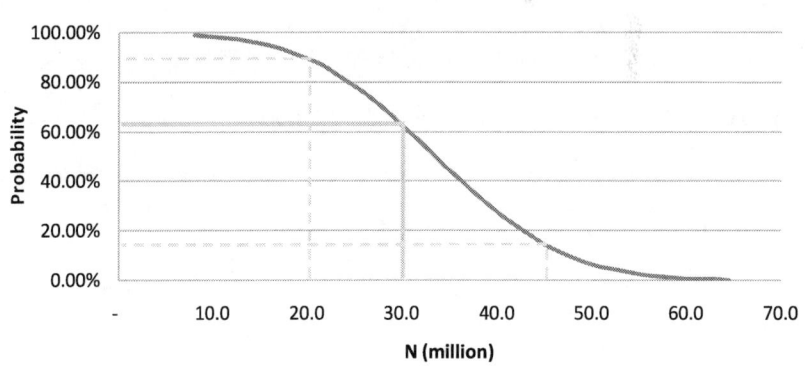

a) **Without Additional AC**

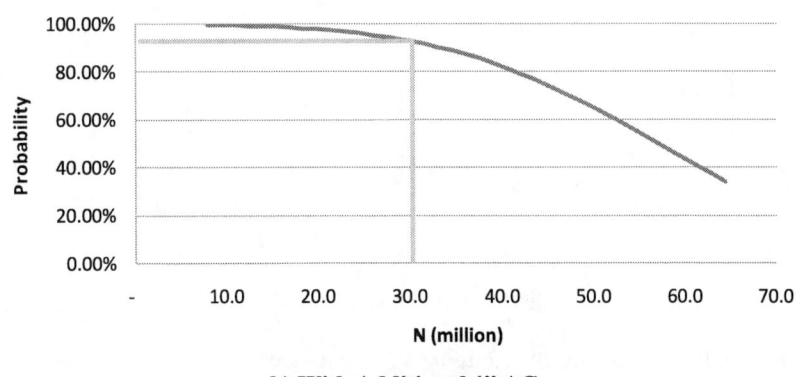

b) With Additional 1" AC

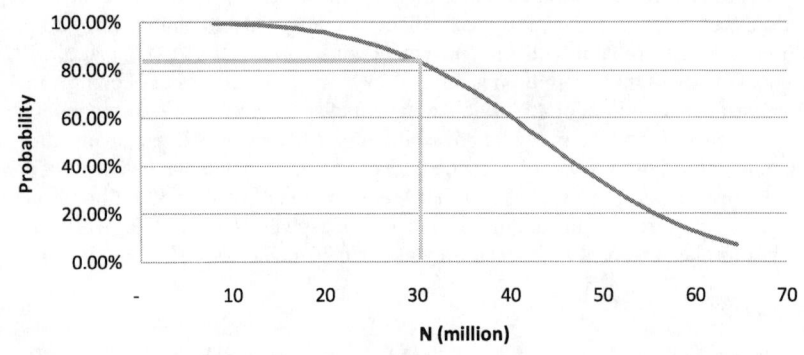

c) With Additional ½" AC

Figure 11: Probability Distribution of N for All Locations (Caltrans Section)

The results of the analysis performed on the 24-hour testing cycles conducted on two test sections, one located in New Jersey and the other one located in California, highlighted that the impact of daily temperature changes is significant and can reduce the probability of the pavement surviving the design traffic. Since these results are site specific, additional analyses were performed by using the ratio between the in-situ structural capacity (in terms of N backcalculated from FWD) and the designed structural capacity (in terms of N used in the design), i.e. N_{FWD}/N_{Design}. Figures 9 and 11a were reproduced using N_{FWD}/N_{Design} and presented in Figures 12a and b.

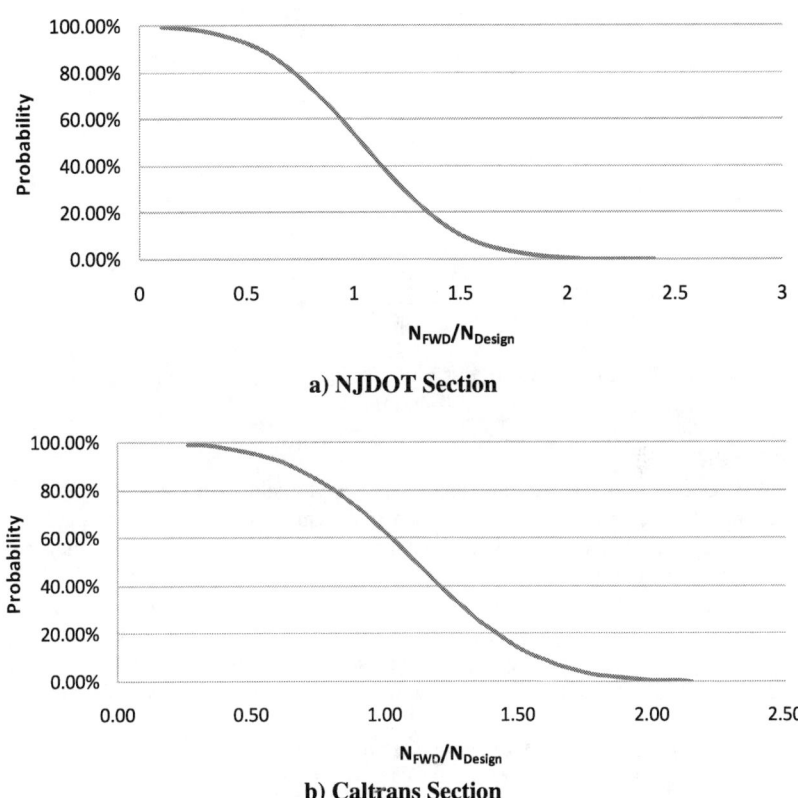

a) NJDOT Section

b) Caltrans Section

Figure 12: Probability Distribution of N_{FWD}/N_{Design} for All Locations

A comparison was made between the results of the NJDOT and Caltrans sections to evaluate the dependency of the results on the environmental region, range of temperature, and pavement structure. As mentioned earlier, the temperature range within the 24 hours of testing was from 8°C to 20°C for the NJDOT section and from 15°C to 30°C for the Caltrans section. Figure 13 shows the results of this comparison. As can be seen, the probability distributions for the two sections are very close despite the fact that these two sections are thousands of miles apart and the temperature ranges they were tested at are significantly different. In addition, the pavement structure and structural capacity of these two sections are significantly different.

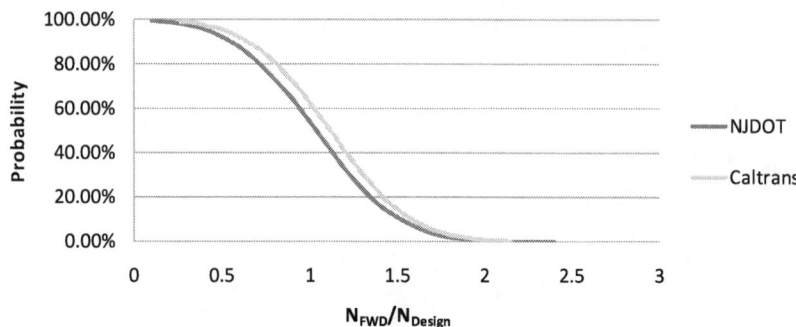

Figure 13: Probability Distribution of N_{FWD}/N_{Design} - NJDOT & Caltrans Sections

DISCUSSION

The analysis results presented in this paper indicated that the daily changes in pavement stiffness could be very significant. The results of the comparison made between the NJDOT and Caltrans sections indicated that although there are major differences in the environmental zone, temperature range, and pavement structure between these two sections, the probability distributions of both sections are very close (Figure 13). The results of this comparison are promising, but not conclusive enough. The analysis results presented in this paper are based on a very limited dataset; therefore more verification and validation is recommended before making generalizations.

After verification and validation, the presented approach could be used as a tool at the design stage to account for daily temperature changes by developing a relationship between the increase in the probability of the pavement surviving the design traffic ($N_{Original\ Design}$) as a result of slightly increasing the design traffic volume ($N_{Revised\ Design}$). An example of this relationship is presented in Figure 14. In this figure, the average of the NJDOT and Caltrans probability distributions (Figure 13) is used to develop a relationship between the change in design traffic, in terms of ($N_{Revised\ Design} - N_{Original\ Design})/N_{Original\ Design}$ and the change in the probability of the pavement surviving the design traffic ($N_{Original\ Design}$). This type of relationship could be used to perform a sensitivity analysis on the designed pavement to investigate the improvement that could be made by slightly changing the pavement structure. In other words, it could be used at the design stage to reduce the chance that the pavement will fail, due to environmental effects, before it reaches the desired age (design traffic).

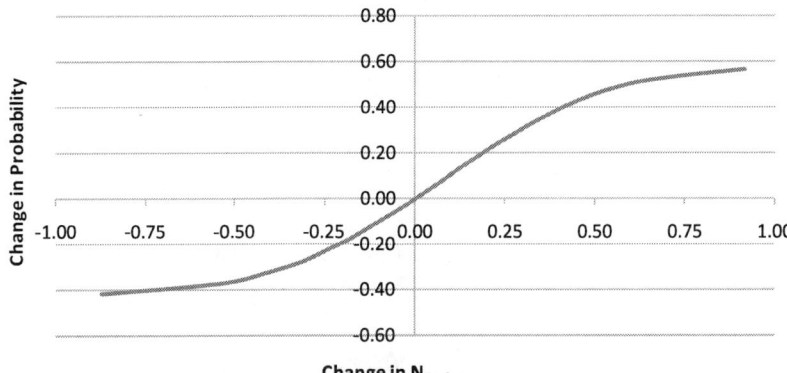

Figure 14: Relationship between Changes in Design Traffic and Probability of Pavement Surviving

SUMMARY, CONCLUSIONS AND RECOMMENDATIONS

The data from two test sections located in New Jersey and California, included in seasonal studies performed for the New Jersey and California Departments of Transportation, was considered in this paper. These two sections were repeatedly tested using the FWD for 24-hour periods to capture the short-term changes in pavement stiffness. In this scenario, FWD tests were repeated at the exact same locations every 1 or 2 hours for a 24-hour period. The same FWD equipment was used, and the same testing protocol followed, in all tests to eliminate the equipment as a source of variation. Backcalculation analysis was performed on the FWD data to determine the in-situ structural capacity in terms of the maximum number of load repetitions the pavement can carry before it reaches a terminal condition (N). The analysis results from both the NJDOT and Caltrans sections indicated that the change in N due to temperature changes can be very significant.

Probability distributions were developed for the NJDOT and Caltrans sections to explain the probability that each section will survive a certain N. These distributions were then used to develop a probability distribution for the ratio between the value of N that a section can survive and the N_{Design}. The results of the comparison made between the probability distributions of NJDOT and Caltrans sections indicated that despite the difference in location, environmental zone, temperature range during testing, and pavement structure, the probability distributions of NJDOT and Caltrans sections were in excellent agreement.

An approach was also presented that could be used as a tool to account for daily temperature changes at the design stage by developing a relationship between the increase in the probability of the pavement surviving a design traffic ($N_{Original\ Design}$) as a result of slightly increasing the design traffic volume ($N_{Revised\ Design}$). It should be noted that although the analysis results presented in this paper are very promising, they are not conclusive enough. The analysis results presented in this paper are based on a very limited dataset; therefore it is not recommended to generalize further without more verification and validation.

REFERENCES

Ahmed, Z., Marukic, I., Zaghloul, S. and Vitillo, N. (2005). "Validation of Enhanced Integrated Climatic Model Predictions Using New Jersey Seasonal Monitoring Data." *TRR*, 1913, 148-161.

Zaghloul, S., Ayed, A., Abd El Halim, A., Vitillo, N. and Sauber, R. (2006). "Investigations of Environmental and Traffic Impacts on MEPDG Predictions." *TRR*, 1967, 148-159.

Zaghloul, S., Helali, K., Ahmed, R., Ahmed, Z. and Jumikis, A. (2005). "Implementation of Reliability-Based Backcalculation Analysis." *TRR*, 1905, 97-106

Speed up Discrete Element Simulation of Asphalt Mixtures with User-written C++ Codes

Yu Liu[1] and Zhanping You [2]

[1] Ph.D. Student and Research Assistant, Michigan Technological University, Department of Civil and Environmental Engineering, Houghton, Michigan 49931, USA. Tel: (906)487-2528, Fax: (906)487-1620, Email: yul@mtu.edu

[2] Tomasini Assistant Professor of Transportation Engineering and Director of the Transportation Materials Research Center, Department of Civil and Environmental Engineering, Michigan Technological University, Houghton, Michigan 49931, USA. Tel: (906)487-1059, Fax: (906)487-1620, Email: zyou@mtu.edu

ABSTRACT: Compared with an embedded programming language "FISH" within the Particle Flow Code (PFC), the user-written C++ code is an optional feature in the PFC. This optional feature is an advanced application and can improve simulation speeds. In this paper, the user-written C++ codes for two and three dimensional models were developed. The discrete element simulation of asphalt mixtures were conducted with user-written C++ codes along with FISH programs. The simulation results from the two methods were compared. It was found that the user-written C++ program can speed up discrete element simulation, where the speed ratios (C++/FISH) are from 2.34 to 4.95 and the averaged value is 3.74.

INTRODUCTION

An asphalt mixture is composed of asphalt, graded aggregates, and air voids. The overall macro-mechanical behaviors of the asphalt mixture are determined by the micromechanics within the cementitious particulate system. Based on the heterogeneous multiphase nature of the asphalt material, it appears that a micromechanical model is more suitable to properly characterize such a material. In recent years, a number of research studies on micromechanical analysis of asphalt mixtures have been conducted using discrete element models. Among the studies, the digital samples of asphalt mixtures were built with the highly idealized model (Collop et al. 2007; Collop et al. 2006; You 2008), randomly created polygon or polyhedron model , and image-based model (Abbas et al. 2007; Buttlar and You 2001; Tian et al. 2007a; Tian et al. 2007b; You and Buttlar 2005; You and Buttlar 2004; You and Buttlar 2006). The discrete element mechanical models included the elastic contact model (Buttlar and You 2001; Collop et al. 2007; Dai and You 2007; You and Buttlar 2005; You and Buttlar 2004; You and Buttlar 2006), the bilinear cohesive model (Kim and Buttlar 2005), and the time-dependent viscoelastic contact model (Collop et al. 2007); (Abbas et al. 2007; Abbas et al. 2005; Chang and Meegoda 1997; Liu et al. 2007) . With the digital samples and the corresponding micromechanical models, the DE modeling was conducted to predict the stiffness (or modulus) (Abbas et al. 2007; Dai and You 2007; Liu and You 2008; You and Buttlar 2005; You and Buttlar 2004) and fatigue properties of asphalt mixtures (Carmona et al. 2007).

Although many studies on discrete element simulation of asphalt mixtures have been very successful, the following issues have to be considered: 1) on one hand, in order to represent the accurate geometrical characteristics of asphalt mixtures, the discrete elements should be as small as possible, which increases the total number of discrete elements and slows down computation speeds, and; 2) on the other hand, in order to simulate the visco-elastic behaviors of asphalt mixtures, the real physical time has to be considered, which makes discrete element simulation very time-consuming. For example, simulation for a typical dynamic modulus test using elastic models can be conducted within several days, while the viscoelastic model simulation needs several months or more. To solve these two issues, the following approaches can be considered: (1) use advanced computers or supercomputers; (2) a newly developed programming method, and; (3) time-temperature superposition. This paper presents a methodology to speed up viscoelastic simulation of asphalt mixtures by developing programs with user-written C++ code which is an optional feature of Particle Flow Codes in two Dimensions and three Dimensions (PFC2D/3D).

DISCRETE ELEMENT PACKAGE: PARTICLE FLOW CODES

Particle Flow Codes in two Dimensions and three Dimensions (PFC2D/3D) was commonly used in the literatures cited above. The internal standard codes in PFC2D/3D are written with the C++ programming language. FISH is a programming language embedded within PFC2D/3D for users to write their own programs. FISH programs are written as a text file and called by PFC2D/3D standard command. Functions and variables in the FISH program can be utilized as the standard commands and variables of PFC2D/3D. Because functions or variables in FISH are not written in the C++ programming language, they have to be interpreted and compiled before they can be executed. Therefore, it is very time-consuming in the situation where many PFC variables and functions must be manipulated and fed back to PFC2D/3D during computation time. In addition to FISH, user-written C++ code is an optional feature in PFC2D/3D, which allows users to write their programs in the C++ language and then create private PFC2D/3D executables. Because user-written C++ functions can have access to many variables or functions in the source program of PFC2D/3D, they can be used instead of FISH program to speed up execution considerably. Before applying the user-written C++ code, however, knowledge on the codes within the PFC2D/3D and C++ programming skills are necessary. Therefore, the user-written code is not commonly used even though it can speed up computation considerably. Simulation of asphalt mixtures are very time-consuming as discussed above: viscoelastic simulations take a few months or more with FISH programs. In order to speed up simulation, this paper develops discrete element simulation programs in the C++ programming language instead of FISH.

USER-WRITTEN C++ CODE FOR DISCRETE ELEMENT MODELING

With the optional feature (user-written C++ code), private PFC2D and PFC3D executables were created and named with PFC2D_user 3.10 and PFC3D_user 3.10. These two executables can be regarded as the user-developed version of PFC2D/3D which are more powerful than PFC2D/3D: they have both user-defined and PFC2D/3D functions. In this research, the following user-developed functions were added in the executables: (1) build digital samples; (2) define material properties, and; (3) apply loading conditions.

Build Digital Samples

The asphalt mixture was regarded as a composite of asphalt mastics (asphalt binder plus fine aggregates), coarse aggregates, and air voids in this study. The microstructures of asphalt mixtures were rebuilt with discrete elements (disks for 2D) or balls for 3D):

(1) Rectangular or cylinder region filled with discrete elements was employed to represent an asphalt mixture sample.

(2) Irregular polygons (2D) or polyhedrons (3D) were randomly created according aggregate gradation.

(3) Discrete elements within the irregular polygons or polyhedrons were bonded together to simulate coarse aggregates, where the interactions between discrete elements were represented with the elastic contact models.

(4) Discrete elements out of the irregular polygons or polyhedrons were bonded together to simulate the two-phase system: asphalt mastic plus air voids. The interactions between discrete elements were simulated with viscoelastic contact models.

(5) Asphalt mastic and air voids were separated by removing discrete elements randomly to form air voids.

The detailed procedure to build the digital sample of asphalt mixtures is included in the previous paper. The final digital samples as shown in Figure 1 were created with the private PFC2D and PFC3D executables: PFC2D_user 3.10 and PFC3D_user 3.10, respectively. Compared with the FISH program, the computation speed was not obviously improved in this case. The reason is that only several functions were manipulated during computation time.

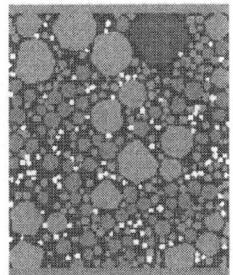 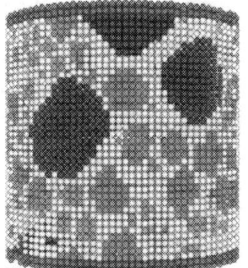

(a) 2D Model from PFC2D_user 3.10   (b) 3D Model from PFC3D_user 3.10

Figure 1. Digital Samples of Asphalt Mixture with User-Written C++ Code

Defining Material Properties

Within the private PFC2D and PFC3D executables, four kinds of interactions within the asphalt mixtures were considered: interactions between aggregates, within aggregates, within asphalt mastics and between aggregates and the mastic. The contact-stiffness model, the contact-bond model, the slip model, and the Burger's contact model (Itasca Consulting Group 2004a; Itasca Consulting Group 2004b; Liu and You 2008) were applied to build the constitutive behaviors at each contact: the contact-stiffness model was employed to represent the interaction between or within aggregates, and the Burger's contact model was employed to characterize the interactions within asphalt mastic or between the mastic and aggregates. The contact-bond model was employed to represent the tensile strength between discrete elements and the slip model was activated when the slip exists at the contact points. Among these four models, Burger's model is an alternative model and the other three models are built-in models (Itasca Consulting Group 2004a; Itasca Consulting Group 2004b). Parameters of built-in models are properties of each discrete element which are constants during simulation. An alternative model associates its parameters directly with the contacts. Therefore, when using an alternative model, such as the Burger's contact model herein, one furthermore consideration is the way in which the model is installed at a new contact. Usually, FISHcall is set so that the creation of a new contact will automatically trigger the calling of a FISH function which installs the Burger's contact model dynamically. In this research, instead of FISHcall, a CPPcall is set to trigger the calling of a user-defined C++ function for installing the Burger's contact model at a new created contact. Both FISHcall and CPPcall are methods in FISH and user-written C++ programs to call a specific function when particular events occur. When a new contact was created in this study, a defined function 'catch-HMAcmodel' was called to update the contact model properties. If new contacts are created frequently during the simulation of an asphalt mixture, the defined function has to be called frequently. The difference between FISHcall and CPPcall is that FISHcall is applied to call a FISH function which has to be interpreted and compiled before execution and CPPcall is to call a C++ function. Therefore, by using CPPcall, the private PFC2D/3D executables can speed up computation, especially in the simulation where the contacts are frequently updated.

Loading Conditions

In the current version of the private PFC2D/3D executables, three loading conditions are included: the uniaxial creep-recovery, ramp, and dynamic modulus tests. As shown in Figure 1, two loading platens were employed to apply loading conditions: on the top platen, the constant or ramp or cyclic load was applied, and the bottom one was fixed in all directions. Within the executables, the loading values and directions of the applied stress are controlled by program and updated at each iterative step by applying CPPcall to trigger the calling of C++ functions just before motion calculation, which can speed up discrete element modeling of asphalt mixtures.

LABORATORY TESTS AND DISCRETE ELEMENT SIMULATION

Laboratory tests developed in previous research work (You and Buttlar 2006) were employed. The asphalt mixture specimens were compacted in a Superpave gyratory compactor to a target air void level of 4% by volume, with a nominal maximum aggregate size of 19mm as shown in Table 1.The sand mastic had a nominal maximum aggregate size of 1.18-mm, which was obtained from the mixture's aggregate gradation by eliminating all the aggregates bigger than 1.18mm except the asphalt. The sand mastic had around 14% asphalt content by weight. The uniaxial compression test was conducted to obtain the dynamic modulus and phase angles for mastics, aggregate (stones), and mixtures. The test procedure and the results of the mastics and mixtures were listed in previous research work (Dai and You 2008; You and Buttlar 2006). The aggregates test procedure and results were listed in a recent paper. Dynamic moduli and phase angles were measured at 0°C with four frequencies, 0.1, 1, 5, and 10Hz, and the results are listed in Table 2. Then the laboratory test results were fitted with Burger's model and the fitted Burger's model parameters are listed in Table 3.

Table 1 Gradation of the asphalt mixture (data source: You and Buttlar 2006)

	Coarse Aggregates						
Size/mm	25-19	19-12.5	12.5-9.5	9.5-4.75	4.75-2.36	2.36-1.18	<1.18
Mass (%)	1.1	21.4	9.1	12.2	10	14.5	31.7
Volume (%)	0.94	18.24	7.76	10.40	8.52	12.36	27.02

	Asphalt Content	Air Voids	Mastic Content	-
Mass (%)	4.80	0	36.50	-
Volume (%)	10.77	4	41.78	-

Table 2: Laboratory test results of aggregates, asphalt mastic and mixture

Loading Frequency (Hz)		0.1	1	5	10
Dynamic Modulus (GPa) (Dai and You 2008; Liu and You 2008; You and Buttlar 2006)	Mastic	4.40	6.19	8.69	9.86
	Mixture	10.62	14.67	17.67	20.08
Phase Angles (°C)	Mastic	38.93	28.80	24.50	28.65
	Mixture	23.24	20.54	18.16	16.87
Young's Modulus (GPa) (Dai and You 2007a)	Aggregates	55.50			

Table 3: Burger's model parameters fitted using lab data

	Frequency (Hz)			
Burger's model parameters	0.1	1	5	10
E_1	901	901	901	901
η_1	607	607	607	607
E_2	3.77	5.64	8.07	8.88
η_2	3.66	0.41	0.011	0.0067

Uniaxial dynamic modulus tests were simulated using the DE model built above and the corresponding loading condition, and the predicted modulus and phase angles were calculated. The predicted results and those from the lab were compared in the previous research, the errors of predicted modulus and phase angles are both less than 10%.

4. Timing Test: Comparisons between C++ and FISH Programs

This section presents a timing test of C++ program (the private PFC2D/3D executables) by comparing with the similar FISH program. Nine different sizes of asphalt mixture digital samples were built, which consist of discrete elements ranging from 890 to 91,668. Discrete element modeling under cyclic loads was conducted on the nine samples, where both C++ and FISH programs were applied. The speed ratios of the C++ program to the FISH program (C++/FISH) were calculated during computation time which is consumed by a computer to finish a specified simulation task. The speed ratio vs. computation time is plotted in Figure 2 where the results of five samples were used and the other four samples have similar results. The average speed ratio of 7 hours' computation time vs. sizes of nine samples is plotted in Figure 3. The conclusions were observed as follows:

(1) Speed ratios (C++/FISH) decrease with computation time increasing. The reason is that discrete elements within the digital samples tend to be stabilized and few contacts need to be updated with computation time increasing. As a result, the FISHcall or CPPcall will not be triggered and the advantage of C++ program is not obvious.

(2) With the sizes of samples (numbers of discrete elements within the samples) increasing, speed ratios decrease and then increase. The reason for the decreasing of the speed ratio is that the smaller samples are not easy to stabilize and new contacts are created frequently. As a result, smaller samples have larger speed ratios. The reason for the speed ratios increasing is that the larger samples have more contacts than the smaller samples, where there are more probabilities to create new contacts in a larger sample. As a result, larger speed ratios were observed when the sample sizes were large enough.

(3) The average speed ratio is 3.74, ranging from 2.34 to 4.95.

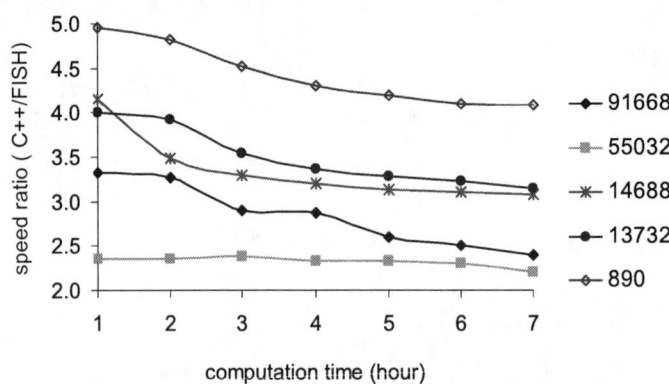

Figure 2. Speed ratio vs. loading time for 5 samples (legend shows the number of discrete elements)

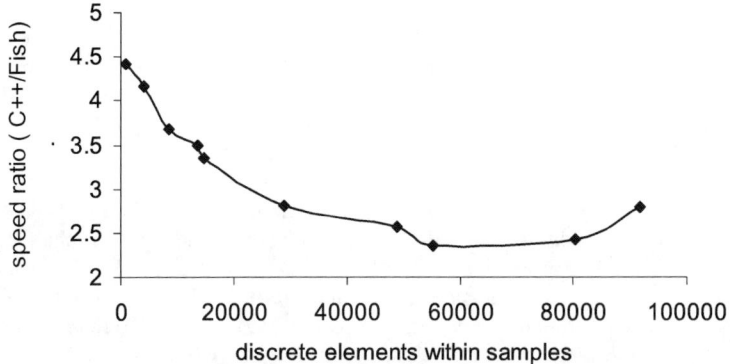

Figure 3. Average speed ratio of 7 hours' loading simulation time vs. sizes of samples

5. Summary and Conclusions

In the discrete element models of asphalt mixtures simulation, the computation speed is a key issue which has to be considered especially for viscoelastic simulation of asphalt mixtures. The objective of this research is to speed up computation by developing discrete element simulation program with an optional feature of PFC2D/3D—user-written C++ code. The private PFC2D/3D executables were developed with the user developed programs in C++ programming language. Timing tests were conducted by running C++ and FISH programs and comparing their results. The following conclusions were observed:

(a) With the user-written C++ codes, discrete element modeling of asphalt mixtures can be sped up, where the speed ratio (C++/FISH) is from 2.34 to 4.95 while the average value is 3.74.

(b) With the computation time increasing, the discrete elements in the digital samples tend to be stabilized and few contacts need to be updated. As a result, the user-defined C++ functions are not triggered frequently, which causes the speed ratios decrease with computation time increasing.

(c) The smaller samples are not easy to stabilize and, new contacts can be created frequently, while larger samples with more discrete elements have more probabilities to create new contacts. Therefore, for models with large number of discrete elements, the speed ratios decrease in general, but increase when the number reaches a certain value.

ACKNOWLEDGEMENTS

This material is based in part upon work supported by the National Science Foundation under Grant CMMI 0701264. Any opinions, findings, and conclusions or recommendations expressed in this material are those of the author's and do not necessarily reflect the views of the National Science Foundation.

REFERENCES

Abbas, A., Masad, E., Papagiannakis, T., and Harman, T. (2007). "Micromechanical modeling of the viscoelastic behavior of asphalt mixtures using the discrete-element method." *International Journal of Geomechanics*, 7(2), 131-139.

Abbas, A., Masad, E., Papagiannakis, T., and Shenoy, A. (2005). "Modelling asphalt mastic stiffness using discrete element analysis and micromechanics-based models." *International Journal of Pavement Engineering*, 6(2), 137-146.

Buttlar, W. G., and You, Z. (2001). "Discrete element modeling of asphalt concrete: Microfabric approach." *Transportation Research Record* (1757), 111-118.

Carmona, H. A., Kun, F., Andrade, J. S., and Herrmann, H. J. (2007). "Computer simulation of fatigue under diametrical compression." *Physical Review E - Statistical, Nonlinear, and Soft Matter Physics*, 75(4), 046115.

Chang, K. N., and Meegoda, J. N. (1997). *Micromechanical simulation of hot mix asphalt*, American Society of Civil Engineers.

Collop, A. C., McDowell, G. R., and Lee, Y. (2007). "On the use of discrete element modelling to simulate the viscoelastic deformation behaviour of an idealized asphalt mixture." *Geomechanics and Geoengineering*, 2(2), 77 - 86.

Collop, A. C., McDowell, G. R., and Lee, Y. W. (2006). "Modelling dilation in an idealised asphalt mixture using discrete element modelling." *Granular Matter*, 8(3-4), 175-184.

Dai, Q., and You, Z. (2007). "Prediction of creep stiffness of asphalt mixture with micromechanical finite-element and discrete-element models." *Journal of Engineering Mechanics*, ASCE, 133(2), 163-173.

Dai, Q., and You, Z. (2008). "Micromechanical finite element framework for predicting viscoelastic properties of asphalt mixtures." *Materials and Structures*, 41 (6), 1025-1037.

Itasca Consulting Group, I. (2004a). "PFC2D Version 3.1." Minneapolis

Itasca Consulting Group, I. (2004b). "PFC3D Version 3.1." Minneapolis.

Kim, H., and Buttlar, W. G. "Micromechanical Fracture Modeling of Asphalt Mixture Using the Discrete Element Method." *Advances in Pavement Engineering (GSP 130)*, Austin, Texas, USA, 17-17.

Liu, Y., Feng, S., and Hu, X. (2007). "Discrete Element Simulation of Asphalt Mastics Based on Burgers Model." *Journal of Southwest Jiaotong University*, 15(1), 20-26.

Liu, Y., and You, Z. (2008). "Simulation of Cyclic Loading Tests for Asphalt Mixtures using User Defined Models within Discrete Element Method." *GeoCongress 2008: Characterization, Monitoring, and Modeling of GeoSystems, Geotechnical Special Publication (GSP 179)*, ASCE, 742-749 .

Tian, L., Liu, Y., Hu, X., and Wang, B. (2007a). "Random Generation Algorithm for Simulation of Polyhedral Particles in Asphalt Mixture Aggregate and Its Program." *China Journal of Highway and Transport* 20(3), 5-10.

Tian, L., Liu, Y., and Wang, B. (2007b). "3D DEM model and digital restructure technique for asphalt mixture simulation." *Journal of Chang'an University(Natural Science Edition)*, 27(4), 23-27.

You, Z., Adhikari, S., and Dai, Q. (2008). "Air void effect on an idealized asphalt mixture using a two-dimensional and three-dimensional discrete element modeling approaches." *ASCE Geotechnical Special Publication: Innovations in the Characterization, Modeling and Simulation of Pavements and Materials*, 55-62.

You, Z., and Buttlar, W. (2005). "Application of Discrete Element Modeling Techniques to Predict the Complex Modulus of Asphalt-Aggregate Hollow Cylinders Subjected to Internal Pressure." *Transportation Research Record*, 1929(-1), 218-226.

You, Z., and Buttlar, W. G. (2004). "Discrete element modeling to predict the modulus of asphalt concrete mixtures." *Journal of Materials in Civil Engineering*, 16(2), 140-146.

You, Z., and Buttlar, W. G. (2006). "Micromechanical Modeling Approach to Predict Compressive Dynamic Moduli of Asphalt Mixture Using the Distinct Element Method." *Transportation Research Record: Journal of the Transportation Research Board, National Research Council, Washington, D.C.*, 1970, 73-83.

Finite Element Modeling of Reflective Cracking under Moving Vehicular Loading: Investigation of the Mechanism of Reflective Cracking in Hot-Mix Asphalt Overlays Reinforced with Interlayer Systems

Jongeun Baek[1] and Imad L. Al-Qadi[2]

[1] Graduate Research Assistant, Department of Civil and Environmental Engineering, University of Illinois at Urbana-Champaign, ATREL, 1611 Titan Dr, Rantoul, IL, 61866; PH (217) 893-0705; FAX (217) 893-0601; email: baek2@illinois.edu

[2] Founder Professor of Engineering, Illinois Center for Transportation Director, Department of Civil and Environmental Engineering, University of Illinois at Urbana-Champaign, 205 N Mathews MC-250, Urbana, IL, 61801; PH (217)265-0427; FAX (217) 333-1924; email: alqadi@illinois.edu

Abstract

The fracture mechanism of reflective cracking in hot-mix asphalt (HMA) overlay was investigated using a three-dimensional finite element model utilizing the cohesive zone model. The effectiveness of interlayer systems to abate reflective cracking in overlays was evaluated. A fractured area due to potential reflective cracking was calculated at a region in which cohesive elements were inserted. A represent fracture area (RFA) is introduced as a weighted average value of degradation to calculate the fractured area. Utilizing global and local RFAs, the mechanism of reflective cracking is investigated for HMA overlay with interlayer systems. For various interlayer systems, this study found that different reflective cracking patterns occurred in terms of quantity and distribution. The sand mix interlayer reduces reflective cracking in leveling binder at the beginning; however, steel netting interlayer retards reflective cracking in wearing surface as well as leveling binder.

Key words: Reflective cracking, FE Modeling, Interlayer, Cohesive zone model, Representative fractured area

Introduction

Reflective cracking is a major structural defect in hot-mix asphalt (HMA) overlay; it results from traffic and environmental loadings when discontinuities exist on underlying pavements such as PCC joints and cracks. To abate reflective cracking, several methods have been utilized, including increasing overlay thickness, sawing (or cracking) and sealing joints, placing a crack-arresting layer underneath overlay, and inserting various interlayer systems (Mukhtar 1994, Cleveland et al. 2002). A certain level of success has been accomplished to delay reflective cracking; but no method has been developed to completely stop the occurrence of reflective cracking (Button and Lytton 2007). Among the successful techniques, interlayer systems have recently been used to successfully and cost-effectively retard reflective cracking (Van Deuren and Esnouf 1996; Buttlar et al. 2000; Steen 2004). Interlayer systems, which are relatively thin layers placed between HMA overlay and existing pavements, have two different mechanisms to delay reflective cracking: 1) by absorbing strain energy which occurs in HMA overlay; and 2) by reinforcing the low tensile strength HMA overlay. However, due to inappropriate installation of interlayer systems as well as a lack of understanding the interlayer system mechanism, the benefit of using interlayer systems has been regarded as questionable or minimal (Peredoehl 1989; Epps et al. 2000).

Reflective cracking analysis using finite element (FE) modeling
Significant efforts have been made to analyze the phenomenon of reflective cracking utilizing a finite element (FE) method. Three-dimensional FE analysis was performed for HMA overlay placed on a jointed PCC pavement (Bozkurt and Buttlar 2002). Reflective cracking induced by temperature variations and traffic loading was evaluated by means of tensile stress on the top and bottom of the overlay. The study evaluated the effectiveness of the interlayer stress absorbing composite (ISAC) in reducing stress around a crack tip. Kuo and Hsu (2003) examined the path of reflective cracking by applying various fatigue criteria on HMA overlay and interfaces. Al-Qadi et al. (2003) and Elseifi and Al-Qadi (2005) utilized the path-independent J integral to compute energy dissipation at a vicinity of a reflective crack tip when steel netting and geocomposite interlayer are used as interlayer systems. Depending on interface conditions, reflective cracking was initiated and propagated in different locations. Theses FE approaches were primarily carried out to analyze either stress or strain based on a continuum mechanics.

Fracture mechanics approaches have been utilized to analyze the fracture behavior of HMA. The cohesive zone model (CZM) proposed in the 1960's has been utilized to model HMA cracking (Jenq and Perng 1991). In their study, various combinations of nonlinear spring and dashpot model were used to represent the viscous behavior of HMA. After that, several research efforts have shown successful applications for modeling a crack in HMA (Soares et al. 2003; Paulino et al. 2004; Song et al. 2004, 2006; Baek and Al-Qadi 2006). Song (2006) inserted the CZM elements into the pre-defined crack path over a PCC joint. He concluded that bottom-up reflective cracking results from traffic loading while top-down reflective cracking is induced by temperature variation. Currently, more applications of the CZM are used in pavement analysis because of the efficiency and the versatility. However, three-dimensional (3D) FE analysis has not been performed to examine the complex behavior of reflective cracking induced by moving traffic loading which can lead to mixed mode fracture in tension and shear.

In this study, a 3D FE model accompanied with cohesive elements was developed for HMA overlay over jointed PCC pavements. The study examines the path of reflective cracking under moving traffic loading. In addition, comparative study to investigate the effectiveness of sand mix and steel netting interlayer systems in retarding reflective cracking was conducted.

FE modeling of hot-mix asphalt overlays
Geometry and boundary conditions
A typical HMA overlaid pavement was modeled using a 3D FE. Special purpose elements, cohesive elements, were used to simulate fracture behavior. Figure 1 illustrates the geometry of the overlay pavement. One lane of HMA overlay over jointed plain PCC pavements is selected (Baek and Al-Qadi 2008). The overlay

pavement consists of 57-mm-thick HMA overlay, 200-mm-thick PCC slab, 150-mm-thick base, and subgrade which has 3.3-m-thick in finite and 6.6-m-thick in infinite domain. The infinite domain is added at a far-field zone in order to achieve zero deformation and to minimize stress wave reflection at the boundary for dynamic analysis. For the finite and infinite zones, 128074 continuum elements with a reduced integration (C3DR) and 3120 infinite elements (CIN3D) are used, respectively. Of the whole PCC slab (6.0-m-long by 3.6-m-wide), one quarter of the slab is used in the model due to the geometrical symmetric condition of the slab. The symmetric condition assumed in this model is not valid for a transverse direction when traffic loading is applied. Nonetheless, the symmetric condition is to be held to achieve computation efficiency since a main region in which critical pavement responses are investigated is limited to around a joint. The horizontal behavior of all interfaces in this model is governed by the Coulomb friction model with friction coefficient of 1.0, and no displacement in a normal direction is allowed simply to exclude complicated interfacial behavior and focus on fracture behavior inside HMA. An implicit dynamic analysis was performed to calculate pavements' responses induced by transient dual-tire assembly loading with non-uniform contact stresses (Yoo et al. 2006). The loading was applied at 8.0km/h traveling 600mm across a joint for 0.27sec.

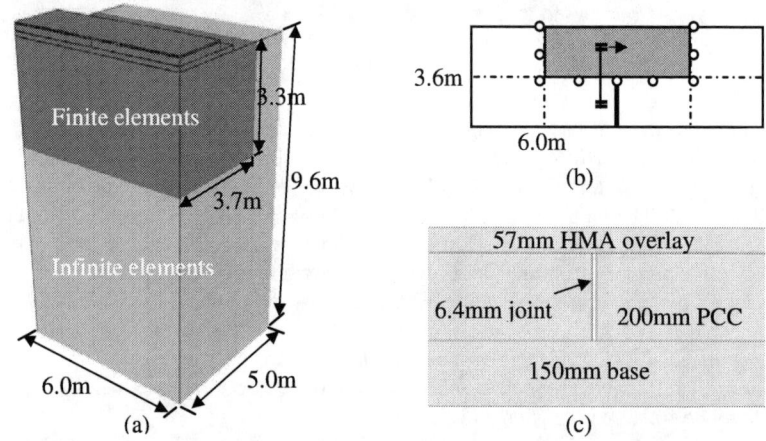

Figure 1. HMA overlay geometry and boundary condition: (a) overall geometry; (b) geometrical symmetry; and (c) pavement profile.

Three overlay designs with two interlayer systems
Three overlaid pavements with two different interlayer systems were modeled in order to investigate the effectiveness of the interlayer systems on retarding reflective cracking. As a control section, design A consists of 38-mm-thick wearing surface and 19-mm-thick leveling binder without an interlayer system. The leveling binder is made of conventional mix, which is nominal 9.5-mm-mixture with PG 64-22 binder. Design B has a sand mix interlayer as a substitute for the leveling binder, which has smaller size aggregates (nominal 4.5mm mix) and PG 78-28 polymer modified binder. Those features give the sand mix interlayer increased toughness to resist cracking.

Finally, design C contains a steel netting interlayer as an additional layer between the leveling binder and PCC surface. The steel reinforcement is thought to compensate for a lack of tensile strength of the leveling binder when properly installed.

Reflective cracking model

Cohesive elements having zero-thickness are used to model an interfacial or cohesive crack (Abaqus 2005). The cohesive elements are governed by a traction separation law (TSL) which describes a transferable force (traction) between two opposite nodes with respect to displacement jump (separation). Among several TSLs, bi-linear TSL was adapted in this study to reduce a compliance problem as proven by Song et al. (2006). Two thousand eighty-five cohesive elements were assigned to a region of HMA overlay where reflective cracking would develop, i.e., right over the middle of a joint (Figure 2).

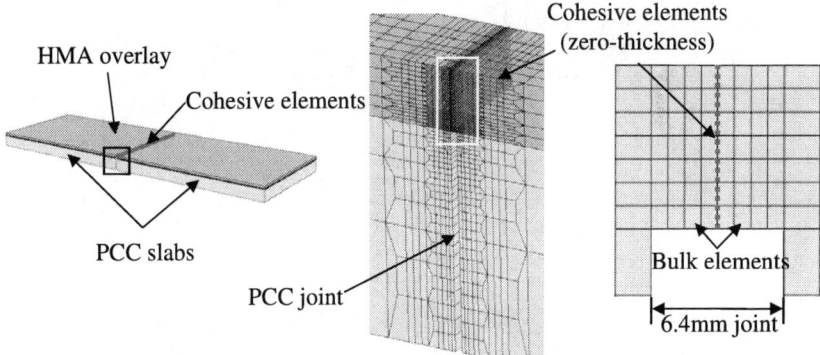

Figure 2. Schematic of cohesive element application to mode reflective cracking in HMA overlay over the joint.

Until a certain amount of force is applied to the cohesive elements, the force transferred to the other side of the cohesive element is proportional to the separation. The maximum traction to be transferred is specified as cohesive strength, t^0, and corresponding separation, δ^0, is also defined (Figure 3). The t^0 and δ^0 are defined by initial stiffness, K. When a separation reaches the δ^0, i.e., applied stress reaches the cohesive strength, the cohesive element starts to lose cohesive strength. The stiffness of the cohesive element is reduced by a function of (1-D) where D is a degradation scalar. For a linear degradation, the degradation is defined as shown in Figure 4 (Davila and Camanho 2003). During a fracture process, the maximum traction is reduced and finally vanishes at a critical separation, δ^{cr}. So, fracture criterion is defined by the degradation scalar, D: no fracture at D<0; micro crack at 0<D<1; and macro crack at D=1.

Material properties

The HMA is the most essential material to govern the overall behavior of the overlay. Since the target temperature of this analysis was -10°C and a short period of moving load was applied at 8.0km/h, HMA was assumed as a linear viscoelastic (LVE) material. The Prony series expansion based on the generalized Maxwell model was utilized to characterize the time-temperature dependent behavior of HMA. Dynamic moduli obtained at -20°C, -10°C, and 0°C under multiple frequencies were converted into shear and bulk relaxation moduli (Baek and Al-Qadi 2008). The Prony series parameters used for the conventional mix leveling binder and sand mix interlayer are shown in Table 1. For the whole loading range of 10^{-4} to 10^3 sec, the relaxation moduli of the conventional mix are higher than those of the sand mix.

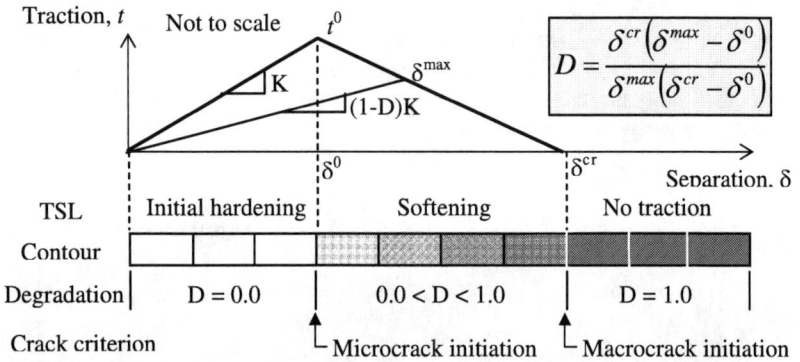

Figure 3. Relationship between a traction-separation law (TSL), degradation, and crack initiation criteria.

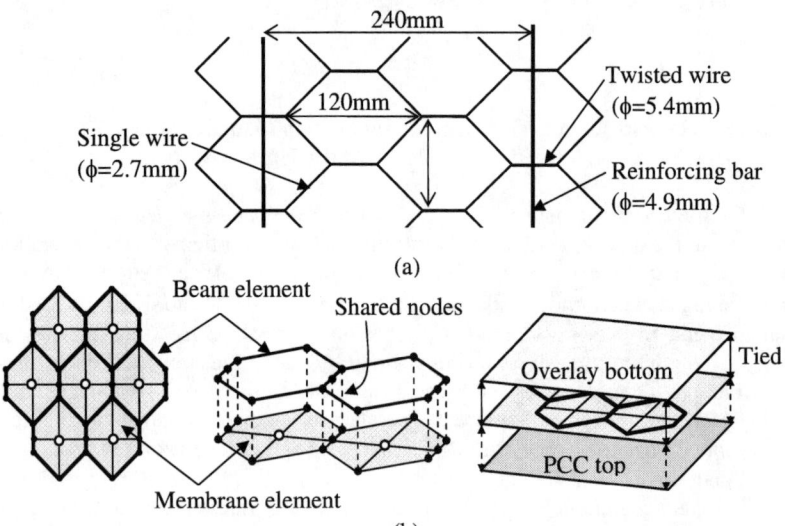

Figure 4. Steel reinforcement modeling: (a) mesh configuration; and (b) details on node connectivity and boundary conditions at the interface.

Since HMA is assumed as a primary material to affect the behavior of reflective cracking, the other layers were simply regarded as homogeneous isotropic linear elastic materials. For PCC, aggregates, and soil, typical values of elastic modulus, Poisson's ratio, and density were selected from the literature (Huang 1993). Table 2 shows elastic material properties used in the pavement model. The elastic modulus of HMA is instantaneous, which is assumed at an extremely low temperature and short loading period.

Table 1. Prony series parameters, g and τ and shear instantaneous modulus, G_0

N		1	2	3	4	5	6	7	8	9	10	G_0
Con.	g_i	0.07	0.10	0.11	0.14	0.12	0.11	0.10	0.08	0.06	0.03	7.0
Mix	τ_i	10^{-4}	10^{-3}	10^{-2}	10^{-1}	10^{0}	10^{1}	10^{2}	10^{3}	10^{4}	10^{5}	GPa
Sand	g_i	0.07	0.14	0.15	0.14	0.13	0.12	0.10	0.08			5.7
mix	τ_i	10^{-4}	10^{-3}	10^{-2}	10^{-1}	10^{0}	10^{1}	10^{2}	10^{3}			GPa

To define the bi-linear TSL in cohesive elements, fracture material properties were obtained from the disk-shape compact tension [DC(T)] test at -10°C with specimens cored from the field (Al-Qadi et al. 2008). The fracture energy is calculated by the area under the load-crack mouth opening displacement (CMOD) curve and ligament area of the specimen. The average of the fracture energy of the conventional mix at -10°C is 274J/m^2 (Baek and Al-Qadi 2008). Also, the average peak load is 1.9kN. Since no test was performed to obtain the tensile strength of HMA, the tensile strength was approximately assumed from the peak load and ligament area of the specimen. The tensile strength of the conventional mix is 0.58MPa. In addition, the fracture energy and tensile strength of the sand mix are 593J/m^2 and 0.79MPa, respectively.

Table 2. Material properties of HMA overlaid pavement model

Material	Elastic modulus (GPa)	Poisson's ratio	Density (ton/m^3)
HMA (Conv. mix)	17.2*	0.22	2.3
HMA (Sand mix)	14.0*	0.22	2.3
PCC	27.5	0.20	2.4
Base	0.3	0.35	1.9
Subgrade	0.14	0.40	1.9
Slurry seal	1.0	0.35	2.0
Steel	200.0	0.28	7.8
	Single wire	Double wire	Reinforcing bar
Diameter (mm)	2.7	5.4	4.9

* Instantaneous modulus

Steel netting modeling
The steel netting has 120mm by 80mm hexagonal openings, which consists of single and double twisted wires and 240mm-interval transverse reinforcing bars (Al-Qadi et al. 2003, Elseifi and Al-Qadi 2005). Figure 4 illustrates the shape of the steel netting interlayer model, node connectivity, and boundary conditions. Each component of the steel netting is modeled with a beam element having a circular cross section. A twisted wire is simply regarded as an equivalent singe wire whose diameter is twofold (5.4mm). In addition, to place the steel reinforcement, slurry seals are applied for

protection and improved bonding between the steel netting and surrounding layers (Elseifi and Al-Qadi 2005). For the slurry seals, a thin layer is added at the interface by means of membrane elements, which carry only in-plane force. By sharing nodes of the membrane layer, the steel mesh is embedded in the slurry seal layer, meaning that no relative displacements occur between the steel mesh and slurry seal as shown in Figure 4(b). It was assumed that friction is infinite between the HMA overlay and slurry seal layer including the steel mesh, i.e., a perfect bonded condition. The interfacial behavior between the PCC slabs and slurry seal layer is governed by the Coulomb friction model. Material properties for the slurry seal and steel netting are presented in Table 2. The slurry seal is assumed as mastic with low stiffness at -10°C. So, from a structure point of view, the slurry seal layer does significantly affect the behavior of the interface; but its function is to connect the steel netting and surrounding layers.

Reflective cracking analysis

The behavior of reflective cracking is examined in the pre-defined potential crack area over a joint in which cohesive elements are inserted. The degradation scalar, D, is utilized to determine reflective crack initiation time. In addition, the quantity of reflective cracking is specified by means of representative fractured area.

Reflective crack initiation

As vehicular loading approaches a joint, shear stress increases first and tensile stress follows at the middle and bottom of the leveling binder, respectively. Once the combined stresses reach the cohesive strength of a cohesive element, a microcrack begins to occur in the cohesive element, i.e., D>0 and reflective cracking is developed if the criterion is met, i.e., D = 1. Figure 5 shows degradation variations by loading time steps (t_d = 0.009sec) at the bottom of the wearing surface (0.35H) and leveling binder (0.01H) for design A. In all locations, degradations jump once and converse to a certain level. The degradations are not decayed since they are calculated with δ^{max}, which is the maximum value a cohesive element has experienced. Under the wheel path (middle of two tires), the degradation jumps to around 0.8 at the second time step ($2t_d$) and converges to 1.0 at $6t_d$ for both locations (0.01H and 0.35H). On the other hand, out of the wheel path (225mm away from the outer edge of one tire), the degradation begins to increase at $5t_d$ and $10t_d$ in 0.01H and 0.35H, respectively. Different from the other locations, the degradation converges around 0.86 at $23t_d$.

When the crack initiation criterion of D=1.0 is applied, no reflective cracking is initiated in the four locations because degradations do not reach 1.0 despite being close to 1.0. For example, the maximum degradation, D_{max}, is 0.9984 at the bottom of the leveling binder under the wheel path. This suggests that one passage of traffic loading is not able to develop a macro reflective cracking. Instead of applying multiple loadings to create reflective cracking, potential reflective cracking is specified by adapting a lower crack criterion as an alternative way to save computational efforts. It is hypothesized that a cohesive element with degradation higher than 0.5 has the potential to develop fracture after more loads are applied. In other words, when the initial stiffness in the bi-linear TSL is reduced by a factor of 0.5, the potential reflective cracking is defined as developed. Thus, the potential reflective cracks are initiated at 1.8, 1.6, 6.6, and $15.0t_d$ for each location marked with "O" in Figure 5.

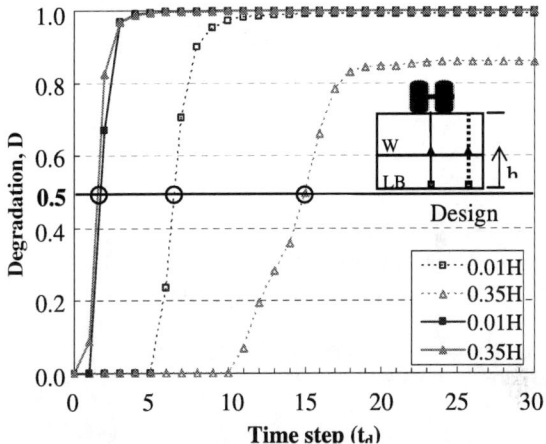

Figure 5. Degradation variations over time steps at various locations for the design A.

Figure 6 illustrates reflective crack initiation time at various locations with respect to a distance from the bottom of the overlay for the design A. The earliest potential reflective cracking occurs in the middle of the leveling binder between 0.08H (1.5mm) and 0.12H (2.3mm) at 0.77 t_d between two tires and at $0.93t_d$ under a tire. Then, the potential reflective cracking propagates upward and downward. This occurs because as a vehicle approaches a joint, coupled shear and tensile stresses induced in the middle is higher than a single tensile stress at the bottom.

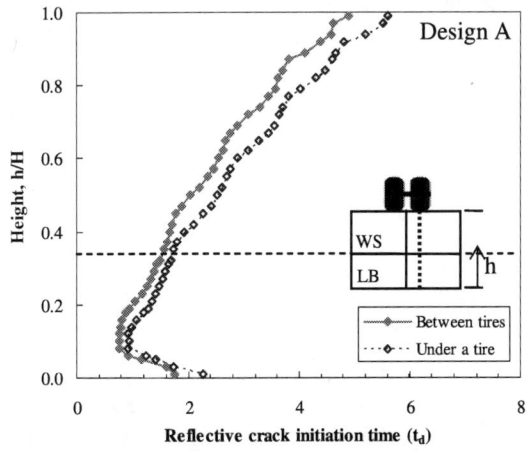

Figure 6. Reflective crack initiation time at various locations for the design A.

Representative fractured area

In order to quantify the fracture characteristic of reflective cracking, a representative fractured area (RFA) is introduced. The RFA describes a degree of progressive fracture as a percentage in a specific in-plane cross-section area of cohesive elements by means of a weighted average value of degradations as follows:

$$RFA_{Z,k} = \frac{\sum_{j}^{M_z}\sum_{i}^{N_z}\left[(A_C)_{ij} \times D_{ij}^k\right]}{A_Z} \times 100 \qquad (1)$$

where,
  RFA is representative fractured area in %;
  $RFA_{Z,k}$ is a RFA for a specific area zone, z, at time step, k;
  A_C is an in-plane cross-section area of a cohesive element at i^{th} row and j^{th} column;
  A_Z is the total area of the zone; and
  N_z and M_z are the number of cohesive elements in row and column.

Compared to the aforementioned lower crack criterion which uses a trigger value of D=0.5, the RFA takes into account all progressive degradations in a specific area linearly. Thus, the RFA ranges from 0 (no degradation in a specific area) to 100 (full degradation). However, it can also stand for various degradation distributions with the same RFA. In order to analyze the distribution of degradation in detail, local RFAs which are computed in each sub-area can be utilized. A global RFA for a certain whole area is decomposed into several local RFAs corresponding to each sub-area. Then a fracture progress in a local area can be evaluated independently.

Fracture mechanism of an interlayer system

The mechanism of interlayer system for retarding reflective cracking is investigated using global and local RFAs for the designs A, B, and C. The global RFA represent an overall quantity of the whole cross section of the overlay; the local RFAs specify localized features in a specific area. The cross-section area of the HMA overlay is divided into four basic sub-areas based on typical patterns shown in degradation contours: zone 1 under the wheel path in the wearing surface; zone 2 under the wheel path in the leveling binder; zone 3 out of the wheel path in the WS; and zone 4 out of the wheel path in the leveling binder. Table 3 lists the global and local RFAs for the three overlay designs. The degradation contours of the designs A, B, and C are obtained at $4t_d$, $4t_d$, and $16t_d$, respectively; no degradation occurs at $4t_d$ for the design C. The degradation contour is presented when fracture is progressed. For each design, unique degradation contours are developed as shown for designs A, B, and C, respectively: Most degradation contours exist in zone 2 in the control section; quite large and small amounts of degradation are spread in zones 1 and 2, respectively; and a narrow strip shape of degradations is distributed at the bottom of the leveling binder (zones 2 and 4). RFA for the whole cross-section area, RFA_{OL}, indicates a total fractured area as a single parameter: 11.9%, 8.4%, and 7.0% for the designs A, B, and C, respectively. Based on the RFA_{OL}s, 29% and 41% of fractured area are reduced by the sand mix and steel netting interlayers, respectively.

Table 3 Typical degradation contours and corresponding RFAs for the three HMA overlay designs

Design	A			B		C	
Degradation Contour	3	1	3				
	4	2	4				
RFA_{OL}	11.9			8.4		7.0	
Subdivision	RFA_{local}	%*		RFA_{local}	%	RFA_{local}	%
Zone 1	12.2	22		31.1	80	0.1	1
Zone 2	78.4	72		14.9	19	33.0	51
Zone 3	0.0	0		0.1	7	0.0	0
Zone 4	3.1	6		0.0	0	15.1	48
WS (1+3)	4.0	22		10.2	81	0.1	1
LB (2+4)	27.7	78		4.9	19	21.0	99
UN (1+2)	34.3	94		25.7	99	11.1	52
OT (3+4)	1.0	6		0.1	1	5.0	48

* A fraction (%) of the RFA_{OL} in each category, RFA_{local} x A_{local} /A_{OL} x 100

In addition, the shape of a degradation contour can be characterized with those local RFAs. For the design A, RFAs for zones 1, 2, and 4 are 0.122, 0.784, and 0.031, respectively. This means that 78.4% of zone 2 area is fractured fully. For the design B, obviously different local RFAs are produced: 0.311 in zone 1 and 0.149 in zone 2. A major fractured area (80%) occurred in zone 1. Compared to the design A, a 1.5 times wider area is fractured in the zone 1; less fracture in zone 2. This suggests that the sand mix interlayer may better reduce reflective cracking. However, reflective cracking may move to the wearing surface because the sand mix has greater fracture energy and tensile strength. The RFA_3 and RFA_4 in designs A and B are insignificant, and those features can be shown in combined RFA in the area away from the wheel, RFA_{OT}. For the design C, RFA in the zones 2 and 4 is 0.330 and 0.151, respectively; RFA_{LB} in both zones 2 and 4 is 0.210 (99%) and RFA_{WS} in both zones 1 and 3 is 0.001 (1%). Thus, all fractured area is distributed only in the leveling binder and no further cracking propagates to the wearing surface. Since steel netting has very high stiffness, reflective cracking, induced by shear in the middle of the leveling binder, is reduced.

Summary

A three-dimensional finite element pavement model was utilized to investigate the fracture mechanism of interlayer systems in hot-mix asphalt (HMA) overlay constructed on jointed PCC pavement. Cohesive elements were utilized to simulate reflective cracking induced by moving traffic loading. Representative fractured area (RFA) was proposed to calculate potential fractured area by reflective cracking as a weighted average value of degradation. Utilizing RFA in global and local zones, the effectiveness of two interlayer systems, sand mix and steel netting, were evaluated. The sand mix and steel netting interlayers reduced the overall fractured area by 29% and 41%, respectively. Compared to a control section, more reflective cracking occurs at an early stage in wearing surface, not in leveling binder, when a sand mix interlayer is placed. Higher fracture toughness of the sand mix results in the delay of reflective cracking in the leveling binder. On the other hand, the steel netting interlayer reduces reflective cracking significantly as shear is controlled by the steel netting interlayer. Also, no reflective cracking is propagated into the wearing surface in the presented case.

Acknowledgments
The authors greatly appreciated the support of our colleagues at the Illinois Center for Transportation (ICT) at UIUC. In addition, this work was partially supported by the National Center for Supercomputing Applications under project #DMS050004N and utilized the IBM p690 Copper machine.

References
ABAQUS (2005) *ABAQUS/Standard User's Manual Version 6.5*, Habbit, Karlsson, & Sorenson, Inc. Pawtucket, RI.
Al-Qadi, I. L., M. Elseifi, and D. Leonard (2003). Development of an Overlay Design Model for Reflective Cracking with and without Steel Reinforcing Nettings, *Journal of the Association of Asphalt Paving Technologists*, Vol. 82, pp. 388-423.
Baek, J. and I. L. Al-Qadi (2006). Finite Element Method Modeling of Reflective Cracking Initiation and Propagation: Investigation of the Effect of Steel Reinforcement Interlayer on Retarding Reflective Cracking in Hot-Mix Asphalt Overlay. *Transportation Research Record 1949*, Transportation Research Board, Washington, D.C., pp. 32-42.
Baek, J. and I. L. Al-Qadi (2008). Mechanism of Overlay Reinforcement to Retard Reflective Cracking under Moving Vehicular Loading. *Proceedings of the Sixth RILEM International Conference on Cracking in Pavements*, Eds. I. L. Al-Qadi, T. Scarpas, and A. Loizos, Chicago, Illinois, USA, pp. 563 – 573.
Bozkurt, D. and W. G. Buttlar (2002). Three-Dimensional Finite Element Modeling to Evaluate Benefits of Interlayer Stress Absorbing Composite for Reflective Crack Mitigation, *Presented for the 2002 Federal Aviation Administration Airport Technology Transfer Conference*.
Buttlar, W. G., D. Bozkurt, and B. J. Dempsey (2000). Cost-Effectiveness of Paving Fabrics Used to Control Reflective Cracking. *Transportation Research Record 1117*, Transportation Research Board, Washington, D.C., pp. 139-149.
Button, J. W., and R. L. Lytton (2007). Guidelines for Using Geosynthetics with Hot Mix Asphalt Overlays to Reduce Reflective Cracking. *Presented at the 86th Annual Meeting of the Transportation Research Board*, CD-ROM, Transportation Research Board, Washington, D. C.
Cleveland, G. S., J. W. Button, and R. L. Lytton (2002). Geosynthetics in Flexible and Rigid Pavement Overlay Systems to Reduce Reflection Cracking. *Report 0-177701*, Texas Transportation Institute, College Station, TX.
Davila, C. G., and P. P. Camanho (2003). Analysis of the Effects of Residual Strains and Defects on Skin/Stiffener Debonding Using Decohesion Elements, *The 44th AIAA/ASME/ASCE/AHS Structures, Structural Dynamics, and Materials Conference*, April 7-10, Norfolk, VA.

Elseifi, M., and I. L. Al-Qadi (2005). Effectiveness of Steel Reinforcing Netting in Combating Fatigue Cracking in New Flexible Pavement Systems. *Journal of Transportation Engineering*, ASCE, Vol. 131, No. 1, 2005, pp. 37-45.

Elseifi, M., and I. L. Al-Qadi (2005). Modeling and Validation of Strain Energy Absorbers for Rehabilitated Cracked Flexible Pavements. *Journal of Transportation Engineering*, ASCE, Vol. 131, No. 9, 2005, pp. 653-661.

Epps, A., J. T. Harvey, Y. R. Kim, and R. Roque (2000). Structural Requirements of Bituminous Paving Mixtures, *Transportation in the New Millennium*, Transportation Research Board, Washington, D.C.

Jenq, Y.-S., and J.-D. Perng (1991). Analysis of Crack Propagation in Asphalt Concrete Using Cohesive Crack Model. *Transportation Research Record 1317*, Transportation Research Board, Washington, D. C., pp. 90-99.

Kuo, C.-M., and T.-R. Hsu (2003) Traffic Induced Reflective Cracking on Pavements with Geogrid-Reinforced Asphalt Concrete Overlay, *Presented at the 82nd Transportation Research Board Annual Meeting*, CD-ROM, Transportation Research Board, Washington D.C.

Huang, Y. H. (1993). *Pavement Analysis and Design*, Prentice-Hall, Inc., New Jersey. pp. 664-718.

Mukhtar, M. T. (1994). Interlayer Stress Absorbing Composite (ISAC) for Mitigating Reflection Cracking in Asphalt Concrete Overlays. *Dissertation*. University of Illinois at Urbana-Champaign, Urbana, IL.

Paulino, G. H., S. H. Song, and W.G. Buttlar (2004). Cohesive Zone Modeling of Fracture in Asphalt Concrete, *Proceedings of the International RILEM Conference, No. 37, Cracking in Pavements – Mitigation, Risk Assessment, and Prevention*, C. Petit, I. L. Al-Qadi, and A. Millien, Eds., Limoges, France.

Predoehl, N. H. (1989). Evaluation of Paving Fabric Test Installation in California, *Final Report*, Draft Report, California Department of Transportation, Translab.

Soares, J. B., F. A. C. de Freitas, and D. H. Allen (2003). Crack Modeling of Asphalt Mixtures Considering Heterogeneity of the Material, *Presented at the 82nd Transportation Research Board Annual Meeting*, CD-ROM, Transportation Research Board, Washington D.C.

Song, S. H., G. H. Paulino, and W. G. Buttlar (2004). Simulation of Mode I and Mixed-Mode Crack propagation in Asphalt Concrete Using a Bilinear Cohesive Zone Model, *Presented at 84th Annual Meeting of the Transportation Research Board*, CD-ROM, Transportation Research Board, Washington, D.C.

Song, S. H., G. H. Paulino, and W. G. Buttlar (2006). A Bilinear Cohesive Zone Model Tailored for Fracture of Asphalt Concrete Considering Viscoelastic Bulk Material, *Engineering Fracture Mechanics*, Vol. 73, pp. 2829-2848.

Steinberg, M. L. (1992). Geogrid as a Rehabilitation Remedy for Asphaltic Concrete Pavements. *Transportation Research Record 1369*, Transportation Research Board, Washington, D. C., pp. 54-62.

Yoo, P. J., I. L. Al-Qadi, M. Elseifi, and I. Janajreh (2006). Flexible Pavement Responses to Different Loading Amplitudes Considering Layer Interface Conditions and Lateral Shear Forces, *International Journal of Pavement Engineering*, Vol. 7-1, pp. 73-86.

Distribution of Permanent Deformations within HMA Layers

Charles W. Schwartz[1] and Regis L. Carvalho[2]

[1]Department of Civil and Environmental Engineering, University of Maryland, College Park, MD 20742; PH (301) 405-1962; FAX (301) 405-2585; email: schwartz@umd.edu
[2]Department of Civil and Environmental Engineering, University of Maryland, College Park, MD 20742; PH (301) 405-8593; FAX (301) 405-2585; email: regis@umd.edu

Abstract

Current mechanistic-empirical rutting models relate permanent strains to resilient strains. However, the influence of layer thickness on the predicted rutting based on these resilient strains is in sharp disagreement with field observations. Thickness or depth correction factors are required to bring predictions into agreement with observations. Nonlinear elastoplastic finite element analyses of representative generic pavement structures are employed to provide qualitative and quantitative insights into this and related issues. Specific findings include: (a) horizontal stresses predicted by the elastoplastic analyses remain compressive at all times—i.e., the material at the bottom of the HMA layer is not in tension as predicted from multilayer elastic analyses but is yielding under triaxial confined compression conditions; (b) the residual strain variations with depth computed from the elastoplastic analyses are qualitatively similar to measurements from field pavements without resort to any empirical depth or thickness correction functions; and (c) the elastoplastic finite element analyses can directly predict surface heave outside the wheel path, which at best is included in mechanistic-empirical rutting models only through the field calibration factors. Concluding suggestions are included for a conceptual approach for enhancing mechanistic-empirical rutting prediction models.

Introduction

Permanent deformation ("rutting") within the wheel path is a common distress in pavements surfaced with hot mix asphalt (HMA). Mechanistic-empirical (M-E) models for HMA rutting couple mechanistic calculation of pavement stress and/or strain response with empirical prediction of accumulated permanent deformation. The mechanistic strains vary with vertical and horizontal location within the HMA layers. The specific variation will be a function of the pavement structure, material properties, load configuration, and other factors. M-E rutting models must rationally account for these strain variations within the HMA and other pavement layers when accumulating the predicted permanent deformation at the surface. Empirical field calibration factors are then employed to bring the predicted surface deformations into better agreement with measured performance over a wide range of pavement conditions.

There are several issues not yet fully resolved in current M-E models for permanent deformations in the HMA layers in flexible and composite pavement systems. First, opinions remain divided over whether rutting is due primarily to axial permanent strains beneath the tire centerline (e.g., NCHRP, 2004) or to shear permanent strains beneath the tire edge (e.g., Deacon *et al*., 2002; Monismith, Popescu, and Harvey, 2006). Second, current M-E rutting models relate permanent strains (axial or shear) to resilient strains computed using multilayer elastic theory. However, in the absence of correction factors, the influence of layer thickness on the rutting predicted from these resilient strains is in sharp disagreement with field observations. Thickness or depth correction factors are required to bring predictions in line with observations. These correction factors further weaken the mechanistic linkage between predicted rutting and computed strains in the M-E approach. Third, current M-E models often assume that the mechanisms and distributions of permanent strains are similar for HMA layers in flexible pavements vs. HMA overlays on rigid pavements. And fourth, current M-E models do not explicitly consider the contribution of heaving at the edge of the wheel paths (Figure 1), although this may be implicitly included in the field calibration corrections.

Nonlinear elastoplastic finite element analyses of flexible and composite pavement structures can provide significant qualitative and quantitative insights into these issues. A series of such analyses and consequent insights is described here.

Field Rutting Measurements

There are many techniques for measuring rutting in the field. One of the simplest approaches is to use a straightedge, as shown in Figure 1. The total measured rutting will be a combination of the settlement in the center and the heave at the edges of the wheel path. Although traffic wander tends to reduce the heave at the edges of the wheel path, it can be a prominent feature in heavily rutted pavements, especially if traffic is channelized.

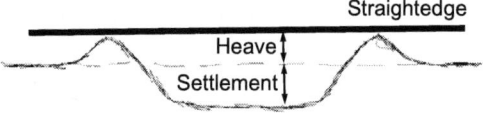

Figure 1. Components of measured surface rutting.

The total rutting measured at the pavement surface is the sum of the permanent deformations in all of the layers. The permanent deformation for each layer in turn is the integration of the vertical permanent strain distribution over the layer thickness. Calibration of M-E rutting models therefore requires as a minimum the allocation of total rutting to the individual layers. Trenches are the preferred approach for determining the permanent deformation in each layer. However, very few field pavement sites available for M-E model calibration have been trenched.

Forensic trenches at the Westrack test facility found that nearly all of the total rutting occurred within the top 150 mm of HMA (Epps et al. 2002; Westrack Forensic Report). Additional detail is provided by trench data from the MnRoad test sections, where permanent deformations were measured for each 40 to 50 mm thick sublayer (Mulvaney and Worel, 2002). Figure 9 summarizes these measurements for several high-volume flexible pavement sections at MnRoad. Permanent deformations occurred primarily in the upper portions of the HMA layers, with 100% of the permanent deformations usually occurring within the top 75 to 125 mm.

M-E Rutting Models for HMA

Current M-E rutting models relate permanent strains (axial or shear) to resilient strains computed using multilayer elastic theory. The Westrack model (Deacon et al., 2002; Monismith, Popescu, and Harvey, 2006) does this in terms of a critical permanent shear strain γ_p that is related empirically to the mechanistically computed linearly elastic shear stress τ and shear strain γ_e at a depth of 50 mm beneath the edge of the tire:

$$\gamma_p = ae^{b\tau}\gamma_e N^c \tag{1}$$

in which a, b, and c are calibration coefficients determined from laboratory constant height repeated shear tests and "tempered" by field observations (Monismith, Popescu, and Harvey, 2006). This critical permanent shear strain is then used to estimate the rut depth in the HMA layer:

$$RD = K\gamma_p \tag{2}$$

in which RD is the total HMA rutting in inches and K is a coefficient relating rut depth to permanent strain. K is a function of the asphalt layer thickness and "ranges from about 5.5 for a 6-in. (150-mm) layer to 10 for a 12-in. (305-mm) thick AC layer" (Monismith, Popescu, and Harvey, 2006). Values for K were determined from plane strain viscoelastic-plastic finite element analyses (Sousa et al., 1994).

The rutting model implemented in mechanistic-empirical pavement design guide (MEPDG) recently adopted by AASHTO is based upon the permanent vertical compressive strains within the HMA (NCHRP, 2004; El-Basyouny, 2004; El-Basyouny, Witczak, and Kaloush, 2005). Since stresses, strains, and temperatures vary through the asphalt thickness, the asphalt layer is divided into n sublayers and the overall rut depth RD for the layer is the accumulation of the permanent deformations in the individual sublayers:

$$RD = \sum_{i=1}^{n} \varepsilon_{pi} h_i \tag{3}$$

in which ε_{pi} is the permanent vertical compressive strain at the center of each sublayer and h_i is the sublayer thickness. Values for ε_{pi} are determined from an empirical strain ratio model:

$$\frac{\varepsilon_p}{\varepsilon_r} = \beta_{\sigma 3}\left[\beta_1 10^{k_1} T^{k_2 \beta_2} N^{k_3 \beta_3}\right] \tag{4}$$

in which the k_i's are laboratory calibration coefficients determined from unconfined uniaxial repeated load permanent deformation tests (Kaloush, 2001), the β_i's are field calibration coefficients, and $\beta_{\sigma 3}$ is depth correction function. During field calibration, the respective k_i and β_i terms were combined into single global calibration values. Final calibrated values (Version 1.000 of the MEPDG software) are $k_1 = -3.35412$, $k_2 = 1.5606$, $k_3 = 0.4791$ with $\beta_1 = \beta_2 = \beta_3 = 1$.

Note that Eq. (4) is based on unconfined laboratory testing. As a consequence, "AC rut depth increased as the thickness of the AC increased. This, obviously, does not conform to known field practice. To overcome this limitation, AC sections that had trenches, in order to obtain the rut depth in each sublayer, were used to correct the predicted rut depth to accurately model the influence of depth and layer thickness upon AC rut depth" (El-Basyouny, 2004). The MnRoad trench data was therefore used to develop a depth correction function $\beta_{\sigma 3}$:

$$\beta_{\sigma 3} = (C_1 + C_2 \cdot depth) \cdot 0.328196^{depth} \tag{5}$$

$$C_1 = -0.1039 \cdot h_{AC}^2 + 2.4868 \cdot h_{AC} - 17.342 \tag{6}$$

$$C_2 = 0.0172 \cdot h_{AC}^2 - 1.7331 \cdot h_{AC} + 27.428 \tag{7}$$

in which *depth* is depth to the point of strain computation, and h_{AC} is the combined thickness of all the asphalt sublayers. Equations (4) through (7) are used for rutting predictions for both new and rehabilitation (e.g., HMA overlay) design scenarios.

Both the Westrack K factor in Eq. (2) and the MEPDG depth function $\beta_{\sigma 3}$ in Eq. (4) can be viewed as attempts to compensate for deficiencies in using linearly elastic stress and strain distributions to estimate permanent deformations. For example, Figure 2 compares the variations of the uncorrected versus corrected permanent strains from the MEPDG model. (Note: This and all subsequent figures adopt the standard mechanics sign convention of + for tension and – for compression.) As dictated by Eq. (4), the uncorrected computed permanent strains are proportional to the mechanistically determined vertical resilient strains, which are largest at the bottom of the HMA layer due to the combination of the direct vertical compression and the compressive Poisson strains induced by the horizontal linearly elastic tensions. As a consequence, the permanent deformations are concentrated in the lower depths of the HMA layer contrary to field experience. The empirical depth correction function distorts both the magnitude and shape of the permanent strain distribution in an attempt to force the majority of the permanent deformations into the upper portions of the HMA layer. The depth correction function dominates the permanent strain values used to compute the total rutting for the layer and, in the process, arguably subverts the "mechanistic" portion of the modeling.

Clearly, the empirical permanent deformation laws in Eqs. (1) and (4) have at best only a very distant relationship to a realistic nonlinear constitutive response of HMA. Multidimensional confinement and plastic flow interactions, which intuitively should strongly influence the permanent deformation response, are treated only in a very approximate way via the empirical thickness/depth corrections in Eqs. (2) and (5). These issues and others can be more rigorously addressed via nonlinear finite element analysis incorporating more realistic constitutive models for the HMA.

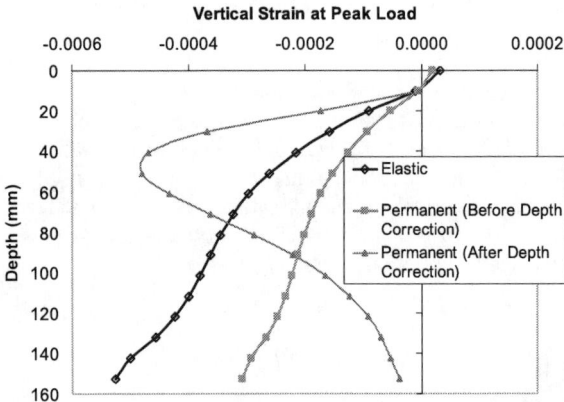

Figure 2. Influence of depth function $\beta_{\sigma 3}$ on calculated permanent vertical strains from the MEPDG model (150 mm HMA layer over crushed stone base, first load cycle).

Finite Element Analyses

Finite element analyses using ABAQUS (2006) were performed to provide a more realistic approximation of the permanent deformation response of an HMA layer in a pavement system. The pavement scenario is very simple: an axially symmetric single tire on a three layer flexible pavement system consisting of 150 mm (6 in) of HMA and 300 mm (12 in) of granular base over natural subgrade. The traffic loading corresponds to a 40 kN wheel with an 830 kPa psi uniform tire pressure over a circular area. Temperature is assumed constant at 38°C throughout the pavement thickness. The HMA layer was modeled both as a linearly elastic material (Table 1) and as a rate-independent elastoplastic material using the Drucker-Prager frictional plasticity model with a linear yield surface and an isotropic piecewise linear hardening law (Table 2). The base and subbase were modeled as linear elastic materials (Table 1). The Drucker-Prager model parameters for the HMA were derived from unconfined and confined triaxial strength test data provided by Kaloush (personal communication) for a typical dense graded HMA mixture at 38°C. Figure 3 compares the predicted vs. measured stress-strain response for the axial strength tests using the calibrated model parameters. Although not a perfect match to the measured data, the elastoplastic finite element (EPFE) predictions are qualitatively acceptable for the types of insights being sought in this study.

Table 1. Linear elastic material properties for all pavement layers.

Layer	Thickness (mm)	Elastic Modulus (MPa)	Poisson's Ratio
HMA (38°C)	150	476	0.35
Base	300	220	0.35
Subgrade	Infinite	57	0.35

Table 2. Drucker-Prager model parameters for the HMA layer (38°C).

Property	Value
Friction angle	43.6°
Initial yield stress	20 kPa
Ultimate yield stress	728 kPa

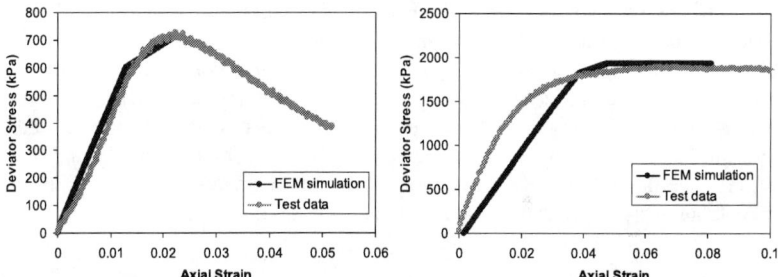

Figure 3. Predicted vs. measured stress-strain response for laboratory triaxial strength tests of typical HMA dense graded mix: (a) unconfined; (b) confined at 276 kPa.

Figure 4 depicts the vertical and horizontal stresses at peak load in the HMA layer along a vertical line beneath the center of the tire as computed in the linear elastic and nonlinear plastic analyses. As might be expected, the vertical stress distributions (Figure 4a) are similar for both cases—i.e., the plastic yielding has a nearly negligible effect. Plastic yielding has a more pronounced influence on the horizontal stresses (Figure 4b), especially in the lower half of the layer where there is clear evidence of yielding as the lateral confinement diminishes. One important observation from Figure 4b is that the horizontal stresses in the nonlinear plastic analysis remain compressive at all times—i.e., the material at the bottom of the layer is yielding under a triaxial confined condition and not under unconfined conditions as commonly employed in laboratory repeated load tests. Horizontal stresses in the lower half of the HMA layer as computed in the EPFE analyses range between approximately 30 to 70 kPa.

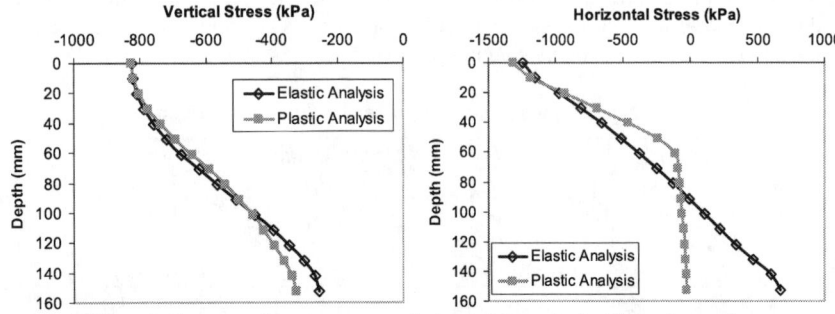

Figure 4. Computed stresses at peak load: (a) vertical; (b) horizontal.

The impact of plastic yielding on the vertical and horizontal total strains in the HMA layer at peak load is significantly more pronounced, as indicated in Figure 5. The vertical compressive strains beneath the tire center (Figure 5a) from the elastic analysis monotonically increase with depth. These are the resilient strains ε_r that are the input for Eq. (4). Consequently, the permanent strains predicted by Eq. (4) will also increase monotonically with depth. The accumulated rutting will therefore be concentrated in the lower portion of the HMA layer, which as described previously is contrary to field experience. The corresponding vertical compressive strains from the EPFE analysis, on the other hand, deviate from the elastic strains in the correct direction, at least qualitatively, with the peak strain occurring near the center of the layer at a depth of about 60 mm. However, rutting in this case is still concentrated at the bottom of the HMA layer. For both analyses, the horizontal strains (Figure 5b) increase monotonically from compression to tension with depth, with the strains from the plastic analysis larger that those from the elastic case, as expected.

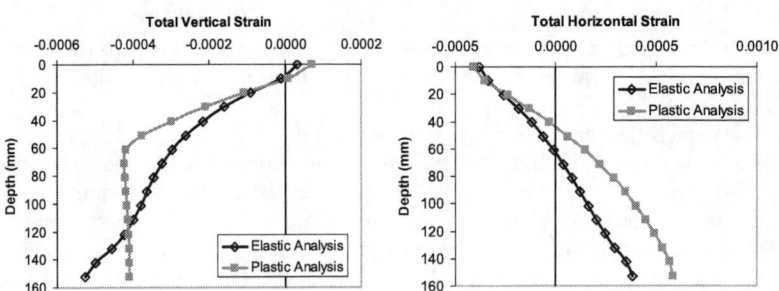

Figure 5. Computed total strains beneath tire center at peak load: (a) vertical; (b) horizontal.

Although not shown here, the differences in the shear stress and strain distributions between the linear elastic and nonlinear plastic analyses are not dramatic, especially in the upper portion of the pavement surrounding the location of the computed shear stress and strain inputs to the Westrack model Eq. (1).

By definition, the elastic stresses and strains at *peak* load are the inputs to the rutting models in the M-E prediction methodology. However, rutting in physical terms is the residual deformation after *removal* of the load. An examination of the residual strains after unloading is therefore instructive. These residual strains for the EPFE analysis are depicted in Figure 6; note that the residual stresses and strains for the elastic analysis are zero by definition. The residual vertical compressive strains after unloading increase with depth until about 100 mm, after which they decrease sharply and eventually become expansive. The yielding at the bottom of the HMA layer under peak load results in residual horizontal compressive stresses after unloading. This residual horizontal compression induces expansive vertical strains—i.e., a decrease in the residual vertical compression--due to the Poisson effect. The residual deformations resulting from this strain distribution are now concentrated in the upper portion of the HMA layer, congruent with field experience. No additional depth correction function is required to bring analysis results into qualitative alignment with physical expectations.

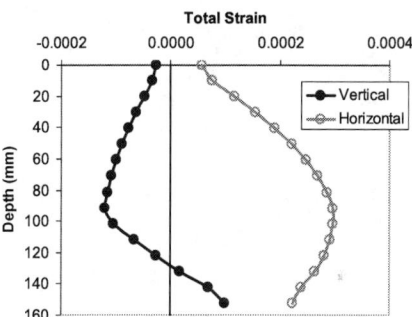

Figure 6. Residual total strains from the EPFE analysis after removal of load.

Another noteworthy feature of the EPFE analysis results is the accumulation of incremental residual deformations over multiple load cycles. Conventional wisdom often purports that a straightforward strain-hardening plasticity analysis of a constant amplitude cyclic loading should produce all permanent deformations in just the first load cycle; since all subsequent load cycles are to the same load magnitude, no additional plastic yielding and/or deformations should develop in the subsequent cycles. As shown in Figure 7, however, this is not the case. Additional plastic deformations develop in each load cycle, with diminishing magnitude in each successive cycle. This matches field observations. The locked-in residual horizontal stresses that are the consequence of plastic yielding at the bottom of the HMA layer introduce stress reversals that produce additional plastic yielding and residual deformation with each successive load-unload cycle.

Direct comparison of rutting predictions from the EPFE analyses and the M-E methods is difficult. The Westrack model requires values for the material parameters a, b, and c in Eq. (1) that are not known for the mixture considered here. The

Westrack (implicitly) and MEPDG (explicitly) models have also been calibrated to observed field performance while the EPFE analyses have not. Consequently, the magnitudes of the predicted strains and HMA rutting are substantially different among the methods. This is illustrated in Figure 8 for the case of the MEPDG; although the shapes of the respective distributions are qualitatively similar, the magnitudes of the residual strains from the EPFE are only about one-fourth of the MEPDG values. A pseudo "calibration" of the EPFE can be performed by scaling the computed residual vertical strains so that the total rutting for the layer matches that predicted by the MEPDG model. These "calibrated" EPFE results are also plotted in Figure 8. Again, it is clear that the distribution of residual strains with depth is quite similar between the "calibrated" EPFE and MEPDG, although the location of the maximum strain is slightly deeper in the EPFE case. The qualitative similarity of the residual strain distributions is also highlighted in Figure 9, which plots the cumulative permanent deformation as a function of depth. The cumulative distributions from the MEPDG and EPFE models are quite similar, and both fit the measured data from MnRoad equally well.

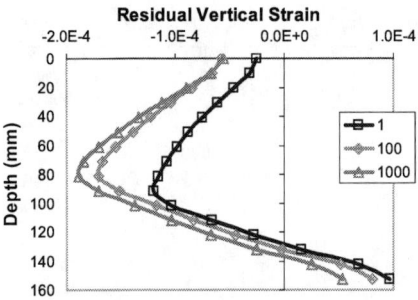

Figure 7. Residual total vertical strains beneath the tire center from EPFE analyses. Legend denotes number of load cycles.

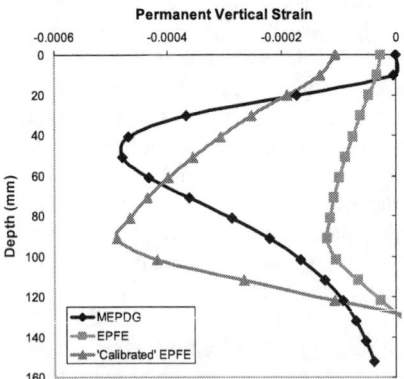

Figure 8. Comparison of uncalibrated and "calibrated" EPFE with MEPDG.

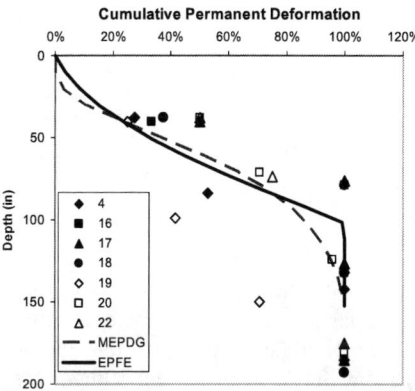

Figure 9. Cumulative permanent deformation distribution as measured at MnRoad and as predicted by the MEPDG and EPFE models. Numbers in legend refer to MnRoad test cells.

Another approach for directly comparing the different rutting models is to examine normalized trends in rutting across different pavement scenarios. Two additional pavement structures were evaluated for this purpose: (a) a full depth asphalt section—i.e., HMA over subgrade, and (b) a composite pavement consisting of a 150 mm HMA layer over a rigid PCC slab. The reference case is the three-layer conventional flexible pavement structure described previously. Properties for all analyses were the same as those given in Tables 1 and 2. Rutting in the wheel path was computed for each of the three models and then normalized by the computed rutting for the conventional flexible pavement structure for that model. Full application of the Westrack model was not possible because values for the material constants a, b, and c in Eq. (1) are unknown. However, it is clear from the form of Eqs. (1) and (2) that rut depth scales linearly with the elastic shear strain γ_e for a given material and HMA layer thickness. Therefore, the normalized rut depths for each pavement structure using the Westrack model will be equal to their respective normalized shear strains.

The normalized rut depths for all three pavement structures as predicted by all three models (MEPDG, Westrack, EPFE) are summarized in Table 3. Only the settlement within the wheel path was considered; as described previously, the M-E rutting models can include surface heave outside the wheel path only indirectly via the field calibration process. The results in Table 3 show some disagreement in trends. The MEPDG model predicts more rutting for the full-depth as compared to the conventional flexible structure and even greater rutting for the composite pavement. The Westrack model also predicts greater rutting for the full-depth structure. However, it deviates from the MEPDG model for the composite pavement, where it predicts less rutting as compared to the conventional flexible structure. The EFPE predictions rank the pavements in the same order as the Westrack model, but quantitatively the predictions are more exaggerated, especially for the composite pavement. More careful validation studies based on full material characterization and measured performance data will be required to resolve these discrepancies.

Table 3. Relative wheelpath settlements for different pavement structures.

Model	Pavement Structure		
	Flexible	Full-Depth	Composite
MEPDG	1.0	3.2	4.9
Westrack	1.0	1.3	0.7
EPFE	1.0	5.1	0.03

Although heave outside the rutted wheel path will often be reduced or eliminated by traffic wander in real pavements, it can still be a factor in heavily rutted sections or when traffic is highly channelized (as in APT tests, for example). The M-E models account for heave only indirectly through the field calibration process. The EPFE model can predict heave directly, however. Figure 10 summarizes the rutting profiles for the conventional flexible, full-depth, and composite structures as predicted by the EPFE analyses. The traffic loading in these analyses corresponds to a fully channelized condition. Since the absolute magnitudes of the surface displacements vary greatly among the three sections, they have been scaled approximately in Figure 10 by factors corresponding to different numbers of loading cycles. In relative terms, the predicted surface heave is a small portion of the total rutting for the full-depth pavement, about half for the conventional flexible section, and nearly all of the total rutting for the composite pavement for the cases considered here.

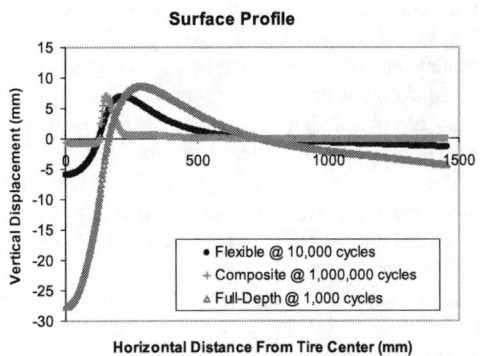

Figure 10. Surface profiles predicted by EPFE analyses.

Summary

Current M-E rutting models relate permanent strains (axial or shear) to resilient strains computed using multilayer elastic theory. However, in the absence of correction factors, the influence of layer thickness on the rutting predictions based on these resilient strains is in sharp disagreement with field observations. Thickness or depth correction factors are required to bring predictions in line with observations.

Thickness correction factors in current models (K in the Westrack model or $\beta_{\sigma3}$ in the MEPDG approach) are the same for all pavement types (e.g., HMA layers in flexible pavements vs. HMA overlays on rigid pavements), unrealistically implying that the mechanisms and distributions of permanent strains are similar. Nonlinear elastoplastic finite element analyses can provide significant qualitative and quantitative insights into these and other issues. Specific observations from the EPFE results presented in this paper include:

- Horizontal stresses predicted by the EPFE analyses remain compressive at all times—i.e., the material at the bottom of the HMA layer is not in tension as predicted by multilayer elastic analyses but is yielding under triaxial confined conditions in contrast to the unconfined conditions commonly employed in laboratory repeated load tests. Horizontal stresses in the lower half of the HMA layer as computed in the EPFE analyses ranged between approximately 30 to 70 kPa.

- M-E predictions of rutting are based on the elastic stresses and strains at *peak* load. In contrast, rutting calculations from the EPFE analyses are based on the more correct physical mechanism of residual deformations after *removal* of the load.

- The EPFE analyses directly predict residual strain variations with depth that are qualitatively similar to measurements from the MnRoad test sections without resort to any empirical depth or thickness correction functions.

- Ordering of relative rut depths in different pavement structures by the different methods is inconsistent. The EPFE ranking is the same as that from the Westrack model but different from the MEPDG ordering. Additional study will be required to resolve these discrepancies.

- The EPFE approach can directly predict heave outside the wheel path. This can be predicted only indirectly by the M-E rutting models via the field calibration process.

Overall, the general approach adopted by current M-E methods is sensible. In conceptual terms it can be expressed as:

$$RD = \beta * S * f(N) \tag{8}$$

in which β is a field calibration coefficient, S is a "structure" factor, and $f(N)$ is a load cycle scaling function. The structure factor S is analogous to K in the Westrack model or $\beta_{\sigma3}$ in the MEPDG (and to $\varepsilon_p(1)$ in the original SHRP models—see Lytton et al., 1993). However, S probably depends upon more than just HMA layer thickness; conceptually, it should also be a function of pavement structure (e.g., conventional flexible vs. composite), applied load (stress) magnitude, available strength at the given pavement temperature, and perhaps other factors. The load scaling function $f(N)$ is analogous to the N^c or $N^{k_3\beta_3}$ terms in the Westrack and MEPDG models, although the exponent terms may perhaps be modestly dependent on relative stress levels, temperature, and/or other factors. Advanced mechanistic modeling approaches like elastoplastic or viscoelastic-viscoplastic finite element analyses can provide insights into the nature of these dependencies for S and $f(N)$.

Acknowledgments

The work reported in this paper was performed as part of National Cooperative Highway Research Program (NCHRP) Project 9-30A. Harold Von Quintus of Applied Research Associates, Inc. is the Principal Investigator and Ed Harrigan is the NCHRP program manager for this project. All views and conclusions expressed in this paper are strictly those of the authors.

References

ABAQUS (2006). *ABAQUS Analysis Users Manual (v6.6)*, HKS, Inc., Providence, RI.

Deacon, J. A., J.T. Harvey, I. Guada, L. Popescu, and C.L. Monismith (2002). "Analytically Based Approach to Rutting Prediction" *Transportation Research Record*, No. 1806, pp 9-18.

El-Basyouny, M.M. (2004). *Calibration and Validation of Asphalt Concrete Pavement Distress Models for the 2002 Design Guide*. Ph.D. dissertation, Arizona State University, Tempe, AZ.

El-Basyouny, Witczak, and Kaloush (2005). "Development of the Permanent Deformation Models for the 2002 Design Guide," *Annual Meetings of the Transportation Research Record*, Washington, DC, January (preprint CD).

Epps, J.A., Hand, A., Seeds, S., Schulz, T., Alavi, S., Ashmore, C., Monismith, C.L., Deacon, J.A., Harvey, J.T., Leahy, R. (2002). *Recommended Performance-Related Specification for Hot-Mix Asphalt Construction: Results of the Westrack Project*. NCHRP Report 455, Transportation Research Board, National Research Council, Washington, DC.

Kaloush, K.E. (2001). "Simple Performance Test for Permanent Deformation of Asphalt Mixtures," Ph.D. dissertation, Arizona State University, Tempe, AZ.

Lytton, R.L., Uzan, J., Fernando, E.G., Roque, R. Hiltunen, D., and Stoffels, S.M. (1993). *Development and Validation of Performance Prediction Models and Specifications for Asphalt Binders and Paving Mixes*, Report No. SHRP-A-357, Strategic Highway Research Program, National Research Council, Washington, DC, October.

Monismith, C.L., L. Popescu, and J.T. Harvey (2006). "Rut Depth Estimation for Mechanistic-Empirical Pavement Design Using Simple Shear Test Results" Association of Asphalt Paving Technologists, Vol. 75, 1294-1338.

Mulvaney, R., and Worel B. (2002). *MnRoad Mainline Rutting Forensic Investigation, Final Report*. Minnesota Road Research Project, Minnesota Department of Transportation, Office of Research Services, Mail Stop 330, St. Paul, MN, October.

NCHRP (2004). "Mechanistic-Empirical Design of New and Rehabilitated Pavement Structures," *NCHRP Project 1-37A Draft Report*, Transportation Research Board, National Research Council, Washington, DC.

Sousa, J. B., J. A. Deacon, S. Weissman, J. T. Harvey, C. L. Monismith, R. B. Leahy, G. Paulsen, and J. S. Coplantz (1994). *Permanent Deformation Response of Asphalt-Aggregate Mixes*. Report No. SHRP-A-415, Strategic Highway Research Program, National Research Council, Washington, D.C.

Westrack Forensic Report, http://www.tfhrc.gov/pavement/pubs/westrack/westrack.htm.

Implications of Complex Axle Loading and Multiple Wheel Load Interaction in Low Volume Roads

Minkwan Kim[1], A. M. ASCE and Erol Tutumluer[2], M. ASCE

Abstract

Various highway vehicle axle/wheel arrangements adopted nowadays to accommodate increasing load levels impact flexible pavement response and performance in accordance with different multiple axle/wheel loading scenarios applied in low volume roads. This paper describes a recent three-dimensional (3D) finite element (FE) pavement modeling research effort focused on investigating effects of complex multiple axle/wheel loading scenarios on mechanistic analyses of critical pavement responses and their implications on low volume pavement designs. For this purpose, realistic nonlinear, stress dependent pavement geomaterial modulus models were employed in the base/subbase and subgrade layers of thinly surfaced asphalt pavements. The low volume road sections were next analyzed using a validated 3D FE structural analysis program. Comparisons made between the single wheel superposition and the full 3D loading results proved the need and importance of 3D FE nonlinear analyses of low volume flexible pavements to properly consider both the stress dependent geomaterial modulus behavior and the implications of multiple wheel loads and their interaction.

Keywords: Low volume roads, multiple wheel loads, superposition, finite element analysis

Introduction

As the demand for heavier wheel loads and number of load applications increases, it becomes especially important to properly characterize the nonlinear behavior of pavement foundation geomaterials, i.e., subgrade soils and base/subbase unbound aggregates, in flexible pavement structures. The resilient or elastic behavior of these geomaterials is highly stress dependent with coarse-grained granular materials showing stress-hardening and fine-grained soils, such as silts and clays, exhibiting stress-softening type response. General-purpose finite element (FE) programs currently being used for two- and three-dimensional (2D and 3D) pavement analyses of multiple axle/wheel loading scenarios often enable the use of available built-in material constitutive models that are not suitable for characterizing such nonlinear resilient behavior of pavement geomaterials. However, to predict accurately pavement response to traffic loading and successfully implement mechanistic-empirical (M-E) pavement design procedures, FE based pavement analyses should properly consider the nonlinear resilient behavior of geomaterials.

In addition, to compute accurate pavement responses, i.e., stress, strain,

---

[1]Indiana Department of Transportation, Greenfield, IN 46140; (PH)317-467-3984, mkim@indot.in.gov
[2]Department of Civil and Environmental Engineering, University of Illinois, Urbana, IL 61801; (PH)217-333-8637, tutumlue@uiuc.edu

deflection, under multiple wheel loads typically applied by typical highway vehicle axle/wheel arrangements, effects of load magnitude and multiple wheel load interaction should also be realistically taken into account in M-E based pavement design. The principle of superposition has commonly been considered as a proper approach to study loading effects of multiple wheels. However, the single wheel load response superposition, especially in the case of nonlinear analysis, may not be adequate to analyze pavements subjected to such multiple wheel load cases.

This paper presents findings from a recent 3D pavement modeling research effort at the University of Illinois focused on the resilient response predictions of mostly low-volume, thinly surfaced pavement structures subjected to complex multiple wheel loading scenarios. The objective has been to obtain more accurately the critical pavement responses from the 3D nonlinear FE analysis and determine implications of multiple axle/wheel load interactions on pavement designs. The 3D FE modeling approach is validated first by predicting field measured pavement responses. Comparisons are then made between the single wheel superposition and full 3D loading results to emphasize the differences between responses of various low volume pavement geometries subjected to single, tandem, and tridem axle multiple wheel loads.

Previous Pavement Studies on Multiple Wheel Load Interaction

To investigate the effect of multiple wheel load interaction, several full-scale test studies on instrumented pavement sections and numerical modeling analyses have been conducted. As part of the early efforts, Chou and Ledbetter (1973) calculated the final pavement responses, such as stress, strain, deflection, equivalent to the summation of the results from each single wheel load case in a Corps of Engineers (COE) study performed in Stockton Airfield followed by the work done related to the multiple wheel heavy gear load tests conducted at the Waterways Experiment Station (WES). The main objective of their study was to investigate the validity of the principle of superposition for airfield flexible pavement analysis. Several loading cases were considered for static and dynamic wheel loads. For the superposition of various single wheel load levels, higher measured deflections were reported when compared to the superposed values, and the superposed stress values tended to be lower than the actual measured stresses for a stress-softening clayey silt section. Yet, the opposite was observed for the stress-hardening sand section. In the end, however, they concluded that when single wheel responses were correctly measured and each wheel had the same load, the superposition for multiple wheel loads was a reasonably valid approach.

Federal Aviation Administration (FAA) constructed National Airport Pavement Test Facility (NAPTF) where the primary objective was to develop new airport pavement design procedures for the next generation aircraft configured with complex and large loading gears (Thompson and Garg, 1999; Hayhoe and Garg, 2002; Gomez-Ramirez, 2002). To quantify load induced responses from aircraft multiple wheel gears, six flexible pavement sections were constructed for the first cycle of testing over low, medium, and high strength subgrades at the NAPTF (Hayhoe and Garg, 2002). The pavement sections were loaded by typical aircraft gear configurations, i.e., dual single, dual tandem, and dual tridem. Vertical subgrade deformations/strains measured from Multi-Depth Deflectometers (MDDs) in the NAPTF first cycle tests showed that accurately predicting pavement responses was extremely difficult (Hayhoe and Garg, 2002). The elastic pavement layer behavior did not well represent significant strain differences between the first and the last peaks of wheel passages. After terminating trafficking on the first cycle tests, the next set of flexible pavements with variable subbase thicknesses were also built over the low strength subgrade to determine the adequacy of subbase thickness designs needed to protect the weak subgrade.

Thompson and Garg (1999) introduced an "Engineering Approach" to determine critical pavement responses under typical multiple wheel aircraft gear loadings and evaluate wheel load interaction effects on the flexible pavement responses. The "Engineering Approach" used average layer modulus values computed from nonlinear axisymmetric ILLI-PAVE FE analysis and these values established the inputs for elastic layered analyses to solve for multiple wheel loading scenarios. The actual modulus distributions were, however, much different from the single modulus assignment into the entire horizontal pavement layer. Based on the findings of the FAA's NAPTF full scale pavement tests, Gomez-Ramirez (2002) also proposed that the principle of superposition could be applied to the design and analysis of airport pavements subjected to aircraft gear loads, if single wheel nonlinear responses were accurately determined. In this approach, the modulus distributions were not realistic in accordance with stress distributions or stress bulbs in the layer. The only way to consider realistic 3D stress dependent modulus distributions under individual wheel loads would be through performing a full 3D structural analysis for pavements subjected to multiple wheel loads.

Nonlinear Stress Dependent Geomaterial Models

The deformation behavior of pavement layers due to repeated loading consists of both elastic or resilient and permanent deformations. If the stress due to applied wheel load is much smaller than the pavement layer material strength and is repeated for a large number of applications, the permanent deformation per each wheel loading becomes negligible. Accordingly, the load-deformation behavior can be considered as elastic or resilient in the mechanistic analysis of flexible pavement structures.

For all the mechanistic FE analyses conducted in this study, the asphalt concrete (AC) layer was considered as linear elastic with only 2 elastic material constants, elastic or resilient modulus (M_R) and Poisson's ratio (ν), used as inputs.

For the fine-grained subgrade soils, the nonlinear behavior was taken into account with a stress dependent resilient modulus, which decreased in proportion to the increasing stress levels to properly exhibit the well known stress-softening behavior. The bilinear model by Thompson and Robnett (1979) was chosen as the resilient modulus model for this the nonlinear subgrade characterization expressed by the following modulus-deviator stress relationship:

$$M_R = K_1 + K_3 \times (K_2 - \sigma_d) \text{ when } \sigma_d \leq K_2$$
$$M_R = K_1 - K_4 \times (\sigma_d - K_2) \text{ when } \sigma_d \geq K_2 \quad (1)$$

where K_1, K_2, K_3, and K_4 are material constants obtained from repeated load triaxial tests and $\sigma_d (=\sigma_1 - \sigma_3)$ is the deviator stress.

Uzan (1985) model properly considers the effects of stress dependency for modeling the nonlinear behavior of base aggregates and are generally suitable for FE programming and practical design use. Especially, this model considers the effects of both confining and deviator stresses and therefore handles very well the modulus or stiffness increase with increasing vertical and horizontal stresses in an unbound aggregate layer. The Uzan (1985) model is given as follows:

$$M_R = K_1 \left(\frac{\theta}{p_o}\right)^{K_2} \left(\frac{\sigma_d}{p_o}\right)^{K_3} \qquad (2)$$

where $\theta = \sigma_1 + \sigma_2 + \sigma_3 = \sigma_1 + 2\sigma_3 =$ bulk stress, $\sigma_d = \sigma_1 - \sigma_3 =$ deviator stress, p_0 is the unit pressure of 1 kPa, and K_1, K_2, and K_3 are multiple regression constants obtained from repeated load triaxial test data on granular materials.

Witczak and Uzan (1988) proposed the use of universal model developed as an improvement over the Uzan model (1985) by replacing the deviator stress term with the octahedral shear stress (τ_{oct}). Since τ_{oct} considers stresses in all three orthogonal directions, i.e., x, y, and z directions, the universal model, given as follows, is more suitable for 3D FE pavement analysis:

$$M_R = K_1 p_a \left(\frac{I_1}{p_a}\right)^{K_2} \left(\frac{\tau_{oct}}{p_a}\right)^{K_3} \qquad (3)$$

where $I_1 = \sigma_1 + \sigma_2 + \sigma_3$, $\tau_{oct} = 1/3\{(\sigma_1-\sigma_2)^2 + (\sigma_1-\sigma_3)^2 + (\sigma_2-\sigma_3)^2\}^{1/2}$, p_a is the atmospheric pressure, and K_1, K_2, and K_3 are multiple regression parameters obtained from repeated load triaxial test data.

To employ nonlinear material models, a user-defined material subroutine, known as a UMAT in ABAQUS™ (ABAQUS, 2005) FE analysis, was successfully utilized in this study. This UMAT material subroutine gave a good convergence of the iterations for computing nonlinear resilient modulus distributions within the base and subgrade layers and the moduli obtained in axisymmetric analyses were in agreement with the distributions predicted by the GT-PAVE nonlinear pavement FE analysis program (Kim and Tutumluer, 2006 & 2007; Tutumluer and Kim, 2006). For further verification, ABAQUS FE pavement responses computed from 3D analyses were found to be very similar to those from the axisymmetric analyses when the same nonlinear universal model was used in the UMAT subroutine under the application of single wheel loading (Kim and Tutumluer, 2007).

Field Validation of the Nonlinear FE Solutions

The developed nonlinear geomaterial models employed in the ABAQUS FE program had to be validated for accurately predicting pavement responses under multiple wheel loads. For this purpose, the field measured responses of the National Airport Pavement Test Facility (NAPTF) flexible pavement test sections were utilized (Gopalakrishnan, 2004). The measured responses were collected using multi depth deflectometers (MDDs) and pressure cells (PCs) installed in the test sections. The field results of medium strength conventional pavement section, herein MFC, were chosen for validation of the distinct effect of nonlinear stress dependent geomaterials. Table 1 lists the MFC section pavement layer thicknesses and material properties including the nonlinear resilient modulus model parameters of the unbound aggregates and subgrade materials used in the NAPTF MFC section (Gopalakrishnan, 2004). The AC surface layer was modeled as linear elastic.

Table 1. Pavement Geometry and Material Properties Used in the Validation Study

Pavement Layer	Thickness (mm)	E or M_R (MPa)	ν	Material Properties		
Asphalt Concrete	127	8,268	0.35	Isotropic and Linear Elastic		
Base	203	Nonlinear	0.38	Nonlinear: $M_R = K_1 (\theta)^{K2}(\sigma_d)^{K3}$		
				K_1 (MPa)	K_2	K_3
				10.3	0.40	0.0
Subbase	305	Nonlinear	0.38	Nonlinear: $M_R = K_1 (\theta)^{K2}(\sigma_d)^{K3}$		
				K_1 (MPa)	K_2	K_3
				6.9	0.64	0.0
Subgrade	-	Nonlinear	0.40	Nonlinear: Bilinear model (Eq.1)		
				$K_1 = E_{Ri}$ (MPa)	$K_2 = \sigma_{di}$ (MPa)	K_3 K_4
				62.8	0.042	420 570

Next, 3D FE analyses were performed to compute the pavement responses under aircraft gear loadings and to compare them with the measured responses of NAPTF. Figure 1 illustrates the modeled pavement geometry and the FE mesh, which consisted of 20-noded hexahedron solid elements. A six-wheel dual-tridem type aircraft gear configuration, similar to that of Boeing 777 aircraft, applied individual wheel loads of 20 metric tonnes with 1372 mm wheel spacing and 1448 mm axle spacing. The single wheels were assumed in the analyses to apply uniform tire pressure of 1.3 MPa over a circular area.

In general, the nonlinear FE mechanistic model predictions were in reasonably good agreement with the measured responses of the NAPTF MFC test section. Figure 2 shows the comparisons from measured and predicted vertical subgrade stresses and surface deflections for the MFC section. Considering the variability of the pressure cell and MDD measurements in the field, predicted ABAQUS nonlinear FE results showed good agreement with measured values. The predicted responses are in the range of measured responses, which validates the adequacy of the subgrade and unbound aggregate base/subbase nonlinear UMAT characterizations.

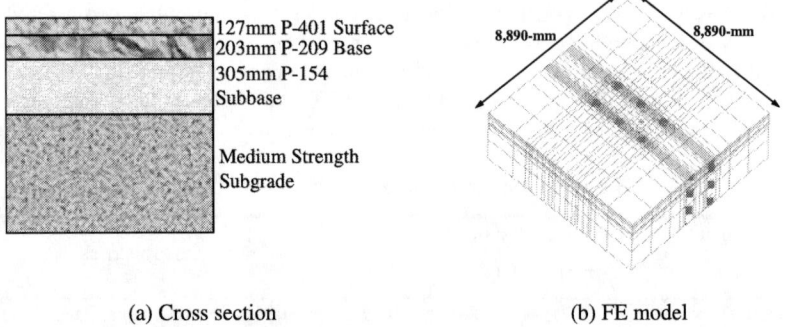

(a) Cross section (b) FE model

Figure 1. Pavement Geometry & FE Mesh used to Analyze NAPTF MFC test section

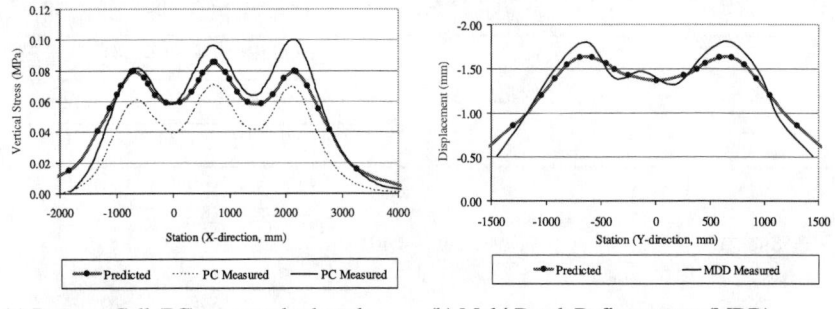

(a) Pressure Cell (PC) measured subgrade stresses

(b) Multi Depth Deflectometer (MDD) measured surface deflections

Figure 2. Comparisons between Measured and Predicted Responses in the NAPTF MFC Pavement Test Section

Full 3D and Single Wheel Superposition Analyses

Figure 3 shows the typical highway vehicle axle/wheel arrangements (Huang, 1993) used to study effects of multiple wheel loads and multiple wheel load interaction through 3D FE pavement structural analyses. Typically, the spacing of 343 mm was considered for each wheel and 1,219 mm for each axle. The 3D FE analyses were performed for all the single, tandem, and tridem axle arrangements applying multiple wheel loads on a typical low volume flexible highway pavement having only 76 mm and 102 mm of AC surfaces.

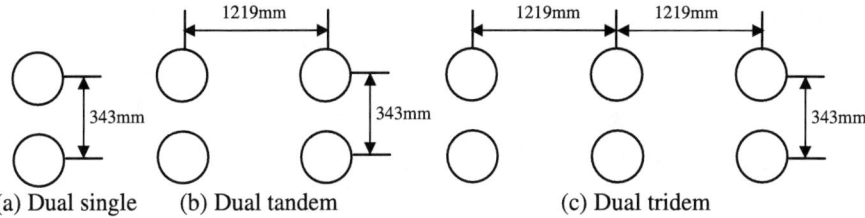

(a) Dual single (b) Dual tandem (c) Dual tridem

Figure 3. Different Circular Contact Areas Associated with Typical Axle/wheel Arrangements

Using the developed ABAQUS UMAT subroutine, conventional flexible pavements analyzed consisted of a linear elastic AC layer underlain by nonlinear elastic unbound base and nonlinear subgrade layers. Table 2 gives material properties and nonlinear model parameters assigned to the pavement layers. The conventional flexible pavement geometries analyzed within four different pavement section cases ranged from relatively thin to substantially thick granular layers listed as follows:

Case (1): 102 mm of AC and 152 mm of aggregate base;
Case (2): 102 mm of AC and 254 mm of aggregate base;
Case (3): 76 mm of AC and 305 mm of aggregate base;
Case (4): 76 mm of AC and 457 mm of aggregate base.

Table 1. Pavement Material Properties Assigned in the 3D FE Analyses

Layer	E (MPa)	ν	Material Properties			
AC	2,759	0.35	Isotropic and Linear Elastic			
Base	138 (initial guess)	0.4	Isotropic and Nonlinear: Universal model by Witczak and Uzan (Eq. 3)			
			K_1 (kPa)	K_2	K_3	
			1,098	0.64	0.065	
Subgrade	41 (initial guess)	0.45	Isotropic and Nonlinear: Bilinear model by Thompson and Robnett (Eq. 1)			
			K_1 (E_{RI}) (kPa)	K_2 (σ_{di}) (kPa)	K_3	K_4
			41,369	41	1,000	200

As shown in Figure 4, the FE structural analyses were then conducted using the first order 8-noded isoparametric linear hexahedron elements in a FE mesh having sizes of 6,096 mm in the horizontal direction and 21,336 mm in the vertical direction. All vertical boundary nodes had roller supports with fixed horizontal boundary nodes used at the bottom of the FE mesh. A uniform pressure of 0.55 MPa was applied over a circular area of 107 mm radius.

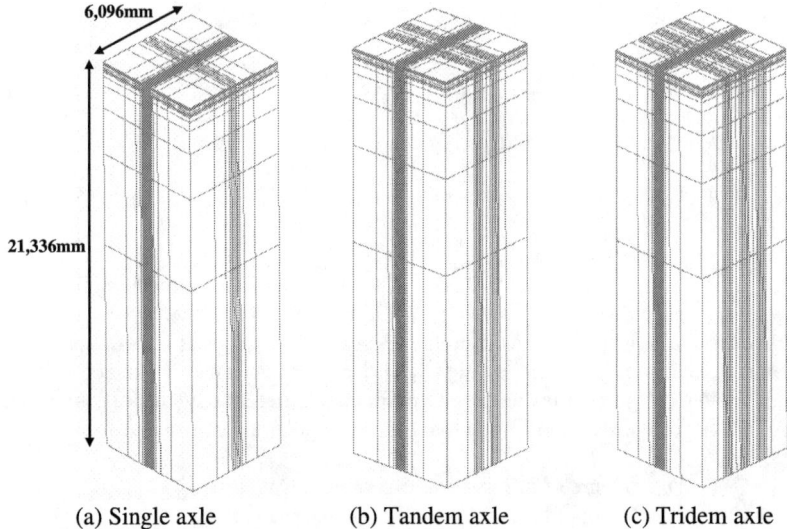

(a) Single axle   (b) Tandem axle   (c) Tridem axle

Figure 4. 3D FE Meshes used in Various Multiple Wheel Loading Cases

In the superposition analysis, the superposed responses determined at a specific pavement loading location were obtained by adding the contribution of each single wheel applied at the corresponding radial offset distance. Accordingly, at that pavement location of interest, the contributions of 2, 4, and 6 wheels were superposed for the single, tandem and tridem axle configurations, respectively (see Figure 3). To quantify the differences, the differences in critical pavement responses between full 3D and superposition analyses were computed using the following equation.

$$\text{Difference}(\%) = \frac{\text{3D Multiple Wheel Response} - \text{Superposed Response}}{\text{3D Multiple Wheel Response}} \times 100 \quad (4)$$

In the case of linear elastic base and subgrade with linear elastic AC material properties, the results from full 3D and superposition analyses do not show any differences as expected. However, when nonlinear pavement geomaterial models are considered, the pavement responses from full 3D loading and superposition from single wheel show differences. Accordingly, Figure 5 summarizes the differences in computed pavement surface deflections and subgrade vertical strains between the full 3D analysis results and the single wheel responses from superposition. Since most pavement responses obtained from single wheel superposition are larger than the full 3D FE analysis results, the differences show the negative sign implying that the superposition is in general more conservative except for the subgrade vertical strains under thin pavements with only 152 to 254 mm base course thicknesses. The thickest pavement (Case 4), which corresponds to 76 mm of AC and 457 mm of aggregate base, results in the maximum differences in computed surface deflections, i.e., 19% difference from single axle, 12% from tandem axle, and 12% from tridem axle. Therefore, such results indicate that the superposition from single wheel loading would not capture adequately effects of multiple wheel load interaction on nonlinear pavement responses. Especially when a high level of material nonlinearity exists and a thick granular layer is considered, large differences will be expected between the results from full 3D and superposition solutions.

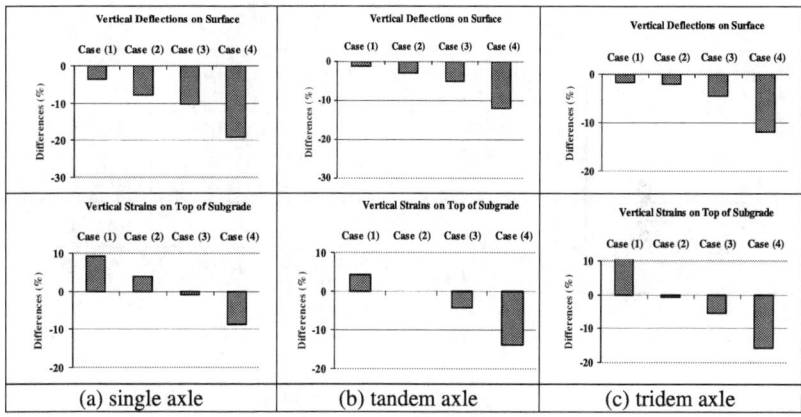

Figure 5. The Differences of Computed Surface Deflections and Subgrade Vertical Strains between the 3D and Superposition Analyses

Multiple Wheel Load Interaction

Using the full 3D FE analysis results, critical pavement responses, i.e., surface deflection ($\delta_{surface}$), horizontal strain at the bottom of AC (ε_h) and vertical strain on top of subgrade (ε_v), were next obtained under the single, tandem and tridem axle load arrangements in an effort to study effects multiple wheel load interaction. To identify the critical pavement response locations under multiple wheel loads, two different conventional flexible pavement geometries, cases (2) and (3), were mainly selected to represent typical low volume roads. In all nonlinear analyses, stress-dependent base and subgrade characterization models were used as given in Table 1 with the linear elastic AC properties.

Table 2 gives detailed comparisons of the pavement responses along the traffic loading direction in accordance with the axle/wheel arrangements shown in Figure 3. The single axle surface deflection responses are indicated as the minimum deflections and the largest responses were obtained from the tridem axle arrangement. While for ε_h at the bottom of AC the critical pavement response location is under the wheel, for $\delta_{surface}$ and ε_v on top of subgrade, the critical pavement response occurs in between the wheels at location. The wheel/axle load interaction was a significant factor affecting pavement responses. Especially, surface displacements were more influenced than others in the nonlinear analyses. This 3D numerical modeling study has utilized realistic nonlinear geomaterial characterizations in the granular base and subgrade and, as a result, clearly indicated the need to better evaluate the locations and magnitudes of critical pavement responses in low volume pavements subjected to various multiple axle/wheel loading scenarios.

Table 2. Comparisons of Predicted Critical Pavement Responses from 3D Nonlinear FE analyses

Case (2): 102-mm of AC and 254-mm of aggregate base				
Pavement response	Location*	Single Axle	Tandem Axle	Tridem Axle
$\delta_{surface}$ (mm)	1	-6.46×10^{-01}	-7.63×10^{-01}	-9.08×10^{-01}
	2	-6.73×10^{-01}	-7.77×10^{-01}	-9.23×10^{-01}
$\varepsilon_{h\ bottom\ of\ AC}$ ($\mu\varepsilon$)	1	204	207	211
	2	-5	0	5
$\varepsilon_{v\ top\ of\ subgrade}$ ($\mu\varepsilon$)	1	-617	-581	-590
	2	-671	-639	-648
Case (3): 76-mm of AC and 305-mm of aggregate base				
Pavement response	Location*	Single Axle	Tandem Axle	Tridem Axle
$\delta_{surface}$ (mm)	1	-7.37×10^{-01}	-8.43×10^{-01}	-9.87×10^{-01}
	2	-7.36×10^{-01}	-8.44×10^{-01}	-9.89×10^{-01}
$\varepsilon_{h\ bottom\ of\ AC}$ ($\mu\varepsilon$)	1	230	230	231
	2	-99	-97	-94
$\varepsilon_{v\ top\ of\ subgrade}$ ($\mu\varepsilon$)	1	-710	-667	-665
	2	-762	-723	-725

*1 and 2 indicate the response locations directly under the wheel (1) and between the wheels (2) showing largest or peak response values, respectively.

Conclusions

This paper aimed to present implications of complex axle loadings and multiple wheel load interaction in the mechanistic analysis of low volume roads. Three-dimensional (3D) finite element (FE) models were developed to analyze thin asphalt concrete (AC) surfaced flexible pavements using the general purpose ABAQUS FE program equipped with a user material UMAT subroutine, which adequately accounted for the nonlinear resilient behavior of pavement geomaterials, e.g., coarse-grained unbound aggregates and fine-gained subgrade soils. To validate the accuracy of nonlinear characterization models, the FE analysis predictions were compared to field measured pavement responses of the FAA's National Airport Pavement Test Facility (NAPTF) test sections. The good agreement between the predicted and field measured pavement responses proved that the ABAQUS UMAT analysis could be reasonably applied to the analysis and design of low volume pavements subjected to complex axle loadings with multiple wheels when the nonlinear behavior of the pavement geomaterials was adequately considered.

This study investigated the applicability of the commonly used superposed single wheel approach subjected to single, tandem and tridem axle loadings in nonlinear analyses of pavements. When superposed responses obtained were compared to the full 3D analysis results with nonlinear base and subgrade properties, important differences in critical pavement responses were found. Further, the multiple wheel/axle load interaction was a significant factor affecting pavement responses. Especially, surface displacements were more influenced than others in the nonlinear analyses. While for horizontal strain at the bottom of AC the critical pavement response location was under the wheel, for surface deflections and subgrade vertical strains, the critical pavement response occurred in between the wheels.

The significant findings from this modeling study have clearly established the need and importance of 3D FE nonlinear analyses of low volume flexible pavements to properly consider both the stress dependent geomaterial modulus behavior and the implications of multiple wheel loads and their interaction.

References

ABAQUS (2005). *ABAQUS/Standard User's Manual*, Version 6.5. Hibbit, Karlsson & Sorensen, Inc., Pawtucket, R.I.
Chou, Y.T., and Ledbetter, R. H. (1973). *The Behavior of Flexible Airfield Pavements under Loads - Theory and Experiments*. M.P. S-73-66, U.S. Army Engineer Waterways Experiment Station, Vicksburg, MS
Gomez-Ramirez, F. (2002). *Characterizing Aircraft Multiple Wheel Load Interaction for Airport Flexible Pavement Design*, Ph.D. Dissertation, University of Illinois, Urbana, IL.
Gopalakrishnan, K. (2004). *Performance Analysis of Airport Flexible Pavements Subjected to New Generation Aircraft*, Ph.D. Dissertation, University of Illinois, Urbana, IL.
Hayhoe, G.F. and Garg, N. (2002). "Subgrade Strains Measured in Full-Scale Traffic Tests with Four- and Six-Wheel Landing Gears." *In Proceedings of the Federal Aviation Administration Airport Technology Transfer Conference*, Atlantic City, NJ
Huang, Y. H. (1993). *Pavement Analysis and Design*, 1st Ed., Pearson Prentice Hall, NJ.
Kim, M. and Tutumluer, E. (2006). "Modeling Nonlinear, Stress Dependent Pavement Foundation Behavior Using A General-Purpose Finite Element Program." ASCE Geotechnical Special Publication, 154, entitled, *Pavement Mechanics and Testing,* Edited by B. Huang, R. Meier, J. Prozzi, and E.

Tutumluer.
Kim, M. and Tutumluer, E. (2007). "Nonlinear Pavement Foundation Modeling for Three-dimensional Finite Element Analysis of Flexible Pavements." In CD-ROM Proceedings of the 86th Annual Meeting of the Transportation Research Board, Washington, D.C.
Raad, L. and Figueroa, J.L. (1980). "Load Response of Transportation Support Systems." *Journal of Transportation Engineering*, ASCE, 106(1), 111-128.
Thompson, M. R. and Robnett, Q. L. (1979). "Resilient Properites of Subgrade Soils. Journal of Transportation Engineering." *Journal of Transportation Engineering*, ASCE, Vol. 105, No.TE1.
Thompson, M.R., and Garg, N. (1999). "Wheel Load Interaction: Critical Airport Pavement Responses." *In Proceedings of the Federal Aviation Administration Airport Technology Transfer Conference*, Atlantic City, NJ.
Tutumluer, E. (1995). *Predicting Behavior of Flexible Pavements with Granular Bases*, Ph.D. Dissertation, Georgia Institute of Technology, GA
Tutumluer, E. and Kim, M. (2007). "Considerations for Nonlinear Analyses of Pavement Foundation Geomaterials in the Finite Element Modeling of Flexible Pavements," In ASCE Geotechnical Special Publication No. 176, entitled, *Analysis of Asphalt Pavement Materials and Systems*, Edited by L. Wang and E. Masad.
Uzan, J. (1985). "Characterization of Granular Material." *Transportation Research Record 1022*, TRB, National Research Council, Washington .D.C., pp. 52-59.
Witczak, M. W., and Uzan, J. (1988). *The Universal Airport Pavement Design System: Granular Material Characterization*, University of Maryland, Department of Civil Engineering, MD.

Measuring the Specific Gravity and Absorption of Steel Slag and Crushed Concrete Coarse Aggregates: A Preliminary Study

Julian Mills-Beale[1] and Zhanping You [2]

[1] Ph.D. Student and Research Assistant, Michigan Technological University, Department of Civil and Environmental Engineering, Houghton, Michigan 49931, USA. Tel: (906)487-2528, Fax: (906)487-1620, Email: jnmillsb@mtu.edu

[2] Tomasini Assistant Professor of Transportation Engineering and Director of the Transportation Materials Research Center (Center of Excellence for Transportation Materials), Michigan Technological University, Department of Civil and Environmental Engineering, Houghton, Michigan 49931, USA. Tel: (906)487-1059, Fax: (906)487-1620, Email: zyou@mtu.edu

ABSTRACT

The research objective is to evaluate a new test procedure for determining the specific gravity and absorption of steel slag and crushed concrete, which are recycled and manufactured coarse aggregates, respectively. Steel slag and crushed concrete, unlike many coarse aggregates possess special physical characteristic such as many effective surface voids. It is hypothesized that the current *AASHTO T 85* (24±4 hours) test method fails to adequately saturate these effective voids on the coarse aggregate surfaces. When the steel slag and crushed concrete were soaked for 36 and 48 hours, it underscored the fact that *AASHTO T 85* underestimates the absorption capacity of special porous coarse aggregates. A new proposed vacuum saturation approach is thus utilized to suck the trapped air within the effective pores while replacing it with water. This method is expected to better determine the specific gravity and absorption of these special aggregates. The steel slag and crushed concrete were tested at 30 minutes of vacuum saturation under 30mm Hg pressure. Statistical analysis indicated that this proposed method was viable in determining the specific gravities and absorption of these aggregates. An extended research is ongoing to expand the scope of coarse aggregates in addition to testing at reduced saturation times to increase the efficiency of the proposed method.

Keywords: Specific Gravity, Absorption, Coarse Aggregates, *AASHTO T 85*, Vacuum Saturation, Steel Slag, Crushed Concrete

1 INTRODUCTION

In the design and construction of airfield and highway pavements, the specific gravity, (Gs) and water absorption capacity (Wa %) of coarse aggregates are key material parameters. They are used in the pavement mixture volumetric analysis, workability determination and freeze thaw estimation. Three types of specific gravity measurements exist – the bulk dry (Gsb, dry), bulk at saturated surface-dry state (Gsb, SSD) and the apparent specific gravity (Gsa). The surface textural characteristics of coarse aggregates are believed to greatly influence the measurement of their specific gravities and absorption. Different types of coarse aggregates exist based on their surface characteristics. While some have minimal surface voids, others have a high number of surface voids. Steel slag and crushed concrete, manufactured and recycled coarse aggregates respectively, are two types of aggregates that have high surface voids.

Many researchers have in the past sought to effectively determine the porosity and its associated water absorptive capacity of coarse aggregates. Boyle's law which relates the pressure applied on a system to its volume was examined to evaluate its feasibility in measuring the effective porosity of coarse aggregate voids by Washburn et al.(1922).

Further research on measuring the effective porosity of coarse aggregate surface voids was conducted by Dolch after modifying Washburn's McLeod gauge porosity (1959) . The approach evaluated the effective void volume by lowering the device head on a dry aggregate sample while it is immersed in mercury. Another method of determining the specific surface of a solid was the sorption method as proposed by Brunauer et al.(1938). This sorption absorption method is used in obtaining a curve for the pore size distribution and ultimately the effective porous voids. A number of other methods that was explored at pore size distribution are small angle – ray scattering, heat of immersion, rate of dissolution, ionic adsorption, and radioactive and electrical methods.

A device known as the mercury porosimeter, which was an enhancement of Washburn's McLeod gauge porosity meter, was designed by Ritter et al. (1945) to improve the penetration of mercury into aggregate pores using high pressure. Drake's high pressure mercury penetration method was used to successfully test PCC aggregates over a range of pore sizes in a research work undertaken by Hiltrop et al. (1960).

The injection of mercury into the pore system of aggregates was used in determining the coarse aggregates absorption by Ritter et al.(1948). The principle was that pores on the surfaces of coarse aggregates existed as a system of circular capillaries.

In another development, Washburn et al. (1922) established that the relationship between the applied pressure, p; radius of the pore, r; surface tension of mercury, σ; contact angle between the mercury and the solid coarse aggregate, θ; is given as $p = -2\sigma \cos\theta$.

The use of vacuum saturation to expel air voids within loose mixtures was first proposed by Rice (1952). Using the Rice chamber and assigned pressure of

30mm Hg, Rice developed a method of removing all trapped air voids within a loose uncompacted HMA mixture. The method became known as the standard test procedure for finding the theoretical maximum density of loose uncompacted HMA mixtures; and also known as the Gmm test.

Attempts at investigating a new, simple and more reproducible method to determine the bulk specific gravity or the saturated surface-dry condition for granular materials was undertaken by Lee et al. (1970). The outcome of this research was the development of a new chemical indicator method to ascertain the saturated surface-dry condition (SSD), in addition to a glass mercury pycnometer which measured the bulk specific gravity of large aggregates.

A little over a decade ago, Vitton et al.(1999) showed how an automated helium pycnometer in coarse aggregate specific gravity analysis. They endeavored to use helium gas, which is more easily absorbed into coarse aggregates pore spaces. The helium pycnometer uses the ideal gas law, $PV=nRT$, where P is the absolute pressure applied; T, the absolute temperature; V, the volume of sample; n, the number of gas molecules; R, the universal gas constant. Vitton also used the automated envelope density analyzer to find the bulk volume or envelope volume of coarse aggregate samples, and finally the specific gravities.

Vacuum sealing coarse aggregates in a plastic bag has been used to measure the specific gravity of coarse aggregates as shown by Hall (2002). Hall examined how this vacuum sealing method compares with the current standard *AASHTO T 85* in terms of specific gravity and absorption determination. Hall focused on individual or single gradation and blended coarse aggregates. The research proved that the vacuum sealing approach had statistically similar results with those of the *AASHTO T 85*.

2 PROBLEM STATEMENT

The accepted standard for testing coarse aggregate specific gravity and absorption is the *AASHTO T 85*. The test method is based on the assumption that 24±4 hours' soaking fully satisfies the full water absorption capacity of coarse aggregates – both low and high absorptive aggregates. The research postulates that when it comes to special aggregates such as those with a high percentage of surface voids like steel slag and crushed concrete, the *AASHTO T 85* fails to satisfy the maximum water absorptive capacity of these coarse aggregates. Underestimating the water absorption of porous coarse aggregates, it is believed, will have an effect on the calculated specific gravity results (bulk dry and saturated-surface dry) due to the high percentage of surface voids involved. The search for an improved method at determining the absorptive potential of porous coarse aggregates is thus crucial in the accurate determination of these specific gravity and absorption calculations.

3 RESEARCH METHODOLOGY

The verification of the hypothesis that *AASHTO T 85* underestimates the absorptive potential of porous coarse aggregates was conducted by increasing the soaking time. 36 and 48 hours were chosen as soaking times for the selected porous coarse aggregates and their absorption values compared with that of 24 hours soaking.

The proposed method of improving the testing of the specific gravity and absorption of the porous coarse aggregates involves the use of Rice's vacuum saturation procedure of HMA theoretical maximum density test *(AASHTO T 209)* to initially saturate all the effective porous voids on the coarse aggregates. 30 minutes of 30mm mercury vacuum saturation is therefore used to remove all entrapped air within the effective porous voids while causing water to penetrate those pores. Vacuum saturating the porous coarse aggregates is expected to be the most suitable method at satisfying the water absorptive capacity of the aggregates. It acts as a replacement for the conventional 24±4 hours soaking which is used to affect water penetration into coarse aggregate pores. Improving the water penetrability of porous aggregates will redefine the current *AASHTO T 85* method of testing the specific gravity and absorption of these special aggregates. After vacuum saturating the coarse aggregates, the standard *AASHTO T 85* test method is followed to determine the specific gravity and absorption of the materials.

4 MATERIALS TESTED

Steel slag and crushed concrete were tested during this research. They represent the scope of coarse aggregates that possess special textural characteristic like voids to surface voids which has immense potential to trap water. The steel slag and crushed concrete were tested at the gradations indicated in Table 1. The steel slag and crushed concrete represented manufactured and recycled coarse aggregates, respectively. Three replicate tests were conducted on the coarse aggregates and the results averaged to represent the specific gravity and absorption.

Table 1. Coarse aggregates and their sieve size fractions

Coarse Aggregate Material	Sieve Size Tested (Retained)
Steel Slag	#4
	3/8"
	1/2"
	3/4"
Crushed Concrete	#4
	3/8"
	1/2"
	3/4"

5 TEST METHOD

5.1 Sample Preparation

The crushed concrete and steel slag coarse aggregates were prepared according to the *AASHTO T 85* specification. The specification required that a minimum mass of the test samples be tested based on the individual sieve size involved. Sample preparation generally involved drying in an electric oven at a temperature of 110 ± 5 °C (230 ± 9 °F) and cooling the sample until it could be handled.

5.2 Vacuum Saturating the Coarse Aggregates

The cooled coarse aggregates are placed in the standard vacuum saturation or Rice chamber. In vacuum saturating the test sample to remove any entrapped air within the mass and forcing water into the porous surface voids, the *AASHTO T 209* test specification procedure was followed. 30 minutes of 30mm mercury vacuum saturation was conducted during this test. After vacuum saturating the sample, the *AASHTO T 85* procedure was then followed for the determination of the saturated-surface dry (SSD) state of the coarse aggregates. Finally, the relevant weights of the material are determined, and used in the calculations of the specific gravities and water absorption results.

6 TEST RESULTS

6.1 Absorptive Behavior at 36 and 48 Hour Soaking

With increased soaking periods from 24 to 36 and 48 hours, water absorption for the steel slag increased by 4.8 and 10.8% for the 36 and 48 hours respectively. For the crushed concrete, increasing the soaking times from 24 to 36 and 48 hours resulted in 3.3 and 6.0% increments for the 36 and 48 hours, respectively, in the calculated water absorption. Fig. 1 and 2 represents the observed difference in the water absorptive behavior of the steel slag and crushed concrete at 24, 36 and 48 hours soaking.

6.2 Specific Gravity and Absorption Analysis

In the analysis of the specific gravity and absorption results, the statistical tools employed were:
1. Linear regression plots
2. t-Test at 95% confidence level
3. Standard Error
4. Coefficient of variation

6.2.1 Bulk Specific Gravity, Gsb (Dry)

The coefficient of linear relationship (R^2) is 0.86. The two results had no statistical difference at the 95% confidence level, a standard error of 0.037, and a coefficient of variation of 0.008. A plot of the linear regression is shown in Fig. 3.

6.2.2 Bulk Specific Gravity (SSD)

The R^2, standard error and coefficient of variation results between the 30 Minutes Vacuum Saturation Test and *AASHTO T85* methods are 0.92, 0.030 and 0.008, respectively. Fig. 4 shows the linear regression relationship between the two methods.

6.2.3 Apparent Specific Gravity (Gsa)

The R^2 relationship is 0.742 and standard error of 0.060. The coefficient of variation is 0.012. No statistical difference exists between the *AASHTO T85* and 10 Minutes Vacuum Saturation Test for Gsa. The linear regression relationship is shown in Fig. 5.

6.2.4 Water Absorption (Wa %)

The R^2 between the two test methods is 0.403, a standard error of 0.063, and a coefficient of variation of 0.657. Results so far indicated a statistical difference between the two methods for absorption determination of the porous aggregates – crushed concrete and steel slag. Fig. 6 shows the linear regression plot between the two methods.

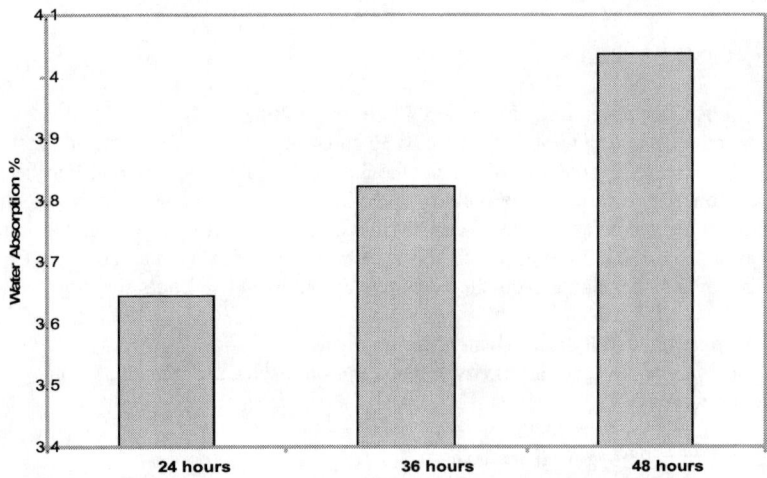

Figure 1. Water absorption trend for porous steel slag (retained No. 4.75mm)

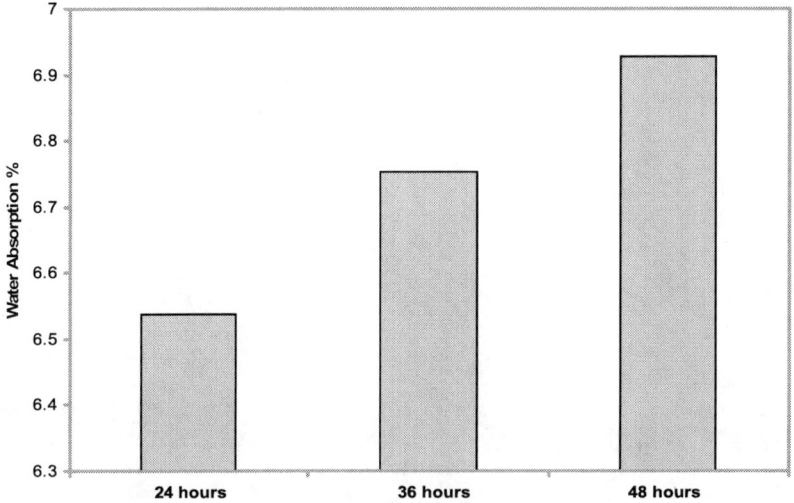

Figure 2. Water absorption trend for crushed concrete (retained 1/2in.)

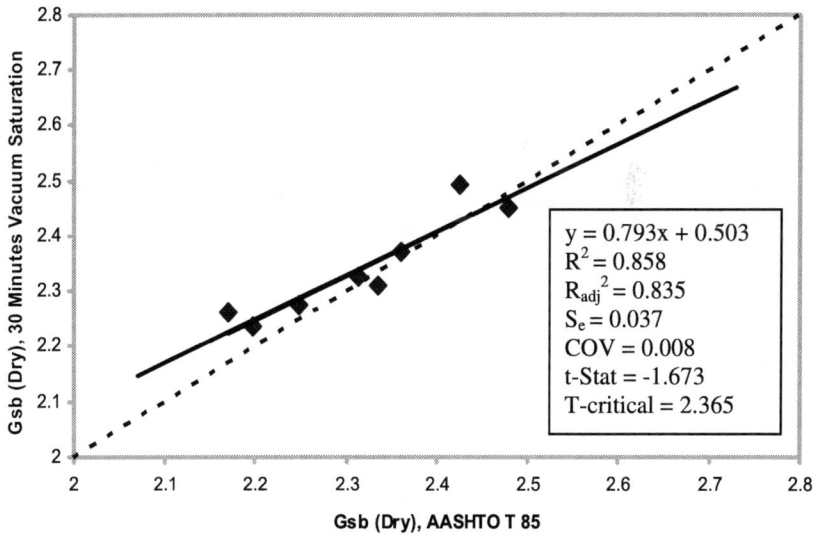

Figure 3. AASHTO T-85 against 30 Minutes Modified Rice Test for Gsb (Dry)

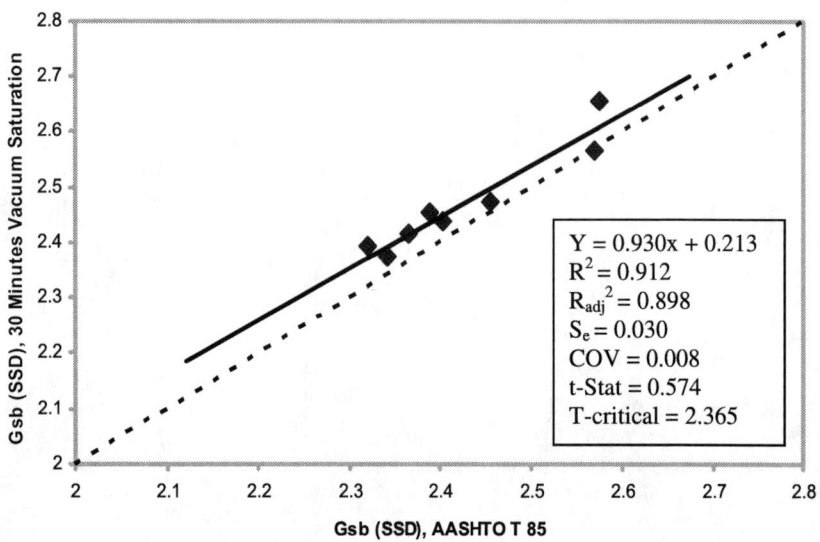

Figure 4. AASHTO T-85 against 30 Minutes Modified Rice Test for Gsb (SSD)

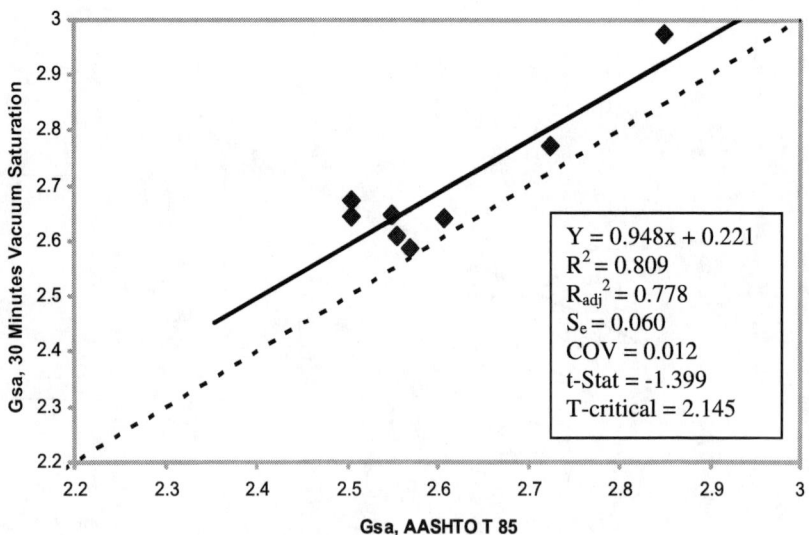

Figure 5. AASHTO T-85 against 30 Minutes Modified Rice Test for Gsa

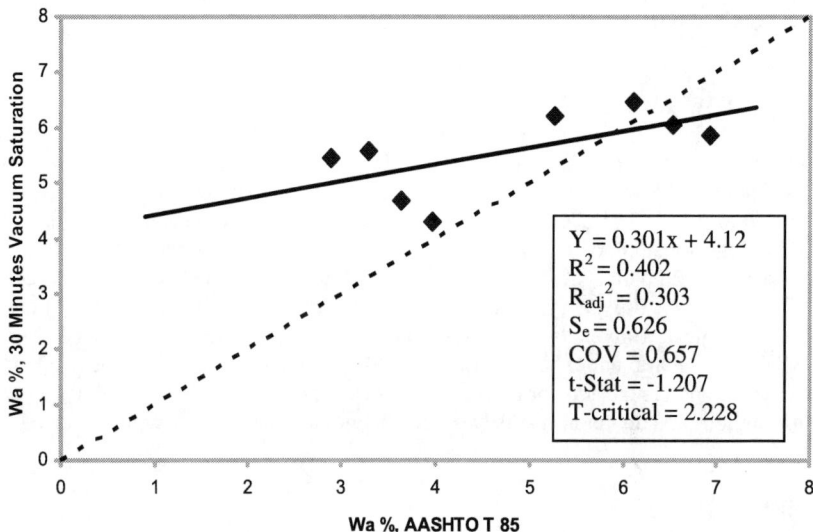

Figure 6. AASHTO T-85 against 30 Minutes Modified Rice Test for Water absorption (Wa %)

7 SUMMARY OF FINDINGS

From the results of the linear regression and statistical analysis, the following deductions can be made:

1. The prolonged soaking times (24 and 36 hours) proved the inadequacy of the current *AASHTO T 85* in satisfying the effective volume capacity of porous coarse aggregates.
2. The vacuum saturation procedure has interesting prospect to be used as a quick method for determining the specific gravity of coarse aggregates.
3. The vacuum saturation method produced higher absorption values than the traditional *AASHTO T 85*. This fact gives credence to the hypothesis that the *AASHTO T 85* may be underestimating the absorptive potential of porous aggregates.
4. No statistical difference occurs between the AASHTO T 85 and the 30 Minutes Modified Rice Test for Gsb (dry), Gsb (SSD) and Gsa.

The proposed method will serve porous aggregates better since the vacuum saturation will tend to force water to satisfy the absorptive capacity of these aggregates. Further research needs to be conducted on more materials to confirm and validate the results of this interesting finding. Additionally, decreased vacuum saturation time (10 and 20 minutes) will be evaluated in this research to quicken the test procedure. This new proposed method is set to save time, and increase significantly the efficiency in QC/QA and construction design.

ACKNOWLEDGEMENTS

The experimental work was completed in the Transportation Materials Research Center (Center of Excellence for Transportation Materials) at Michigan Technological University, which maintains the AASHTO Materials Reference Laboratory (AMRL) accreditation on asphalt and hot mix asphalt, aggregates, and Portland cement concrete. The authors wish to express their sincere gratitude to Jim Vivian and Ed Tulpo in the Transportation Materials Research Center at Michigan Technological University, undergraduate student Kelly Heidbrier of Michigan Technological University. The research work was partially sponsored by the Michigan Department of Transportation (Federal Pass Through funds). Any opinions, findings, and conclusions or recommendations expressed in this material are those of the author's and do not necessarily reflect the views of any government agency.

REFERENCES

1. AASHTO (2004). "Standard Method of Test for the Specific Gravity and Absorption of Coarse Aggregates." T 85.
2. ASTM (1993). "Standard Method of Test for Specific Gravity and Absorption of Coarse Aggregate." ASTM Designation C 127-88.
3. Washburn, E. W. and E. N. Bunting (1922). "The Determination of the Porosity of Highly Vitrified Bodies." Journal of the American Ceramic Society **Volume 5** (Issue 8): Page 527-537.
4. Dolch, W. L. (1959). "Studies of Limestone Aggregates by Fluid Flow Methods." ASTM Special Technical Publication **Vol. 59**: pp. 1204 -1215
5. Brunauer, S., P. H. Emmett, et al. (1938). "Adsorption of gases in multimolecular layers." Journal of American Chemical Soceity **vol. 60**: p. 309-319.
6. Ritter, H. L. and L. C. Drake (1945). "Pore size distribution in porous materials." Industrial Engineering and Chemical Analysis **Edition 17**: pp. 782–786.
7. Hiltrop, C. L. and J., Lemish (1960). "The Relationship of Pore-Size Distribution and Other Rock Properties to the Serviceability of Some Carbonate Rocks." Highway Research Board Bulletin **239** 1-23.
8. Ritter, H. L. and L. C. Erich (1948). Journal Of Analytical Chemistry **20**(7): 665.
9. Washburn, E. W. and E. N. Bunting (1922). "The Determination of the Porosity of Highly Vitrified Bodies." Journal of the American Ceramic Society **Volume 5** (Issue 8): Page 527-537.
10. Rice, J. M. (1952). "The Measurement of Voids in Bituminous Mixtures by Pressure Methods " Assoc Asphalt Paving Technol Proceedings **Vol 21**: pp 92-119.

11. Lee, D. Y. and P. S. Kandhal (1970). "An Evaluation of the Bulk Specific Gravity for Granular Materials." Highway Research Record, National Research Council,Washington, D.C. **307**: 44-55.
12. Vitton, S. J., M. A. Lehman, et al. (1999). "Automated Soil Particle Specific Gravity Analysis Using Bulk Flow and Helium Pycnometry." American Soceity for Testing and Materials **STP 1350**: 3-13.
13. Hall, K. D. (2002). "A Comparison of Fine and Coarse Aggregate Specific Gravity Results and Variability Using Vacuum Sealing and SSDetect Methods Preprint." Transportation Research Board, National Academy of Sciences, Washington, DC.

Optimizing Low Density Concrete Behavior for Soft Ground Arrestor Systems
E. Heymsfield, P.E.,[1] W.M. Hale, P.E.,[2] and T.L. Halsey[3]

[1]University of Arkansas, Department of Civil Engineering, 4190 Bell Engineering Center, Fayetteville, AR 72701; PH (479) 575-7586; FAX (479) 575-7168; email: ernie@uark.edu
[2]University of Arkansas, Department of Civil Engineering, 4190 Bell Engineering Center, Fayetteville, AR 72701; PH (479) 575-6348; FAX (479) 575-7168; email: micah@uark.edu
[3]Tatum Smith Engineers, Inc., Rogers, AR 72756; PH (479) 621-6128; FAX (479) 621-6985; email: tlh@tatumsmith.com

Abstract

There are approximately 10 aircraft overruns per year that occur within the United States. Overruns are more common during landing than during takeoff and more likely to occur in wet conditions. Aircraft overruns can result in passenger fatalities, injuries, and extensive aircraft damage. The majority of aircrafts involved in an overrun stop within 305 m (1000-ft.) of the runway threshold. To reduce overrun hazards, the Federal Aviation Administration (FAA) requires airports to have a 305 m (1000 ft) runway safety area beyond the design runway length. However, many airports are restricted from extending their runway because of either natural or man-made barriers. As an alternative to creating a 305 m (1000-ft) runway safety area through runway extension, the FAA allows for the installation of an engineered material arrestor system (EMAS). Arrestor systems are designed as passive systems to reduce aircraft stopping distance by inducing drag forces on aircraft landing gear. An EMAS is typically constructed using a low-density cementitious material. Previous work by the authors has studied the sensitivity of aircraft stopping distance to aircraft characteristics considering two types of arrestor bed materials: phenolic foam and low-density concrete. This article investigates a much larger suite of possible low-density concrete mixes. Three arrestor bed configurations along with twenty-six low-density concrete mixtures were studied to evaluate aircraft stopping distance behavior as a function of low-density concrete mixture. Two aircrafts were considered, B727 and B747. The aircrafts represent different landing gear configurations and weights. Stopping distance was determined for each low-density concrete material considering each aircraft with and without reverse thrust. Over four hundred stopping distance analyses were conducted during the study using the FAA ARRESTOR computer code. In this paper, plots are used to summarize the study results.

Introduction

An aircraft overrun occurs when a plane is unable to stop within the design runway length. Overruns can occur during landing or takeoff; however, are more common during landing. Approximately three times as many overruns occur during landing than occur during takeoff (Kirkland & Caves, 2002). Between 1978 and 1987, thirty-three accident/incident overruns involving Part 121/129 (23 overruns), and Part 135 operators (10 overruns) occurred within the U.S. (David, 1990). Between 1980 and 1998, 180 jet and turboprop overruns occurred within the English-speaking world: U.S. (133), Canada (16), United Kingdom (26), and Australia (5) (Kirkland & Caves, 2002). In a more recent study, twelve overruns occurred between October 2004 and March 2006 internationally involving 34 deaths and 7 destroyed aircrafts (Edwards, 2006). Since 1983, there have been 45 overruns at U.S. airports with fatalities (Lautenberg, 2006). Overruns can be the result of runway conditions; overruns are more likely to occur during wet runway conditions. Overruns can also result from aircraft malfunctions: brakes or reverse thrust, or pilot error.

Of the thirty-three overruns in the David study conducted for the FAA, thirty-one stopped within 305 m (1000-ft) of the design runway end (David, 1990). In response, to promote aircraft safety during an overrun the FAA requires a 305 m (1000-ft) runway safety area. However, due to natural or man-made barriers, many existing airports are unable to comply with this FAA requirement without reducing runway length and therefore restricting aircraft usage. As an alternative to the 305 m (1000-ft) runway safety area, the FAA allows an airport operator to instead install an engineered material arrestor system (EMAS). An EMAS is a system installed at the end of the design runway length that significantly reduces an aircraft's stopping distance. An EMAS is a passive system made of a crushable material. Drag forces are induced on the aircraft landing gear wheels as the aircraft wheels traverse the arrestor bed. Currently, there are EMAS systems installed at thirty-five runways at twenty-one airports (FAA, 2008). Still, 325 major U.S. airports have runways with at least one runway that does not satisfy FAA runway safety area requirements (Lautenberg, 2006). At the time of the Lautenberg report, there were 507 out of 1017 runways at major US commercial airports that had deficient runway safety areas either not satisfying the 305 m length or having an installed EMAS.

An EMAS design is a function of aircraft fleet usage, arrestor bed material, and available airport runway safety area. Preliminary guidelines for EMAS bed length are available in the FAA AC 150 / 5220A (FAA, 2005). However, in-depth studies describing aircraft stopping distance as a function of EMAS properties is limited. Previous work by the authors has investigated the sensitivity of aircraft stopping distance in an EMAS as a function of aircraft parameters. Conversely, this paper presents stopping distance as a function of material strength behavior.

EMAS Description

An EMAS is a passive system used to provide passenger safety with corresponding minimal aircraft structural damage during an aircraft overrun. An EMAS is positioned within the runway safety area with a setback to protect the arrestor material from jet blast, Figure 1. An EMAS consists of an asphalt ramp and an arrestor bed. The inclined asphalt ramp increases wheel penetration at bed intrusion and protects against water collecting at the bed entrance interface. Bed dimensions and material are dependent on minimizing aircraft stopping distance within limits of available runway safety area space, passenger safety, and aircraft landing gear strength, a function of the specific aircraft. Since landing gears are aircraft type dependent, operating aircraft fleets need to be considered during an EMAS design. An EMAS is designed as a passive system and therefore dissipates aircraft energy through drag forces. As the aircraft landing gears traverse the EMAS bed, vertical force equilibrium is established through wheel and arrestor material deformation. F_{DRAG} forces are also created through wheel and arrestor material deformation. Aircraft deceleration is calculated through force equilibrium in the horizontal direction of the aircraft considering all aircraft landing gears. Varying vertical forces applied to the landing gears result in changes in landing gear stroke, the landing gear piston movement within the strut. Consequently, differential changes in stroke at the aircraft landing gears results in angular acceleration about the pitch axis. As the aircraft traverses the arrestor bed, force equilibrium and pitch moment equilibrium dictate the reorientation of the aircraft. A crushable material is used for the arrestor bed to induce significant drag forces. Low-density concrete is most typically used as an arrestor bed material with a unit weight less than 4.72 kN/m^3 (30 pcf). This translates to a material that is highly crushable; however, is capable of supporting snow removal and emergency vehicles. Conversely, a disadvantage to using a weak material is durability, which is an issue at EMAS installations.

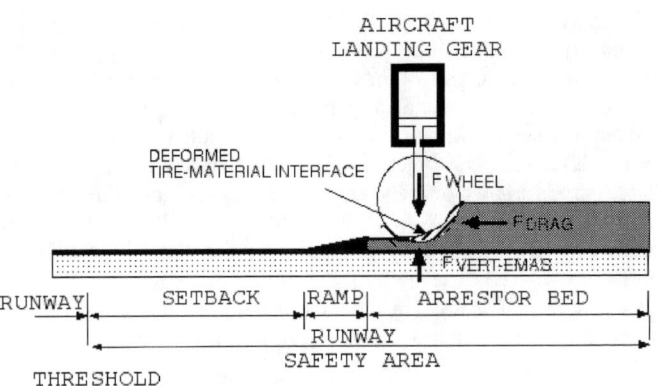

Figure 1. Landing gear – arrestor bed interaction.

An installed EMAS at the Little Rock, AR airport is shown in Figure 2. The background in Figure 2 shows a river, a natural barrier to runway extension. The EMAS includes a sloped profile from bed entry to its maximum bed depth. Sides of the EMAS include steps for easy emergency vehicle entry. An EMAS is configured using 1.22 m x 1.22 m (4-ft x 4-ft) low-density concrete panels with varying thickness. Panels are positioned to be adjoining, Figure 3.

Runway Safety Area with EMAS   Side View of EMAS

Figure 2. Installed EMAS.

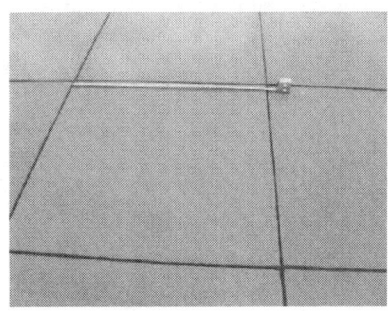

In-place EMAS panels.   Stacked 4-ft x4-ft EMAS panels

Figure 3. EMAS Panels 1.22 m x 1.22 m (4-ft x 4-ft).

ARRESTOR Code

ARRESTOR is a computer code developed for the Federal Aviation Agency to approximate aircraft stopping distances as a function of aircraft type and arrestor bed characteristics (Cook et al., 1995). A summary of code input and output is shown in Figure 4. Required input includes aircraft parameters (weight, center of gravity location, mass moment of inertia, and wheel friction), arrestor bed geometry, and arrestor bed material characteristics (unit weight and stress-strain behavior). Aircraft behavior is calculated at incremental time steps through force and moment equilibrium. Aircraft response at each recorded time increment includes aircraft velocity, acceleration, and landing gear forces. In addition, ARRESTOR output includes material response in terms of material deformation.

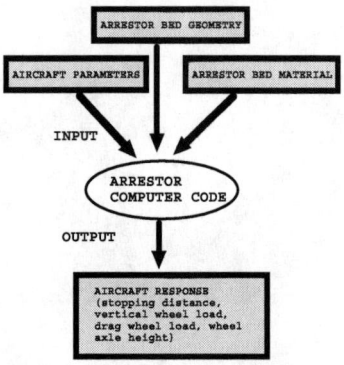

Figure 4. ARRESTOR code flowchart.

Samples of nose gear acceleration and velocity plots from an ARRESTOR code analysis are shown in Figure 5. Stopping distance in this paper is identified as the nose-gear distance from the start of the arrestor material to the point that the aircraft velocity is 0. Jumps in nose gear acceleration occur at points where the aircraft's landing gears enter the arrestor bed.

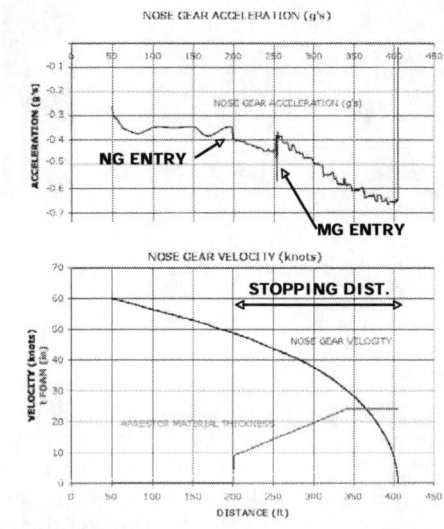

Figure 5. ARRESTOR code output.

Aircraft Types

The FAA ARRESTOR code is limited to analyzing stopping distances for three aircrafts: B707, B727, and B747. However, only two aircraft types are considered in this study, B727 and B747. Although a B707 is available for analysis in the FAA ARRESTOR computer code, it was not considered because of the aircraft's very limited usage. The two aircrafts in this study represent commercial aircrafts with significant weight differences, length dimensions, and landing gear configurations. Weights of the B727 and B747 are 752 kN (169,000-lb) and 2804 kN (630,000-lb), respectively. The B727 has a total of two main landing gears with dual wheel at each main landing gear. Conversely, the B747 uses two wing main landing gears and two body main landing gears. Each B747 main landing gear includes a dual-tandem wheel arrangement.

Arrestor Bed Geometry

The arrestor bed geometry shown in Figure 6 is used as a basis for stopping distance calculations presented in this paper (Cook et al., 1995). The EMAS is comprised of two arrestor beds to allow for the option of a stiffer material in Bed 2. An asphalt ramp is positioned 30.5 m (100-ft) beyond the runway safety area entry and inclined. The arrestor material begins at 61 m (200-ft) from the runway threshold to protect against jet blast. The thickness of the bed varies from 229 mm (9-in) to 610 mm (24-in) over a 42.7 m (140-ft) distance. The maximum thickness of the EMAS, 762 mm (30-in), occurs in Bed 2. The bed geometry is designed so as to promote minimum stopping distance considering passenger safety in conjunction with minimal aircraft damage.

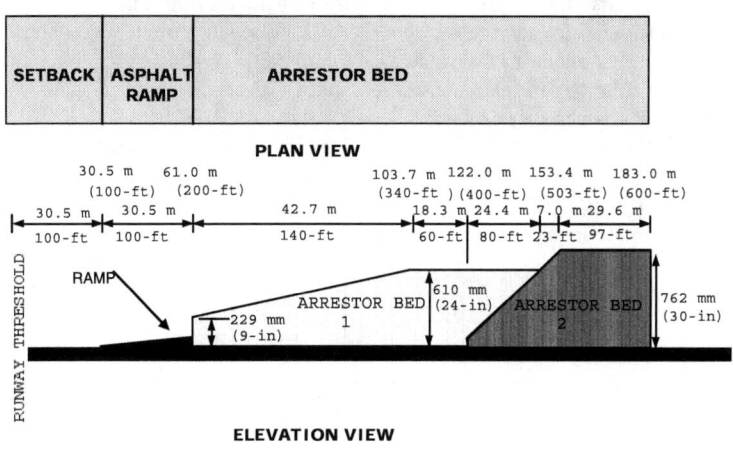

Figure 6. Base EMAS profile.

Material Mix Description

Low-density concrete mixtures were developed during the study and compared to examine the sensitivity of stopping distance to material stress-strain behavior. A compressive strength of approximately 207 kPa (30 psi) at 28 days of age was used as a target to develop a highly crushable material. Over 100 different concrete mixture proportions were developed and tested throughout the project's duration. Preliminary test results showed a direct correlation between the concrete's compressive strength and the unit weight of the fresh concrete. To attain a compressive strength of approximately 207 kPa (30 psi), a concrete mixture with a fresh unit weight of approximately 4.40 kN/m^3 (28 pcf) was needed. Therefore, the focus of many of the early batches was unit weight only.

Throughout the study, the lightweight aggregate type and content were varied to attain a target unit weight of 4.40 kN/m^3 (28 pcf). For mixtures containing aggregates, two lightweight aggregate types were considered, vermiculite and perlite. For mixtures without aggregates, several cellular concrete mixtures were batched. Cellular concrete mixtures contained a cement binder, water, and foam. Calcium aluminate cement (CAC), portland cement or combinations of both were tried as cement binders. A foam generator and a foaming agent were used to introduce the foam into the concrete.

It was desired to have a mixture attain 207 kPa (30 psi) at an early age (< 3 days) for production purposes. Consequently, several mixtures were batched with CAC because of its fast setting times and high early strength. However, because of characteristic strength regression associated with CAC mixtures, other mixtures were included consisting of combinations of CAC and portland cement, and mixtures containing only portland cement. The use of lightweight aggregates alone was not enough to achieve the targeted unit weight of 4.40 kN/m^3 (28 pcf) and therefore, chemical admixtures were also used. Additional unit weight reduction was attained using an air-entraining agent (AEA). Both dry and liquid AEAs were examined. High range water reducers (HRWR) and viscosity modifying admixtures (VMAS) were also considered to increase the concrete's fluidity and prevent segregation. The final admixture examined in the study was a non-chloride accelerator. Non-chloride concrete accelerators were also considered as a CAC replacement to increase early age strength and reducing setting times.

Stress-strain material behavior is required input for the ARRESTOR code. Stress-strain behavior for each of the low-density concrete mixtures was developed with load testing equipment using a 305 mm (12-in) diameter by 305 mm (12-in) deep concrete material sample subjected to a 25 mm (1-in) diameter ram, Figure 7.

Figure 7. Stress-strain material testing.

An ARRESTOR code analysis was conducted on twenty-eight potential arrestor bed materials. Two of these materials, low-density concrete and phenolic foam were previously discussed in earlier papers (Heymsfield, et al, 2007 and Heymsfield & Halsey, 2008). Of the twenty-six newly considered low-density concrete mixtures (LDC), results of twelve mixtures were considered viable and therefore are presented in this paper. The stress-strain behavior of the two previously documented materials (LDC-BASE 1, LDC-BASE 2; and PHENOLIC FOAM) and the twelve new low-density concrete samples (LDC- mix #) are shown in Figure 8. Unit weights of the mixtures fall within 3.14 kN/m^3 – 4.72 kN/m^3 (20 – 30 pcf). For each of the presented materials, large strains accompany relatively low stress conditions. However, when the material is fully crushed, significant stress increase accompanies nominal strain changes. Consequently, the material is crushable identifying it as a potential arrestor bed material that incurs significant wheel penetration during aircraft intrusion. Since, drag forces are a function of pressure between the aircraft wheel and arrestor material, larger drag forces are induced with greater wheel penetration.

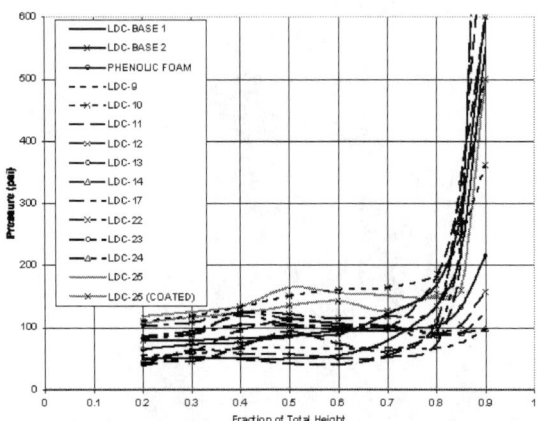

Figure 8. Material stress-strain behavior.

Stopping Distance as a Function of Low Density Concrete Mix

Stopping distances for the low-density concrete mixtures are determined using the EMAS profile in Figure 6 as a basis and measured from the start of the arrestor bed. Values are presented for two aircraft types, B727 and B747, considering 0 reverse thrust, Figure 9, and full reverse thrust, Figure 10. Both aircrafts are assumed to enter the runway safety area at 60 knots. Stopping distances exceeding 122 m (400-ft) identify scenarios in which the aircraft does not stop within the arrestor bed. As the aircraft traverses the arrestor bed, large vertical loads due to angular acceleration about the pitch axis, and drag loads due to wheel-material interaction are induced on the landing gears. Cases in which gear loads are exceeded on the B727 aircraft are identified on these figures with a "*". The ARRESTOR code assumes that B747 aircraft maximum loads are not exceeded because of the heavy aircraft design. Both figures show the intrinsic problem in an arrestor bed design to minimize stopping distance while not exceeding maximum aircraft gear loads.

Of aircraft parameters, gross aircraft weight has the greatest impact on stopping distance. Consequently, stopping distances are typically less for the B727 aircraft than the B747. However, since aircraft deceleration due to drag is a function of wheel penetration, a few mixtures were found to cause a greater stopping distance for the lighter aircraft. For the majority of the mixtures, some minimum reverse thrust needs to be applied for the B747 to stop within the 121.9 m (400-ft) total arrestor bed length. With full reverse thrust, except for one mixture, both aircrafts stopped within the arrestor bed length.

Although the mixtures in Figure 8 have similar behavior, stopping distance results in Figure 9 and 10, show high sensitivity of an aircraft's stopping distance to a material's low strain strength. During this preliminary investigation, emphasis has been placed on developing a low-density concrete mixture to minimize stopping distance while considering passenger safety and landing gear strength. However, the material weakness makes it susceptible to degradation through moisture and freeze/thaw cycling. Consequently, future work will need to investigate sealing the arrestor bed material against water intrusion.

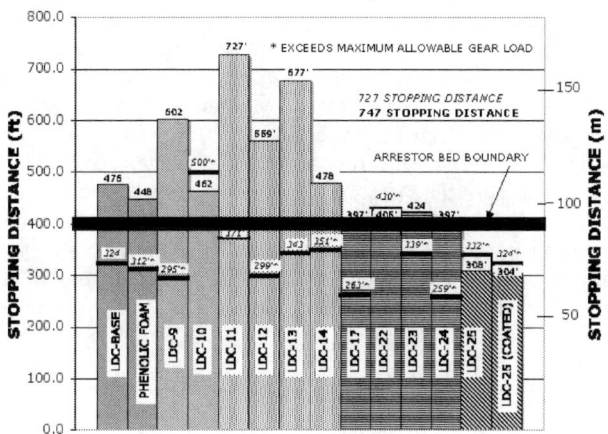

Figure 9. Stopping distance as a function of mixture design, 0 reverse thrust.

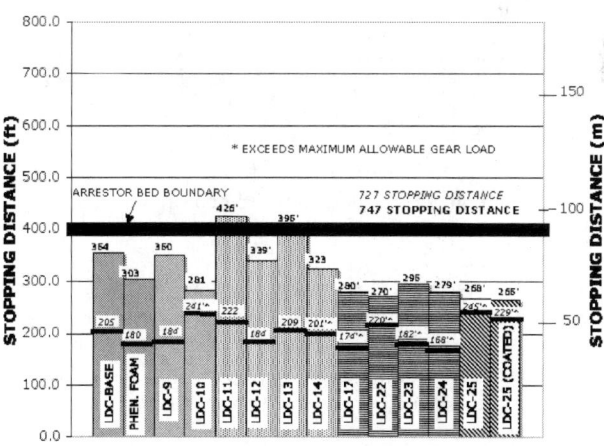

Figure 10. Stopping distance as a function of mixture design, full reverse thrust.

Stopping Distance as a Function of Arrestor Bed Geometry

Aircraft stopping distance is influenced by arrestor bed geometry. In addition to the basis arrestor bed shown in Figure 6, two additional arrestor bed geometries are examined. Design 1 in Figure 11 represents the basis geometry discussed in the previous sections of this paper. Design 2 represents an increase in material thickness by continuing the bed slope to attain a 152 mm (6-in.) greater bed thickness. Conversely, Design 3 is derived by increasing the Design 1 Bed thickness by a constant 152 mm (6-in.) throughout the bed profile. Considering a total 122 mm (400-ft.) length for each of the three beds, Design 2 and Design 3 represent material volume increases of 13% and 26%, respectively. Minimum stopping distances are superimposed on Figure 11 for each Design considering the twenty-eight mixtures using the two aircrafts with 0 reverse thrust and 60 knots entry speed. The three designs showed insignificant differences in stopping distance for the considered aircrafts and mixtures. However, the results illustrate that the most economical arrestor bed design should consist of a two-bed system. The first bed of the system is used to stop lighter planes without incurring significant landing gear damage, and the second bed, comprised of a stiffer material and greater thickness, to stop heavier planes.

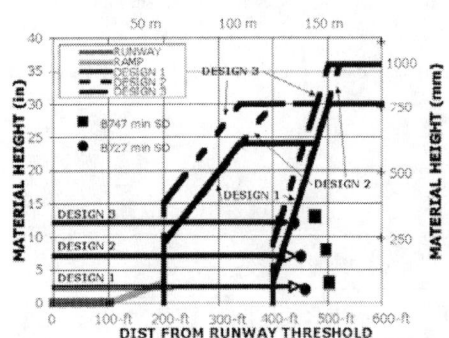

Figure 11. Stopping as a function of bed geometry.

Conclusions

Previous research has investigated aircraft parameters and bed geometry while limiting the considered number of potential arrestor bed materials. Conversely, this paper investigates stopping distance as a function of material behavior. An ARRESTOR code analysis was conducted on twenty-eight potential arrestor bed materials. In addition to two previously documented materials, stopping distance results for twelve of the twenty-six newly considered mixtures were presented in this paper. All mixtures exhibited crushable behavior, large deformation accompanying low constant stress. However, study results indicate that stopping distance is highly sensitive to material behavior before the material fully crushes. Aircraft reverse thrust significantly reduces aircraft stopping distance. The most viable arrestor bed system consists of two beds, one bed to stop lighter aircrafts and a second bed of stiffer material and greater thickness to stop heavier aircrafts. Because of weak material strength, arrestor bed material durability is a major concern and is paramount in any future study.

Acknowledgements

The research presented in this article was funded by the Mack-Blackwell Transportation Center. The authors are appreciative to this agency for its financial support of this research work.

References

Cook, R.F., Teubert, C.A., and Hayhoe, G., (1995). "Soft Ground Arrestor Design Program," Technical Report # DOT/FAA/CT-95, National Technical Information Service, Springfield, VA

David, R.E. (1990). "Location of Commercial Aircraft Accidents / Incidents Relative to Runways," Federal Aviation Administration, Washington, DC.

Edwards, D., (2006). "Engineered Material Arresting Systems," Air Line Pilots Association, International (ALPA) 52nd Annual Air Safety Forum, Washington DC

Federal Aviation Administration, (2005). "Advisory Circular 150/5220-22A, Engineered Materials Arresting System (EMAS) for Aircraft Overruns".

Federal Aviation Administration, (2008), "Fact Sheet, FAA, Engineered Material Arresting System (Jan. 31, 2008).

Heymsfield, E., Hale, M.W., and Halsey T.L., (2007) "A Parametric Sensitivity Analysis of Soft Ground Arrestor Systems," Aviation A World of Growth, Proceedings of the 29th International Air Transport Conference, Irving, TX, ASCE

Heymsfield E., and Halsey, T.L., (2008). "Sensitivity Analysis of Engineered Material Arrestor Systems to Aircraft and Arrestor Material Characteristics," Transportation Research Record, accepted for journal publication

Kirkland, I.D. & Caves, R.E., "New Aircraft Overrun Database, 1980-1998," Transportation Research Record, Washington, DC, #1788, pp 93-100, 2002

Lautenberg, F.R., "An Accident Waiting to Happen: Over Half of America's Airports Have Runways that Fail to Meet FAA Standards," Report by the Office of Senator Frank R. Lautenberg, 2006

Mechanistic Characteristics of Moisture Damaged Asphalt Matrix and Hot Mix Asphalt Mixtures

Mohammad J. Khattak[1], A.M., ASCE, P.E. and Vikram Kyatham[2]

[1]Associate Professor, Department of Civil Engineering, University of Louisiana at Lafayette, Lafayette, LA 70504-2291, Phone # (337) 482-5356, Fax # (337) 482-6688, mxk0940@louisiana.edu
[2]Graduate Research Assistant, Department of Civil Engineering, University of Louisiana at Lafayette, Lafayette, LA 70504-2291, Phone # (337) 482-5356, Fax # (337) 482-6688, kyatham_vikram@hotmail.com

ABSTRACT: The Hot mix asphalt (HMA) can be characterized as a combination of three phases: air voids, coarse aggregate, and asphalt matrix (AM), which includes binder and fine aggregate. The coarse aggregate phase is stiffer than the matrix phase, and is elastic in nature. The matrix phase makes the HMA a visco-elastic material due to the time dependent strain response of the binder. The matrix phase is considerably weaker than the coarse aggregate phase, and more susceptible to moisture damage. This study focuses on the characterization of the mechanistic responses of the AM and the HMA mixtures under dry and moisture damaged conditions. The mixtures were made using limestone aggregates and two asphalt type: neat and hydrated lime-modified. The dynamic shear rheometer was utilized to determine fatigue damage at 25°C and complex shear modulus (G^*) of AM mixtures for a range of temperatures and loading frequencies. An indirect tensile load test was conducted to characterize tensile strength, fatigue life, elastic and permanent deformation characteristics of HMA mixtures. The G^*-master curve analysis revealed that the hydrated lime-modified AM significantly improves the visco-elastic response of the moisture damaged AM. The fatigue life of AM mixtures increased due to the increase in G^* values as a result of hydrated lime addition. Moreover, the fatigue life and permanent deformation characteristics of the HMA improved due to decreases in the rates of accumulation of tensile and compressive plastic deformation, respectively. Finally, relationships were established between various HMA and AM mechanistic properties.

INTRODUCTION AND BACKGROUND

Moisture typically reduces the stiffness of the asphalt binder (AC) through moisture diffusion and degrades the adhesive bonding between the aggregates and the AC. This causes moisture related distress in flexible pavements, commonly called as stripping. Several mechanisms for stripping have been reported by researchers. In general, an interaction of the various mechanisms is often presented as the best explanation for the moisture damage process in the hot mix asphalt (HMA) mixtures (Roque et al. 2005). Anti-stripping additives are used to reduce the moisture damage susceptibility of HMA. Previous research has shown the beneficial use of hydrated lime (HL) as a multifunctional additive that helps in reducing stripping, increasing the resistance to fatigue and permanent damage of HMA (Lesueur et al. 1999; Little et al. 2001; Sebaaly et al. 2007).

HMA is defined as three phase material: air voids (AV), coarse aggregate, and an asphalt matrix (AM), which is a combination of AC, and fine aggregate passing No. 4 sieve. The three phases of HMA can also be defined as a combination of aggregate, AV, and mastic, which is a combination of AC and mineral filler (Anderson et al. 2000; Kim et al. 2004). The coarse aggregate phase is considered elastic in nature and the AM phase as visco-elastic material due to the time dependent strain response of the AC.

In this study, an attempt was made to first identify the fundamental AM properties that affect and control the moisture damage. The mechanical behavior of AM and HMA mixtures in moisture conditioned and unconditioned states was compared to investigate if the fundamental properties can identify moisture related damage. Next, the effect of HL modification on the mechanistic characteristics of the moisture damaged AM and HMA is investigated, and finally, an effort is made to correlate the fundamental material properties of HMA to various AM properties.

RESEARCH OBJECTIVES

The objectives of this study are to evaluate:
1. The moisture damage susceptibility of HMA and AM mixtures.
2. Relationship between various mechanical properties of HMA and AM mixtures.

LABORATORY INVESTIGATION

A brief description of the test materials, sample preparation, and testing is provided below. It should be noted that triplicate specimens were tested for each type of test.

Test Materials and Sample Preparation

Both HMA and AM mixtures were made using Limestone aggregate and PG64-22 AC obtained from the local contractor. The AC was modified with 20 percent of HL (Kanitpong et al. 2007). Since the HL was used as a binder modifier, it was not accounted for as aggregate filler.

Superpave® design procedure was used to make HMA mixtures. The nominal maximum aggregate size of 25 mm was used with an optimum AC content of 4.4%. The cylindrical HMA samples of 150 mm diameter were compacted by the Superpave® Gyratory Compactor, and then sliced using the diamond saw to obtain 50.8 mm thick specimens. The average AV for the test specimens were $7\pm1.0\%$. Table 1 shows the aggregate gradations for both the AM and HMA mixtures.

The moisture conditioning of the HMA specimens was conducted in accordance with ASTM D 4867/D 4867M-04. Briefly, the procedure consists of partially saturating the specimens using a vacuum chamber at room temperature and then soaking in water at 60±1.0°C for 24 hours before testing.

Table 1 Aggregate Gradation Data.

Sieve Sizes	US	1.5"	1"	¾"	½"	3/8"	No. 4	No. 8	No. 16	No. 30	No. 50	No. 100	No. 200
	Metric	37.5	25	19	12.5	9.5	4.75	2.36	1.18	0.6	0.3	0.15	0.075
% Passing	HMA	100	96.6	89.1	67.9	56.8	32.5	21.2	15.6	11.7	7.5	5.2	4.2
	AM	-	-	-	-	-	-	100	74	55	35	25	20

The AM mixtures were made using 10% AC and fine aggregate passing No. 8 sieve. The aggregate gradation for AM was extracted from HMA mixture gradation (Table 1). The AC and the aggregate were mixed at 167°C and compacted at 147°C. In order to prepare the AM specimen about 54 grams of loose mix was poured into 25.4 mm diameter and 50.8 mm high mold. The amount of AM to be poured into the mold was calculated based on a target AV content of 10% in the sample. The sample was cored from the middle to produce specimens of 15.86 mm diameter. The heights of specimens were 15.86 mm and 22.2 mm for frequency sweep and fatigue tests, respectively.

For moisture conditioning, the specimens were placed in a vacuum chamber and vacuum was applied for 15 minutes. The water was then released into the chamber and the specimens were allowed to submerge in water. The vacuum was applied for another 15 minutes, and the specimen was soaked for 12 hours in water at room temperature. It should be noted that specially fabricated environmental chambers were used to facilitate the testing on DSR under moisture conditioning.

HMA-Indirect Tensile Cyclic Load Test (ITCLT)

The ITCLT was conducted using the Materials Testing System. The peak load of 1,111 N was used with a sustained load of 111 N. The deformations were measured in the vertical and horizontal directions in the middle 25.4 mm of the specimen, using linear variable differential transducers. All the mixtures were tested at 25°C and a test frequency of 1 Hz with a loading-unloading time of 0.1s and rest period of 0.9s. The test was conducted on the mixture for a period of 24 hours, or untill the total vertical deformation reached 12.5 mm. The ITCLT was conducted to determine the fatigue and permanent deformation characteristics of HMA mixtures. The fatigue life was defined as the number of load cycles where the rate of tensile plastic deformation (TPD) just starts to increase (Khattak et al. 2001). This represents the rate at which the damage is induced into the mixtures, which is ultimately responsible for initiating tensile cracks or fatigue cracks in the mixture. Vertical plastic deformation as a function of the number of load cycles was obtained from the ITCLT to determine the rutting resistance of the HMA mixtures.

HMA-Indirect Tensile Strength (ITS) Test

The ITS test was conducted on the HMA mixtures to determine stress-strain characteristics including ITS, fracture energy (FE) and equivalent modulus (E_q). In a typical stress-strain curve of a HMA under indirect tensile loading mode, with increasing stress the resulting strain would increase, non-linearly. The stress will reach a peak value and then starts to drop off. The peak stress is defined as ITS. The area under the stress-strain curve to the ITS is the FE. It is defined as the total energy required for complete failure of the HMA. The E_q is defined as the slope of the stress-strain curve to one-half of the ITS. It represents the stiffness of the HMA. Higher the modulus higher is the resistance to fatigue and permanent deformation (Khattak et al. 2001).

AM-Frequency Sweep Test

The Bohlin's Dynamic Shear Rheometer (DSR) was used to determine the complex shear modulus (G*) of AM at high and intermediate temperatures. The frequency sweep test was conducted at frequency levels of 1 Hz through 60 Hz at log-increments at temperatures of 15, 20, 25, 34, 46, 58 and 64°C. The test was conducted under the controlled stress mode. The applied shear stress at each temperature was selected such that it was within the linear visco-elastic range. The specimen was conditioned by applying 100 cycles at 10 Hz at one-half the stress level that was used in the test.

AM-Fatigue Test

The test was conducted under stress controlled mode, and a single frequency of loading of 1.58 Hz was applied at 25°C. Shear stress level of 14,000 Pa was applied on the AM. The test was terminated at load repetitions of 10,000 cycles. The pre-conditioning of the AM was done by applying 200 cycles of loading at a frequency of 10 Hz before conducting the actual fatigue test. The G* was plotted as a function of the number of load repetitions to identify the fatigue failure in the mixture.

RESULTS & DISCUSSION

Mechanistic Characteristics of AM Mixtures

Complex Shear Modulus (G) Master Curve of AM*

The master curve was constructed using the principle of time-temperature superposition; that is, the response of the mixture at low temperatures remains the same, as that under short or fast loading rates, and at high temperatures, remains the same under long or slow loading rates. The G* master curve was constructed at a reference of 25°C and represented by a Sigmoidal function as:

$$\log|G^*| = a + \frac{b}{1 + \dfrac{1}{\exp^{d+e(\log f_R)}}} \quad (1)$$

Where G^* = complex shear modulus; f_R = frequency of loading at reference temperature; a and b= fitting parameters for a given set of data; a represents the minimum value of G^* and $a+b$ represents the maximum value of G^*; and d and e = parameters describing the shape of the function.

Fig 1 shows the comparison of G*-master curve at 25°C for unconditioned and conditioned, neat and HL-modified AM mixtures made from limestone aggregate. The master curve indicates that unconditioned HL-modified AM exhibited significantly higher G* values under all loading frequencies as compared to unconditioned neat AM.

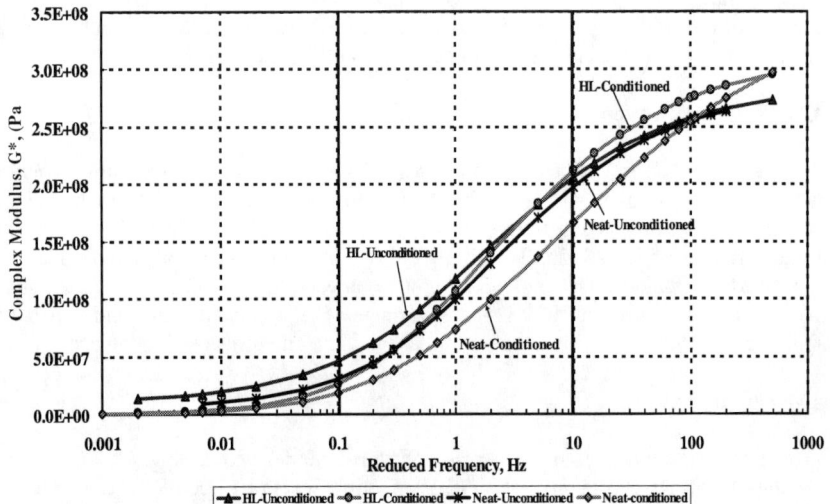

FIG 1. The G*-master curve for neat and HL-modified limestone AM at 25°C.

At lower frequencies (0.1 Hz and lower) the effect of moisture conditioning is evident as the average G* value for the unconditioned neat AM is about 3 times greater than the conditioned neat AM. Similarly, the unconditioned HL-modified AM showed 2.5 times greater G* values than the conditioned ones. For moisture damaged specimens, the average G* of HL-modified AM exhibited twice the G* of neat AM mixture.

For medium frequency range (between 0.1Hz and 10Hz), the improvement in the average G* values due to HL modification is evident from the fact that the average G* value for the condition HL-modified AM is 1.3 times higher than the conditioned neat AM. This indicates lower moisture susceptibility of the AM with HL modification.

At high frequencies (higher than 10Hz), it can be observed that there is hardly any difference between the average G* values for all kinds of AM. The effect of HL modification on the resistance to moisture damage seems to be marginal, with only an 8 percent increase in the average G* values for the conditioned HL-modified and conditioned neat AM mixtures. It was also found that the improvement in G* for the AM due to HL modification was only for the conditioned AM, and not for the unconditioned ones.

Fatigue Characteristics of AM

Fatigue testing was conducted to evaluate the effect of moisture damage and HL on the fatigue damage susceptibility of the AM mixtures. For effective comparison amongst various AM mixtures it was essential to identify a laboratory fatigue failure criteria. Fig

2 illustrates G* values as function of the logarithm of the number of loading cycles for conditioned and unconditioned AM. It can be seen that the slope of the curve changes at two different locations, which were identified as inflection points. The G* were almost constant until a certain number of loading cycles and then started to decrease at a constant rate; this was identified as the first inflection point. It was assumed that the damage to the mixture initiated at this point, and hence, was named the "fatigue damage initiation point". After a certain number of load cycles, it was observed that the G* starts to level off with increasing numbers of loading cycles; this was the second inflection point, and was labeled the "fatigue failure point'. It was assumed that the fatigue damage had caused full crack growth, and hence, completes failure of the AM.

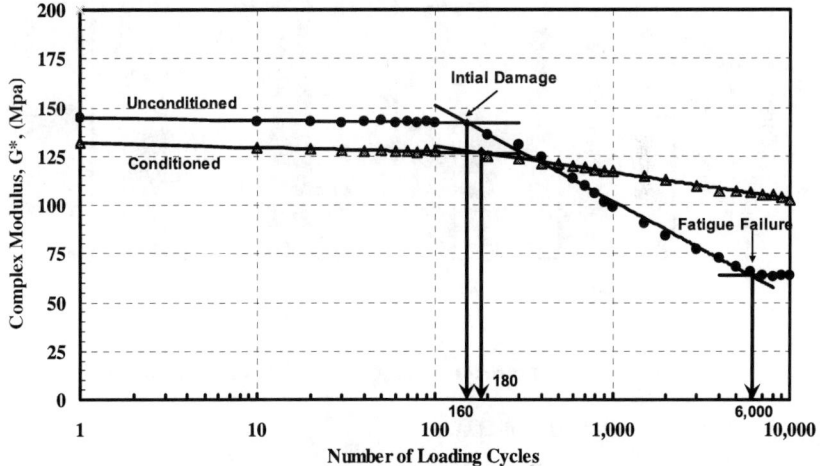

FIG 2. Identification of fatigue failure in AM mixture.

Fig 2 also depicts the initial G* value of the conditioned AM is lower than the unconditioned AM, which indicates moisture damage that was induced in AM due to vacuum conditioning. However, it is interesting to note that the rate of damage accumulation in the unconditioned AM is higher than the conditioned mixture. Moreover, the G* values of the unconditioned AM at the completion of the test (10,000 load cycles) are lower than the conditioned AM. The conditioned AM did not reach the fatigue failure at the end of testing. This trend was observed for all the different types of AM mixture tested in this study. This behavior may be due to the presence of water in the AV of the conditioned AM mixtures due to vacuum saturation during initial conditioning. Therefore, during the course of the testing, pore water pressure might have developed that stiffened the conditioned AM (pseudo stiffness). Since the fatigue test was conducted under stress controlled mode this increase in stiffness due to pore water pressure resulted in low damage due to number of load cycles.

It was found that the AM specimens were tested at different AV content. Therefore, the fatigue life of AM mixtures was normalized to a target AV content of 10 percent. Figure 3 shows the fatigue life of AM mixtures with and without AV normalization. The improvement in the fatigue damage resistance due to HL-modification is clearly represented by the increase in number of loading cycles to fatigue failure. The HL modified AM exhibited 2 and 1.5 times higher fatigue lives, relative to neat unconditioned and conditioned mixtures, respectively.

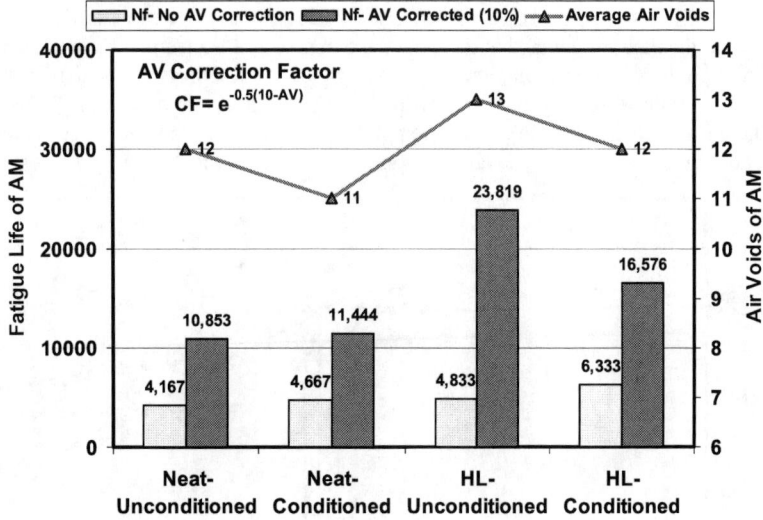

FIG. 3. Comparison of fatigue life of AM mixtures.

Mechanistic Characteristics of HMA

HMA-Fatigue Life (N_f)

The laboratory fatigue failure of an HMA specimen was defined as the number of load cycles at which the rate of accumulation of tensile plastic deformation (TPD) increases rapidly (Khattak et al. 2001). The number of load cycles to fatigue failure of the specimen can be determined by plotting the normalized slope of accumulation of TPD as a function of the number of load cycles, as shown in Fig 3. It is obvious from the Fig 3 that the normalized slope of the curve decreases first, reaches a valley, and then starts to increase. The increase in slope indicates the crack initiation and propagation in the specimen, and hence, the fatigue life is defined as the number of load cycles at which the slope of TPD just starts to increase (Khattak et al. 2001).

Based on the aforementioned criteria, the fatigue life of the HMA mixtures were determined and plotted as shown in Fig 4. The figure also depicts the rate of TPD for various mixtures. The addition of HL exhibited an increase in N_f of about 35% for conditioned HL-modified HMA mixtures. The increase in fatigue life of the HL-modified mixtures is due to the resistance to TPD (see, Fig 4). Fatigue failure in an HMA pavement is due to the tensile stresses and strains at the bottom of the HMA layer. The indirect tensile cyclic load test, which simulates the tensile stresses and strain conditions in the test specimens in the horizontal direction, is a good indicator of the mixture performance under a cyclic load.

AIRFIELD AND HIGHWAY PAVEMENTS 2008   141

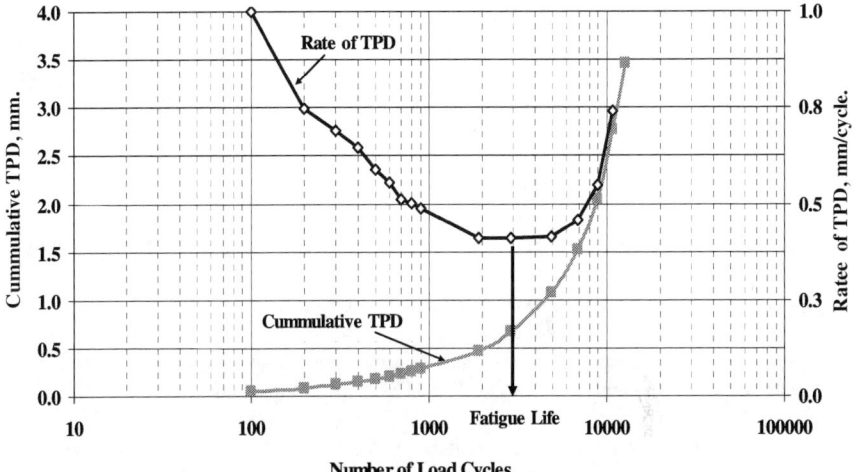

FIG. 3. Typical cumulative TDP and slope of TPD as a function of number of load cycles.

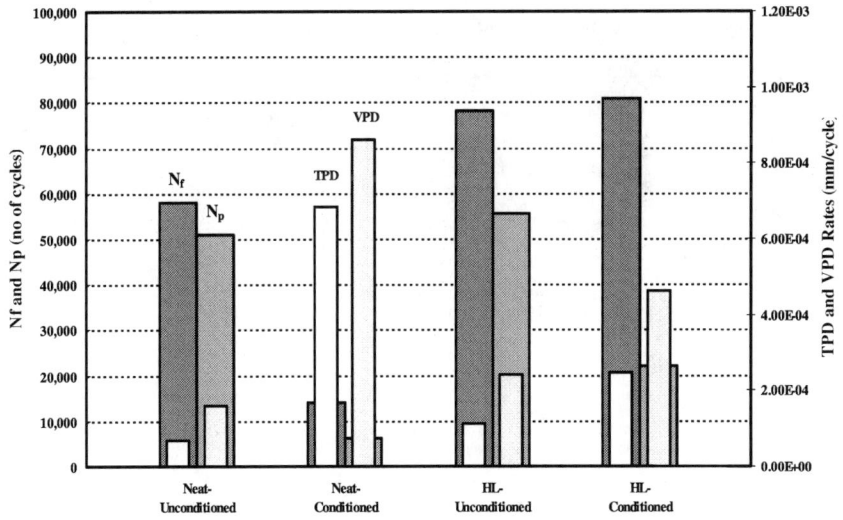

FIG. 4. Nf, Np, VPD, TDP rate and VPD rate of HMA mixtures.

HMA-Permanent Deformation

The number of loading repetitions (N_p) to obtain a vertical permanent deformation (VPD) of 0.25 mm in the HMA mixture was considered to compare various HMA mixtures. The Fig 4 also illustrates Np and initial rate of VPD for neat and HL-modified HMA mixture. A comparison of the Np and VPD rate of conditioned HL-modified and conditioned neat HMA mixtures indicate significant improvements in the resistance to moisture damage. The HL-conditioned mixtures exhibited 4 time higher Np values than the conditioned neat mixtures. On the other hand, there was no significant difference in the Np values of unconditioned HL-modified and neat HMA mixtures. Similar to N_f, the Np improvements were due to decreases in the VPD rates as a result of HL addition.

HMA-Stress Strain Characteristics Under Static Loading

The stress strain response of the various limestone mixtures under static loading is illustrated in Fig 5, and Table 2 lists the ITS, FE and E_q for all the HMA mixtures. A comparison between the stress-strain curves for unconditioned and conditioned neat mixtures indicated higher strength values for the unconditioned mixtures. Initially, for the same stress level the resulting strain of the unconditioned neat mixtures was lower, as compared to the conditioned neat mixtures. However, it is interesting to note that the unconditioned neat mixtures failed at a much lower strain level as compared to the conditioned mixtures, which sustained the stress for a longer duration and resulted in higher strain level at failure. This indicates more viscous response from the conditioned mixtures, as compared to the unconditioned mixtures.

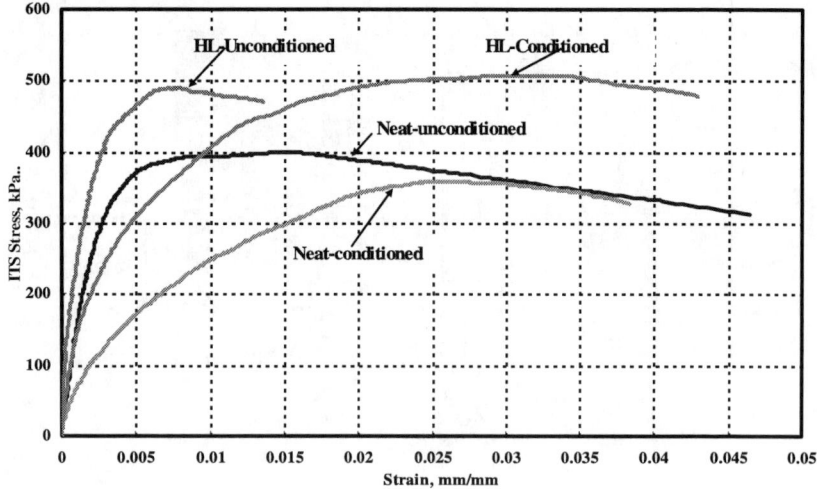

FIG. 5. Typical stress-strain curves for all limestone HMA mixtures.

The FE indicates the effect of moisture conditioning, as the energy to cause failure in the unconditioned mixtures is greater than the conditioned mixtures. A similar comparison between the conditioned neat and the HL-modified indicates the effect of HL modification on the strength and fracture energy. It can be observed from Fig 5 that both the conditioned neat HL-modified mixtures show a similar stress-strain response. The effect of HL is evident from the fact that the conditioned mixtures take more stress as compared to the conditioned neat mixtures for the same resulting strain, which is reflected by the higher FE for the HL-conditioned.

It is interesting to note from Fig 5 that the unconditioned HL-modified mixture shows the highest strength value among all the mixtures, but does not sustain a high strain level, which is the opposite of the trend shown by the conditioned mixture. Similarly, the average E_q for conditioned HL-modified is twice the E_q of the neat mixtures. The stress-strain curve analysis indicates that the moisture conditioning reduces the elastic component of the HMA response, and the HL modification does indicate significant improvement in the strength characteristics of conditioned HMA mixtures.

Table 2. Summary of stress-strain characteristics of limestone HMA mixtures.

Material Property	Neat-Unconditioned	Neat-Conditioned	HL-Unconditioned	HL-Conditioned
ITS, (Kpa)	468	394	517	554
FE, (Kpa)	4.55	3.42	1.83	5.72
E_q, (Kpa)	3,089	1,457	5,549	3,262

Relationships between HMA and AM Mechanistic Properties

The various fundamental material characteristics of AM mixtures were correlated with the HMA N_f and N_p. It was found that N_f and N_p of HMA mixtures were a function of G*, G*/sinδ and tanδ (ratio of loss modulus (G') to storage modulus (G")) of AM at 1.58 and 10 Hz, as shown in Fig 6. It can be observed from Fig 6a that with an increase in G* of AM, the N_f and N_p of HMA increases. It is important to note that all the tests were run in a stress-controlled mode and the high stiffness or G* of AM corresponds to greater resistance to stress, and hence, higher fatigue life of the mixture.

The parameter to determine the permanent deformation resistance of asphalt binder is given as G*/sinδ by Superpave®. It can be seen from Fig 6b that as G*/sinδ of AM increases, the N_p also increases. This can be explained by the fact that increasing G* of AM enhances the resistance to rutting, likewise decreasing the δ of the AM makes it more elastic, and hence, more resistant to rutting.

The parameter to determine the fatigue damage resistance for binders under controlled strain mode as given by Superpave® is G*sinδ. As mentioned earlier, all the testing in this study was conducted in the stress controlled mode, and hence, the parameter G*/sinδ was used instead of G*sinδ. It can be seen from Fig 6b that as G*/sinδ increases the N_f of HMA increases. This phenomenon can be explained by understanding the stress controlled test using the stress-strain relationship (recall Fig 5). It is clear from the figure that for the same stress levels the material with higher modulus value experiences lower strain relative to the one with lower values. Hence, with repeated load test the accumulation of strain for high modulus values will be lower, and therefore require a high number of load cycles to cause failure, thereby exhibiting higher fatigue life.

(a) Relationship between N_f and N_p of HMA and G^* of AM

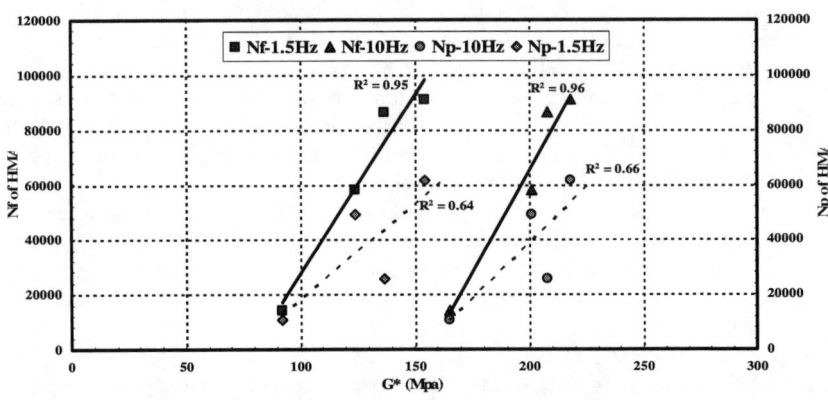

(b) Relationship between N_f and N_p of HMA and $G^*/\sin\delta$ of AM

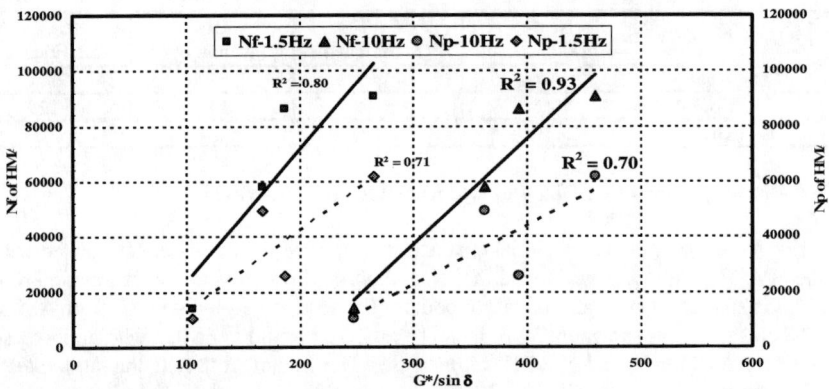

(c) Relationship between N_f and N_p of HMA and $\tan\delta$ of AM

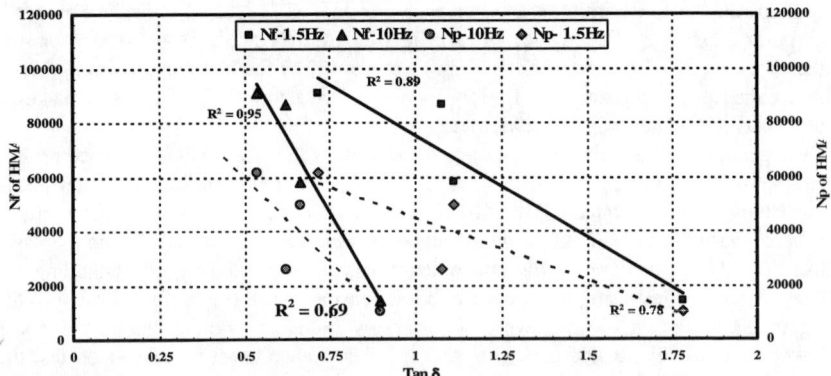

FIG. 6. N_f and N_p of HMA as a function of various AM properties.

Hence, for the stress controlled test, the parameter $G^*/\sin\delta$ of AM mixture exhibited a good relationship with the N_f and N_p of HMA mixtures.

Fig 6c depicts the relationship between N_f and N_p of HMA, and $\tan\delta$ of AM. It can be observed that as the $\tan\delta$ of AM increases, N_f decreases. Since $\tan\delta$ is the ratio of G" to G', higher $\tan\delta$ values represents greater viscous component of the G^* indicating more viscous behavior, and hence lower fatigue life. Similarly, as $\tan\delta$ of AM increases, N_p of HMA decreases. The increase in $\tan\delta$ signifies decreasing elasticity of AM mixtures, which leads to permanent deformation.

The data in Figure 7 shows that as the N_f of AM increases the N_f of HMA also increases. It should be noted that although the correlation is not good between the two, the trend is acceptable (see, Fig 7). This is because the fatigue tests for both AM and HMA mixtures were conducted at different AV content and moisture conditioning.

The above observations clearly indicate that by improving the AM properties the mixtures performance can be improved. In addition, any effect that is caused by moisture damage can also be identified by AM properties.

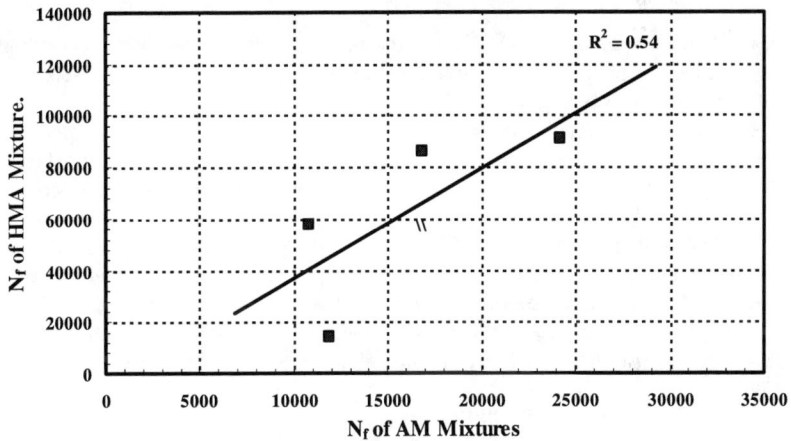

FIG. 7. Relationship between fatigue life of HMA and fatigue life of AM.

CONCLUSIONS AND RECOMMENDATIONS

Based on the laboratory test results for the AM and HMA mixtures, the following conclusions and recommendations were drawn:

- The moisture damage and the subsequent improvement in moisture damage resistance of AM mixtures as a result of HL modification was evident from the G^* master curve analysis.
- Improvement in fatigue damage resistance of HMA mixtures was observed as indicated by the higher fatigue lives, and lower TPD rates, due to HL addition.
- The stress-strain characteristics of moisture damaged HMA mixtures exhibited improved mixture behavior due to HL modification.

- The N_f and N_p of HMA showed good relationship with the G^*, $G^*/\sin\delta$ and $\tan\delta$ of AM mixtures at 1.58 and 10 Hz.
- There was no good relationship between the N_f of HMA and N_f of AM mixtures, however, the trend was reasonable.

It is recommended that the research study should be expanded to different types of aggregate, aggregate gradation and angularity, and types of asphalt, to identify key AM parameters that can facilitate the understanding of moisture damage in the HMA mixtures.

ACKNOWLEDGMENTS

The authors wish to express their sincere thanks to the University of Louisiana at Lafayette and the Louisiana Board of Regents for their financial support. A special thank you is extended to Mark Leblanc for his assistance in the laboratory experimentation.

REFERENCES

Anderson D.A., Marasteanu M.O (2000). "Establishing Linear Viscoelastic Conditions for Asphalt Binders." *Transportation Research Record 1728*, Transportation Research Board, Washington, D.C.

Kantipong K., Atud J., Tebid M. W. (2007). "Laboratory evaluation of hydrated lime application process in asphalt mixture for moisture damage and rutting resistance." *87thTransportation Research Board Annual Meeting* (CD-ROM) Transportation Research Board, Washington, D.C.

Khattak M. J., Baladi G. Y. (2001). "Fatigue and permanent deformation models for polymer-modified asphalt mixtures." *81thTransportation Research Board Annual Meeting* (CD-ROM) Transportation Research Board, Washington, D.C.

Kim Y. R., Little N. D. (2004). "Linear Viscoelastic Analysis of Asphalt Mastics." *Journal of Materials in Civil Engineering*, Vol. 16, No. 2. ISSN 0899-1561/2004/2-122-132.

Lesueur D., Little N. D. (1998). "Effect of Hydrated Lime on the Rheology, Fracture and Aging of Bitumen." *78thTransportation Research Board Annual Meeting* (CD-ROM) Transportation Research Board, Washington, D.C.

Little N. D., Epps A. J. (2001). "The benefits of Hydrated Lime in Hot Mix Asphalt." *National Lime Association*. Arlington, Virginia.

Roque R., Bjorn B., Mang T., and Masad E. (2005). "Development and Evaluation of Test Methods to Evaluate Water Damage and Effectiveness of Anti-stripping Agents." *Report No. New UF # 00026632*, Final Report, Florida Department of Transportation, University of Florida.

Sebaaly E P., Little N. D., Hajj Y. E., Bhasin A. (2007). "Impact of Lime and Liquid Antistrip on the Properties of an Idaho Mixture." *87thTransportation Research Board Annual Meeting* (CD-ROM) Transportation Research Board, Washington, D.C.

A Comparative Study of Laboratory Measured and Predicted Dynamic Modulus for Characterizing Florida Asphalt Mixtures

W. Virgil Ping[1], and Yuan Xiao[2]

[1]Professor, Department of Civil and Environmental Engineering, FAM-FSU College of Engineering, Florida State University, 2525 Pottsdamer Street, Tallahassee, FL 32310, USA; PH 1-850-410-6129, FAX 1-850-410-6142; e-mail: ping@eng.fsu.edu

[2]Graduate Research Assistant, Department of Civil and Environmental Engineering, FAM-FSU College of Engineering, Florida State University, 2525 Pottsdamer Street, Tallahassee, FL 32310, USA; e-mail: xiaoyu@eng.fsu.edu

ABSTRACT: Fatigue cracking under repeated load is one of the primary factors to be considered for flexible pavement design. Dynamic modulus test appeared to be a promising simple performance test for linking fundamental material properties to fatigue cracking observed in the field. The dynamic modulus concept was introduced in the AASHTO mechanistic-empirical pavement design guide as one of the critical inputs for characterization of asphalt concrete mixture, and it is employed as a key parameter in most commonly used models to predict fatigue life of bituminous mixtures. This paper presents a laboratory experimental program to evaluate the dynamic modulus of Florida Superpave asphalt mixtures including granite and limestone materials. The dynamic modulus master curves were constructed using the time-temperature superposition principle. The results indicated that the master curves for same type of materials were similar in shape and close to each other. The Witczak prediction model was used to perform a comparison between the measured and predicted dynamic modulus. The comparison showed that the Witczak prediction model provided a good estimation and could be adopted for estimating the modulus characteristics of commonly used Florida asphalt mixtures.

1 INTRODUCTION

Fatigue cracking is one of the principal types of distress in flexible pavement design. It is important to evaluate the fatigue resistance of asphalt mixtures for the design of pavement structures in order to achieve improved performance. Fatigue test is a destructive and usually time consuming test compared with the non-destructive stiffness test such as the dynamic (complex) modulus test. A number of fatigue life prediction models were developed to estimate the hot mix asphalt (HMA) fatigue resistance. In the National Cooperative Highway Research Program (NCHRP) mechanistic-empirical design guide for pavement structures (NCHRP 2004), the most commonly used Asphalt Institute's fatigue equation (Asphalt Institute 1982) was adopted to predict fatigue cracking of flexible pavement, in which the dynamic modulus functions as a key parameter. In addition, some other models also used the dynamic modulus in the fatigue prediction equation, such as the fatigue cracking model developed by Shell Oil Company (Bonnaure et al. 1980).

The dynamic complex modulus is the primary stiffness property of interest defining the fatigue cracking response of HMA mixtures and strongly influences the performance of asphalt pavement. The dynamic modulus ($|E^*|$) test (DMT) is used to characterize the modulus response of asphalt concrete mixtures by the mechanistic-empirical (M-E) design guide for pavement structures recently developed in the NCHRP project 1-37A. This is a significant change compared with the AASHTO Design Guide (1993), in which the indirect diametral resilient modulus (M_r) test (IDT) was employed as the material property to define characteristic of HMA layers. The transition from the resilient modulus to the use of dynamic modulus for design of flexible pavement structures has hardly been smooth. The potential impact of adopting the dynamic modulus for implementing the new AASHTO M-E Design Guide is tremendous for state transportation agencies, such as the Florida Department of Transportation (FDOT). This paper presents a laboratory experimental program to characterize the Florida HMA mixtures using the dynamic complex modulus. The experimental program is described as follows.

2 MATERIALS AND LABORATORY TESTING PROGRAM

A summary of the 18 Superpave mix designs and their gradations and volumetric properties are presented in Table 1 and 2 respectively. These mix designs were selected as they are commonly used FDOT gradations and are known to perform well in the field. The nominal maximum aggregate sizes for the mixtures tested are 19.0 mm, 12.5 mm, and 9.5 mm, respectively. The types of aggregates used include granite and limestone.

Table 1. Summary of Superpave Mix Designs

Series	L1	L2	L3	L4	G1	G2	G3	G4	G5
Mixture	LD-00 2502A	LD-02 2529A	SP-03 2610A	SP-03 2627A	SP-02 2180A	SP-03 2921A	SP-03 2922A	SP-04 3034A	SP-04 3225A
Size (mm)	12.5	12.5	12.5	12.5	9.5	9.5	19.0	12.5	12.5
Type	Fine	Coarse	Fine	Fine	Fine	Coarse	Coarse	Coarse	Fine
Load Level	D	D	C	C	B	D	D	D	C
Aggregate*	CFL	CFL	AL	SFL	Ga553	Ga553	Ga553	Ga553	Ga553 Ga206

Table 1. Summary of Superpave Mix Designs (Cont.)

Series	G6	G7	G8	G9	G10	G11	G12	G13	G14
Mixture	SP-02 2194A	SP-03 2452A	SP-03 2941A	SP-03 2351A	SP-05 4015A	SPM-05 4044A	SPM-05 4051A	SP-05 4100A	SP-02 2052A
Size (mm)	12.5	12.5	19.0	12.5	12.5	9.5	9.5	12.5	12.5
Type	Coarse	Coarse	Fine	Fine	Fine	Fine	Fine	Fine	FC
Load Level	D	D	C	C	B	D	D	D	C
Aggregate*	NS	Ga553	Ga553	Ga553	Ga553	Ga553	Ga553	Ga553	Ga553

*CFL: Central Florida Limestone; AL: Alabama Limestone; SFL: South Florida Limestone; Ga553 & Ga206: Georgia Granite; NS: Nova Scotia Granite.

Table 2. Gradations and Volumetrics of Mixtures Used

Sieve (mm)	L1	L2	L3	L4	G1	G2	G3	G4	G5
19	100	100	100	100	100	100	100	100	100
12.5	93	94	96	96	100	100	90	95	98
9.5	89	89	90	88	100	100	77	84	90
4.75	71	56	72	69	74	71	51	52	57
2.36	53	30	52	54	48	42	32	32	40
1.18	42	20	34	38	39	28	22	21	34
0.6	35	15	24	27	28	18	16	15	28
0.3	22	10	11	19	16	13	12	9	16
0.15	9	6	6	12	7	9	9	6	4
0.075	4.5	4.3	4.5	4.5	4.5	6.9	6.4	5.2	4.5
Volumetrics									
G_{mm}	2.276	2.253	2.494	2.313	2.550	2.563	2.603	2.589	2.554
G_b	1.035	1.035	1.035	1.035	1.035	1.035	1.035	1.035	1.035
G_{mb}	2.185	2.162	2.393	2.220	2.448	2.460	2.499	2.485	2.452
P_b	8.2	8.2	5.7	7.5	5.3	5.8	4.5	5.0	5.3
G_{sb}	2.346	2.311	2.689	2.389	2.745	2.776	2.781	2.775	2.729
G_{se}	2.549	2.518	2.726	2.570	2.778	2.819	2.803	2.811	2.783
P_{ba}	3.514	3.676	0.527	3.056	0.442	0.572	0.293	0.480	0.730
P_{be}	4.97	4.8	5.2	4.68	4.9	5.2	4.2	4.5	4.61
VMA	14.5	14.1	16.1	14.0	15.5	16.5	14.2	14.9	14.9
V_a	4.0	4.0	4.0	4.0	4.0	4.0	4.0	4.0	4.0
VFA	72	72	75	71	74	76	72	73	73
D/A	0.9	0.9	0.9	1.0	0.9	1.3	1.5	1.2	1.0

Table 2. Gradations and Volumetrics of Mixtures Used (Cont.)

Sieve (mm)	G6	G7	G8	G9	G10	G11	G12	G13	G14
19	100	100	100	100	100	100	100	100	100
12.5	98	99	90	100	99	100	100	100	98
9.5	89	75	79	90	90	100	100	90	90
4.75	58	44	61	55	61	74	75	60	59
2.36	38	29	44	40	42	50	48	43	40
1.18	24	19	35	34	33	39	40	34	34
0.6	16	13	26	28	26	31	30	27	26
0.3	10	9	18	16	18	23	16	19	11
0.15	5	6	8	4	7	9	6	7	4
0.075	4.5	4.5	4.4	2.9	2.8	5.6	3.0	3.0	3.5
Volumetrics									
G_{mm}	2.420	2.555	2.571	2.570	2.557	2.550	2.535	2.567	2.539
G_b	1.035	1.035	1.035	1.035	1.035	1.035	1.035	1.035	1.035
G_{mb}	2.322	2.454	2.468	2.467	2.455	2.448	2.434	2.464	2.438
P_b	6.0	5.4	5.0	4.8	5.0	5.2	6.0	5.0	5.3
G_{sb}	2.604	2.701	2.763	2.752	2.764	2.750	2.748	2.756	2.757
G_{se}	2.646	2.789	2.789	2.778	2.772	2.773	2.793	2.784	2.764
P_{ba}	0.631	1.206	0.347	0.348	0.101	0.307	0.612	0.376	0.092
P_{be}	5.41	4.26	4.6	4.4	4.9	4.9	5.4	4.6	5.2
VMA	16.2	14.1	15.1	14.7	15.6	15.6	16.7	15.1	16.3
V_a	4.0	4.0	4.0	4.0	4.0	4.0	4.0	4.0	4.0
VFA	75	72	74	73	74	74	76	74	75
D/A	0.8	1.1	1.0	0.7	0.6	1.2	0.8	0.7	0.7

The sample preparation for the dynamic modulus test was based on the conclusions of an extensive study on sample geometry and aggregate size conducted during NCHRP Project 9-19. The HMA mixtures were compacted in the laboratory with the targeted optimal air voids content (4%) using a Servopac gyratory compactor. Then the specimens were cut and cored to a nominal diameter of 100 mm (4.0 in.) and a height of 150 mm (5.9 in.).

The dynamic moduli and phase angles were measured by applying compressive sinusoidal (haversine) load generally following the Simple Performance Test System developed in NCHRP Project 9-29. Tests were performed at temperatures of 5, 25, and 40 °C and frequencies of 25, 10, 5, 1, and 0.5 Hz. In general, for each test series, three DMT specimens were tested and averaged to determine the dynamic modulus of each mix design.

3 EXPERIMENTAL RESULTS

The complex modulus is a viscoelastic response of asphalt concrete mixture under sinusoidal loading conditions at different test temperatures and loading frequencies, which accounts for both elastic and viscous effects of the material. The magnitude of complex modulus, commonly referred to as dynamic modulus $|E^*|$, is determined as the ratio of the amplitude of applied stress to the amplitude of measured strain response experimentally. The real part of complex modulus represents the elastic stiffness and the imaginary part characterizes the internal damping of the material. The dynamic modulus test is typically conducted on cylindrical specimens subjected to a compressive loading.

The applied stress and strain response are measured to calculate the dynamic modulus and phase lag of the specimen. The fundamental relationships are presented in the following equations:

$$E^* = |E^*| \cdot e^{i\phi} = \frac{\sigma_0}{\varepsilon_0} \cdot e^{i\phi} \tag{1}$$

$$\sigma = \sigma_0 \cdot \sin(\omega \cdot t) \tag{2}$$

$$\varepsilon = \varepsilon_0 \cdot \sin(\omega \cdot t - \phi) \tag{3}$$

$$\phi = \frac{t_{lag}}{t_p} \cdot (360°) = t_{lag} \cdot f \cdot (360°) \tag{4}$$

Where,

- σ_0 = stress amplitude
- ε_0 = strain amplitude
- ω = angular frequency (rad/sec)
- φ = the phase lag (degrees)
- t_{lag} = time lag between a cycle of sinusoidal stress and a cycle of strain
- t_p = time period of a stress cycle (seconds)

The average dynamic modulus values of replicate specimens for each asphalt mixture design series are summarized in Table 3. As expected as well as in accordance with other research studies, the dynamic modulus increased as the test frequency increased under a certain testing temperature; the dynamic modulus increased with a decrease in test temperature under a certain loading frequency. The phase angle increased as the test temperature increased. The phase angle had a decreasing trend with increasing load frequency under a certain temperature. A more complex behavior of the phase angle as a function of the loading frequency was observed at higher temperatures. The phase angles were grouped and averaged by aggregate type shown in Figure 1 for granite and limestone materials respectively, at each testing temperature.

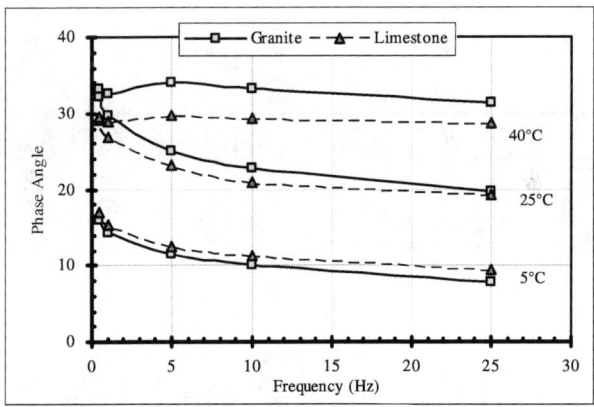

Figure 1. Average phase angles for different materials

Table 3. Summary of Dynamic Modulus Testing Results

Mixture	Temperature (°C)	Dynamic Modulus (MPa) at Frequency (Hz)				
		25 Hz	10 Hz	5 Hz	1 Hz	0.5 Hz
L1	5	15135	14092	13298	11332	10433
	25	5777	4716	4127	2848	2542
	40	2555	2410	1900	1082	899
L2	5	11482	10505	9538	7183	6389
	25	7004	5793	4888	3253	2712
	40	3309	2506	1996	1176	962
L3	5	14522	13093	11803	8800	7541
	25	5501	4519	3590	2060	1697
	40	1908	1370	1093	655	510
L4	5	15717	14657	13541	10783	9592
	25	7803	6458	5482	3487	2741
	40	2973	2260	1758	1101	896
G1	5	11641	10446	9486	7295	6366
	25	5230	4268	3583	2265	1890
	40	2215	1650	1308	860	706

Table 3. Summary of Dynamic Modulus Testing Results (Continued)

Mixture	Temperature (°C)	Dynamic Modulus (MPa) at Frequency (Hz)				
		25 Hz	10 Hz	5 Hz	1 Hz	0.5 Hz
G2	5	13381	12505	11722	9816	8956
	25	7218	6202	5412	3781	3166
	40	3014	2294	1833	1074	887
G3	5	22200	19906	18453	14937	13354
	25	9671	8039	6661	4093	3166
	40	3280	2545	1913	996	869
G4	5	18322	16858	15546	12352	10335
	25	7842	6613	5470	3256	2474
	40	2924	2081	1568	922	719
G5	5	19111	17174	15763	12469	11045
	25	7696	6266	5151	3050	2330
	40	2693	1887	1405	841	656
G6	5	16962	15282	13879	10759	9426
	25	6980	5682	4635	2707	2113
	40	2223	1573	1225	725	574
G7	5	16213	14416	13025	9845	8507
	25	6394	5087	4059	2241	1820
	40	1923	1272	1052	582	466
G8	5	20764	19431	18095	14733	13266
	25	9158	7758	6575	4223	3348
	40	3855	2894	2260	1295	1023
G9	5	18812	17234	15996	13029	11747
	25	8403	7176	6085	3940	3145
	40	3822	2858	2235	1305	986
G10	5	17736	16514	15284	12239	10925
	25	7056	5812	4810	3149	2418
	40	2759	1969	1481	915	704
G11	5	18462	17163	15711	12603	11267
	25	7693	6512	5428	3349	2607
	40	2658	1942	1519	904	698
G12	5	17418	15522	14234	11313	10077
	25	7285	6247	5312	3451	2767
	40	2826	2148	1710	1056	829
G13	5	17330	16165	14999	12061	10788
	25	7888	6452	5372	3214	2468
	40	3044	2212	1683	979	748
G14	5	18889	17653	16316	13086	11651
	25	7229	5811	4800	2806	2131
	40	2691	1906	1432	831	661

4 ANALYSES OF EXPERIMENTAL RESULTS

4.1 *Master Curve Construction*

In the AASHTO M-E Pavement Design Guide, the dynamic modulus of HMA, at all levels of temperature and loading frequency, is determined from a master curve constructed by fitting a sigmoidal curve to the measured dynamic modulus test data, which describes the time dependency of the material. The mathematical model for master curve construction developed by Pellinen and Witczak (2002) was used for obtaining predicted master curves for all mixtures in this study.

$$\log(|E^*|) = \delta + \frac{\alpha}{1+e^{\beta+\gamma \log f_r}} \quad (5)$$

Where:

$Log(|E^*|)$ = log of dynamic modulus
f_r = reduced frequency
δ = minimum modulus value
α = span of modulus value
β, γ = shape parameters

$$a(T) = \frac{f}{f_r} \quad or \quad \log[a(T)] = \log(f) - \log(f_r) \quad (6)$$

Where:

f = testing frequency at desired temperature
f_r = reduced frequency
T = temperature of interest

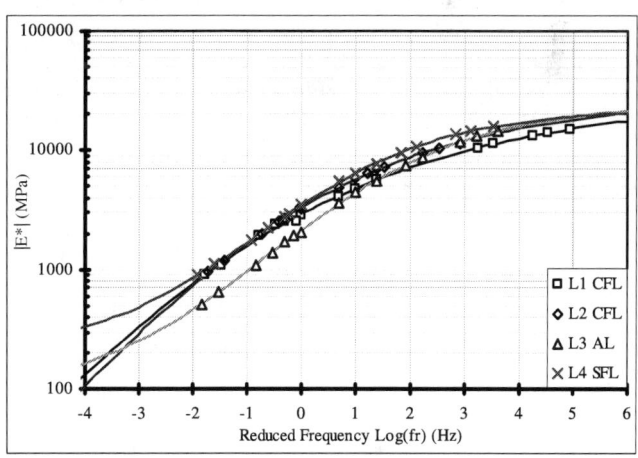

Figure 2. Master curves for limestone materials

In all master curve constructions, the reference temperature was taken as 25°C (77°F). The shifting factors were obtained simultaneously with the coefficients of the sigmoidal function through nonlinear regression. The master curves for limestone and granite materials were grouped together and shown in Figures 2 and 3, respectively.

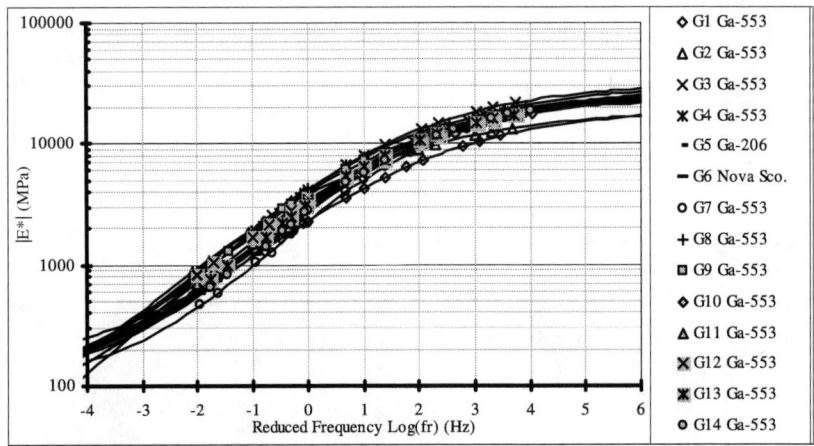

Figure 3. Master curves for granite materials

4.2 *Witczak Prediction Model*

Efforts were made by asphalt pavement researchers to develop regression equations to estimate the dynamic modulus for a specific mix design. One of the most comprehensive mixture dynamic modulus models is the Witczak et al. prediction model (NCHRP 2004). It was proposed in the AASHTO M-E design guide and the calculations were based on the volumetric properties of a given mixture. In this model, the parameter η (bitumen viscosity) for each dynamic modulus test temperature is determined by:

$$\log(\log \eta) = A + VTS \cdot \log T \qquad (7)$$

Where A is the regression intercept, T is Rankine temperature and VTS is the slope of log-log viscosity versus temperature relationship. A and VTS are functions of binder type and material characteristics, and they are determined by regression using experiment data of binder viscosity versus temperature T. All HMA in this study used the asphalt binder PG 67-22 (AC-30). The input binder viscosity was obtained from two sources:

- Mix/Laydown conditions for PG 67-22 asphalt binder (Witczak & Fonseca 1996):

    $A = 10.6768$, $VTS = -3.56455$

- Brookfield rotational viscometer results on short-term Rotational Thin Film Oven (RTFO) aged PG 67-22 specimens (Birgisson et al. 2004):

    $A = 10.407$, $VTS = -3.4655$

Witczak's prediction equation is presented as follows:

$$\log|E^*| = -1.249937 + 0.029232P_{200} - 0.001767(P_{200})^2$$
$$- 0.002841P_4 - 0.058097V_a - 0.802208\frac{V_{beff}}{(V_{beff} + V_a)}$$
$$+ \frac{[3.871977 - 0.0021P_4 + 0.003958P_{38} - 0.000017(P_{38})^2 + 0.00547P_{34}]}{1 + e^{(-0.603313 - 0.313351\log f - 0.393532\log \eta)}} \quad (8)$$

Where
$|E^*|$ = dynamic modulus, in 10^5 psi
η = bituminous viscosity, in 10^6 poise (at any temperature, degree of aging)
f = load frequency, in Hz
V_a = percent air voids content, by volume
V_{beff} = percent effective bitumen content, by volume
P_{34} = percent retained on 19-mm sieve, by total aggregate weight (cumulative)
P_{38} = percent retained on 9.5-mm sieve, by total aggregate weight (cumulative)
P_4 = percent retained on 4.75-mm sieve, by total aggregate weight (cumulative)
P_{200} = percent passing 0.75-mm sieve, by total aggregate weight (cumulative)

4.3 Predicted Versus Measured Dynamic Modulus

A few comparative studies were conducted to evaluate the effect of Witczak prediction model since it was proposed in the M-E design guide. Loulizi et al. (2006) performed the dynamic modulus and resilient modulus tests on two typical mixes used in the Commonwealth of Virginia. Witczak's prediction equation was used to generate the dynamic modulus master curves for both mixes. The results of analysis showed that the predicted dynamic modulus values were on the same order of magnitude as the measured ones. Birgisson et al. (2004) presented an evaluation of how well the Witczak predictive modulus equation worked for mixtures typical to Florida. They found that the prediction resulted in a slight bias for mixtures commonly used in Florida, which allowed for a correction of the bias between predicted and measured dynamic modulus. Schwartz (2005) evaluated the accuracy and robustness of the Witczak predictive model. The study showed that the Witczak predictive model provides sufficiently accurate and reasonably robust estimates of complex modulus for use in the M-E performance prediction and design, although it has more limited ability to make fine distinctions between the performance of different mixtures at the same temperature and other design conditions. King et al. (2005) performed a series of dynamic modulus testing of Superpave mixtures commonly used in North Carolina. Prediction accuracy of Witczak's equation was evaluated by comparing the measured and predicted dynamic moduli values. The study showed that the Witczak prediction at cooler temperatures was better than that at warmer temperatures.

In this study, the comparisons between the predicted and measured dynamic moduli for all mixture series under the two binder conditions are presented in Figure 4. Linear regression with zero intercept was performed for the data analysis. Since the comparison was made by using measured dynamic modulus as horizontal x-values, the points above the line of equality indicated a prediction that is not conservative, in which the predicted dynamic modulus was higher than the measured one. The comparison indicated a fair prediction for the mixtures tested in this study.

Since only one type of asphalt binder (PG 67-22) was used for all mix designs, the differences of stiffness properties between tested mixtures were primarily due to the different types of aggregates used for each mix design. Therefore, comparisons were made by types of materials used in the mixtures. The comparisons of predicted and measured dynamic modulus for limestone and granite materials are shown in Figures 5-6, respectively, under the two binder conditions. As shown in the figures, the prediction model provides conservative estimates for granite materials. However, the regression analysis provided a slope that was higher than 1 (1.22 for the mix-laydown condition, 1.20 for the RTFO aged condition) for limestone materials, which meant a prediction that was not conservative. That was likely because the stiffness of granite materials was much higher than that of limestone in the actual testing.

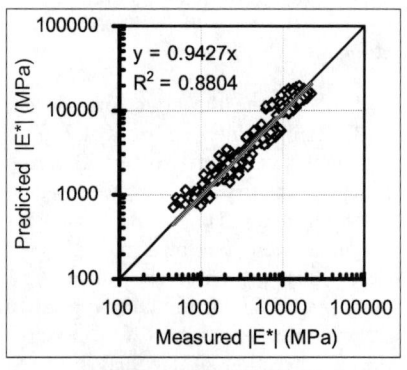
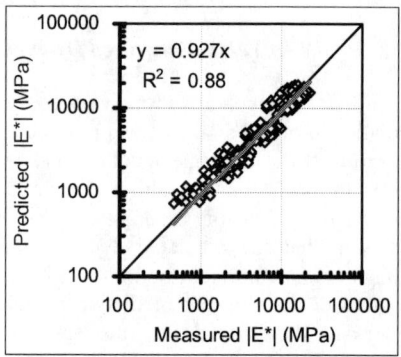

(a) Mix-laydown condition (b) RTFO aged condition

Figure 4. Measured vs. predicted values of dynamic modulus for all mixtures

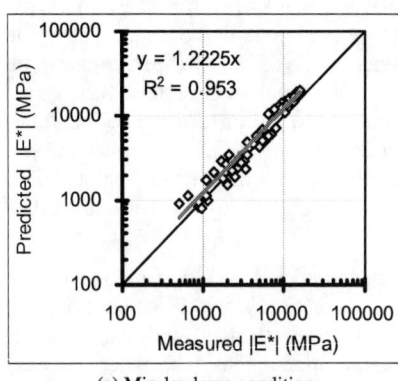
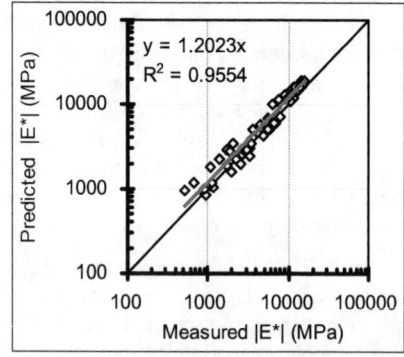

(a) Mix-laydown condition (b) RTFO aged condition

Figure 5. Measured vs. predicted values of dynamic modulus for limestone materials

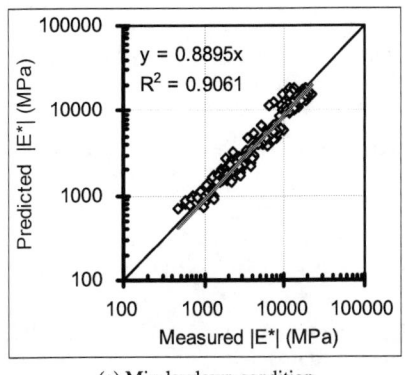
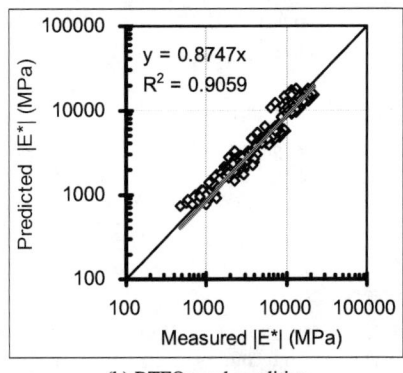

(a) Mix-laydown condition   (b) RTFO aged condition

Figure 6. Measured vs. predicted values of dynamic modulus for granite materials

5 SUMMARY AND CONCLUSIONS

All 18 Superpave asphalt mixtures were tested at three temperature levels (5, 25, and 40 °C) and at the following frequencies: 25, 10, 5, 1, and 0.5 Hz. The dynamic modulus decreased with an increase in the testing temperature at a specific loading frequency. At a constant testing temperature, the dynamic modulus increased with an increase in the loading frequency. These trends are in agreement with other studies. At testing temperatures of 5°C and 25°C, the phase angle decreased with an increase in frequency. At the temperature of 40°C, with an increase in frequency, the phase angle had a tendency to increase at low frequencies (below 5 Hz) and decrease at higher frequencies, possibly due to a combined effect of the softer binder and more aggregates contribution. At higher frequencies (5 Hz and above), the phase angle increased with an increase in testing temperature; at lower frequencies, the phase angle as a function of the temperature was more complicated. No relationship between aggregate type and phase angle was observed.

The master curves for same type of materials were similar in shape and close to each other; due possibly to only one binder type (PG 67-22) that was used for the mixtures. The Witczak prediction model provided a good estimation of the dynamic modulus values for the selected asphalt mixtures common to Florida in this study. The over all prediction was balanced with a regression coefficient close to 0.94. The comparisons showed that the prediction model provided a slightly conservative estimate for the granite mixtures and a minor overestimate for the limestone mixtures, which appeared to depend on the nature of aggregate type.

6 REFERENCES

American Association of State Highway and Transportation Official (AASHTO) (1993). *AASHTO Design Guide*, AASHTO, Washington, D.C

Asphalt Institute (1982). Research and Development of the Asphalt Institute's Thickness Design Manual (MS-1), 9th edition, Research Report 82-2, 1982.

Birgisson, B., R. Roque, J. Kim and L. V. Pham (2004). "The Use of Complex Modulus to Characterize the Performance of Asphalt Mixtures and Pavements in Florida." *Final report*, The Florida Department of Transportation, BC 354, RPWO #22, Gainesville, FL.

Bonaquist, R. and D. W. Christensen (2005). "Practical Procedure for Developing Dynamic Modulus Master Curves for Pavement Structural Design." *Transportation Research Record: Journal of the Transportation Research Board*, No. 1929: 208-217

Bonnaure, F., A. Gravois and J. Udron (1980). "A New Method of Predicting the Fatigue Life of Bituminous Mixes." *Journal of the Association of Asphalt Paving Technologists*, Volume 49: 499-529.

El-Basyouny, M. and M. Witczak (2005). "Development of the Fatigue Cracking Models for the 2002 Design Guide." *TRB*, 2005 Annual Meeting CD-ROM.

Fonseca, O. A. and M. W. Witczak (1996). "A Prediction Methodology for the Dynamic Modulus of In-Place Aged Asphalt Mixtures." *Journal of the Association of Asphalt Paving Technologists*, Vol. 65: 532-559.

King, M., M. Momen and Y. R. Kim (2005). "Typical Dynamic Moduli Values of Hot Mix Asphalt in North Carolina and Their Prediction." *TRB*, 2005 Annual Meeting CD-ROM.

Loulizi, A., G. W. Flintsch, I. L. Al-Qadi and D. Mokarem (2006). "Comparing Resilient Modulus and Dynamic Modulus of Hot-Mix Asphalt as Material Properties for Flexible Pavement Design." *Journal of the Transportation Research Board*, TRB No. 1970: 161-170.

National Cooperative Highway Research Program (2004). 1-37A-2002 Design Guide (Draft), NCHRP, Washington, DC.

Pellinen, T. K. and M. W. Witczak (2002). "Stress Dependent Master Curve Construction for Dynamic (Complex) Modulus." *Journal of the Association of Asphalt Paving Technologists*, Vol. 71: 281-309.

Ping, W. V. and Y. Xiao (2007). "Evaluation of the Dynamic Complex Modulus Test and Indirect Diametral Test for Implementing the AASHTO 2002 Design Guide for Pavement Structures in Florida." *Final Report*, Florida Department of Transportation, BC-352-12, Tallahassee, FL.

Romanoschi, S., N. Dumitru, O. Dumitru and G. Fager (2006). "Dynamic Resilient Modulus and the Fatigue Properties of Superpave HMA Mixes used in the Base Layer of Kansas Flexible Pavements." *TRB*, 2006 Annual Meeting CD-ROM.

Schwartz, C. W. (2005). "Evaluation of the Witczak Dynamic Modulus Prediction Model." *TRB*, 2005 Annual Meeting CD-ROM.

Witczak, M. W. and O. A. Fonseca (1996). "Revised Prediction Model for Dynamic (Complex) Modulus of Asphalt Mixtures." *Journal of the Transportation Research Board*, TRB, No. 1540: 15-23.

Witczak, M. W., T. K. Pellinen and M. M. El-Basyouny (2002). "Pursuit of the Simple Performance Test for Asphalt Concrete Fracture/Cracking." *Journal of the Association of Asphalt Paving Technologists*, Vol. 71: 767-778.

Determination of the Elastic Modulus and Poisson's Ratio of Asphaltic Mixtures Using Uniaxial Creep Recovery Tests

H. Taherkhani[1] and A.C. Collop[2]

[1]Assistant professor, University of Zanjan, Zanjan, Iran, email: htaherkhanik@yahoo.com
[2]Professor of Civil Engineering, University of Nottingham, Nottingham, UK, email: Andrew.collop@nottnigham.ac.uk

ABSTRACT

Elastic modulus and Poisson's ratio are the main parameters in elastic analysis of pavements, and also used in viscoelastic and elasto-visco-plastic analysis. These are the main parameters in elastic analysis of pavements, and also used for viscoelastic and elasto-visco-plastic analysis. The values of these parameters for asphaltic mixtures are determined using empirical values, formulations or experiments specifically designed for their determination. This describes the determination of these parameters using uniaxial single and repeated creep recovery tests by measuring the radial and axial strain of specimen. The tests were carried out, over a range of test conditions, on two types of asphaltic mixtures used in the UK pavements, namely a 10 mm DBM and HRA30/10. The variation of the parameters with temperature, stress level and accumulated strain level has been studied for the mixtures. It is found that the elastic modulus of asphaltic mixtures decreases with increasing damage in the mixtures. A formulation is developed for prediction of the elastic modulus of the mixtures as a function of stress level, temperature and the accumulated damage in the mixture.

Key Words: Asphaltic Mixtures, Elastic modulus, Poisson's Ratio, Creep Recovery

1. Introduction

For a successful analysis and design of a pavement, it is necessary to determine a functional relationship between stress and strain for the pavement materials at different working conditions. Elastic, viscoelastic and elasto-visco-plastic theories have been used for the analysis of flexible pavements. Elastic theory was used first by Westergaard [1927] for analysis of Portland cement concrete pavements, after which the principles of elasticity were widely used in the analysis and design of flexible and rigid pavements. Burmister [1943] developed elastic layer theory for a two-layer pavement, which later was used for three and more layers. In this theory, it is usually assumed that the materials comprising the pavement layers are homogeneous, isotropic and linear elastic and characterized by time-independent constants of proportionality between stress and strain [Huang, 1967].

For a linear elastic material, the proportionality between stress and strain follows Hooke's Law. For three-dimensional bodies the relationship between stress and strain can be written as follows [Sluys *et al.*, 2002]:

$$\{\sigma\} = [K]\{\varepsilon\} \quad (1)$$

where $\{\sigma\}$ and $\{\varepsilon\}$ are the stress and strain tensors, respectively, and $[K]$ is the stiffness matrix of elasticity. For a homogenous linear elastic isotropic material, the stiffness matrix is defined as:

$$[K] = \frac{E}{(1+v)(1-2v)} \begin{bmatrix} 1-v & v & v & 0 & 0 & 0 \\ v & 1-v & v & 0 & 0 & 0 \\ v & v & 1-v & 0 & 0 & 0 \\ 0 & 0 & 0 & 0.5-v & 0 & 0 \\ 0 & 0 & 0 & 0 & 0.5-v & 0 \\ 0 & 0 & 0 & 0 & 0 & 0.5-v \end{bmatrix} \quad (2)$$

where E = Young's modulus
       v = Poisson's ratio.

Linear viscoelastic models have been used in many studies for characterisation of asphaltic mixtures, analysis of stresses and strains, and prediction of rutting in flexible pavements e.g. [Papazian, 1962; Pagen, 1964, 1968 & 1972; Sayegh, 1967; Huang, 1967; Thrower, 1975; Huschek, 1977; Kenis, 1977; Nunn, 1986; Hopman et al. 1992 and Blab et al. 2002]. Some researchers have extended linear viscoelastic models to include non-linear effects e.g. [Fitzgerald et al., 1973; Lai et al., 1973 and Vakili, 1983, Collop et al., 1995 and Collop et al., 2003].

Under a load application, asphaltic materials usually show elastic, plastic, viscoelastic (delayed elastic) and viscoplastic (viscous) deformations. An elasto-visco-plastic constitutive model usually comprises formulations for each of these components [Abdulshafi et al. 1968; Uzan, 1996; and Chehab et al. 2003].

In both the viscoelastic and elasto-visco-plastic models, the elastic modulus is usually used for simulation of the elastic component of the deformation [Collop et al., 2003].

Collop et al. [2003] developed a non-linear viscoelastic model for simulation of the deformation behaviour of asphaltic mixtures. The approach was based around the Burger's mechanical model (Figure 1) comprising elastic, delayed elastic and viscous elements. The creep and recovery behaviour of an asphaltic mixture can be simulated by the Burger's mechanical model. This mechanical model is comprised of a number of Voigt elements in series with a viscous dashpot and an elastic spring. When a stress σ is applied to this combination, the spring in series deforms instantaneously in proportion to the applied stress and the inverse of the elastic modulus of the spring E_0. However, neither the dashpot in series nor the dashpots in the Voigt elements can move in the same short period of time. When the stress is held constant, the Voigt elements deform in proportion to the applied stress σ, elastic modulus of the springs and the viscosity of the dashpots in the Voigt elements, and the dashpot in series deforms gradually proportional to the applied stress and the viscosity of the dashpot. Upon application of a stress, the spring in series will experience an instant strain that is recovered instantly after releasing the stress. Therefore, when the behaviour of asphaltic mixtures is elastic the elastic modulus of the spring in series can be assumed as the elastic modulus of the mixtures and used in elastic analysis of pavements.

This paper explains determination of the elastic modulus of the spring in Burger's mechanical model and Poisson's ratio of asphaltic mixtures using uniaxial creep recovery tests. The variation of these parameters with stress level, temperature, type of mixture and accumulated strain in the mixture is discussed.

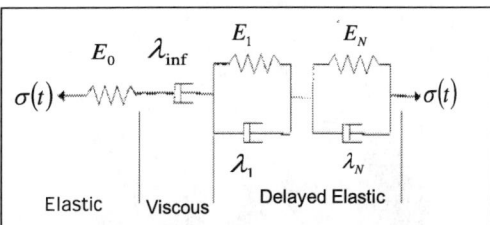

Figure 1. Mechanical arrangement of the model.

2. Materials

Two generic types of asphaltic mixture were chosen for this study; a 10mm Dense Bitumen Macadam (DBM) (British Standards Institute, 2003a) and a 30/10 Hot Rolled Asphalt (HRA) (British Standards Institute, 2003b). These mixtures were chosen because different types of behaviour were anticipated. The DBM is a densely graded mixture that relies primarily on aggregate interlock for its strength, whereas the HRA is a gap graded mixture that relies more on the properties of the bitumen/sand/filler mortar. Granite aggregates and a 70/100 penetration grade of bitumen were used to produce both mixtures. The target air void content was chosen to be 4% for both mixtures and binder contents of 5.5% and 8% were chosen for the DBM and HRA respectively.

Cylindrical specimens, 100mm in diameter and 100mm in height, were manufactured for the compression testing programme. The mixtures were compacted in a 150mm diameter gyratory compactor at a temperature between 150°C and 156°C. After cooling, the mixtures were extruded from the mould and a 100mm diameter core was taken from the centre of the specimen. Both ends of the specimen were trimmed and the air void content was measured and only specimens with an air void content between 3% and 5% were selected for testing. The specimens were stored in a cold room at 5°C until they were required for testing.

3. Test Equipment and Instrumentation

An Instron 1332 loading frame with a temperature-controlled cabinet (–5°C to 50°C) and a servo-hydraulic actuator with a load capacity of ±100 kN and ±50mm axial stroke was used for the testing programme. Figure 2 shows the experimental set-up for the compressive uniaxial testing undertaken in this study. In this test, the specimen is placed between two polished chrome plates. A friction reduction system, comprising a layer of plastic film sandwiched between two layers of soap, was used to minimise lateral confinement due to friction between the platens and the specimen. Two LVDTs, positioned on the top platen were used to measure the axial deformation of the specimen and an LVDT mounted on the collar was used to measure radial deformation at the mid-height of the specimen. The radial LVDT was used for measuring the instant radial strains to be used for determination of the Poisson's ratio.

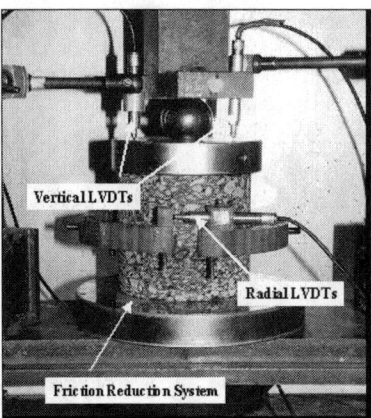
Figure 2. Uniaxial test set-up.

4. Testing Procedure

Uniaxial creep recovery tests were performed over a range of temperatures (10°C to 40°C) and stress levels (from 400 to 2500kPa). To ensure uniformity of temperature, the specimens were stored in the temperature-controlled cabinet for at least 12 hours prior to testing. The specimen was instrumented and a small compressive pre-load was applied to take up any slack in the system and allow the friction reduction system to deform in order to minimise subsequent measurement errors. In a single creep recovery test, after preloading, the target stress σ was immediately applied on the specimen over a period of 30 to 40 milliseconds. The load was then released over a short time equivalent to the time taken to apply the target load. After removing the load the specimen was allowed to recover until the rate of recovered deformation was almost zero. During the creep and recovery stages, the axial and radial deformation and the load were measured and logged by a computer connected to the equipment. The data were converted to ASCII files and were transferred for data analysis.

In repeated creep recovery tests, for each cycle, the axial stress σ was applied on the specimen and was held constant for a time period of ΔT after which the stress was released and the specimen was allowed to recover for a time period of ΔR. The rate of loading and unloading was chosen such that the target load be applied and removed at a short time between 30 to 40 milliseconds. The tests were continued by repeating the cycles until specimen failure. The axial strain and load were logged during the tests by the computer connected to the equipment. For each mixture the compression tests were carried out at one stress level (1000kPa) and one temperature (20°C) and three different unloading times (30, 60 and 120Sec).

5. Test Results

Figure 3 shows a typical compressive creep recovery test result for the HRA mixture at 20°C for an axial stress level of 1000kPa where the axial strain is plotted as a function of time. Similar behaviour was observed for the DBM, which for brevity is not shown. It can be seen from Figure 3 that, when the load is applied, an instantaneous (time-independent) strain appears in the specimen followed by gradually increasing (time-dependent) strains. When the load is removed, there is an instantaneous recovery (elastic strain) followed by time-dependent elastic strain (delayed elastic strain) recovery. Irrecoverable deformation remains in the specimen as a permanent strain, which is comprised of plastic and viscous components.

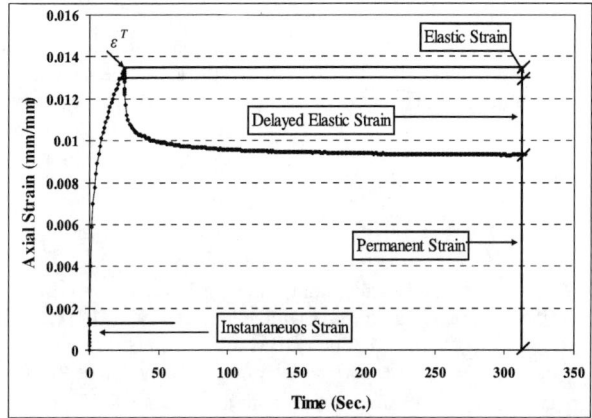

Figure 3. A typical creep recovery test result in compression for the HRA30/10 at 20°C (Axial stress level of 1000 kPa).

5.1. Elastic Modulus, E_0

In an idealised creep recovery test on the Burger's mechanical model, the instantaneous strain after load application will be equal to the instantaneous recovered strain after load removal. However, for the asphalt mixtures the instantaneous strain after load application was seen to be higher than the instantaneous strain recovered after releasing the load (see Figure 3). This difference is thought to be due to the plastic strain caused by aggregate movement, densification, the deformation of friction reduction system and the viscous deformation occurred in the short period in which the load varies from zero to the target load. The instantaneous recovered strain after releasing the load is the time-independent elastic strain and used for calculation of the elastic modulus. Therefore, the elastic modulus of the spring E_0 was calculated from the unloading path in the creep recovery test results. As shown in Figure 4, the elastic modulus was taken as the slope of the straight line from a plot of axial strain versus stress during the short period of time in which the stress varies from the target stress to zero in the creep recovery tests. In this figure the strain prior to unloading has been taken as zero.

Ideally, to obtain the elastic modulus of the spring E_0, the unloading time should be short enough so that the time dependent recovery deformation (delayed elastic) is not significant. Achieving a target load at a very short time depends on the accuracy of the loading system and the software which controls the machine. Based on the data logging software used in the tests and the data points required for calculation of the elastic modulus, the shortest loading and unloading time was estimated to be around 40 milliseconds for which the behaviour of the mixtures can be assumed to be elastic.

Figure 5 shows the elastic modulus E_0 for the DBM mixture at different temperatures and stress levels. As can be seen in the figure, over the range of stress levels utilized in these experiments, the elastic modulus of the mixture is independent of the stress level and only depends on the temperature. Similar behaviour was observed for the HRA mixture, which for brevity is not presented here.

Figure 6 shows the elastic modulus E_0 as a function of temperature. The solid lines represent the WLF equation (Eq. 3) fitted to the experimental data. As can be seen, the temperature dependency of the parameter for the mixtures is well captured by the WLF equation.

$$E_0 = A\exp\frac{-2.303.c_s^1.(T-T_s)}{c_s^2 + (T-T_s)} \quad (3)$$

where
- T = temperature in Celsius degree
- E_0 = the elastic modulus in GPa
- A = material constant with values of 17400 GPa and 21000 GPa for the HRA and DBM mixtures, respectively
- T_s = the reference temperature with a value of -60°C
- c_s^1 and c_s^2 = universal constants with the values 8.86 & 101.6

Results from the single creep recovery tests, unloaded at different strain levels, along with the repeated creep recovery test results, revealed that the elastic modulus of the mixtures decreases with increasing damage in the mixtures. Figure 7 shows the elastic modulus E_0 as a function of the accumulated strain in the mixtures. As can be seen, up to a certain level of strain, the elastic modulus remains almost constant, after which it reduces with increasing accumulated strain or damage in the mixtures. The figure clearly shows that the reduction of the elastic modulus for the DBM mixture occurs at a lower strain level than that for the HRA. Moreover, the reduction of the elastic modulus with accumulated strain is linear with approximately the same slope for both mixtures. The reduction of the elastic modulus is thought to be due to the development of cracks and internal damage in the mixtures. Because of continuous grading and thinner binder films in the DBM mixture the internal cracks develop at a lower axial strain level than in the HRA. The following equations can be used for determination of the elastic modulus E_0 (in GPa) for the mixtures at 20°C. Developing an equation as a function of temperature and damage needs further experimental work at different temperatures.

$$E_0^{DBM} = 2.97 \qquad \varepsilon_t \leq 0.03$$
$$= -21.3\varepsilon_t + 3.6 \qquad \varepsilon_t > 0.03 \quad (4)$$

$$E_0^{HRA} = 1.9 \qquad \varepsilon_t \leq 0.043$$
$$= -20.7\varepsilon_t + 2.79 \qquad \varepsilon_t > 0.043 \quad (5)$$

where ε_t is the total accumulated strain in the mixtures.

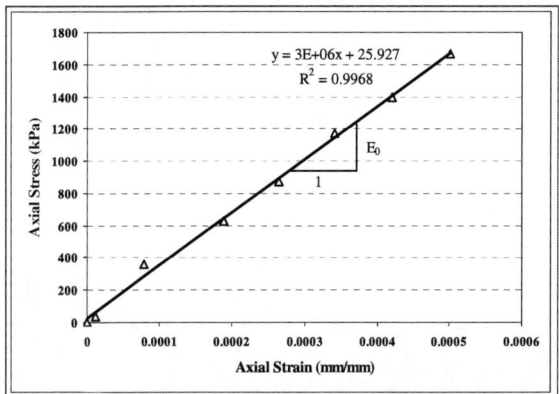

Figure 4. Calculation of the elastic modulus E_0 for the HRA at 20°C (stress level of 2000 kPa).

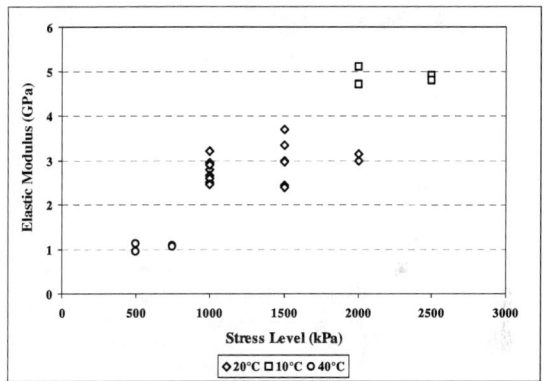

Figure 5. Elastic modulus E_0 for the 10 mm DBM.

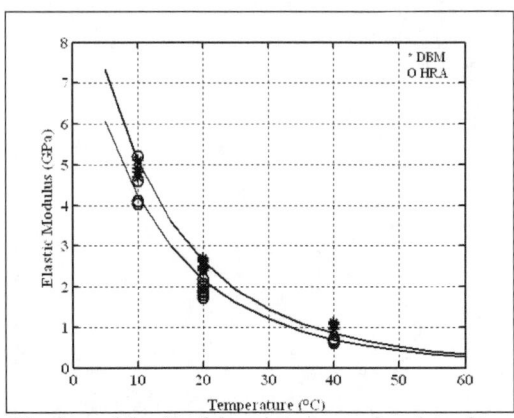

Figure 6. Elastic modulus E_0 of the mixtures as a function of temperature.

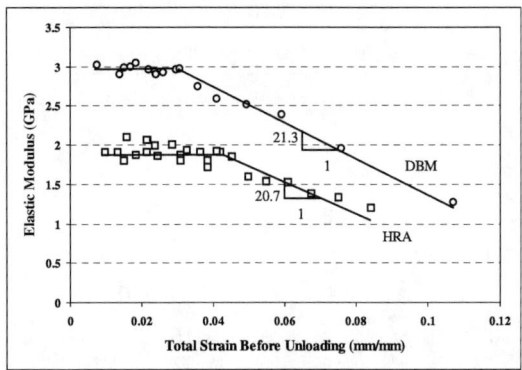

Figure 7. Elastic modulus E_0 of the mixtures as a function of accumulated strain in the mixtures.

5.2. Poisson's Ratio

Poisson's ratio is defined as the negative ratio of the transverse extension strain to the longitudinal contraction strain in the direction of loading in the elastic region of behaviour, as given in Eq. 6. As tensile deformation is considered positive and compressive deformation is considered negative, therefore, Poisson's ratio is a positive value.

$$\upsilon = -\frac{\varepsilon_r}{\varepsilon_a} \quad (6)$$

Poisson's ratio for all elastic materials must fall in the range of 0 to 0.5. Materials with a Poisson's ratio of 0.5 are incompressible, since the sum of all their strains leads to a zero volume change. Poisson's ratio in a viscoelastic material is time dependent. Poisson's ratio values for asphaltic mixtures have been found to lie somewhere in the range 0.1 to 0.45 [Read, 1996]. Table 1 presents the temperature dependent values for the Poisson's ratio of asphalt mixtures recommended by TRL [Nunn, 1995] and Dunill [2002].

Poisson's ratio of the mixtures was determined from the data at the beginning of the uniaxial creep recovery tests. In the uniaxial creep recovery tests, Poisson's ratio was determined as the average ratio of the radial strain to the axial strain over the short time period in the beginning of the test during which the load increases from zero to the target load (Figure 8). Figure 9 shows Poisson's ratio of the mixtures at 20°C. The scatter in the values is caused by the experimental errors in measurement of the small radial and axial strains by which the Poisson's ratio was calculated. As can be seen, Poisson's ratio of both mixtures is independent of the applied stress. A summary of the Poisson's ratio for the mixtures is presented in Figure 9 as a function of temperature. As can be seen in the figure, Poisson's ratio increases with increasing temperature and is almost the same for both mixtures. Also in the figure the superimposed line refers to the values of Poisson's ratio recommended by the TRL [Nunn, 1995], as given in Table 1. As can be seen, the values determined in this study are consistent with the values recommended by TRL, and therefore, those values can be used as Poisson's ratio of the mixtures in the constitutive model.

Table 1. Values for Poisson's ratio recommended by the TRL and proposed by Dunhill [2002]

Temperature (°C)	Poisson's ratio recommended by the TRL	Poisson's ratio proposed by Dunhill [2002]
0 and less	0.25	0.3
20	0.35	0.35
30 and more	0.45	0.4

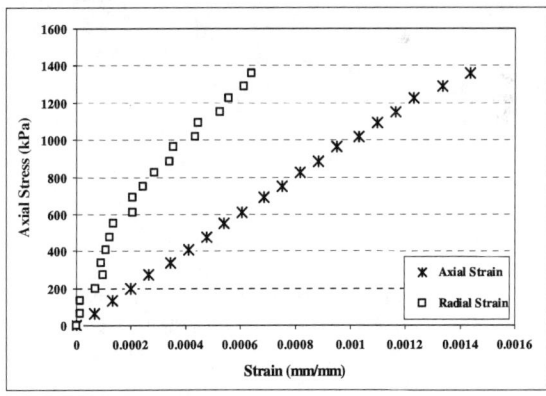

Figure 8. The axial and radial strain at the short period of load application for the HRA30/10 (20°C, 1500 kPa).

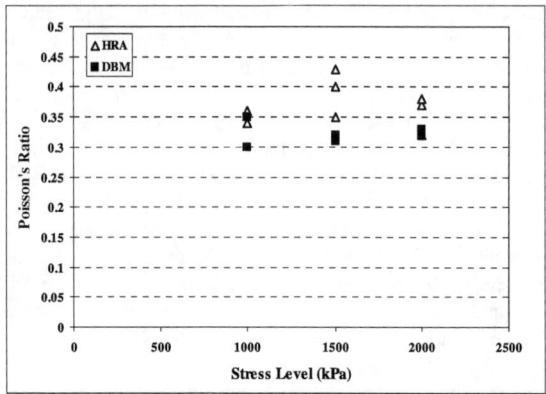

Figure 9.   Poisson's ratio of the mixtures as a function of stress level at 20°C.

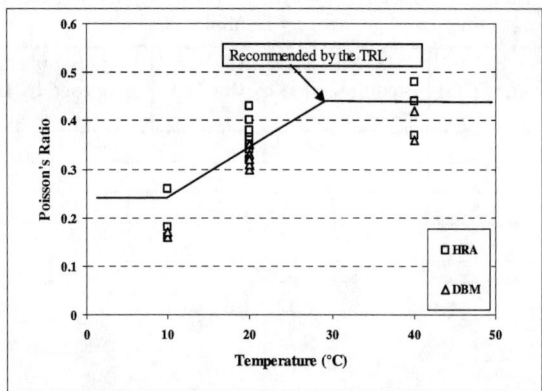

Figure 10.   Poisson's ratio of the mixtures as a function of temperature.

6. Conclusions

The following conclusions can be drawn from this paper:
- The uniaxial creep recovery tests can be used for determination of the elastic parameters of elastic modulus and Poisson's ratio.
- Over the stress ranges utilised in this research, the elastic modulus of asphaltic mixtures was found to be independent of stress level and decreases with increasing temperature.
- The temperature dependency of elastic modulus of asphaltic mixtures can be well captured by WLF equation.
- The elastic modulus of asphaltic mixtures decreases with increasing damage in the mixtures.
- Poisson's ratio of asphaltic mixtures is independent of stress level and increases with increasing temperature.

References

Abdulshafi, A., Majidzadeh, K. (1968) 'Combo viscoelastic-plastic modelling and rutting of asphalt mixtures.' Transportation Research Record 968, Asphalt Mixtures and Performance, TRB, pp 19-31.

Blab, R. and Harvey, J. T. (2002) 'Viscoelastic rutting with improved loading assumptions.' Proc. 9th International Conference on Asphalt Pavements, Copenhagen, Denmark, Volume I, Paper 1.

British Standards Institution (2003a) 'Coated macadams for roads and other paved areas.' BS 4987: Part1, London.

British Standards Institution (2003b) 'Hot rolled asphalt for roads and other paved areas. ' BS 594:Part1, London.

Burmister, D. M. (1943) 'The theory of stresses and displacements in layered systems and applications to the design of airport runways.' Highway Research Board, No. 23, pp 126-144.

Chehab, G. R., Kim, Y. R., Schapery, R. A., Witzack, M., Bonaquist, R. (2003) 'Characterisation of asphalt concrete in uniaxial tension using a viscoelastoplastic model.' Journal of Asphalt Paving Technologists, Vol.72.

Collop, A. C., Cebon, D. and Hardy, M. S. (1995) 'Viscoelastic approach to rutting in flexible pavements.' Journal of Transportation Engineering Vol. 121, pp 88-29.

Collop, A. C., Scarpas, A., Kasbergen, C. and de Bondt, A. (2003) 'Development and finite element implementation of a stress dependent elasto-visco-plastic constitutive model with damage for asphalt.' Proc. of 82nd TRB Annual Meeting, Washington D.C., U.S.

Dunhill, S. (2002) 'Quasi-static characterisation of asphalt mixtures.' Ph.D. Thesis, University of Nottingham, Nottingham.

Fitzgerald, J. E. and Vakili, J. (1973) 'Non-linear Characterisation of Sand Asphalt Concrete by Means of Pavement-memory Norms. 'Experimental Mechanics, Vol. 13, pp 479-493.

Hopman, P. C., Pronk, A. C., Kunst, P. A. J. C., Molenaar, A. A. A. and Molenaar, J. M. M. (1992) 'Application of the viscoelastic properties of asphalt concrete.' Proc. 7th Int. Conf. on the Structural Design of Asphalt Pavements, Nottingham, UK, pp 73-88.

Huang, Y. H. (1967) 'Stresses and displacements in viscoelastic layered systems under circular loaded areas.' Proc. 2nd International Conference on the Structural Design of Asphalt Pavements, Ann Arbor, 1967, pp 225-244.

Huschek, S. (1977) 'Evaluation of rutting due to viscous flow in asphalt pavements.' Proc. 4th Int. Conf. on the Structural Design of Asphalt Pavements, Ann Arbor, pp 497-508.

Kenis, W. J. (1977) 'Predictive design procedures - A design method for flexible pavements using the VESYS structural subsystem.' Proc. 4th Int. Conf. on the Structural Design of Asphalt Pavements, Ann Arbor, pp 100-110.

Lai, J. S. and Anderson, D.(1973) 'Irrecoverable and Recoverable Nonlinear Viscoelastic Properties of Asphalt Concrete. ' Highway Research Record, 468, pp 73-88.

Nunn, M. E. (1986) 'Prediction of permanent deformation in bituminous pavement layers.' Transport and Road Research Laboratory. Report nr 29.

Nunn, M. E. (1995) 'The Characterisation of Bituminous Macadams by Indirect Tensile Stiffness Modulus.' TRL Project Report.

Pagen, C. A. (1964) 'Rheological response of bituminous concrete.' Highway Research Record, No. 67.

Pagen, C. A. (1968) 'Size and thermological relationships of asphaltic concrete.' Proceedings of the Association of Asphalt Paving Technologists, p 228.

Pagen, C. A. (1972) 'Dynamic structural properties of asphalt pavement mixtures.' Proceedings of the 3rd International Conference on Structural Design of Asphalt Pavements, pp 290-316.

Papazian, H. S. (1962) 'The response of linear viscoelastic materials in the frequency domain with emphasis on asphalt concrete.' Proc. 1st Int. Conference on the Structural Design of Asphalt Pavements, Ann Arbor, pp 454-463.

Read, J. M. (1996) 'Fatigue cracking of bituminous paving mixtures.' Ph.D. Thesis, University of Nottingham, Nottingham, UK.

Sayegh, G. (1967) 'Viscoelastic properties of bituminous mixes.' Proc. the 2nd International Conference on Structural Design of Asphalt Pavements, pp 743-755.

Sluys, L. J. and de Borst, R. (2002) 'Computational Methods in Nom-Linear Solid Mechanics.' Delft, Netherlands, Delft University of Technology: Faculty of Civil Engineering and Geosciences.

Thrower, E. N. (1977) 'Methods of predicting deformation in road pavements.' Proc. 4th Int. Conf. on the Structural Design of Asphalt Pavements, Ann Arbor, pp 540-554.

Uzan, J. (1996) 'Asphalt concrete characterisation for pavement performance prediction.' J. of Association of Asphalt Paving Technologists, Vol. 65, pp 573-607.

Vakili, J. (1983) 'An Experimental study of Asphalt Concrete Based on a Multiple Integral Representation of Constitutive Equation of a Non-linear Viscoelastic Solid. .' Journal of Rheology, 27(3), pp 211-222.

Westergaard, H. M. (1927) 'Theory of concrete pavement design.' Proceedings of Highway Research Board, Vol. 1, Part1, pp 175-181.

Properties of Asphalt Mixtures with RAP in the Mechanistic-Empirical Pavement Design of Flexible Pavements: A Preliminary Investigation

Shu Wei Goh[1] and Zhanping You[2]

[1] Ph.D. Student and Research Assistant, Department of Civil and Environmental Engineering, Michigan Technological University, Houghton, Michigan 49931, USA. Tel: (906)487-2528, Fax: (906)487-1620, Email: sgoh@mtu.edu

[2] Tomasini Assistant Professor of Transportation Engineering and Director of the Transportation Materials Research Center, Department of Civil and Environmental Engineering, Michigan Technological University, Houghton, Michigan 49931, USA. Tel: (906)487-1059, Fax: (906)487-1620, Email: zyou@mtu.edu

ABSTRACT

The Reclaimed Asphalt Pavement (RAP) can save money and save energy when recycling is done on the side. The use of RAP significantly reduced the usage of natural resources (i.e., aggregate and petroleum product), and assisted local governments to meet the global reducing disposal standard. In this study, a 15% RAP was used in Superpave mixtures and a mixture without RAP was used as the control mixture. Comprehensive laboratory tests such as dynamic modulus and rutting tests were conducted to evaluate the performance of the mixtures with RAP. The results show that the difference in E* for control and RAP mixture compacted at 7% air void is not significant. However, it was found that the additional 15% RAP decreased the rut depth significantly after 8000 cycles using the Asphalt Pavement Analyzer (APA). The dynamic modules of asphalt mixtures were used in the AASHTO Mechanistic-Empirical Pavement Design Guide (MEPDG) analysis for a given pavement structure as well. In the preliminary study, it was found that an additional 15% of RAP reduced the rut depth up to 13% over a 20-year period in MEPDG analysis.

Keywords: Reclaimed Asphalt Pavement, Dynamic Modulus, Mechanistic-Empirical Design Guide, Asphalt Mixes, Asphalt Pavement Analyzer, Flexible Pavements

BACKGROUND

Dwindling sources of traditional aggregate, increasing haulage distance, and increasing asphalt unit price were the primary reasons that leading to the development of the Reclaim Asphalt Pavement (RAP). Previous research indicated that the life cycle cost for a pavement is lower if the pavement is maintained at an acceptable level of service (ARRA 2001; Button et al. 1995).

RAP has been developed for many years. During the 1930's, the Hot In-Place Recycling (HIR) technology was first discovered in the asphalt recycling area (ARRA 2001). During the 1970's, two events – the petroleum crisis of the early 1970's, and the development of large scale cold planning equipment and tungsten carbide milling tools – led to an interest in asphalt recycling technology. Since then, paving contractors have been making extensive use of RAP and various kinds of research were conducted intensively to evaluate its performance.

Typically, asphalt will become stiffer, often referred to as aging, with time. Researchers (Peterson et al. 2000) reported that different kind of solvents, extraction and recovery method resulted in a significant variability on the properties of asphalt binder. Researchers also studied the aging effect of the binder and investigated the effect of each composition on asphalt recycling agents. Lewandowski et al. (1992) were trying to simulate the aging effect of the binder through microwaving and studied its performance by using the Gel Permeation Chromatography (GPC), Fourier-Transform Infrared Spectroscopy (FTIR), and Dynamic Mechanical Analysis (DMA) (Lewandowski et al. 1992). The results indicated there was a large change in molecular size when a recycling agent was incorporated into the asphalt. In addition, the shear modulus, G* was found to increase during microwaving which correlated with the measured decrease in penetration (increase in viscosity). Similar results were also found from the FTIR test. Peterson et al. (1994) investigated the effect of metals, asphaltenes, and paraffins content on the properties of recycled aged asphalts (Peterson et al. 1994). The composition of asphalt (asphaltenes, aromatics, oils, and waxes) was separated by supercritical fractionation. Peterson et al. indicated that asphaltenes increase the hardening rate but not the oxidation rate and the effect of saturation depended on the asphaltene content. Wax doesn't show any significant effect toward the hardening of asphalt and asphalt shows robust performance when highly aromatic recycling agents were added. Researchers (Chaffin et al. 1997) also studied the fractions of asphalt by GPC, high-performance liquid chromatography, and viscosity to determined its aging effect. The main objective of this study was utilizing part of the asphalt fraction as recycling agents. The results indicated that all the RAPs tested have superior aging index compared to the original asphalt. Researchers (Chaffin et al. 1997; Terrel and Fritchen 1978) also indicated that the RAP will harden more slowly than the original asphalt and the hardening degree was highly correlated with the total saturated content in the RAP.

Several laboratory tests were also conducted on RAP to investigate the mixture characteristics and its performances. Because RAP is stiffer than a new asphalt mixture, researchers (Yamada 1984; Yamada et al. 1987) studied the compactability of the mixture containing RAP. It was found out that the mixture can be compacted as easily as a general mixture at a new compacting temperature estimated from penetration-index. However, Daniel and Lachance (2005) indicated that the void in mineral aggregate (VMA) and void filled with asphalt (VFA) increased when different percentages of RAP were added (Daniel and Lachance 2005). McDaniel et al. (2007) studied the properties of Plant-produced RAP mixture using Dynamic Modulus (E*) test, G* test, low temperature creep compliance test and indirect tensile strength test (McDaniel et al. 2007). They found out that there are no statistical differences in mean strength and E* for mixtures with 15% and 25% RAP level. However, the E* for the mixture with 40% RAP was found to be significantly different (higher E*) at the high test temperature based on statistical analysis (pair-t test). Chehab and Daniel (2006) evaluated the sensitivity of the predicted performance of the RAP mixture and found that the MEPDG predicted IRI was not sensitive to the RAP content (Chehab and Daniel 2006).

The specifications of the asphalt binder in the United States are usually based on Superpave binder criteria. However, for RAP binder, Kandhal and Foo reported that Dynamic Shear Rheometer (DSR), one of the Superpave Binder Tests, was not recommended for RAP binder because it is too liberal (Kandhal and Foo 1997). A series of recommended guidelines and specifications for RAP to be used in the field were developed under the National Cooperative Highway Research Program (NCHRP) Report 9-12 (McDaniel and Anderson 2001; McDaniel and Soleymani 2000; McDaniel et al. 2000). This report mainly discussed how the RAP acted through the "black rock study", examined the effect in asphalt binder through "binder effect study," and evaluated the effect of additional RAP in the asphalt mixture.

Several RAP projects were constructed in the United States and Canada to investigate its field performance and RAP up to 50% was used in the pavement (Emery 1993; Hossain et al. 1993; Hossain and Scofield 1992; Kandhal et al. 1995; Paul 1995). These field projects show the performance of the recycled pavements containing RAP have similar or better performance, in some cases, compared to the virgin asphalt pavement.

SCOPE AND OBJECTIVES

In this paper, an asphalt mixture with 15% RAP was studied. The RAP mixture was sampled from the job site and the control mixture (mixture without RAP) was batched and mixed through the bucket mixer in lab. The control mixture was heated in the oven for two hours to simulate the asphalt mixture's short-term aging effect during the field production. Two different gyration numbers were used during the test. 86 gyrations were used to compact the mixture to 4% air void level and 30 gyrations for 7% air void level. A total of three replicates were applied to each test in order to meet the Superpave criteria. In addition, both RAP and control mixtures were evaluated and compared using the same compaction temperature (142°C).

The objectives of this study are to: 1) compare the mixture volumetric properties of RAP and control mixtures under the same gyration and compaction temperature; 2) evaluate the performance of RAP mixture through dynamic modulus (E*) and Asphalt Pavement Analyzer (APA) rutting test; 3) use the results obtained from the laboratory tests to analyze the RAP mixture through the Mechanical-Empirical Pavement Design Guide (MEPDG); and 4) compare the rutting distress predicted from MEPDG with the APA result.

EFFECT OF RAP ON VOLUMETRIC CHARACTERISTIC

Table 1 Gradation of Control and RAP mixture

Aggregate Type:	1/2 x 3/8	Manufacture Sand	Fine Manufacture Sand	Blend Sand	Bag House	RAP
Blend %:	20%	13%	30%	20%	2%	15%

Binder Grade: PG 64-28

	Percent Passing	
Sieve Size	Control Mixture	RAP Mixture
19mm	100.0%	100.0%
12.5mm	95.9%	98.7%
9.5mm	89.1%	86.6%
4.75mm	70.1%	71.8%
2.36mm	54.1%	51.4%
1.18mm	42.6%	38.1%
0.80mm	33.3%	25.5%
0.30mm	20.1%	14.7%
0.1mm	10.5%	7.7%
0.075mm	6.7%	5.4%
Asphalt Content:	5.57%	5.57% (4.9% of new binder)
Compaction temperature:	142°C	142°C
Total Gyration Number:	86	86
Average Air Void Level:	3.56%	3.90%
Total Gyration Number:	30	30
Average Air Void Level:	6.20%	6.73%

The gradations for both control and RAP mixtures are shown in **Table 1**. A total of twelve mixtures (six mixtures for 4% air void and another six for 7% air void) were used in this evaluation. Under the same compaction temperature (142°C) and gyration number, it was expected that the RAP mixture will have a higher air void level. When RAP was added, it increases the amount of asphaltene in the original asphalt mixture and hence increases the mixture's viscosity. This study investigates the suitability and the effect of the mixture with 15% RAP compacted at the same gyration number and same compaction temperature as the control mixture. It was found that for RAP mixture compacted at 86 gyrations and 30 gyrations, the air void level increased by 0.34% and 0.53%, respectively. In addition, an average of 0.44% increase in air void level was found for both cases. The maximum air void difference between the control and RAP mixtures was found to be 1.27%. The reason for the increased air void is mainly because RAP is stiffer than the original asphalt and need additional compaction or a higher compaction temperature to achieve the desired air void level. However, the additional 15% RAP didn't significantly affect the air void level in this study.

DYNAMIC MODULUS

The dynamic modulus (E*) is the ratio of stress to strain under haversine (or sinusoidal) loading conditions and is used as one of the material characterization inputs in the MEPDG to model pavement performance. In this study, control and RAP mixtures compacting at 4% and 7% air void levels were evaluated. A total of three replicate samples were used for each air void level and the E* test in this study was performed based on the AASHTO TP62-03 Standard. Four different temperatures were used (i.e. -5°C, 4°C, 13°C and 21.3°C) and the frequencies ranged from 0.1 Hz to 25 Hz in this test. The E* test results were then combined and regressed into a single Sigmoidal Master curve using the time-temperature superposition principle. During the formation of the Sigmoidal Master curve, -5°C was used as the reference temperature. The results of the master curve for mixtures at 4% and 7% air void level are shown in **FIG. 1** and **FIG. 2**, respectively.

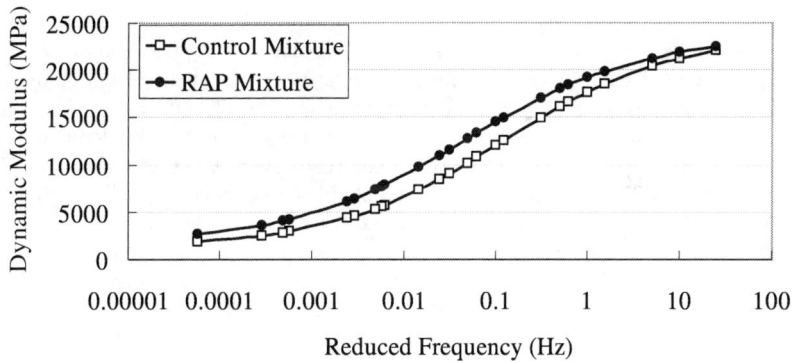

FIG. 1 Comparison of Dynamic Modulus Master Curve with the Reference Temperature of -5°C for Control and 15% RAP mixture (4% Air Void)

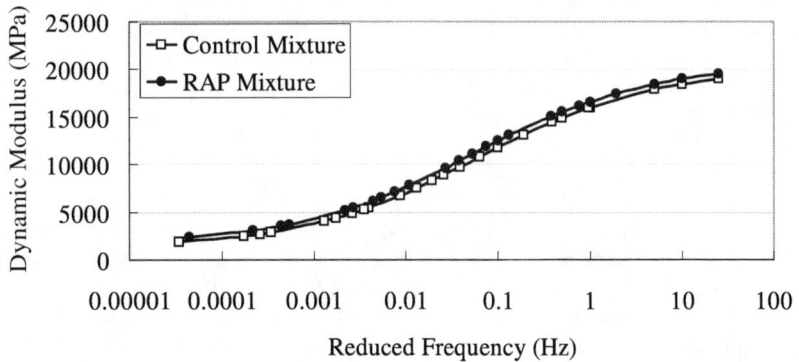

FIG. 2 Comparison of Dynamic Modulus Master Curve with the Reference Temperature of -5°C for Control and 15% RAP mixture (7% Air Void)

MECHANISTIC-EMPIRICAL PAVEMENT DESIGN

The Mechanistic-Empirical Design Guide (MEPDG) is being developed under the National Cooperative Highway Research Program (NCHRP) Project 1-37A and is designed to be adopted by the American Association of State Highway and Transportation Officials (AASHTO) for use as the future pavement design guide for the public and private sectors. The development of the MEPDG is based on the collective experience of pavement experts, data from road tests, calculation of pavement response, and mechanistic and empirical pavement performance models (Mulandi et al. 2006; Priest et al. 2005). The MEPDG performance prediction model consists of four major sub-models. These sub-models are environmental effect, pavement response, material characteristics, and performance predictions. The MEPDG software is able to predict the development and propagation of various kinds of pavement distress, including rutting and fatigue cracking, using input data on asphalt mixture characteristics obtained from laboratory testing. There are three hierarchical levels in the MEPDG, Level 1, Level 2, and Level 3, with the accuracy of prediction increasing from Level 3 to Level 1. In this study, a Level 1 design was used with the measured dynamic modulus as shown in the previous discussions. The design pavement life was set at 20 years and assumed values for creep compliance were used for the RAP and control mixtures. The creep compliance will most dramatically impact the prediction of thermal cracking and the dynamic modulus results are for the rutting prediction. This study focuses exclusively on the development and propagation of rutting.

One of the features in the MEPDG is that it allows the user to input very specific climatic data and traffic information. In this study, the climate data for Lansing, Michigan obtained from the MEPDG climate database was used. Default input parameters in MEPDG for traffic information and vehicle distribution were used as well.

Previously, it was mentioned that the addition of RAP increases the stiffness of an asphalt mixture and has a lower rutting potential based on the results of the E* test. In this section, a 20-year total rut depth for both mixtures was predicted using the MEPDG for comparison. FIG. 3 and FIG. 4 show the results of the predicted rutting depth over 20 years using MEPDG software version 1.0 for both control and RAP mixtures at 4% and 7% air void level. Observation indicated that the depth of rutting increases rapidly during the first 20 months with a decreasing rutting rate thereafter. FIG. 3 shows that the predicted rutting depth for the RAP mixture at the 4% air void level is significantly lower than the control mixture. The largest difference was 13% after 20 years, which shows that the addition of RAP did not have much effect on rutting distress. For 7% mixtures, no significant differences were observed based on results shown in FIG. 4. A pair t-test was used (shown in Table 2) and the results were located at p= (0.031, 0.034) and p= (0.0151, 0.0164) for 4% and 7% air void, respectively. This predicted that the RAP mixture will have a significant rutting depth for both the 4% and 7% mixtures after 20 years.

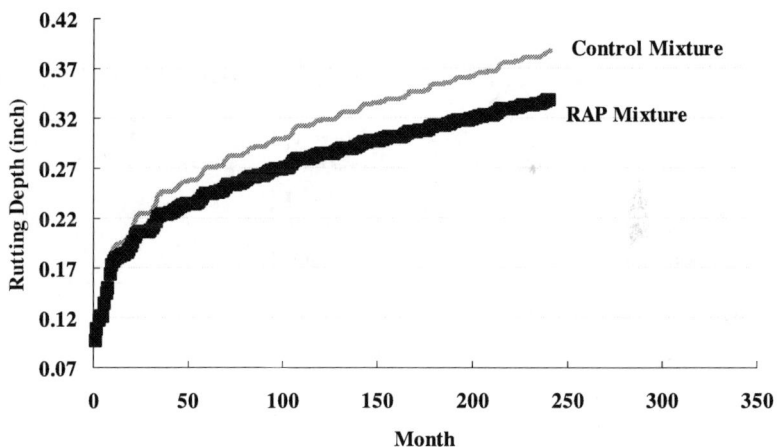

FIG. 3 Prediction of Rutting Depth over 20 years using MEPDG Software Version 1.0 for Control and RAP mixture (4% Air Void Level)

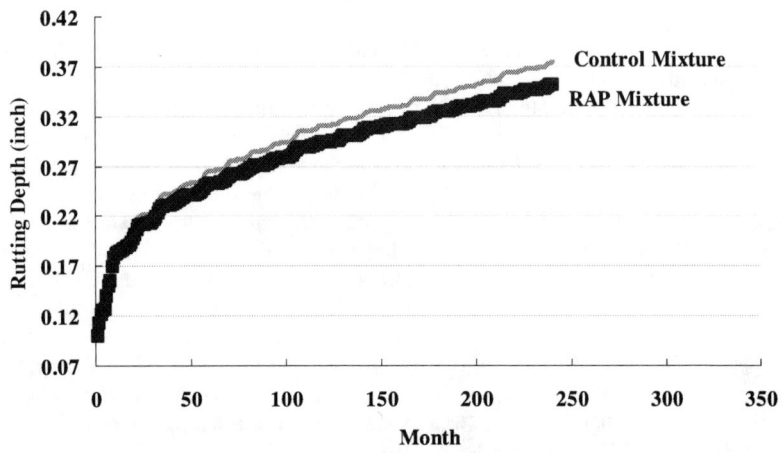

FIG. 4 Prediction of Rutting Depth over 20 years using MEPDG Software Version 1.0 for Control and RAP mixture (7% Air Void Level)

Table 2 Statistical Analysis of Rutting Performance for Control and RAP Mixtures

Sample	Mean	Variance	Average Standard Error	Mean Difference at 95% Confidence Level
Control 4% Air Void	0.3046	0.06125	0.00395	0.031, 0.034
RAP 4% Air Void	0.2718	0.05020	0.00324	
Control 7% Air Void	0.2974	0.05767	0.00372	0.0151, 0.0164
RAP 7% Air Void	0.2817	0.05281	0.00341	

APA RUTTING

The rutting tests were conducted through the Asphalt Pavement Analyzer (APA) device based on AASHTO TP 63-03 at 64°C (147°F). The purpose of this test was to determine the rut resistance for RAP mixture and compare the results with the control mixture. In this test, only mixtures with 7% air void level were used. The results of the APA test are presented in **FIG. 5**.

Based on the test conducted, it was found that RAP mixture has a significantly lower rutting depth compared to the control mixture. This is most likely due to the aging of the RAP binder (RAP binder is stiffer). The final total depth after 8000 cycles for control and RAP mixture are 5.05mm and 3.81mm, respectively. It was found that the additional 15% RAP decrease the total rutting depth by 24%.

The APA test measured the mixture's permanent deformation by applying the load directly to the mixture surface. It shows the rutting potential for the control and RAP mixture under the same temperature and workload. For the MPEDG, the rutting depth was predicted using different models developed by the researchers. The model included climate effect, pavement structure, pavement properties and pavement response. Hence, it can be concluded that the additional 15% RAP reduced the mixture's rutting potential significantly under the same workload and temperature. However, when the climate, traffic volume, pavement structure and properties, and mixture volumetric characteristic are taken into consideration, the additional 15% RAP has a lower decrease in rutting potential than in the APA testing.

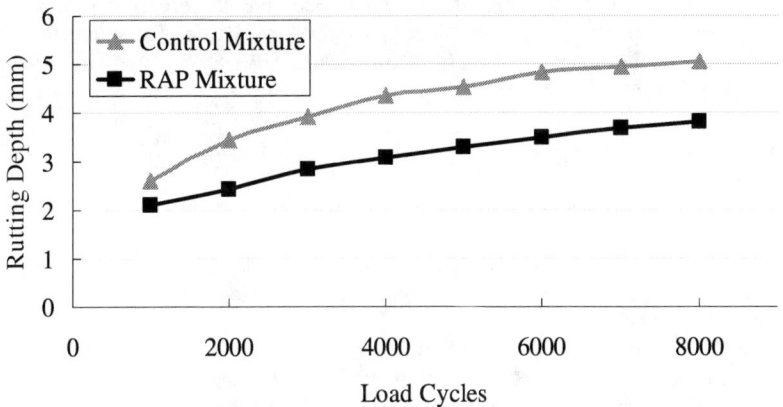

FIG. 5 APA Rutting Result for Control and RAP mixture at 64°C (7% Air Void)

SUMMARY AND CONCLUSIONS

This paper discusses the effect of 15% RAP on HMA mixture number properties and performance, and its rutting potential at the same gyration and compaction temperature. The performance of the mixture was evaluated using the dynamic modulus test, and the rutting potential was analyzed using the MEPDG and the APA rutting test.

Based on the volumetric analysis, the additional 15% RAP increased the average air void by 0.44% when same gyration number and compaction temperature were applied. In order to achieve the desired air void level, a higher compaction temperature or gyration is needed. For the dynamic modulus test, the additional 15% RAP increased the E*. However, the difference in E* for control and RAP mixture compacted at 7% air void is not significant. The E* was used in the MEPDG analysis to predict the rutting potential of the control and the RAP mixture. Through the MEPDG analysis, it was found that the additional 15% RAP decreased the rutting depth significantly based on the pair t-test statistical analysis. However, the largest difference was 13% after 20 years (for a 4% air void mixture) which shows that the addition of RAP did not give much effect on rutting distress. The results from APA test were presented in this study as well. Based on the results, it was found that the additional 15% RAP decreased the rut depth significantly after 8000 cycles. In addition, a decrease of 24% rutting depth was observed at the end of the test. The actual rutting depth results from the APA were compared with the predicted values from MEPDG level 1. The additional 15% RAP reduced the mixture's rutting potential significantly under the same workload and temperature. However, when the climate, traffic volume, pavement structure and properties, and mixture volumetric characteristics were taken into consideration, the additional 15% RAP had a lower decrease in rutting potential.

The literature reviews indicated that thermal cracking and fatigue cracking are always the major concern for RAP mixture. Hence, it is important to evaluate the severity and the potential of these distresses for RAP mixture. The evaluation and investigation of the potential of these distresses using the tensile strength ratio and four point beam fatigue test are ongoing.

REFERENCES

ARRA. (2001). Basic Asphalt Recycling Manual, Asphalt Recycling and Reclaiming Association.

Button, J. W., Estakhri, C. K., and Little, D. N. (1995). "Performance and cost of selected hot in-place recycling projects." *Transportation Research Record*, (1507), 51-66.

Chaffin, J. M., Liu, M., Davison, R. R., Glover, C. J., and Bullin, J. A. (1997). "Supercritical fractions as asphalt recycling agents and preliminary aging studies on recycled asphalts." *Industrial & Engineering Chemistry Research*, 36 (3), 656-666.

Chehab, G. R., and Daniel, J. S. (2006). "Evaluating recycled asphalt pavement mixtures with mechanistic-empirical pavement design guide level 3 analysis." *Transportation Research Record*, (1962), 90-100.

Daniel, J. S., and Lachance, A. (2005). "Mechanistic and Volumetric Properties of Asphalt Mixtures with Recycled Asphalt Pavement." Transportation Research Board, pp 28-36.

Emery, J. J. (1993). "Asphalt concrete recycling in Canada." *Transportation Research Record*, (1427), 38-46.

Hossain, M., Metcalf, D. G., and Scofield, L. A. (1993). "Performance of recycled asphalt concrete overlays in Southwestern Arizona." Report No. *0361-1981*.

Hossain, M., and Scofield, L. A. "Performance of Recycled Asphalt Concrete Materials in an Arid Climate." Atlanta, GA, USA, 415-427.

Kandhal, P. S., and Foo, K. Y. "Designing recycled hot mix asphalt mixtures using Superpave technology." New Orleans, LA, USA, 101-117.

Kandhal, P. S., Rao, S. S., Watson, D. E., and Young, B. (1995). "Performance of recycled hot-mix asphalt mixtures in Georgia." *Transportation Research Record*, (1507), 67-77.

Lewandowski, L. H., Graham, R., and Shoenberger, J. "Physicochemical and rheological properties of microwave recycled asphalt binders." Atlanta, GA, USA, 449-461.

McDaniel, R., and Anderson, R. M. (2001). Recommended use of Reclaimed Asphalt Pavement in the Superpave Mix Design Method: Technician's Manual, Transportation Research Board, North Central Superpave Center.

McDaniel, R. S., Shah, A., Huber, G. A., and Gallivan, V. L. (2007). Investigation of Properties of Plant-Produced RAP Mixtures, Transportation Research Board.

McDaniel, R. S., and Soleymani, H. (2000). Superpave RAP Mixtures, University of Texas, Austin, Texas A&M University, College Station, Texas.

McDaniel, R. S., Soleymani, H., Anderson, R. M., Turner, P., and Peterson, R. (2000). Recommended use of Reclaimed Asphalt Pavement in the Superpave Mix Design Method, National Cooperative Highway Research Program, North Central Superpave Center, IN.

Mulandi, J., Khanum, T., Hossain, M., and Schieber, G. "Comparison of pavement design using AASHTO 1993 and NCHRP Mechanistic- Empirical Pavement Design Guides." Atlanta, GA, United States, 912-923.

Paul, H. R. (1995). Evaluation of Recycled Projects for Performance, Louisiana Department of Transportation and Development, Louisiana Transportation Research Center, Federal Highway Administration.

Peterson, G. D., Davison, R. R., Glover, C. J., and Bullin, J. A. (1994). "Effect of composition on asphalt recycling agent performance." *Transportation Research Record*, (1436), 38-46.

Peterson, R. L., Soleymani, H. R., Anderson, R. M., and McDaniel, R. S. (2000). Recovery and Testing of RAP Binders from Recycled Asphalt Pavements, Association of Asphalt Paving Technologists.

Priest, A. L., Timm, D. H., Solaimanian, M., Gibson, N., and Marasteanu, M. "A full-scale pavement structural study for mechanistic-empirical pavement design." Long Beach, CA, United States, 519-556.

Terrel, R. L., and Fritchen, D. R. (1978). "Laboratory Performance of Recycle Asphalt Concrete." *ASTM Special Technical Publication*, 104-12.

Yamada, M. (1984). "Characterization of Recycled Asphalt Mixes and their Pavement Performance." *Doboku Gakkai Rombun-Hokokushu/Proceedings of the Japan Society of Civil Engineers*, (348), 51-60.

Yamada, M., Ninomiya, T., and Mise, T. (1987). "Recycled asphalt mixtures in Osaka and their performance." *Memoirs of the Faculty of Engineering, Osaka City University*, 28, 197-201.

Laboratory Evaluation of Warm Asphalt Properties and Performance

Amy Hearon, E.I.T.[1] and Stacey Diefenderfer, P.E., Associate Member[2]

[1] Graduate Research Assistant, Department of Civil and Environmental Engineering, University of Virginia, 351 McCormick Road, PO Box 400742, Charlottesville, VA 22904; Phone (434) 293-1976; Fax (434) 293-1990; email: ajh7d@virginia.edu

[2] Research Scientist, Virginia Transportation Research Council, 530 Edgemont Road, Charlottesville, VA 22903; Phone (434) 293-1933; Fax (434) 293-1990; email: stacey.diefenderfer@vdot.virginia.gov

Rising energy costs and increased environmental awareness have brought attention to the potential benefits of warm asphalt in the United States. Warm mix asphalt (WMA) is produced by incorporating additives into asphalt mixtures to allow production and placement of the mix when heated to temperatures well below the 150°C+ temperatures of conventional hot mix asphalt (HMA). Potential benefits such as reduced plant emissions, improved compaction in the field, extension of the paving season into colder weather, and reduced energy consumption at the plant may be realized with different applications.

Trial installations of WMA, including two sections using the Sasobit WMA additive, have been investigated in Virginia. This study presents the results of laboratory testing to evaluate the performance of the mixtures used in the two trial sections. The evaluation included comparisons of volumetric properties, moisture susceptibility, rutting resistance, and fatigue performance between the HMA and WMA mixtures used in each section. Few differences were found. In addition, the findings indicated that the performance of HMA and WMA should be equal when proper construction methods are used.

Keywords: warm mix asphalt, hot mix asphalt, moisture susceptibility, rutting resistance, fatigue performance, tensile strength ratio

Introduction

The focus in construction has recently shifted toward sustainability and environmentally friendly practices. New technologies in asphalt paving have emerged that may save fuel and lower emissions as well as provide other benefits to contractors and agencies. These technologies have been grouped together under the name "warm mix asphalt" (WMA). Conventional hot mix asphalt (HMA) is typically produced and compacted at temperatures between 140°C to 170°C; cold mix asphalt is compacted at ambient temperatures (20°C to 50°C). WMA falls between the two and is generally defined as asphalt mixtures produced at temperatures between 100°C and 135°C. There are several WMA technologies available, including the following:
- *LEA*, Advanced Concepts Engineering Co.
- *CECABASE RT*, Arkema Group
- *Double Barrel Green System*, Astec Industries
- *Evotherm*, MeadWestvaco Asphalt Innovations
- *Advera WMA*, PQ Corporation
- *Terex*, Terex Roadbuilding
- *Sasobit*, Sasol Wax Americas, Inc. (National Asphalt Paving Association [NAPA], 2007).

When asphalt is produced at lower temperatures, there are many benefits such as reduced emissions and energy consumption and increased worker safety. WMA technologies also allow asphalt to be placed at cooler ambient temperatures and to be hauled farther. The lower production temperatures result in less oxidation during production and laydown, which may lead to greater fatigue resistance. Drawbacks of the technology potentially include an increased susceptibility to moisture damage since the lower production temperatures may lead to the aggregate not being sufficiently dried before mixing. Additional concerns include an increased potential for rutting, possibly because of less aging (stiffening) of the binder or compaction issues at the lower placement temperatures. The potential for increased curing times has also been reported, which could mean delays in opening roads to traffic (Newcomb, 2007). However, the answers regarding most of these issues, both positive and negative, have not yet been determined.

The Virginia Department of Transportation (VDOT) has evaluated HMA and WMA mixtures used during two trial sections paved in Virginia in 2006 (Diefenderfer et al., 2007). The WMA was produced with the Sasobit technology. Sasobit is a wax byproduct of the Fischer-Tropsch process of natural gas and coal gasification. It comes in pellets or flakes and is combined with the binder to lower its viscosity (Sasol Wax, 2004). This paper discusses the effects of lower temperature production on the compactibility, volumetrics, moisture susceptibility, rutting potential, and fatigue resistance of the two mixtures used during VDOT's field installations.

Materials

Mixture A was a Superpave 9.5-mm nominal maximum aggregate size (NMAS) surface mix produced using PG64-22 binder. Morelife 3300 antistrip was used at a dosage rate of 0.5% by weight of the binder. The aggregate was a mix of granite and siltstone. A summary of the mix properties is provided in Table 1. This was a typical 9.5 mm HMA surface mix used by the contractor; the only adjustment made to the mix design to produce WMA was the addition of Sasobit and the reduction of the production temperature. Sasobit was added at a dosage rate of 1.5% by weight of the binder. The binder content used for the WMA was the same as that used for the HMA.

Mixture B was a 12.5-mm NMAS Superpave surface mix produced using PG64-22 binder. Hydrated lime was used as an anti-stripping agent. The aggregate was a mix of limestone and gravel. A summary of the properties of this mix is provided in Table 1. Again, this was a typical surface mix used by the contractor. The mix design was the same as for the WMA and HMA mix, except for the addition of 1.5% Sasobit by weight of the binder and the reduction of production temperatures.

Table 1. Properties of Mixtures A and B

Properties	Mixture A	Mixture B
Mixture type	9.5 mm surface mix	12.5 mm surface mix
Design gyrations	65	65
Cumulative percent passing		
19.0 mm	100	100
12.5 mm	100	96
9.5 mm	92	86
4.75 mm	60	-
2.36 mm	43	34
75 µm	5.7	6.0
Binder content	5.50%	5.20%
Anti-stripping agent	Morelife 3300, 0.5% by weight of asphalt	Hydrated lime, 1.0% by weight of asphalt
Recycled asphalt pavement	20%	10%
Sasobit	1.5% by weight of asphalt	1.5% by weight of asphalt
Production temperature	HMA, 150°C; WMA, 120°C	HMA, 163-165°C; WMA, 150°C

Testing Program

The laboratory study included mix produced at the plant and mix produced in the laboratory. During construction of the two trial sections in 2006, HMA and WMA were sampled at each plant. In addition, for Mixture A, HMA was produced in the laboratory using the plant production temperatures and WMA was produced in the laboratory at temperatures of 110°C, 130°C, and 150°C. In all cases, the mixing and compaction temperatures were the same. For Mixture B, only plant-produced HMA and WMA were tested. The testing matrix for the mixtures is presented in Table 2.

Table 2. Laboratory Testing Matrix for Mixtures A and B

Laboratory-Mixed Material: Mixture A

Temperature	Control HMA, 150°C	WMA, 150°C	WMA, 130°C	WMA, 110°C
Asphalt content	%AC from job mix formula			
Volumetrics				
No. of gyrations	Design, 65 gyrations			
Moisture Susceptibility				
TSR	Aging states: none, short term, long term[a]			
Hamburg	7% air voids; 50°C; wheel load, 158 lb			
Rutting				
APA	7% air voids; 64°C; hose pressure, 120 psi; wheel load, 120 lb			
Fatigue				
Flexural beams	Test to 30% stiffness; 10 Hz haversine load; 20°C			

Plant-Mixed Material: Mixtures A and B

Temperature	Control HMA 150°C: Mixture A 165°C: Mixture B	WMA 120°C: Mixture A 150°C: Mixture B
Volumetrics		
No of gyrations	Design, 65 gyrations	
Moisture susceptibility		
TSR	Modified AASHTO T283	
Hamburg	7% air voids; 50°C; wheel load,158 lb	
Rutting		
APA	7% air voids; 64°C; hose pressure, 120 psi; wheel load, 120 lb	
Fatigue		
Flexural beams	Test to 30% stiffness; 10 Hz haversine load; 20°C	

[a]Short-term aging: loose mix was aged for 4 days in a forced draft oven at 85°C. Long-term aging: loose mix was aged for 8 days in a forced draft over at 85°C (Bell et al., 1994).

Volumetrics

Volumetric analyses were performed to determine fundamental mixture properties. Properties included maximum theoretical specific gravity (G_{mm}), bulk specific gravity (G_{sb}), voids in total mix (VTM), voids filled with asphalt (VFA), and voids in mineral aggregate (VMA). Specimens were compacted using the Superpave Gyratory Compactor (SGC) in accordance with AASHTO T312, Preparing and Determining the Density of Hot-Mix Asphalt (HMA) Specimens by Means of the Superpave Gyratory Compactor, and the bulk specific gravity was measured using AASHTO T166, Bulk Specific Gravity of Compacted Hot-Mix Asphalt Using Saturated Surface Dry Specimens (American Association of State Highway and Transportation Officials [AASHTO], 2007). The binder content and gradation were determined using AASHTO T308, Determining the Asphalt Binder Content of Hot-Mix Asphalt by the Ignition Method, and AASHTO T30, Mechanical Analysis of Extracted Aggregate (AASHTO, 2007).

Volumetric results for Mixtures A and B are shown in Tables 3 and 4, respectively. *Plant* specimens refer to gyratory specimens that were compacted on site at the contractor's plant to eliminate differences in volumetrics or other properties that might be affected because of reheating. *Lab* specimens were made post-construction from loose mixture samples to evaluate the effects of reheating. Both plant and lab gyratory specimens were made for Mixture A HMA and WMA; however, for Mixture B, plant specimens were produced only for the WMA.

In the case of the plant-produced Mixture A WMA, there was a difference in total air voids between the plant-compacted and lab-compacted specimens; this may have been influenced by the presence of moisture in the plant-produced mix attributable to damp stockpiles. Otherwise, the air voids for all specimens of both trials were similar.

Table 3. Mixture A Volumetric Results

Property	Plant-Produced Mix				Lab-Produced Mix			
	HMA		WMA		HMA		WMA	
	Plant[a]	Lab[b]	Plant	Lab	150°C	110°C	130°C	105°C
% AC	5.5	5.9	5.9	5.8	5.4	5.7	5.6	5.6
Rice SG (G_{mm})	2.504	2.501	2.498	2.502	2.508	2.507	2.508	2.509
% VTM	2.8	3.1	2.7	4.5	4.2	5.1	5.1	4.1
% VMA	14.7	15.7	15.3	16.8	15.9	17.3	17.1	16.2
% VFA	80.7	80.4	82.2	73.3	73.6	70.4	70.2	74.5
Bulk SG (G_{mb})	2.433	2.424	2.430	2.390	2.402	2.379	2.380	2.406
% Density @ N_{ini}	89.5	89.5	89.6	88.1	88.3	87.2	87.2	88.3
Sieve	Percent Passing							
¾ in (19.0 mm)	100	100	100	100	100	100	100	100
½ in (12.5 mm)	99.6	100	100	99.8	99.6	99.7	99.4	99.5
³/₈ in (9.5 mm)	94.3	93.5	94.6	93.5	94.1	93.3	91.5	92.2
No. 4 (4.75 mm)	62.0	62.9	61.1	62.0	62.6	65.9	60.5	62.1
No. 8 (2.36 mm)	43.9	44.0	42.9	44.0	45.0	48.2	44.1	45.1
No. 200 (75 µm)	6.1	6.3	5.5	6.1	6.5	6.5	6.8	6.8

[a] *Plant* indicates gyratory specimens compacted at the plant during production.
[b] *Lab* indicates gyratory specimens compacted after construction from loose mixture samples.

Table 4. Mixture B Volumetric Results

Property	Plant-Produced Mix			
	HMA		WMA	
	Plant[a]	Lab[b]	Plant	Lab
% AC	No specimens compacted at plant	5.4	5.6	5.8
Rice SG (G_{mm})		2.604	2.571	2.597
% VTM		3.3	2.3	2.9
% VMA		15.8	15.3	16.5
% VFA		79.1	85.2	82.5
Bulk SG (G_{mb})		2.518	2.513	2.522
% Density @ N_{ini}		86.7	87.8	87.0
Sieve	Percent Passing			
¾ in (19.0 mm)	No specimens compacted at plant	100	100	100
½ in (12.5 mm)		95.8	97.3	97.0
⅜ in (9.5 mm)		84.1	84.1	85.2
No. 4 (4.75 mm)		48.3	49.9	51.0
No. 8 (2.36 mm)		32.7	33.0	33.4
No. 200 (75 μm)		6.5	5.9	6.3

[a]*Plant* indicates gyratory specimens compacted at the plant during production.
[b]*Lab* indicates gyratory specimens compacted after construction from loose mixture samples.

Moisture Susceptibility

The moisture susceptibility was measured using a modified version of AASHTO T283, Resistance of Compacted Asphalt Mixtures to Moisture-Induced Damage (AASHTO, 2007). The 16-hour curing time and 24-hour storage time are waived in Virginia. The rest of the method is followed; the specimens are subjected to vacuum saturation before undergoing one freeze-thaw cycle. It was hypothesized that WMA may continue to increase in tensile strength over time. To investigate this, additional specimen sets were subjected to short-term and long-term aging before being tested. Short-term and long-term aging consisted of holding the compacted specimens in a forced draft oven at a temperature of 85°C for 4 days and 8 days, respectively, prior to testing them in accordance with AASHTO T283.

Tensile strength ratio (TSR) results were mixed, as evident in Tables 5 and 6. The HMA specimens were not improved after short-term aging or by the addition of an anti-stripping agent; however, results were improved after long-term aging. The addition of an anti-stripping agent improved all of the WMA samples. The long-term aging improved the samples produced at 130°C and 150°C. An analysis of variance (ANOVA) of the strength data showed that both temperature and aging had an effect on the WMA dry and wet strengths. The strengths increased with increased aging and with increased temperature. The t-test was used to compare wet and dry strengths of the materials at 150°C both with and without an anti-stripping agent. There was no significant difference in the wet strengths of the WMA or HMA with or without an anti-stripping agent. Likewise, there was no significant difference between the dry strengths of the WMA and HMA without an anti-stripping agent. There was a difference in the dry strengths of the HMA and WMA with an anti-stripping agent.

Table 5. Tensile Strength Ratio for Laboratory-Produced Mixes, Mixture A

Mix	HMA 150°F	WMA 150°F	WMA 130°C	WMA 110°F
Unaged specimens, no anti-strip	0.78	0.82	0.72	0.48
4-day oven-aged specimens, no anti-strip	0.66	0.83	0.90	0.84
8-day oven-aged specimens, no anti-strip	0.80	0.93	0.92	0.74
Unaged specimens, with anti-strip	0.76	0.86	0.92	0.72

Table 6. Tensile Strength Ratio for Plant-Produced Mixes

Mix	Mixture A	Mixture B
HMA, plant-compacted	0.82	0.85
WMA, plant-compacted	0.69	0.90
WMA, lab-compacted[a]	0.75	-
[a]Sample was collected during production and reheated for compaction.		

The plant-produced mixes generated mixed results. The TSR for the Mixture A plant-compacted WMA was particularly low, so a loose mix sample was taken back to the lab and reheated and compacted for additional testing. It rained the day before production, so the stockpiles for Mixture A were wet. This may explain the low TSR and the improvement of the WMA after reheating. Both materials for Mixture B performed well, having TSR values greater than the 0.80 specification.

The Hamburg wheel-track test was also used to evaluate moisture susceptibility. The test method used was AASHTO T324, Hamburg Wheel-Track Testing of Compacted Hot-Mix Asphalt (HMA) (AASHTO, 2007). Testing was performed using an Asphalt Pavement Analyzer (APA) that was modified to perform Hamburg testing. Specimens were compacted using the SGC and saw cut to fit into the high-density polyethylene molds as illustrated in Fig. 1. The specimens were compacted to 7% air voids. The water temperature was maintained at 50°C during the test, and the applied wheel load was 158 lb. The plant-produced mixtures were tested; a summary of the results is presented in Table 7.

Fig. 1. Hamburg test setup (after draining)

Table 7. Summary of Hamburg Results

Mix	Average Air Voids (%)	Standard Deviation (Air Voids)	Rut Depth at 20,000 Cycles (mm)
Mixture A HMA	7.6	0.49	2.11
Mixture A WMA	7.8	0.17	2.13
Mixture B HMA	7.4	0.50	2.44
Mixture B WMA	7.1	0.27	2.07

The maximum allowed deformation at 20,000 cycles is not specified in the AASHTO method, but a maximum of 10 mm after 20,000 cycles is specified in Colorado (Federal Highway Administration [FHWA], 2006). The measured rut depths of both samples for both trials were well below the 10 mm criteria. It can be seen in Figs. 2 and 3 that the specimens experienced only plastic deformation and had not yet reached the stripping inflection point. The Hamburg machine malfunctioned with the Mixture A HMA sample, and data for the first 9,000 passes were not recorded. Based on the depth of rut after 20,000 passes, it can be assumed that this sample also experienced only plastic deformation. From these observations, it can be concluded that all mixes should be resistant to stripping.

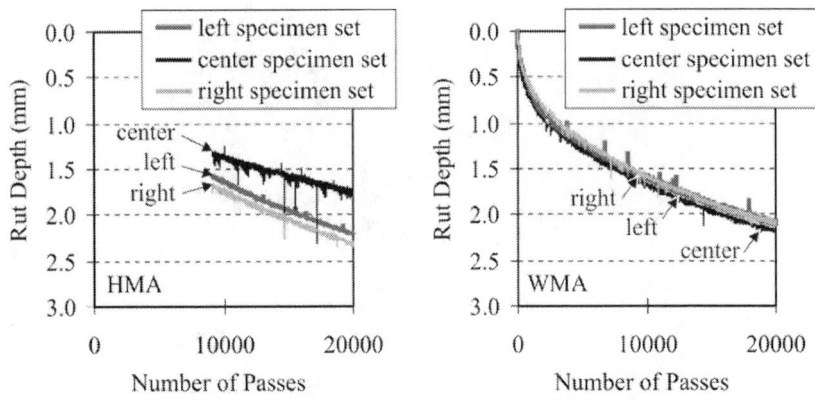

Fig. 2. Hamburg rut depths for Mixture A

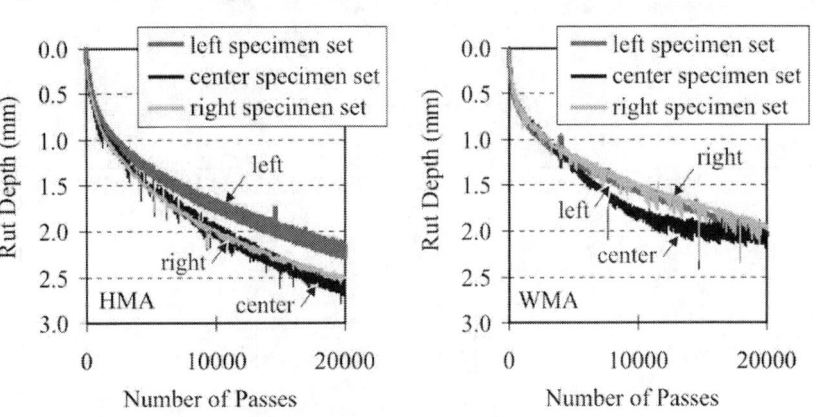

Fig. 3. Hamburg rut depths for Mixture B

Rutting

Testing was performed on gyratory samples compacted from plant-produced material using the APA to evaluate the rutting resistance of each mixture in accordance with Virginia Test Method 110, Method of Test for Determining Rutting Susceptibility Using the Asphalt Pavement Analyzer (VDOT, 2007). Testing was performed at a temperature of 64°C and a hose pressure of 120 psi. A vertical load of 120 lb was applied for 8,000 cycles, and the resulting rut depth was measured. Hand average values were calculated from measurements of rutting taken prior to testing and after the completion of 8,000 cycles of loading; automatic average values were automatically detected and collected by the APA.

Table 8 shows the average values for each sample. Previous research (Maupin and Mokarem, 2006) indicated that hand and automatic measurements are highly correlated; thus, either is acceptable. Both HMA and WMA specimen sets for

Mixture A and B were found to be acceptable in rutting resistance; the specification limit for these mixtures is 7.0 mm of rutting. For Mixture A, the WMA specimen set was shown to have an average of more than 0.5 mm less rutting than the control mixture; however, this was not found statistically significant using the t-test. The difference in performance is likely due to the stiffening influence of Sasobit at temperatures below the additive's melting point, which has been promoted as a benefit of the technology (Butz et al., 2001). However, both Mixture B specimen sets had similar rutting results; the t-test confirmed that the rutting resistance values were not significantly different.

Table 8. APA Rut Measurements for Mixtures A and B

Mix	HMA Lab Specimens			WMA Lab Specimens		
	Hand Average (mm)	Automated Average (mm)	Average Voids (%)	Hand Average (mm)	Automated Average (mm)	Average Voids (%)
Mixture A						
Average	5.23	4.39	6.97	4.32	3.81	7.03
Standard deviation	0.59	0.43	0.15	0.35	0.48	0.06
Mixture B						
Average	3.24	2.74	8.53	2.83	2.72	8.53
Standard deviation	0.04	0.11	0.06	0.53	0.25	0.25

Fatigue

The fatigue resistance of Mixtures A and B was evaluated in accordance with AASHTO T321, Determining the Fatigue Life of Compacted Hot-Mix Asphalt (HMA) Subjected to Repeated Flexural Bending (AASHTO, 2007). Plant-produced mix was collected and compacted in the laboratory as the contractors did not have any means for compacting beam specimens. The test specimens were cut from larger beam samples compacted in a vibratory compactor. The beams were tested in the controlled strain mode until the measured stiffness was reduced to 30% of the initial stiffness. The predicted cycles to failure, N_f, were taken at 50% of the initial stiffness. As can be seen in Figs. 4 and 5, the WMA has a slightly lower fatigue resistance. However, as the applied strain increased, the difference in fatigue life between HMA and WMA became less.

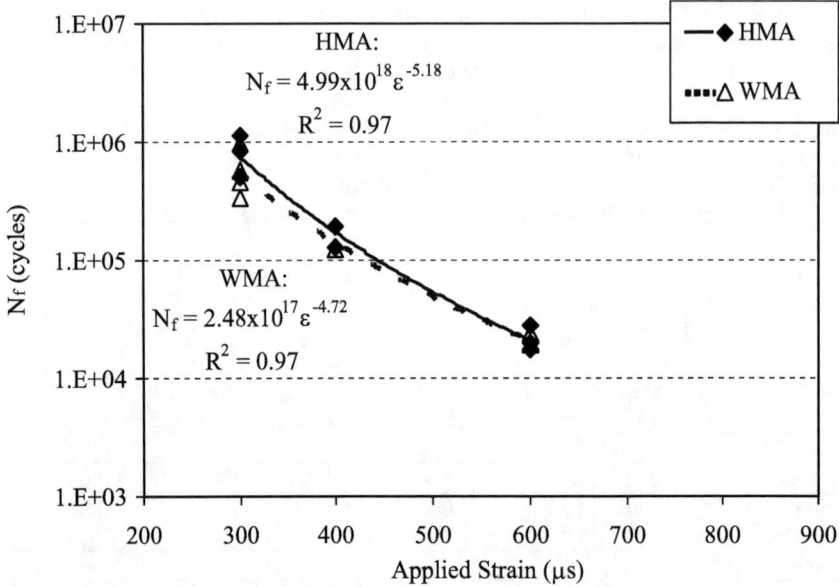

Fig. 4. Fatigue results for Mixture A

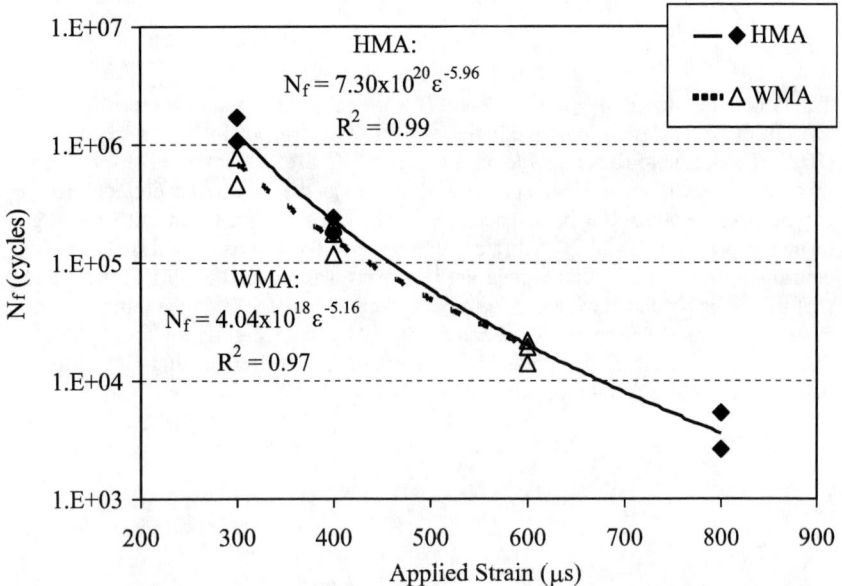

Fig. 5. Fatigue results for Mixture B

Summary and Conclusions

Based on the two mixtures considered in this study, HMA and WMA should have equal performance when properly constructed.
- The volumetrics of the HMA and Sasobit WMA produced at different temperatures were not significantly different.
- The TSR results were mostly inconclusive. There did, however, appear to be a positive effect from aging in the WMA. Sasobit may have the ability to stiffen the binder at temperatures below its melting point by forming a lattice structure (Butz et al., 2001). A further investigation of the characteristics of this lattice structure is needed.
- The HMA and WMA appear to be resistant to moisture susceptibility; there was little difference between the performance of HMA and WMA as measured by the Hamburg test.
- The HMA and WMA performed similarly in the APA. Both mixes met the criterion for rutting resistance.
- Based on the flexural fatigue test, the HMA performed slightly better at lower strains than the WMA; however, the performance of the mixes appeared nearly equal at higher strains.

Further research is needed to validate fully these findings for mixtures with different binders and aggregate structures. The performance of WMA in base or intermediate pavement layers should also be investigated.

References

AASHTO. (2007). *Standard specifications for transportation and methods of sampling and testing,* 27th ed. Washington, DC.

Bell, C.A., Wieder, A.J., and Fellin, M.J. (1994). "Laboratory aging of asphalt-aggregate mixtures: Field validation." SHRP-A-390. National Research Council, Washington, DC.

Butz, T., Rahimian, I., and Hildebrand, G. (2001). "Modifications of road bitumins with the Fischer-Tropsch Paraffin Sasobit®." *Journal of Applied Asphalt Binder Technology,* (1)2, 70-86.

Diefenderfer, S.D., McGhee, K.K., and Donaldson, B.M. (2007). "Installation of warm mix asphalt projects in Virginia." VTRC 07-R25. Virginia Transportation Research Council, Charlottesville.

FHWA. (2006). "Asphalt pavement technology, bituminous mixtures laboratory (BML), equipment Hamburg wheel tracking device." Washington, DC, <http://www.fhwa.dot.gov/pavement/asphalt/labs/mixtures/hamburg.cfm> (March 3, 2008).

Maupin, G.W., Jr., and Mokarem, D.W. (2006). "Investigation of proposed AASHTO rut test procedure using the Asphalt Pavement Analyzer." VTRC 07-R11. Virginia Transportation Research Council, Charlottesville.

Newcomb, D. (2007) "An introduction to warm-mix asphalt." National Asphalt Pavement Associations, Lanham, MD, <http://fs1.hotmix.org/mbc/Introduction_to_Warm-mix_Asphalt.pdf > (March 3, 2008).

National Asphalt Pavement Association. (2007). "WMA technologies." Lanham, MD <http://www.warmmixasphalt.com/WmaTechnologies.aspx> (March 3, 2008).

Sasol Wax. (2004). "Product Information 124." Hamburg, Germany, <http://www.sasolwax.com/data/sasolwax_/Bitumin%20Modification/Sasobit%20Since%201997.pdf> (May 26, 2006).

Virginia Department of Transportation, Materials Division. (2007). *Virginia test methods*. Richmond, <http://www.virginiadot.org/business/resources/bu-mat-VTMs070704.pdf> (Nov. 7, 2007).

Acknowledgments

The authors thank the Virginia Transportation Research Council and the Federal Highway Administration for their support of this research, especially Troy Deeds, Donnie Dodds, Ken Elliton, and the VTRC Asphalt Laboratory staff for the collection and testing of samples. This study would not have been possible without the support of the personnel from the Virginia Department of Transportation, Materials Division, Culpeper and Staunton District Materials Offices, Warrenton and Verona Residencies, and Rappahannock and Monterey Area Headquarters. In particular, the assistance of Trenton Clark is acknowledged. Appreciation is also extended to the staff of Superior Paving Corp. and B&S Construction, Inc., for their assistance and cooperation during construction. Additional thanks are due to personnel from Hi-Tech Asphalt Solutions, Inc., and to Larry Michael for their assistance and support. Finally, the authors thank Linda Evans of VTRC for her editorial comments.

Laboratory Simulation of Warm Mix Asphalt (WMA) Binder Aging Characteristics

Tejash Gandhi[1] Serji Amirkhanian[2]

Abstract

As warm asphalt has been gaining increasing popularity in the recent years, there are still several characteristics about warm asphalt that are unknown. While several studies have been conducted to study the performance of warm asphalt mixtures, aging characteristics of warm asphalt binders are not known in great detail as the technology is relatively new, and there are no old pavements to study the aging behavior of warm asphalt. This paper presents the results of a study conducted to simulate the aging of warm asphalt binder in the laboratory by preparing asphalt mixtures containing two different binder sources and three different warm asphalt additives (Control, Asphamin® and Sasobit®). The mixtures were artificially aged in the oven, and the binders were extracted for testing. The binders extracted from freshly prepared samples were considered being short term aged binders and binders extracted from oven aged mixtures were considered being long term aged binders.

Results of several tests (e.g., viscosity, high and low temperature properties, Gel Permeation Chromatography, etc.) are presented in this study. The results indicated that the binders extracted from the warm mix asphalt (WMA) had significantly lower aging index (ratio of the viscosity of extracted binders to original binders) compared to the binders extracted from control hot mix asphalt (HMA). It was also observed that the binders extracted from the WMA had aged significantly lower compared to binders extracted from control HMA. The results indicated that the warm asphalt additives did not have any significant effect on the fatigue cracking parameter ($G^*.\sin \delta$) or the creep stiffness of the binders. However, Asphamin® significantly increased the m-value of the binders.

Keywords: Short term aging, long term aging, warm mix asphalt binder aging.

---

[1] Graduate Research Assistant, Department of Civil Engineering, Clemson University, Clemson SC 29634; eMail: tgandhi@clemson.edu

[2] Professor, Department of Civil Engineering, Clemson University, Clemson SC 29643; eMail: kcdoc@clemson.edu

Introduction

Rising energy prices, global warming, and more stringent environmental regulations have resulted in an interest in warm mix asphalt (WMA) technologies as a means to decrease the energy consumption and emissions associated with conventional hot mix asphalt (HMA) production. The asphalt industry has been experimenting with warm and cold asphalt for decades now in order to reduce energy requirements and for environmental benefits. However, in many cases, most of the cold products have been inferior to hot mix asphalt, and the extra costs of the cold and warm asphalts are not offset by the savings in energy.

The warm asphalt technologies allow asphalt mixes to be produced at lower mixing and compaction temperatures, addressing the prominent environmental and economic factors currently faced by the industry. There are also technical benefits to the use of warm mixes; namely extension of the construction season and reduced aging of the asphalt binder. Reduction of the short term aging (oxidation and volatilization) of the asphalt binder during conventional construction could potentially enhance pavement performance through reduced thermal and fatigue cracking, thus improving the life cycle cost of the pavement. Apart from the technical advantages, there are several other advantages like lower wear and tear of the plants, safer working environments for the workers, ability of opening the site to traffic sooner, etc. With the availability of several proprietary chemicals and processes to produce warm asphalt, it is now possible to produce warm asphalt without affecting the properties of the mix. Some of the most common processes / chemicals available are described in previous publications *(Biro, et Al., 2007; Gandhi and Amirkhanian, 2007)*.

The concept driving warm mix technologies is the reduction in asphalt binder viscosity, which allows the asphalt to attain suitable viscosity for coating of the aggregate and compaction of the mix at lower temperatures. The implementation of warm mix technology as a viable option for paving operations is a promising concept. However, further investigation of the effects of the aforementioned additives on the constituent materials of asphalt mixtures and pavement performance is needed. While several studies have been conducted to study the performance of warm asphalt mixtures *(Hurley and Prowell, 2006; Barthel, et Al.; Hurley and Prowell, 2005)*, the aging properties of binder containing the warm asphalt additives have not been studied in great detail. This paper presents the results of some comparative studies conducted on binders extracted from laboratory prepared and aged HMA and WMA samples. The binders extracted from freshly prepared samples were considered being short term aged binders and binders extracted from oven aged mixtures were considered being long term aged binders.

Materials and Procedures

Two different binder sources were used in this study. The first binder used was from a blend of different crude sources, the second was a Venezuelan crude source. The properties of the binders are as shown in Table 1.

Table 1: Binder properties.

Property	Binder 1	Binder 2
Original Binder		
Viscosity, Pa-s (135°C)	0.405	0.626
G*/sin δ, kPa (64°C)	1.207	1.801
RTFO Residue		
Mass Loss, % (163°C)	-0.02	-0.24
G*/sin δ, kPa (64°C)	2.815	4.608
PAV Residue		
G*sin δ, kPa (25°C)	2970	2420
Stiffness (60), MPa (-12°C)	183	129
m-Value (60) (-12°C)	0.311	0.345
PG Grade	64 -22	64 -22
Mixing Temp.[+], °C	150–155	163–170
Compaction Temp.[+], °C	139-144	150–155

[+]*Information provided by the suppliers*

WMA was prepared using two of the available technologies, Asphamin® and Sasobit®. Asphamin® is Sodium – Aluminum – Silicate, hydro-thermally crystallized into a fine powder. The crystals contain about 21% water, which induces a fine spray in the binder causing a volume expansion, thereby increasing the workability and compactibility of the mixture at lower temperatures. It has been reported, by the manufacturer, that a reduction of about 10 to 15 °C is possible *(Eurovia Services)*. Sasobit® is a long chain aliphatic wax (chain lengths of 40 – 115 carbon atoms) obtained from coal gasification using the Fischer – Tropsch process. Sasobit® melts in the asphalt binder at temperatures of 85 – 115 °C, causing a marked reduction in the viscosity of the binder. The manufacturer reports a reduction in mixing and handling temperatures of 10 – 30 °C *(Sasol Wax)*.

Laboratory compacted HMA and WMA samples were prepared with the two binder sources and one aggregate source. The HMA was compacted in the temperature range of 150 – 154 °C, and did not contain any warm asphalt additive. The WMA was compacted in the range of 120 – 124 °C, and contained Asphamin® or Sasobit®. Asphamin® was added to the mix at the rate of 0.3% by weight of the mix, just as the binder was mixed with the aggregate, and Sasobit® was first blended with the binder at the rate of 1.5% by weight of the binder, and the modified binder was mixed with the aggregate.

The asphalt binders were then extracted and evaluated from the HMA and WMA using a rotovapor as per ASTM D2172, *Standard Test Methods for Quantitative Extraction of Bitumen From Bituminous Paving Mixtures, (ASTM Standards)*, and ASTM D5404, *Standard Practice for Recovery of Asphalt from Solution Using the Rotary Evaporator, (ASTM Standards)*. Binder was extracted from the HMA and WMA before and after subjecting the mixes to long term oven aging as per AASHTO TP30, *Standard Practice for Mixture Conditioning of Hot Mix Asphalt (AASHTO Standards)*.

Results and Discussions

In this research, Binders 1 and 2 extracted from the WMA were compared with Binders 1 and 2 extracted from the HMA. Statistical analysis systems (SAS) software was used to perform the analysis of variance (ANOVA) with the null hypothesis (H_0) that the means are equal, with a level of significance of 0.05. If the F – value obtained from the ANOVA table is greater than the F_{crit} value (which depends on the level of significance and the degrees of freedom), H_0 is rejected, which means that the sample means between different treatments are not equal. If it was determined using ANOVA that the sample means between different treatments were different, the least significant difference (LSD) was calculated. If the difference between two means is greater than or equal to the LSD value, the two means are said to be significantly different. Thus, the LSDs were calculated for all pairs of means within different treatments, and compared with the pairs of sample means to determine which pairs were significantly similar and which pairs were significantly different. The differences in the means are indicated on the graphs by the means of alphabet letters (e.g., A or B) on the bars. Bars with the same letters indicate that there were no statistically significant differences in the means.

The extracted binders were evaluated to determine the effects of several properties on the binders including viscosity, rutting and fatigue parameters, and the stiffness and m-values. The results of the study are discussed in the following sections.

Effects on Viscosity

The viscosities of the binders extracted from the WMA and HMA after short term aging were measured and normalized to the viscosity of the virgin binder without any warm asphalt additive. The binders extracted form the short term aged mixtures are considered to be similar to binders aged in the rolling thin film oven (RTFO), as both aging procedures simulate the aging of the binders during the mixing process. All the viscosities were measured at 135 °C using a rotational viscometer as per AASHTO T316, *Viscosity Determination of Asphalt Binder Using Rotational Viscometer, (AASHTO Standards)*. The normalized viscosities are as shown in Figure 1.

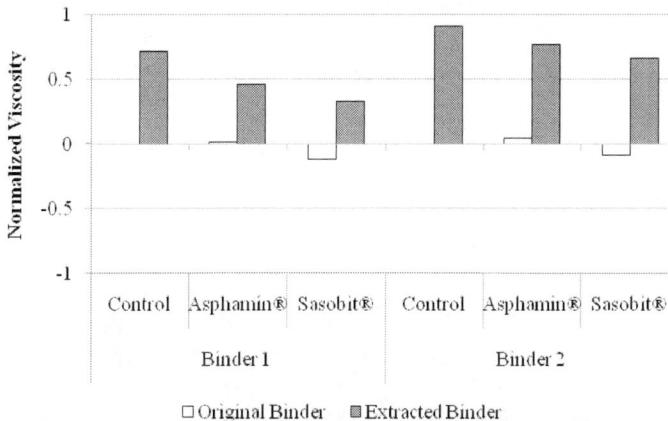

Figure 1: Normalized viscosities of original and extracted binders with and without the warm asphalt additives.

The results indicated that in case of original binders, binders containing Asphamin® have significantly similar viscosities compared to the control binders, whereas binders containing Sasobit® have significantly lower viscosities compared to control binders and binders containing Asphamin®. The results of this study are consistent with a previous study *(Gandhi and Amirkhanian, 2007)*.

Similarly, when the extracted binders were compared, it was observed that the binders with no warm asphalt additives had the highest viscosities, followed by the binders containing Asphamin® having significantly lower viscosities, followed by binders containing Sasobit®. This shows that the binders in the warm mix asphalt undergo lesser aging as a result of reduced mixing and compaction temperatures. It is hypothesized that the viscosity of the binders containing Asphamin® is significantly more than the viscosity of the binders containing Sasobit® as a result of the mineral filling effect of the Asphamin® zeolites in the binder.

Effects on Rutting Parameter ($G^*/\sin \delta$)

After the binders were extracted form the short term aged HMA and WMA mixtures, the rutting parameter ($G^*/\sin \delta$) of the binders were measured using a dynamic shear rheometer (DSR). The $G^*/\sin \delta$ values are plotted in Figure 2.

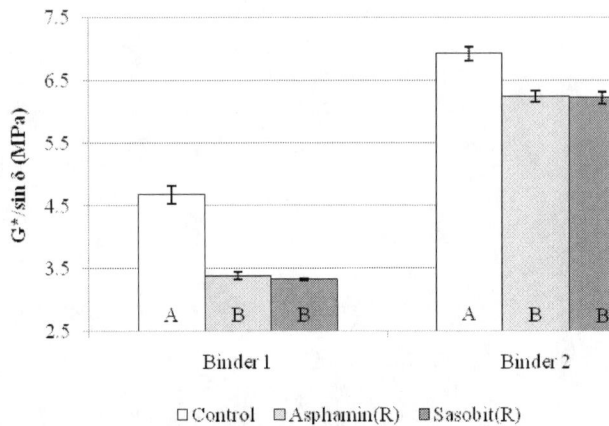

Figure 2: G*/sin δ values of the extracted binders with and without warm asphalt additives.

As seen with the viscosities of the binders, it was observed that the G*/sin δ values of the binders extracted from the WMA were significantly lower than the G*/sin δ values of the binders extracted from the HMA. Additionally, there were no significant differences in the G*/sin δ values of the binders containing Asphamin® and Sasobit®. Thus it could be concluded that the reduced mixing and compaction temperatures of the WMA reduce the aging of the binders significantly.

Effects on Fatigue Parameter (G*.sin δ)

To evaluate the fatigue parameters (G*.sin δ) of the binders, the binders were extracted from long term oven aged HMA and WMA mixtures. This binder is considered to be equivalent to pressure aging vessel (PAV) aged binder. The G*.sin δ of the binders extracted from the long term aged mixes were measured and the results are shown in Figure 3.

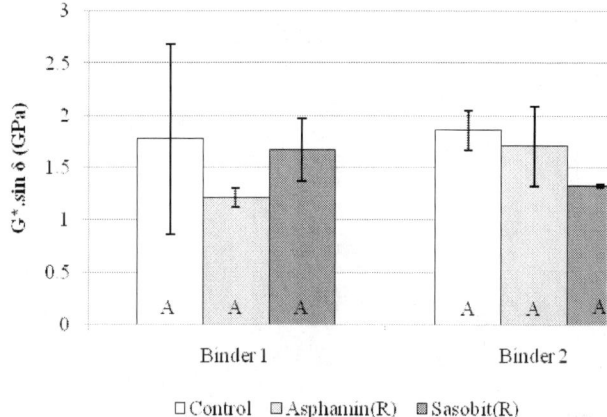

Figure 3: G*/sin δ values of the extracted binders with and without warm asphalt additives.

The results indicated that the binders extracted from the HMA and from the WMA had significantly similar G*.sin δ values. This shows that the binders in the HMA and WMA have similar tendencies towards fatigue cracking, and that the addition of the warm asphalt additives does not have any significant effect on the fatigue cracking tendencies of the binders.

Effects on Stiffness and m-value

The binders extracted form the long term aged HMA and WMA mixes were also tested for flexural creep stiffness using a bending beam rheometer (BBR) to determine the low temperature cracking tendencies of the mixes. The stiffness values and the m-values are plotted in Figures 4 and 5 respectively.

From the figures, it can be observed that the creep stiffness of the binders extracted form the HMA and WMA are significantly similar. Thus it could be concluded the addition of the warm asphalt additives does not significantly affect the low temperature creep stiffness of the binders tested in this research.

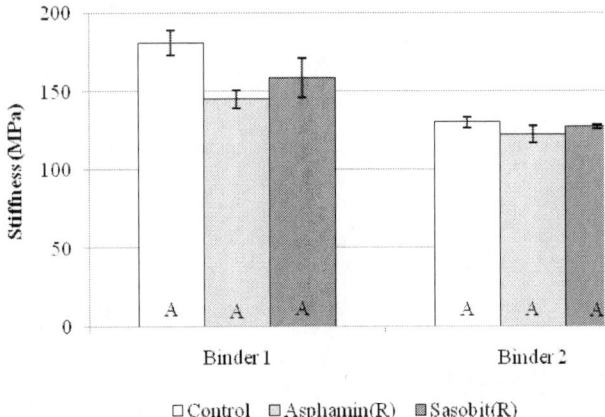

Figure 4: Stiffness values of the extracted binders with and without warm asphalt additives.

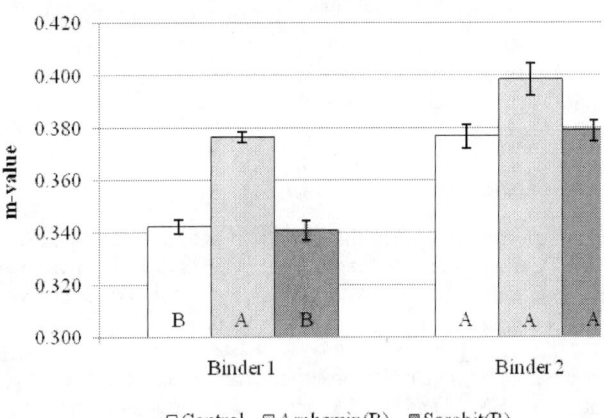

Figure 5: m- values of the extracted binders with and without warm asphalt additives.

When the m-values were compared, it was observed that the binders containing Asphamin® generally had significantly higher m-values compared to binders containing Sasobit® and the control binder (no warm asphalt additive). However, in a previous study *(Gandhi, et al., 2008)*, it was observed that the binders containing the warm asphalt additives had significantly higher stiffness values and significantly lower m-values compared to the control binders, when the binders had aged under similar conditions. That would mean that when the binders undergo similar aging, binders with the warm asphalt additives show higher tendencies towards cracking at low temperatures.

The results of this study indicate that the reduced mixing and compaction temperatures of the mixes containing Asphamin® improve the m-values of the binders and thus reduce the low temperature cracking tendencies of the mixes.

Conclusions

From this limited study of the properties of the binders extracted from HMA and WMA mixtures used in this study, the following can be concluded.

- The aging characteristics of warm mix asphalt binder including viscosity, rutting parameter ($G^*/\sin \delta$), fatigue parameter ($G^*.\sin \delta$), and stiffness and m-value are studied through experiment research in this paper. It can be concluded that for the original binders, the binders containing Asphamin® and the control binders had significantly similar viscosities, and binders containing Sasobit® had significantly lower viscosities compared to binders containing Asphamin® and the control binders; for the extracted binders, the viscosities of the binders extracted from WMA were significantly lower than the viscosities of the binders extracted from HMA.

- Similarly, the binders extracted from WMA showed significantly lower $G^*/\sin \delta$ values compared to binders extracted form HMA. Thus, it could be concluded that the lower mixing and compaction temperatures of the WMA significantly lower the aging in the binders.

- Furthermore, lower mixing and compaction temperatures of the WMA significantly lowered the aging in the binders. WMA mixes used in this study were not more prone to cracking at low temperatures. The binders containing Asphamin® show improved resistance to low temperature cracking compared to HMA binders and WMA binders containing Sasobit®.

References

"Annual Book of ASTM Standards: Road and Paving Materials; Traveled Surface Characteristics", American Society for Testing and Materials, 2005, Vol. 04.03.

"Standard Specifications for Transportation Materials and Methods of Sampling and Testing", American Association of State Highway Transportation Officials, 2004.

Barthel, W., J.P. Marchand, M. Von Devivere. "Warm asphalt mixes by adding a synthetic zeolite", Eurovia. www.asphamin.com.

Biro, S., Gandhi, T.S., and Amirkhanian, S.N., (2007), "Mid Range Temperature Rheological Properties of Warm Asphalt Binders", Submitted for Review for

Publication in the Journal of Materials in Civil Engineering, American Society of Civil Engineers.

Eurovia Services Website : http://www.eurovia.com/en/produit/135.aspx, accessed January 2007.

Gandhi, T.S. and Amirkhanian, S.N., (2007), "Laboratory Evaluation of Warm Asphalt Binder Properties – A Preliminary Analysis", 5th International Conference of Maintenance and Rehabilitation of Pavements and Technological Control, Park City, Utah, pp 475-480.

Gandhi, T.S., Akisetty, K., and Amirkhanian, S.N., (2008), "Laboratory Evaluation of Warm Asphalt Binder Aging Characteristics", International Journal of Pavement Engineering, in press.

Hurley, G., and Prowell, B., (2005), "Evaluation of Sasobit® for use in Warm Mix Asphalt", NCAT Report 05-06, Auburn.

Hurley, G., and Prowell, B., (2006), "Evaluation of Evotherm® for use in Warm Mix Asphalt", NCAT Report 06-02, Auburn.

Hurley, G., and Prowell, B., (2006), "Evaluation of Potential Process for use in Warm Mix Asphalt", Journal of the Association of Asphalt Paving Technologist, Volume 75, pp 41 – 90.

Sasol Wax Website : http://www.sasolwax.com/Sasobit_Technology.html, accessed January 2007.

Volumetric Properties of Warm Rubberized Mixes Depending on Compaction Temperature

Chandra K. Akisetty[1], Soon-Jae Lee[2], and Serji N. Amirkhanian[3]

[1]Department of Civil Engineering, Clemson University, Clemson, SC 29634; e-mail: cakiset@clemson.edu
[2]Department of Technology, Texas State University – San Marcos, San Marcos, TX 78666; e-mail: SL31@txstate.edu
[3]Department of Civil Engineering, Clemson University, Clemson, SC 29634; e-mail: kcdoc@clemson.edu

ABSTRACT

The hot mix asphalt (HMA) industry is commencing on a program to substantially decrease mix production temperatures. Reduced mix production and paving temperatures would decrease the energy required to make HMA, reduce emissions and odors from plants, and improve the working conditions at the plant and paving site. With regard to rubberized asphalt mixtures, they are produced and compacted at higher temperatures than conventional mixtures. If the technologies of warm mix asphalt are incorporated, it is expected to reduce the mixing and compaction temperatures of rubberized asphalt mixtures to those of conventional mixtures. This study was initiated to investigate the effect of compaction temperature on warm rubberized mixes. For this, two Superpave mix designs for two asphalt binders and one aggregate size (12.5mm) were conducted to determine the optimum asphalt contents (OAC). Warm rubberized mixes were produced using two of the available processes. A total of 96 specimens (4 mix types: *control mix, rubberized mix, warm rubberized mix 1 and warm rubberized mix 2* * 4 compaction temperatures: *97, 116, 135, and 154°C* * 6 repetitions) were fabricated using Superpave gyratory compactor. Volumetric properties of the specimens were evaluated. The results showed that the warm mix processes were effective to improve the volumetric properties of rubberized mixes at a certain range of compaction temperatures.

INTRODUCTION

The "warm mix asphalt" (WMA) refers to technologies which allow a considerable reduction of mixing and compaction temperatures of asphalt mixes through lowering the viscosity of asphalt binders by use of chemical additives. Reduced mix production and paving temperatures would decrease the energy needed to produce HMA, reduce emissions and odors from plants, and make better working conditions at the plant and paving site (*Hurley and Prowell 2005 a; Hurley and Prowell 2005 b; Hurley and Prowell 2006; Gandhi and Amirkhanian 2007*).

Rubberized asphalt mixes are generally compacted at a higher temperature than conventional mixes, based on the field experience (*Amirkhanian and Corley 2004*). With lower compaction temperatures, the rubberized mixes might result in several problems such as inadequate volumetric properties and poor short-term and long-term performance. If the technologies of warm mix asphalt are incorporated into the mixes, optimum mixing and compaction temperatures of the rubberized mixes are expected to decrease and be comparable to those of conventional mixes. Among a number of warm mix additives, this study evaluated two additives, Aspha-min® and Sasobit®. Information regarding the two additives can be found in other reports (*Hurley and Prowell 2005 a; Hurley and Prowell 2005 b*).

The objective of this study was to investigate the volumetric properties of rubberized asphalt mixtures containing warm mix additives as a function of compaction temperature using the Superpave Gyratory Compactor (SGC).

MATERIALS AND TEST PROGRAM

Two binders (control PG 64-22 and crumb rubber modified (CRM) binders) were used in this study. One type of rubber, which was produced by mechanical shredding at ambient temperature, was used with a gradation as shown in Table 1. CRM binders were made by adding a specified amount of rubber (-40 mesh) to the control binder, mixing with a stirrer (700 rpm) at 177°C for 30 minutes (*Shen et al. 2006*). This mixing condition matches the field practices used in South Carolina to produce field CRM mixtures. One aggregate source was used for preparing samples. Hydrated lime, used as an anti-strip additive, was added at a rate of 1% by dry mass of aggregate according to the SC DOT specifications. The experimental flow chart of this study and test combinations are shown in Figure 1.

Table 1. Crumb rubber gradation.

Sieve No. (µm)	Ambient CRM	
	% Retained	% Cumulative Retained
30 (600)	0	0
40 (425)	9.0	9.0
50 (300)	31.9	40.9
80 (180)	32.9	73.8
100 (150)	7.6	81.4
200 (75)	18.6	100.0

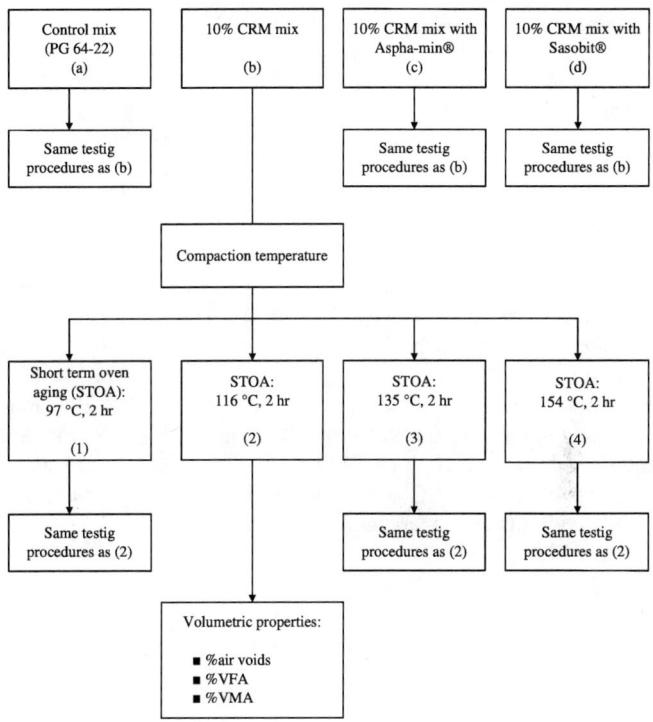

Figure 1. Flow chart of experimental design procedures.

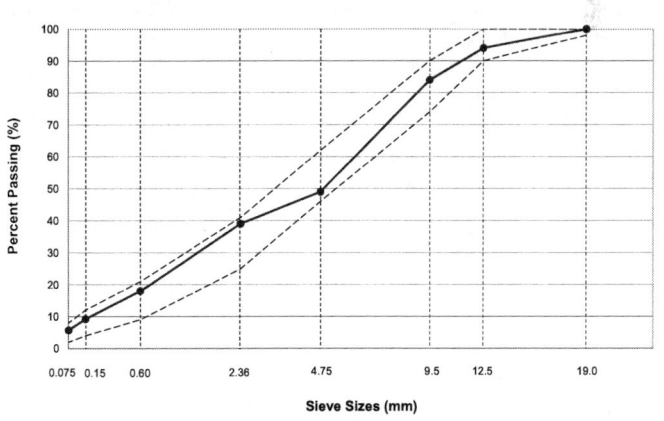

Figure 2. Gradation chart of 12.5mm asphalt mixture.

A nominal maximum size 12.5mm Superpave mixture was used for the mix design in this study. The procedures described in AASHTO T 312 regarding the preparation of HMA specimens were followed. All mixtures used an identical aggregate structure to distinguish the influence of the binders and the warm mix additives (Figure 2). The optimum asphalt contents were obtained and used to produce specimens at four different compaction temperatures.

The mixing of the aggregates with the asphalt binders was conducted at temperatures (control mix: 154°C; CRM mix: 175°C; CRM mix with Aspha-min®: 150°C; CRM mix with Sasobit®: 150°C) recommended by the manufacturers of asphalt binder and warm mix additives. The loose asphalt-aggregate mixtures were oven aged at the compaction temperatures for 2 hours prior to the compaction. The four compaction temperatures used were 97°C, 116°C, 135°C, and 154°C. This range was selected based on the temperatures (135°C and 154°C) which are commonly used as short-term oven aging temperatures in the laboratory to simulate binder aging and absorption during the construction of HMA pavements (*Asphalt Institute 2003*). The compaction temperature of 97°C was chosen to evaluate the effect of warm mix additives at relatively lower temperature.

The specimens were manufactured to the target air void content of 4±1% using 75 gyrations of SGC. Each specimen was 150 mm in diameter and 110±5 mm in height. A total of 96 specimens (4 binder types * 4 compaction temperatures * 6 repetitions) were prepared and tested.

RESULTS AND DISCUSSIONS

The optimum asphalt contents for control mix and 10% CRM mix were found to be 4.2% and 5.0%, respectively. The mix design result of CRM mix was used for CRM mixes containing warm mix additives also.

Bulk and maximum specific gravities were measured and the air void contents of 96 specimens fabricated at four compaction temperatures were calculated. Figure 3 shows the air void contents of the mixtures as a function of the compaction temperature. Similar to the previous research (*Lee et al. 2007*), the air void contents of CRM mixtures significantly decreased with an increase in the compaction temperature. Both the warm mix additives of Aspha-min® and Sasobit® were found to have an effect to decrease the air void content of CRM mixtures at each compaction temperature used in this study. However, specimens made with control binder of PG 64-22 had almost the same air void content over a range of compaction temperatures (116°C to 154°C). This result is also consistent to the previous studies (*Azari et al. 2003; Bahia 2000; Stuart 2000*).

For the aggregate source used in this study, the CRM mixtures showed that the compaction temperature could be decreased to 139°C and 133°C for Aspha-min® and Sasobit®, respectively. In general, the warm mix additives resulted in approximately

20°C to 30°C reduction of compaction temperature required for the target air void content, indicating that the compaction temperatures of CRM mixtures containing the additives can be reduced to those of conventional control mixtures.

Figures 4 and 5 illustrate the change of %VFA and %VMA of the specimens with an increase in the compaction temperature from 97°C to 154°C, respectively. Similar to the air void contents, as expected, the %VFA and %VMA of specimens produced with control PG 64-22 binders were found to be almost the same values over the compaction temperatures, except for the lowest temperature of 97°C. In terms of the CRM mixtures, the general trends of %VFA and %VMA were also similar to the change in the air void contents of the CRM mixtures. Still, the %VMA values of the CRM mixtures were relatively higher than those of the control mixtures with the same air void contents. This is thought to be associated to the higher OAC of the CRM mixtures, increasing the effective asphalt contents of the mixtures.

In general, the warm mix additives used in this study were observed to have an effect to increase the %VFA values and decrease the %VMA values for all compaction temperatures, compared to the conventional CRM mixtures. On the other hand, Aspha-min® and Sasobit® were generally found to have insignificantly different influences on the CRM mixtures regarding the %VFA and the %VMA.

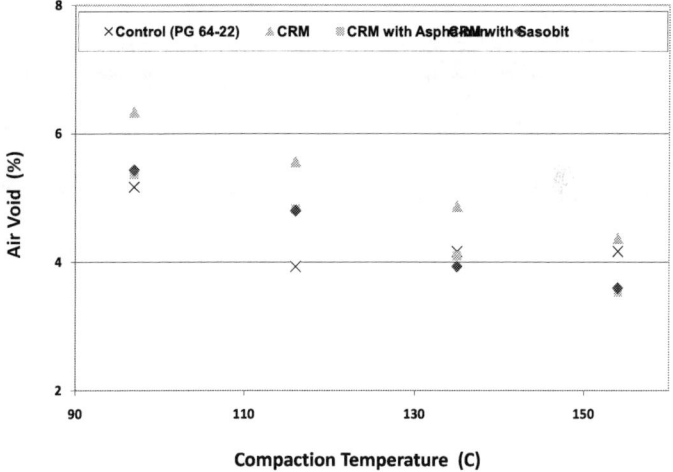

Figure 3. Change in %air voids as a function of compaction temperature.

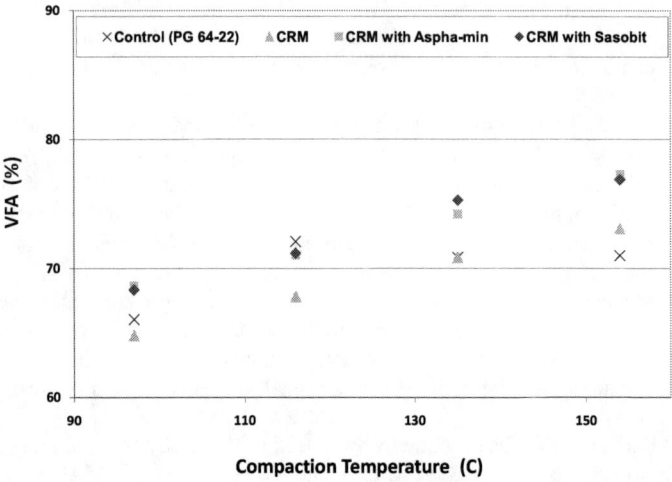

Figure 4. Change in %VFA as a function of compaction temperature.

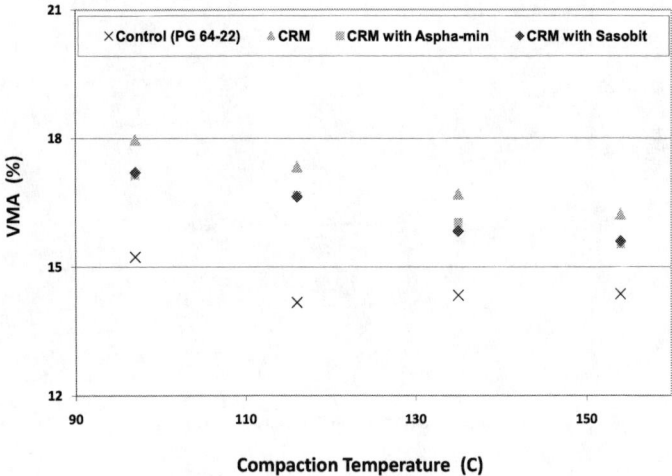

Figure 5. Change in %VMA as a function of compaction temperature.

CONCLUSIONS

1) Regardless of the warm mix additives, the air void contents of the mixtures with CRM binders decreased as the compaction temperature increased from one temperature to the next consecutive temperature.
2) In general, the compaction temperatures of CRM mixtures containing the warm mix additives can be decreased to those of the control mixtures, with the target air void contents satisfied.
3) Irrespective of compaction temperature, the addition of warm mix additives into CRM mixtures resulted in the increase of %VFA values and decreased the %VMA values.
4) More testing should be performed to characterize and evaluate performance properties of warm rubberized mixes such as rutting resistance, fatigue crack and thermal crack resistance and long term durability.

ACKNOWLEDGEMENT

The authors wish to acknowledge and thank South Carolina's Department of Health and Environmental Control (DHEC) for their financial support of this project.

REFERENCES

Amirkhanian, S. and Corley, M. (2004). "Utilization of Rubberized Asphalt in the United States - An Overview." *Proc., Advanced Technologies in Asphalt Pavements*, South Korea, 3-13.

Azari, H., McCuen, R. H., and Stuart, K. D. (2003). *"Optimum Compaction Temperature for Modified Binders." Journal of Transportation Engineering*, ASCE, Vol. 129, No. 5, 531-537.

Bahia, H. U. (2000). *Recommendations for Mixing and Compaction Temperatures of Modified Binders*, Draft Topical Report for NCHRP study No. 9-10, National Cooperative Highway Research Program, Washington, D.C.

Gandhi, T. and Amirkhanian, S., (2007). Laboratory Investigation of Warm Asphalt Binder Properties – A Preliminary Investigation, MAIREPAV5 Proceedings, Vol. 5, 475-480, Park City, Utah.

Hurley, G., and Prowell, B.,(2005 a). "Evaluation of Aspha-Min® for use in Warm Mix Asphalt", NCAT Report 05-04, Auburn.

Hurley, G., and Prowell, B.,(2005 b). "Evaluation of Sasobit® for use in Warm Mix Asphalt", NCAT Report 05-06, Auburn.

Hurley, G., and Prowell, B.,(2006). "Evaluation of Evotherm® for use in Warm Mix Asphalt", NCAT Report 06-02, Auburn.

Lee, S.-J., Amirkhanian, S., Putman, B.J., and Kim, K.W. (2007). Laboratory Study of the Effects of Compaction on the Volumetric and Rutting Properties of CRM Asphalt Mixtures, *Journal of Materials in Civil Engineering, ASCE*, Vol. 19, No. 12, 1079-1089.

Shen, J., Amirkhanian, S., Lee, S.-J., and Putman, B.J. (2006). Recycling of Laboratory-Prepared RAP Mixtures Containing Crumb Rubber Modified Binders in HMA. *Transportation Research Record*, 1962, 71-78.

Stuart, K. D. (2000). *Methodology for Determining Compaction Temperatures for Modified Asphalt Binders*, Draft FHWA Report, Federal Highway Administration, McLean, Va.

The Asphalt Institute (2003). *Performance Graded Asphalt Binder Specification and Testing, SP-1*. The Asphalt Institute, Lexington, KY.

Impacts of Laboratory Curing Condition on Indirect Tensile Strength of Cold In-place Recycling Mixtures using Foamed Asphalt

Hosin "David" Lee[1], Soohyok Im[2], and Yongjoo Kim[3]

[1] Associate Professor, Public Policy Center, Department of Civil and Environmental Engineering, University of Iowa, Iowa City, IA, 52242-1527; PH (319) 335-6818; FAX (319) 335-6801; email: hlee@engineering.uiowa.edu
[2] Research Assistant, Public Policy Center, University of Iowa, Iowa City, IA, 52242-1527, PH (319) 335-2957; FAX (319) 335-6801; email: sooim@engineering.uiowa.edu
[3] Research Associate, Public Policy Center, University of Iowa, Iowa City, IA, 52242-1527; PH (319) 335-2957; FAX (319) 335-6801; email: yongjkim@engineering.uiowa.edu

Abstrct:

Cold in-place recycling (CIR) layer is normally covered by a wearing surface in order to protect it from water ingress and traffic abrasion while obtaining the required pavement structure and texture. Currently, various agencies have differing moisture content requirements prior to placement of the wearing surface based either on the total moisture content in the mixture or the increase in moisture content from the pavement prior to recycling. The industry standard for this curing time is 10 to 14 days or a maximum moisture content of 1.5 percent. The main objective of this research is to develop technically sound methods to identify minimum in-place CIR properties necessary to permit placement of the HMA overlay through the laboratory curing process. Indirect tensile strength and moisture content of CIR mixtures were measured from the specimens with two different curing procedures: uncovered and semi-covered. Based upon the limited test results, both the length of the curing time and the moisture content significantly affect the indirect tensile strength of the CIR mixtures. It should be noted that, given a similar moisture level, the longer curing period would produce the higher indirect tensile strength.

Key Words: cold in-place recycling, curing condition, moisture content, indirect tensile strength, foamed asphalt.

INTRODUCTION

The cold in-place recycling (CIR) has become one of the most popular methods in rehabilitating the existing asphalt pavements due to its cost-effectiveness and the conservation of paving materials. Particularly, in Iowa, the CIR has been used widely in rehabilitating the rural highways because it significantly increases the service life of the existing pavement. To place CIR-foam mixtures with specific engineering properties, the laboratory test procedure was modified to improve the consistency of the CIR-foam mix design process (Kim and Lee 2006)

A CIR layer is normally covered by a hot mix asphalt (HMA) overlay in order to protect it from water ingress and traffic abrasion and obtain the required pavement structure and texture. Curing is the term currently used for a time period that a CIR layer must remain exposed in the air for drying before an HMA overlay is placed. The curing period depends on several factors, which include day and night-time temperatures, humidity levels and rainfall activity, wind, layer thickness, type of stabilizing agent used, moisture content of CIR mixtures before and after recycling, the level of compaction, in-place voids, and the drainage characteristics of the material below the CIR layer and the shoulders. Overlaying the CIR surface prior to adequate moisture loss through a proper curing may result in a premature failure of the CIR and/or HMA overlay (ARRA 2001). The Asphalt Institute (1998) reported that the inadequate curing can produce high retained moisture contents that would increase the possibility of asphalt stripping and slow rates of strength development after HMA overlay is placed.

The current practice in Iowa simply controls the maximum moisture content of the CIR, keeping it to 1.5 percent. The main objective of this study is to identify the minimum CIR properties necessary to permit placement of HMA overlay through the laboratory curing process. This paper discusses impacts of the laboratory curing condition on the development of the indirect tensile strength of CIR mixtures using foamed asphalt.

LITERATURE REVIEW

AIPCR and PIARC (2002) recommended that the application of the HMA overlay should be delayed until the residual water has largely evaporated. This duration should not only depend on the climatic conditions following CIR construction, but also on the traffic level that the CIR layer could support after completion of the pavement construction. In most European countries, the residual moisture content is used to determine the timing of placing the HMA overlay in the ranges between 1.0% and 1.5%. Particularly, in Spain, it is recommended to place the HMA overlay only after the moisture content in the CIR layer has become less than 1.0% for at least seven days or when the materials can be extracted from the CIR pavement by coring. Recently, Lee and Kim (2007) reported that the curing temperature and period significantly affect the engineering properties of the CIR-foam mixtures such that CIR-foam mixtures cured in the oven at 60°C for two days consistently exhibited significantly higher indirect tensile strengths than those cured in the oven at 40°C for three days.

Based on the FHWA's survey (2005), each state has differing moisture content or curing period requirements prior to placement of the wearing surface. Arizona, Iowa, South Dakota, Vermont and Washington recommend that the CIR layer shall be allowed to cure until the moisture of the CIR mixes is reduced to 1.5% or less. Colorado recommends the moisture content of 1.0 % and Kansas recommends 2.0 %. Delaware, Idaho, Maine, Maryland, Nebraska, Nevada, New Hampshire, New York, Ohio, Ontario and Pennsylvania require a curing period between 4 and 45 days.

Although most agencies specified the same curing requirement for both CIR-foam and CIR-emulsion, AIPCR and PIARC (2002) allows a higher moisture content for CIR-foam than CIR-emulsion. They recommend between 1.0% and 1.5% moisture content for CIR-emulsion layer before an HMA overlay whereas at least 2.0% below the optimum moisture content (OMC) of the CIR-foam layer. Assuming a typical OMC value of RAP materials between 4.0% and 4.5%, it allows the moisture content between 2.0% and 2.5% for CIR-foam (2.0% below OMC). Particularly, in the United Kingdom, the minimum curing period of CIR-foam before the HMA layer is specified as just 36 hours.

LABORATORY SPECIMEN PREPARATION

The purpose of this experiment is to determine the effects of curing period and moisture content on indirect tensile strength in the laboratory. Indirect tensile strengths and moisture contents of the specimens cured for various curing periods were measured.

Curing Condition

To represent the curing process of CIR pavement in the field, two different laboratory curing procedures were examined: uncovered and semi-covered. As shown in Figure 1 (a) and (b), respectively, the uncovered specimen is placed on two wooden sticks to allow ventilation for effective curing in the air and the semi-covered specimen is covered with the plastic mold to allow water to evaporate through its top surface.

(a) Uncovered   (b) Semi-covered
Figure 1. Pictures of CIR specimens under Two Different Curing Conditions

Sample Preparation

As shown in Figure 2, RAP materials were collected from six different CIR project sites across Iowa. Overall, the RAP material from Muscatine County was the coarsest, those from Montgomery and Wapello Counties were coarse, and those from

Hardin and Lee Counties were rather densely graded.

In order to study the influence of moisture content and curing time on the indirect tensile strength (ITS) of CIR-foam mixtures, the test specimens were prepared using foamed asphalt (PG 52-34). The CIR-foam mixtures were prepared with 2.0% foamed asphalt content and 4.0% moisture content. Uncovered and semi-covered specimens were compacted using gyratory compactor at 25-gyration, which would give a similar density as the specimen with 75-blow of the Marshall hammer (Kim and Lee 2007). Uncovered CIR-foam specimens were prepared using RAP materials from six counties and semi-covered CIR-foam specimens were prepared using RAP materials from three counties.

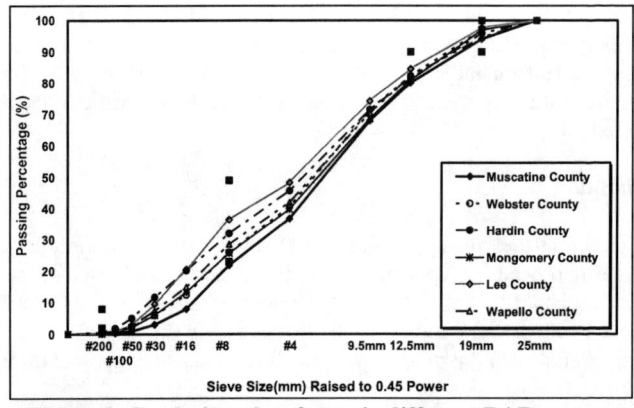

Figure 2. Gradation plots from six different RAP sources

INDIRECT TENSILE STRENGTH GAIN OVER TIME AND MOISTURE

Uncovered Curing Condition

Uncovered CIR-foam specimens were cured in the air at 25°C to allow water to evaporate. Twelve uncovered CIR-foam specimens were prepared to measure indirect tensile strengths and moisture contents at four target moisture contents for each of six different RAP sources. The indirect tensile strength was measured from three specimens at each of four target moisture contents of 2.0%, 1.5%, 1.0% and 0.5%. Moisture content was measured every hour until the uncovered CIR-foam specimens would achieve the target moisture content assuming the initial moisture content of 3.0%. After indirect tensile strength was measured from twelve CIR-foam specimens at four target moisture contents, all CIR-foam specimens were dried in the oven at 40°C for three days to determine the initial moisture content. If the measured initial moisture content is different from the assumed moisture content of 3.0%, all measured moisture contents were adjusted accordingly. For each of six RAP sources, Figure 3 shows plots of the indirect tensile strengths against the curing periods. A different curing period was needed to achieve the target moisture contents depending on its RAP source. Most RAP materials were cured very quickly and reached the moisture content of about 1.0% within ten hours. It should be noted that there is a significant variation in moisture contents among specimens with the same curing time. The indirect tensile strength did not increase for the first 10 hours of curing, but increased when the curing time increased up to 50 hours.

It is interesting to note that RAP materials from Lee County, which lost moisture the fastest reaching the moisture content from 2.24% (in 2 hours) to 0.83% (in 22 hours), exhibited a slight increase in the indirect tensile strength from 105.9kPa to 134.2kPa. However, when RAP materials from Wapello County, which lost moisture slowly from 2.23% (in 2 hours) to 0.73% (in 50 hours), exhibited a significant increase in the indirect tensile strength from 122.5kPa to 222.0kPa. A similar trend was observed from the RAP materials from Webster County such that, when the moisture decreased from 2.23% (in 2.5 hours) to 0.72% (in 50 hours), the indirect tensile strength increased by a similar amount from 121.8kPa to 222.5kPa. This result indicates that, for the given moisture content between 0.7% and 0.8%, the specimen with more curing time (50 hours vs. 22 hours) produces the higher tensile strength. RAP materials from Montgomery, Muscatine and Hardin Counties, which were cured fast in the early stage to 1.64% (in 2 hours), 1.65% (2.5 hours) and 1.62% (in 3.5 hours), respectively, exhibited the relatively high initial tensile strengths of 193.1kPa, 211.0kPa and 148.5kPa, respectively, and, when cured for 50 hours, their indirect tensile strengths continued to increase to 224.8kPa, 292.1kPa and 240.4kPa, respectively. This result indicates that, for the given curing time between 2.0 and 3.5 hours, the specimen with lower moisture content (1.6% vs. 2.2%) produces the higher tensile strength.

Overall, the indirect tensile strength increased as the moisture content decreased. It is interesting to note that the indirect tensile strength of all uncovered CIR-foam specimens did not increase significantly when the moisture content changed from 2.0% to 1.0% whereas the indirect tensile strength increased significantly when the moisture content changed from 1.0% to 0.5%.

Semi-covered Curing Condition

To simulate the curing condition of the CIR pavement in the field, semi-covered CIR-foam specimens were prepared by allowing water to evaporate through their top surfaces while being cured. Gyratory compacted CIR-foam specimens were placed into the plastic mold and cured in the oven (forced-air circulation) at 25°C while only top of the specimen was exposed to the air. Twenty-one CIR-foam specimens were prepared using RAP materials from Hardin, Webster and Muscatine counties to measure indirect tensile strengths and moisture contents after 3 hrs, 6 hrs, 12 hrs, 24 hrs, 48 hrs, 168 hrs and 336 hrs.

Figure 4 shows relationships between indirect tensile strength and curing period for semi-covered CIR-foam specimens from three different RAP sources. It should be noted that there is a significant variation in moisture contents among CIR specimens with the same curing period. Due to the slower moisture loss, the indirect tensile strength gain of semi-covered CIR specimens was much less than that of uncovered CIR specimens. Similar to the uncovered specimens, Up to 12 hours of curing, the indirect tensile strength did not increase but it increased when curing time increased from 12 hours to 336 hrs (moisture content changed from 2.0% to 0.5%). Given the similar moisture content, the indirect tensile strength of some specimens cured for 14-day (336 hrs) was higher than those cured for 7-day (168 hrs). This result indicates, given the similar moisture level, the longer curing time would produce the higher tensile strength.

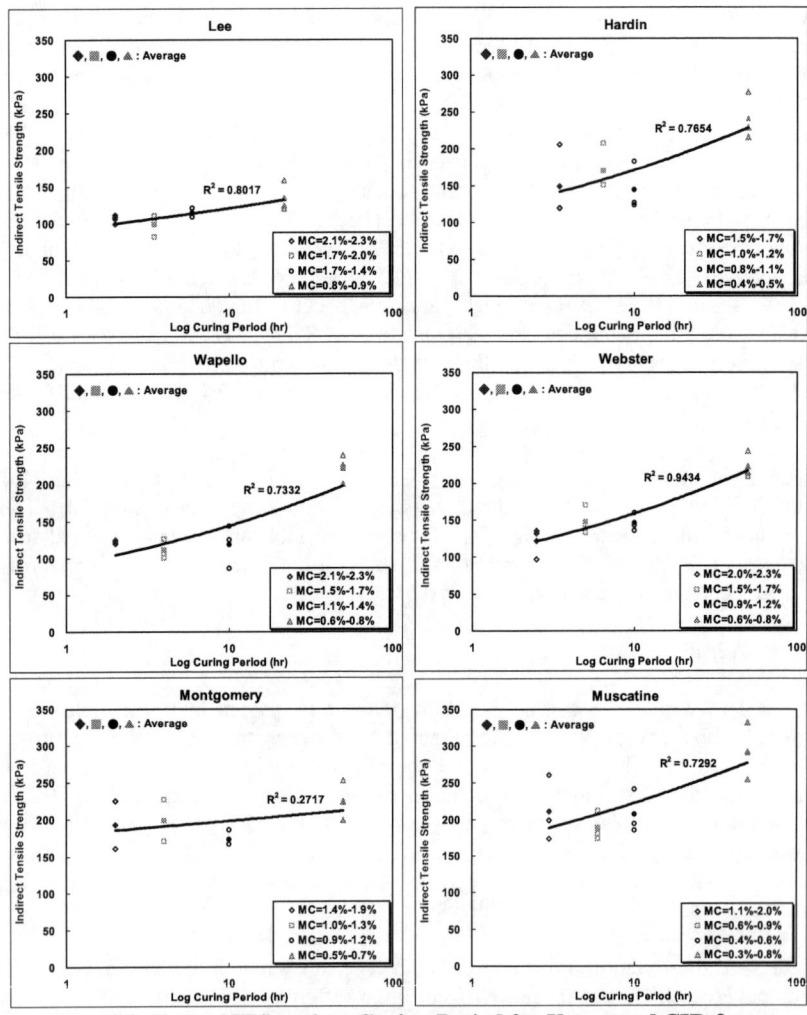

Figure 3. Plots of ITS against Curing Period for Uncovered CIR-foam Specimens from Six Different RAP Sources

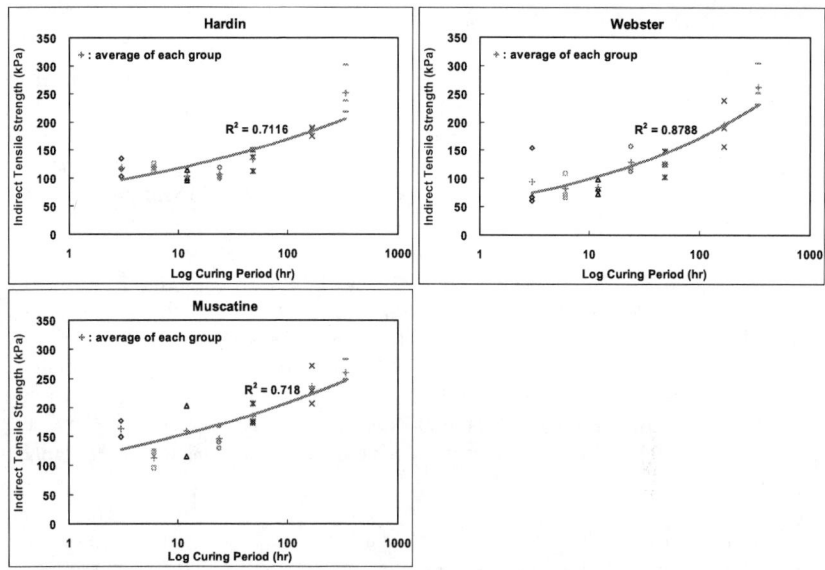

Figure 4. Plots of ITS against Moisture Content for Semi-covered CIR-foam Specimens from Three Different RAP Sources

Impacts of Moisture Contents on Indirect Tensile Strength

Figure 5 (a) and (b) show plots of indirect tensile strengths against moisture contents for uncovered and semi-covered CIR-foam specimens, respectively. The uncovered and semi-covered CIR-foam specimens exhibited similar relationships between the indirect tensile strength and moisture content such that the indirect tensile strength increased as moisture content decreased.

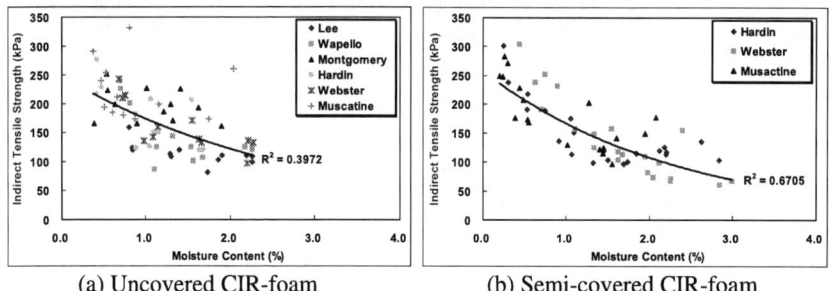

(a) Uncovered CIR-foam   (b) Semi-covered CIR-foam

Figure 5. Plots of ITS against Moisture Content for Uncovered and Semi-covered CIR Specimens

SUMMARY AND CONCLUSIONS

A CIR layer is normally covered by a wearing course to protect it from water ingress and traffic abrasion and obtain the required pavement structure and texture. Currently, the industry standard for measuring this curing period is 7 to 14 days or maximum moisture content of 1.5 percent whereas numerous projects with some CIR projects, struggling with unfavorable climate, have been overlaid with higher levels of moisture.

Based on the limited laboratory test results, the indirect tensile strengths of both uncovered and semi-covered CIR-foam specimens increased as the curing period increased and the moisture content decreased. Most RAP materials were cured very quickly and reached the moisture content of about 1.0% within ten hours. Indirect tensile strength did not increase during the early stage of curing but increased during a later stage of curing, usually when the moisture content fell below 1.5%. Given the similar moisture level, the longer curing period would produce the higher indirect tensile strength.

In the future, to develop the CIR curing index, the moisture content should be measured from various depths of a CIR layer in the field. CIR-foam project sites should be identified for the installation of the moisture meter, which can measure both moisture and temperature in the field. The moisture meter should be installed at the bottom of the CIR layer and monitored for several weeks. The area measuring feature of the moisture meter will be very valuable given the inherent variability of moisture within a CIR layer caused by a non-uniformity of water application during construction.

ACKNOWLEDGMENTS

The authors would like to thank the financial support provided by the Iowa Highway Research Board (IHRB) and the members of the steering committee for their guidance throughout the project.

REFERENCES

TAI (1998). *"A Basic Asphalt Emulsion Manual."* Manual Series No. 19. Asphalt Institute, Lexington, Kentucky.
ARRA (2001). *"Basic Asphalt Recycling Manual."* Asphalt Recycling & Reclaiming Association, Maryland.
AIPCR and PIARC (2002) *"Cold In-place Recycling of Pavements with Emulsion or Foamed Bitumen."* Technical committee 7/8 Road Pavements.
FHWA (2005) "Cold In-place Recycling State of Practice Review.",
<http://www.fhwa.dot.gov/Pavement/recycling/cir> (March 12, 2008)

Kim, Y. and Lee, H. (2006). "Development of Mix Design Procedure for Cold In-Place Recycling with Foamed Asphalt." J. Mater. Civ. Eng., 18(1), 166–124.

Kim, Y. and Lee, H. (2007). "Validation of New Mix Design Procedure for Cold In-place Recycling with Foamed Asphalt." J. Mater. Civ. Eng., 19(11), 1000–1010.

Lee, H. and Kim, Y. (2007). "Validation of the Mix Design Process for Cold In-Place Rehabilitation Using Foamed Asphalt." Iowa Highway Research Board (IHRB), Iowa DOT.

Experimental Study on Gilsonite-modified Asphalt
Juanyu Liu[1] and Peng Li[2]

[1] Assistant Professor, Department of Civil and Environmental Engineering, University of Alaska, Fairbanks, AK 99775-5900, Phone: 907-474-5764, Fax: 907-474-6030, Email: ffjl@uaf.edu

[2] Graduate Research Assistant, Alaska University Transportation Center, University of Alaska, Fairbanks, AK 99775-5900, Phone: 907-474-5054, Email: fspl4@uaf.edu

Abstract: This paper presents an experimental study on investigating the properties of Alaskan asphalt binder modified with different percentages of gilsonite over a wide range of climatic conditions by addressing the performance grade (PG) of gilsonite-modified binders according to Superpave specifications. The base asphalt used in this study was the neat asphalt with PG 52-28, and 5 different percentages of gilsonite (0, 3%, 6%, 9%, and 12% of total binder content) were introduced. With the increase of gilsonite content from 0% to 12%, the PG high temperature increased from 52°C to 70°C. However, the PG low temperature also increased from -28°C to -22°C. The results indicated that the addition of gilsonite tends to improve the rutting resistance of asphalt binders, however, increases the tendency for fatigue cracking and low temperature cracking. Adding low content of gilsonite (i.e. 3% within the scope of this study), the modified binder presents improved rutting resistance without any compromise of resistance to low temperature cracking.

Key words: Gilsonite, asphalt, rutting, fatigue cracking, low temperature cracking

Introduction

The use of additives to improve the performance of hot mixes continues to generate worldwide interest and attention, and various asphalt binder modifiers have been used to enhance the properties of asphalt binder, such as styrenebutadiene-styrene (SBS), ethylene-vinyl-acetate (EVA), styrene-butadiene-rubber (SBR), carbon black, gilsonite, etc (Button 1992, Huang et al. 1995, Tia et al. 1997, Raad et al. 1997). Among many modifiers, gilsonite is a naturally occurring glossy black asphaltic, solid hydrocarbon resin with a low specific gravity. It occurs in its very pure natural state in a mineral called Uintaite. Gilsonite is known for its easy use and good affinity with asphalt. Due to the fact that gilsonite is also a kind of asphalt binder in nature, it can be quickly dissolved into asphalt binder and coat aggregate particles during the mixing process. Thus, adding gilsonite into hot mix asphalt (HMA) mixture does not cause any problems to blending, mixing and compacting of HMA mixture that other asphalt binder modifiers usually cause. Two procedures have been reported on applying gilsonite into HMA mixtures (Gaughan 1990). Most often, gilsonite is directly applied into asphalt to make a stiffer binder for the mixtures. Another way is to coat the gilsonite onto the aggregates prior to mixing them with conventional asphalt binder. These two different application procedures will lead to different microstructures of HMA mixture and eventually result in different properties and performance of HMA mixture and flexible pavements.

Gilsonite has been used for many years by various industries worldwide as an additive in carbon black dispersing agents, hard resin printing inks for newspapers and magazines, asphalt modifying agent for road paving, and as an additive in sand molds used by the foundry industry (Davis and Gilbreath 2002). Gilsonite has been reported to be successfully used in various areas ranging from high stress areas in the City of Oslo, Norway, toll booth approaches on the New Jersey Turnpike in the United States, and major city streets and highways in Australia, Singapore, Indonesia, Japan, France and Germany (SealMaster 2007). It was found that gilsonite-modified HMA mixtures extend the expected pavement life, improve stripping resistance, and significantly reduce shoving and rutting. Gilsonite-modified asphalts exhibit generally significantly improved high temperature properties; however, the low and intermediate temperature properties may potentially be adversely affected due to the changes of the oil-to-asphaltene content with the addition of gilsonite (Tia et al. 1997). A major research interest in this area is how to achieve and extend the performance-range of the asphalt by increasing the stability without compromising other properties.

As the largest state in the United States in land area, Alaska covers different climatic zones. Alaskan pavements experience the extreme temperature conditions that range, in some instances, from about -50 °C in winter to 40°C in summer. To explore the application of gilsonite to improvement of Alaskan pavements performance, this paper reported an experimental study on investigating the properties of asphalt binder modified with different percentages of gilsonite over a wide range of climatic conditions through the determination of performance grade (PG) of gilsonite-modified binders according to Superpave criteria.

Experimental Work

This study conducted by the University of Alaska Fairbanks (UAF) determined the PG of the gilsonite-modified binders to assess the high and low temperature limits for adequate performance (i.e. resistances to rutting, fatigue cracking and thermal cracking) according to Superpave criteria. The base asphalt used in this study was the neat asphalt with PG 52-28, and the granulated gilsonite used was provided by SealMaster Company. Five different percentages of gilsonite (0, 3%, 6%, 9%, and 12% of total binder content) were introduced to produce modified asphalt binders, and effect of gilsonite content on the properties of modified asphalt binders were evaluated.

Table 1 summarizes the tests conducted in the laboratory. For each test, three replicates were provided for each temperature measured. All tests were performed according to AASHTO R29 (Standard Practice for Grading or Verifying Performance Grade of an Asphalt Binder).

Table 1. Superpave Tests for Gilsonite-modified Binders

Binder aging	Tests	Performance parameter	Test equipment/model
Original binder	Flash point (AASHTO T48)	safety	Gilson/PT-6
	Rotational viscosity (ASTM D4402)	Handling and pumping	Brookfield/RV DV-III
	DSR test (AASHTO T315)	$G^*/\sin\delta$ - resistance to rutting	Rheometric Scientific/ARES-RAA
Short term aging (RTFO)	DSR test (AASHTO T315)	$G^*/\sin\delta$ - resistance to rutting	Rheometric Scientific/ARES-RAA
	Mass loss	the amount of volatiles evaporating during the mixing and construction process	Mettler Toledo Balance
Long term aging (PAV residue)	DSR test (AASHTO T315)	$G^*(\sin\delta)$ - resistance to fatigue cracking	Rheometric Scientific/ARES-RAA
	BBR test (AASHTO T313)	Creep stiffness and m-value – resistance to thermal cracking	Cannon Instrument/TE-BBR
	Direct tension (AASHTO T314)	Resistance to thermal cracking	Instron/BTI-3

Test Results and Discussion

Results of Superpave binder tests are summarized in Table 2. The direct tension test (DTT) strain data were not available, because based on the specification, the DTT strain is only necessary when m-value requirement is satisfied but the creep stiffness requirement is not. The DTT is not required if the creep stiffness is less than 300 MPa. As shown in Table 2, for different gilsonite additions, the PG high temperature ranges from 52°C to 70°C, while low temperature ranges from -28°C to -22°C. According to Superpave specification, a PG 52-28 grade is intended for use in an environment where an average seven-day maximum pavement temperature of 52°C and a minimum pavement design temperature of -28°C, are likely to be experienced. A PG 70-22 grade is intended for use in an environment where an average seven-day maximum pavement temperature of 70°C and a minimum pavement design temperature of -22°C, are likely to be experienced, etc.

Figure 1 illustrates PG grading of gilsonite-modified asphalt binders. Since the Superpave asphalt binder specification is meant to be performance based, it addresses three primary performance parameters of asphalt pavements: permanent deformation (rutting), fatigue cracking, and low temperature (thermal) cracking.

Table 2. Summary of Superpave Binder Test Results

Gilsonite content, %	Flash point, °C	Mass loss, %	Viscosity @ 135°C	Grade Temp. at which specified criterion is satisfied, °C						PG grade	
				DSR origin.	DSR RTFO	DSR PAV	BBR S	BBR m-value	DTT strain	High	Low
0	310	0.59	0.2168	52	52	16	-18	-18	-	52	-28
3	316	0.61	0.2555	58	58	19	-18	-18	-	58	-28
6	310	0.21	0.3400	58	58	22	-12	-12	-	58	-22
9	316	0.24	0.3842	64	64	25	-12	-12	-	64	-22
12	316	0.23	0.5208	70	70	25	-12	-12	-	70	-22

Effects of Gilsonite on Rutting Performance of Asphalt Binders

It can be seen from Table 2 that the addition of gilsonite increases the viscosity of an asphalt at high pavement temperatures. This effect increases as the level of gilsonite addition increases. The Superpave binder specification uses a rutting factor, $G^*/\sin\delta$, which is a measure of asphalt binder's stiffness or rut resistance at high pavement service temperatures. Figure 2 illustrates $G^*/\sin\delta$ measured at 58°C for both original binder and RTFO aged binders. For both conditions, rutting factor increases with the increase of gilsonite addition. The high temperature of PG grade is determined based on that the rutting factor must be at least 1.00 kPa for the original asphalt binder and a minimum of 2.20 kPa for the RTFO aged asphalt binder when tested by DSR. According to the PG grading, with the increase of gilsonite addition from 0 to 12% of total binder weight, the PG high temperature increases from 52°C to 70°C, indicating a higher rutting resistance of the binder.

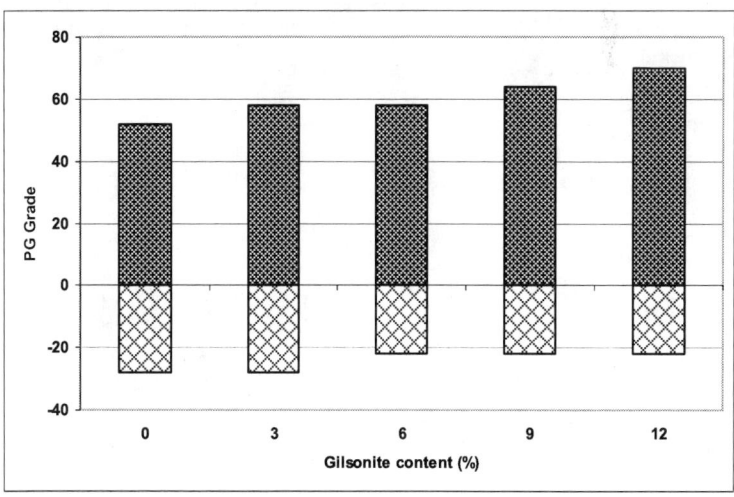

Figure 1. PG Grading of Gilsonite-modified binders

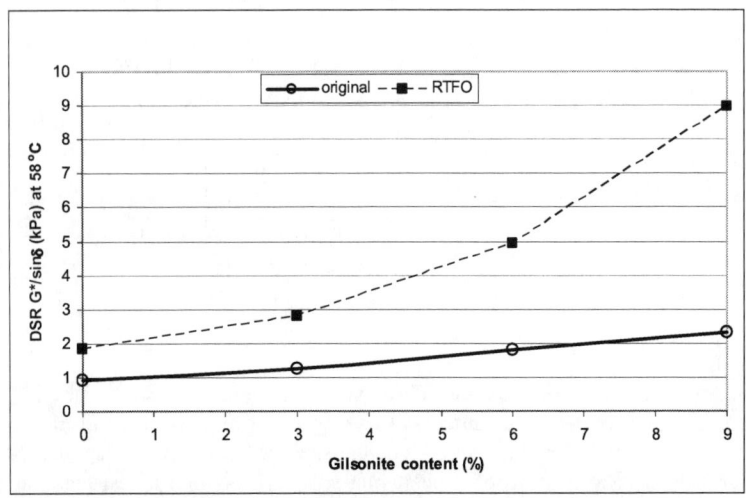

Figure 2. Effects of Gilsonite Addition on Rutting Factor

Effects of Gilsonite on Fatigue Resistance of Asphalt Binders

The specification uses a fatigue factor, $G^*\sin\delta$, which represents asphalt binder's resistance to fatigue cracking. The specification has a maximum limit of 5000 kPa for $G^*\sin\delta$ for the binder subjected to PAV aging, and tested at intermediate pavement service temperature. As shown in Figure 3, the intermediate pavement service temperature increases with the increase of gilsonite addition, which implies higher fatigue factor at same intermediate temperature, and associated reduced resistance to fatigue cracking.

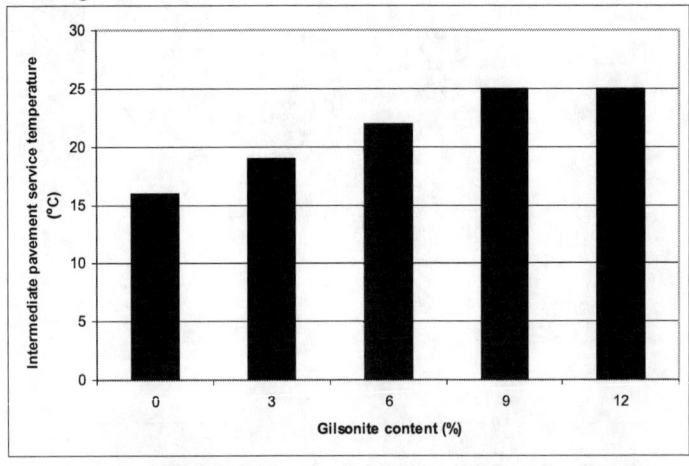

Figure 3. Effects of Gilsonite Addition on Fatigue Resistance

Effects of Gilsonite on Resistance to Low Temperature Cracking of Asphalt Binders

A lower creep stiffness and a higher m-value of PAV aged binders at a low temperature usually mean a higher resistance to low temperature cracking of pavement materials. Figure 4 illustrates creep stiffness and m-value of PAV age binders with different additions of gilsonite measured at -18°C by BBR test. With the increase of gilsonite content, creep stiffness increases while m-value decreases. In addition, according to the PG grading (Table 2 and Figure 1), the PG low temperature increases from -28°C to -22°C, indicating that gilsonite addition increases the tendency for low temperature cracking of the base asphalts.

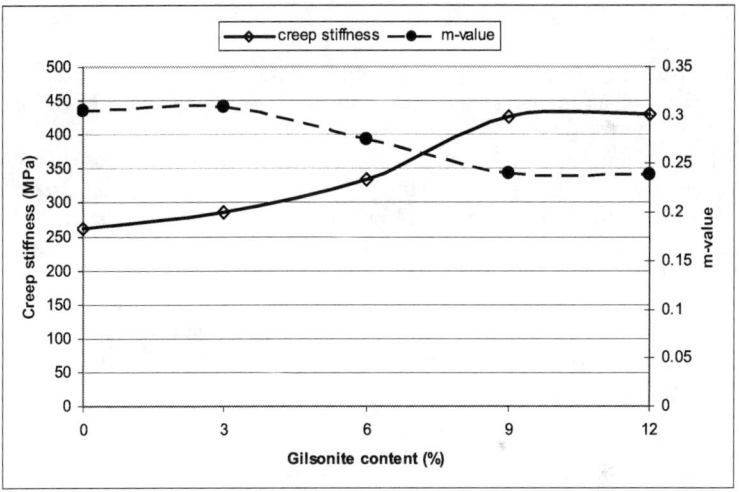

Figure 4. Effects of Gilsonite Addition on Creep Stiffness and m-value (-18°C)

Conclusions

In this study, the properties of asphalt binder modified with different additions of gilsonite over a wide range of climatic conditions were investigated by determining the PG of the gilsonite-modified binders according to Superpave specifications. The base asphalt used in this study was the neat asphalt with PG 52-28, and 5 different percentages of gilsonite (0, 3%, 6%, 9%, and 12% of total binder content) were introduced.

With the increase of gilsonite content from 0% to 12%, the PG high temperature increased from 52°C to 70°C. However, the PG low temperature also increased from -28°C to -22°C. The results indicate that the addition of gilsonite tends to improve the rutting resistance of asphalt binders, however, increases the tendency for fatigue cracking and low temperature cracking. Adding low content of gilsonite (i.e. 3% within the scope of this study), the modified binder presents improved rutting resistance without any compromise of resistance to low temperature cracking. Tests on gilsonite-modified HMA mixtures are suggested in the future study to further explore the application of gilsonite to the asphalt paving in Alaska.

References

Button, J. (1992). "Summary of Asphalt Additive Performance at Selected Sites." *Transportation Research Record 1342*, 67-75.

Davis, N., and Gilbreath, K. (2002). "The Many Facets of Gilsonite." *Proceeding of AADE 2002 Technology Conference on Drilling & Completion Fluids and Waste Management*, Houston, TX.

Gaughan, R.L. (1990). "Development of High Strength Asphalt Mixes." *Proceedings of Conference of the Australian Road Research Board*, Pavements and Materials, 41-52.

Huang, S.-C., Tia, M., and Ruth, B. E. (1995). "Evaluation of Aging Characteristics of Modified Asphalt Mixtures." *ASTM Special Technical Publication*, No. 1265, 128-145.

Raad, L., Saboundjian, S., Sebaaly, P., Epps, J., Carnilli,B., and Bush, D. (1997). "Low Temperature Cracking of Modified AC Mixes in Alaska." *Research Report INE/TRC 97.05*, Institute of Northern Engineering, University of Alaska Fairbanks, Fairbanks, AK.

SealMaster Company. (2007) Personal communication.

Tia, M., Ruth, B. E., Roque, R., Shih, C. T., Upshaw, P., and Chou, C. P. (1997). "Evaluation of Asphalt Additives for Improved Cracking and Rutting Resistance of Asphalt Paving Mixtures." *Research Report #0510646*, University of Florida, Gainesville, FL.

Experimental Research on Preparating Conductive SMA Doped Graphite

Qingjun Ding[1], Xuewei Wu[2], Xinquan Liu[3], Shuguang Hu[4]

School of Materials Science and Engineering, Wuhan University of Technology, Wuhan 430070, Hubei, P.R China;
[1]dingqj@whut.edu.cn,   [2]wuxw1014@163.com,   [3]liuxq_523@163.com,
[4]hsg@whut.edu.cn

Abstract: Conductive stone mastic asphalt (SMA) mixture is a kind of asphalt pavement material with excellent electrical performance, produced through adding conductive paste material to the mixture. It was produced by using graphite as conductive particle, and a kind of resistance test method of embedding showerhead electrode was designed in the present study. It was found that graphite can improve electrical performance of SMA mixture effectively. Fractal theory finds the reason why different graphite have various effect in improving electrical performance of mixture is their model parameters are diverse; electrical performance of conductive SMA mixture containing graphite is steady, and will not change with time; with the increase of graphite content, temperature sensitivity of conductive SMA decreases gradually and its linear V-I characteristic is more obvious; the present research shows: the conductive mechanism of conductive SMA is the result of combined action of conductive pathway, tunneling effect and field emission.

Keywords: Conductive SMA; Resistivity; Graphite; Electrical Performance

1 Introduction

Conductive asphalt mixture may be defined as a bituminous-base composite contained a certain amount of electronically conductive paste materials to attain stable and relatively high electrical conductivity, and can be used as bituminous pavement materials. Conductive bituminous pavement can self-heat when connected with power supply, it has a profound effect of deicing or snow-melting in winter, and solves traffic problems caused by icing road at cold zone in winter. When conductive asphalt mixture was used as pavement materials, information of changes in resistivity of concrete can be obtained. Corresponding information of road status would be collected, and then it can also be applied as self- monitoring asphalt concrete for evaluation on bearing capacity of pavement structure, structure health monitoring, fatigue life estimation, and intelligent management, etc. The self-heat characteristic can also be used to prevent low temperature cracking at the time when temperature suddenly drops. Electrical conductive asphalt composite is also valuable as a coating made to provide a shielded structure. Electrical conductive asphalt mixtures can not only be used as a kind of pavement materials, but also can be used as new building functional materials, which has broad application foreground.

A certain amount of graphite was added into the asphalt mixture to prepare conductive SMA, and a method to test electrical performance of conductive SMA was introduced in the present study. Mechanism of conductive SMA was discussed by experimental research, micro measurement, and theoretical analysis. Effect of graphite type on improving electrical performances of asphalt mixtures was investigated, and the change direction of resistivity of conductive SMA mixtures doped graphite with time, temperature and input voltage was studied in present research.

2 Materials and experiment scheme

2.1 Materials

SBS modified asphalt was produced by Hubei Guochuang High Technology Material, the characteristics of asphalt binder is presented on Table 1. Aggregates are basalt crushed stone, mineral powder is limestone grinding powder, and fibers are polyester fiber. Three types of graphite were used in the present study, middle flake graphite LZ100-92, high flake graphite LG150-94 and high flake graphite LG150-96. The characteristics of three kinds of graphite are shown in Table 2.

Table 1 Characteristics of SBS modified asphalt binder

Test	Results
Penetration (25°C, 100g, 5s)/0.1mm	46.7
Softening Point /°C	75
Ductility (5°C, 5cm/min)/cm	35
Viscosity (135°C) /Pa·s	1.9
Flash Point /°C	348
Elastic Restitution /%	95
Ductility after TRFOT(5°C, 5cm/min) /cm	33
Penetration after TRFOT /0.1mm	38.8

Table 2 Characteristics of graphite

Index	LZ100-92	LG150-94	LG150-96
Density(g/cm^3)	2.18	2.26	2.30
Particle size(μm)	150	105	105
Fixed-carbon(%)	≥92.0	≥94.0	≥96.0
Moisture content (%)	≤0.50	≤1.0	≤0.5
Volatile content (%)	≤1.0	≤1.5	≤1.2

2.2 Preparation and test method

2.2.1 Preparation of conductive SMA doping graphite

The conductive SMA was prepared using a certain amount of graphite to substitute a part of mineral powder, on the basis of proportion design of ordinary SMA-13. Graphite was used as a kind of filler to be added in asphalt mixtures (by volume of asphalt, particles passing No. 200 sieve in the mix gradation), to guarantee the void in mineral aggregate (VMA) in bituminous mixture and ensure interlocking conditions of SMA coarse aggregates. Optimal asphalt content (OAC) was determined by marshal method, and bitumen mixtures must be mixed sufficiently to ensure graphite can well disperse in mixtures.

2.2.2 Electrode design and test method

A kind of resistance test method of embedding showerhead electrode was designed as shown in Figure 1, to decrease contact resistance effectively and to make sure the accuracy and feasibility. Electrodes are conductive silk screen, region I and II are filled of compacted conductive SMA mixtures. The electrodes must be sure of erecting in the mixtures in the process of forming. Desktop multimeter ESCORT3146A was used to measure the resistance of specimens after being molded.

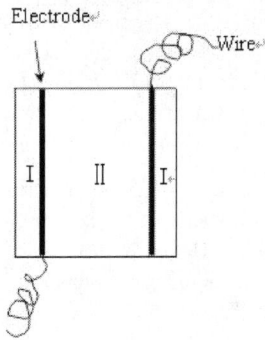

Figure1. Schematic diagram of molding conductive specimen

2.2.3 Moisture susceptivity test

The moisture susceptibility of the asphalt mixtures was evaluated according to "Standard test for effect of moisture on asphalt concrete paving mixtures" (ASTM D4867-04). A minimum TSR of 80% is recommended for this test method. The samples to be used in the moisture susceptibility tests were compacted using a Marshall compactor.

2.2.4 Wheel tracking test

Wheel tracking test was taken to evaluate the rutting resistance property of the asphalt mixture. The specimen for wheel tracking test is 300mm length, 300mm width and 50mm thickness, which is compacted by the slab compactor. The specimen is immersed in dry atmosphere at 60°C for no less than 5 hours, and then a contact wheel pressure of 0.7MPa is loaded on the specimen to test for 60min under 60°C circumstance. The wheel traveling distance of the wheel is 230±10mm at a speed of 42±1 cycles/min. Dynamic stability (DS) was calculated as

$$DS = \frac{15N}{d_{60} - d_{45}} = \frac{42 \times 15}{d_{60} - d_{45}} \qquad \text{Equation 1}$$

Where: N is the wheel traveling speed, generally, N=42 cycles/min, d_{60} and d_{45} are the deflection at 45 and 60 min, respectively.

3 Results and discussion

3.1 graphite factor

Table 2 shows the resistivity of conductive SMA with different graphite content. Figure 2 illustrates the logarithm to base 10 of resistivity with the change of graphite content. The resistivity of conductive SMA decreased continuously with the increase of graphite content, and the change trend of electrical performance was shown in the Figure 2. With the example of graphite LZ100-92, the resistivity of conductive SMA did not decrease remarkably, when the content of graphite was below 10%. But when graphite content changed from 10% to 20%, the resistivity of specimen decreased significantly, it reduced from $1.82 \times 10^9 \Omega \cdot m$ to $676.5 \Omega \cdot m$. If the graphite content continued to increase, the resistivity changed relatively weakly. The resistivity of conductive SMA decreases to $10\Omega \cdot m$ with the content of graphite is about 40%.

Figure 2 shows that these three kinds of graphite have the same general tendency in improving electrical performance of conductive SMA, but the improve effects are different, i.e., graphite LG150-96 is the best of the three. According to Pitchumani&Yao theory, Zhangmingqiu proposed fractal theory to explain the formation of conductive path. From the theory, electrical conductivity of system can be expressed as:

σ=f (V, σ_0, distribution of conductive paste materials)   Equation 2

Where: σ is electrical conductivity of composite,

V is volume fraction of conductive paste materials,

σ_0 is conductivity of conductive particles.

In two-phase system,

$$\sigma = \sigma_p^G \cdot \sigma_s^{1-G} \qquad \text{Equation 3}$$

Where: σ_p is conductivity in parallel limit condition,

$\sigma_p = V\sigma_0 - (1-V)\sigma_m$,

σ_s is conductivity in series limit condition,

$\sigma_s = [V/\sigma_0 + (1-V)/\sigma_m]^{-1}$,

G is distribution function, corresponding to the appeared probability of conductive pathway, $G=V^{aL'}$, $(L'=(\pi/4V)^{1/2}D'$, $D'= (\pi/4V)^{d/(2d-2)}$, a is model parameters, d is fractal dimension). Model parameters have two types: one is geometric parameters (size and geometrical condition of conductive paste materials, distribution of filled density, etc.), and the other is materials parameters (conductive materials, matrix properties, compatibility, etc). Fractal theory explains the main reason why graphite LG150-94 is batter than graphite LZ100-92 is their particle size are different, so their geometric parameters are different. The key reason why graphite LG150-96 is batter than graphite LG150-94 is the difference of carbon content of graphite, so conductivity of conductive particles are different.

Table 2 Resistivity of conductive SMA in different graphite content

content/	Conductivity ρ /$\Omega \cdot$m		
%	LZ100-92	LG150-94	LG150-96
5	5.28×10^9	4.27×10^9	3.80×10^9
10	1.82×10^9	1.42×10^9	1.01×10^9
15	5.96×10^6	3.98×10^5	6.03×10^4
20	676.5	277.1	57.4
25	105.0	61.7	17.0
30	35.5	23.1	8.6
35	14.9	10.2	3.6
40	10.1	3.9	2.6

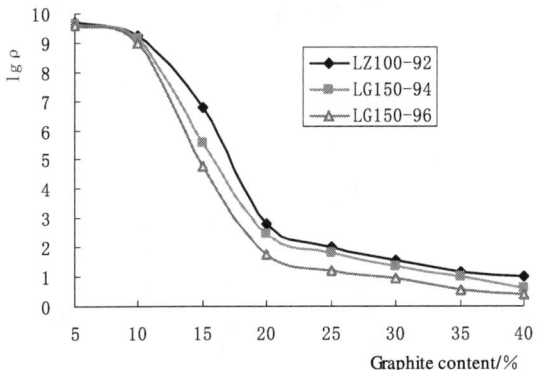

Figure 2 Relationship between lgρ and graphite content

3.2 External factors

The electrical performance of conductive SMA with changes of external factors must be understood before it was used in engineering practice, the changes of resistivity of conductive SMA mixtures doped graphite with time, temperature and input voltage are discussed below.

3.2.1 Time stability

Two conductive specimens were used for long-term observation with two LZ100-92 graphite content, that is, 20% and 35%. Figure 3 shows resistivity of specimens with the change of time, and it was observed that resistivity of specimens did not change strikingly with time process, i.e.,variation range of resistivity was less than 5%. Graphite is an excellent conductive material with fine physical chemistry stability, which assures that conductive SMA mixtures has stable electrical performance.

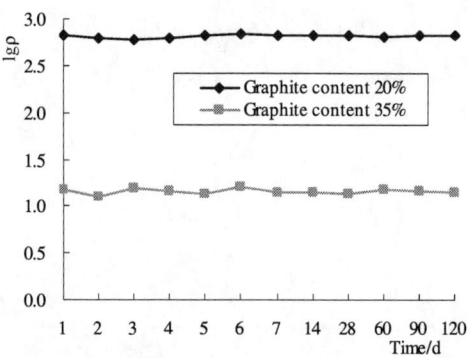

Figure 3 Relationship between lgρ and time

3.2.2 Temperature sensitivity

Relationship between change rates of resistivity and temperature was investgated with 20°C as the base. As can be found from Figure 4, resistivity decreased with temperature reduced. When temperature reduce from 20°C to -30°C, resistivity of specimen with 20% graphite content decreased by 78%, and resistivity of specimen with 30% graphite content decreased about 50%. Comparing two curves, the one corresponding to 30% graphite content is smoother than the one corresponding to 20% graphite content. When graphite content is different, the sensitivity of resistivity of relative conductive SMA to temperature is various. The higher graphite content was, the lower influence of temperature on resistivity of conductive SMA would have. Thus when conductive SMA wound be used as pavement material for deicing or snow-melting in winter, output powder need to be adjusted appropriately according to different circumstances.

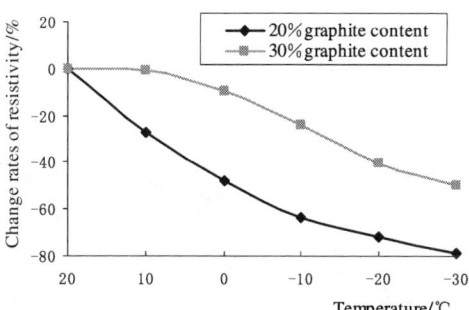

Figure 4 Curves for Variation Rate of Resistivity with Temperature

3.2.3 V-I characteristic

It is very important to control output powder in engineering practice. Through research on V-I characteristic of conductive SMA, change law of current in conductive SMA at various alternative voltages can be concluded. Figure 5 shows current in conductive SMA at various alternative voltages with different graphite content.

The diversity of V-I characteristic of conductive SMA with different graphite content is very prominent. V-I characteristic of conductive SMA with low graphite content is chaos. Linear relationship between V and I becomes more and more significant with the graphite content increase, when graphite content increases to 30%, linear relationship of V-I characteristic curve is remarkable, i.e., R^2 is above 0.99, relationship between V and I coincide ohms law.

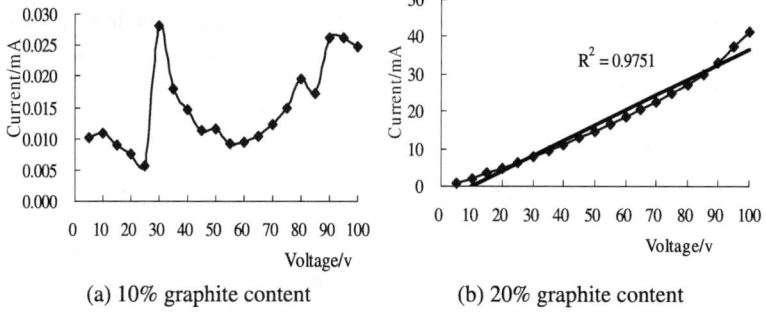

(a) 10% graphite content   (b) 20% graphite content

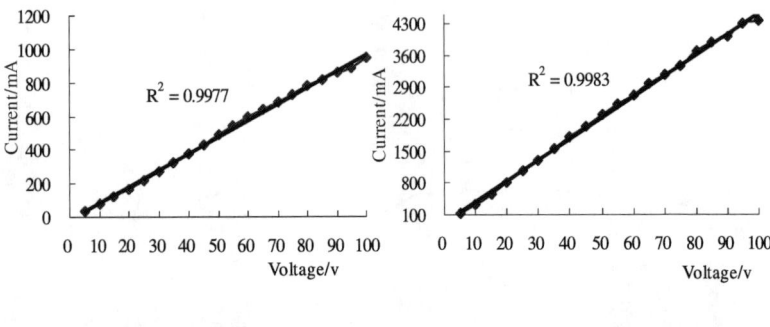

(c) 30% graphite content (d) 40% graphite content

Figure 5 V-I Characteristic of Conductive SMA with Various Graphite Content

3.3 Conductive mechanism research

Some theories are used to explain conductive mechanism of composite materials. Conductive pathway theory suggests that when conductive particles contact each other and form chain network, or gaps between conductive particles are very small (<100nm), current pass forms. The more conductive particles contacting each other were, the more perfect conductive network was, and the higher the conductivity of composite was. Tunneling effect explains conductive phenomena in composite materials, when the distance between conductive particles is long (>100nm), and the particles did not form chain network. So tunneling effect holds that composite do not only depend on particles contacting each other to form conductive chains, but electrons transfer caused by thermal vibration between conductive particles also helps materials become conductive, and proposes threshold value Pc. Conductive particle is different, relative threshold value is various. Beek found that because of the existence of interface effect in composite, when voltage increases to a certain value, strong electric field between conductive particles causes emission electric field, electrons pass over potential barrier and produce current, the result is the increase of current departs from linear relationship. It is field emission theory.

The micro-photo in Figure 6 shows that the graphite in the present study was sheet particles. There were three conditions in conductive mechanism of SMA, when graphite was used as the sole conductive paste material. The first one is some conductive particles form chain conductive pathway though contacting each other, this grantee conductive SMA can be conductive. The second is a part of particles distribute in SMA mixtures with the form of isolated particles or small surfactant aggregates. These particles hardly participate in conduction. When the distance between these isolated particles or small surfactant aggregates is very close, that they are just separated by a layer of asphalt membrane, electrons activated by thermal vibration pass through the potential barrier of asphalt membrane, and transit to adjacent particle to form tunneling current. Also, when internal electric field among particles is strong, electrons have great probability to cross the potential barrier of asphalt membrane, and transit to adjacent particle to form emission current. So the conductivity of conductive SMA is the result of combined action of these three kinds of conductive mechanisms. When graphite content is small and voltage is low, the probability of forming conductive pathway is low, tunneling effect is the main conductive mechanism. The proportion of conductive pathway forming through contact of conductive particles increases with graphite content, conductive mechanism of SMA transits from tunneling effect to conductive pathway theory gradually. When graphite content exceeds a certain value which is threshold value Pc, the effect of conductive pathway is more significantly. When graphite content is near by threshold value Pc, these two conductive mechanisms are both exiting. When external voltage is very strong, field emission also influences the conductivity of SMA.

Figure 6 micro-photo of graphite

The conductive mechanism can explain variation characteristics of resistivity of conductive SMA caused by external factors effectively. In the research of temperature sensitivity, when graphite content is 20%, tunneling effect and conductive pathway theory are both exiting, and when the temperature of mixtures decrease, asphalt shrinks faintly. Thus the distance between graphite particles decreases. This is useful for the formation of conductive network and it can promote the formation of conductive pathway theory. When graphite content is 30%, the main mechanism in SMA is conductive pathway theory, three-dimensional conductive network basically formed, conductive network would be more perfect with temperature decrease. But it has relatively weak influence on resistivity, the change range of resistivity of specimen increases comparatively lower. Non-linear V-I characteristic of conductive SMA can be explained that when graphite content is low, graphite particles is segregated by asphalt membrane. The distance between conductive particles is long, tunneling effect is the main mechanism, and so conductive behavior is non-linear. Probability of the formation of conductive pathway increases with graphite content. The more graphite content is, the more significant conductive pathway is, so linear conductive behavior is more obvious.

4 Road performances

The indoor tests according to the Standard Test Methods of Bitumen and Bituminous Mixtures for Highway Engineering (JTJ 052-2000, the P. R. China) were carried out to verify the performances of asphalt mixtures. In order to investigate the affection done by graphite on SMA mixtures, Electrically conductive SMA with various LG150-96 graphite content, i.e., 0%, 10% ,20%,30% and 40% were under tested. The test results are presented in Table 3.

Table 3 Road Performance of Asphalt Mixtures

Test	0%	10%	20%	30%	40%
OAC/%	6.1	6.8	7.5	8.2	8.9
Air-Void /%	3.78	3.65	3.53	3.40	3.32
Marshall Stability /kN	10.74	9.73	9.36	8.85	8.31
Moisture Susceptibility (TSR) /%	96.8	88.9	93.1	89.8	88.4
Dynamic Stability / (cycles/mm)	8875	9713	9825	8003	8700

It can be seen from Table 3 that the OAC of asphalt mixtures increased with the rising of adding amount of graphite. The OAC reached 8.9% when the dosage of graphite is 40%, which is 2.8% higher than that of SMA mixture without graphite. This increasing of OAC can be mainly due to the higher specific surface area of graphite than limestone powder. But the more graphite content, the lower marshal stability of conductive SMA mixtures. The TSR values obtained from all the five mixes is higher than 80%, which satisfied the specification. Dynamic stability (DS) was calculated to evaluate the permanent deformation resistance of the mixes. As shown in Table 3, the results of DS were higher than 8000cycle/mm. Compared to ordinary SMA mixtures, partial performance of conductive SMA mixtures decrease slightly, but still keep on a high level.

5 Conclusions

1. Graphite added as a kind of conductive material can enhance the electrical performance of SMA mixtures. There is a toboggan process of resistivity of conductive SMA mixtures, when graphite content exceeds a certain value. Conductive SMA mixture can be produced by choosing fitting kind of graphite and content, in order to meet to various resistivity need. Fractal theory suggests that the main factors influence conductive performances of SMA mixtures are geometric

parameters and materials parameters of conductive materials.

2. External factors such as time, temperature, and voltage have different influence on electrical performance of conductive SMA. Resistivity of conductive SMA mixtures doesn't change acutely with time, but impacted by temperature and voltage significant, temperature sensitivity and V-I characteristic of conductive SMA various with graphite content. When graphite content is low, V-I characteristic of conductive SMA is confused, when graphite content increases to a certain value, the curve of V-I characteristic of conductive SMA is linear.

3. The conductivity of conductive SMA is the result of combined action of three conductive mechanisms. The contribution of these mechanisms to electrical performance of SMA is different with graphite content. Threshold value and change law of resistivity influenced by external factors of conductive SMA can be explained by conductive mechanisms effectively.

4.With the increaseing of graphite content, the OAC of conductive SMA would increase and marshal stability would decrease, but other performance can keep on a high level, which give security for it application as pavement materials.

Acknowledgement

This work is supported by National High Technology Research and Development Program of China (2006AA11Z117).

References

1 F.Carmonna. "Conducting filled polymers."R. Solid State commu. 1989, 157(1):461-469
2 P. Xie, J. J. Beaudoin. "Electrically Conductive Concrete and Its Application in Deicing."R. Advances in Concrete Technology, Proceedings, Second CANMET/ACI International Symposium, SP-154, American Concrete Institute, Farmington Hills, Mich., 1995:399-417
3 Rajagopal, M.Stayam. "Studies on electrical conductivity of insulator-conductor composites."J. Appl.phys.1978,49(11):461-469
4 Sherif Yehia, Christopher Y Tuan, David Ferdon et al. "Conductive Concrete Overlay for Bridge Deck Deicing : Mixture Proportioning , Optimization, and Properties."J. ACI Materials Journal, 2000,97(2): 172-181
5 Sherif Yehia, Christopher Y Tuan. "Conductive Concrete Overlay forBridge Deck Deicing."J. ACI Materials Journal, 1999,96(3): 382-390
6 S Wen and D D L Chung. "Seebeck Effect in Carbon Fiber Reinforced Cement."J. Cement and Concrete Research,1998,23(4):837-841
7 S. Wen, D.D L.Chung. "Pitch-matrix composites for electrical, electromagnetic and strain-sensing applications."J. Journal of materials science, 40(2005)3897-3903.
8 TANG Zu-quan, LI Zhou-qiu, HOU Zuo-fu, XU Dong-liang. "Properties analysis

on electrically conductive concrete road material and conductive additive selection."J.Concrete, 2002,4:28-31.
9   Wu Shaopeng, Liu Xiaoming,Mo Liantong, Ye Qunshan. "Research of self-monitoring mechanism of electrically conductive asphalt-based composite." J.Key engineering Materials, Vols.326-328(2006)pp.1499-1502.
10 Xie Ping, Gu Ping, Beaudoin J J. "Electrical Percolation Phenomena in Cement Composites Containing Conductive Fibres"J. Journal of Materials Science, 1996,31(15): 4093-4097.

Research and Optimization on Flame-retarding Asphalt System Based on ATH

Qingjun Ding[1], Fan Shen[2], Xinquan Liu[3], Shuguang Hu[4]

School of Materials Science and Engineering, Wuhan University of Technology, Wuhan 430070, Hubei, P.R China;
[1]dingqj@whut.edu.cn,   [2]wwwsf5227@163.com,   [3]liuxq_523@163.com,
[4]hsg@whut.edu.cn

Abstracts: With the development of engineering construction and transport, the mileage of tunnels built by countries around the world have greatly increased. Asphalt Pavement has become the main form in the pavement of highways and bridges. Tunnel is a semi-closed space that when an accident and firing happen in it, gasoline and other flammable liquids flowdown and will burn quickly and wildly. The fire hazard seriously threaten the human and vehicles' security in tunnel, and cause great casualties and economic losses. which have occurred in many countries of the world. Because of the flammability of asphalt, the asphalt pavement appliances are severely limited in the tunnel project. However, the use of asphalt pavement in the tunnel can be promoted through the approach to flame-retarding treating and changing the material's property of combustion. Based on the flame-retarding system of ATH(aluminum hydroxide), this paper selected the suitable ATH ,tested the flame-retarding performance of MH (magnesium hydroxide) and improved ATH flame-retardant properties by Zeolite powder, then designed a reasonable program about ATH flame-retardant technology. Based on researches on the flame-retarding performance of ATH & MH and the accelerating effect of zeolite powder through tested of limiting oxygen index and flash point, the optimum mixture ratio of compound flame-retarding asphalt can be presented. Results showed the limiting oxygen index of compound flame-retarding asphalt could climb to more than 29 percent, and the flash point came to about 420°C, the flame-retarding performance of asphalt was improved greatly. Its mechanism and theory also have been discussed and explored through DSC-TG analysis and SEM observation.

Keywords: Tunnel Pavement; Flame-retarding Asphalt; limiting Oxygen Index; Flash Point; ATH; DSC-TG Analysis

1 Introduction

The flammability of asphalt has been severely limiting the application of asphalt pavement in the tunnel project for a long time, but it can be greatly improved through flame-retarding treating and changing the material's combustion property. At present, most of organic flame retardants (such as org anic bromine category, phosphorus flame retardant), which are commonly used in the chemical industry, are toxic and tend to volatilize toxic and irritant gases at high temperature. It had not only raised very high demand for the production and construction of Hot Mix Asphalt mixture, but also made a bad effect on the workers' health. Moreover, the organic flame retardants cost us too much in order to be suitable for large-scale use in the asphalt pavement. Aluminum hydroxide (ATH) is a cheap and common used inorganic flame retardant, which can decompose and volatilize steams at 230°C to 350°C without generating harmful gases. However, ATH flame retardant efficiency is low, certain technical means should be taken to improve the efficiency of the flame retardant so that ATH flame retardant can be used in the asphalt road. The paper selected the suitable ATH, tested the flame-retarding performance of MH (magnesium hydroxide) and was improved ATH flame-retardant properties by Zeolite powder, then designed a reasonable program about ATH flame-retardant technology, and discussed the design mechanism about the System of ATH flame-retardant.

2Test Methods and Raw Materials

2.1 Limiting oxygen index method and the flash point test

A combustion reaction has three elements: Combustible Material, Oxygen and Temperature. The combustion properties of asphalt materials can be tested and evaluated through tests of limiting oxygen index (LOI) and flash point (FP). LOI method tests the limits of combustion oxygen concentration, and FP tests the limiting ignition temperature of asphalt combustion.

The flash point was tested according to the "highway projects asphalt and asphalt mixture pilot order" JTJ 052-2000 Cleveland Open Cup flash point test method.

LOI is the minimum concentration of oxygen which is required by the sample to maintain combustion in oxygen and nitrogen mixture under certain conditions, that is, the volume percent of oxygen in gaseous mixture. The formula is:

$$LOI = \frac{[O]}{[N]+[O]} \times 100\% \qquad \text{Equation 1}$$

In equation 1: LOI is limiting oxygen index, %;[O] is the volume flow-rate of oxygen at critical concentration of oxygen, L/min;[N] is the volume flow-rate of nitrogen at critical concentration of oxygen, L/min

At present, it has not constituted any relevant standard about the test method of LOI on asphalt flame-retardant properties in China, so relative testing method according to GB10707-1989, GB2406-1993 has been consulted. The sample is a bar mixed by glass fibre cloth and asphalt which is 3 to 5 mm thick, 8 to 10 mm wide. And Japan JISK 7201 defines: LOI more than 30% is the fireproof grade 1; LOI in the 27 ~ 30% is the fireproof grade 2; LOI in the 24 ~ 27% is the fireproof grade 3; LOI in the 21 ~ 24% is the fireproof grade 4; LOI less than 21% is fireproof grade 5, which is flammable material.

2.2 Equipment

HC-2 oxygen index testing instrument and SYD-3536 Cleveland open cup testing instrument was used for asphalt combustion performance test in this paper.

2.3 Materials

1 asphalt

The research object is I-D modified asphalt manufactured by Guochuang High Technology Material INC. in China Hubei, specific performance indexes of which are shown in Table 1:

2 ATH, MH and Zeolite accelerator

The reagent-ATH has a fineness of 100 mesh and the flame retardant industrial ATH has finenesses of 325 mesh and 1250 mesh respectively, all of which are manufactured by Tianlong fire-retardant material Co.Ltd. in China Henan.

As to MH, the reagent-Mg (OH)2 and superfine brucite powder whose fineness is 1250 mesh were chosen: the content of MgO is more than 70% , both of them are manufactured by Tianlong fire-retardant material Co.Ltd.in China Henan. And grade 4-A molecular sieves original Zeolite powde was selected to be experimental Zeoliter.

Table 1. Performance of I-D Modified Asphalt

Test	Results
Penetration (25°C, 100g, 5s) / 0.1mm	46
Softening Point /°C	75
Ductility (5°C、5cm/min) / cm	35
135°C Viscosity / Pa·s	1.9
Flash Point (FP) /°C	324
Limiting Oxygen Index (LOI) / %	20.8
Elastic Restitution /%	95
Ductility after TRFOT (5°C, 5cm/min)/cm	33

3 Results and Discussion

3.1 The Effect of ATH on Flame Retardant Properties of Asphalt

Tested the LOI and FP of asphalt which added three kinds of ATH respectively with different content additions. The test results are shown in Table 2 and Figure 1:

The results show that the asphalt added ATH flame-retardant materials could improve its fire-retardant properties, and its LOI and the FP temperature rose too. Added ATH with different finenesses, the flame retardant's properties are different. That is, the finer the fineness is, the higher the LOI would be. Added 30% ATH whose fineness is 1250 mesh, the asphalt LOI reached to 29 percent, which was close to the fireproof grade 1 technical requirements in the definition of Japan JISK 7201. In the FP temperature testing, the asphalt's surface generated a mass of inflated bubbles when the temperature was above 250°C, which indicated that ATH released moisture during its decomposition. The FP temperature of asphalt raised to more than 350°C after adding 30% ATH to the asphalt whose performance of high-temperature combustion resistance improved remarkably.

In summary, the addition of ATH can greatly restrain asphalt combustion, and increase the LOI and the FP temperature. But if used as the only modifier, ATH should be used in large quantity and superfine size, in order to obtain the good fire-retardant properties.

Table 2. Flame-retarding Performance of Asphalt adding ATH

ATH Addition Rate /%	Reagent		325 mesh		1250 mesh	
	LOI /%	FP /°C	LOI /%	FP /°C	LOI /%	FP /°C
0	20.8	324	20.8	324	20.8	324
10	22.3	329	22.6	336	24.6	342
20	23.7	338	25.0	343	28.4	349
30	25.4	344	26.8	348	29.6	351

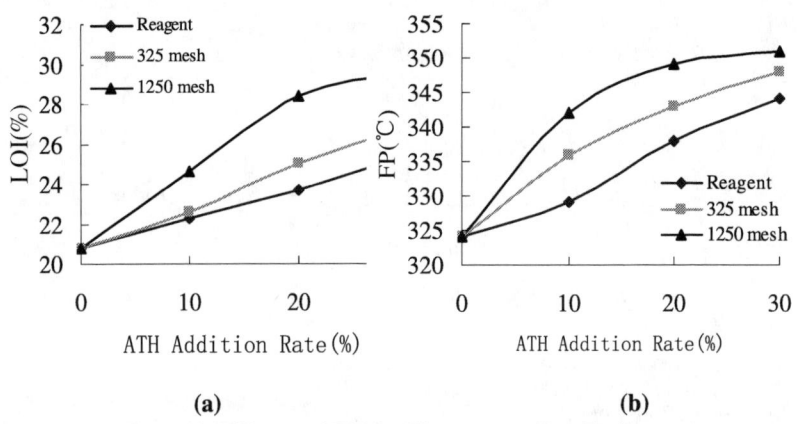

Figure 1. Effect on ATH for Flame-retarding Performance

3.2 ATH Synergies with MH Flame Retardant

Based on the LOI and FP temperature of asphalts with different content of MH, and the gas situation when the sample combusted vertically, the effect of different contents of MH on flame retardant properties of asphalt can be evaluated. The test results are shown in Table 3 and Figure 2 :

The LOI testing indicates that the flame-retardant properties of MH are better than that of ATH. Added 10% MH whose fineness is 1250 mesh to asphalt, the LOI of MH flame retardant asphalt reached 25.2% which exceeded that of ATH with the same fineness for 0.6%. Flame-retardant effect of MH whose fineness is 1250 mesh, are obvious better than that of the reagent Mg $(OH)_2$. Through the testing, we observed that flue gas volume had significantly reduced with the addition of MH. However, the addition of MH was unable to raise the FP temperature of the asphalt effectively, and didn't generate many bubbles during the test. So, using MH as the only flame retardant can not effectively improve the asphalt combustible temperature too.

Taken 5% MH whose fineness is 1250 mesh as the synergy of ATH flame retardant, and added it to the flame retardant asphalt which contained 20% ATH with a fineness of 325 mesh, the LOI and the FP of asphalt and the smoke conditions can be tested and shown in Table 4.

Table 3 Flame-retarding Performance of Asphalt adding MH

MH Addition Rate /%	Reagent		1250 mesh	
	LOI /%	FP /°C	LOI /%	FP /°C
0	20.8	324	20.8	324
5	22.6	326	24.4	327
10	23.4	327	25.2	329
15	24.2	326	26.1	330

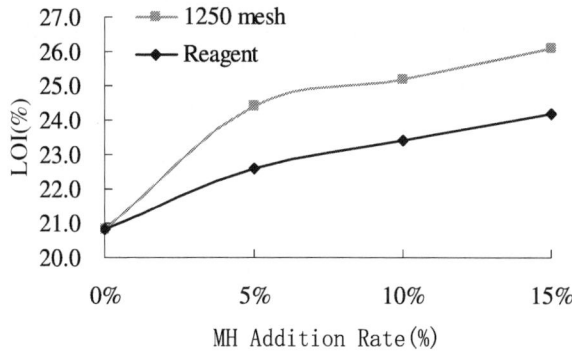

Figure 2. Effect of MH for Flame-retarding Performance

Table 4. The Cooperation of ATH & MH

ATH Addition Rate /%	MH Addition Rate /%	LOI /%	FP /°C	Smoke Conditions
20	0	25.0	343	Black Smoke
0	5	24.4	327	Black Smoke
20	5	27.4	415	A little Smoke

When MH and ATH were combinedly used as the fire retardant materials, the fire-retardant properties of asphalt could be effectively improved, the FP temperature rised to 415°C, the LOI reached over 27 percent, and only a small amount of smoke was produced. The effect was much better than when using ATH or MH alone.

3.3 Zeolite flame retardant accelerator

Zeolite is a kind of original zeolite powder used for the preparation of 4A molecular sieves. Because it has the property of selective adsorption and can be permeated by the particles, when used as a flame retardant accelerator, Zeolite can not only effectively improve the efficiency of the flame retardant, but also restrain smokes well, which is a prominent technology in the flame-retarded field. However, if 4 A molecular sieve is used directly as a flame retardant accelerator, the technical difficulties are great and economic costs are high, and it is unsuitable for large-scale use. Accordingly, original zeolite powder is chosen as the flame retardant accelerator besides 20% ATH, which were added into asphalt to prepare the flame-retardant asphalt. Research results about the accelerating effect of zetlite on the ATH flame retardant performance are shown in the Table 5.

The efficiency of ATH flame retardant improved when Zeolite was added. When the addition of Zeolite was 3%, although the FP temperature remained nearly unchanged, the LOI increased by 1.4%, and flue gas got a noticeable improvement that only a small amount of smoke was generated, so the effect of restraining smoke are clearly better than that of ATH or MH. By increasing the zeolite addition, the LOI increased correspondingly, and the content of flue gas also became slightly lower.

On the basis of these studies, flame-retarding asphalt system based on ATH is designed by using ATH, MH and Zeolite (shown in Table 6). When adding 20% ATH, 5% MH and 3% Zeolite into asphalt, the LOI of flame-retarding asphalt system is 29.2%, reaching the fireproof grade 2 technical requirements accroding to the definition in Japan JISK 7201, the FP temperature was 419°C, and flue gas is little. It is completely appropriate to be used as the inorganic flame retardant composition of flame retardant asphalt material, which can be used in the tunnel.

Table 5. The Effect of Zeolite for ATH Modified Flame-retarding Asphalt

ATH Addition Rate /%	Zeolite Addition Rate /%	LOI /%	FP /°C	Smoke Conditions
20	0	25.0	343	Black Smoke
	3	26.4	345	A little Smoke
	6	27.6	344	A little Smoke
	9	28.2	346	A little Smoke

Table 6. Compounding of ATH Modified Flame-retarding Asphalt Sysytem

ATH Addition Rate /%	MH Addition Rate /%	Zeolite Addition Rate /%	LOI /%	FP /%	Smoke Conditions
20	5	3	29.2	419	A little Smog

4 Discussions

1 the flame-retardant mechanism of ATH

The main flame retardant mechanism of ATH is that ATH decomposed and unwatered at the temperature range of 230°C to 350°C, and absorbed a large amount of heat. The differential thermal analysis curve (Figure 3a) indicates the following analysis: Firstly, ATH had endothermic peaks at 250°C and 315°C, which can absorb a large amount of calories, and restrain the temperature rise of organic polymer. Furthermore, large amounts of water vapor released from the ATH heating can dilute the concentration of flammable gases and prevent burning. Secondly, the weight losses of ATH amounted to 22.97% at the temperature range of 270°C to 350°C, that means, there was masses of water vapor generated and volatilized. Thirdly, ATH dehydration generated Al_2O_3 protective film with stable chemical property, which can isolate the air to prevent burning. Therefore, the flame-retardant mechanism of ATH can be summarized as the effect of dehydrated decalescence, water vapor's dilution and the isolation of Al_2O_3 protective film.

In Figure 3a, the decomposition peak of ATH is at 315°C, lower than the FP temperature of asphalt. In the process of FP testing, while the asphalt temperature was lower than 340°C, ATH decomposed and released so much water that it generated bubble expansion, which prevented asphalt from combusting and improved FP Temperature. But while the asphalt temperature increased to above 340°C, ATH decomposition would complete, and the FP temperature was unable to improve with the addition increase.

2 MH and ATH Synergies

MH and ATH have the same flame retardant mechanism, but the decomposition temperature of MH is higher, as shown in Figure 3b. The decomposition temperature of

MH ranged from 340°C to 430°C, and its mass loss was more than 29%, slightly better than that of ATH. MH decomposing temperature was higher than 340°C, whereas the FP of asphalt was only 324°C, and the asphalt was already in a combustible state before MH thermal decomposition. So, the only addition of MH into asphalt can dramatically enhance its LOI, but there is a hidden danger of FP temperature. When both MH and ATH were added into asphalt, the decomposition temperature range of flame retardant material expanded, which made up for the later flame retardant performance of ATH, and also resolved this incipient fault of FP temperature mentioned above. From the Figure 3c, in the mixture of MH and ATH, MH immediately started decomposing and absorbing heat after the ATH endothermic decomposition, and the decomposition temperature of the flame retardant ranged from 230°C to 430°C. Therefore the efficiency of fire-retardant had been greatly improved that LOI increased by 2.4%, and the FP temperature raised to 415°C.

3 Zeolite roles in acceleration

From comparative analysis of Figure 3c and Figure 3d, the addition of zeolite has not apparently accelerated the heat decomposition of system. The role of flame-retardant acceleration and smoke suppression of Zeolite owe to its unique microstructure. Zeolite is a kind of hydrous tectosilicate mineral with three-dimensional structure of silicon (aluminum) oxygen skeleton, its basic unit is silicon-oxy tetrahedron (SiO_4) which has silicon as the center and four oxygen ions rank around, as shown in Figure 4. Silicon-oxy tetrahedron connect with aluminum-oxy tetrahedron via angular point, so they shaped the three-dimensional structure of silicon (aluminum) oxygen skeleton, formed a large number of multi-cavities and channels that zeolite powder looked like sponge in the micro-structure (shown in Figure 5). These structure feature permit different particles get through selectively: the water vapor generated by ATH and MH heat decomposition can volatilize through this zeolite layer, while the smoke generated by asphalt combustion and outer oxygen can not pass the zeolite layer. Because the cation imbalance of aluminum-oxy tetrahedron result in the holistic negative charge, this structure has the ability of electrostatic adherence, so the smoke produced by asphalt combustion can be adsorbed and form a covering layer when it passes through the zeolite layer. Therefore, the addition of Zeolite can effectively ameliorate the performance of flame retardant system, and smoke suppression effect has also been significant improved.

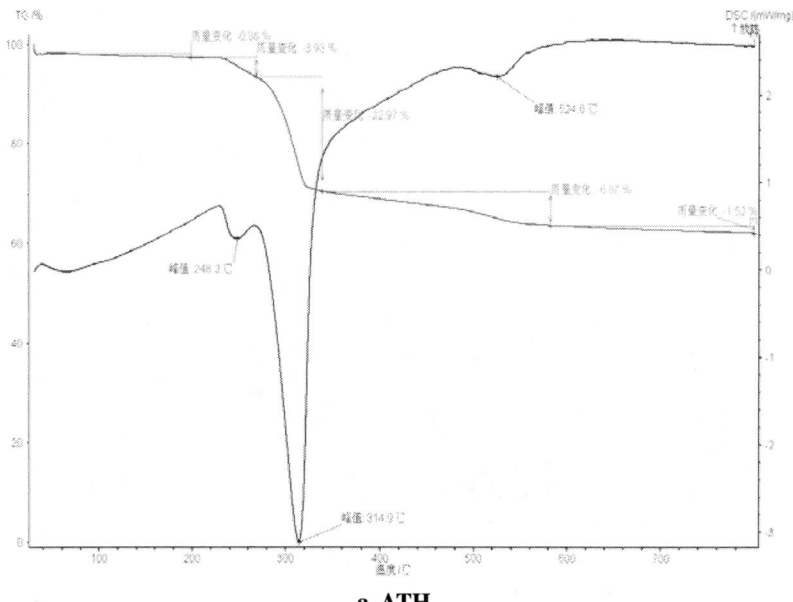

a. ATH

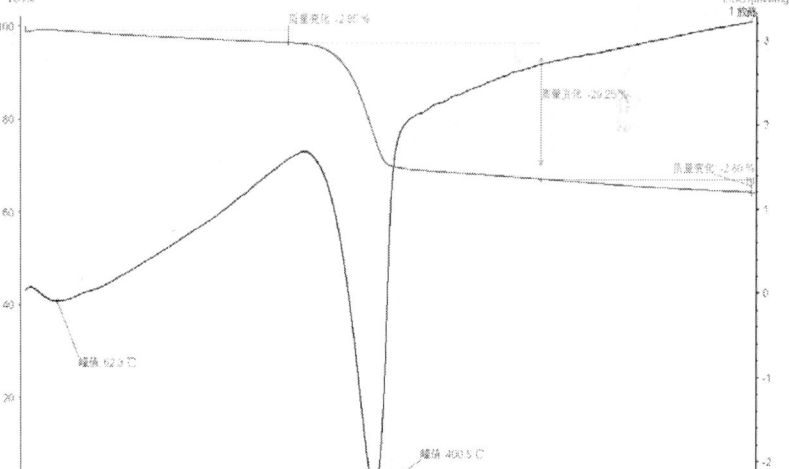

b. MH

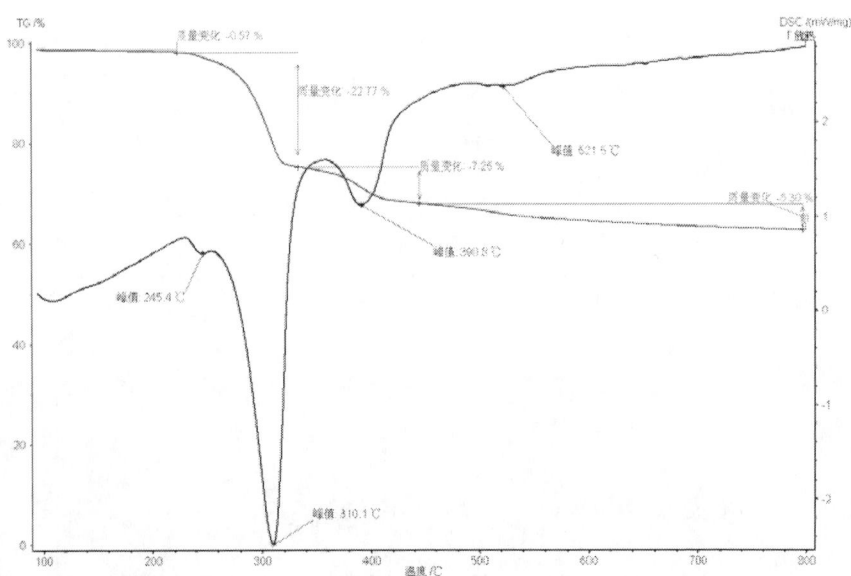

c. ATH&MH Mix

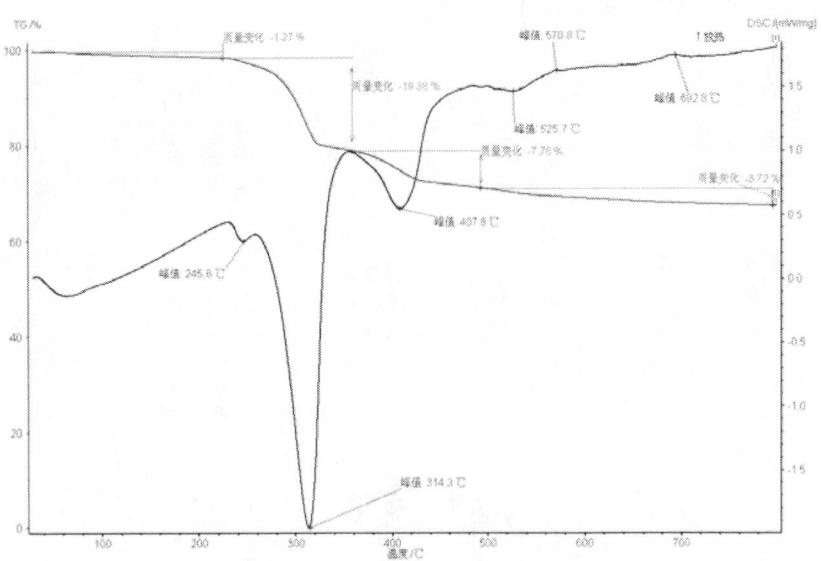

d. ATH, MH and Zeolite Mix
Figure 3. The Analysis of DSC-TG

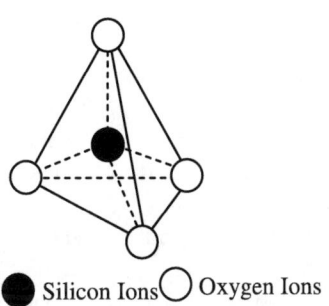 

Figure 4. Construction Unit of Zeolite   Figure 5. SEM Analysis of Zeolite

5 Conclusions

Based on this study, following conclusions were made:

1 The addition of ATH can effectively improve the property of fire-retardant. When 30% ATH whose fineness is 1250 mesh was added into asphalt, the LOI reach 29.6%, and the FP temperature increased to 350°C, the flame-retardant performance is excellent. But when 20% ATH with a fineness of 325 mesh was added, the LOI is only 25%. Thus a large quantity of fine powder must be added if ATH is used alone;

2 The FP temperature of asphalt is lower than MH decomposition temperature, so it can't be elevated by the addition of MH, and there would be certain hidden danger if MH was used alone. As a kind of synergetic flame retardant of ATH, MH can benefit ATH well in flame retardant property. When 5% MH was added in asphalt which contains 20% ATH, the LOI could increase to 27.6 percent, the FP temperature improve to 415°C, and the flame-retarding performance of ATH-asphalt system is greatly enhanced.

3 Zeolite formed molecular sieves on the surface of burning things, which can selectively be passed by water vapour and impede the spread of oxygen and smoke. It can significantly accelerate the flame-retarding property of ATH-asphalt system and improve flue gas situation. With the addition of ATH, MH and Zeolite into asphalt, the LOI reach above 29 percent, the FP temperature is close to 420°C, and the asphalt system generate little smoke and toxicity, so it is suitable for using as an indispensable component of flame retardant asphalt.

Acknowledgement

This work is supported by National High Technology Research and Development Program of China(2006AA11Z117).

Reference:
1 Feng naiqian. "Application of Natural Zeolite Concrete."M. Bingjing: China Architecture&Building Press, 1996.
2 FAN Jun, YANG Qun. "Test Method Investigation of Asphalt Oxygen-index." J. Journal of PLA University of Science and Technology(Natural Science, 2004(6):30~32.
3 LI Zu wei,CHEN Hui qiang,MU Jian bo, CHEN Shi zhou. "Study on Technology to Improve Flameresistant SBS Modified Bitumen and Its Mechanism ."J. Journal of Changsha Communications University, 2002(4):44~47.
4 Luo Xiaofeng,Yu Jianying,Wu Shaopeng ,Cong Peiliang. "Preparation and Properties of Flame-retardant Asphalts. "J. Petroleum Asphalt, 2005(4): 11~13.
5 GB 9343-1988. "Test method for flammability of plastics Flash-ignition temperature and self-ignition temperature test."S.
6 GB 10707-1989. "Rubber-Determination of flammability by oxygen index ."S.
7 GUO Jin-cun, LIAO Ke-jian, DAI Yue-ling. "Development of Asphalt for Flame Retardancy." J. Journal of Liaoning University of Petroleum & Chemical Technology, 2005(2):5~8.
8 JISK 7201-1997."Japanese Industrial Standards. "S.
9 Wang Yunbo ,Tan Wanchun. "Study on the Structure Character of Zeolite and its Application in Water Treatment." J. Water Purification Technology, 2007(2): 21~24.
10 Yasuo I. "Covered Highway Structure with Means for Easy and Quick Access to Tunnel Interior." U.S Patent, 752 632. 2004.

On the Mechanical Modeling of Asphalt Matrix and Hot Mix Asphalt Mixtures

Mohammad J. Khattak[1] Zhanping You[2] and Vikram Kyatham[3]

[1]Mohammad J. Khattak, Associate Professor, Department of Civil Engineering, University of Louisiana at Lafayette, Lafayette, LA 70504-2291. Phone #: (337) 482-5356, Fax #: (337) 482-6688, Email: mxk0940@louisiana.edu
[2]Assistant Professor and Director of Transportation Materials Research Center (Center of Excellence for Transportation Materials), Department of Civil and Environmental Engineering, Michigan Technological University, Houghton, Michigan, 49931-1295, Fax: 906-487-1620, Office phone: 906-487-1059, Email: zyou@mtu.edu
[3]Graduate Research Assistant, Department of Civil Engineering, University of Louisiana at Lafayette, Lafayette, LA 70504-2291. Phone #: (337) 482-5356, Fax #: (337) 482-6688, Email: kyatham_vikram@hotmail.com

ABSTRACT: This paper discusses the mechanical modeling of the asphalt matrix (AM) and hot mix asphalt (HMA) mixtures for two types of aggregates and one asphalt grade. The AM gradation was extracted from the HMA gradation passing No. 8-sieve. Hydrated lime was used as a binder additive to investigate the improvement in mixture properties. A dynamic shear rheometer (DSR) was used to conduct frequency sweep and creep tests at various temperatures on AM specimens. An indirect tensile load test was conducted to determine the resilient modulus and the creep compliance of HMA mixtures. The master curves for complex shear moduli (G^*) and creep compliance of AM were constructed using the data obtained at different temperatures from a DSR test. The G^*-master curve data indicated improvements due to the hydrated lime addition for a range of frequencies, and were a function of aggregate type. The results showed that the generalized mechanical model (i.e., a combination of a Maxwel model and Kelvein models) can effectively characterize the creep response of both the AM and HMA mixtures. The study also focused on the application of a clustered distinct element two- dimensional (2-D) modeling (DEM) to predict the uniaxial compression modulus using a synthetic, heterogeneous microstructure reconstructed from scanned images of actual test specimens. The HMA 2-D microstructure was obtained by scanning HMA test specimens using a high-resolution scanner to obtain grayscale images. Bulk properties of the AM and aggregate structure were assigned and virtual compressive test simulations were conducted. The results indicated that the DEM simulation produced repeatable HMA moduli for a range of AM moduli, and can model the aggregate interactions, which represent some effects of the aggregate interlock in asphalt mixture. Furthermore, the micromechanical model using the DEM approach provides a reasonable physical portrayal of the force chains developed in the aggregate skeleton, which are known to be a critical aspect of HMA micromechanical modeling.

Key Words: Mechanical modeling, Creep compliance, Asphalt matrix, Distinct element modeling, complex shear modulus.

INTRODUCTION AND BACKGROUND

Hot mix asphalt (HMA) is a particulate composite material consisting of three phases: air voids, coarse aggregate and asphalt matrix (AM), which is a combination of asphalt binder (AC) and fine aggregate passing No. 4 sieve (Khattak et al. 2008). The three phases can also be defined as a combination of aggregates, air voids, and mastic, which is a combination of AC and mineral filler (Anderson et al. 2000). The coarse aggregate phase is stiffer than the AM phase and is basically elastic in nature. The AM phase makes the HMA a visco-elastic material due to the time dependent strain response of the binder. Efforts have been directed towards relating the mastic behavior to pavement performance. Two approaches (mirco-mechanical based and rheology based models) have been used in the literature for modeling the dynamic mechanical behavior of asphalt mastic.

Little and Kim (2004) studied the suitability of different micromechanical and rheology based models to characterize the dynamic behavior of asphalt mastics. Two distinct, compositionally different asphalt binder types (AAD-1 and AAM-1) and two mineral fillers (limestone and hydrated lime), were used in this study. The elastic-viscoelastic correspondence principle was used to modify elastic micromechanical models to viscoelastic models to account for the time and temperature dependence of mastic. In the case of the rheology based models the Nielsen model was selected. It was found that micromechanical models showed good agreement with testing data at low particle volume concentrations. The rheological model was successful in predicting the stiffening effect of limestone filler when added up to 25% by volume. However, the behavior of mastics modified with hydrated lime (HL) was suggested to be controlled by physiochemical interaction, and thus highly binder specific.

Buttlar et al. (1999) used a generalized self consistent scheme model to characterize the dynamic mechanical behavior of the mastic. Three aggregate related reinforcement mechanisms (volume filling reinforcement mechanism, physiochemical mechanism, and particle interaction mechanism) resulting in the increased stiffness of asphalt materials were discussed. It was suggested that volume filling reinforcement was evident in mastics, and the particle interaction mechanism was more pronounced in asphalt mixtures, where aggregates form a stone skeleton through which stresses are transferred.

The microfabric discrete element modeling MDEM approach was used by You and Buttlar (2004 and 2006) to predict the HMA complex modulus across a range of test temperatures and load frequencies. The MDEM approach developed in the two papers allowed many different constitutive models to be employed to describe particle and interface properties, such as normal and shear stiffness and strength. In the study of MDEM, a program called 'Particle Flow in Two Dimensions,' or PFC2D was employed. The MDEM was able to model and simulate the HMA The model provided better modulus estimates across a range of test temperatures and load frequencies as described by You and Buttlar (2004 and 2006).

Chang et al (1999) discussed the mechanical behavior and internal structure of HMA due to changes in temperature. A micromechanical model, ASBAL, was used to simulate the stress-strain behavior and variation of internal structure of HMA. The simulation showed the ability to predict the temperature susceptibility of HMA assembly. Therefore, they concluded that this micromechanical model can be used as a tool to understand the mechanism of high temperature rutting.

RESEARCH OBJECTIVES

The objectives of this study are to evaluate:
1. The effectiveness of the Generalized Creep Compliance model in characterizing the creep response of AM and HMA mixtures.
2. The application of a clustered distinct element modeling to predict the uniaxial compression modulus using a synthetic, heterogeneous microstructure reconstructed from scanned images of actual test specimens.

TESTING PROGRAM

Test Materials and Sample Preparation

Both HMA and AM mixtures were made using one PG64-22 AC and two aggregates types: limestone (LS) and gravel (GR), obtained from the local contractor. The AC was modified with 20 percent of Hydrated Lime (HL) (Kanitpong et al. 2007). The HL was used as a binder modifier and therefore, it was not accounted for as aggregate filler.

Superpave® design procedure was used to make HMA mixtures. The nominal maximum aggregate size of 25 mm was used with an optimum AC content of 4.4%. The cylindrical HMA samples of 150 mm diameter were compacted by the Superpave® Gyratory Compactor, and then sliced using the diamond saw to obtain 50.8 mm thick specimens. The average AV for the test specimens were $7\pm1.0\%$. Table 1 shows the aggregate gradations for both the AM and HMA mixtures.

The AM mixtures were made using 10% AC and fine aggregate passing No. 8 sieve. The aggregate gradation for AM was extracted from HMA mixture gradation (Table 1). The sample dimensions were 15.86 mm in diameter, and 15.86 mm in height. A detailed description of AM sample preparation and molding can be found elsewhere (Khattak, et al. 2008).

Table 1. Aggregate Gradation Data.

Sieve Sizes	US	1.5"	1"	¾"	½"	3/8"	No. 4	No. 8	No. 16	No. 30	No. 50	No. 100	No. 200
	Metric	37.5	25	19	12.5	9.5	4.75	2.36	1.18	0.6	0.3	0.15	0.075
% Passing	HMA	100	96.6	89.1	67.9	56.8	32.5	21.2	15.6	11.7	7.5	5.2	4.2
	AM	-	-	-	-	-	-	100	74	55	35	25	20

Asphalt Matrix (AM) Testing

AM-Frequency Sweep Test

The Bohlin's Dynamic Shear Rheometer (DSR) was used to determine the complex shear modulus (G^*) of AM at various temperatures. The frequency sweep test was conducted at frequency levels of 1 Hz through 60 Hz at log-increments at temperatures of

15, 20, 25, 34, 46, 58 and 64°C. The test was conducted under the controlled stress mode. The applied shear stress at each temperature was selected such that it was within the linear visco-elastic range. The specimen was conditioned by applying 100 cycles at 10 Hz at one-half the stress level that was used in the test (Khattak, et al. 2008).

AM-Creep Compliance [J(t)] Test

The $J(t)$ of the AM under shear load is defined as:

$$J(t) = \frac{\gamma(t)}{\tau} \tag{1}$$

Where, τ= applied shear stress, and $\gamma(t)$ = shear strain at time t.

The τ was applied for 2 min, then followed by an unload time of 2 min. The τ at each temperature was selected such that it was within the linear visco-elastic range. Angular rotation of the specimen was measured, and $\gamma(t)$ was calculated. The $J(t)$ was determined by taking the inverse of shear modulus. The $J(t)$ was obtained at temperatures of 15, 20, 25, 34, 46, 58 and 64°C. It should be noted that the same specimen as that used for the frequency sweep test was used for the creep test. The creep test was conducted after frequency sweep test. A rest period of 5 to10 minutes was used to allow for specimen recovery for each test temperature (Khattak, et al. 2008).

Hot Mix Asphalt (HMA) Testing

HMA-Creep Compliance [D(t)] Test

The D(t) of the HMA was determined in the indirect tension loading mode at a temperature of 25°C. The specimen was first conditioned under a cyclic load of 445 N for 100 cycles followed by a zero load period of 4 minutes. The creep load was then applied on the specimen for a duration of 2 minutes followed by an unloading period of 4 minutes (Khattak, et al. 2008).

The deformations in the horizontal and the vertical directions were measured at the middle 25.4 mm of the specimen, using linear variable differential transducers. The creep load applied on the specimen was such that it produced a horizontal strain between 38 and 89μ for the duration of the loading. These limits prevent both non-linear response, characterized by exceeding the upper limit, and problems associated with noise and drift inherent in sensors when violating the lower strain limit. The D(t) was calculated using the equation derived by Kim et al.(2004).

$$D(t) = \frac{\pi.a.d}{2.P} [\frac{\gamma_2.D_V - \beta_2.D_H}{\beta_1.\gamma_2 - \beta_2.\gamma_1}] \tag{2}$$

Where $D(t)$= creep compliance at time t (1/Pa); P= applied load (N); a= loading strip width, (m); d= thickness of specimen, (m); D_V= vertical deformation (m); D_H= horizontal deformation, (m); and β_1, β_2, γ_1, and γ_2 are coefficients that are equal to -0.009731, -0.003153, 0.003043, and 0.009376, respectively. These values are based on the loading strip width of 19 mm, specimen diameter of 150 mm, and gauge length of 25.4 mm.

HMA-Resilient Modulus (M_R) Test

The resilient modulus (M_R) test was conducted in indirect tension loading mode at 25°C. A cyclic load of 1,000 N, with a sustained load of 111 N, was applied to the specimen. The deformations were measured in the vertical and horizontal directions in the middle 25.4 mm of the specimen. The test frequency was one cycle per second with a load and unload times of 0.1 second and a rest period of 0.9 second. The equation to determine the dynamic modulus in indirect tension loading mode derived by Kim et al. (2004) was modified to determine the M_R of the specimens. The M_R was calculated using the following equation:

$$M_R = \frac{2.P}{\pi.a.d}\left[\frac{\beta_1.\gamma_2 - \beta_2.\gamma_1}{\gamma_2.D_{VE} - \beta_2.D_{HE}}\right] \tag{3}$$

Where D_{VE} = vertical elastic deformation (m); D_{HE} = horizontal elastic deformation, (m); and rest of the parameters are previously defined.

RESULTS AND ANALYSIS

Asphalt Matrix (AM)

AM-Frequency Sweep Master Curve Analysis

The master curve was constructed using the principle of time-temperature superposition; that is, the response of the mixture at low temperatures is the same as that under short or fast loading rates, and the response of the mixture at high temperatures is the same as that under long or slow loading rates. The G* master curve was constructed at a reference of 25°C as shown in Figure 1. It was found that the G*master curve of AM mixtures can be modeled using the Sigmoidal function:

$$\log|G^*| = a + \frac{b}{1 + \dfrac{1}{\exp^{d+e(\log f_R)}}} \tag{4}$$

Where G^* = complex shear modulus; f_R = frequency of loading at reference temperature; a and b = fitting parameters for a given set of data; a represents the minimum value of G^* and $a+b$ represents the maximum value of G^*; and d and e = parameters describing the shape of the function.

Fig. 1 shows the comparison of G*-master curve at 25°C for neat and HL modified AM mixtures made with LS and GR aggregates. The master curve indicates that the GR AM specimens exhibited significantly higher G* values under all loading frequencies as compared to LS AM specimens.

At lower frequencies (0.1Hz and lower) the effect of HL is very evident as the average G* value for the LS-HL AM is about 1.5 times greater than the LS-neat AM. However, in the case of GR AM specimens, the average G* values for HL AM were 20% higher than the neat-AM. For medium frequency range (between 0.1Hz and 10Hz), the improvement in the average G* values for LS specimens due to HL modification is evident from the fact that the average G* value for the HL-modified AM is 1.13 times higher than the neat AM. For GR AM specimens though no improvement was observed due to the HL modification. Throughout it can be observed (Fig. 1) that the G* values of the GR AM specimens are higher than LS specimens over the complete range of frequencies.

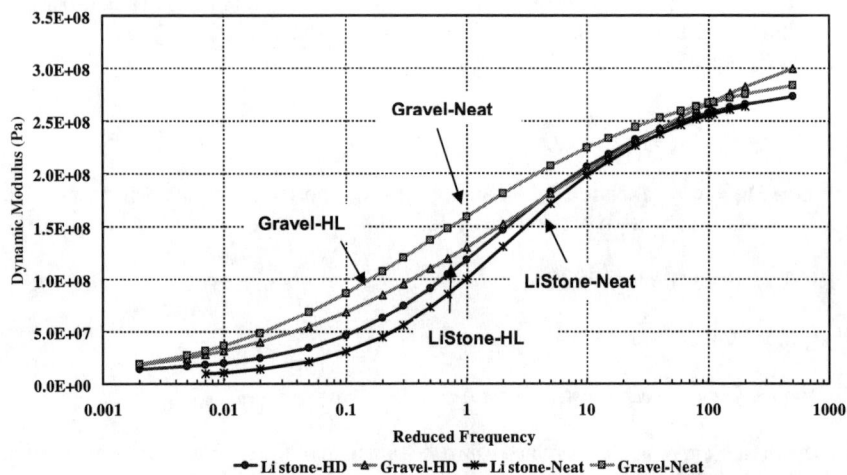

FIG 1. G*-master curves for AM with limestone and gravel aggregate at 25°C.

AM-Creep Compliance Master Curve Analysis

The $J(t)$ master curve for AM mixtures was generated by shifting the data for all the temperatures at a reference temperature of 25°C. Similar to the G* master curve the $J(t)$ master curves were also developed using the time-temperature equivalence principle. The high temperature data shifted the master curve further with respect to time, which can be used for predicting the creep modulus values for longer durations.

Fig. 2 shows the $J(t)$ master curves of the LS and GR, HL modified and neat AM specimens. It can be observed that the trend followed by the $J(t)$ curves is similar to the G* master curves. The improved effect of HL modification is evident in the LS AM specimens as indicated by the lower $J(t)$ values for LS-HL specimens.

Hot Mix Asphalt (HMA)

HMA-Resilient Modulus (M_R)

The M_R and the total modulus (M_T), which includes a combination of resilient and visco-elastic strain, was considered in the analysis to determine the effect of HL modification on the visco-elastic behavior of HMA.

The data in Fig. 3 indicates that the moduli increased by 4% with HL modification for the LS-HMA mixtures. However, no improvement was observed in the GR-HMA mixtures. The M_R of LS-neat is almost 20% higher when compared to GR-neat mixtures. Similarly, the M_R of LS-HL is 36% greater than GR-HL mixtures. This indicates that LS-HMA mixtures are performing better than GR-HMA mixtures. Similar results were observed for Total Modulus (M_T).

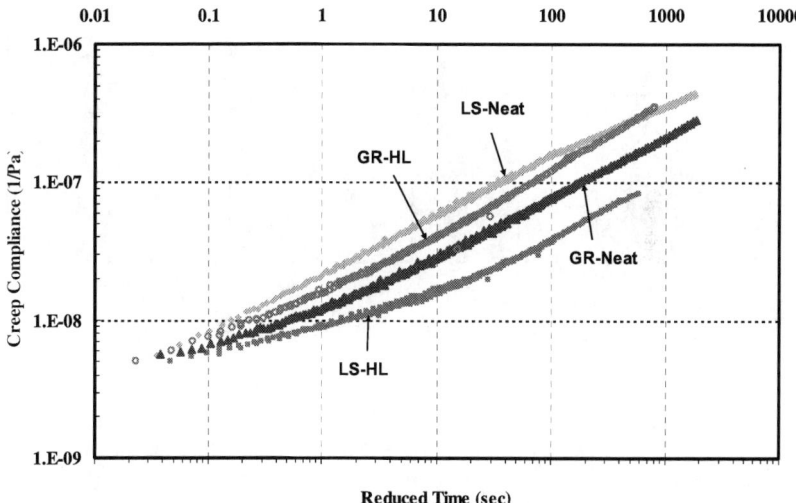

FIG. 2. J[t] Master Curves for the AM with Limestone and Gravel at 25°C.

Mechanical Modeling

AM- Creep Compliance [J(t)] Mechanistic Models

The time dependent stress-strain behavior of visco-elastic AM can be characterized using the generalized model. The generalized model is the combination of the Maxwell and Kelvin model. The generalized model was fitted to the *J(t)* master curve data for the AM using "the method of successive residuals". Under a constant stress, the creep compliance for the generalized model is given as:

$$J(t) = \frac{1}{Go}(1+\frac{t}{To}) + \sum_{i=1}^{n}\frac{1}{G_i}[1-\exp(\frac{-t}{T_i})] \qquad (5)$$

The Maxwell model is commonly described using its relaxation time (T_o), which is defined as "the time required for the stress to reduce to 36.8% of the original value". The Kelvin model is commonly described using the retardation time (T_1), which is defined as "the time required for the strain to reach 63.2% of the retardation time". The AM mixtures exhibit both stress relaxation and strain retardation characteristics. Therefore, the stress-strain behavior of AM under a constant creep load can be characterized using a generalized model, which is a combination of both Maxwell and Kelvin models.

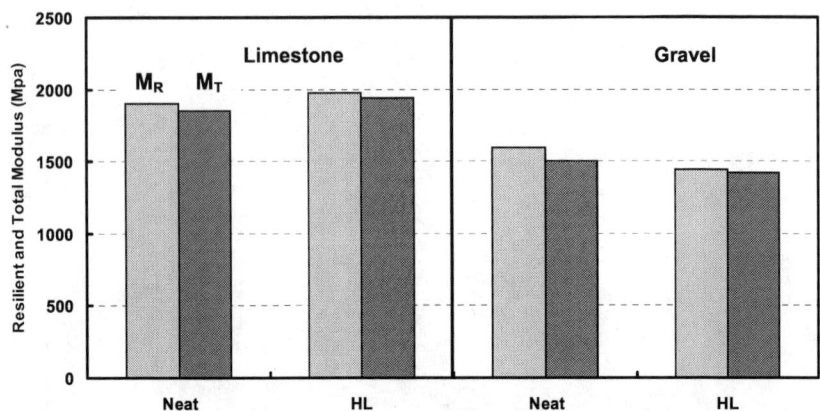

FIG. 3. Resilient and total modulus of HMA mixtures.

A graphical description of these mechanical models is shown in Fig 4. This model explains the effect of load duration on pavement responses. Under a single load application, the instantaneous and the retarded elastic strains predominate, and the viscous strain is negligible. However, under a large number of load repetitions, the accumulation of viscous strains is the cause of permanent deformation.

The generalized model was used to express the visco-elastic behavior of the AM mixtures. Fig 5 shows a typical measured $J(t)$ master curve and predicted values using the generalized model of AM at 25°C. In general, all the AM required one Maxwell component and three Kelvin components. The model parameters for all the AM are listed in Table 2. The G_o for the LS-Neat AM is lower than the corresponding value for the LS-HL AM mixtures. This indicates that the LS-neat AM requires more time for the stress to relax and the response is more viscous in nature or indicates improvements in the elastic properties due to the addition of HL. However, no improvement in the elastic properties was observed in the case of GR AM mixtures. A comparison of mechanistic model parameters between LS and GR AM mixtures indicates that LS AM mixtures exhibit lower creep compliance, and hence, improved elastic response of the mixtures.

HMA-Creep Compliance (D[t])Mechanistic Model

A generalized model was fitted to the HMA $D[t]$ data. The model components for all types of mixtures are shown in Table 2. In general, one Maxwell and one Kelvin model were sufficient to characterize the creep response of HMA mixtures. It was found that the E_o and T_o increased with the addition of HL, suggesting higher resistance to creep damage. However, no change was observed in the case of the Kelvin model parameters, as a result of HL modification.

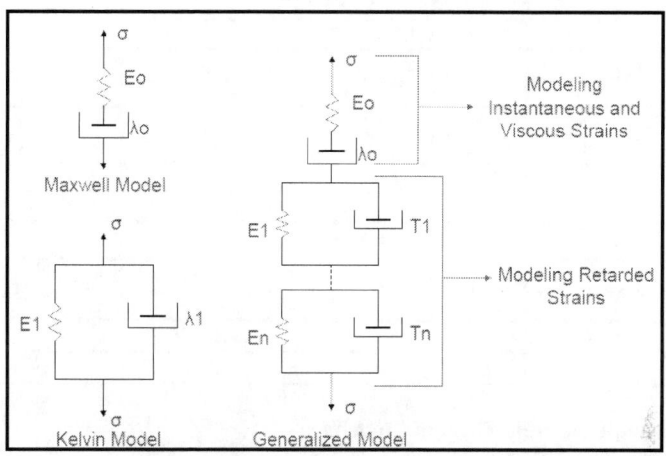

FIG. 4. Mechanical models for visco-elastic materials.

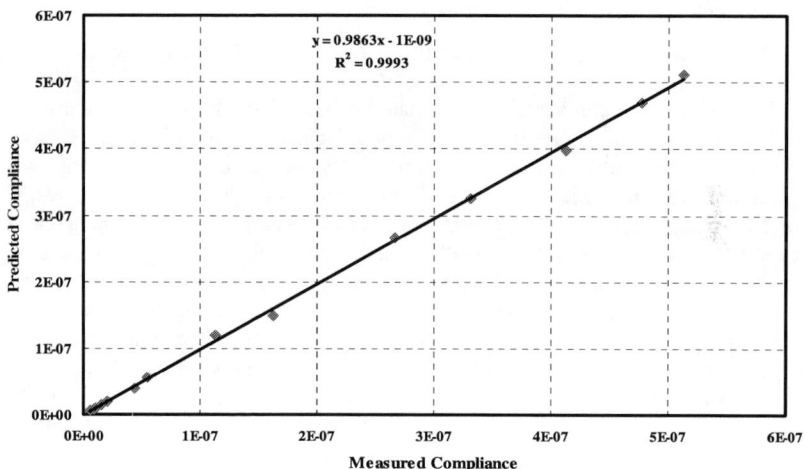

FIG. 5. Predicted Vs measured compliance of AM mixtures.

Table 2. Generalized Model Parameters for AM and HMA mixtures.

Asphalt Matrix									
		Maxwell Model				Kelvin Model			
Aggregate Type	Sample Type	G_0(Mpa)	T_0(sec)	G_1	T_1	G_2	T_2	G_3	T_3
Gravel	Neat	195	67.75	9.58	258.53	45.7	8.39	225	0.16
	HL	164	22.39	10.1	98.14	50.3	4.64	182	0.12
Limestone	Neat	179	82.56	6.06	291.21	13.2	20.24	77.9	0.72
	HL	205	93.5	26.1	159.01	118	3.87	365	0.09
HMA									
Aggregate Type	Sample Type	E_0(Mpa)	T_0(sec)	E_1	T_1	E_2	T_2	E_3	T_3
Limestone	Neat	1242.12	15.15	344.5	20.33	-	-	-	-
	HL	1418.8	23.26	344.5	20.49	-	-	-	-

Micromechanic Discrete Element Modeling

A clustered DEM modeling approach was applied to predict the uniaxial compression modulus of a synthetic, heterogeneous microstructure reconstructed from scanned images of actual test specimens. The current modeling approach was limited to two dimensions (2-D).

DEM Prediction of Compressive Moduli Based Upon the 2D Microstructure

In the distinct element approach for material modeling, the complex constitutive behavior of a material can be simulated using a detailed morphological reconstruction of the material microstructure, along with relatively simple constitutive descriptions of material stiffness and contact behavior. The commercial code called 'Particle Flow in Two Dimensions,' or PFC^{2D} is employed. In this study, the asphalt concrete 2-D microstructure was obtained by optically scanning smoothly sawn asphalt concrete test specimens using a high-resolution scanner to obtain grayscale images from the sections. Fig 6 shows one of the slices of the GR mixture using an optical scanner. Several image processing techniques were then applied to process and analyze the images. For the 2-D image processing in this study, the average of the polygon diameter was chosen as a threshold to determine which aggregates would be retained on a given sieve. The polygons used then were filtered as coarse aggregate. Then, the outlines of the aggregates were converted into many sided polygons. The resulting polygons were mapped onto a sheet of uniformly sized disks, then the bulk material and interface properties of the aggregates and mastic were assigned in the PFC^{2D} DEM models. The procedure used in this study is the same as used by Buttlar and You (2001), and You and Buttlar (2004 and 2006)

DEM Simulation

The current DEM virtual test was limited to 2-D analysis techniques and involved the simulation of uniaxial cylinder specimens. The simulated specimens contained up to

100,000 disk-shaped particles, depending on the size of the specimens. Fig 7a shows the micromechanical distinct models for the mastic, where the elements for the microstructure of aggregate are not shown in the figure. Fig 7b shows the enlarged part of the mastic. After a certain of uniaxial compressive loading (continuous incremental load or sinusoidal load), the response of each aggregate and each piece of the mastic can be monitored. Fig 7c demonstrates the compressive and tensile contact force chains of the enlargement of Fig 7b. Even in a compressive test, due to the high heterogeneity of the material, local tensile force (stress) may occur. The DEM simulation can model the aggregate interactions, which represent some effects of the aggregate interlock in asphalt mixture.

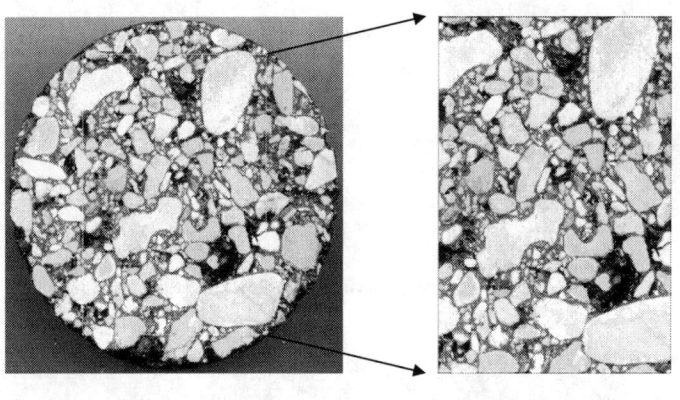

(a). Raw Image is converted to black and white color picture

(b). Change into rectangular size of image with a height of 126.67 mm and a width of 86.67 mm

FIG. 6. Scanned HMA image for DEM.

Uniaxial Compression Test Prediction for Different Cross-sections of Specimens

Four DEM models from four images were analyzed to evaluate the repeatability of the model, and also to evaluate the individual model's accuracy in predictions. Simple statistical analysis was employed to compare the prediction to the measurements. Fig 8 also shows the DEM prediction results. It seems that the four images provided similar results for a range of AM stiffness. The M_R and M_T values obtained from indirect tensile test of HMA mixtures were compared with the simulation results and it was found that the measured values were slightly lower than the predicted values. One problem of the simulation was that at really low asphalt mastic modulus level, the simulation was difficult to be completed due to the instability problem of the microstructure.

Further understanding of the PCF 2D program and constitutive models for the HMA is currently being studied at UL Lafayette with the aid of Michigan Technological University. In addition, the G* measurement of the AM will be measured at wider temperature range (in particular lower temperatures) in order to fully assess the applicability of the DEM prediction.

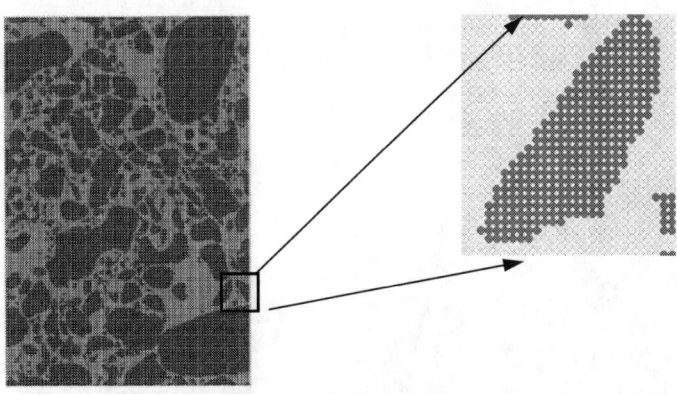

(a). The discrete element for the mixture.

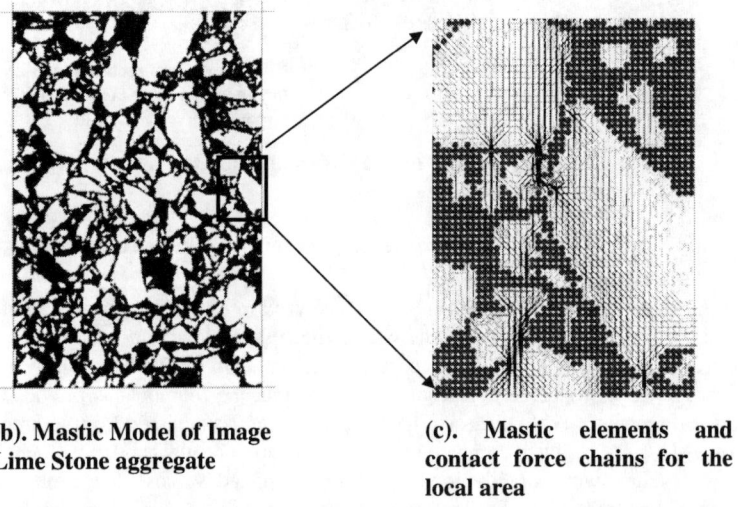

(b). Mastic Model of Image Lime Stone aggregate

(c). Mastic elements and contact force chains for the local area

FIG. 7. Discrete element modeling and contact for chain for each discrete element.

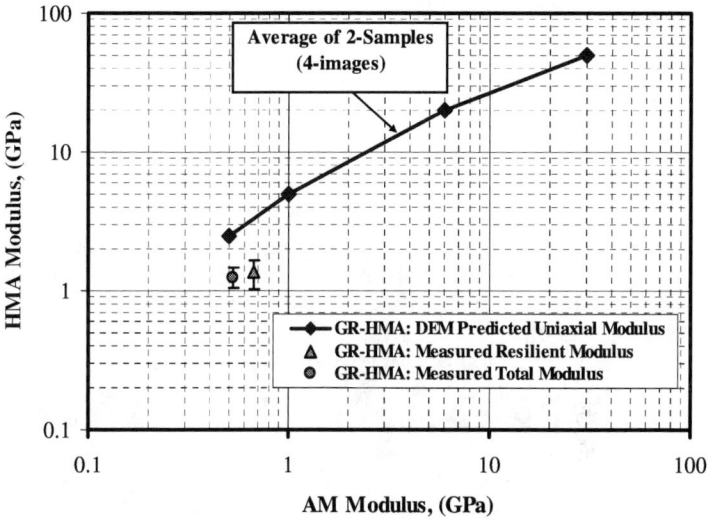

FIG. 8. Simulation results for GR-neat HMA mixtures.

CONCLUSIONS

Based on the laboratory test results for the AM and HMA mixtures, the following conclusions were made:
- The G^*-master curve of the AM mixtures can be characterized by using the sigmoidal curve, and hence, can identify the effectiveness of HL modification for a range of frequencies.
- The HL modification is effective in improving the mechanistic characteristics of the AM and HMA mixtures made from limestone aggregate.
- The generalized model can be used to effectively characterize the creep behavior of AM and HMA mixtures made with limestone and gravel aggregates.
- The micromechanical model using DEM approach provides a reasonable physical portrayal of the force chains developed in the aggregate skeleton, which is known to be a critical aspect of HMA micromechanical modeling.
- The predicted HMA modulus using the discrete element approach exhibited reasonable trends for a range of the AM moduli. The resilient and total moduli of HMA were slightly lower than the predicted moduli indicating that a calibration factor is required for a better prediction.

ACKNOWLEDGMENTS

The authors wish to express their sincere thanks to the University of Louisiana at Lafayette and the Louisiana Board of Regents for their financial support. A special thank you is extended to Mark Leblanc for his assistance in the laboratory experimentation.

REFERENCES

Anderson D.A., Marasteanu M.O (2000). "Establishing Linear Viscoelastic Conditions for Asphalt Binders." *Transportation Research Record 1728*, Transportation Research Board, Washington, D.C.

Buttlar W.G., Bozkurt D., Al-Khateeb G., and Waldhoff A.S. (1999). "Understanding Asphalt Mastic Behavior through Micromechanics." *Transportation Research Record 1681*, Transportation Research Board, Washington, D.C.

Buttlar, W.G., and You, Z. (2001). "Discrete Element Modeling of Asphalt Concrete: A Micro-Fabric Approach." Journal of the Transportation Research Board, National Research Council, the National Academies, Washington, D.C., No. 1757, pp. 111-118.

Chang G.K., Meegoda J.N. (1999). " Micro-mechanical Model for Temperature Effects of Hot Mix Asphalt Concrete." *79th Transportation Research Board Annual Meeting* (CD-ROM), Transportation Research Board, Washington, D.C.

Kantipong K., Atud J., Tebid M. W. (2007). "Laboratory evaluation of hydrated lime application process in asphalt mixture for moisture damage and rutting resistance." *87th Transportation Research Board Annual Meeting* (CD-ROM) Transportation Research Board, Washington, D.C.

Khattak M.J., Kyatham V. (2008). " Visco-elastic Behavior of Hydrated-Lime Modified Asphalt Matrix & HMA under Moisture Damage Condition." *88th Transportation Research Board Annual Meeting* (CD-ROM), Transportation Research Board, Washington D.C.

Kim Y. R, Y. Seo, and M. King. (2004). Dynamic Modulus testing of asphalt concrete in indirect tension mode. *84th Transportation Research Board Annual Meeting (CD-ROM),* Transportation Research Board, Washington D.C.

Little N. D., Kim Y.R. (2004). "Linear Viscoelastic Analysis of Asphalt Mastics." *Journal of Materials in Civil Engineering*, Vol. 16, No. 2. 122-132.

You, Z., and W.G. Buttlar, (2004). "Discrete Element Modeling to Predict the Modulus of Asphalt Concrete Mixtures." *Journal of materials in Civil Engineering*, ASCE, March-Apri, 2004.

You, Z. and Buttlar, W.G. (2006), Micromechanical Modeling Approach to Predict Compressive Dynamic Moduli of Asphalt Mixture Using the Distinct Element Method, *Journal of the Transportation Research Board*, National Research Council, Washington, D.C., No. 1970, pp 73-83.

Effect of Loading and Temperature on Dynamic Modulus of Hot Mix Asphalt Tested under MMLS3

Sudip Bhattacharjee[1], Rajib B. Mallick[2] and Jo Sias Daniel[3]

[1] Assistant Professor, Alabama A & M University, Normal, AL, 35762, Email: sudip.bhattacharjee@aamu.edu, Phone: 256-372-4148, Fax: 256-372-5909, Corresponding Author

[2] Associate Professor, Civil Engineering, Worcester Polytechnic Institute, Worcester, MA, 01609, Email: rajib@wpi.edu, Phone: 508-831-5289, Fax: 508-831-5808

[3] Associate Professor, University of New Hampshire, Durham, NH, 03824, Email: jo.daniel@unh.edu, Phone: 603-862-3277, Fax: 603-862-2364

Abstract: In this study, the accelerated loading equipment MMLS3, which applies scaled wheel load on a scaled pavement structure, has been used to investigate the effect of loading and temperature on the dynamic modulus ($|E^*|$) of hot mix asphalt (HMA). Several HMA test slabs have been prepared in the laboratory by compacting mixes using vibratory roller compactor. The test slabs, instrumented with strain gauges and thermocouples, have been subjected to repeated wheel load using MMLS3 under controlled temperature condition. The strain and temperature data have been acquired continuously during loading with data acquisition system. The analysis of data indicated that strain signals are composed of several dominant frequencies. The constitutive equation of linear viscoelasticity in time domain was then transformed to frequency domain using the Fast Fourier Transform technique at steady state condition and the complex $|E^*|$ values have been determined continuously during the loading.

The test results indicated that $|E^*|$ is significantly affected by the combined effect of loading and temperature. A general trend of reduction of $|E^*|$ values due to these factors has been observed in the study. A simple model has been proposed which can be used to determine $|E^*|$ under moving wheel load and varying temperature condition. A comparison of test result with the MEPDG model of dynamic modulus indicates that the MEPDG model represents the condition at the beginning of loading when the damage is minimal.

INTRODUCTION

The dynamic modulus ($|E^*|$) of hot mix asphalt (HMA) is one of the important material property inputs of HMA used in the Mechanistic Empirical Pavement Design Guide (MEPDG) (NCHRP 2004a) to determine the elastic response of pavement under wheel load. As outlined in MEPDG, $|E^*|$ is predicted at any temperature and frequency using the $|E^*|$ master curve prepared at the reference temperature, the reduced frequencies and the shift factors.

In the current protocol (NCHRP 2004b), the master curve of $|E^*|$ is determined based on an *undamaged* specimen. Then the effect of temperature is incorporated by applying the master curve to the individual incremental segment of the temperature profile and the corresponding strain is calculated. However, none of the levels of design input considers explicitly how the dynamic modulus changes during loading due to the damage development. On the other hand, it is a well known fact that stiffness of the mix decreases due to the damage development under the combined action of loading and temperature. The damage development in HMA has been investigated thoroughly by Kim et al in numerous studies (Kim et al 1997) using the principle of continuum damage mechanics. Therefore, it is important that the combined effect of loading and temperature on $|E^*|$ should be investigated.

In order to find the effect of loading and temperature on $|E^*|$, two important criteria should be fulfilled during a fatigue test: (a) a proper test condition similar to the actual field condition should be employed and (b) the continuous determination of $|E^*|$ is required during the entire fatigue test until failure. The proper test condition is obtained when the method of specimen preparation (compaction) followed in the laboratory is similar to the actual field compaction and the type of loading is similar to the actual pavement loading in the field. The SHRP study (Sousa et al 1991) recommended that roller compactors should be used to produce laboratory specimens for performance testing. Since the distribution of air voids in gyratory compacted specimens is different from the distribution in roller compacted (and hence actual pavements) specimens (Cominsky et al 1994; Tashman et al 2002) and the aggregate orientations can also be different among the field and laboratory specimens (Masad et al 2002), it can be concluded that the vibratory roller compaction is a better method of compaction in the laboratory. This method of compaction can produce air void distribution similar to the actual field condition.

The previous studies on continuous damage development in HMA were performed with either uniaxial or beam specimens under either controlled stress or controlled strain condition. The present literature does not provide results of any study where continuous measurement of dynamic modulus under actual wheel loading, support and environmental condition has been performed.

Considering the importance of simulation of actual field conditions as closely as possible in the laboratory, fatigue testing of HMA has been performed in the laboratory using an accelerated loading equipment, called the Model Mobile Load Simulator (MMLS3). The equipment applies scaled loading on a layered pavement structure constructed in the laboratory, which corresponds to full scale loading in the field. Some of the important facts related to MMLS3 established by previous studies (Bhattacharjee et al 2004a) are: (1) the stress distribution under the MMLS3 is similar to the stress distribution observed in the field under the full scale loading, (2) MMLS3 applies unidirectional loading and (3) same strains and stresses are observed under the HMA slab as in the field under full scale loading. These facts and the analysis of the results of fatigue performance of HMA under MMLS3 loading (Bhattacharjee et al 2004a,b) showed that the methodology of testing scaled pavement structure in the laboratory using MMLS3 may lead to a better interpretation of fatigue performance of HMA than using other conventional fatigue test methods.

OBJECTIVES AND SCOPE

The objectives of the study were to:
1. Develop data acquisition system to acquire and simultaneous real time analysis of data to determine $|E^*|$ continuously during loading and
2. Determine how $|E^*|$ is affected by loading and temperature.

The scope of the work reported in this paper consists of real time analysis of data obtained from strain gauges and thermocouples during fatigue testing of HMA using the accelerated loading equipment MMLS3.

TEST EQUIPMENT AND SETUP

The test equipment MMLS3 is a laboratory accelerated loading equipment which applies scaled loading to a layered pavement structure with a running wheel load of 2.7 kN and 690 kPa tire pressure. The previous dimensional analysis has shown (Kim et al 1998) that if N is the scaling factor applied between the prototype and the model, the loading on model pavement will result same strains as in the prototype pavement provided the thickness is scaled down by 1:N and the load magnitude by $1:N^2$ and keeping 1:1 scale for material properties. Therefore the 2.7kN load corresponds to the full scaled single axle single wheel load of 40kN with a scale factor of 3.85. The load is applied unidirectionally at the rate of 7200 load applications per hour under controlled environmental condition. Figure 1 shows a schematic diagram of the sequence of the work performed with the accelerated loading equipment.

Vibratory roller compactor was used to compact loose mix in a mold and four test slabs of thickness 36 mm and dimension 1295 mm x 495 mm were prepared by saw cutting each compacted mix. Four transverse and four longitudinal stain gauges were glued under each test slab and four thermocouples were used along the length of the slabs. A typical test slab was placed on flexible neoprene layers followed by a steel plate and sand foundation. The number of neoprene layers was varied to provide different support conditions to achieve different strain levels. The neoprene layers provided the condition for flexible support to the slabs and the rigid base below the neoprene layers helped minimizing rutting. The data from strain gauges and thermocouples was recorded continuously during the loading using NI DAQ data acquisition software and later post processed using MATLAB. The mix consisted of a PG 64-28 asphalt binder and a 9.5 mm NMAS aggregate gradation used by Maine Department of Transportation.

More details of the test setup, the principles of MMLS3 and the usefulness of this equipment have been reported by Bhattacharjee et al (2004a,b).

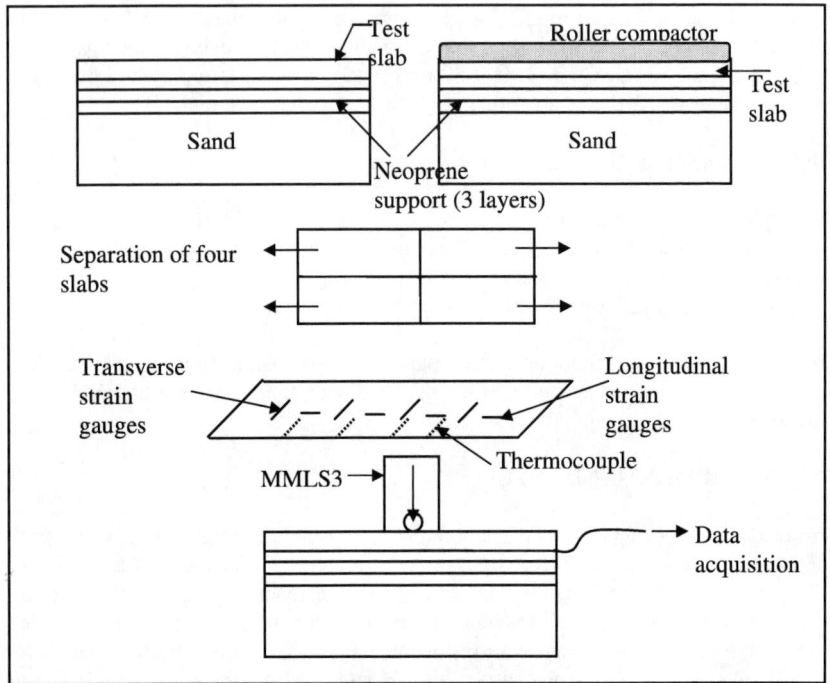

Figure 1. Sequence of work in accelerated loading and testing with MMLS3

FATIGUE RESPONSE OF HMA UNDER MMLS3

The strain histories recorded by the longitudinal and transverse strain gauges are shown in Figure 2 as well as the final cracked surface. It is evident from Figure 2 that the longitudinal strains consist of both compression and tension whereas the transverse strains show tension only (and of relatively greater magnitude than longitudinal strains), which is similar to the phenomenon observed in full scale testing. Since the transverse strain gauges were under tension only, they showed higher permanent strains compared to longitudinal strain gauges. Figure 3 shows the final test setup without the environment chamber and the instrumented test slab.

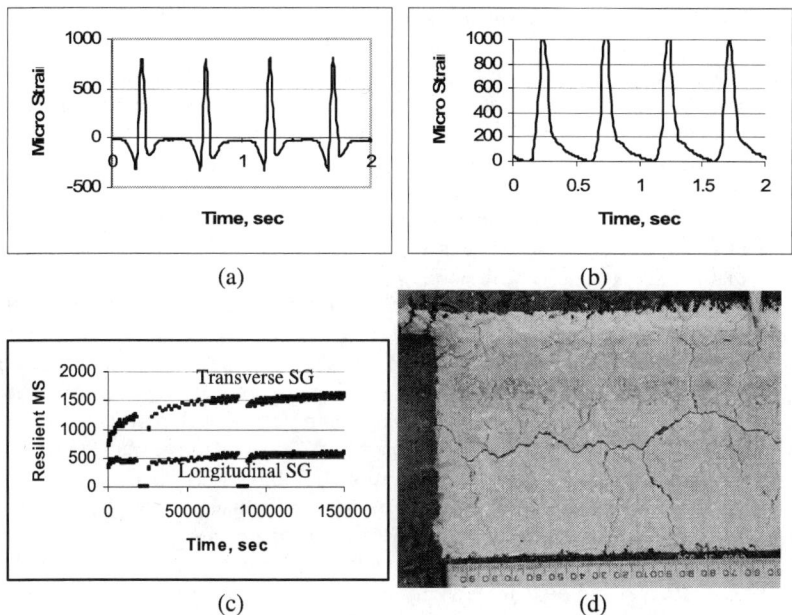

Figure 2. (a) longitudinal strain, (b) transverse strain, (c) resilient strain history, (d) cracked model pavement surface after the test

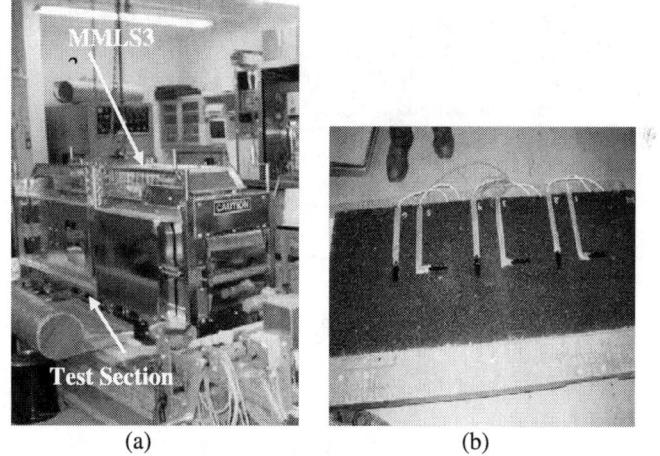

Figure 3. (a) Final test setup with MMLS3 and the model pavement structure (without the environment chamber), (b) instrumentation

It was observed that the transverse strain gauges were more affected by the permanent strains than the longitudinal strain gauges and therefore only the data from longitudinal strain gauges was used in the analysis. It is to be noted that the recoverable strains increased continuously during testing (Figure 2(c)), which was the indication of damage development in the pavement. The damage development was characterized by the reduction of the values of dynamic modulus. To investigate this situation, the test slabs with uniform temperature variation and same type of foundation supports were selected. The following sections explain the analysis performed on the strain history measured from each strain gauge.

Frequency Composition of Strain Signals

One of the major advantages of using MMLS3 as loading equipment is that it is capable of producing strain histories in the laboratory similar to the ones observed under actual traffic. By applying the principle of Fourier Transform (FT), the strain signal in time domain can be transformed to frequency domain, thus allowing us to observe the frequency components of the signal. Figure 4 shows the frequency composition (Fourier magnitude spectrum) of a typical strain signal from the longitudinal strain gauge shown in Figure 2. It should be noted from Figure 4 that even at constant speed, the strain signal is composed of different frequencies. Although the MMLS3 was run at constant speed of 7200 loads per hour, the strains are composed of several dominant frequencies, 2, 4, 6, 8, 10, 12, 14, 16 Hz; the rest of the frequencies constitute the noise. The spectrum of frequencies arises due to the interaction of rolling wheel load with the layered structure of the pavement. Each of these frequencies has a corresponding complex dynamic modulus value and a complex stress value. The total strain is the summation of the individual components at various frequencies.

In the following analysis, the magnitudes of strains at different frequencies are considered at various time instances during the loading. Once the dominant frequencies are determined for individual strain gauges, the corresponding dynamic modulus values can be calculated by converting strains and stresses in frequency domain. In the next section the basic principle is outlined first and then the calculation process is explained in more detail.

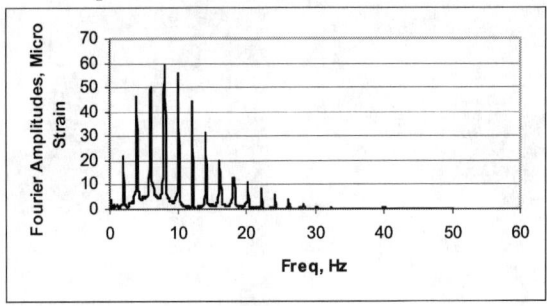

Figure 4. Fourier amplitude spectrum of strain signal

EFFECT OF LOADING AND TEMPERATURE ON DYNAMIC MODULUS

The Basic Principle

Dynamic modulus is defined as the ratio of the stress amplitude to the strain amplitude at a particular frequency and temperature calculated at the steady state. The elastic strain under the test slab in longitudinal direction, $\varepsilon_y(t)$, can be expressed as:

$$\varepsilon_y(t) = \frac{1}{E}\left(\sigma_y(t) - v\sigma_x(t)\right) = \frac{1}{E}g(t) \qquad (1)$$

where, $\sigma_x(t)$ and $\sigma_y(t)$ are stress functions in transverse and longitudinal direction respectively, $g(t)$ is the combined stress time function, v is the Poisson's ratio and E is the elastic modulus. Application of the correspondence principle (Christensen 2003) to Eq. 1 leads to the following form of the viscoelastic strain in longitudinal direction at any time during the loading:

$$\varepsilon_{y,VE}(t) = \int_0^t D(t-\tau)\frac{dg(\tau)}{d\tau}d\tau \qquad (2)$$

where, $D(t)$ is the creep compliance. Under steady state condition, the both sides of Eq. 2 can be transformed to frequency domain by Fourier Transform (FT) as follows:

$$F[\varepsilon(t)] = F[D(t)]i\omega F[g(t)] \qquad (3)$$

where, $F[D(t)]$ is the FT of $D(t)$, $i\omega F[g(t)]$ is the FT of derivative of $g(t)$ and ω is the angular frequency in radians. From theory of viscoelasticity (Christensen 2003) we can write:

$$D^*(\omega) = i\omega F[D(t)] \qquad (4)$$

where, $D^*(\omega)$ is the complex compliance. Substituting Eq. 4 into Eq. 3 results:

$$F[\varepsilon(t)] = D^*(\omega)F[g(t)] \qquad (5)$$

Therefore, at steady state condition, complex creep compliance at any frequency is equal to the ratio between Fourier transform of strain and Fourier transform of stress function at the same frequency. Once $D^*(\omega)$ is determined, the complex modulus is obtained as:

$$|E^*(\omega)| = \frac{1}{|D^*(\omega)|} \qquad (6)$$

Using the above equations, it is possible to calculate the complex modulus continuously during the loading by converting the strains and stresses in frequency domain. The steps of performing the analysis are explained below.

Calculation Steps
The flow chart of calculation steps is shown in Figure 5 and is explained below.

Step 1 find stead state time: The time instance of the steady state condition was first determined. Figure 6 shows a typical variation of slope of longitudinal strain over

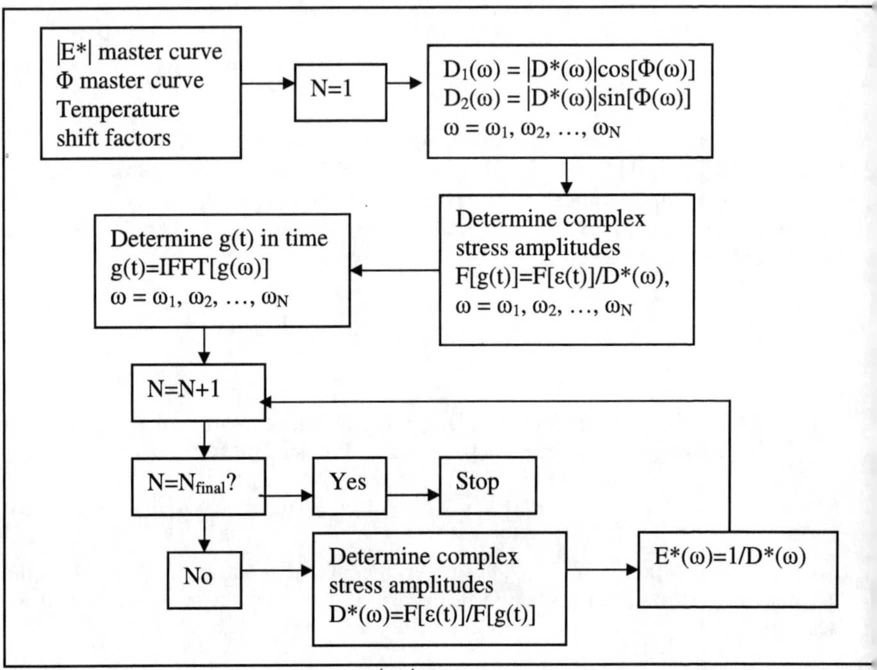

Figure 5. Flow chart of calculation of $|E^*|$ during testing. N represents the number of acquisition files analyzed.

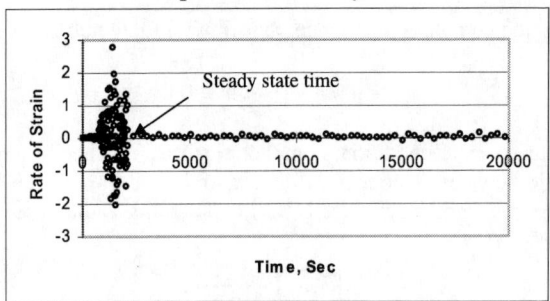

Figure 6. The steady state time

time, which indicates that the steady state condition was reached within a few hundred seconds of load applications.

Step 2 find complex strain: After the steady state condition was reached, the strain history (each acquisition, consisting of ten seconds of strain data) in time domain was considered. The average strain in each history was subtracted from the total strain history to remove any permanent strain that was present. Then the strain in time domain was transformed to frequency domain by Fast Fourier Transform (FFT). This resulted Fourier amplitudes of strains at a series of frequencies.

Step 3 find complex stress: The complex Fourier amplitudes of stress function $g(t)$ were determined at these frequencies by taking the ratio of complex strain amplitudes and complex creep compliances (Eq. 5). The complex creep compliance values were calculated using the master curve of dynamic modulus and phase angle of specimens tested in the laboratory. The stress function $g(t)$ was then determined by Inverse Fast Fourier Transform (IFFT) of the complex stress amplitudes of $g(t)$. Figure 7 shows the stress function in frequency and time domain. Once the stress function $g(t)$ was determined for the first strain history, it was applied in subsequent strain histories considering the fact that the tests were run at constant stress condition and the amount of rutting was not significant to affect the applied stress (Bhattacharjee et al. 2004a,b).

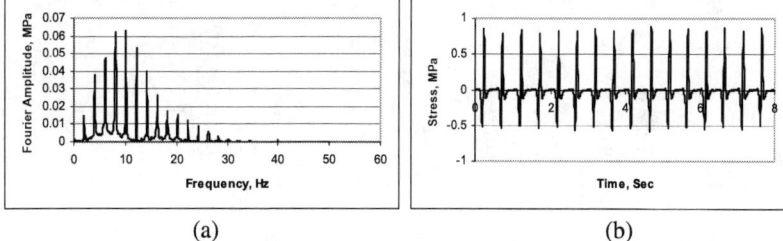

(a) (b)

Figure 7. (a) Fourier amplitude spectrum of stress function $g(t)$, (b) the stress function g(t)

Step 4 go to next acquisition and continue: In the subsequent instances of strain histories, the complex compliances were calculated by taking the ratios between FFT amplitudes of $\varepsilon(t)$ and $g(t)$.

Step 5 find dynamic modulus: The complex creep compliance was then obtained by dividing complex strain amplitudes by the corresponding complex stress amplitudes (Eq. 5) and then the complex dynamic modulus is obtained by taking inverse of the complex creep compliance (Eq. 6). The dynamic modulus in real time was then calculated by taking absolute value of complex dynamic modulus (Eq. 6).

Changes in Dynamic Modulus Due to Loading

Following the above procedure, the dynamic moduli of the longitudinal strain gauges were calculated throughout the length of the loading time for each test slab under consideration. Figure 8(a) shows the plot of the variation of normalized $|E^*|$ for entire length of the test at an average steady temperature of 20C. The normalization was done on the basis of the initial $|E^*|$ value. Figure 8(b) shows the same thing at average 15C steady temperature. Several points can be noted from Figure 8:
1. The values of $|E^*|$ reduced as the loading continued.
2. The amount and rate of decrease of $|E^*|$ are dependent on temperature.
3. The general form of $|E^*|$ can be expressed by various decreasing time functions. A simple linear decreasing time function of the following form has been adopted for the present study:

$$|E^*(\omega)|(t) = |E^*(\omega)|_0 - \alpha t \qquad (7)$$

where, $|E^*|_0$ is the initial value of modulus and α is the rate of decrease. In several cases it was found that good exponential fit could be achieved. Since there was some scatter in the data it was not possible to fit exponential curve through all strain gauges, although the authors believe that exponential form will be the best curve representing the $|E^*|$ as function of time. However, the study of the values of α at various temperatures will serve the purpose of demonstrating the effect of loading on $|E^*|$.

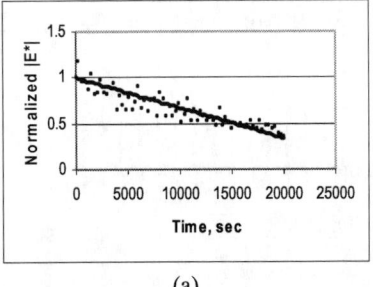
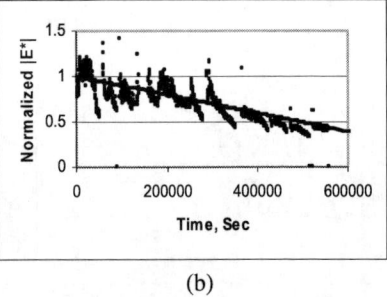

(a) (b)

Figure 8. Normalized $|E^*|$ at (a) 20C and (b) 15C versus time during loading

The effect of temperature on the values of α has been studied and is shown in the Figure 9 where values of α has been plotted against temperature for various frequencies. It can be seen from Figure 9 that α is dependent on temperature; it increases as temperature increases. Therefore, it can be concluded that rate of decrease of $|E^*|$ increases as the temperature increases. This is consistent with the common understanding of the damage development in HMA under repeated loading.

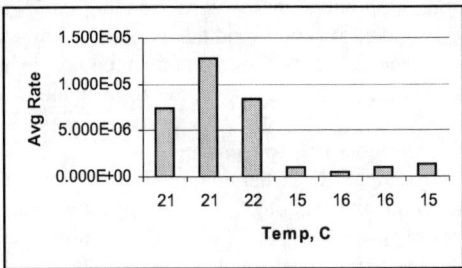

Figure 9. Effect of temperature on the rate of decrease of $|E^*|$

COMPARISON WITH THE MEPDG EQUATION FOR $|E^*|$

In the MEPDG, the design period of the pavement structure is divided into small intervals depending on local climatic conditions and the dynamic modulus of HMA is calculated at every time interval using the equation provided in the guide. The calculated dynamic modulus is then used in layered elastic analysis to predict strain at that interval of time and this leads to the calculation of the damage index, the cumulative value of which over all time intervals provides the total damage anticipated during the pavement structure. Therefore, it is very important to accurately predict the dynamic modulus over length of time. In this study, |E*| values calculated throughout the loading time have been compared to the values predicted by the MEPDG equation. The comparison for a typical test slab is shown in Figure 10, which indicates that during the initial stages of loading the differences of |E*| values obtained from the two different methods are smaller compared to the later stages of loading. This indicates that at initial stages of loading the |E*| values calculated from MMLS3 tests are close to the values obtained from undamaged tests and as the damage increases over time, the difference between them increases. This is possibly due to the fact that the MEPDG considers only the effect of temperature on the |E*|, but not the change of |E*| over time due to damage development.

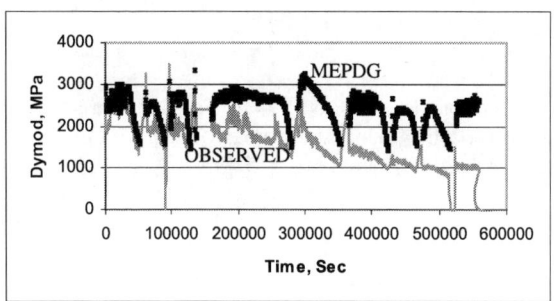

Figure 10. Comparison of observed |E*| values with the values obtained from MEPDG equation

CONCLUSIONS

Based on the analysis of results, following conclusions can be drawn:

1. |E*| is affected by traffic loading; it decreases over time due to loading.
2. The decrease of |E*| is strongly dependent on temperature. The rate of decrease is dependent on the temperature.
3. The comparison of the predicted |E*| values with the actual values obtained during the loading indicates that simple form of Eq. 7 can be used predict |E*|.
4. The accelerated pavement testing equipment MMLS3 can be used to determine the change in |E*| values which can be used along with the MEPDG equation to better predict the values of |E*| at different stages of loading.

REFERENCES

Bhattacharjee S., Gould J., Mallick R. B. and Hugo F. (2004a). "An Evaluation of Use of Accelerated Loading Equipment for Determination of Fatigue Response of Asphalt Pavement in Laboratory." *International Journal of Pavement Engineering*, Vol5, No2, pp61-79.

Bhattacharjee S., Gould J., Mallick R.B. and Hugo F. (2004b). "Use of MMLS3 Scaled Accelerated Loading for Fatigue Characterization of Hot Mix Asphalt in the Laboratory." *Proc. of the Second International Conference on Accelerated Pavement Testing* (CD-ROM), Minneapolis, MN.

Cominsky R., Leahy R. B. and Harrign E. T. (1994). "Level One Mix Design: Material Selection, Compaction and Conditioning." *Strategic Highway Research Program* (SHRP), Project SHRP A-408, National Research Council, Washington, DC.

Christensen R. M. 2003. "Theory of Viscoelasticity." *Second Edition, Dover Publication.*

Kim S. M., Hugo F. and Roesset J. M. (1998). "Small Scale Accelerated Pavement Testing." *ASCE Journal of Transportation Engineering*, Vol 124, No 2, pp 117-122.

Kim Y. R., Lee H. J. and Little D. N. 1997. "Fatigue Characterization of Asphalt Concrete Using Viscoelastic Continuum Damage Theory." *Journal of Association of Asphalt Pavement Technologists*. Vol 66, pp 520-545.

Masad E., Jandhyala V. K., Dasgupta N. and Somadevan N. (2002). "Characterization of Air Void Distribution in Asphalt Mixes Using X-Ray Computed Tomography." *ASCE Journal of Materials in Civil Engineering*, Vol 14, No 2, pp 122-129.

National Cooperative Highway Research Program (NCHRP) Project 1-37A. (2004). "Guide for Mechanistic-Empirical Design of New and Rehabilitated Pavement Structures."

National Cooperative Highway Research Program (NCHRP) Project 1-37A, Provisional Test Method DM-1 (2004). "Standard Test Method for Dynamic Modulus of Asphalt Concrete Mixtures."

Sousa, J. B., Harvey J, Painter L, Deacon J. A., and Monismith C. L. (1991). "Evaluation of Laboratory Procedures for Compacting Asphalt-Aggregate Mixtures." *Strategic Highway Research Program* (SHRP), Project SHRP-A-003A, National Research Council, Washington, DC.

Tashman L., Masad E., D'Angelo J., Bukowski J. and Harman T. (2002). "X-Ray Tomography to Characterize Air Void Distribution in Superpave Gyratory Compacted Specimens." *International Journal of Pavement Engineering*, Vol 3(1), pp 19-28.

Using PCI Data to Define Major Rehabilitation Projects at Washington Dulles International Airport

Richard G. Thuma, P.E., M. ASCE[1]
Gary K. Fuselier, M. ASCE[2]
Peter K. Yip, P.E.; M. ASCE[3]

ABSTRACT

Since 2002, Washington Dulles International Airport has taken an aggressive approach to effectively managing its airfield pavement system. Each year, roughly one third of the total airfield pavement network is inspected based upon the FAA procedures, Advisory Circular 150/5380-6B, "Guidelines and Procedures for Maintenance of Airport Pavement," to develop the Pavement Condition Index (PCI). The PCI data is used for two purposes: first, to plan for in-house annual preventive and stop-gap maintenance requirements; and second, to define major rehabilitation projects. The airport has effectively used the process to develop successful, high-impact projects to rehabilitate the original airfield pavements that are 40 to 45 years old. During the process, each major project is defined into a constructible package, with cost and duration estimates developed for each. With all projects defined, they are prioritized and fit into the operational landscape and integrated with other airfield projects. This paper explains the project definition process and components and shows how PCI inspection data can be turned into a key tool for developing an airport's plan to maintain and improve its pavement infrastructure.

INTRODUCTION

Washington Dulles International Airport (IAD) first implemented a Pavement Management System (PMS) in 1979, although it was called a Pavement Monitoring Program at the time. This implementation included a visual inspection of the

---

[1] Project Engineer, Crawford, Murphy & Tilly, 2750 W. Washington Street, Springfield, IL 62702; PH (217)787-8050; email: rthuma@cmtengr.com
[2] Design Project Manager, Metropolitan Washington Airports Authority, West Building, Room 155, Ronald Reagan Washington National Airport, Washington, D.C. 20001; PH (703)417-8189; email: Gary.Fuselier@MWAA.com
[3] President, Roy D. McQueen & Associates, 22863 Bryant Court, Suite 101, Sterling, VA 20166; PH (703)709-2540; email: pckyip@rdmcqueen.com

pavement distress following the newly-developed Pavement Condition Index (PCI) system, and tabulating distress data by hand. Since that time, the airport has been re-inspected five times through the 1980's and 1990's and the inspection data has been migrated into the MicroPAVER program to calculate PCI's, predict deterioration rates, and determine repair budget needs. Although the Pavement Management System throughout these years was very good at identifying the current condition of the pavements and determining maintenance budgets, the data was not used to develop requirements for major repair projects.

NEED FOR A PLAN

During the 1990's, many of the original Dulles pavements started showing signs of significant distress. The airfield was originally built in 1960-1961, with nominally 15" Portland Cement Concrete pavement on 9" aggregate base course. Pavements were constructed in 25-ft wide lanes, with transverse spacing at 20 or 25 feet. As aircraft traffic increased throughout the 1980's, more and more of the slabs began cracking into two or more pieces. By the late 1990's, the PCI for many of the 40 year-old pavement areas at Dulles was approaching the critical PCI value, where stopgap maintenance was no longer cost effective and major rehabilitation efforts were needed.

In 2001, the Metropolitan Washington Airports Authority (Airports Authority) recognized the looming need for major rehabilitation of its aging pavement infrastructure. At the same time, IAD was in the midst of the Dulles Development (d^2) capital development program – a $3.4 billion program that would impact all parts of the airport, including parking garages, an Air Traffic Control Tower, an underground train to connect current and future terminals, a new fourth runway complex, and extensive updates to utility systems across the airport. The Airports Authority knew the impacts of the d^2 construction program on airport operations would be extensive, and they recognized that major pavement rehabilitation work at the same time would only compound the impacts. Therefore, the Airports Authority decided they needed a clear plan for integrating the pavement rehabilitation construction demands with the d^2 program. The result was a new approach to pavement management and repair project development, as described below.

TWO DELIVERABLES

When first establishing this program in 2002, the Airports Authority requested two distinct deliverables – first, a complete PCI inspection and update to the pavement management system, and second, a separate "Project Definition" report. The PCI inspection and update to the pavement management system included the traditional pavement management functions – visual PCI inspection, transferring field data into the MicroPAVER program to calculate the PCI, predicting deterioration rates, and developing maintenance budget requirements. The Project Definition Report uses the PCI data to identify the pavement areas that are in need of major reconstruction/rehabilitation efforts, and group those requirements into logical construction packages that fit within the constraints of airfield operations.

Pavement Management System Updates

For the purpose of pavement management, the Dulles Airfield was divided into four separate complexes: Runway 1L-19R, Runway 1R-19L, Runway 12-30 and Midfield. These four areas are illustrated in Figure 1. Each year, one or two of the four complexes is inspected so that all pavements are inspected within a three-year period. This approach is advantageous because it keeps the data as current as possible, minimizes the impacts to airport operations, and keeps inspectors on the airfield every year, which allows for easy re-inspection of problem areas and initial inspection of newly constructed pavements.

Field inspection, in accordance with FAA Advisory Circular 150/5380-06B, "Guidelines and Procedures for Maintenance of Airport Pavement," includes 100% PCI survey of the Runway Complexes and development of a detailed distress map. The Midfield Complex is only inspected at a 25% rate because rehabilitation needs have not been as acute on this complex. However, as the other complexes have been reconstructed and repaired, this approach may change in the future.

MicroPAVER is used to manage the pavement system. Field inspection data is entered into the database and then MicroPAVER calculates the PCI. PCI statistics are developed for the network, branch and sections.

Family Condition Forecast Models are developed in MicroPAVER and used to predict the rate of pavement deterioration. Models were developed based upon pavement use and type using data sets from 2000 and 2002 through 2008. For example, the high-speed exit taxiways only get trafficked by lighter, arriving aircraft as opposed to parallel taxiways that get used by both arriving and departing aircraft. It follows that these two areas would deteriorate at different rates.

Policies for Stop-Gap (Safety) and Preventive Maintenance actions were developed according to the repair practices currently used at IAD. Unit prices for the repairs are included in the models, and then MicroPAVER is used to develop maintenance budgets for multiple funding scenarios. Ultimately, an in-house maintenance budget is developed for each complex to maintain the PCI at the current level.

In addition, a detailed deterioration analysis is performed for each pavement section and the information is summarized on a single page that shows the past, present and predicted future pavement condition as well as the extrapolated distresses found within the section. An example summary sheet is shown in Figure 2.

Figure 1 – Four Complexes at Washington Dulles International Airport

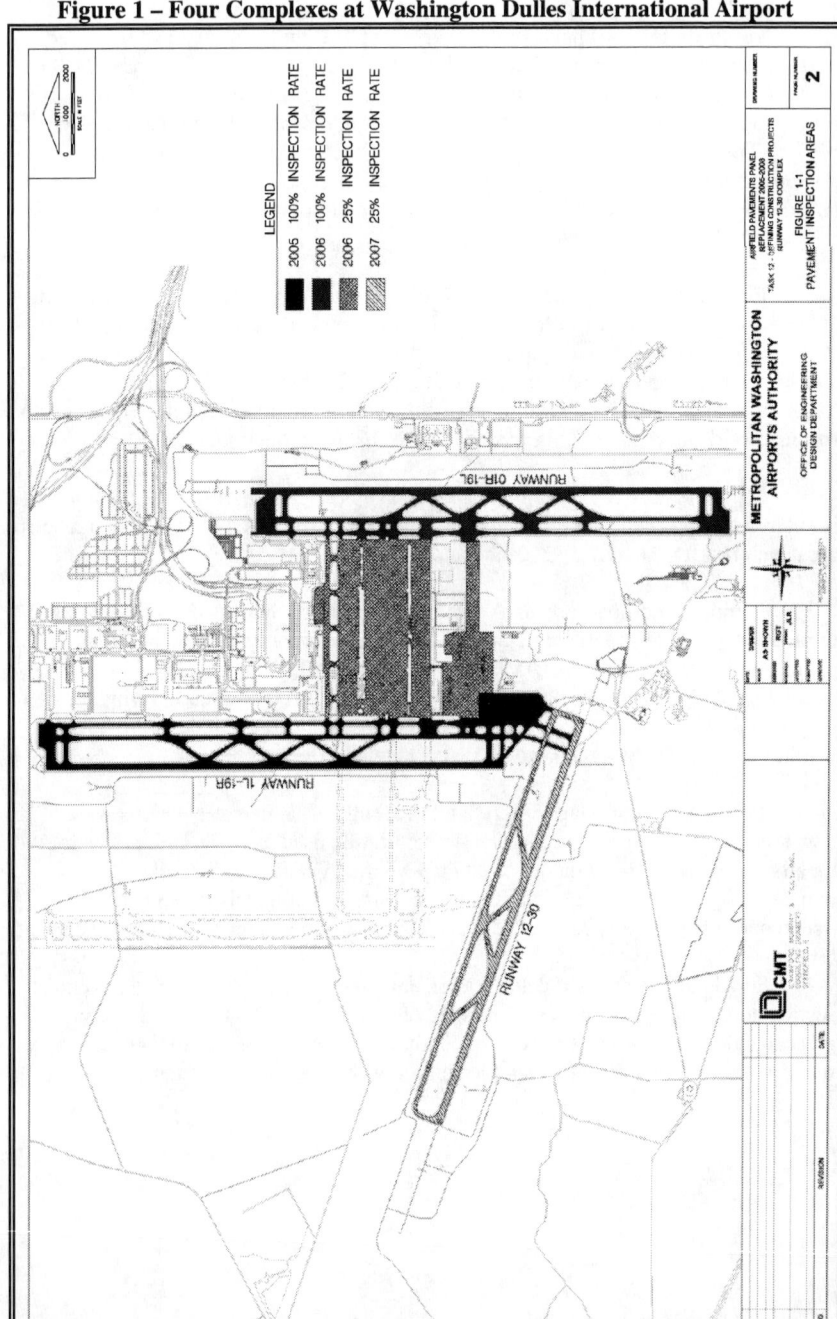

Figure 2 – Example Section Condition Analysis Summary

WASHINGTON DULLES INTERNATIONAL AIRPORT – PAVEMENT MANAGEMENT SYSTEM

Pavement Condition Analysis

Branch Name	Taxiway J	Previous PCI	(2003)	49
Branch ID	TWJ	Current PCI	(2006)	37
Section ID	03	Predicted PCI	(2011)	29
Pavement Section	15"PCC/9"CA	Last Construction Date	1/1/1959	

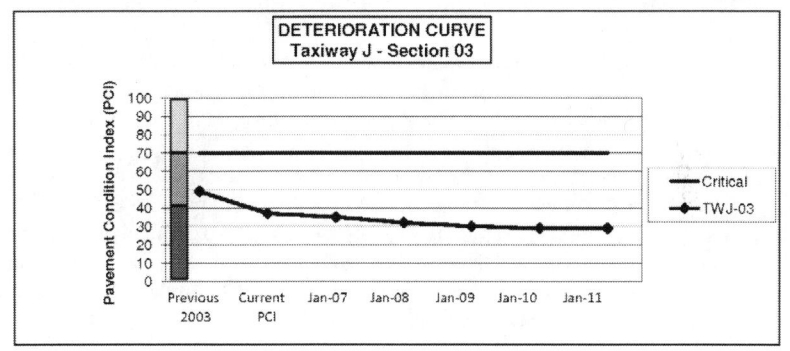

INSPECTION SUMMARY – EXTRAPOLATED DISTRESS QUANTITIES

Distress	Description	Sev	Quantity	Units	Density %	Deduct
62	CORNER BREAK	LOW	1.1	Slabs	0.19	0.7
62	CORNER BREAK	HIGH	2.3	Slabs	0.37	3
63	LINEAR CR	HIGH	38.8	Slabs	6.37	18.77
63	LINEAR CR	LOW	157.4	Slabs	25.84	15.69
63	LINEAR CR	MEDIUM	66.1	Slabs	10.86	19.68
64	DURABIL. CR	HIGH	3.4	Slabs	0.56	2
64	DURABIL. CR	LOW	1.1	Slabs	0.19	0.5
64	DURABIL. CR	MEDIUM	2.3	Slabs	0.37	1
65	JT SEAL DMG	HIGH	51.3	Slabs	8.43	12
65	JT SEAL DMG	LOW	502.9	Slabs	82.58	2
65	JT SEAL DMG	MEDIUM	54.7	Slabs	8.99	7
66	SMALL PATCH	MEDIUM	39.9	Slabs	6.55	3.62
66	SMALL PATCH	HIGH	6.8	Slabs	1.12	2.4
66	SMALL PATCH	LOW	131.2	Slabs	21.54	2.89
67	LARGE PATCH	LOW	152.8	Slabs	25.09	12.04
67	LARGE PATCH	MEDIUM	44.5	Slabs	7.3	13.6
67	LARGE PATCH	HIGH	4.6	Slabs	0.75	4
69	PUMPING	NONE	25.1	Slabs	4.12	3.93
70	SCALING	LOW	21.7	Slabs	3.56	1.54
72	SHAT. SLAB	LOW	1.1	Slabs	0.19	2.5
73	SHRINKAGE CR	NONE	3.4	Slabs	0.56	0.6
74	JOINT SPALL	HIGH	17.1	Slabs	2.81	8.76
74	JOINT SPALL	LOW	38.8	Slabs	6.37	2.49
74	JOINT SPALL	MEDIUM	17.1	Slabs	2.81	3.1
75	CORNER SPALL	MEDIUM	10.3	Slabs	1.69	1.12
75	CORNER SPALL	HIGH	11.4	Slabs	1.87	2.91
75	CORNER SPALL	LOW	46.8	Slabs	7.68	2.85

Load Related Distress (%)	Climate/Durability Related Distress (%)	Other Related Distress (%)
40	16	44

Project Definition Report

After the PCI data has been collected, the data is analyzed to develop a list of major rehabilitation projects that are needed in the next five years. Based upon the functional and structural conditions, designers develop rehabilitation methods, such as partial-depth patches, full-depth patches, slab replacement or complete reconstruction to repair distressed areas. The approach is selected according to structural condition, the type of pavement distress and the feasibility of closing the pavement for an extended period. If only nightly closures are acceptable, then only partial-depth repairs with fast-setting materials can be used. If longer closures are permitted, then full slab replacement or complete reconstruction can be undertaken.

Pavement sections with PCIs below about 55 become candidates for complete reconstruction. However, the selected approach can depend on the type of distress within the section, and also on the amount of patching that has already been accomplished within the section.

Projects are developed keeping constructability and impacts to airfield operations in mind. For instance, at taxiway intersections all pavements within the area that must be closed are evaluated for reconstruction need. In addition, the areas designated for reconstruction are checked to see how they fit into the ultimate buildout of the airport, and pavements are either added to or deleted from the project.

Each year, projects are developed for each complex and prioritized according to condition and operational value. For each project, concept-level cost and construction duration estimates are developed. This information is summarized and listed in priority order on a single sheet, as shown in Figure 3. To account for accelerated construction schedules, complex phasing, and nights-only work, factors are applied to the duration and cost estimates. The construction duration is increased by a factor of 1.1 to 1.15 for projects with constrained access, and for projects that only allow work at night the duration is increased by 1.25 to account for the shorter 6-hr allowable work shift. Only one factor is used to adjust the conceptual cost estimate. A 25% increase is applied for the areas of the project that will require accelerated construction techniques (special materials and/or 24 hr per day schedule). Because all projects completed in the last six years have included complex phasing and nights only work areas, the past bid history includes the contractor's additional costs for those work scenarios. Therefore, additional factors for night work and/or complex phasing are not applied to the conceptual cost estimate.

After all the projects within the complex under study that year are summarized, the projects from all the complexes are combined into a single list to develop a prioritized list of needed projects for the entire airfield, as illustrated in Figure 4. This presentation of the repair requirements has proven to be a very effective tool for planning upcoming projects and coordinating the project sequence with other d^2 projects on the airfield. Note that Pavement Condition Indices and estimated project costs are not shown in the Figure in accordance with Airports Authority policy.

Figure 3 – Example Project Definition Cost and Duration Summary

PROJECT TITLE: RECONSTRUCT Taxiway J, Section 3
Total PCC Pavement Area: 54,000 SY
Nightwork Area: - SY

SKETCH

PROJECT COST ESTIMATE

Work Item	Unit	Unit Cost	Quantity	Item Cost
PCC Pavement Removal (36" Total Removal)	SY		54000	$ -
Bituminous Shoulder Removal	SY		4650	$ -
Unclassified Excavation	CY		1650	$ -
Unsuitable Subgrade Undercut Excavation	CY		9000	$ -
Soil Stabilization Fabric	SY		54000	$ -
Undercut Backfill Stone	TON		40000	$ -
Aggregate Base Course	TON		22500	$ -
Cement Treated Base Course (6")	SY		58650	$ -
Bituminous Shoulder Pavement (5")	SY		4650	$ -
17" PCC Pavement	SY		54000	$ -
PCC Test Batch	EA		1	$ -
Pavement Marking	SF		49000	$ -
Inpavement Taxiway Centerline Light	EA		0	$ -
Miscellaneous Elec (Cable, Ducts, etc.)	LS		1	$ -
Maintenance and Protection of Air Traffic	LS		1	$ -
Beam Barricade	EA		200	$ -
Bucket Barricade	EA		75	$ -
Class A Barricade	EA		12	$ -
Erosion Control Plan	LS		1	$ -
Partial Depth Spall Repair - Nights	SF		0	$ -
Bituminous Linear/Crack Patch - Nights	SF		0	$ -
Joint Resealing - Nights	LF		0	$ -
6" PVC Underdrain Pipe	LF		10000	$ -
Underdrain Cleanout/Connection	EA		30	$ -
			Subtotal	$ -
		Mobilization (6%)	Subtotal	$ -
		Contingency (15%)		$ -
			Total Cost	$ -

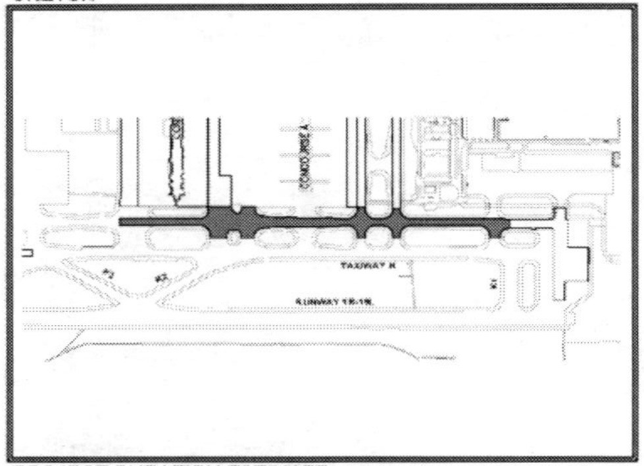

PROJECT DURATION ESTIMATE

Work Item	Unit	Daily Rate	Quantity	Work Days
Traffic Control Setup	DAY	1	1	1
Pavement Removal	SY	2000	54000	27
Excavation & Undercut	CY	1200	10650	9
Subgrade Preparation	SY	5000	54000	11
6" Cement Treated Base	SY	5000	58650	12
In-Pavement Light Placement	EA	4	0	0
17" PCC Pavement	SY	3000	54000	18
Pavement Marking	SF	5000	1	1
Pavement Cleaning	SY	5000	54000	11
Shoulder Repair	SY	600	4650	8
Traffic Control Removal	DAY	1	1	1
			Subtotal	99
Phasing Complexity Factor		1.1	Additional Days	10
% Area Nightwork		0%	Additional Days	0
			Subtotal	109
			Contingency Days (10%)	11
			Total Workdays	120

Assume 17 Workdays per 30 Calendar Days

Total Calendar Days	212
Total Months	7.1

286　　　　　　　　AIRFIELD AND HIGHWAY PAVEMENTS 2008

Figure 4 – Prioritized Project List

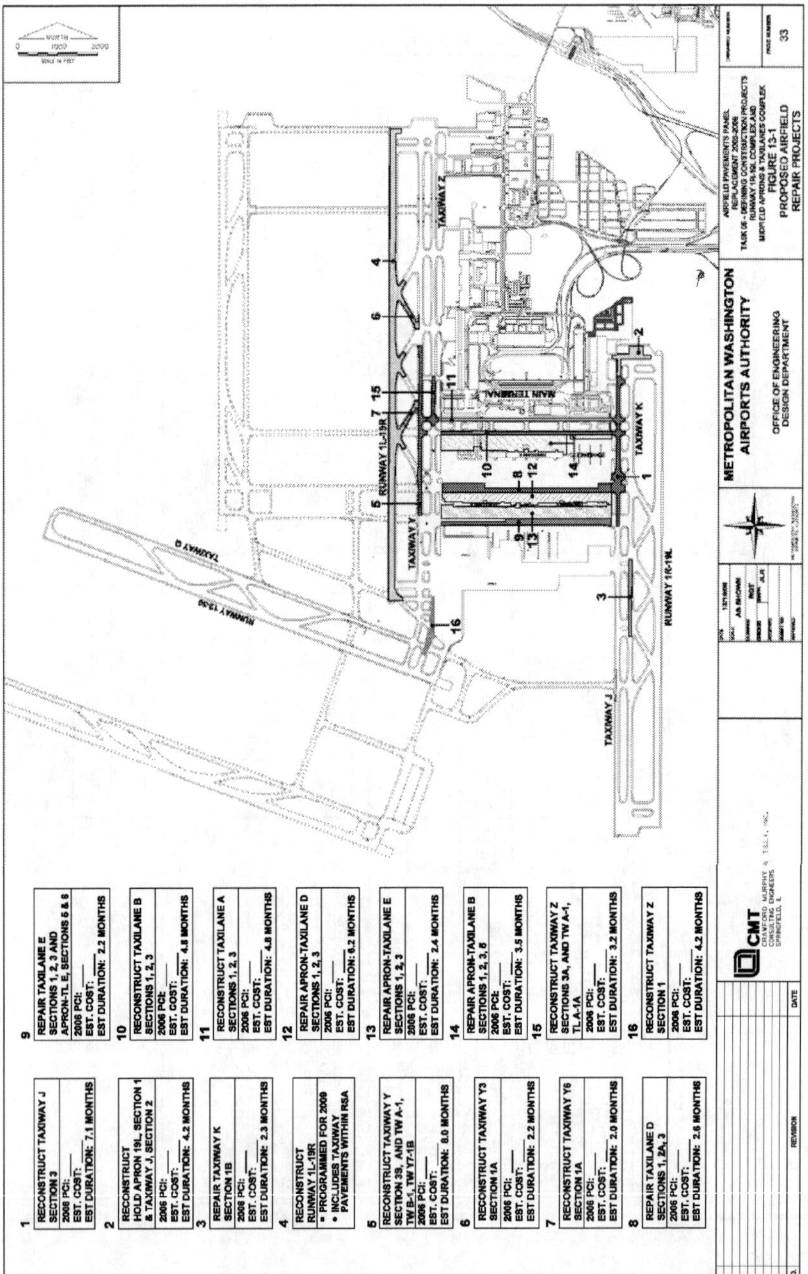

PCI IMPROVEMENTS DUE TO REPAIR AND RECONSTRUCTION

This system of pavement management and project definition reports has been in place for six years. The results of targeted major repairs can be seen in the area-weighted PCI of the airfield. Table 1 lists the projects completed during this time frame and the reconstructed or repaired area. In reconstructed areas, the PCC taxiway was demolished and a new full-depth pavement section was reconstructed. In repaired areas, spalls and cracks were fixed using partial-depth repair procedures.

Table 1. Pavement Repair and Rehabilitation Projects, 2003 - 2008

Construction Year	Complex	Project Title	Reconstructed Area (SF)	Repaired Area (SF)
2003	Runway 1L-19R	Airfield Pavement Panel Replacement 2003	346,570	
2004	Midfield Aprons & Taxilanes	Taxilane E Rehabilitation		300,000
2004	Runway 12-30	Runway 12-30 Reconstruction	1,500,000	
2005	Runway 1R-19L	Reconstruct Taxiways K, K2 & K3	278,122	
2006	Runway 1R-19L	Reconstruct Taxiways K & K7	345,500	
2007	Runway 12-30	Taxiway Q Rehabilitation		1,390,200
2007	Runway 1R-19L	Hold Apron 19L Reconstruction	144,000	
2008	Runway 1R-19L	Taxiway J Reconstruction	514,350	

As projects were completed, the PCI for those areas improved from a substandard level to an acceptable level. Although the projects were significant, they did not necessarily impact a large percentage of the total area in the complex. Table 2 lists each complex, its approximate area, and the percentage of the area that has been rehabilitated by either full-depth reconstruction or by partial depth repairs. The PCI improvement due to repair and reconstruction was calculated by comparing the actual PCI versus the forecast PCI that would occur without the major rehabilitation projects. Note that in each complex there have been substantial improvements in the PCI, considering the relatively small area that has been repaired or reconstructed. This illustrates the value of an active pavement management system that helps the owner target repair projects on the areas that will provide the most benefit for the available funds.

Table 2. Pavement Repair and Rehabilitation Projects, 2003 - 2008

Airfield Complex	Approximate Total Area (SF)	Reconstructed Area (%)	Repaired Area (%)	PCI Net Gain
Runway 1L-19R (West Side)	7,100,000	4.9%	6.0%	+ 2.7
Runway 1R-19L (East Side)	6,000,000	19.5%		+ 11.0
Midfield Aprons & Taxilanes	9,000,000	0.2%	14.0%	+ 3.2
Runway 12-30	2,800,000	50.2%	44.3%	+ 25.1

SUMMARY

Washington Dulles International Airport initiated an aggressive airfield pavement system in 2002 to begin planning for major reconstructions and repairs while massive new development was occurring on the airport. The pavement management approach consists of two deliverables: a traditional pavement management system developed with the aid of MicroPAVER, and a project definition report. The project definition report focuses on the major repair and reconstruction projects that are developed from the Pavement Condition Index inspection data. A detailed cost and duration estimate is developed for each major project, and then the projects for the entire airfield area prioritized according to condition, operational value, and construction constraints. This project definition report has become an invaluable tool for the Airports Authority to develop (and annually revise) their 5-year capital improvement program.

After six years since implementing this program, the results can be seen in the improved average pavement condition for each complex on the airfield.

Federal Aviation Administration (2007). *Advisory Circular 150/5380-6B, Guidelines and Procedures for Maintenance of Airport Pavements*, US Department of Transportation, Washington, D.C.

Fast-Track Construction of Runway 14-32 Pavement Rehabilitation at the Sarasota-Bradenton International Airport

E.M. Vélez-Vega, P.E.[1] and D.R. Bardt, P.E.[2]

[1] Kimley-Horn and Associates, Inc., Aviation Group, 4431 Embarcadero Dr., West Palm Beach, FL 33407; PH (561) 845-0665; FAX (561) 863-8175; Email: eileen.velez@kimley-horn.com

[2] Kimley-Horn and Associates, Inc., Aviation Group, 4431 Embarcadero Dr., West Palm Beach, FL 33407; PH (561) 845-0665; FAX (561) 863-8175; Email: dave.bardt@kimley-horn.com

Abstract

The Sarasota Manatee Airport Authority (SMAA) sponsored the Runway 14-32 Pavement Rehabilitation at the Sarasota Bradenton International Airport (SRQ) in Sarasota, Florida. Runway 14-32 is the primary runway at SRQ and serves commercial jet and general aviation traffic. Runway 14-32 is a 150-foot-wide asphalt runway, which was extended in 2002 to its present length of 9,500 feet, with published precision instrument approaches at both ends. This project consisted of a pavement rehabilitation of the center 7,000-foot section of Runway 14-32. The project consisted of milling two inches off of the existing runway pavement and overlaying the runway with two inches of new P-401 asphalt. Due to the continuous flight operations at SRQ, the project had a strictly enforced seven-hour construction window which consisted of three hours for paving before installing temporary pavement markings. Reopening the runway on time was critical to the Airport and for the Contractor to avoid financial penalties. This paper will present the construction coordination practices and schedule conducted during the fast-track pavement rehabilitation of Runway 14-32. The final construction was completed within the contract timeframe and below the contract amount.

Introduction

The Sarasota Bradenton International Airport (SRQ) is located on the border of Sarasota County and Manatee County on Florida's west coast. Initially developed by the Sarasota Manatee Joint Airport Authority, the aviation facilities were constructed in 1942. The airport was then leased to the Army Air Corps for use as a fighter pilot training base during World War II. The facility was transferred back to civilian use in 1947.

The airport has two runways, Runway 4-22 and Runway 14-32. Runway 4-22 is currently 5,004 feet with an Aircraft Design Group (ADG)[2] classification of B-II, due to the length of its runway safety areas at both ends of the runway. It is currently used primarily for general aviation aircraft. Runway 14-32, currently at 9,500 feet with displaced thresholds, is the main air carrier runway at SRQ. Originally constructed in 1942, Runway 14-32 was overlaid and a parallel taxiway, Taxiway Alpha, was added in 1963 and then extended to 7,000 feet in length in the early 1970s. In 2002, the runway and parallel Taxiway Alpha were extended to its current length of 9,500 feet. Only the extensions were constructed during that project and the center 7,000 feet of both the runway and parallel taxiway were left untouched. In June of 2006, Kimley-Horn and Associates, Inc. completed the design for the rehabilitation of the center 7,000 feet of Runway 14-32 and SRQ advertised for bids in July 2006. Construction was completed in May, 2007.

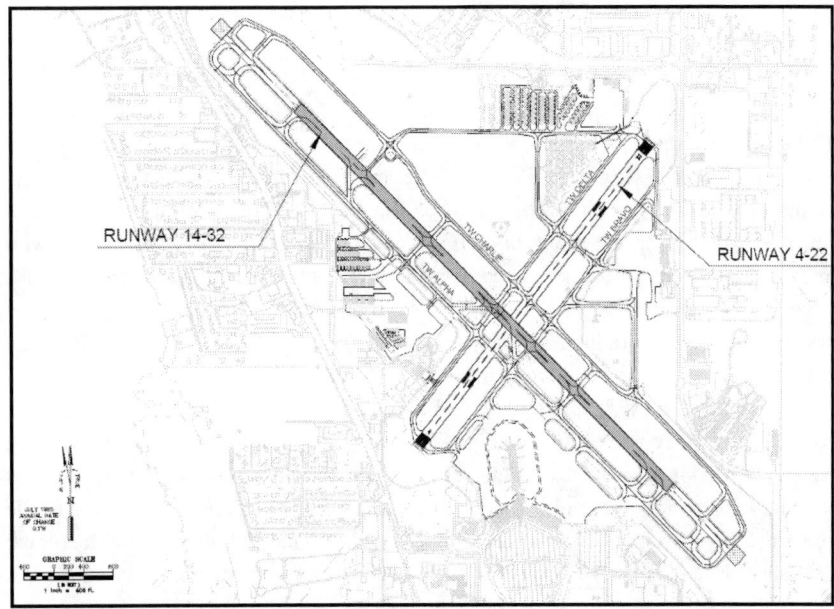

Figure 1. Sarasota Bradenton International Airport - Plan View

Due to the classification, length, and runway safety area restrictions on Runway 4-22, Runway 14-32 was the only runway available for use by the thirteen commercial air carriers with service to SRQ. As with most small Federal Aviation Administration (FAA) Part 139 airports, air carrier service is a fragile commodity and the Airport's management team has made the protection of that service its number one priority. The directive to the design team was that construction phasing should accommodate the airline schedules in a manner that would not result in any delayed, diverted, or canceled flights. With no other runway available, this became a major design consideration.

The first challenge of this project was to quantify the condition of the pavement, its structural capacity versus traffic conditions, and determine the appropriate design section to serve the traffic forecast. In order to accomplish this, both a visual pavement condition survey and non-destructive testing, using a Heavy Falling Weight Deflectometer (HWD), were undertaken. Pavement core samples were taken to confirm the existing pavement sections and based on the structural analysis and the traffic forecasts, options for the pavement rehabilitation were developed.

Data Collection

Pavement Condition Survey and NDT Data

Using procedures detailed in the FAA Advisory Circular (AC) 150/5380-6B[8], a pavement condition survey was performed on Runway 14-32 during January 2006 by Roy D. McQueen and Associates, Ltd (RDM). The pavement condition survey consisted of both a Pavement Condition Index (PCI) survey and non-destructive testing (NDT). Once the condition survey was completed the data was loaded into the MicroPAVER program for functional analysis. The pavements were visually inspected for distresses. The pavements were divided into sections and sample units, generally with an area of 5,000 square feet, for the asphalt concrete (AC) surfaced runway pavements. The results from the pavement condition survey determined that the keel section of the runway had an average PCI of 56 or "fair." The average PCI for the side section of the runway was 60 or "fair." The intersection area between Runway 4-22 and Runway 14-32 had an average PCI of 57 or "fair." The overall average PCI for Runway 14-32 was 58 or "fair", which is below the critical PCI of 70 for runway pavements. Current civil and military policy recommends a minimum PCI of 70 for runways; 60 for aprons and taxiways; and 50 for other pavements. Generally, a PCI value of 55 to 70 indicates pavement areas that are showing initial signs of deterioration. The PCI survey indicated that most of the distresses were primarily related to non-load related mechanisms, such as extensive longitudinal and transverse cracking, patching, and weathering of the AC surface, which showed a consistency with the age of the pavement surface. Although some load-related distresses such as alligator cracking with rutting were observed during the inspection, their severity levels were generally low. From the PCI survey, it was clear that the functional condition of the runway indicated a need for pavement rehabilitation. Before any rehabilitation options were identified, the structural condition of the pavements was evaluated by NDT methods. Approximately 265 NDT's were conducted on Runway 14-32 in accordance with FAA AC 150/5370-11A[7]. NDT's were generally located on a 50-foot grid in trafficked areas and a 100-foot grid in side sections between Station 100+00 and 170+00. The primary purpose of the NDT was to measure the structural properties of the pavement systems. All tests were conducted under an impulse forcing function at nominal amplitude of 20,000 lbs. RDM's HWD machine was used for the testing program. For the conventional and layered elastic design procedures used for the pavement analysis, the primary strength characterization was the elastic modulus (E) and the California Bearing Ratio (CBR) of the subgrade. The NDT data for the various pavement areas were reduced to sets of elastic moduli using the WESDEF back-calculation computer program, developed by the U.S. Army Engineer Waterways Experiment Station in Vicksburg, Mississippi, for the flexible runway pavements.

RDM also identified a geotechnical testing pattern on the pavement components and the subgrade soil CBR tests which was subsequently conducted by Nodarse & Assoc. The investigation which consisted of 12 pavement cores taken from Runway 14-32 at the locations established by the NDT testing were mainly full-depth AC pavements on sand subgrade. For the full-depth AC pavements, the thickness began with about 20 inches on the Runway 14 end and decreased gradually to 13 inches. There were approximately 700 feet on the Runway 32 end containing a limerock base course. The laboratory CBR on the sand subgrade soil ranged from 19 to 51, which indicated a strong subgrade support. This CBR range usually applies to well-graded sand. The subgrade modulus ranged between approximately 18,000 to 42,000 psi along Runway 14-32 with an average of 26,000 psi for the keel section and 24,000 psi for the side section. The average CBR for the runway was 17 with a range from 12 to 18 for the keel section and 13 to 23 for the side section. Based on FAA procedure, statistical data processing indicated that a reasonable value for pavement analysis was E = 21,000 psi or CBR = 14%. However, the CBR values from the NDT back-calculation indicated lower CBR values that do not necessarily apply to well-graded sand but to natural sand. As the CBR samples were taken adjacent to the runway and the NDT data represented the actual runway conditions, the back-calculated CBR value of 14 was used for pavement design.

Geotechnical Investigation Data

Nodarse & Associates, Inc. conducted a geotechnical investigation along the Runway 14-32 in February 2006. Geotechnical investigations in the vicinity of Runway 14-32 found sub-grade soils to be sand, and base course material to be limerock. The geotechnical report dated February 17, 2006, described the field investigation results of 12 pavement core samples at specific locations on Runway 14-32. Four bulk soil samples were also collected at various locations adjacent to the runway for CBR testing. CBR test samples were collected at approximate depths of one to two feet below the existing ground surface of Runway 14-32. The testing of the pavement cores was performed at various locations on the runway. The locations were selected in order to verify the results from the NDT conducted by RDM on Runway 14-32, which is explained in the previous section. The pavement cores were collected using a four-inch diameter hollow core barrel and the pavement surface course and base course samples were extracted, classified, and measured for thickness. The pavement cores encountered approximately 8.6 to 20.4 inches of asphalt and were used to confirm NDT results. A limerock base course with approximate thicknesses of 12.0 to 12.4 inches was encountered beneath the asphalt in four of the pavement cores.

Traffic Data

Traffic forecast information for Runway 14-32 was provided by the SRQ staff. Monthly aircraft operations for Runway 14-32 in 2005 were summarized to obtain a design traffic mix. The growth factor considered for the next 20 years was three to five percent. The traffic forecast data showed an imbalance between the arrival and departing traffic. The imbalance could not be reconciled by assuming that some of the air carrier traffic was using an alternative runway. Therefore, the 2005 baseline operations were balanced by assuming fourteen (14) Boeing 767-400 departures and arrivals. This adjustment was used with the three and five percent growth scenarios for structural evaluation and design. A combination of the FAA Layered Elastic Design[3] (LEDFAA) methods for airport pavements program and conventional LED analyses were used to compute requirements for any needed pavement strengthening. Table 1 shows the results of the structural capacity analysis for Runway 14-32 which indicates that Runway 14-32 is generally structurally adequate and strengthening is not required. It also indicates that Runway 14-32 has more than 10 years of structural life left based on the future traffic growth scenario. However, the Runway overall PCI of 58 indicated that a functional rehabilitation was needed.

Although the runway pavement was structurally adequate for the current traffic design, one of the pavement core samples indicated that the pavement was deficient in asphalt thickness by approximately two inches. A mill and replace of the existing asphalt surface was recommended to address the extensive non-load related distresses. A two-inch net thickness increase was recommended to address the pavement thickness variability (from 13 inches to 15 inches) and traffic usage. Therefore the initial recommended rehabilitation consisted of milling two inches of the existing asphalt pavement and replacing it with four inches of new asphalt pavement. The two inch milling depth was set to remove the top lift of asphalt without leaving a thin wedge that could delaminate.

Table 1 Structural Capacity Analysis for Runway 14-32

Thickness	Design CBR	Remaining Life	
		3% Growth	5% Growth
13" AC	14	< 9 years	< 8 years
15" AC	14	> 10 years	< 10 years
9" AC/12" Limerock	14	> 10 years	> 10 years

Runway Construction Phasing

Although the recommended design section was to mill two inches of the AC surface and replace it with four inches of new AC overlay for minor strengthening and thickness variability corrections, funding and construction costs were a concern and the final construction plans included a base bid mill and the replacement of two inches of the existing AC surface with an add alternate to place the second two- inch lift of asphalt. Based on final bid cost, only the base bid was accepted. Phasing options were developed for the project to minimize construction impacts on normal airport operations. The phasing needed to account for the airline schedules which had the last flight arriving at around 11 PM and the first flight out in the morning at 6 AM. The resulting seven-hour construction window was a major constraint on the project given the directive that no flights were to be canceled, delayed, or diverted. Construction sequences with time allowances for each step were developed for the mill and overlay phases and for the overlay-only phases.

These sequence delineations outlined specific issues of concern for each step in the process. Based on the time window available for actual paving, including three hours for mill and overlay phases, and 4.5 hours for overlay-only, production rates were estimated at 500 tons of asphalt a night and an overall construction schedule was developed. The limits of nightly construction were also evaluated and the schedule was based on a 1,500-foot long, 25-foot wide nightly area for milling and paving.

The final phasing plan included six phases of paving on the main runway, and a final phase, after curing, for runway grooving and permanent pavement markings. The runway phasing included a phase for each side of the intersecting runway with the third phase being the runway intersection for the mill and overlay, and a repeat of that phasing for the overlay. The phasing was setup to give the contractor the longest possible paving runs to minimize lateral joints. The contractor was required to use two paving machines in tandem to minimize cold joints between paving lanes. Final phasing plans in the construction documents included multiple plan sheets of phasing and schedules. Once the phasing options and production rates had been refined to a point that they could be articulated to the stakeholders, communication was initiated by the engineer with the various airport users that could be affected by the construction. Specifically, phone calls were initiated to the chief pilots of eleven airlines serving SRQ at the time, to explain the project and discuss any concerns they may have had; none of the airline representatives had conflicts with the proposed schedule or construction sequences.

Fast-Track Rehabilitation Requirements

The Engineer was aware of how critical the construction sequencing and planning for contingencies were before beginning the design process. Careful attention was paid to ensuring the contractor had adequate resources on site and that an asphalt plant or equipment breakdown would not prevent the runway from reopening each morning on time. Detailed schedules were included in the bid documents and reviewed at the pre-bid meeting for the bidders to understand how the project was to be phased. Sequences such as the one shown in Table 2, on page 8, were developed for all phases that included a runway closure. Paving production rates were vigilantly evaluated to ensure adequate days were included in each phase. Due to concerns of operating on fresh, hot asphalt, a cool down period was added to the construction sequence. This would minimize the possibility of burned paint and asphalt shoving from landing aircraft. The designated tack coat material for this project was an Emulsified Asphalt, RS-1 or approved equivalent which was approved as a modification to standards to the FAA P-603 Tack Coat specification[6]. The tack coat material was chosen because it provided a very fast break that would not delay paving operations. Once the design section and phasing sequence were established, the contract documents were developed in a manner that would allow the enforcement of the schedule and the specific time frames for each night's work. The final documents included specification requirements aimed at ensuring the runway would open each morning on time. In order to protect against asphalt plant problems once the runway was milled, the contractor was required to have the asphalt for each night's production produced and in the silos before the runway was closed. To protect against equipment breakdown, the contractor was required to have a spare of each major piece of equipment on site. The contractor was not allowed to use that equipment for any purpose other than to replace a machine that had broken down. To emphasize the importance of finishing on time, the contract documents included a provision for "runway rental fees." Under that provision, if the contractor failed to complete the night's work and reopen the runway by 6 AM, the contractor paid rental fees to the airport. Specifically, if work was not inspected and cleared by operations personnel and barricades removed and aircraft able to land, the contractor was charged $1,500 for any delay up to 15 minutes. After 15 minutes, the contractor was charged an additional $50 per minute. In addition, if the contractor failed to complete the overall project in the total time frame, liquidated damages accrued at a rate of $1,000 per day. The engineer also recognized that there was an impact on the contractor if they were unable to get the work area on time from the airport due to late aircraft or other delays outside their control and a bid item was added for "reverse runway rental." This was an amount that SRQ would pay the contractor for each night it turned the runway over to them later than 15 minutes after the scheduled time. The value of this item was set by the bidder and a quantity of 10 days was included in the bid form. The quantity was set high enough so that if the bidder put more than real costs in this item, it could have enough impact so that they could lose the bid.

Table 2   Runway 14-32 Nightly Work Sequence

Step 1 – Daytime Preparation
Make daily GO/NO GO decision, based on weather forecast
Confirm daily NOTAM with FAA
Pre-batch each night's total production asphalt and store in hot silos
Queue and/or service all equipment on site
Perform survey layout during traffic gaps
Perform cores on previous nights asphalt

Step 2 – Nightly Runway Closure and Mobilization (30 Minutes)
Verify radio contact and procedures with ATCT; close runway(s)
Turn off runway and taxiway lights
Turn off affected NAVAIDS
Install lighted "X"s
Install lighted barricades (if Runway 4-22 is open)
Set up perimeter gate/security queue
Mark nightly travel routes/work limits
Set up mobile light units
Perform final survey/layout control

Step 3 – Perform Milling Operations (2 Hours)
Mill 2" x 25' wide x 1,750' long lane using two milling machines
Minimum one "extra" milling machine on site for "breakdowns"
Haul millings to on site or offsite milling spoil area
Provide continuous trucking resources(no waiting for trucks to cycle)
Initiate brooming and vacuuming immediately behind milling
Stop milling at nightly phase limits; remove equipment
Clean up all excess FOD

Step 4 – Perform Asphalt Lay down (3 Hours including overlap with Step 3)
Final broom/vacuum pavement
Spray P-603 tack coat (use rapid curing tack)
Wait for tack to "break" prior to paving
Pave 2" x 25' wide x 1,750' long paving lane using two pavers in tandem
Sweep and clean up all spillage; remove equipment

Step 5 – Asphalt Cool down, Temporary Paint Application (2 Hours)
Perform artificial cooling of asphalt to lower its temperature to 150°F
Layout temporary markings for that night's 25'-wide work lane
Paint temporary markings after asphalt temperature is acceptable (150°F)
Allow paint to dry prior to operating on it

Step 6 – Final Cleanup and Runway Reopening (30 Minutes)
Perform final sweep/vacuum/clean up of work limits and all haul routes
Test runway lighting and NAVAID circuits for operability
Inspect all work limits with SRQ Operations staff
Remove lighted barricades
Remove lighted "X"s
Advise ATCT that Runway 14-32 is available

The SMAA had a history of a limited numbers of bidders for previous airfield projects. In order to overcome this, a site was identified on airport property where the successful bidder could locate a temporary asphalt plant, and the project was bid concurrently with a service road paving project. The final pavement rehabilitation option was divided into the base bid consisting of a two-inch asphalt milling and a two-inch asphalt overlay for the runway only as well as striping and grooving. Add alternates were included which consisted of the remaining work on the taxiway connector pavements; the additional two-inch runway overlay and the Runway 4-22 intersection; and signage, lighting, and electrical modifications to remove Taxiway A6 from service. Unfortunately, the SRQ only received two bids from the same two asphalt companies that had bid on all the other paving projects for the Airport. Additionally, bids were received during a period when contractors had heavy workloads and oil prices were increasing and were highly unpredictable. Those factors, combined with a very limited daily production rate, resulted in total bid prices well in excess of available funding.

In order to salvage the project, meetings were held with the apparent low bidder to evaluate costs in order to reach an agreement on a reduced scope that would be fundable. In the process, the $30,000 line item for reverse runway rental was evaluated and found to be a fair representation of costs if the contractor lost a work night at the last minute. It included not only the contractor's costs for manpower and equipment that were idle, but wasted asphalt that was in the silo and costs for subcontractors, primarily the milling subcontractor that was paid $10,000 per night they were on site, whether or not they were able to work. The total number of days for the reverse runway rental was reduced in the contract from 10 to one to preserve the unit price, while reducing the contract price by $270,000, recognizing that if we delayed the Contractor more than one night we would have to add it back in. At the end of the project, no "runway rental" days were needed which resulted in an additional saving of $30,000. An additional $175,000 in savings by allowing the contractor to keep the standby equipment in their yard rather than on site, and allowing them to work from only one staging area instead of the two shown on the plans. This arrangement was possible since the contractor's yard was in close proximity to SRQ and standby equipment could be on site in approximately fifteen minutes. The second bidder's location was not as close. The decision to award the base bid only also resulted in a shorter construction period, reducing cost for the owners construction phase services, construction observation and material testing.

Construction

By the time construction began, flight schedules had changed and the last arriving flight at SRQ was at 11:23 PM and the first departing flight was at 7:00 AM. The runway was closed every night at 12:00 AM and reopened at 7:00 AM. Due to the hours of operation for this project and the fact that the Air Traffic Control Tower (ATCT) closed at 11 PM, the SRQ provided an operations aid on site while work was performed to monitor aircraft traffic and control crossings of Runway 4-22. Table 2, on page 8, shows the nightly construction sequence for Runway 14-32.

Figure 2. Tandem Paving Machines

Figure 3. Asphalt Pavement Overlay Compaction

Conclusions

The final construction was completed within the contract timeframe and below the contract amount. However, the Owner's directive of no delayed, diverted, or canceled flights, and the lack of a second runway that would accommodate the air carrier traffic, put major restrictions on the construction phasing and production rates which in turn resulted in costs well above the original estimates. The Engineer was not able to accurately estimate the impact that the low production levels, high contractor workload, and high asphalt costs would have on bid prices. The inclusion of a "reverse runway rental" did have the positive effect of avoiding buried contingent cost of delays in the bid price and allowing an accurate representation of the real cost of delaying runway closure each night. This showed airport personnel the real cost of a late flight and they were able to work with the airlines to avoid any late closures of the runway.

Finally, close coordination between the contractor, owner, and engineer on a daily basis made it easier to make early decisions on possible weather issues so work nights would be canceled and extra days allowed before asphalt was produced and costs incurred. This helped avoid paving operations being undertaken under questionable weather conditions and helped keep quality high. Figure 3 shows an aerial photograph of the completed Runway 14-32 pavement rehabilitation project.

Figure 3. Aerial Photo of the Completed Runway 14-32 Project

Acknowledgments

The authors would like to acknowledge the Kimley-Horn and Associates, Inc. West Palm Beach and Sarasota offices for their design efforts and construction phase services. Special thanks to Mr. Raymond White from the Sarasota Manatee Airport Authority Planning and Engineering Department and his team as well as Mr. Rick Piccolo, SRQ CEO. Other project personnel that contributed to the successful completion of this project were Roy D. McQueen and Associates, Ltd., Nodarse and Associates Inc., and the contractor, APAC Southeast Inc., Southern Florida Division. The careful coordination between the owner, engineers, contractor, and subcontractors resulted in a well defined and flawless runway rehabilitation project.

References

1. Airport Ground Service Guide Manual, Safety 1st, Aviation Training Institute, 2006, 2007 Ed.

2. FAA Advisory Circular 150/5300-13 Change 12, Airport Design

3. FAA Advisory Circular 150/5320-6D Change 4, Airport Pavement Design and Evaluation

4. FAA Advisory Circular 150/5340-1J, Standards for Airport Markings

5. FAA Advisory Circular 150/5370-2E, Operational Safety on Airports During Construction

6. FAA Advisory Circular 150/5370-10C, Standards for Specifying Construction of Airports

7. FAA Advisory Circular 150/5370-11A, Use of Nondestructive Testing in the Evaluation of Airport Pavements

8. FAA Advisory Circular 150/5380-6B, Guidelines and Procedures for Maintenance of Airport Pavements

9. Runway 14-32 Pavement Rehabilitation Report at the Sarasota-Bradenton International Airport, Florida, Roy D. McQueen & Associates, Ltd., March 2006

10. Sarasota Bradenton International Airport website: http://www.srq-airport.com/

Materials and Pavement Evaluation for the New Doha International Airport Using Mechanistic-Empirical Technology

R.B. Leahy[1], L. Popescu[2], C. Dedmon[3], and C.L. Monismith[2]

Abstract

The paper discusses the use of mechanistic-empirical analyses to predict the performance of the air-side asphalt concrete pavements at the New Doha International Airport. The materials and pavement layer geometry were designed in accordance with Federal Aviation Administration (FAA) requirements. To check the adequacy of the 75-blow Marshall mix design, moisture sensitivity and permanent deformation testing was conducted. For moisture sensitivity studies, both the AASHTO T-283 (Lottman) and AASHTO T-324 (Hamburg wheel tracking) tests were performed. While the AAHTO T-324 procedure can be used to evaluate the mix in a "wet" or "dry" state, it does not provide a realistic estimate of in situ rutting potential under actual traffic loading. Thus, to estimate the rutting performance of the materials in the pavement section subjected to the design traffic, data from the repeated load simple shear test at constant height (RSST-CH), AASHTO T-320, and mechanistic pavement analyses were utilized.

Mixes at the design binder content, i.e., "optimum" and "optimum + 0.5%" were tested in repeated loading at 122°F (50°C). These test results were used in a rut depth estimation procedure developed during the WesTrack accelerated pavement test program conducted from 1995-2000. In this approach calculated stresses and strains within the multilayer elastic system and RSST-CH results are used to estimate rutting in the asphalt-bound layer for the anticipated traffic and site conditions, i.e., pavement temperatures. To accumulate permanent strains resulting from the anticipated aircraft mix, a time-hardening cumulative damage hypothesis was utilized. Results of these analyses suggest that the proposed mix design should provide adequate rutting resistance. However, the results also suggest that the production tolerance for binder content should be limited to ± 0.25 percent.

Using subgrade strain criteria suggested by Shell researchers, an estimate of rutting contributed by the unbound layers was also determined for the anticipated traffic and site conditions. Results of this analysis indicate that the overall pavement thickness is adequate. A brief description of the rut prediction methodology is included. RSST-CH results were compared to rutting criteria established for San Francisco International Airport subjected to stop-and-go movements of B747-400 aircraft. This comparison supports the conclusion reached from the mechanistic-empirical permanent deformation analyses for the asphalt bound layer.

---

1. Nichols Consulting Engineers, Chtd., Sacramento, CA
2. Pavement Research Center, ITS, University of California, Berkeley, CA
3. Bechtel Corporation

In general, the results of these analyses indicate that the proposed materials and pavement structure should perform satisfactorily from a rutting perspective in this environment when subjected to the anticipated aircraft traffic. The results also indicate that good construction practices are required to achieve the desired performance.

1 Introduction

The New Doha International Airport (NDIA) is a key project in Qatar's national development strategy, positioning the country as a leading regional aviation hub and supporting the continued dynamic growth of its national airline. The Overseas Bechtel Incorporated (OBI) contract includes the design, construction management, and project management of this greenfield airport covering a site of approximately 2,200 hectares.

Situated in the Arabian Peninsula, Doha temperatures average over 104°F (40°F) from May to September. Rainfall is scarce, (average 2¾ in (70 mm) per year), falling mostly between October to March.

Expected to open in 2010, NDIA will be the first airport in the world to accommodate the new A380 aircraft during all stages of planning, design and construction. The greenfield airport will comprise two parallel runways, a 40-gate passenger terminal accommodating 24 million annual passengers, a full range of airport support facilities, and extensive commercial developments.

NDIA will be built partially on reclaimed land, requiring the placement of approximately 63 million cubic meters of land fill, surrounded by a 14 kilometer dike. Airfield pavements consist of flexible asphalt pavement for all runways and taxiways and of rigid Portland Cement Concrete pavement for aircraft parking aprons. When complete, approximately 1.8 million cubic meters of asphalt airfield pavement (approximately 4.5 million tons of hot mix asphalt) will have been constructed at NDIA.

Selecting and proportioning materials for flexible pavements is described in FAA Advisory Circular 150/5370, Item P-401, *Plant Mix Bituminous Pavements*, and is based on the Marshall method of mix design method. This method focuses on the selection of an "optimum" asphalt content which corresponds to the equilibrium or ultimate air void content of the as-built pavement. The overall goal is to determine (within the limits of the project specifications) a cost-effective blend of aggregate and asphalt that yields a mix with sufficient asphalt to ensure a durable pavement and sufficient stability to satisfy the demands of traffic with minimal distortion or displacement. OBI elected to conduct independent testing that exceeds FAA standard procedures to verify the asphalt concrete mix design and assess the rut resistance of the airfield asphalt pavements. The purpose of this paper is to illustrate how recent developments in testing and analysis can provide greater insight into long-term performance.

As shown in Table 1, a series of laboratory tests were used to evaluate the polymer modified bituminous paving materials proposed for NDIA. The objectives of the testing were threefold:
- to verify the PG (performance grade) of the polymer modified binder (PMB);
- to verify the mix design, i.e., job mix formula (JMF) of the asphalt concrete; and
- to assess the rutting susceptibility of the hot mix asphalt (HMA) concrete which will be made with the aforementioned JMF.

Only the results from the rutting tests are addressed in this paper.

To quantify the HMA susceptibility to rutting (i.e. permanent deformation), the following permanent deformation tests were undertaken:
- the Hamburg Wheel Tracking Device (HWTD)
- the repeated load simple shear test at constant height (RSST-CH)

Table 1 – Proposed Materials Testing

Objective	Method
PG Verification	AASHTO M320-05, "Standard Specification for Performance-Graded Asphalt Binder"
JMF Verification (75-blow Marshall)	Asphalt Institute MS-2, "Mix Design Methods for Asphalt Concrete and Other Hot-Mix Types, 6th Edition"
Rutting Susceptibility (Permanent Deformation Testing)	AASHTO T324-04, "Standard Method of Test for Hamburg Wheel-Track Testing of Compacted Hot-Mix Asphalt (HMA)"
	AASHTO T320-03, "Standard Method of Test for Determining the Permanent Shear Strain and Stiffness of Asphalt Mixtures Using the Superpave Shear Tester, (SST)"

2   Laboratory Testing

Of particular interest for airfield flexible pavements is deformation resistance or rutting. It is a key performance parameter that depends largely on the HMA job mix formula, i.e., aggregate gradation and asphalt content. Rutting is defined as the accumulation of small amounts of unrecoverable strain resulting from applied wheel loads to the HMA pavement. This deformation is caused by compaction (volume reduction) or shear deformation (without volume change), or both, of the HMA under traffic. Shear failure (lateral movement) in a HMA pavement generally occurs in the top 100 mm (4 in) of the HMA structure. Rutting not only decreases the useful life of a pavement but also creates a safety hazard. Since rutting of asphalt concrete is sensitive to asphalt content, specimens were tested at not only the design or "optimum" asphalt content, but also at "optimum + 0.5%." This testing would demonstrate the potential adverse effects of excess asphalt that can (and frequently) occur during full-scale production. The additional 0.5% asphalt cement is reasonable as the "action limit" specified by FAA P-401 is 0.45%.

2.1 Loaded Wheel Testers

Standardized laboratory equipment and test procedures that predict field-rutting potential would be of great benefit to the HMA industry. Currently, the most common type of laboratory equipment of this nature is a loaded wheel tester (LWT). Laboratory wheel-tracking devices that measure rutting do so by rolling a small loaded wheel device repeatedly across a prepared HMA specimen. To identify asphalt concrete mixes that may be prone to rutting, many agencies use LWTs to supplement their mix design procedure. The LWTs allow for an accelerated evaluation of rutting potential in the designed mixes.

<u>Hamburg Wheel Tracking Device</u>
The Hamburg Wheel Tracking Device (HWTD) shown in Figure 1 was used to assess the rut resistance of pavement at NDIA. The HWTD device was selected as it allows testing at elevated temperatures – comparable to what are anticipated at the NDIA site – in both "wet" and "dry" conditions. Also, the HWTD has been shown to distinguish between "good" and "poor" performing asphalt concrete mixes with respect to both moisture damage and rut resistance. Furthermore, HWTD data have been shown to correlate with field performance. Several state DOTs (Department of Transportation) – e.g., Colorado, Texas and Utah – have established criteria for material acceptance based on HWTD data.

The HWTD captures the combined effects of rutting and moisture damage by rolling a steel wheel across the surface of an asphalt concrete slab that is immersed in hot water. Typically, specimens are compacted to 7±1 percent air voids for dense-graded HMA. The most commonly used test temperature is 122°F (50°C), although 104° F (40°C) has been used when testing certain base course mixes.

Test samples are typically "loaded" for 20,000 passes or until 0.8 in (20 mm) of deformation occurs. Various agencies have established maximum allowable rut depths based on correlations with field data; e.g., both Colorado and Utah DOTs recommend maximum allowable rut depths for dense-graded mixes of 10 mm at 20,000 wheel passes.

2.2 Mechanistic-Empirical Analysis and Design

Although wheel tracking test data correlate well with actual field performance, they cannot be used in mechanistic pavement analyses and cannot be used to determine fundamental engineering properties of the material, i.e., modulus. The inference space is limited to the materials and loading conditions of the test. Thus, one cannot extrapolate these results to other materials and/or loading conditions. Hence, to supplement this "torture" testing, a mechanistic-empirical approach using repeated load and modulus testing was undertaken.

Figure 1 – Photo of Hamburg Wheel Track Device

A mechanistic approach explains phenomena by reference to physical causes. In pavement design, the phenomena are the stresses, strains and deflections within the pavement structure and the physical causes are the loads and material properties of the pavement structure. The relationship between these phenomena and their physical causes is typically described using a mathematical model. Various mathematical models can be (and are) used, the most common of which is a layered elastic model. Along with this mechanistic approach, empirical elements are used when defining what value of the calculated stresses, strains and deflections result in pavement failure. The relationship between physical phenomena and pavement failure is described by empirically derived equations that compute the number of loading cycles to failure.

The laboratory testing and a methodology for estimating rutting in the asphalt concrete pavement structure are summarized herein. The procedure utilizes stiffness and plastic strain versus repetitions data obtained from the repeated load simple shear test. In this approach, the pavement was assumed to behave as a multilayer elastic system for determination of key stresses and strains to permit rutting estimates to be made at the pavement surface. Figure 2 illustrates the idealization of a pavement structure which was used to estimate the accumulation of plastic strain.

To reflect the temperature distributions with depth in the asphalt bound layer, h_{AC}, the layer was subdivided into a number of sublayers as seen in Figure 2. Temperature distributions were determined using the Enhanced Integrated Climate Model (EICM) on an hour-by-hour basis for a representative day for each month of the year (3). The temperature at mid-depth in each of the sublayer was then used to select the mix stiffness for that sublayer for an appropriate time of loading associated with the anticipated traffic. In addition to the determination of τ and γ^e in the asphalt concrete layer, the vertical compressive strain, ε_v, at the surface of the subgrade may also be determined. The general framework used to determine the accumulation of plastic strain is shown in Figure 3.

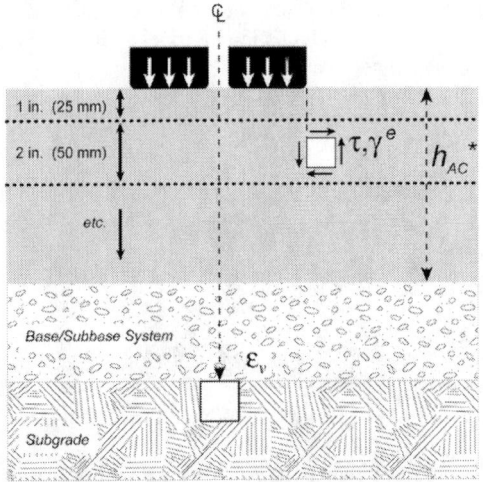

* h_{AC} can be subdivided into 3 or more layers depending on the layered elastic analysis program utilized.

Figure 2 – Pavement idealization for accumulation of plastic strain (2)

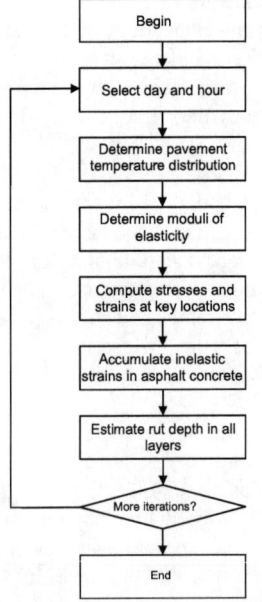

Figure 3 – Framework for rut depth estimates (2)

Rutting in the asphalt concrete is assumed to be controlled by shear deformations. Accordingly, the computed values for τ and γ^e at 2 in (50 mm) depth beneath the edge of the tire were used for the estimates of plastic strain, as shown in Figure 2. In simple loading, permanent shear strain in the asphalt concrete is assumed to accumulate according to the following expression:

$$\gamma^i = a' \cdot \exp(b'\tau)\gamma^e n^c \quad \text{(Equation 1)}$$

where:
- γ^i = permanent (inelastic) shear strain at a 2 in. (50 mm) depth
- τ = shear stress determined at this depth using elastic analysis
- γ^e = corresponding elastic shear strain
- n = number of axle load repetitions
- a', b', c = experimentally determined coefficients

Plastic (or inelastic) strain in the asphalt concrete layer due to shear deformation was computed using a time-hardening principle, illustrated schematically in Figure 4. To estimate the contribution to deformation from base and subgrade, a modification to the Asphalt Institute subgrade strain criterion was utilized. This approach permits prediction of plastic strain as a function of traffic and environment as well as a function of mix parameters. As noted above, data needed to conduct this analysis are generated from two laboratory tests: a frequency sweep test to determine modulus and the repeated shear test at constant height. A brief overview of these tests follows (4). Photos of the shear test device and samples are shown in Figure 5.

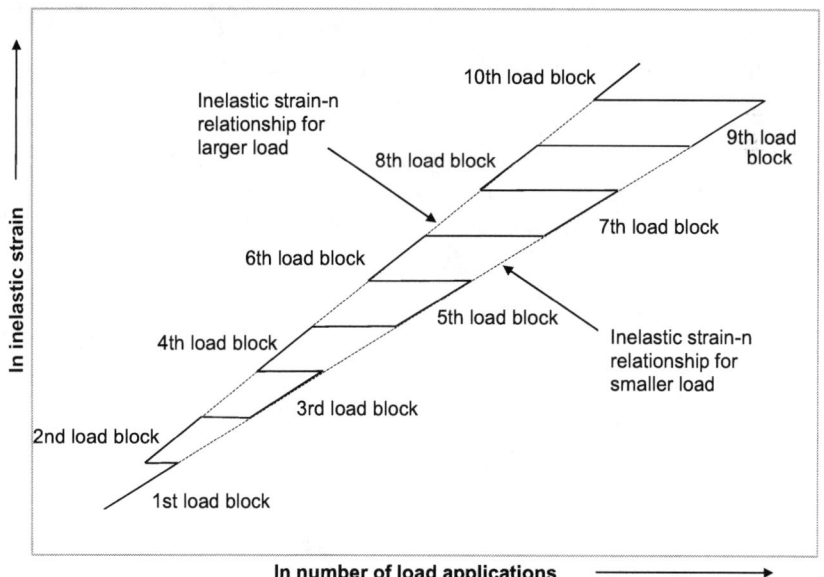

Figure 4 – Time-hardening procedure for accumulation of plastic strain (2)

Superpave Shear Tester (SST)    Loading Chamber

Prepared Sample   Prepared Sample (left) and Sample After Test (middle and right)

Figure 5 – Superpave Shear Test apparatus and samples (4)

Performance Testing – Modulus and Shear
Repeated Load Test in Shear – The Superpave shear tester (SST), developed as part of the Strategic Highway Research Program, can perform a repeated load test in shear. In this test, known as the repeated simple shear at constant height (RSST-CH) test, a repeated haversine shear stress is applied to the specimen. As noted in Table 1, the test procedure is described in AASHTO T320-3.

Dynamic Modulus Test in Shear – The dynamic modulus test in shear is known as the frequency sweep at constant height (FSCH) test. Horizontal strain is applied at a range of frequencies (from 10 to 0.10Hz) using a haversine loading pattern, while the specimen height is maintained constant by compressing or pulling it vertically as required.

3 Test Results and Discussion

Materials, polymer modified binder (PMB) and aggregate, were shipped directly from the project site in Doha to Advanced Asphalt Technologies, where all testing was accomplished.

3.1 Rutting Susceptibility– Hamburg Wheel Track (HWT) Testing

Shown in Figure 6 are test results from the Hamburg Wheel Track (HWT) testing. Duplicate specimens were fabricated (design asphalt content, gyratory compaction to air void content of 7%±1%) and tested in accordance with AASTHO T324 at 122°F (50°C). Specimens were tested both "DRY" and "WET." From Figures 6a and 6b, note that the average maximum rut depths recorded after 20,000 wheel passes were approximately 0.14 in (3.6 mm) and 0.16 in (4.1 mm) for the dry and wet testing, respectively. These rut depths are significantly less than the 0.8 in (20 mm) limit specified by Utah DOT. These data indicate that the mix is both moisture and rut resistant.

3.2 Rutting Susceptibility– Frequency Sweep and Repeated Shear Testing

To address the anticipated performance of the materials as a function of the environment, pavement layer geometry and aircraft traffic, both frequency sweep and repeated shear tests were conducted in accordance with AASHTO TP-7. Triplicate specimens were fabricated (gyratory compaction to air void content of 7%±1%) at two asphalt contents: the "design" or "optimum" asphalt content (4.7%), and "optimum + 0.5%" (5.2%). Excess asphalt tends to make a mix more susceptible to rutting (or permanent deformation), hence the testing at "optimum + 0.5%." Frequency sweep tests were conducted at 3 temperatures 39, 68, and 115 °F (4, 20 and 46°C) and 3 frequencies (0.1, 1 and 10 Hz). Repeated shear testing was conducted at 122°F (50°C) for 5000 load cycles.

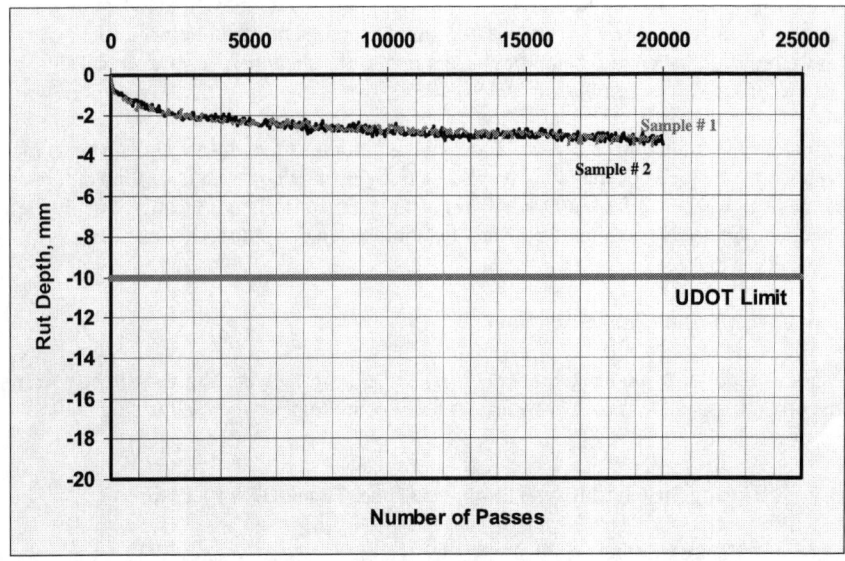

Figure 6a – Hamburg Wheel Track Data (DRY)

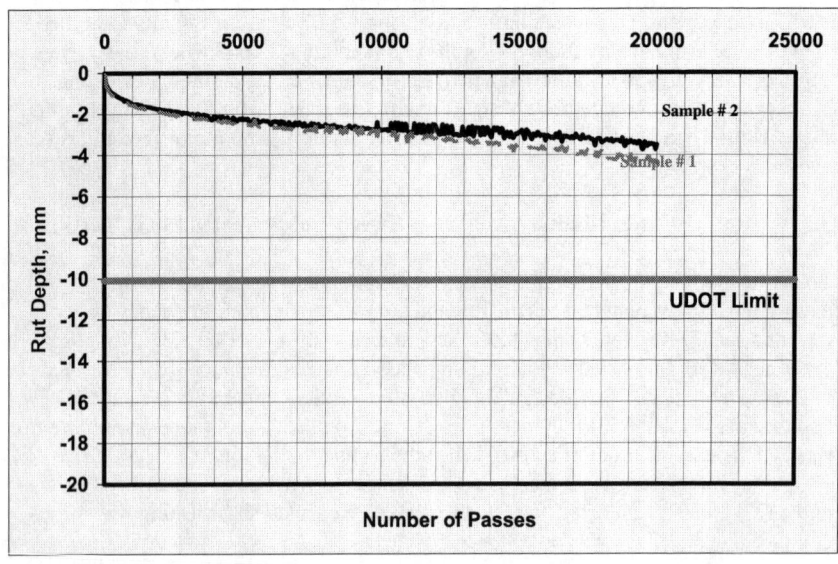

Figure 6b – Hamburg Wheel Track Data (WET)

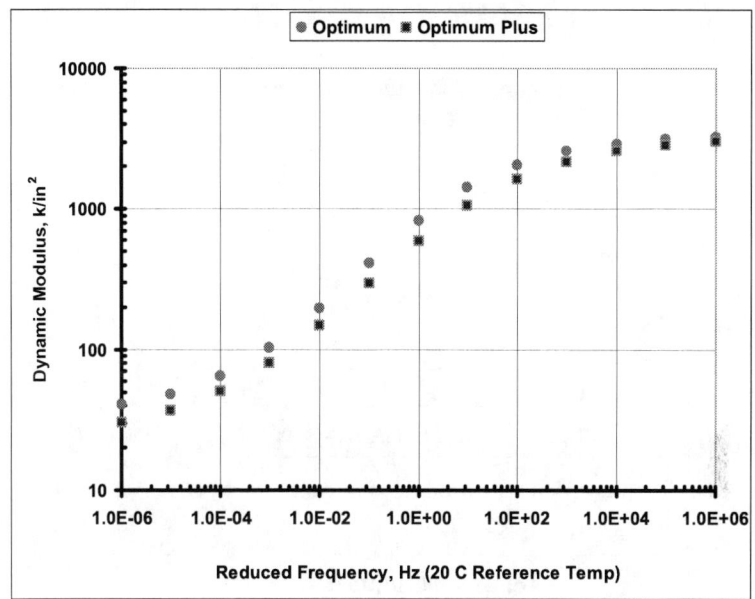

Figure 7a — Comparison of Asphalt Concrete Modulus for "Optimum" and "Optimum Plus" Mixes

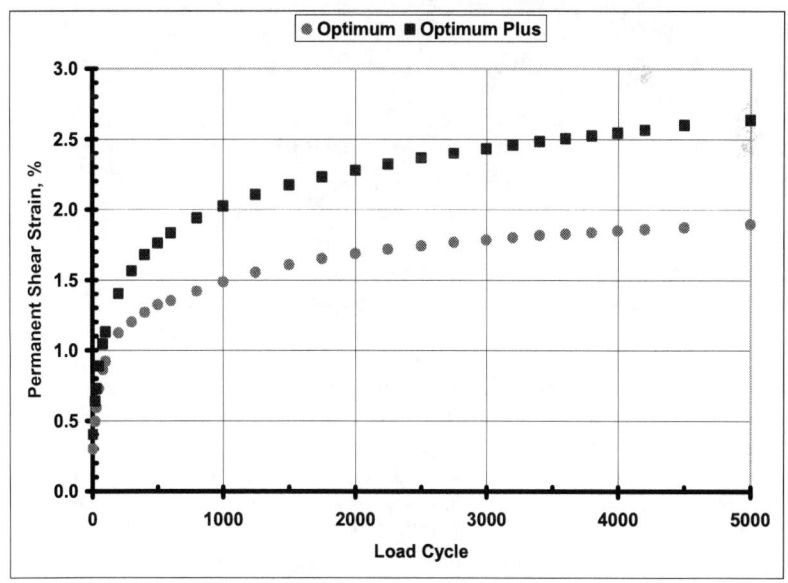

Figure 7b — Comparison of Permanent Shear Strain for "Optimum" and "Optimum Plus" Mixes

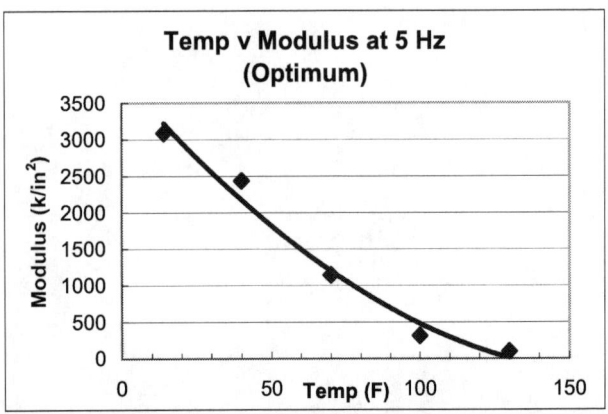

Figure 8a — Asphalt Concrete Modulus v Temperature (4.7% Asphalt Content)

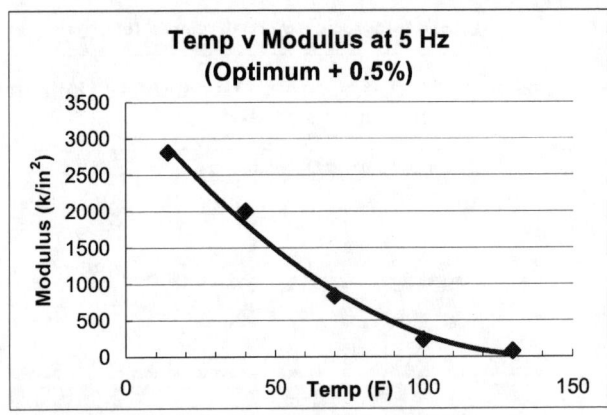

Figure 8b — Asphalt Concrete Modulus v Temperature (5.2% Asphalt Content)

Shown in Figure 7 is a comparison of the modulus and permanent shear strain for the "optimum" and "optimum plus" mixes. Shown in Figure 8 is the relationship between temperature and modulus for the "optimum" and "optimum plus" mixes. The effect of asphalt content on mix properties may be observed from Figures 7 and Figure 8. From Figure 7b note that the permanent shear strains for the "optimum" and "optimum plus" mixes are approximately 1.9% and 2.6%, respectively. From Figures 8a and 8b note that the moduli at 77°F (25 °C) for the "optimum" and "optimum plus" mixes are approximately 1,000 k/in^2 and 750 k/in^2, respectively. On the "plus" side of the optimum asphalt content, modulus decreases and permanent shear strain increases. These data underscore the effect of asphalt content on mix susceptibility to permanent deformation and, hence, the importance of HMA plant process control.

3.3 Rutting Susceptibility– Mechanistic-Empirical Analysis

Having determined the asphalt concrete modulus as a function of temperature and frequency, an analysis of the NDIA pavement section with the anticipated traffic was undertaken. The actual pavement layer geometry and that idealized for the analysis are shown in Table 2. Material properties required for the multilayer elastic analysis include modulus and Poisson's ratio. The modulus of the asphalt concrete used in the analysis was based on the results of the frequency sweep testing. Poisson's ratio and moduli for the aggregate base and subgrade were assumed, i.e., taken from the published literature (7).

Table 2 – Pavement Layer Geometry and Material Properties

Pavement Layer	Actual	Idealized for Analysis	Material Properties
Asphalt Concrete	8 inches	20 inches Asphalt Concrete	E = f(temperature) $46,775 - 733,800$ lb/in^2 ν = f(temperature) $0.40 - 0.45$
Asphalt Treated Base	12 inches		
Aggregate Base	15 inches	15 inches Aggregate Base	$E = 30,000$ lb/in^2 $\nu = 0.30$
Subgrade (CBR = 10)			$E = 1500$ lb/in^2 $\nu = 0.30$

Aircraft operations data, provided by Bechtel, and used for the thickness design of pavement at NDIA, are shown in Table 3. The data were sorted by category and aircraft type in descending order of the number of operations. As may be seen from Table 3, passenger aircraft account for nearly 90% of the operations. Accordingly, the analysis focused on the weight and main gear configurations of the Airbus and Boeing aircrafts. A review of FAA, Boeing and Airbus technical documents which describe gear configurations revealed that load per tire on the main gear ranged from 36,000 to 65,000 lb/tire. Tire pressures ranged from 175 to 235 lb/in^2 (8, 9, 10).

Because of time constraints, the analyses of critical stresses and strains in the pavement section were limited to two gear configurations: a four-tire configuration for the Airbus; and 6-tire configuration for the Boeing. The load per tire and tire pressure for the Airbus analysis were 56,000 lb and 220 lb/in^2, respectively. The load per tire and tire pressure for the Boeing analysis were 51,250 lb and 215 lb/in^2, respectively. A schematic of the main gear configurations used in the CIRCLY analyses is shown in Figure 9. The Airbus 330-200/300 (equivalent to the Boeing 767-400ER) and Boeing 777-300 dimensions were used to represent the aircraft operations. As shown in Table 4, results of the CIRCLY analyses indicated that the stresses and strains at the critical locations were comparable.

Table 3 — Aircraft Operations: Scenario 2a Forecast (2009 - 2028)

TYPE OF OPERATION	CODE	Aircraft Type	Total Operation	Average Annual			
Passenger	E	A350-800	862,867	21,572			
	E	A350-900	712,837	17,821			
	E	A330-200/300	441,594	11,040			
	E	B777-200	431,052	10,776			
	C	B737-700	368,602	9,215			
	E	A340-600	326,163	8,154			
	E	B777-300	227,787	5,695			
	C	A319-21	137,929	3,448			
	F	A380-800	117,196	2,930			
	E	B777-800	60,151	1,504			
	D	A300-600	24,000	600		% of annual operations	
	D	A310	6,400	160			
	D	B767-300	6,400	160	93,074	88.90	passenger
Cargo	B	B1900F	135,603	3,390			
	D	ABF	98,057	2,451			
	E	B777F	89,669	2,242			
	E	B74F	22,903	573			
	C	B70F	17,738	443			
	D	A310F	17,404	435	9,534	9.11	cargo
GA	B	Gulfstream IV	69,844	1,746	1,746	1.67	GA
Emiri	C	A319	5,483	137			
	C	CL6	5,483	137			
	E	A340-300	2,741	69	343	0.33	Emiri
		total →	4,187,900	104,698	104,698	100	

Table 4 – Summary of CIRCLY Analysis Output

	Airbus	Boeing
Shear stress at tire edge (lb/in^2)	66 - 68	63 - 65
Shear strain at tire edge	0.0004 – 0.0006	0.0004 – 0.0006
Vertical compressive stress at top of subgrade (lb/in^2)	9 - 13	11 - 12
Vertical compressive strain at top of subgrade	0.0007 – 0.0008	0.0006 – 0.0007

As noted in previously, the accumulation of plastic strain in the asphalt concrete may be computed using Equation 1. In addition to the stresses and strains determined from the CIRCLY analyses, reasonable assumptions with respect to temperatures at the project site and aircraft operations were required to compute the plastic strain. The assumptions were as follows:

- aircraft operations uniformly distributed throughout the year
- plastic strain accumulated during the warmest months, i.e., April – November
- plastic strain accumulated 8 hours/day
- one-half the aircraft operations at maximum weight
- conservative value of shear modulus, 11,000 lb/in^2 at 122°F (50°C)
- no aircraft wander was assumed

Shown in Figure 10 is the predicted accumulation of plastic strain for the first 5 years. Several conclusions may be drawn from the data presented in Figure 10. For a well designed mix, the accumulation of plastic strain tends to level off very rapidly with time because of the inevitable aging (i.e., hardening) of the asphalt cement with exposure to UV radiation. Mirroring the laboratory data shown in Figure 7b, the "optimum plus" mix shows considerably greater accumulation of plastic strain than does the "optimum" mix. From the laboratory test data presented in Figure 7b, the "optimum plus" mix sustains approximately 1½ times the plastic strain of the "optimum" mix. From the results presented in Figure 10, the "optimum plus" mix sustains approximately 2½ times the plastic strain of the "optimum" mix. The greater accumulation of predicted plastic strain is likely the result of the assumptions that were made (i.e., temperature distribution at the site, time over which plastic strain was accumulated, aircraft operations, and the use of model coefficients that were developed for highway loading conditions). This approach, though calibrated for highway loads, yields reasonable results for the aircraft considered and is consistent with the laboratory test results. It demonstrates the adverse effect of excess asphalt cement on permanent deformation.

As another check of the rut resistance of the proposed pavement structure, the vertical compressive strain at the top of the subgrade (shown in Figure 11) generated from the CIRCLY analysis were used to estimate the influence of the Airbus and Boeing aircraft operations on permanent deformations in the unbound base and subgrade. An analysis based on the Shell criterion – load repetitions to ¾ inch rutting – suggests that after 5 years of aircraft operations, contribution of the unbound materials to surface rutting would be minimal. This analysis provides additional confirmation as to the adequacy of the structural section and materials. The results and assumptions made in the analysis are shown in Table 4.

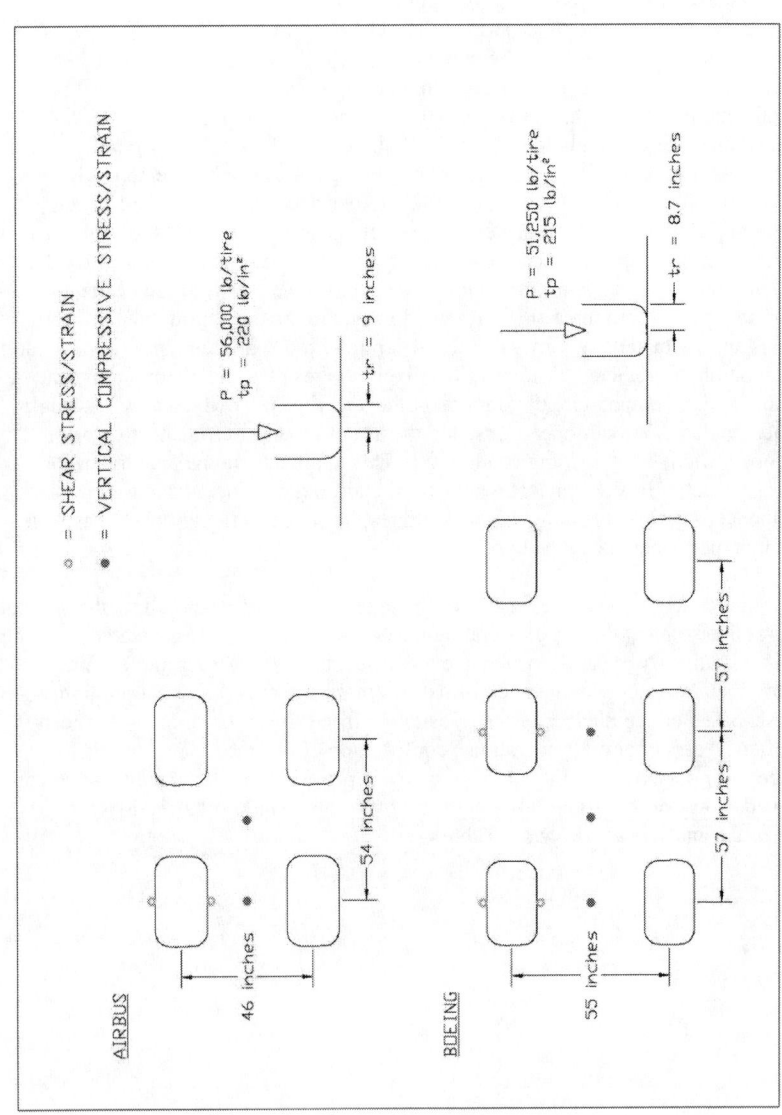

Figure 9 – Gear Configurations and Stress/Strain Locations for CIRCLY Analysis

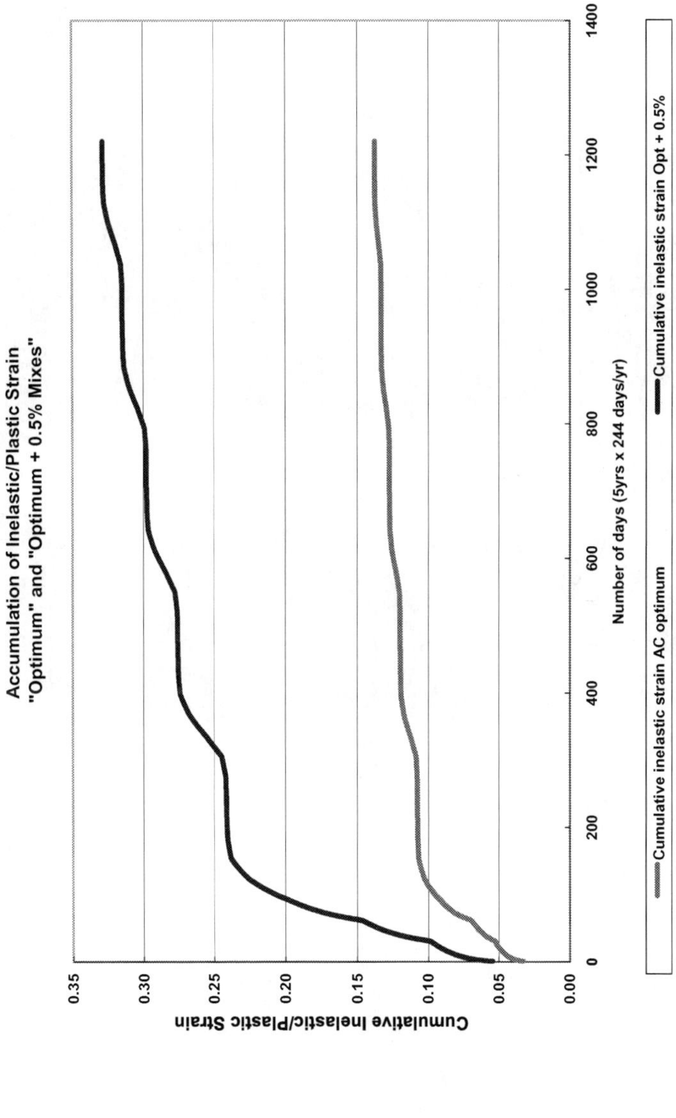

Figure 10 – Predicted Accumulation of Plastic Strain

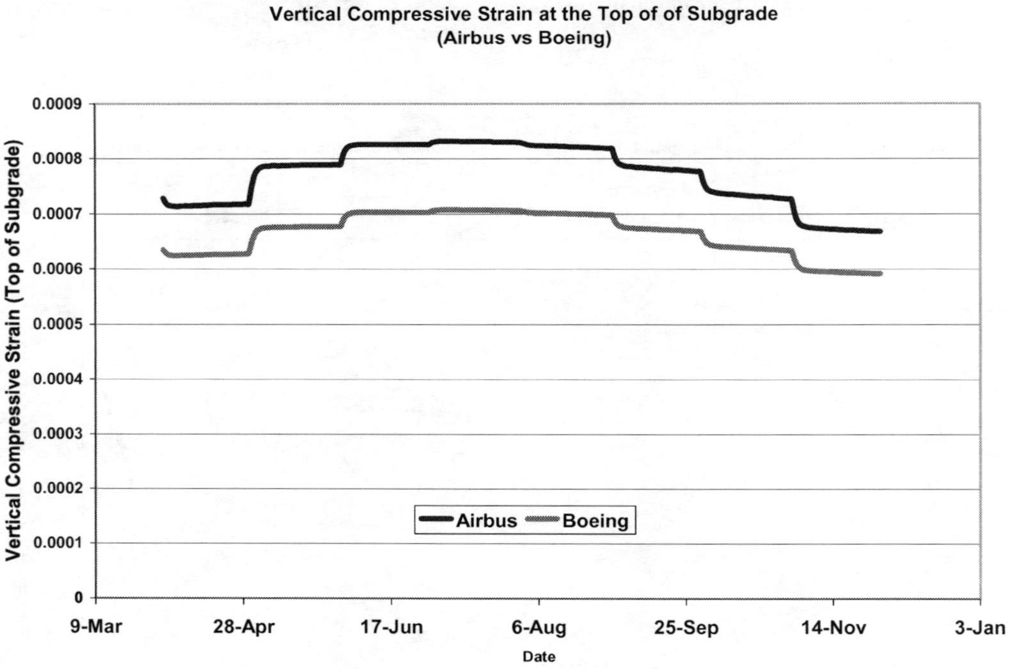

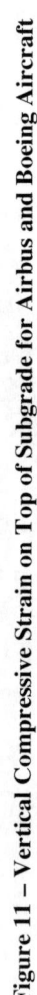

Figure 11 – Vertical Compressive Strain on Top of Subgrade for Airbus and Boeing Aircraft

Table 4 – Results of Permanent Deformation Analysis Based on Shell Criterion (12)

	ε_v * (micro-strain)	Load Repetitions per Month	Load Repetitions (to ¾ inch rutting)	Damage (in 5 years)
		Airbus	N **	$n_{i/N}$
April	720	4,118	2,287,189	0.0090
May	780	4,118	1,660,557	0.0124
Jun	830	4,118	1,295,149	0.0159
Jul	830	4,118	1,295,149	0.0159
Aug	830	4,118	1,295,149	0.0159
Sep	770	4,118	1,748,514	0.0118
Oct	740	4,118	2,049,771	0.0100
Nov	670	4,118	3,050,232	0.0068
		Boeing		
April	630	1124	3,901,844	0.0014
May	670	1,124	3,050,232	0.0018
Jun	700	1,124	2,560,000	0.0022
Jul	700	1,124	2,560,000	0.0022
Aug	700	1,124	2,560,000	0.0022
Sep	670	1,124	3,050,232	0.0018
Oct	640	1,124	3,663,635	0.0015
Nov	600	1,124	4,742,716	0.0012

* From Figure 10   total damage → 0.1121
** $\varepsilon_v = 2.8 \times 10^{-2} \times N^{-0.25}$

Assumptions:
1) Aircraft Operations Per Year (Airbus: 65,885; Boeing: 17,975)
2) ¾ of Aircraft Operations Are at Maximum Weight
3) Rutting Occurs ONLY During April - November
4) Rutting Occurs ONLY 8 hours/day

4   Conclusions

The following conclusions are noteworthy:
- Hamburg Wheel Track testing of specimens both "wet" and "dry" yielded rut depths were less than 5 mm. This is significantly less than the 20 mm limit specified by Utah DOT. These data indicate that the mix is both moisture and rut resistant.
- Frequency sweep and repeated shear testing, similarly, yielded modulus and permanent strain values that compare favorably with typical values reported in

the literature. These data do, however, underscore the effect of asphalt content on the rut resistance of the mix. The repeated shear test data indicate that the "optimum plus" mix sustains approximately 1½ times the plastic strain of the "optimum" mix. It is likely that an asphalt concrete mix with excess asphalt would sustain proportionally greater rutting in the field; hence, the importance of plant process control. The contractor is urged to impose an asphalt content tolerance of ± 0.25%.

- Accumulation of plastic strain based on laboratory testing does not account for long-term aging, i.e., binder hardening, and hence, increased rut resistance that occurs in the field. Accordingly, the laboratory plastic strain data are conservative measures of the likely deformation that would occur in the as-built pavement.
- Shown in Figure 3.3b is the percent shear strain (γ) as a function of load repetitions (N). From a log (γ) – log (N) regression for "optimum" and "optimum-plus" mixes one can then compute the percent strain at any number of load repetitions. A computation of percent strain at 25,000 load repetitions for the "optimum" and "optimum-plus" mixes yielded values of 2.5 percent and 3.4 percent, respectively. These compare favorably to the criterion used for San Francisco International Airport of 5 percent shear strain at 25,000 repetitions, as noted in reference 12.
- An analysis based on the Shell criterion for subgrade strain – load repetitions to ¾ inch rutting – indicated that after 5 years of aircraft operations, predicted minimal contribution to surface rutting from the unbound base and subgrade. This analysis provides additional confirmation as to the adequacy of the structural section and materials.

References

1. Cooley, L; Kandhal, P; Buchanan, M; Fee, F; Epps, A; NCAT Report 00-04, *Loaded Wheel Tests in the United States: State of the Practice*, July 2000.

2. Monismith, C; Popescu, L; Harvey, J; *Rut Depth Estimation for Mechanistic-Empirical Design Using Simple Shear Test Results*, Journal of the Association of Asphalt Paving Technologists, Volume 75, 2006.

3. Dempsey, B., W. Herlache, and A. Patel. *Environmental Effects on Pavements—Theory Manual, Volume 3*. FHWA/RD-84/115, University of Illinois at Urbana-Champaign, 1985. [N.B. The model has been recently updated: Federal Highway Administration, Integrated Climate Model (ICM Release version 2.0.0), prepared for FHWA by University of Illinois, October 1997.

4. *Washington DOT Pavement Guide Interactive*, web site http://training.ce.washington.edu/WSDOT

5. Brown, R R, and Cross, S A, *A Study of In-Place Rutting of Asphalt Pavements*, Association of Asphalt Paving Technologists, v58, 1989.

6. Ford, M C, and Hensley, M J, *Asphalt Mixture Characteristics and Related Pavement Performance,* Association of Asphalt Paving Technologists, v57, 1988.

7. *AASHTO Guide for Design of Pavement Structures*, AASHTO, 1993.

8. *Impact of New Large Aircraft on Airport Design*, DOT/FAA/AR97/26, March 1998.

9. *Airplane Characteristics for Airport Planning – A340-500/-600, AIRBUS,* January 2003.

10. *Airplane Characteristics for Airport Planning – A380-500/-600, AIRBUS,* January 2004.

11. *Shell Pavement Design Manual*, Shell International Petroleum Company, Limited, London, 1985

12. Monismith, C., B.A. Vallerga, J.T. Harvey, , F. Long, and A. Jew, *Asphalt Mix Studies - San Francisco International Airport,"* Proceedings, 26th International Air Transportation Conference, San Francisco, CA, June 2000, American Society of Civil Engineers, 2000.

25 YEARS OF NDT ANALYSIS RUNWAY 1L-19R AT WASHINGTON DULLES INTERNATIONAL AIRPORT

Stanley M. Herrin[1], P.E., Roy D. McQueen[2], P.E., Michael J. Darter[3], P.E. Gary Fuselier[4] and Joseph S. Grubbs[5], P.E.

[1]Manager, Springfield Aviation Group Military & Special Projects, for Crawford, Murphy & Tilly, Inc., 2750 W. Washington St., Springfield, IL 62702; PH 217-787-8050; email sherrin@cmtengr.com.

[2]Vice President, R. D. McQueen & Associates, 22863 Bryant Court, Suite 101, Dulles, VA 20166; PH 703-709-2540; email rdmcqueen@rdmcqueen.com

[3]Principal Engineer, Applied Research Associates, Inc., 100 trade Centre Drive, Suite 100, Champaign, IL 61820; PH 217-369-4500; email:mdarter@ara.com

[4]Project Design Manager for Metropolitan Washington Airports Authority, Ronald Reagan Washington National Airport, Washington DC 20001-4901; PH 703-417-8189; email: gary.fuselier@mwaa.com.

[5]Senior Aviation Engineer, CH2M Hill, 200 Corporate Center Drive, Suite 150 Moon Township, PA 15108; PH 412-604-4043' email: jgrubbs@ch2m.com

Abstract

In 1982 the Federal Aviation Administration, owners of Washington Dulles International Airport, undertook a Pavement Management program at the airport that included the Nondestructive Testing (NDT) of Runway 1L-19R. NDT analysis was obtained with the US Army Corps of Engineers' WES Vibrator and Falling Weight Deflectometer. The data was used to determine modulus of subgrade reaction, k, uniformity of the subgrade support, center and joint deflections and joint load transfer efficiency.

Between 1982 and 2007, maintenance to the PCC runway pavement was limited to select panel replacement, joint sealing, crack repair and spall repair.

In 2007, as part of a project to replace Runway 1L-19R, NDT tests were conducted on the runway to determine subgrade support values.

This paper describes the results of this unique opportunity to compare NDT values taken on the same pavement 25 years apart, with similar equipment.

Introduction

In 1982 the Federal Aviation Administration (FAA) undertook a comprehensive Pavement Management (PM) study at Washington Dulles International Airport. The FAA, owners of Washington Dulles International Airport, had previously undertaken an evaluation of the airfield pavements in 1978 that included a visual survey and nondestructive testing. However, the 1982 study included a Pavement Condition Index (PCI) survey, and nondestructive testing using the falling weight deflectometer.

The PCI method of determining the pavement condition was developed by the Corps of Engineers in the mid 1970's and adopted by the FAA in 1982. Dynatest Consulting, Inc introduced the Falling Weight Deflectometer to North America in the late 1970's. Thus, this 1982 study was one of the first PM studies at an airport using techniques that are standard investigative procedures today.

Two types of Nondestructive Testing (NDT) equipment were used in 1982 to evaluate the pavements at Washington Dulles International Airport. The first NDT equipment was the WES Vibrator, developed by the U.S. Army Engineer Corps Waterways Experiment Station (WES). The WES Vibrator consists of a loading plate, a 15-kip dead load, and a force generator mounted in a semi-truck trailer (Figure 1). The second NDT equipment was the Falling Weight Deflectometer (FWD), introduced to North America by Dynatest Consulting, Inc. in the late 1970's. The FWD consists of a loading plate, a mass which can be lifted to a predetermined height and dropped and a spring shock system onto which the mass is dropped. The applied load to the pavement was measured as approximately 15,000 pounds.

The 1982, Washington Dulles International Airport consisted of three runways, three air carrier ramps and connecting taxiways. All of the airfield pavements were included in the 1982 Pavement Management study, but only the most heavily trafficked pavements were subjected to a detailed NDT survey.

1982 NDT Program

Runway 1L-19R was included in the detailed NDT survey. NDT tests were taken in the center 50 feet of the runway, alternating the testing locations between the slabs located left and right of centerline.

The FWD tests were conducted at the center of slab, at the transverse joint plus supplementary tests at slab corners, at longitudinal joints and at cracks. Center of slab and transverse joint test locations were usually the same locations as tested using the WES Vibrator.

Figure 1. US Army COE WES Vibrator

2007 Program

Between 1982 and 2007, repairs to Runway 1L-19R consisted of Panel Removal and Replacement for new Touchdown Zone Lights, and a repair program in 2003 consisting of the removal and replacement of approximately 38 panels, crack and spall repair.

In 2007 the Metropolitan Washington Airports Authority began a program to reconstruct Runway 1L-19R, to begin upon completion of the new north-south runway being constructed at the airport. The runway is to be referred to as 1C-19C beginning June 2008.

As part of the reconstruction of Runway 1L-19R, a HWD survey of the runway was conducted to determine the uniformity of the pavement and the subgrade support. The applied load was approximately 32,000 pounds.

Summary of Paper

This paper compares results of NDT surveys taken 25 years apart on the same pavement with similar equipment.

1982 Data Analysis

The original NDT has been lost over the years; however a summary of the NDT results was presented in the Pavement Evaluation Study[1] report, and presented here. All of the deflections were normalized to an applied load of 15,000 pounds.

Deflection Profile – A profile of the center of slab deflection is shown in Figure 2. The average value of all tests was 2.24 mils. The highest value was 3.2 mils, and the lowest value was 1.25 mils. Most of the values fall between 1.7 and 2.5 mils.

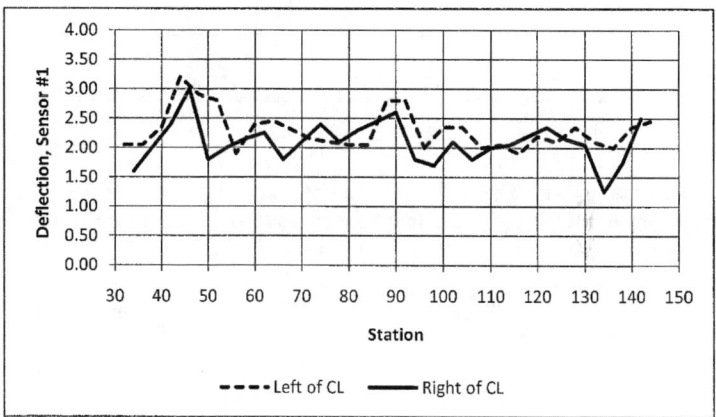

Figure 2. 1982 FWD Deflection Profile for 15,000 pound load

Uniform Deflection Pavement Section – The 1982 NDT data collected using the WES Vibrator was analyzed and divided into pavement sections according to criteria given in FAA Advisory Circular AC 150/5370-11. These pavement sections have similar deflection values and are statistically different than adjacent sections. The average deflection value for each section was calculated.

The 1982 NDT data for Runway 1L-19R was very consistent, and only one pavement section was found to exist on the runway. However, the 1978 NDT data from the WES Vibratory found three distinct pavement sections. Table 1 summarizes the NDT data from 1978 and 1982.

The average FWD deflection for Runway 1L-19R was 2.24 mils (2.24 x 10^{-3} inches) in 1982. The average WES Vibrator deflection for Runway 1L-19R was 2.87 mils (2.24 x 10^{-3} inches) in 1982.

Table 1. Average Deflection Runway 1L-19R 1978 and 1982

Pavement Element	Location	Stationing	Equipment	Testing Year	Sensor #1 Deflection (x10⁻³ Inches)
1L-19R	All Slabs	30-145	WES	1978	2.32
	All Slabs	30-145	WES	1982	2.87
	Slabs Left of Centerline	30-145	WES	1982	2.89
	Slabs Right of Centerline	30-145	WES	1982	2.87
	All Slabs	30-145	FWD	1982	2.24
	All Slabs	30-70	WES	1978	2.19
		70-90	WES	1978	2.42
		90-145	WES	1978	2.38
	All Slabs	30-70	WES	1982	2.74
		70-90	WES	1982	3.07
		90-145	WES	1982	2.88

Deflection Data – The deflection data obtained during the 1982 testing data is presented in table 2. The data has been grouped by PCI Sample units, and the stationing for these sample units has been provided.

The average center deflection for the WES Vibratory is approximately 28% higher than the average center deflection for the FWD. These are very different types of loadings, so this was not expected even for the same relative load magnitude of 15,000 pounds.

Table 2. 1982 Deflection Data

Pavement Element	Feature	PCI Sample Units	Stationing	Equipment	Average Center Deflection (x10⁻³ Inches)	Average Transverse Joint Deflection (x10⁻³ Inches)	Average Corner Deflection (x10⁻³ Inches)	Average Longitudinal Joint Deflection (x10⁻³ Inches)	Average Joint-to-Midspan Ratio
1L-19R	R1LB1	1-13B	30-55	WES	2.83	4.38			1.58
	R1LB2	14-42B	58-114	WES	2.91	5.39			1.9
	R1LB3	43-57B	115-145	WES	2.79	4.16			1.53
	R1LB1	1-13B	30-55	FWD	2.22	4.26	9.55	5.63	2.03
	R1LB2	14-42B	58-114	FWD	2.26	6.29	15.51	6.4	2.81
	R1LB3	43-57B	115-145	FWD	2.23	5.94	12.5	7.68	2.78

Deflection vs PCI and Subsurface Conditions – As part of the 1982 analysis, the average deflection values were plotted along with the PCI sections and cut/fill profile, as shown in Figure 3. This analysis determined that for Runway 1L-19R, the deflection values are 1) not related to the pavement condition, measured by the PCI (the runway ends had more cracking) and 2) not related to cut and fill locations along the runway.

2007 Data Analysis

In the spring of 2007, an NDT survey using a 'Heavy' HWD was conducted on Runway 1L-19R. The primary purpose of the 2007 NDT program was to obtain an estimate of the in-situ modulus of subgrade reaction, k. The applied load was approximately 32,000 pounds. The data were normalized to a load of 15,000 to allow comparison of 2007 data with 1982 data.

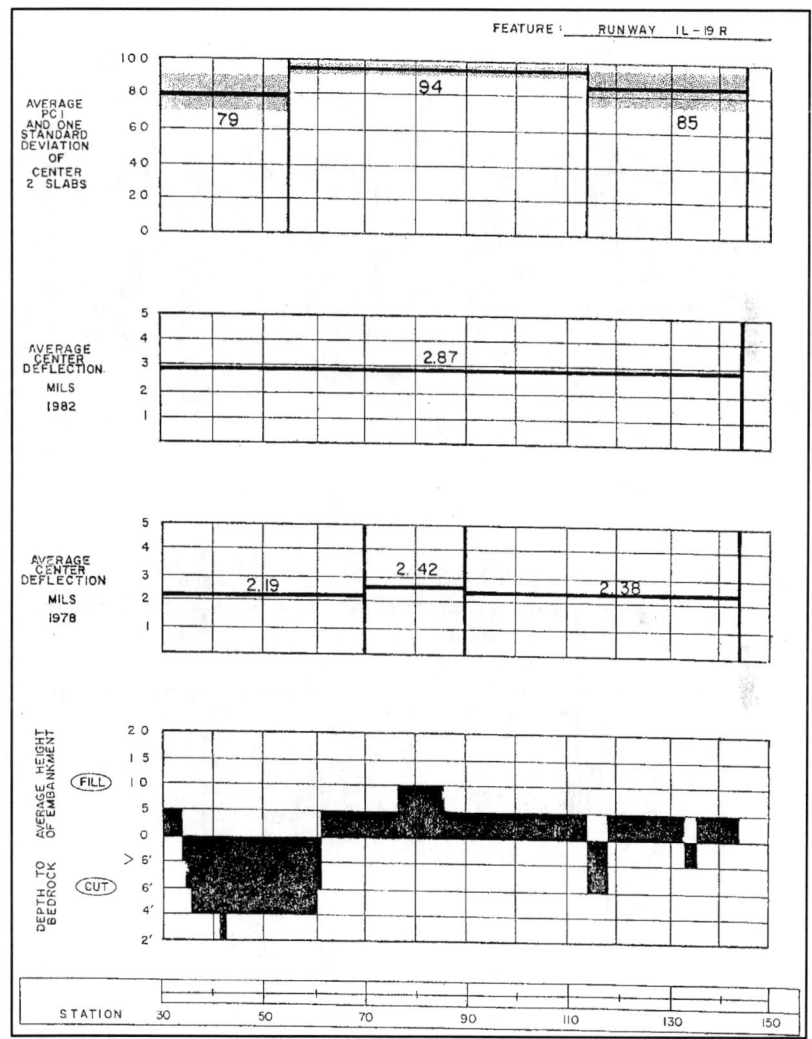

Figure 3. Relationship between PCI, NDT data and cut/fill on Runway 1L-19R

Deflection Profile – A profile of the center of slab deflection is shown in Figure 4. The average deflection for the runway is 2.4 mils. The highest value was 6.3 mils, and the lowest value was 1.1 mils. Most of the values fall between 2.0 and 3.0 mils.

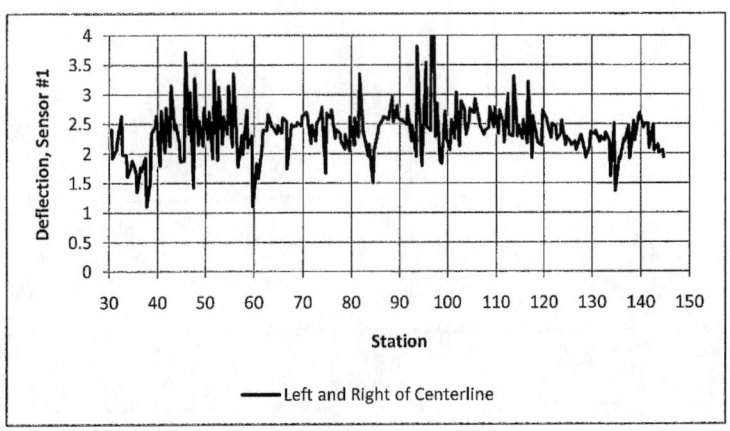

Figure 4. 2007 FWD Deflection Profile

As a comparison, the 1982 FWD deflection data, from both the left and right of centerline, was superimposed on the 2007 data, and is shown in Figure 5. Visually, there is a close comparison of the two sets of data.

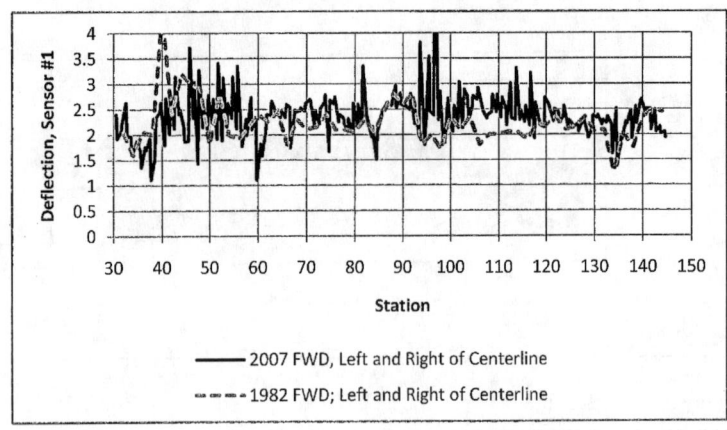

Figure 5. Comparison of 1982 and 2007 FWD Deflection Profile

Uniform Deflection Pavement Section – The 2007 FWD data was analyzed and divided into uniform pavement sections.

While the 1982 NDT data for Runway 1L-19R was very consistent, the 2007 data indicated there are three pavement sections, shown in Table 3.

Table 3. 2007 Pavement Sections

Station	Average Deflection (mils)
30+00 to 38+60	1.85
38+60 to 124+20	2.44
124+20 to 145+00	2.20

Although the 1978 NDT survey found three distinct pavement sections, as shown in Figure 3, the limits of these sections do not correspond to the 2007 sections.

It is interesting to note in Table 3, that the ends of the runway (last 1,000 to 2,000 feet) had lower deflections than the center of the runway. This trend was not observed previously, perhaps due to the reduced number of deflection points taken in 1978 and 1982.

Deflection Data – The deflection data obtained during the 2007 testing program is summarized in table 3. Since the purpose of the 2007 testing program was to obtain an estimate of in-situ k, mot of the tests were conducted at mid-slab. The limited joint tests that were obtained were not sufficient for detailed analysis.

Deflection vs PCI and Subsurface Conditions – The relationship of 2007 PCI values and deflection values will be discussed as part of another paper.

The relationship between the 2007 deflection values and subsurface conditions does not appear to be related. Although the pavement section from station 30+00 to 38+60 is lower than the rest of the runway, and in a cut section, as shown in Figure 3, the cut section extends approximately 2,000 feet beyond this pavement section. The pavement section from 124+20 to 145+00 is primarily on embankment. Again, as in 1982, there does not appear to be a relationship between deflection and cut and fill locations along the runway. This may be due to significant mixing of soil materials along the runway to achieve a uniform support throughout.

1982 Determination of Subgrade Support

In 1982, the collected FWD data was used to "backcalculate" the dynamic k-value immediately below the PCC pavement using the following procedure:

First, the deflection basin was measured by the FWD at several center slab locations on the airfield.

Second, the "area" of the deflection basin was calculated from this data. The "area" is based on deflection values under the load, 12, 24 and 36 inches from the load, as shown in Figure 6.

Third, the computer program "ILLISLAB" was used to calculate theoretical deflection and area over a wide range of support values immediately under the PCC pavement and modules of elasticity for the PCC pavement. The results were plotted, as shown in Figure 6. This curve was developed for the Tower Ramp, where the pavement was 9-inches in thickness. Similar curves were developed for the 15-inch runway and taxiway pavements.

Finally, using the PCC modulus of elasticity value determined in 1978 (5.4 million psi) as the PCC slab dynamic modulus of elasticity, the measured deflection and the calculated area, the K-value was determined. The k-value is the support value immediately under the PCC pavement, at the top of the base course.

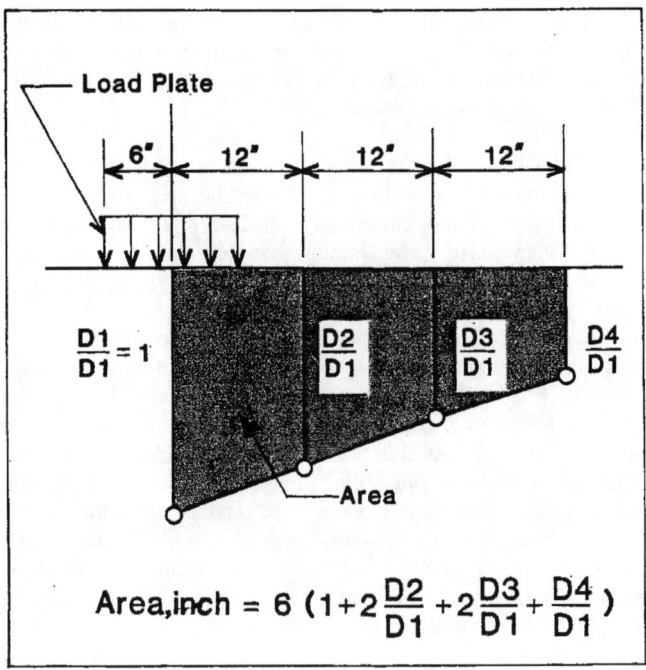

Figure 6. Calculation of Deflection "Area"

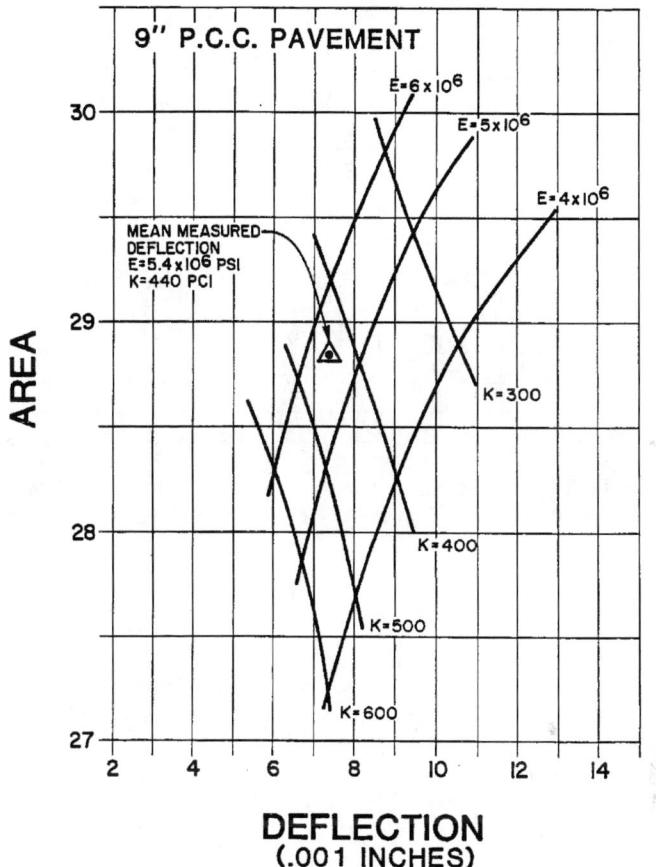

Figure 7. k-Value Determination based on Computer Modeling.

Since the k-value appeared to vary considerably throughout the airport due to drainage, distance to bedrock and other unknown factors, the measured deflection and calculated area for each feature was used to determine the dynamic k-value of each section.

For Runway 1L-19R, the dynamic k-values, immediately under the PCC pavements, as determined during the 1982 investigation, are shown in Table 4.

In 1982, the k-value at the top of the subgrade was determined by using Figure 2-4 of FAA AC 150/5320-6D, the airfield pavement design procedure for 1982. Based on thickness of subbase and k-value at the top of the subbase, the subgrade k-value is extrapolated. This value is also shown in Table 4, termed Extrapolated Dynamic K-value top of Subgrade.

The backcalculated k-value is a dynamic k-value since the k-value was computed from a dynamic load. In 1982, the static value was estimated by dividing these values by approximately two.

Table 4.   1982 Support Values

Feature	Dynamic k-Value, below PCC pavement, psi/in	Extrapolated Dynamic k-Value, Top of Subgrade, psi/in	Equivalent Static k-Value, Top of Subgrade, Psi/in
R1LB1	350	195	98
R1LB2	450	330	165
R1LB3	450	330	165
Average			143

This 1982 procedure for estimating the subgrade support values was one of the first attempts to do so on an airfield. The AC 150/5320-6E procedure should be used today.

2007 Determination of Subgrade Support

As part of the reconstruction of 1L-19R in 2007, a subsurface investigation and a NDT survey were conducted. The NDT survey was conducted with a Heavy Weight Deflectometer and the NDT tests were performed in accordance with FAA AC 150/5370-11A. Both the geotechnical and NDT results were considered in formulating the input for subgrade k.

The NDT data were reduced using closed form back-calculation methods based on the AREA method as described in the advisory circular. For the runway, the mean minus one standard deviation for k (design k) at the top of the existing aggregate base is 248 psi/in. This equates to a design k of approximately 150 psi/in. at the top of the subgrade.

The k-values determined from the 1982 and 2007 NDT surveys are of the same magnitude (150 versus 143 psi/in). Differences between the two values may be due to 1) the 2007 value is based on more data points and 2) season affects.

Conclusion and Summary

These conclusions and summary are limited to the changes in data between 1982 and 2007, and subgrade support analysis. Although tempting, conclusions and discussion on pavement performance during the 25 years has been reserved for others.

1. There has not been a significant change in center-of-slab deflection on Runway 1L-19R at Washington Dulles International Airport during the past 25 years. The slight change may be due to one of more reasons, including:

a. The condition of the concrete slabs deteriorated between 1982 and 2007.
b. The original 1982 data is not available. Had the data been available, the conversion between 2007 applied load and 1982 applied loads may have changed.
c. The surveys were conducted at different times of the year. The 2007 survey was conducted in the spring, and the 1982 survey was conducted in the fall.
d. Variability between testing equipment.
e. Changes in moisture content of the subgrade over 25 years would result in changes in deflection.

2. Although the center-of-slab deflection has increased, the increase would not have significantly affected the 1982 investigation and conclusions.

3. If the nondestructive test data from 1982 and 2007 were used to determine required pavement thicknesses for Runway 1L-19R, for the same traffic conditions, the thickness would be very similar.

4. The method used to determine k-value in 1982 was one of the first attempts to do so. The results of this effort produced a k-value very similar to season effects from the spring vs. fall testing performed in 2007 and 1982, respectively.

References

1. Crawford, Murphy & Tilly, Inc.; ERES Consultants, Inc.; Barenberg, E. J.; and Shahin, M. Y. (1984) *1982 Report on Evaluation of Airfield Pavements at Dulles International Airport.*

2. Federal Aviation Administration; AC 150/5370-11A *Use of Nondestructive Testing Devices in the Evaluation of Airport Pavements.*

3. Federal Aviation Administration; (1995) AC 150/5320-6D *Airport Pavement Design and Evaluation*, p 15.

4. Howard, Needles, Tammen and Bergendoff; ARE, Inc. (1979) *Evaluation of Airfield Pavements at Dulles International Airport.*

THE LIFE CYCLE OF A RUNWAY PAVEMENT
A CASE STUDY OF RUNWAY 1L-19R AT THE WASHINGTON DULLES INTERNATIONAL AIRPORT

Gary K. Fuselier, M. ASCE [1]

Joseph S. Grubbs, PE; M.ASCE [2],

Roy D. McQueen, PE, M.ASCE [3]

Abstract

In 1961, the original construction of Runway 1L-19R was completed and the pavement remained idle for approximately one year until the opening of the Washington Dulles International Airport in 1962. The pavement was designed for a 25 year life and will have been in continuous use for nearly 50 years at the time when it is scheduled for reconstruction in 2009.

This paper will examine the original design parameters and the history of loading (traffic growth) over the life of the pavement. It will also discuss the active pavement management system implemented by the Authority and the subsequent maintenance activities used to extend life of this pavement.

Finally, the paper will discuss the pavement design associated with the upcoming reconstruction project as the Authority attempts to match the history of performance of the original pavement life.

Airport Overview

In 1962, the Washington Dulles International Airport (IAD) opened for service. IAD is located in Loudoun and Fairfax Counties in northeastern Virginia, approximately 26 miles northwest of Washington, D.C. The Metropolitan Washington Airports Authority (the Authority) owns and operates the facility.

---

[1] Project Design Manager for Metropolitan Washington Airports Authority, Ronald Reagan Washington National Airport, Washington, DC 20001-4901; PH (703) 417-8189; email: gary.fuselier@mwaa.com

[2] Senior Aviation Engineer for CH2M HILL, 200 Corporate Center Drive, Suite 150, Moon Township, PA 15108; PH (412) 604-4043; email: jgrubbs@ch2m.com

[3] Vice President for Roy D. McQueen & Associates, 22863 Bryant Count, Suite 101, Sterling, VA 20166; PH (703) 709-2540; email: rdmcqueen@rdmcqueen.com

Currently, IAD is designated as a large hub primary commercial service airport in the FAA's National Plan of Integrated Airport Systems (NPIAS). IAD serves as the "growth" airport in the metropolitan Washington Region, since Ronald Reagan Washington National Airport is constrained by Federal Legislation and surrounding land uses. Baltimore Washington International Airport (BWI) also serves the greater Washington DC region but is similarly encumbered by land constraints.

The original runway and taxiway system was completed approximately one year prior to the airport opening. As originally constructed in 1961, three Portland cement concrete (PCC) runways continue to serve the airport. These runways are designated 1L-19R, 1R-19L and 12-30. Runways 1L-19R and 1R-19L are each 11,500 feet long by 150 feet wide. Runway 12-30 is 10,500 feet long by 150 feet wide. IAD is scheduled to open a third parallel runway in November of this year. This new PCC runway will be 9,400 feet long by 150 feet wide and will be designated Runway 1L-19R. The original Runway 1L-19R will be re-designated as Runway 1C-19C.

Figure 1 – Aerial photograph of the Washington Dulles International Airport. Runway 1R-19L is in the foreground with its parallel, Runway 1L-19R, and the cross wind, Runway 12-30, in the background.

Original Pavement Design

In 1958, the original pavement design was prepared for the Civil Aeronautics Administration (CAA), which was later incorporated into the Federal Aviation Agency (FAA). The CAA (FAA), which operated Dulles Airport at that time, decided on the use of Portland cement concrete pavement over the use of asphaltic concrete for the entire airport, formally known then as the Washington International Airport.

The pavement design was prepared in accordance with the CAA (FAA) design methodology and compared to the Portland Cement Association (PCA) / Westergaard and Corps of Engineers (COE) methodologies. The design assumed the following:

1. Equivalent Single Wheel Load of 100,000 lbs.
2. Using the PCA and COE methodologies, an airplane whose gross load is 500,000 lbs with a DC-8 gear configuration and a tire pressure of 200 psi was assumed equal to the 100,000 lbs equivalent single wheel load.
3. The soil was classified as E-8, in accordance with the CA (FAA) soil classification system.
4. Portland cement concrete - Mr = 750 @ 28 days and 850 @ 90 days
5. Effective Subgrade Modulus at top of aggregate subbase, k = 200 psi
6. Factor of Safety - 1.7 for Critical Areas and 1.25 for Non-critical Areas

The volume of aircraft operations anticipated at the time this design was forecast based on the 1958 operations at other major airports across the United States including New York (LaGuardia and Idlewild); Chicago (O'Hare and Midway); Los Angeles; and San Francisco. The original forecasted aircraft operations were 214,100 for 1965; 252,000 for 1970; and 396,000 for 1975. The way the airfield was intended to be operated, it was assumed that 33 percent of the operation were planned to take place on Runway 1L-19R.

Using the CAA pavement design method with the aforementioned design assumptions and the forecasted traffic, the critical area pavement requirements were 15 inches of un-reinforced concrete on 9 inches of crushed aggregate base course material. The pavement requirements for the non-critical areas were determined to be 12 inches of un-reinforced concrete on 9 inches of crusher run base course material. The PCA pavement design methodology yielded the same critical area pavement section. The COE design methodology resulted in a much thicker critical area pavement, 21 inches of concrete with the same base course.

In 1960, construction of Runway 1L-19R with a 15-inch Portland cement concrete (PCC) un-reinforced pavement on a crushed aggregate base course pavement section began. The typical slab size was 20 feet by 25 feet. The transverse joints were primarily aggregate interlock contraction (dummy) joints. In the final three joints at each end of the runway, these contraction joints included a tie bar. The typical pavement section, including longitudinal joint types, is shown in Figure 2.

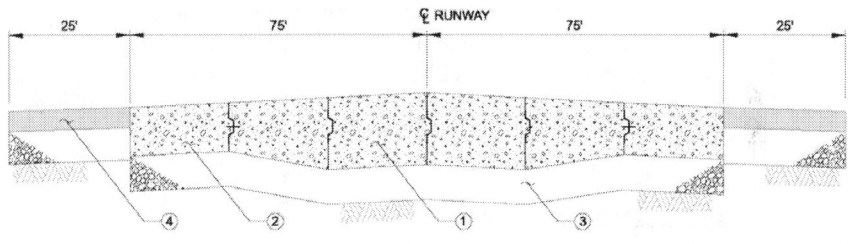

Legend
1. PCC Pavement, 15 inch depth
2. PCC Pavement, 12 inch depth
3. Aggregate Base, 9 inch depth
4. Bituminous Concrete

Figure 2 – Original Runway Pavement Section

It should be noted that at the runway/taxiway intersections, the PCC pavement was maintained at a constant 15 inches.

The original pavement section did not include pavement base drains or underdrains, even though it is apparent that water would be trapped in the aggregate base of the keel section. However, given the pavement's 47 year old age, this trapped water, which commonly would surface through the in-pavement electrical fixtures, had no measurable ill effects on the pavement. The pavement longevity, in spite of the presence of subsurface water, may be attributed to high PCC flexural strength as well as the low volume of initial traffic.

Traffic Growth

Traffic growth did not increase as originally forecast. Table 1 summarized the historical operations of Washington Dulles International Airport.

Table 1 - Historical IAD Operations

Year	Total Operations	Year	Total Operations	Year	Total Operations
1962	8,016	1980	165,420	2000	456,436
1965	158,883	1985	208,333	2005	509,652
1970	184,226	1990	242,209	2007	382,939
1975	177,673	1995	308,144		

Though the overall volume of traffic did initially grow as originally forecast, the past twenty years has seen a substantial increase in large aircraft operations. This growth in large aircraft traffic spurred changes throughout the years in the Maintenance and Repair (M&R) activities. It should be noted that the drop in operations between 2005 and 2007 was due to the cessation of a based regional low cost carrier. This led to the elimination of an estimated 600 daily regional jet flights.

Pavement Management Activities

Although the Runway 1L-19R pavements were constructed in the early 1960's, Dulles did not experience heavy traffic until the mid 1980's. Up to that time, M&R activities consisted mainly of routine preventive maintenance measures, such as joint re-sealing. However, as the traffic increased in the 1980's, pavement deterioration, and consequently M&R requirements and funding, increased. To keep the aging pavements serviceable, more emphasis was necessarily placed on crack sealing, spall repairs and isolated slab replacements. Figure 3 depicts slab replacement activities that were performed during a series of nightly runway closures.

Figure 3 - Photo of Isolated Slab Replacement

To keep pace with ever increasing maintenance requirements, the Airports Authority instituted a formal pavement management system (PMS) in 1982 to better forecast and budget necessary M&R. The initial implementation of PMS was in 1982 and updates are currently performed yearly, with all pavements re-inspected on a maximum 3-year cycle. PMS activities consist of visual condition survey at 100% frequency (i.e., all slabs are inspected), preparation of distress repair maps for in-house and contract maintenance programs, use of MicroPAVER for storing and organizing data and generating annual M&R budget estimates, selective nondestructive testing and structural analysis, and development of discrete projects. An example deterioration curve for a section of Runway 1L-19R keel prepared in 2005 is shown in Figure 4.

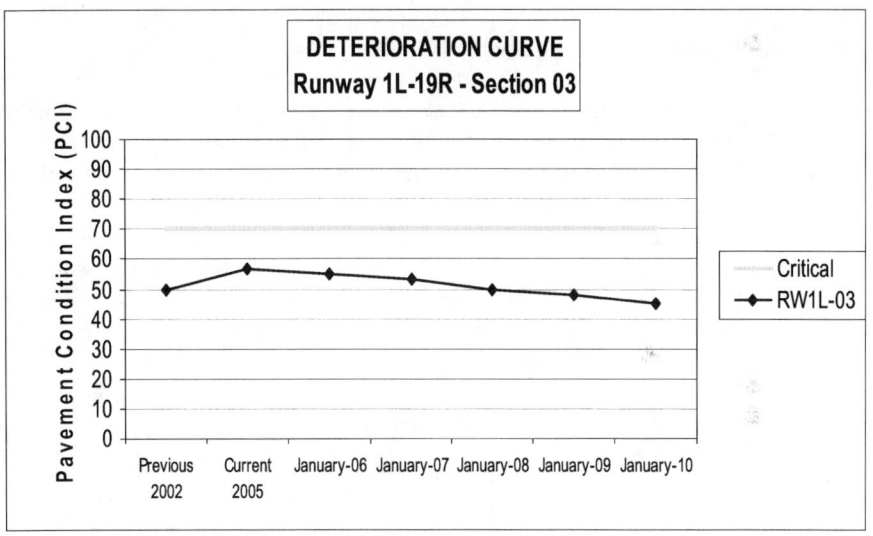

Figure 4 - Runway 1L-19R Deterioration Curve

As shown, the application of PMS and initiation of timely M&R activities, primarily isolated slab replacement brought the PCI up to 58 in 2005. However, it is evident that without a major rehabilitation, the downward general trend in the condition of the 47 year old pavement, as depicted in Figure 4, would continue to worsen. Further, approximately 50% of the recorded distresses on the 15-inch concrete pavement were attributed to load related mechanisms. More details on the PMS program at Dulles can be found in *"Using PCI Data to Define Major Rehabilitation Projects at Washington Dulles International Airport"* by Thuma, Fuselier and Yip.

Reconstruction of Runway 1L-19R

Order of Reconstruction. Utilizing the tools of the PMS, the airport has established a systematic approach to reconstruction of the original runway pavements. Based on the PMS condition rankings, Runway 12-30 was the first feature to be reconstructed. Runway 12-30 was reconstructed in 121 calendar days during the summer of 2004. The expedited construction was necessitated by the operational impact of only having two active runways. Runway 1L-19R is scheduled for reconstruction in 2009 with Runway 1R-19L to follow in the not too distant future.

Coordinated Reconstruction of Runway 1L-19R. To mitigate the operational need to expedite the reconstruction of Runway 1L-19R (soon to be re-designated as Runway 1C-19C), the reconstruction has been coordinated with the opening of the third parallel runway, the new Runway 1L-19R. The new Runway 1L-19R will lessen the operational impact, thereby permitting a more moderate timeframe for the Runway 1C-19C reconstruction.

Reconstruction Pavement Design. The pavement design for Runway 1C-19C was performed in accordance with FAA Advisory Circular 150/5320-6D. A fundamental requirement for pavement thickness design is a reasonable estimate of aircraft demand forecast. As a starting point, the airport's 2003 report on "Updated Activity Forecasts and Simulation" was consulted. Based on sensitivity studies performed during the pavement design, it was assumed that the B777 aircraft that will use Runway 1C will consist of 25% B777-300ER and 75% of B777-300. The traffic forecast used to support the pavement design is summarized on Table 2.

Table 2 – Forecast Traffic for the Pavement Design

Aircraft Type	Weight	Gear	Total Departures 2010	Total Departures 2020	2020 RW 1C-19C Departures 25%
Business Jets & Turbo	30,000	D	35,150	51,565	12,891
C130	155,000	ST	1,550	2,274	568
K135	335,000	2D	2,550	3,741	935
Commuters	45,000	D	22,783	33,422	8,355
Regional Jets	75,000	D	111,287	163,258	40,815
A-319, A-320, B-737 & MD-80	160,000	D	27,529	40,385	10,096
A321	187,500	D	13,559	19,891	4,973
Aircraft Type	Weight	Gear	Total Departures		2020 RW 1C-19C Departures

			2010	2020	25%
B-717	121,000	D	3,947	5,790	1,447
B-727	210,000	D	1,240	1,819	455
B-737-800	175,000	D	8,859	12,995	3,249
B-757	250,000	2D	20,312	29,798	7,449
DC-9S	125,000	D	4,276	6,272	1,568
DC-8-63	358,000	D	260	381	95
B-767-200	335,000	2D	1,818	2,666	667
B-767-300,400	409,000	2D	1,981	2,906	727
B-777-300ER	777,000	3D	2,690	3,946	987
B777-300	662,000	3D	8,069	11,837	2,959
B-747-Series	873,000	2D/2D2	1,944	2,852	713
MD-11/DC-10	621,000	2D/D1	960	1,408	352
A-310	331,000	2D	1,040	1,526	381
A-330	459,000	2D	1,745	2,559	640
A-340	567,000	2D/D1	2,155	3,161	790
A380	1,300,000	2D/3D2	1,000	1,467	367
Total			276,701	405,920	101,480

Table 2 Cont'd

For the design analyses, FAA's current and proposed mechanistic-empirical pavement design methods were utilized. The LEDFAA (v.1.3) program was used for the pavement design and sensitivity analysis studies. The finite element method using the Beta version of the FAARFIELD program was used for the final design sequence and compared to the LEDFAA results. Generally, the correspondence between FAARFIELD and LEDFAA was quite good.

To gauge the sensitivity of input variations on the thickness designs, the following primary design inputs in LEDFAA were varied within a reasonable range of the expected value:

- Runway Traffic Frequency: 25%, 33%, 50% (25% baseline)
- Concrete Flexural Strength (MR): 650 < MR < 750 (675 psi baseline)
- Modulus of Subgrade Reaction (k): 150 < k < 250 (200 psi/in baseline)

The results of the sensitivity analysis for the variable traffic frequency at the variable flexural strengths are as shown in Table 3.

Table 3 – PCC Thickness (on 6 inches of CTB with k = 200 psi/in) Sensitivity to MR

Flexural Strength, MR (psi)	PCC Thickness (inches)		
	25% Traffic	33% Traffic	50% Traffic
650	17.6	17.9	18.4
675	17.0	17.3	17.8
700	16.4	16.7	17.2
725	15.9	16.2	16.6
750	15.5	15.8	16.1

The results of the sensitivity analysis for the potential varying modulus of subgrade reaction at the variable traffic percentages are shown in Table 4.

Table 4 – PCC Thickness (on 6 inches of CTB, MR = 675 psi) Sensitivity to k

Traffic Percentage	PCC Thickness (inches)		
	150 psi/in	200 psi/in	250 psi/in
25%	17.9	17.0	16.6
33%	18.2	17.3	16.9
50%	18.7	17.8	17.3

Minimizing Subgade Variability. Based on the geotechnical evaluation, the existing subgrade is not conducive to serving as a construction platform without improvements. Therefore, the following subgrade improvements were considered to provide adequate support for construction traffic, and, ultimately, for the rigid pavement:

- The first method is to treat the subgrade material with cement, which is a common practice at Dulles. Typically, the depth of soil cement stabilization is 12 inches.

- The second treatment method is to remove subgrade soils to a depth of between 12 to 16 inches, and replace with a well blended large aggregate material. To manufacture this aggregate, crushing and reusing the existing PCC pavement was evaluated.

Based on the cost comparisons, cement stabilization was selected for subgrade improvement. However, as part of a sustainable design, the existing PCC pavement will be recycled for other uses on the project.

After evaluating all data sources, the modulus (k) of the in-situ was conservatively estimated at 116 psi/in. This is less than the in-situ k that was measured by nondestructive testing (125 to 150 psi/in), yet slightly higher than that estimated in the geotechnical report (90 psi/in) from CBR correlations. This value also corresponds to the historical value of 120 psi/in used for the original pavement design.

With the k-value on the unimproved subgrade conservatively estimated at 116 psi/in., the FAA correlations in Advisory Circular 150/5320-6D were used to estimate the effective k at the top of the soil cement layer. This resulted in an effective k ranging from 215 psi/in. to 350 psi/in., depending on how the cement stabilized subgrade is characterized. For the thickness designs an effective k at the top of the soil cement layer was conservatively estimated at 200 psi/in, about what would be expected from a 12-inch layer of CBR=20 subbase. The use of a design effective k of 200 psi/in. at top of soil cement was also supported by plate load tests on a test strip constructed for the newly constructed Runway 1L-19R at Dulles. So, it appears that an effective k of 200 psi/in. at the top of soil cement is reasonably conservative based on the NDT results, recent plate load tests and application of FAA correlations. From the sensitivity study, other design inputs included design flexural strength of 675 psi and application of 25% of airport wide traffic.

Using the aforementioned baseline parameters (MR = 675 psi, k = 200 psi/in and traffic = 25%), the pavement structure using the LEDFAA program was determined to be 17 inches of Portland cement concrete (PCC) pavement (Item P-501) with 6 inches of Cement Treated Base (CTB) Course (Item P-304) on 12 inches of Soil Cement Stabilized (SCS) Subgrade (Item P-301). The pavement structure using the FAARFIELD program was determined to be a comparable 17.2 inches of PCC pavement with 6 inches of CTBC on 12 inches of SCSS.

Given the potential variables evaluated as part of the sensitivity analysis, the recommended pavement section for the runway reconstruction was 18 inches of PCC pavement with 6 inch CTB course on 12 inches of SCS subgrade as shown in Figure 5. This pavement section will provide the airport with the greatest operational flexibility. The runway pavement would be capable of handling up to 50% of the airport's departures.

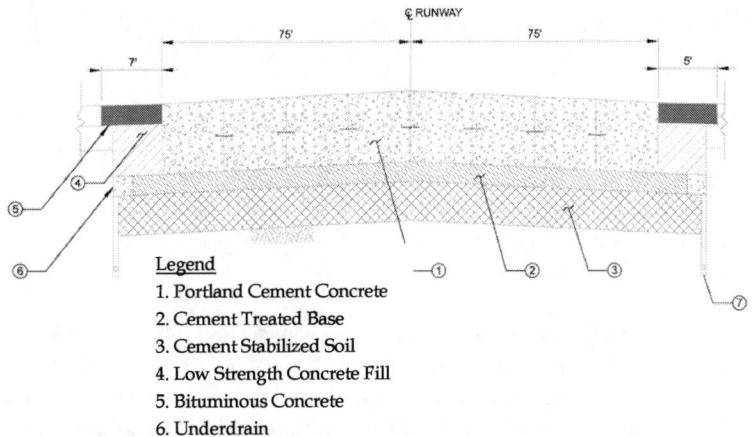

Legend
1. Portland Cement Concrete
2. Cement Treated Base
3. Cement Stabilized Soil
4. Low Strength Concrete Fill
5. Bituminous Concrete
6. Underdrain

Figure 5 – Proposed Pavement Section for the Runway Reconstruction

Unlike the original runway pavement, the replacement pavement section will include underdrains to collect subsurface water. Aiding in this collection will be periodic drains of various airfield electrical system items.

Although the design computations can also support a runway slab thickness of 17 inches of PCC pavement, due to the uncertainties in aircraft utilization (e.g., B777 split and future B787 use), as well as possible degradation of the soil cement layer over time, the use of an 18-inch slab thickness for the runway is prudent. It will also result in a design life somewhat in excess of

20-years, whereas the theoretical design life of the 17-inch slab is 20-years for the assumed inputs. Therefore, the additional inch of concrete will provide reserve capacity to accommodate unexpected increases in aircraft weights and/or frequency over the pavement design life.

Summary and Conclusion

Why did the original pavement last so long? The pavement longevity can be attributed to two main factors. The first factor is the slower than expected traffic growth. The original forecast traffic anticipated that the 396,000 annual operations would be reached by 1975. This number of annual operations was not exceeded until 1999, twenty four (24) years later than originally anticipated. The second factor contributing to the longevity of the runway pavement is the implementation of the airfield wide Pavement Management System. The PMS led to timely maintenance and repair of the runway pavement, which maintained its serviceability throughout the years.

Expectation of new pavement section. The new pavement section is expected to likewise exceed the standard twenty year design life. This expectation is based on several factors. The first factor is the selection of conservative inputs for the pavement design procedures which led to the selection of an 18-inch PCC pavement section. The second factor is the sophistication of an established and fully implemented Pavement Management System with a history of successful M&R activities. Lastly, the project employs many of the Innovative Pavement Research Foundation (IPRF) Best Practices for PCC pavement construction. Using the current Advisory Circular 150/5370-10c, this P-501 specification was tailored to include years of historical paving experience at IAD. Additionally, this specification was further tailored to include sections of the supplemental specification P-50X.

With the conservative pavement design parameters, an effective PMS and M&R program and the specifying of PCC Best Practices, there is a good probability that the new runway pavement life will exceed the twenty-year design life. Whether this new pavement section can match the performance of the original pavement section, only time will tell.

ACCURACY OF PAVEMENT MANAGEMENT PREDICTIONS A CASE STUDY AT WASHINGTON DULLES INTERNATIONAL AIRPORT

Stanley M. Herrin[1], and Gary Fuselier[2]

[1]Manager, Springfield Aviation Group Military & Special Projects, for Crawford, Murphy & Tilly, Inc., 2750 W. Washington St., Springfield, IL 62702; PH 217-787-8050; email sherrin@cmtengr.com.

[2]Project Design Manager for Metropolitan Washington Airports Authority, Ronald Reagan Washington National Airport, Washington DC 20001-4901; PH 703-417-8189; email: gary.fuselier@mwaa.com.

Abstract

In 1982 a Pavement Management program was undertaken on the airfield pavements at Washington Dulles International Airport to establish rate of pavement deterioration, predict remaining service life, formulate a pavement rehabilitation program and estimate rehabilitation cost.

Based on three prediction models, two levels of pavement quality and the time phase for rehabilitation was predicted. The prediction models took into account the Pavement Condition Index (PCI), slab cracking and aircraft traffic.

During the 25 years since the Pavement Management program, the airport has continued maintenance programs, including a several panel replacement programs and complete reconstruction of several airfield pavements.

This paper describes the predictions methods and compares the time period predicted for reconstruction with the actual year the pavements were reconstructed.

Introduction

In 1982 the Federal Aviation Administration (FAA) undertook a comprehensive Pavement Management (PM) study at Washington Dulles International Airport. The FAA, owners of Washington Dulles International Airport, had previously undertaken an evaluation of the airfield pavements in 1978, but the 1982 study included a Pavement Condition Index (PCI) survey and non-destructive testing using the falling weight deflectometer.

The PCI method of determining the pavement condition was developed by the Corps of Engineers in the mid 1970's and adopted by the FAA in 1982. Dynatest Consulting, Inc introduced the Falling Weight Deflectometer to North America in the late 1970's. Thus, this was one of the first PM studies at an airport using techniques that are standard investigative procedures today.

Two of the PM study objectives were:
- To establish the rate of deterioration and predict the remaining service live of the airfield pavements
- To formulate a pavement rehabilitation and maintenance program including an estimated timetable and cost estimate.

Three methods were used to predict the end of service life. Based on these three prediction models, for each pavement feature the "best" estimate of the remaining service life and time period for rehabilitation was identified.

This paper presents the three prediction models, their predicted periods of rehabilitation and the "best" estimate of remaining life of those pavements. The actual rehabilitation date of those pavements during the 25 years since the PM study and dates of major rehabilitation are presented.

Washington Dulles International Airport In 1982

In 1982, Washington Dulles International Airport consisted of three runways, three air carrier ramps and connecting taxiways, as shown in Figure 1.

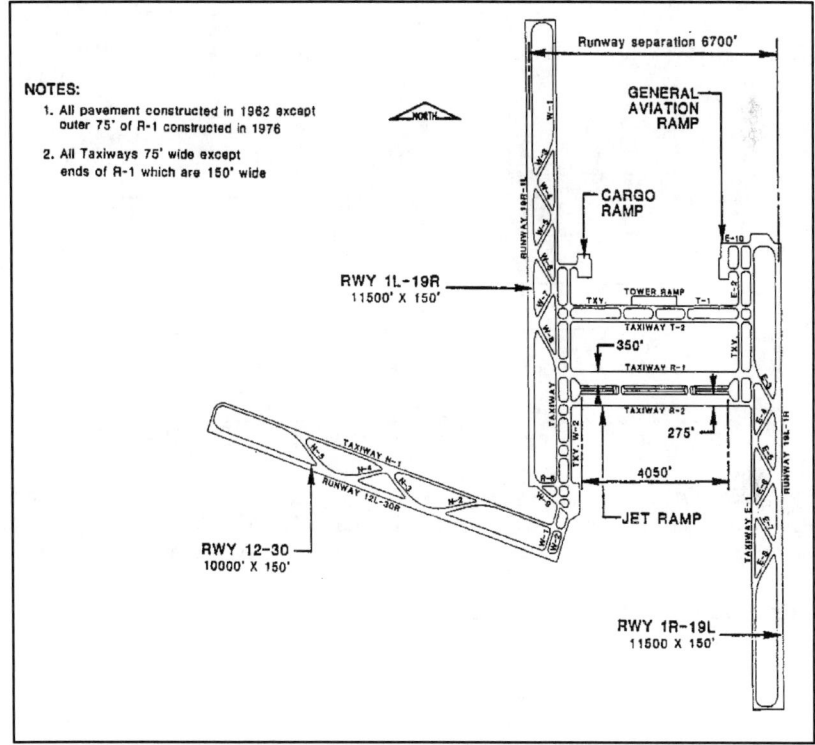

Figure 1. Washington Dulles International Airport in 1982

Construction of Washington Dulles International Airport was begun in 1958 and the airport was opened to traffic in 1962. Between 1962 and 1982, when this PM study was undertaken, the only major improvement to the airfield was the expansion of the jet ramp in 1976 to accommodate larger aircraft. In 1982, none of the pavements constructed in 1962 had been reconstructed; however several panel replacement projects had occurred.

Essentially all of the pavements included in the 1982 Pavement Management study at Washington Dulles International Airport were constructed at the same time, by the same contractor, to the same design details, and with the same materials. These pavement details are shown in Figure 2.

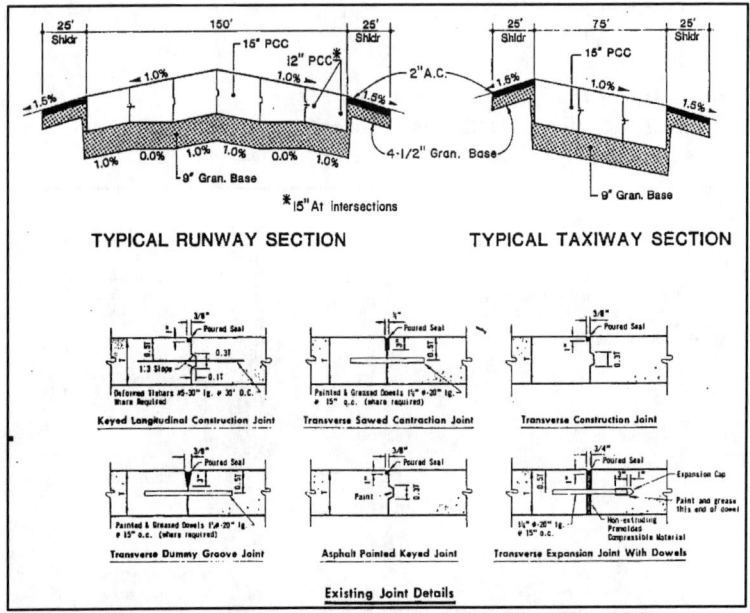

Figure 2. Paving Details, Pavements Constructed in 1958

All of the airfield pavements were surveyed in 1982, but only the most heavily trafficked pavements were subjected to a detailed PCI and NDT surveys. The high-speed taxiways and several connecting taxiways were only visually surveyed.

Pavement Life Prediction Methods

Three approaches were used to estimate the remaining life of each feature:
- Straight-Line PCI Projection
- PCI-Fatigue Analysis
- Cracking-Fatigue Analysis

Figure 3 presents the pavement life prediction approaches in graphical format.

Straight-Line PCI Projection - Straight-Line PCI Projection assumes that all pavements have a PCI of 100 upon completion of construction. Based on the PCI of each pavement feature, an average annual PCI point loss was calculated and the future annual loss of PCI was extrapolated at the same rate.

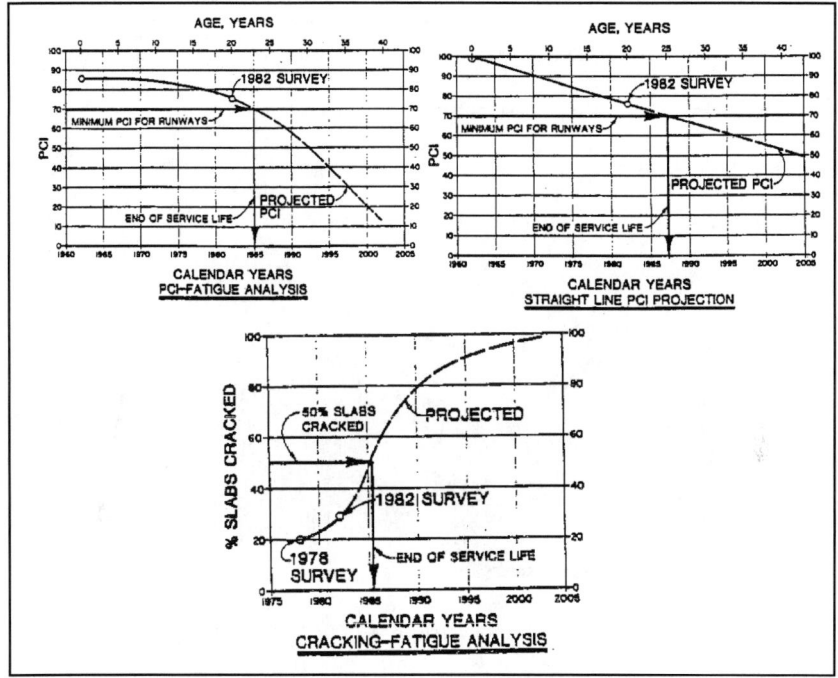

Figure 3. Graphically Representation of the Prediction Models

Straight line extrapolation is a simplistic procedure, valid only for high PCI and non-traffic load pavements. PCI deterioration is curve-linear. Since all of the pavements were of the same age, a "PCI family curve", (the PCI of similar pavements are plotted by age) could not be developed.

PCI-Fatigue Analysis – PCI-Fatigue Analysis relates pavement deterioration (PCI) to traffic loadings. The current PCI and traffic history were determined for each pavement feature. Mathematical models were developed by regression techniques to closely approximate the historic rate of pavement deterioration. The procedure that best fits actual conditions included separate equations for runways, taxiways and ramps.

Cracking-Fatigue Analysis - Cracking-Fatigue Analysis predicts the percentage of structurally cracked pavement panels over the analysis period based upon the Miner's fatigue damage number.

Laboratory and field pavement studies have shown that pavement panel cracking and the cumulative fatigue damage number can be represented by an S-shaped curve on a log-normal plot.

The shape of the curve indicates that cracking begins slowly but then accumulates rapidly before slowing down after a large percentage of slabs are cracked. The shape of the S-curve was determined for the Dulles pavements by analyzing the wide range of past traffic and corresponding observed cracking of the different pavement features.

The models developed for pavement life prediction are:

Runways

$$PCI = 86 - 0.000224 \sum \left(\frac{\sigma}{M_r}\right)^{4.0} \cdot N_{departures} \quad (1)$$

Taxiways

$$PCI = 79 - 0.000477 \sum \left(\frac{\sigma}{M_r}\right)^{3.8} \cdot N_{departures} \quad (2)$$

Aprons

$$PCI = 93 - 0.00529 \sum \left(\frac{\sigma}{M_r}\right)^{4.0} \cdot N_{departures} \quad (3)$$

σ = Critical stress induced by each individual aircraft, psi
M_r = Modulus of rupture of concrete of pavement feature, psi
departures = Total departures by each individual aircraft.

Failure Criteria

Failure criteria was established to determine the useful life of a pavement. For the two pavement life prediction models based on PCI values, two PCI values were selected to define end of service life. The two levels represent two different levels of pavement condition. A pavement with PCI above the High Pavement Quality level only preventative maintenance is required. A pavement with a PCI below the Acceptable Pavement Quality normally requires complete rehabilitation. A pavement with a PCI in-between these two levels requires concentrated maintenance effort to slow the rate of deterioration. If the rate cannot be slowed, rehabilitation is required.

The PCI minimum values for end of service life were:

Location	PCI Level to Maintain High Pavement Quality	PCI Level to Maintain Acceptable Pavement Quality
Runways	70	55
Taxiways	60	40
Ramps	50	40

For the fatigue cracking model, end of service life was represented by 50 percent of the slabs as being cracked. This is recognized in AC 150/5370-11, Appendix 1, page 11 paragraph 6.1.6, Step 6(2) that says "If more than 50 percent of the slabs show load-induced cracking, the pavement should be considered failed."

If only low severity cracks were found within the feature with 50% of the slabs being cracked, the equivalent PCI would be 78. If one assumes that 30% of the panels in the feature have low severity, 15% of the panels have medium severity cracks and 5% of the panels have high severity cracks, the equivalent PCI would be 46. This assumes no other distress is found in the panel.

Remaining Service Life Prediction

The remaining service life, predicted by each method, varies slightly between the models. To account for this variability, the predicted remaining service life were tabulated and an "end of service life time phase" was estimated. The time phases for reconstruction were:

Immediate	1984 – 1985
Short Range	1985 – 1988
Medium Range	1989 – 1993
Long Range	1994 – 2003
Beyond 2003	Reconstruction not predicted within 20 years

The End of Service Life Time Phase for each feature is presented in Table 1.

Table 1. End of Service Life for Each Feature

Location	Pavement Element	Pavement Feature	HIGH QUALITY				ACCEPTABLE QUALITY		
			Cracking Fatigue Analysis	Straight Line PCI Analysis	PCI-Fatigue Analysis	End of Service Life Phase	Straight Line PCI Analysis	PCI-Fatigue Analysis	End of Service Life Phase
Runways	1L-19R	B1	1987	1990	1987	1986-1988	>2003	1998	1989-1993
		B2	>2003	>2003	No Traffic	1994-2002	>2003	No Traffic	>2003
		B3	1991	2002	1991	1989-1993	>2003	>2003	1994-2002
	1R-19L	B1	>2003	>2003	>2003	1994-2002	>2003	>2003	>2003
		B2	>2003	>2003	No Traffic	1994-2002	>2003	No Traffic	>2003
		B3	>2003	2002	>2003	1994-2002	>2003	>2003	>2003
	12-30	B1	1982	1982	1982	1984-1985	1988	1988	1986-1988
		B2	1994	>2003	>2003	1994-2002	>2003	>2003	1994-2002
		B3	>2003	>2003	>2003	1994-2002	>2003	>2003	>2003
		B4	1991	1982	>2003	1989-1993	>2003	>2003	1989-1993
Taxiways	W-1	W11	1996	>2003	>2003	1994-2002	>2003	>2003	1994-2002
		W12	>2003	1987	2001	1986-1988	2000	>2003	1994-2002
		W13	>2003	>2003	>2003	>2003	>2003	>2003	>2003
	W-2	W21	1998	1990	1995	1986-1988	>2003	>2003	1989-1993
		W22	1982	1982	1982	1984-1985	1982	1982	1984-1985
		W23	1991	>2003	>2003	1989-1993	>2003	>2003	1989-1993
	E-1	E11	1991	1998	>2003	1989-1993	>2003	>2003	1989-1993
		E12	1988	1982	1982	1984-1985	1989	>2003	1986-1988
	E-2	E21	1984	1987	>2003	1986-1988	2000	>2003	1986-1988
	N-1	N11	1997	1982	>2003	1989-1993	>2003	>2003	1994-2002
		N12	>2003	>2003	>2003	>2003	>2003	>2003	>2003
	T-1	T11	1988	1995	>2003	1989-1993	>2003	>2003	1989-1993
		T12	1989	>2003	>2003	>2003	1986	>2003	>2003
		T13	1999	1994	>2003	1994-2002	>2003	>2003	1994-2002
	T-2	T21	>2003	>2003	>2003	>2003	>2003	>2003	>2003
		T22	>2003	>2003	>2003	>2003	1987	>2003	>2003
		T23	1997	>2003	>2003	1994-2002	>2003	>2003	1994-2002
Aprons	R-1	A1A1	1997	>2003	1989	1994-2002	2002	1991	1994-2002
		A1B1	1987	1989	1988	1989-1993	1994	1993	1989-1993
		A1C1	>2003	1985	1984	>2003	1989	1992	>2003
		A1DE1	>2003	>2003	1997	1994-2002	>2003	1998	1994-2002
	R-2	A2AC1	2002	>2003	1994	1994-2002	>2003	1998	1994-2002
		A2D1	1982	1982	1982	1984-1985	1982	1982	1984-1985
		A2D2	1982	1982	1982	1984-1985	1983	1983	1984-1985
	Tower	ATWRP	>2003	>2003	>2003	>2003	>2003	>2003	>2003

The years included in the time phase for reconstruction increases with future years, reflecting the accuracy of the approaches and changes in traffic, environmental factors and unknown factors.

Anomalies that were noted include:

> Subsurface Moisture – Pavement stains were observed during the PCI field surveys and noted as pumping. However since no debris was found and the stains were found where aircraft infrequently travel, it was concluded that this was not pumping. The End of Service life predicted by the Cracking Fatigue Analysis was favored.

> Key-Way Failure – One feature had substantial key-way failure, although cracked panels were not observed. The End of Service life predicted by the Straight Line PCI and PCI-Fatigue Analysis were favored.

> Cracked Slabs – The Cracking-Fatigue Analysis was based on changes in cracking between 1978 and 1982. Features with a large increase in the number of cracked slabs were examined in more detail, and the End of Service Life was adjusted accordingly.

> Replaced Slabs – Had the PCI approach included replaced slabs, the PCI value would have been lower and end of service life reduced. The Cracking-Fatigue Analysis includes replaced panels in the prediction model. The End of Service Life for features with a high number of replaced panels is more in line with the Cracking-Fatigue Analysis.

Pavement Repair and Performance, 1982 – Present

Washington Dulles International Airport has a maintenance staff that does temporary repairs, such as filling spalls and cracks, on a nightly basis. However, the crew does not have the money nor the equipment to perform permanent repairs such as permanent spall repair and replace PCC panels.

During the past 25 years, the permanent spall repairs have been contracted out. These repairs have been included with other projects, such as new taxiways, and included in projects that consist primarily of removal and replacement of cracked panels.

The determination of the repair projects and their effectiveness in increase the airfield pavement condition is the subject of another paper.

However, the extent and year of the removal and replacement of the PCC panels can be used as an indicator of End of Service Life – High Quality pavement condition and End of Service Life – Acceptable Quality pavement condition.

Between 1982 and 2007 there were approximately 16 projects that included panel removal and replacement projects. Nine of those projects included complete reconstruction of the pavement feature, including subbase, subgrade and portions of the shoulder.

The dates of the panel replacement and reconstruction projects, along with the projected End of Service Life (presented in Table 1) are shown in Table 2.

AIRFIELD AND HIGHWAY PAVEMENTS 2008   353

The dates of the panel replacement and reconstruction projects, along with the estimated remaining service life predicted by the three pavement life prediction models are shown in Table 3.

Comparison – Time Period for Rehabilitation to Rehabilitation Date

Table 2 presents a comparison of the End of Service Life predictions with actual rehabilitation date. The End of Service Life was predicted for some pavements 'within 5 years' and '20 years or greater'. Repairs predicted to occur in the '5 to 20 year' time frame did not often occur.

One feature on Runway 12-30 and two features on Taxiway W2 (now Taxiway Y) were required immediately repairs and were repaired. These features also had low PCI values and a high number of cracks.

Pavement repairs that occurred after 17 years, correlated better with the time period predicted for Acceptable Pavement Quality than with High Pavement Quality.

The Time Period for Rehabilitation, an estimate based on three prediction models, did not consistently predict rehabilitation in the 5 to 15 year range.

Table 2. Time Period Predicted Performance Compared to Rehabilitation Date

Projected End of Service Life	Actual Panel Replacement	Pavement Element	Feature Designation		1982 Pavement Management Study		Rehabilitation Date	
			1982 Pavement Management Study	Present Pavement Management System	End of Service Life Phase	End of Service Life Phase	Date of Panel Replacement Projects	Date of Complete Reconstrcution
Within 5 Years	Within 5 Years	12-30	B1	R12-2	1984-1985	1986-1988	1987, 1991	2004
		W-2	W21	TWZ-01	1986-1988	1989-1993	1987, 1993, 1999	2003
		W-2	W22	TWZ-02 &	1984-1985	1984-1985	1987, 1993	2003
5 <	> 20	E-1	E12	TWK-02A	1984-1985	1986-1988		2005
5 <	> 20	E-1	E12	TWK-02B	1984-1985	1986-1988		2002
5 <	> 20	E2	E21	TWJ-3	1986-1988	1986-1988		2007
5 <	9	R-2	A2D1	TL-E	1984-1985	1984-1985	1991, 2004	
5 <	9	R-2	A2D2	TL-D	1984-1985	1984-1985	1991, 2004	
5-10	20	1L-19R	B1	RW1L-01	1986-1988	1989-1993	2003	2009 (e)
5 - 10	> 20	12-30	B4	R12-4	1989-1993	1989-1993		2004
5 - 10	> 20	W-2	W23	TWZ-03A	1989-1993	1989-1993		
5 - 10	> 20	E-1	E11	TWK-01	1989-1993	1989-1993		2006
5 - 10	> 20	T-1	T11	TL-A	1989-1993	1989-1993		
5 - 10	17	R-1	A1B1	TL-D2	1989-1993	1989-1993	1999	2001
5 - 20	> 20	1L-19R	B3	RW1L-03	1989-1993	1994-2002		2009 (e)
10 - 20	11	1R-19L	B1	RW1R-01	1994-2002	>2003	1993	
10 - 20	11	1R-19L	B2	RW1R-02	1994-2002	>2003	1993	
10 - 20	11	1R-19L	B3	RW1R-03	1994-2002	>2003	1993	
20 <	> 20	N-1	N11	TWQ-2	1989-1993	1994-2002	2007	
20 <	> 20	T-1	T13	TL-A	1994-2002	1994-2002		
20 <	> 20	T-2	T23	TL-B	1994-2002	1994-2002		
20 <	> 20	R-2	A2AC1	TL-E	1994-2002	1994-2002		
> 20	17	R-1	A1C1	TL-D	>2003	>2003	1999	2001
5 - 20	18	W-1	W12	TWY-02	1986-1988	1994-2002		2000
10 - 20	20	1L-19R	B2	RW1L-02	1994-2002	>2003	2003	2009 (e)
10 - 20	22	12-30	B2	R12-3	1994-2002	1994-2002		2004
10 - 20	22	12-30	B3	R12-3	1994-2002	>2003		2004
10 - 20	18	W-1	W11	TWY-01	1994-2002	1994-2002		2000
10 - 20	17	R-1	A1A1	TL-D	1994-2002	1994-2002	1999	2001
10 - 20	17	R-1	A1DE1	TL-D	1994-2002	1994-2002	1999	
> 20	> 20	W-1	W13	TWY-03S & TWY-03N	>2003	>2003		
> 20	25	N-1	N12	TWQ-3	>2003	>2003	2007	
> 20	> 20	T-1	T12	TL-A	>2003	>2003		
> 20	> 20	T-2	T21	TL-B	>2003	>2003		
> 20	> 20	T-2	T22	TL-B	>2003	>2003		
> 20	> 20	Tower	ATWRP		>2003	>2003		

Comparison – Straight Line PCI Analysis to Rehabilitation Date

Table 3 compares predicted Straight Line PCI Analysis End of Service Life with actual rehabilitation date.

The Straight Line PCI Analysis predicted two pavements would need immediate repairs and they were repaired within five years. However two features projected to need immediate repairs were not repaired within six years and three features needing immediate repairs were not repaired for over 17 years.

The Straight Line PCI Analysis predicted 27 pavement features would need repairs in the 20 year range. 23 of the features were repaired after 17 years. Four of the features were repaired after 11 years.

Ten features were not repaired within four years of the year predicted by the Straight Line PCI Analysis.

Straight Line PCI Analysis did not consistently predict rehabilitation in the 5 to 15 year range.

Table 3. St. Line PCI Predicted Performance Compared to Rehabilitation Date

Location	Pavement Element	Feature Designation		1982 Predicted Performance					Rehabilitation Date	
				Cracking Fatigue Analysis	High Quality		Acceptable Quality			
		1982 Pavement Management Study	Present Pavement Management System		Straight Line PCI Analysis	PCI-Fatigue Analysis	Straight Line PCI Analysis	PCI-Fatigue Analysis	Date of Panel Replacement Projects	Date of Complete Reconstrcution
Runways	1L-19R	B1	RW1L-01	1987	1990	1987	>2003	1998	2003	2009 (e)
		B2	RW1L-02	>2003	>2003	No Traffic	>2003	No Traffic	2003	2009 (e)
		B3	RW1L-03	1991	2002	1991	>2003	>2003		2009 (e)
	1R-19L	B1	RW1R-01	>2003	>2003	>2003	>2003	>2003	1993	
		B2	RW1R-02	>2003	>2003	No Traffic	>2003	No Traffic	1993	
		B3	RW1R-03	>2003	2002	>2003	>2003	>2003	1993	
	12-30	B1	R12-2	1982	1982	1982	1988	1988	1987, 1991	2004
		B2	R12-3	1994	>2003	>2003	>2003	>2003		2004
		B3	R12-3	>2003	>2003	>2003	>2003	>2003		2004
		B4	R12-4	1991	1982	>2003	>2003	>2003		2004
Taxiways	W-1	W11	TWY-01	1996	>2003	>2003	>2003	>2003		2000
		W12	TWY-02	>2003	1987	2001	2000	>2003		2000
		W13	TWY-03S &TWY-03N	>2003	>2003	>2003	>2003	>2003		
	W-2	W21	TWZ-01	1998	1990	1995	>2003	>2003	1987, 1993, 1999	2003
		W22	TWZ-02 & TWZ-03	1982	1982	1982	1982	1982	1987, 1993	2003
		W23	TWZ-03A	1991	>2003	>2003	>2003	>2003		
	E-1	E11	TWK-01	1991	1998	>2003	>2003	>2003		2006
			TWK-02A							2005
		E12	TWK-02B	1988	1982	1982	1989	>2003		2002
	E2	E21	TWJ-3	1984	1987	>2003	2000	>2003		2007
	N-1	N11	TWQ-2	1997	1982	>2003	>2003	>2003	2007	
		N12	TWQ-3	>2003	>2003	>2003	>2003	>2003	2007	
	T-1	T11	TL-A	1988	1995	>2003	>2003	>2003		
		T12	TL-A	1989	>2003	>2003	1986	>2003		
		T13	TL-A	1999	1994	>2003	>2003	>2003		
	T-2	T21	TL-B	>2003	>2003	>2003	>2003	>2003		
		T22	TL-B	>2003	>2003	>2003	1987	>2003		
		T23	TL-B	1997	>2003	>2003	>2003	>2003		
Aprons	R-1	A1A1	TL-D	1997	>2003	1989	2002	1991	1999	2001
		A1B1	TL-D2	1987	1989	1988	1994	1993	1999	2001
		A1C1	TL-D	>2003	1985	1984	1989	1992	1999	2001
		A1DE1	TL-D	>2003	>2003	1997	>2003	1998	1999	
	R-2	A2AC1	TL-E	2002	>2003	1994	>2003	1998		
		A2D1	TL-E	1982	1982	1982	1982	1982	1991, 2004	
		A2D2	TL-D	1982	1982	1982	1983	1983	1991, 2004	
	Tower	ATWRP		>2003	>2003	>2003	>2003	>2003		

Comparison – PCI-Fatigue Analysis to Rehabilitation Date

Table 4 compares predicted PCI-Fatigue Analysis End of Service Life with actual rehabilitation date.

The PCI-Fatigue Analysis predicted two pavements would need immediate repairs and they were repaired within five years. However two features projected to need immediate repairs were not repaired within nine years and one was not repaired for 17 years.

The PCI-Fatigue Analysis predicted 23 pavement features would need repairs in the 20 year range. 21 of the features were repaired after 16 years. Two of the features were repaired after 11 years.

Eleven features were not repaired within four years of the year predicted by the PCI-Fatigue Analysis.

PCI-Fatigue Analysis did not consistently predict rehabilitation in the 5 to 15 year range.

Table 4. PCI-Fatigue Predicted Performance Compared to Rehabilitation Date

				Feature Designation		Predicted		Rehabilitation Date	
Projected End of Service Life - High Quality (Years)	Projected End of Service Life - Acceptable Quality (Years)	Actual Panel Replacement (Years)	Pavement Element	1982 Pavement Management Study	Present Pavement Management System	PCI-Fatigue Analysis - High Quality	PCI-Fatigue Analysis - Acceptable Quality	Date of Panel Replacement Projects	Date of Complete Reconstrcution
No Traffic	No Traffic	21	1L-19R	B2	RW1L-02	No Traffic	No Traffic	2003	2009 (e)
No Traffic	No Traffic	11	1R-19L	B2	RW1R-02	No Traffic	No Traffic	1993	
0	0	5	W-2	W22	TWZ-02 & TWZ-03	1982	1982	1987, 1993	2003
0	6	5	12-30	B1	R12-2	1982	1988	1987, 1991	2004
0	0	9	R-2	A2D1	TL-E	1982	1982	1991, 2004	
0	1	9	R-2	A2D2	TL-D	1982	1983	1991, 2004	
2	10	17	R-1	A1C1	TL-D	1984	1992	1999	2001
5	16	21	1L-19R	B1	RW1L-01	1987	1998	2003	2009 (e)
6	11	17	R-1	A1B1	TL-D2	1988	1993	1999	2001
7	9	17	R-1	A1A1	TL-D	1989	1991	1999	2001
9	> 20	27	1L-19R	B3	RW1L-03	1991	>2003		2009 (e)
13	> 20	5	W-2	W21	TWZ-01	1995	>2003	1987, 1993, 1999	2003
12	16	> 20	R-2	A2AC1	TL-E	1994	1998		
> 20	> 20	11	1R-19L	B3	RW1R-03	>2003	>2003	1993	
> 20	> 20	11	1R-19L	B1	RW1R-01	>2003	>2003	1993	
0	> 20	23	E-1	E12	TWK-02A	1982	>2003		2005
0	> 20	20	E-1	E12	TWK-02B	1982	>2003		2002
15	16	17	R-1	A1DE1	TL-D	1997	1998	1999	
19	> 20	22	W-1	W12	TWY-02	2001	>2003		2000
> 20	> 20	22	12-30	B2	R12-3	>2003	>2003		2004
> 20	> 20	22	12-30	B3	R12-3	>2003	>2003		2004
> 20	> 20	22	12-30	B4	R12-4	>2003	>2003		2004
> 20	> 20	22	W-1	W11	TWY-01	>2003	>2003		2000
> 20	> 20	24	E-1	E11	TWK-01	>2003	>2003		2006
> 20	> 20	25	E2	E21	TWJ-3	>2003	>2003		2007
> 20	> 20	25	N-1	N11	TWQ-2	>2003	>2003	2007	
> 20	> 20	25	N-1	N12	TWQ-3	>2003	>2003	2007	
> 20	> 20	> 20	W-1	W13	&TWY-03N	>2003	>2003		
> 20	> 20	> 20	W-2	W23	TWZ-03A	>2003	>2003		
> 20	> 20	> 20	T-1	T11	TL-A	>2003	>2003		
> 20	> 20	> 20	T-1	T12	TL-A	>2003	>2003		
> 20	> 20	> 20	T-1	T13	TL-A	>2003	>2003		
> 20	> 20	> 20	T-2	T21	TL-B	>2003	>2003		
> 20	> 20	> 20	T-2	T22	TL-B	>2003	>2003		
> 20	> 20	> 20	T-2	T23	TL-B	>2003	>2003		
> 20	> 20	-1982	Tower	ATWRP		>2003	>2003		

Comparison – Cracking Fatigue Analysis to Rehabilitation Date

Table 5 compares predicted Cracking Fatigue Analysis End of Service Life with actual rehabilitation date. Two pavement features were predicted to need immediate repairs and they were repaired within five years. However, two features predicted to need immediate repairs were not repaired for nine years.

The Cracking Fatigue Analysis predicted nineteen features would need repairs after 15 years; fifteen of these features were repaired after 15 years. Three of the features were repaired after 11 years and one after five years.

Nineteen features were not repaired within four years of the year predicted by the Cracking Fatigue Analysis.

Cracking Fatigue Analysis did not consistently predict rehabilitation in the 5 to 15 year range.

Table 5. Cracking Fatigue Predicted Performance Compared to Rehabilitation Date

Projected End of Service Life - High Quality (Years)	Actual Panel Replacement (Years)	Pavement Element	1982 Pavement Management Study	Present Pavement Management System	Cracking Fatigue Analysis	Date of Panel Replacement Projects	Date of Complete Reconstrcution
0	5	12-30	B1	R12-2	1982	1987, 1991	2004
0	5	W-2	W22	TWZ-02 & TWZ-03	1982	1987, 1993	2003
0	9	R-2	A2D1	TL-E	1982	1991, 2004	
0	9	R-2	A2D2	TL-D	1982	1991, 2004	
2	25	E2	E21	TWJ-3	1984		2007
5	17	R-1	A1B1	TL-D2	1987	1999	2001
5	21	1L-19R	B1	RW1L-01	1987	2003	2009 (e)
6	20	E-1	E12	TWK-02B	1988		2002
6	23	E-1	E12	TWK-02A	1988		2005
6	> 20	T-1	T11	TL-A	1988		
7	> 20	T-1	T12	TL-A	1989		
9	22	12-30	B4	R12-4	1991		2004
9	24	E-1	E11	TWK-01	1991		2006
9	27	1L-19R	B3	RW1L-03	1991		2009 (e)
9	> 20	W-2	W23	TWZ-03A	1991		
12	22	12-30	B2	R12-3	1994		2004
14	18	W-1	W11	TWY-01	1996		2000
16	5	W-2	W21	TWZ-01	1998	1987, 1993, 1999	2003
> 20	11	1R-19L	B1	RW1R-01	>2003	1993	
> 20	11	1R-19L	B2	RW1R-02	>2003	1993	
> 20	11	1R-19L	B3	RW1R-03	>2003	1993	
15	17	R-1	A1A1	TL-D	1997	1999	2001
15	25	N-1	N11	TWQ-2	1997	2007	
15	> 20	T-2	T23	TL-B	1997		
17	> 20	T-1	T13	TL-A	1999		
20	> 20	R-2	A2AC1	TL-E	2002		
> 20	17	R-1	A1DE1	TL-D	>2003	1999	
> 20	17	R-1	A1C1	TL-D	>2003	1999	2001
> 20	18	W-1	W12	TWY-02	>2003		2000
> 20	21	1L-19R	B2	RW1L-02	>2003	2003	2009 (e)
> 20	22	12-30	B3	R12-3	>2003		2004
> 20	25	N-1	N12	TWQ-3	>2003	2007	
> 20	> 20	W-1	W13	TWY-03S &TWY-03N	>2003		
> 20	> 20	T-2	T21	TL-B	>2003		
> 20	> 20	T-2	T22	TL-B	>2003		
> 20	> 20	Tower	ATWRP		>2003		

Summary and Discussion

Based on the predicted performance compared to actual rehabilitation date, the three End of Service Life prediction models all produced similar results.
- The prediction models can predict where pavement repairs were required immediately, though not consistently
- The prediction models were capable of predicting when repairs would not be required for over 15 years.
- The prediction models did not consistently predict rehabilitation in the 5 to 15 year range.

Repairs to pavements with a predicted service life between 5 and 15 years may not have been required due to:
- Change in aircraft traffic
- Change in traffic patterns
- New pavement construction
- The Airport's maintenance effort

Likewise, a pavement with a predicted 20 years service life may not last for 20 years due to these same reasons. Actual rehabilitation date correlated with the Acceptable Pavement Quality, as predicted by the Straight Line PCI (Table 4) and PCI-Fatigue (Table 5) prediction models.

Concluding Remarks

At Washington Dulles International Airport many variables have affected pavement performance. Attempts at predicting remaining pavement life based on pavement condition, slab cracking and traffic have had limited success due to changes in traffic and maintenance.

The three prediction models described herein appear capable of predicting pavements that need immediate repairs and predicting pavements that will last over 15 years. Attempts to predicting pavements that would require repairs in 5 to 15 did not produce consistent results.

One of the products of the 1982 Pavement Management Study was an estimated budget of pavement repair and rehabilitation costs. The PCI survey results were used to estimate a repair budget. Based on the airport's experience with the prediction models for immediately repair, the airport has confidence in a repair budget based on PCI surveys. Presently, Washington Dulles International Airport uses annual PCI surveys to define rehabilitation projects. This program is discussed in further detail in a paper presented at this conference.

In-house maintenance has been successful in extending the service life of a pavement, but this success is limited. Eventually the pavement will need to be reconstructed.

A PCI in the range between High Quality and Acceptable Quality indicates a pavement where in-house maintenance should concentrate its effort and provisions for eventual rehabilitation should be planned and budgeted.

Based on the rehabilitation date compared to the three prediction models and Time Period for Rehabilitation, the time period for rehabilitation indicates need and the associated rehabilitation cost is just a planning budget. Frequent PCI surveys track the pavement performance and alerts the airport when the pavement condition has fallen into the need to be repaired within five years.

The time periods for rehabilitation are thresholds set by the airport. Pavement rehabilitation may not occur during these time periods since pavement condition is not the controlling factor that determines when rehabilitation actually occurs.

References

1. Crawford, Murphy & Tilly, Inc.; ERES Consultants, Inc.; Barenberg, E. J.; and Shahin, M. Y. (1984) *1982 Report on Evaluation of Airfield Pavements at Dulles International Airport.*

2. Thuma, R. G., Fuselier, G. K. and Yip, P. K. *Using PCI Data to Define Major Rehabilitation Projects at Washington Dulles International Airport.*

Remaining Service Life Analysis of Concrete Airfield Pavements at Denver International Airport using the FACS Method

Michael T. McNerney[1], PhD, PE, M ASCE

Abstract

Denver International Airport (DEN) has completed a comprehensive pavement evaluation of the entire airfield using the Geospatial Airfield Pavement Evaluation and Management System (GAPEMS). The comprehensive use of TabletPCs and GPS for distress mapping, photographic documentation, core sampling, structural evaluation, and mapping of legacy construction data has provided the pavement engineer with more data and a clearer big picture of the pavement condition. Traditional declining Pavement Condition Index (PCI) value was found to be ineffective as a means of forecasting remaining life for several reasons including the grouping of slabs unrelated to type of distress. The averaging of slabs in groups of 20 slabs for concrete is an inherent flaw in the PCI method when averaging slabs of highly variable distresses.

A new pavement remaining life calculation method called the "FACS Method" was developed that used the acronym from four separate failure modes of concrete pavement: **Fatigue**, Alkali Silica Reaction (**ASR**), **Cracking**, and **Spalling**. Using geospatial mapping of all distresses and the additional data including geospatial photography, the team was able to develop a network forecast of remaining life based on each of the four failure modes by evaluating data at the individual slab level. This improved method of remaining life analysis was implemented at DEN and resulted in significant savings in the slab replacement program.

GAPEMS Data Collection

Denver International Airport (DEN) has completed a comprehensive pavement evaluation of the entire airfield using the Geospatial Airfield Pavement Evaluation and Management System (GAPEMS) (McNerney and Kelley 2007). The pavement evaluation included approximately 125 miles of continuous deflection data of the Rolling Dynamic Deflectometer (RDD) to evaluate the structural capacity of the pavements (Bay and Stokoe, 1998). Pavement evaluation also included 200 core locations and 128 cores tested for alkali silica reaction (ASR) by petrography. Archived construction records were geospatially coded for date of concrete placement, lot number, mix design, and method of construction. Forensic analyses using the distress and construction data found there is a significant difference in remaining service life because of differences in the coarse aggregates.

---

[1] DMJM Aviation, 1200 Summit Ave, Suite 320, Fort Worth, TX 76102, TEL 817 698-6800, FAX 817 698-6802, email: mike.mcnerney@dmjmaviation.com

Deficiency of Traditional Remaining Service Life Analysis

The traditional method of forecasting remaining life by estimating the decline of the pavement condition index (PCI) is not an effective method for airports with significant aircraft loading (McNerney and McCullough, 2000). The PCI was developed for the US Air Force in the 1970's primarily as a maintenance tool to assist in the allocation of maintenance funds between competing Air Force bases for a limited amount of maintenance funding. The PCI is only an indication of what distress is visible on the surface and the index does not take into account structural integrity or carrying capacity of the pavements measured by deflections. The PCI calculation method was never intended to be used as a method to forecast remaining life, but the majority of non pavement engineers view it as the primary indication of the need for reconstruction or rehabilitation.

Identification of distress types associated with the inspection of pavement is a valuable tool and was enhanced for this application at Denver International Airport. Even though the PCI forecast of remaining life is an extremely poor methodology for Denver International Airport, the actual mapping of the individual distress types and their severity is valuable data when available to an experienced pavement engineer.

Too often pavement evaluation reports and airport mangers are making decisions to rehabilitate pavements based upon a low PCI number without knowing the cause of pavement deterioration and often misleading PCI numbers. The FACS Method of determining remaining life was developed to provide a pavement engineer the data to estimate remaining life based upon the causes of the distresses and functional deterioration of the airfield pavements. The basis for remaining life analysis using GAPEMS data is the mapping of all distresses supplemented by deflection data, analyses of historic construction data, aircraft and operations, pavement deicing operations, and the petrographic analyses of cores. At DEN, additional laboratory testing was conducted on core samples to determine material properties and material durability to be included in the remaining life analysis.

The PCI method of pavement management has several deficiencies that have been overcome by using GAPEMS. The first deficiency of ASTM D5340 method of PCI inspection is that it is very labor intensive and therefore, usually only 10 to 20 percent of the PCI sample units are rated to develop a statistical sample of the pavement. The GAPEMS method provides a distress rating for a 100 percent sample and maps each distress location for future repairs. This computer mapping capability was not available when the PCI method was developed so all distresses were reduced to a single index number.

The second inherent deficiency of the PCI method is the requirement to group concrete slabs into sample units of 20 slabs within a range of plus or minus 4 slabs. The problem comes in the selection of the PCI sample units. Slab replacements are not necessarily made in the same 20 slab sample units. In the example of DEN Taxiway M near Runway 8-26, the airport selected PCI sample units along the full width of the taxiway. As shown in the left side of Figure 1, the traditional method indicated no pavement problems because mostly good slabs were included in the sample. However, the right side of the figure also shows that, by using the remaining life analysis method on a slab basis, only the slabs in the center of Taxiway M need immediate replacement.

Improvements to ASTM D5340

Denver International Airport was known to have much surface cracking on the airfield pavements. It was hoped that some of the cracking would be an indication of potential alkali silica reaction (ASR). Therefore, in GAPEMS some improvements to the identification of the surface cracking distress were made to the procedure defined in ASTM D5340 and the equivalent Air Force methods. Figure 2 shows the examples of the surface cracking.

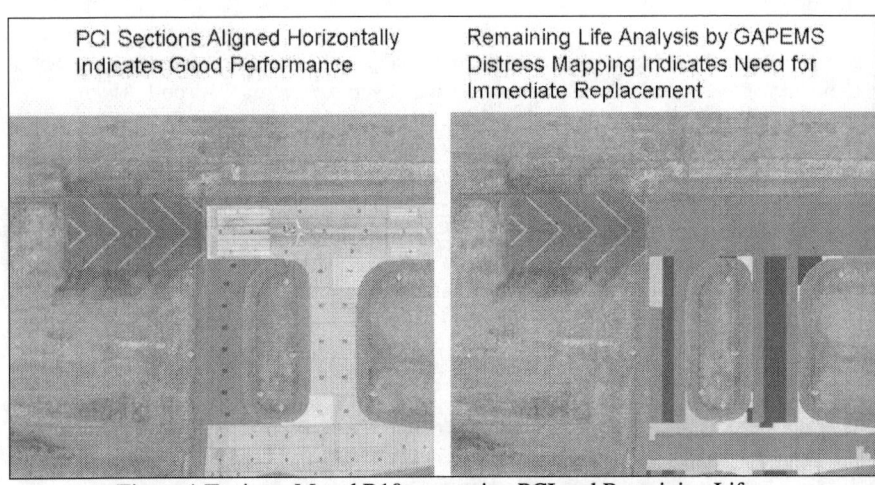

Figure 1 Taxiway M and R10 comparing PCI and Remaining Life

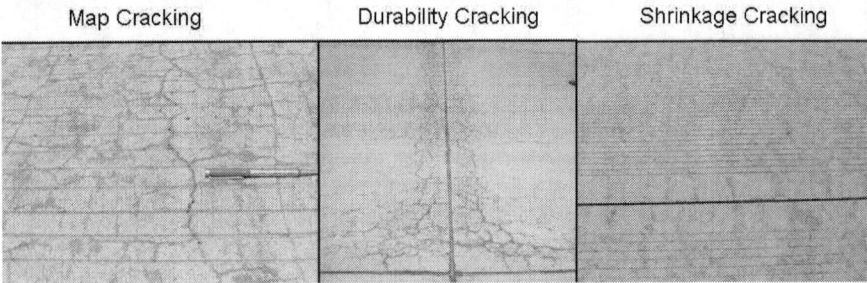

Figure 2 Three Different Types of Surface Cracking at Denver International Airport

Distress number 70 - Map Cracking (also called crazing and scaling) has a low medium and high severity rating in the ASTM and Air Force procedures. However, medium and high severity ratings are given when the cracking has progressed to the point that scaling has already taken place. At most commercial service airports, pavements would be replaced whenever scaling is occurring because of the foreign object damage potential. Therefore, map cracking severity was modified in GAPEMS to provide severity levels of 1 through 7 as shown in Table 1. Level 6 severity corresponds with medium severity and level 7 corresponds with high severity in the ASTM scale.

Table 1 GAPEMS Map Cracking Severity Scale

GAPEMS Severity	Description	ASTM Equivalent Severity
1	Hairline cracking is visible and some discoloration is noted over 50% of the panel	none
2	Map cracking is occurring in over 50% of the panel but no areas have resolved into blocks of less than 6 inches	Low
3	Map cracking is occurring in over 50% of the panel and more than 10% of the panel has resolved into blocks of less than 6 inches	Low
4	Map cracking is occurring in over 90% of the panel and over 50% of the panel has blocks of less than 6 inches	Low
5	Map cracking is occurring in over 90% of the panel and has blocks of less than 6 inches or cracking has a white substance or has become very wide.	Low
6	Scaling is occurring in over 5% of the panel and is causing some FOD potential	Medium
7	Severe scaling is occurring and is causing high FOD potential	High

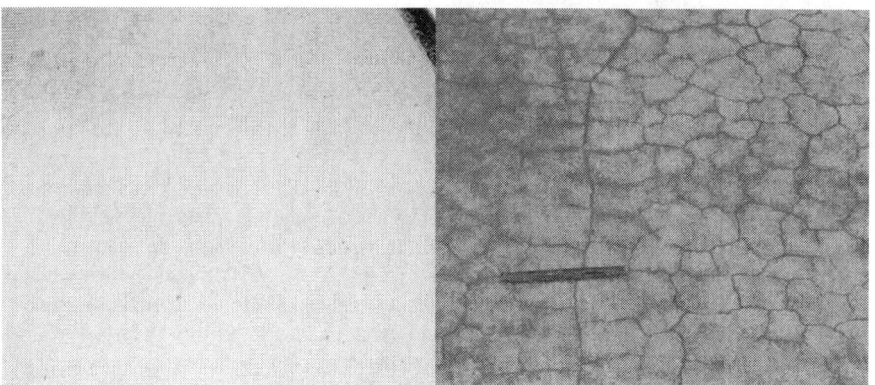

Figure 3 Examples of Map Cracking Severity Levels 1 and 5

Distress number 64 – Durability Cracking is defined by the Air Force as a pattern of cracking running parallel to a joint or crack usually with a dark coloring. This distress was identified because of concrete that has a poor freeze thaw resistance. In contemporary concrete pavement construction practices with adequate minimum air entrained values, freeze thaw aggregate damage to airfield pavements is not often observed.

At DEN, there was a type of map cracking that follows along the joints only and often has dark coloration. It was decided that this observed distress would be called durability cracking rather than map cracking because it follows along the joints rather than being present in the whole slab. Even though this cracking is not freeze thaw related and the cracks are not linear as in classic D-cracking, this distress was recorded differently than map cracking for analysis purposes. After petrographic analysis it appears that this type of cracking has a higher correlation to the presence of ASR than either shrinkage or lower levels of severity of map cracking. Although, the conclusion from petrographic analysis was that ASR is not related to the surface cracking observed at DEN.

The ASTM procedure was used to rate all other distresses. Shrinkage cracking has no severity levels in the ASTM procedure. This refinement to durability cracking and map cracking proved to be very useful in discriminating among the surface cracking distresses to determine remaining service life. Without these modifications to the ASTM distress identification and severity rating, the remaining life would have not been as specific.

FACS Method Remaining Service Life Theory

The concept of remaining service life as applied to Denver International Airport is to determine when it is necessary to replace any given slab or panel of concrete pavement. There are over 6 million square yards of concrete pavement in over 140,000 slabs at DEN, most of which were constructed in 1992 and 1993. Remaining service life can, and should, be extended by maintenance actions whenever possible. When service life has ended, then it is time to replace the concrete slab either because the slab has become a (FOD) object damage threat to aircraft, it has cracked into pieces, it has multiple spalls, it has been patched too much or too poorly, or it is the most convenient or cheapest time to replace it by grouping it with other slabs.

The traditional declining PCI method assumes when the index has been reduced from 100 to 50, or another selected number, the pavement should be replaced or overlaid. However, with GAPEMS there is much more data with which to base an informed decision as to when a slab should be replaced. Using the GAPEMS data, which include historic construction data, date of concrete placement, evaporation rate, mix design and type of coarse aggregate as well as structural capacity and distress information, the author has identified at least different four primary ways in which a concrete pavement can reach failure or end of service life which is applicable to all airport concrete pavements.

Four Modes of Concrete Pavement Failure

The author first applied the notion of different remaining life for different pavement failure modes at Dallas/Fort Worth Airport (McNerney and McCullough, 1997). Based upon the observation of distress and the operations and maintenance of the airfield pavements the author has identified very four different modes of failure in which a pavement can reach the end of its service life at DEN. The author has used the acronym FACS to label these four different failure modes:

1. Fatigue - The pavement can fail because the action of repeated loading or other repeated stresses that result in the fracture and cracking in the pavement. When it reaches a point that full depth cracks are severely spalled or the slab is shattered into pieces the only solution is to replace the slab.

2. ASR - Alkali silica reaction is a chemical reaction between the cement and aggregate causing expansion when a gel is formed from the chemical reaction. Determined by petrographic analysis of core samples there is no alternative than to replace the concrete panels when the ASR has reached certain levels of severity. Petrographic analysis of 128 cores at DEN has shown that the ASR mode of failure is unaffected by surface application of pavement deicer.

3. Cracking, Crazing, and Scaling - This failure mode is a surface deterioration of concrete that leads to cracking and eventually scaling on the surface. The mechanism might be accelerated by freezing and thawing, drying shrinkage or the application of pavement deicer.

4. Spalling - This deterioration mode is a result of spalling of the concrete usually either along a joint or corner of a slab. Normally joint spalling and corner spalling can be repaired with partial depth patching. However, when the spalling becomes systemic and there are multiple spalls either from poor construction practices or poor performing joints, it may be economically justified to replace slabs, rather than continue patching.

There may be other modes in which a concrete slab will fail at Denver International Airport and will need to be replaced such as elevation change, faulting, or profile roughness. However, the author has decided only to concentrate on these four different modes of failure as necessary for the GAPEMS remaining service life analysis. These four modes should cover more than 98 percent of predicted slab failures.

Evaluation Factors for the Four Failure Modes

There are many different factors that can be used to evaluate the remaining service life for each mode of failure. Different distresses indicate each failure mode. Table 2 is a list of the potential factors than can be evaluated for determining remaining service life for each different failure mode.

Fatigue – As shown in Table 2 there are several factors that can affect the remaining life analysis for the fatigue mode of failure. Some of these factors are similar for the entire airport and some are very specific for each individual slab. At DEN the pavement cross section thickness was standardized for construction. Although this is one of the most important evaluation factors for the fatigue failure mode, it is not significant for DEN.

Table 2 Failure Modes and Factors that Affect Remaining Life

Failure Mode	Distress Code	Factors Affecting Remaining Service Life
F: Fatigue	63	Longitudinal/Transverse Cracking
	66	Patching (small)
	67	Patching (large)
	69	Pumping
	72	Shattered Slabs
		HWD/RDD Deflection Data
		Traffic Loading
		Pavement Construction Date/Lane
A: ASR		Petrography
		Expansion Test
		Aggregate Type
		Mix Design (Fly Ash, Cement)
	64	D Cracking
	70	Map Cracking
		Potassium Acetate Deicer
		Traffic Loading
		Pavement Construction Date/Lane
C: Cracking Crazing and Scaling	70	Map Cracking
	73	Shrinkage Cracking
		Aggregate Type
		Mix Design (Fly Ash, Cement)
	64	D Cracking
		Potassium Acetate Deicer
		Traffic Loading
		Pavement Construction Date/Lane
S: Spalling	63	Longitudinal/Transverse Cracking
	64	D Cracking
	66	Patching (small)
	67	Patching (large)
	74	Joint Spalling
	75	Corner Spalling
		Aggregate Type
		Potassium Acetate Deicer
		Traffic Loading
		Pavement Construction Date/Lane

The thickness design for DEN was based upon a theoretical pavement life of 30 years rather than the FAA standard of 20 years of traffic loading. Consequently, the pavement was constructed 1 to 2 inches thicker than if they had used a 20 year life. This leads to a relatively stiff pavement. If fatigue were the only actual mode of failure, then this thickness design would be the ideal method of remaining life prediction.

In general, the stiffness of the runways and taxiways at DEN is greater than any other airfield pavement DMJM Aviation has measured with the Rolling Dynamic Deflectometer and among the stiffest DMJM Aviation has measured with the Heavy Falling Weight Deflectometer. As a consequence, DMJM Aviation has seen little or no evidence of fatigue related distress on the runways and taxiways.

The GAPEMS remaining life theory separates out fatigue mode of failure as only one potential failure mode. Indications of areas where the fatigue failure mode may be occurring are areas that would have a loss of subgrade support, extremely high aircraft or heavy vehicle traffic, or areas in which high environmental stresses such as curling or warping are additive to loading stresses.

When high stresses begin to exceed design stress the concrete begins to fracture and crack. Distresses that would be indicators of reduced service from a fatigue failure mode would include any full depth cracking, shattered slabs, or pumping which is an indication of loss of subgrade support. Longitudinal/transverse cracking can be a significant indication of reduced service life in the fatigue or fracture mode of failure.

The mapping of distress indicated there were severity levels of low, medium and high present on the airfield. A high severity crack (longitudinal, transverse or diagonal) is defined as a filled or nonfilled crack that is severely spalled creating a FOD potential or a nonfilled crack that has a mean width greater than 1 inch creating tire damage potential. High severity cracking was almost never observed on the runways and was seldom observed on the taxiways. However, high severity cracking was frequently reported as a distress condition in concourse aircraft parking aprons A, B and C where aircraft loads are applied over a longer time per application and subgrade support has been compromised.

ASR – The failure mode of ASR is based upon a chemical reaction between the alkali in the cement and the silica of the aggregates that results in expansion of the concrete matrix. This expansion from within the concrete matrix can manifest itself as cracking, expansion, and spalling. There are different levels of ASR that are occurring in the DEN pavements. Some pavements have a little degradation from ASR because, although it is present, there is not enough to cause any damage to the matrix. In other locations, the ASR is widespread throughout the depth of the concrete and with microscopic examination cracks are observed with gel deposits that have cracked the coarse aggregate.

There is no known mitigation measure that is effective in the field on concrete pavements to stop ASR. The use of lithium nitrate has been proposed but tests indicate that it is not likely to penetrate below the surface of pavement to stop the reaction.

It is very difficult to predict when pavement might fail due to ASR damage. The biggest factor is the alkalinity of the cement and reactiveness of the aggregate. At DEN, DMJM Aviation has found significant differences in the three types of aggregates that were used for construction.

The brown granite coarse aggregate, which was labeled Type A, has the most ASR indications in petrographic analysis. The pink granite coarse aggregate, which was labeled Type B, had some ASR indications. The blue grey coarse aggregate, which was labeled Type C, had no indications of ASR.

The analysis of remaining life of the slabs was based upon the 128 petrographic cores and the laboratory testing conducted. Petrographic examination and laboratory testing did not confirm any correlation between ASR and observed surface cracking or field applied potassium acetate deicer. It is well documented that concrete specimens in the laboratory soaked in potassium acetate show an accelerated degradation from the ASR process. However, our analysis of the 128 cores from both treated and untreated core samples did not show this effect in the field performance.

The Denver airport was constructed primarily with class C fly ash added as a cement replacement in the mix design at a rate of about 20-25 percent. After the airport was built, the industry has learned that class F fly ash is better for ASR prevention and that class C fly ash in that quantity is probably worse than no fly ash for reducing ASR degradation. Recent laboratory testing has indicated that there can be a degradation of concrete from the interaction of potassium acetate deicer and fly ash in concrete that could easily be confused with ASR degradation, but it can occur without ASR present (Thomas, 2007).

Cracking, Crazing, and Scaling – The failure mode of Cracking, Crazing and Scaling is a surface deterioration of the concrete pavement. This is a completely different mechanism than the ASR failure mode that is a chemical attack of the concrete pavement and is present within the full depth of the pavement. However, ASR can also exhibit surface cracking distress. The durability of the concrete surface is a difficult mechanism to predict because there are many factors that influence surface durability.

The DEN concrete is a matrix of cement and class C fly ash binder that hold together fine and coarse aggregates. Mix design properties such as percent of air entrainment, percent and type of fly ash, type and setting properties of the cement, and type of aggregate used can influence the durability of the concrete surface. Construction practices such as the rate of evaporation and the finishing and curing procedures also affect the properties of the final concrete pavement. Too much water on the surface, over finishing, high evaporation rates, and insufficient curing methods relative to the evaporation rate can all lead to surface cracking. The surface durability is also affected by aircraft loading, environmental conditions, and the use of pavement deicing.

Not all surface cracking leads to slab replacement. Much of the surface cracking observed at DEN is caused by drying shrinkage and it is only on the surface. Generally, the surface cracking is less than a few mm in depth leaving the underlying concrete in good condition. As long as surface cracking is prevented from leading to scaling, the concrete panels will not have to be replaced. Much of the observed surface cracking is very tight cracking and is more visible or only visible when the surface has recently been wetted. The application of potassium acetate deicer is thought to stain the surface to make the surface cracking more visible as well.

It is recommended that in areas with surface cracking and the application of pavement deicers that the life of the concrete panels can be extended or preserved with the application of a sealer or polymer composite micro-overlay.

Spalling – The failure mode of spalling is when there is a systemic pattern of spalling occurring from poor performance either of a joint or poor construction technique. Spalling in concrete pavement usually occurs from expansion around stress points in the concrete. The most common areas are in corners or along joints. Usually the normal repair method is partial depth patching of the localized spall areas. However, spalling as a failure mode is considered when there is a systemic problem.

Random spalling is unpredictable. However, mapping and tracking of spall locations can lead to identifying systemic problems usually along certain high stress joints. There are a couple of areas at DEN that might be candidates for slab replacement because of the spalling mode of failure.

GAPEMS Remaining Service Life Calculation using the FACS Method

The GAPEMS remaining service life calculation methodology which was customized for Denver International Airport is a network level analysis based upon estimation of the time to slab replacement using GAPEMS data at the individual slab level. This is significantly different from the PCI methodology which uses only distress data for a group of 20 or more concrete slabs. It is suggested that the remaining life analysis need only be reaccomplished after a major GAPEMS data collection process has been completed such as once every 3-4 years.

It is important to realize that the remaining life analysis, although very detailed in that it calculates remaining life to the square foot of replacement, is only a network level analysis of the entire airfield. Additional analysis is required to begin project level slab replacement planning. In order to develop a specific project level slab replacement program for the upcoming construction season, the existing data should be augmented with a project level inspection. That project level inspection should use GAPEMS pavement evaluation techniques including photographs. The primary purpose of the project level inspection is to determine if the distress is as severe as was reported or if a maintenance action can extend the life of the pavement for a period of about 5-6 years. The secondary purpose is to determine the best strategy for combining and grouping panels to be replaced into a coordinated and efficient project.

The actual remaining life methodology is a manual process in which a pavement engineer sits at the computer screen with the GAPEMS data and analyzes the distresses, construction history, cores, and structural data and assigns a range of remaining life to each of the concrete slabs as listed in Table 3. Although, this seems like a daunting task, the GIS data query capability makes it a much easier task than one would think. However, the enormous size of Denver International Airport coupled with the construction variables and operational differences does make it a time consuming task. The time to complete the remaining life analysis at DEN for the 140,000 slabs was about 7 work days. However, for the future, a series of approximately 30 rules and guidelines were developed that could help to partially automate and speed this process for the next airport wide remaining life analysis.

Table 3 Remaining Life Categories

Remaining Life Category	Construction Improvement Program Years
0-2 years	2007, 2008
2-5 years	2009, 2010, 2011
5-7 years	2012, 2013
7-10 years	2014, 2015, 2016
10-15 years	2017 - 2021
15+ years	2022+

It is very important to note that the rules and guidelines that were developed for remaining life analysis are specific to Denver International Airport based upon the current aircraft loading, the environmental conditions, and the construction history data that was mined from records at the airport. The rules developed have these conditions in the assessment as applied by the pavement engineer and are not applicable to any other airport other than Denver International. Although, the remaining life analysis is the result of a single pavement engineer's assessment, the methodologies and the rules developed were reviewed and approved by the entire project team. Also the final work product was reviewed by a separate pavement engineer for quality control purposes.

The methodology employed to analyze and classify slabs for remaining life was as follows:
1. Select an area such as Zone 4.
2. Review all construction data for the zone including dates of construction, aggregate type, contractor, and cement type.
3. Review all structural data including HWD and RDD deflections.
4. Review all core sample petrographic data and expansion tests in the Zone.
5. Analyze the pattern and distribution of the type and severity of distresses.
6. Review geo-located photographs of distresses to see what the raters observed in the field.
7. Select one of the four pavement failure modes such as fatigue.
8. Select those slabs which have the lowest remaining life (0-2 years) for that failure mode.
9. Repeat step 8 for the next higher remaining life category for that mode of failure up to 10-15 years category
10. After completing one failure mode repeat steps 7, 8 and 9 for the next mode of failure such as ASR
11. After completing all four failure modes, review your work for slabs you may have missed, Select all remaining slabs in the Zone and assign to remaining life category 15+.
12. Select a new Zone and repeat steps 2-11 until the airport is completed.

Conclusions

The FACS method of remaining life analysis was reviewed by other engineers in DMJM Aviation and was accepted by the Denver International Airport. The results were implemented immediately and the slab by slab network level analysis was easily converted into projects using the GAPEMS Data. The analysis resulted in the shift from perceived remaining life failures due to ASR on runways to fatigue failures mostly of the apron. The analysis clearly identified potential Type A aggregate failures likely on taxiways. Using the GIS data, an immediate change was made to upcoming slab replacement program which properly prioritizes the greatest needs which were different than those the airport had planned.

The GAPEMS pavement management system using a FACS method of remaining life provides a cost savings to DEN and can be applied to other hub airports. The GAPEMS method of pavement evaluation should be used on all airports with commercial service and has been used successfully at Fort Worth Alliance Airport.

Acknowledgements

This work was conducted at Denver International Airport as part of an FAA discretionary grant and the author was the project manager and wishes to thank Denver International Airport including Mike Steffens the project officer, pavement engineers Pete Stokowski and Rudy Amiscaray, and Mark Kelley the DMJM Aviation Project Principal at Denver.

References

AC 150/5380-6B, Guidelines and Procedures for Maintenance of Airport Pavements, Federal Aviation Administration, Washington DC, September 28, 2007

ASTM D5430-Standard Test Method for Airport Pavement Condition Index Surveys, Philadelphia, PA

Bay, James A., Kenneth H. Stokoe, II, Michael T. McNerney and B. F. McCullough, (1998) Continuous Profiling of Runway and Taxiway Pavements With the Rolling Dynamic Deflectometer (RDD) at the Dallas-Fort Worth International Airport, Transportation Research Record No. 1639, 1998, pp. 102-111.

McNerney, Michael T., B. Frank McCullough, Kenneth H. Stokoe, Ngar-kok Lee, James Bay and Jim Wilde, (1997) "Prediction of Remaining Life on Airport Pavements," Airfield Pavement Conference, Seattle, Washington, August 17-20, 1997, published by ASCE, Reston, VA. pp. 77-93.

McNerney, Michael T. and B. Frank McCullough (2000), "Fatigue Cracking in Rigid Airfield Pavements at Large Commercial Service Airports," Transportation Research Record 1703, 2000, pp 65-71.

McNerney, Michael T. and Mark E. Kelley (2007), "The Use of TabletPCs and Geospatial Technologies for Pavement Evaluation and Management at Denver International Airport," 2007 FAA Worldwide Airport Technology Transfer Conference, Atlantic City, NJ, April 2007.

Thomas, Michael D. A., Laboratory tests performed in 2007 at the University of New Brunswick, unpublished.

Pavement Surface Mixture, Texture and Skid Resistance: A Factorial Analysis

M. Alauddin Ahammed,[1] and Susan L. Tighe[2]

ABSTRACT. Skidding on wet pavements contributes to a substantial portion of highway crashes. The resistance to skidding however depends on the microtexture and macrotexture available on pavement surfaces. Several past studies have focused this aspect but with inadequate or inconsistent conclusions. The uniqueness of this study is that surface texture performance has been evaluated controlling the variability in aggregate mineralogy, environmental condition, construction and service. Portland cement concrete (PCC) specimens were prepared from a single mixture to evaluate the true effect of various surface textures on friction properties. Asphalt concrete (AC) surfaces with the same construction record were tested to examine the true effect of mix properties on surface texture and friction. Pavement surface texture was measured using the sand patch method and Automated Road Analyzer (ARAN) while the skid resistance was measured using the British Pendulum and skid trailer.

Analysis has shown that a Mean Texture Depth (*MTD*) of about 1.8 mm is the optimum macrotexture for maximum surface friction on textured concrete surfaces. Exposed aggregate concrete may not be a preferred texture because of the benefit of sand microtexture is lost with washing out of surface mortar. AC surfaces with complex macrotexture have shown to perform differently from PCC surfaces with simple macrotexture pattern. The hypothesis that British Pendulum Number (*BPN*) is dependent only on surface microtexture and represents low speed friction has appeared to be invalid. The skid number-speed gradient is not something universal but varies from mix to mix. Several statistically significant models have also been developed correlating the skid resistance with the asphalt mix grading composition and surface macrotexture and with concrete surface macrotexture.

Keywords: Asphalt pavement, concrete pavement, surface texture, sand patch method, automated road analyzer, surface friction, British Pendulum, skid trailer, skid number-speed gradient, regression model.

---

[1]Ph.D. Candidate, Department of Civil and Environmental Engineering, University of Waterloo, 200 University Avenue West, Waterloo, Ontario, Canada, N2L 3G1; PH (519) 888-4567 x 33872; FAX (519) 888-4300; E-mail: maahamme@engmail.uwaterloo.ca

[2]Canada Research Chair in Pavement & Infrastructure Management, Associate Director of CPATT and Associate Professor, Department of Civil and Environmental Engineering, University of Waterloo, 200 University Avenue West, Waterloo, Ontario, Canada, N2L 3G1; PH (519) 888-4567 x 33152; FAX (519) 888-4300; e-mail: sltighe@uwaterloo.ca

INTRODUCTION

According to Transport Canada (2004), 2,766 people died and another 222,455 were injured on Canadian roads in 2003 from a total of 156,904 police reported crashes. In 2000, 42,643 fatalities and about 2.9 million injuries were reported from about 6.3 million traffic crashes in the United States (Noyce et al. 2005). According to Hoerner and Smith (2002), uncontrolled skidding due to inadequate surface friction contributes to 15 to 35 percent of wet weather accidents. Study has shown that an improvement of average friction by 0.1 can reduce the wet-accident rate by 13% (Kennedy et al. 1990 and Hosking 1987). The skid resistance therefore represents one of the major functional performances of the pavement identified as smoothness, safety and comfort in ride.

The surface friction or resistance to skidding is obtained from microtexture and macrotexture on the pavement surface. Textures on pavement surfaces provide a retarding force at the tire-pavement interface that resist sliding when a braking force is applied (Dahir and Gramling 1990). This facilitates controlled driving maneuvers, especially on wet surfaces. In addition, the splash and spray, which contributes to 10 percent of the wet weather accidents due to poor visibility, can also be minimized with deeper textures on the pavement surface (Hoerner and Smith 2002). Alternatively, deeper textures may contribute to surface roughness, and thereby travelers discomfort in terms of increased noise and vibration in addition to vehicle wear and extra fuel consumption. Selection of appropriate surface mix for AC pavement or appropriate texture for PCC pavement is important to achieve the desired surface performances.

The objectives of this paper are: 1) to provide a brief review of surface texture and skid resistance, 2) to determine the contribution of AC mix properties and PCC pavement various surface textures on surface macrotexture and friction independent of variation in aggregate mineralogy and great variability in construction practices, service and environment, and 3) to develop correlations of surface mix properties, macrotexture and friction.

FUNDAMENTALS OF SURFACE TEXTURE AND SKID RESISTANCE

Pavement surface friction provides the necessary grip at the tire-pavement interface and facilitates the driver to perform controlled maneuvers during running, turning and stopping of a vehicle. A good correlation was found between the wet/dry pavement accident ratio and the ribbed tire skid number (SN) at 64 km/h (40 mph) (SN_{40}). The wet/dry pavement accident ratio was shown to increase sharply from 0.23 to about 0.7 as the SN_{40} has dropped below 41 (Rizenbergs et al. 1976). Pavement surface friction therefore is an essential component of safety on the roadway.

The friction or resistance to skidding however is available from different shape of textures on the pavement surface. Pavement surface textures that contribute to beneficial or favorable effect in resistance to skidding are classified as surface microtexture with texture amplitude/wavelength of <0.5 mm and macrotexture with texture amplitude/wavelength in the order of 0.5 to 50 mm (Sandberg and Ejsmont 2002). Microtexture provides an adhesive force as the texture asperities of aggregate surface contact the rolling tire. Macrotexture, on the other hand, causes an energy loss (known as hysteresis) and produces a retarding force when the tire deforms around the coarse aggregate particles (Goodman et al. 2006). According to Forster (1989), microtexture penetrates the thin film of water on wet pavement surface to maintain intimate tire-pavement contact whereas macrotexture prevents build up of water and reduces hydroplaning as it contributes to drainage of water accumulated on the pavement surface.

AC pavement surface texture and friction is generally controlled by the choice of asphalt mixes including aggregate gradation, type (polish resistant) and size. Coarse aggregate with hard and angular fine particles and/or harsh fine aggregate bonded on asphalt surface provide beneficial microtexture (Balmer and Hegmon 1980). Alternatively, macrotexture are larger irregularities associated with the gaps between the stone particles on the pavement surface. Macrotexture depends on size, shape and gradation of coarse aggregate in the paving mixture and the construction practices for the surface layer (Noyce et al. 2005).

Microtexture on concrete pavement surfaces is mainly contributed by the fine aggregate in the mortar (ACI Committee 325, 1988). As mentioned by Balmer and Hegmon (1980), hard and angular sands on the concrete surface can provide excellent microtexture for PCC pavements. Macrotexture on concrete surface are formed using some special texturization tools. These include small surface channel, indentations or grooves on fresh concrete or cut on hardened concrete (Hoerner et al. 2003). The contributions of microtexture and macrotexture to offered friction on PCC pavement surfaces are similar to those for AC pavement surfaces provided that there is no bleeding on AC surfaces and the tire can contact the aggregates in the mix.

RELEVANT PAST RESEARCH

After four years of traffic exposure, a 19 mm spaced (3.2 mm wide and 3.2 mm deep) transversely tined PCC pavement surface was shown to exhibit a bald tire SN_{40} of 47. A similar longitudinally tined and an exposed aggregate surfaces were shown to exhibit bald tire SN_{40} of 33 (Mahone et al. 1977). The reason for such large difference in SN between the longitudinally and transversely tined surfaces was not clarified. Burlap dragged surface was shown to exhibit a bald tire SN_{40} of 35 on passing lane and 20 on traffic lane.

Grady and Chamberlin (1981) found that grooved PCC surface with MTD of 1.70 mm (0.067 in.) and 0.41 mm (0.016 in.) provide ribbed tire SN_{40} of 62 and 35, respectively. For the same pavement, the SN was shown to vary by 9 to 22 points for a difference in MTD of 0.10 mm (0.004 in.) to 0.23 mm (0.009 in.). Although such variation seems to be very large, no clarification was provided. Overall, a mean groove depth (MGD) of 1.27 mm (0.05 in.) was found to be necessary for a minimum acceptable ribbed tire SN_{40} of 32. Equation 1 and Equation 2 show the developed models for ribbed tire and smooth tire SN, respectively. Equation 1 shows a consistent decrease in SN with an increase in MGD whereas Equation 2 shows the reverse trend. These two models suggest that either pavement should be constructed with no surface texture (smooth surface) to avail the maximum ribbed tire SN or the vehicle should use smooth tire for safe driving on textured surfaces.

$$SN_{40}^{R} = 44.49 - 4.88(\log MGD) \tag{1}$$

$$SN_{40}^{B} = 18.33 + 17.15(\log MGD) \tag{2}$$

Where, $SN^R_{40} = SN$ at 40 mph using ribbed tire, $SN^B_{40} = SN$ at 40 mph using smooth (bald) tire, and MGD = Mean Groove Depth (in.).

Yager and Buhlmann (1982) found no correlation between the BPN and macrotexture whereas Olek et al. (2004) found a linear relationship between the BPN and macrotexture on PCC surfaces. Corley-Lay (1998) also found that the variation in BPN from section to section resembles the variation in SN. In Corley-Lay's study, large-stone mix with 95% passing the 19.1 mm (3/4'') sieve and Heavy Duty Surface (HDS) course with 95% passing the 9.5 mm (3/8'') sieve were shown to provide SN_{40} of 52.2 (BPN = 62.9) with MTD of 0.822 mm and SN_{40} of 53.1 (BPN = 64.3) with MTD of 0.78 mm, respectively. Despite lower MTD and smaller maximum aggregate size, HDS mix has shown to exhibit greater friction than the large-stone mix. Rubber or fiber modified stone mastic asphalt with 95% passing the 12.5 mm (1/2'') sieve was shown to provide the lowest friction where the SN_{40} ranged from 47.1 (BPN = 60.5) with MTD of 0.89 mm to 50.9 (BPN = 67.8) with MTD of 1.07 mm.

Wambold et al. (2004) found a good correlation (Equation 3) of BPN and the international friction index (IFI) in the international PIARC experiment.

$$F_{60} = 0.0079 BPN + 0.0778 \tag{3}$$

Where, F_{60} = IFI at 60 km/h

Henry and Saito (1983) presented a large number of regression models, with good coefficient of correlation (r) values, for the prediction of SN, MTD and BPN. The proposed models however used several highly interdependent i.e. correlated predictor variables in the same model which may be considered statistically erroneous (Ahammed and Tighe 2008). An example of the developed model is given by Equation 4 (r = 0.905) in which SN^B_{64} and SN^R_{64} (highly correlated variables) were used together for the prediction of BPN.

$$BPN = 20.0 + 0.405 SN^R_{64} + 0.039 SN^B_{64} \tag{4}$$

Where, BPN = British Pendulum Number, SN^B_{64} = SN at 64 km/h using blank tire, and SN^R_{64} = SN at 64 km/h using ribbed tire.

Wambold (1988) refined some of the models in Henry and Saito (1983) including some additional data but without altering the model forms. Forster (1989) attempted to correlate the BPN or SN_{40} with texture shape factor but ended with detection of problems in both texture and skid data.

SMA mixes with 0/10 and 0/15 aggregates were shown to exhibit a BPN of 60 to 72 and 58 to 68, respectively i.e. higher friction on mix with smaller (10 mm) aggregate than the mix with larger (15 mm) aggregate (Boscaino et al. 2004). This further indicates that higher texture (beyond a level) do not necessarily mean a greater friction. Ergun et al. (2005) developed a complex model for the prediction of surface friction (measured with Odoliograph) from the macrotexture and microtexture (determined through image analysis). The authors themselves however doubted about the acceptance of the method because of its imprecision in addition to the complexity.

Fwa et al. (2003) studied the effect of asphalt pavement aggregate spacing and portland cement mortar groove configurations on the tire-pavement contact and skid resistance measured with the British Pendulum. Analysis has shown that the choice of aggregate grading i.e. the asphalt mix design directly influences the skid resistance of an asphalt pavement surface. However, the paper indicated that the correlations are not simple and that further research is needed to better understand the texture-skid resistance relationship.

RESEARCH APPROACH AND METHODOLOGY

Preparation of PCC Specimens

This study has incorporated a controlled laboratory experiment of PCC pavement various surface textures as well as laboratory and field experiments of asphalt pavements various surface mixes. A standard 30 MPa ready mix concrete with 20 mm nominal maximum size aggregate was used to prepare the PCC specimens. All the specimens were prepared at the same time, using 152 mm (6 in.) diameter cylindrical plastic molds, in the laboratory of Centre for Pavement and Transportation Technology (CPATT) at the University of Waterloo (UW), Ontario. Specimen preparation in such a manner has enabled to evaluate the true effect of various surface texture on skid resistance properties, controlling the effect of varying materials/mixes, temperature, aggregate gradation and age/uses. A new PCC pavement test section with same mix was also constructed at CPATT test track.

Texturization and Testing of PCC Specimens

The PCC specimens were surface textured in different configurations with three specimens in each configuration. These include: screed finish, burlap, corn broom and plastic turf drag, exposed aggregate, and various tining of 3.2 mm wide and 4 mm deep tines having rectangular and triangular (45°) tips. Two exposed aggregate surface textures were produced by spraying different rates of surface retarder. The tines were spaced uniformly at 16 mm c/c or randomly at 10 to 22 mm c/c. Although two different tips were used, the resulting grooves after concrete hardening showed no noticeable difference probably due to texturization timing. The field section received a combination of longitudinal burlap drag and uniform 16 mm c/c spaced longitudinal tine texture. Figure 1 shows the list of sixteen surface textures that were tested ("F" indicates field section) in this study.

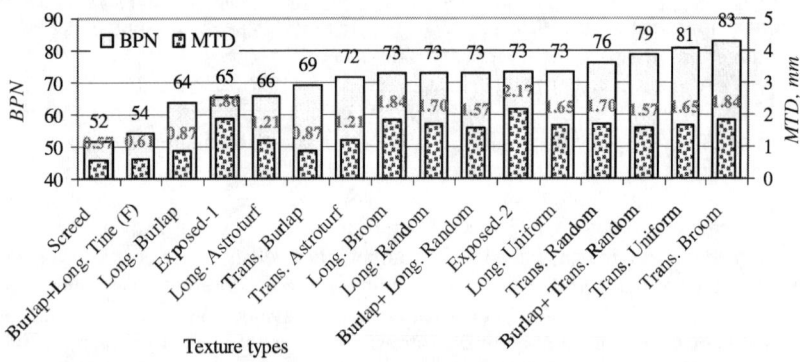

Figure 1. Comparison of *MTD* and *BPN* for various surface textures of PCC.

The surface texture (*MTD*) was measured using the sand patch while the friction (*BPN*) was measured using the British Pendulum. As the degree of saturation has shown remarkable difference in measured *BPN* of PCC surfaces, all the specimens were soaked in water for 24 hours prior to friction testing to be consistent in moisture content.

AC Pavement Sections

The tested AC pavements are located at two sites in the Region of Waterloo (ROW). The first test site, called CPATT test track, is located at Waterloo landfill (LF) station and consists of five test sections each about 140 m long. The surface mixes are: conventional dense HL 3 (two control sections), Polymer Modified HL 3 Asphalt (PMA), Stone Mastic Asphalt (SMA) and Superpave (SP). The other site, named as ROW/CPATT quiet pavement site, is located on regional Road 11 at Crosshill (CH). Four AC sections, each 600 m long, at this site include: HL 3, SMA, Rubber-modified Open Friction Course (ROFC), and Rubber-modified Open Graded Course (ROGC). Each of the surface courses contain 16 mm maximum size aggregates from same source with similar Micro-Deval abrasion (17% to 18%) that enabled direct evaluation of the effect of aggregate gradation independent of variation in aggregate size and mineralogy. However, on both sites, the SP, SMA and ROFC contain premium (100% crushed coarse and fine) and all other sections contain conventional aggregates. All test sections were constructed in 2004 utilizing the same contractor which has eliminated or reduced the variability in construction practices.

Testing of AC Pavements

For laboratory testing, fifteen 152 mm diameter specimens were obtained by coring from five sections (three cores from each section) at landfill site and were tested for the *BPN* and sand patch *MTD*. The *SN* of each of these sections was measured at 64 km/h and 80 km/h at about 25 m interval through multiple passes of ASTM skid trailer (ASTM E 274) mounted with standard ribbed tire (ASTM E501). In addition, the surface macrotexture of these sections were measured using the laser profiler mounted with the Automated Road Analyzer (ARAN). Alternatively, twelve cores were obtained from four sections at quiet pavement site and tested only for sand patch *MTD* because of smaller diameter than that required for *BPN* measurement. However, the *SN* of each section was measured at 64 km/h, 80 km/h, 90 km/h and 100 km/h on the same day (in October 2007) of skid testing at landfill site.

ANALYSIS AND RESULTS

MTD and Skid Resistance of PCC Surfaces

Figure 1 compares the average *MTD* and *BPN* based on three replicate specimens or test points for various surfaces of PCC. Among the studied surfaces, the greatest skid resistance (*BPN* = 83) was provided by the deep transverse broom surface with a *MTD* of 1.84 mm. Smooth surface with trowel finish has exhibited the lowest skid resistance with the lowest *MTD* of 0.52 mm. Longitudinal randomly spaced tining with a *MTD* of 1.70 mm has provided a *BPN* of 73. A similar tining with a burlap drag in advance has exhibited a similar *BPN* (73) but with a lower *MTD*

(1.57 mm) showing the added benefit of cross texturing. Transverse texture has shown to exhibit 7% to 14% higher skid resistance as compared to longitudinal texture with same texture configuration.

The greatest *MTD* (1.86 mm and 2.17 mm) were available for two exposed aggregate surfaces but they have exhibited lower friction as compared to other surfaces with similar or lower *MTD*. Loss of microtexture due to washing out of sand in the surface mortar has probably resulted in the reduced friction. Therefore, exposed aggregate surface may not be a preferred texturing method because of reduction in skid resistance in addition to difficulty in construction and associated extra cost. A variation in *MTD* of 0.31 mm between two exposed aggregate surfaces has resulted in a variation in *BPN* of eight points. All these together indicate that when referring a texture configuration it is important to indicate the actual texture depth in addition to the texturization method, direction and shape.

For the field section, although the tining specification (spacing/depth/width) was similar to the laboratory specimen with uniformly spaced longitudinal tining (*MTD* = 1.65 mm and *BPN* = 73), the *MTD* was shown to be just 0.61 mm with a *BPN* of 54. It indicates that construction variation is a big factor in achieving the desired texture and friction.

MTD and *BPN* Relationship for PCC Surfaces

Figure 2 shows the variation of *BPN* with the variation in *MTD* for the PCC surfaces included in this study. As shown in the figure, the skid resistance increases to the maximum value for a *MTD* value of about 1.8 mm and decreases thereafter as the *MTD* further increases. This can be justified from the fact that for higher *MTD*, with deeper textures, the net tire-pavement contact area does not increase (rather decrease) because of a constant tire deformation beyond a certain depth of surface undulation. This indicates that PCC texturing should be specified to attain a maximum *MTD* of about 1.8 mm as the available friction does not increase beyond such texture level whereas driver's discomfort and tire-pavement noise are likely to increase.

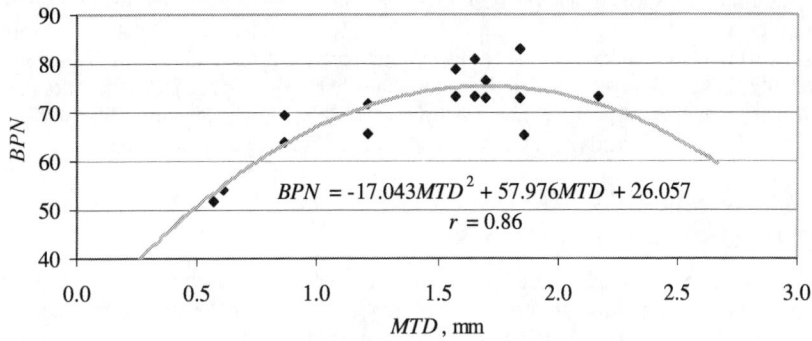

Figure 2. Variation of *BPN* with *MTD* on PCC Pavements.

Wambold (1988) has recommended a *BPN* of 55 as the desired minimum surface friction for adequate resistance to skidding on wet surface. Rizenbergs et al.

(1976) indicated a SN_{40} of 41 as the critical value in recommending the desired minimum surface friction. A *MTD* of 0.7 mm can exhibit such level of skid resistance. However, considering the variability in skid and texture measurements, and surface wear over time, the desired minimum *MTD* should be substantially higher depending on traffic volume and pattern, concrete mix materials and local environment. A *MTD* of maximum 1.8 mm would probably be adequate for every circumstance. The optimum or desired maximum macrotexture should however be verified through friction measurement using the skid trailer although a justified variation of *BPN* with the *MTD* was found in this study. The correlation was also statistically significant at a 5% level of significance with a good correlation coefficient.

Skid Resistance Prediction Model for PCC Pavement Surfaces

SPSS (Version 16.0), a popular software for statistical analysis, was used for developing all the regression models. A quadratic relationship between the *BPN* and *MTD* has exhibited the best correlation in terms of model predictability, statistical significance and logical sense. In the first model, the average *BPN* and *MTD* for sixteen different texture configurations were used. An indicator variable was included in the model to account for the variation in texture directions. The summary of the resulting model is presented in Table 1 and given by Equation 5. The model showed a good coefficient of determination (R^2) of 0.906 with all the predictor variables statistically significant at 5% significance level (*p*-values of 0.000). Equation 5 shows that for a given *MTD*, transverse texture will exhibit about eight points higher *BPN* as compared to other textures.

$$BPN_{Avg} = 34.057 + 39.507 MTD_{Avg} - 10.171 MTD_{Avg}^{2} + 7.609 TD \quad (5)$$

Where, BPN_{Avg} = average *BPN* for each texture configuration, MTD_{Avg} = average *MTD* (mm) for each texture configuration, and *TD* = Texture Direction code (transverse texture = 1, other texture = 0).

Table 1. Skid resistance models for PCC surfaces

Model	IVs	N	b	Std. Error	t	p-value	R^2
Equation 5 (BPN_{Avg})	(Constant)		34.0566	9.170	5.503	0.000	0.906
	MTD, mm	16	39.507	6.188	3.722	0.003	
	MTD^2	16	-10.171	10.614	-2.536	0.026	
	TD	16	7.609	4.010	4.600	0.001	
Equation 6 (BPN_{Avg})	(Constant)		34.752	5.732	6.063	0.000	0.958
	MTD, mm	14	37.829	10.564	3.581	0.005	
	MTD^2	14	-8.993	4.277	-2.103	0.062	
	TD	14	6.697	1.215	5.510	0.000	

As the exposed aggregate texture was shown to behave differently from other surfaces because of loss of sand microtexture with washed mortar, further attempt was made to develop skid resistance prediction model excluding the two exposed aggregate surfaces. The resulting model is also summarized in Table 1 and given by Equation 6. All the regression coefficients were statistically significant at a 5% level of significance with improved R^2 values of 0.958, except the square term which is statistically significant at a 6% level of significance. For a given *MTD*, transverse texture is expected to provide about seven points higher *BPN* as compared to longitudinally textured surfaces.

$$BPN_{Avg} = 34.752 + 37.829 MTD_{Avg} - 8.993 MTD_{Avg}^2 + 6.697 TD \quad (6)$$

Where, BPN_{Avg} = Average *BPN* of each texture, except the exposed aggregate.

MTD and Skid Resistance of Various AC Pavement Surfaces

The comparison of *MTD* and *SN* at 64 km/h (SN_{64}) for nine AC surfaces at two test sites is shown in Figure 3. HL3-1 and HL3-2 (same control mix at LF) have shown to exhibit a SN_{64} of 48 and 44 with a *MTD* of 0.87 mm and 0.76 mm, respectively i.e. a four point difference in *SN* for a difference in *MTD* of 0.11 mm on dense AC mix. Alternatively, the HL3 at CH has provided a SN_{64} of 49 with a *MTD* of 0.56 mm. The PMA mix that contains same aggregate gradation as of HL3-1 and HL3-2 has exhibited SN_{64} of 53 with a *MTD* of 0.92 mm. Two SMA mixes have shown to exhibit SN_{64} of 59 and 57 with the greatest *MTD* of 1.75 mm and 1.53 mm, respectively where a variation in *MTD* of 0.22 mm has resulted in a two point difference in *SN*. The superpave mix has provided the greatest SN_{64} of 61 but with a relatively low *MTD* of 0.91 mm while the ROFC was shown to exhibit a good SN_{64} of 58 but with a *MTD* of 1.19 mm. The ROGC mix with similar gradation of ROFC has exhibited the lowest SN_{64} of 44 with a good *MTD* of 1.15 mm. All these together indicate that AC surface friction is a complex function of many factors including surface texture level, aggregate and mix properties and net contact at the tire/ pavement interface.

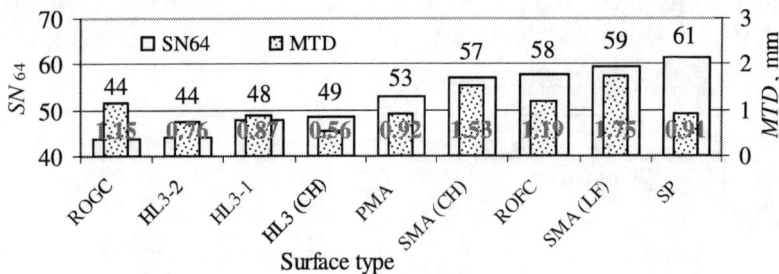

Figure 3. Comparison of *MTD* and SN_{64} for various surface mixes of AC pavement.

SN_{64} versus BPN of Asphalt Surfaces

Figure 4 shows the comparison of *BPN* and SN_{64} for five surfaces at landfill site. In this figure, the variation of *BPN* has shown to fairly resemble the variation of SN_{64} rejecting the hypothesis that *BPN* is dependent on surface microtexture only and represents low speed friction. In other words, it further establishes the fact that *BPN* is dependent on both surface microtexture and macrotexture and simulates well with other dynamic testing method.

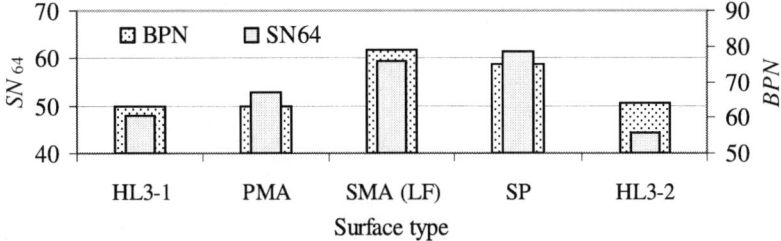

Figure 4. SN_{64} versus BPN of Asphalt Surfaces.

The correlation between BPN and $SN_{64,}$ as given by Equation 7, shows that SN_{64} is fairly 77% of the BPN value. With this correlation, a BPN of 55 would mean a SN_{64} of 42 which agrees well with previous recommendations despite small number of data points used in this study. The correlation is statistically significant at 95% confidence level with t-value of 30.65, p-value of 0.000 and good correlation coefficient (r) of 0.85.

$$SN_{64} = 0.7734 BPN \tag{7}$$

Mix Properties versus MTD and SN of AC Surfaces

Both SN_{64} and MTD were shown to go increasing linearly with an increase in the ratio of $C.A.$ to $F.A.$ percentages (by weight). The MTD and SN_{64} however were shown to decrease with an increase in air void (AV) content in the mixture. Such trends have appeared to be counterintuitive to previous knowledge. The SN_{64} was also shown to decrease with an increase in Voids in Mineral Aggregates (VMA) although the MTD was shown to slightly increase with an increase in VMA. None of these relationships however were statistically significant at 95% (even 90%) confidence level. In fact, both the AV and VMA represent the voids within the asphalt/aggregate mixture and are interior properties of the asphalt mixture. The surface microtexture/ macrotexture and frictional performance are the properties of uncoated aggregate textures and gap/depression between exposed aggregates. This supports the finding in this study.

Variation of Skid Resistance with Surface Texture of AC Pavements

Figure 5 shows the variation of SN_{64} with macrotexture i.e. MTD of AC pavement surfaces. The trend for the variation of skid resistance with MTD has appeared to be different from that of PCC pavement surfaces. This is probably due to different scenarios of the surface macrotexture of two pavement types where the asphalt pavements macrotexture generally is more complex with multi-directional variation as compared to the unidirectional macrotexture on PCC surfaces. Furthermore, the maximum MTD value was 1.75 mm for the AC mixes used in this analysis. However, the trend in Figure 5 shows that a MTD of 1.8 mm on AC surface can provide a SN_{64} of 60 which is similar to transverse texturing on PCC pavement surface with *BPN* of 78. A *MTD* of 1.8 mm therefore may also be taken as the desired maximum texture level for all pavement surfaces to minimize the roughness, tire wear, fuel consumption and tire/pavement noise.

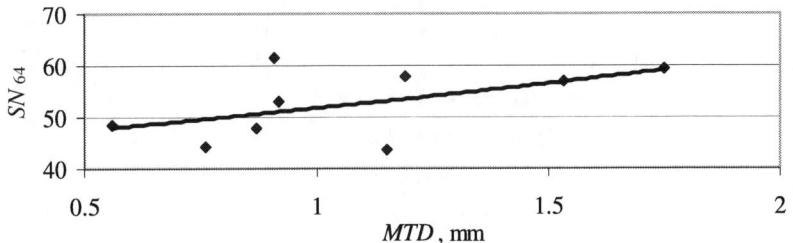

Figure 5. Variation of skid number at 64 km/h with AC surface macrotexture.

Variation of Skid Resistance with Speed

The variation of *SN* for four different surfaces at Crosshill site with the speed of the vehicle is shown in Figure 6. As shown in the figure, *SN* decreases linearly with an increase in vehicle's speed. The correlation for individual mix is very strong with *r* of 0.96 and over but with varying slopes. When combing all mixes for a single trend line the *r* value stands to be 0.47 only. These indicate that the gradient is not something universal for all mixes rather skid number-speed gradient should be developed for each mix separately for best utilization of the available friction.

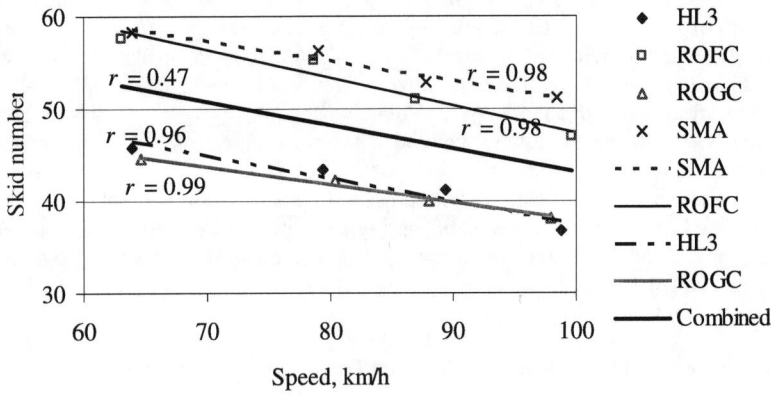

Figure 6. Vehicle's speed versus skid number.

Models for AC Pavement Macrotexture and Friction Prediction

Preliminary model for macrotexture of AC surface has shown that the *C.A./F.A.* ratio and *VMA* (%) are statistically significant, however, the counterintuitive (negative) correlation with *VMA* (%) appeared to be meaningless. Therefore, a model for *MTD* was developed correlating the *C.A./F.A.* ratio as given by Equation 8 and presented in Table 2. The model was statistically significant at 5% significance level with good R^2 value of 0.843.

$$MTD = 0.501 + 0.296(C.A./F.A.) \quad (8)$$

Where, MTD = Mean Texture Depth (mm), and $C.A./F.A.$ = Coarse to fine aggregates ratio in the AC mix (by weight).

Since the MTD and $C.A./F.A.$ are highly correlated, both of them were not included as independent variables in the same model for skid number prediction. Two binary codes were incorporated to distinguish between the premium and normal aggregates and to capture the difference between rubber/polymer modified mixes and conventional mixes (no rubber/polymer). Outliers with standardized residual value >|2| were filtered out as part of regression diagnostics. The models for the SN were then developed as given by Equation 9 and Equation 10 and summarized in Table 2.

$$SN_V = 60.703 + 12.959(PREMIUM) - 0.233V \quad (9)$$

$$SN_V = 57.032 + 11.435MTD - 0.258V \quad (10)$$

Where, SN_V = Skid Number at speed V, V = vehicle's speed (km/h), MTD = Mean Texture Depth (mm), and $PREMIUM$ = code for aggregate crushing (1 for 100% crushed aggregates, 0 for normal aggregates).

Table 2 shows that all the parameters are statistically significant at 5% level of significance. The model for SN incorporating aggregate quality (premium versus normal) and speed has shown a very good R^2 of 0.976. Alternatively, the model for SN as related to MTD and speed has shown a fair R^2 value of 0.558. The code for rubber or polymer in the mix was not statistically significant. These models and preceding analysis indicate that quality of aggregates is most important in achieving good friction.

Table 2. MTD and skid number prediction models for AC surfaces

Model	IVs	N	b	Std. Error	t	p-value	R^2
Equation 9	(Constant)		0.501	0.059	8.519	0.000	0.843
(MTD)	C.A./F.A.	26	0.296	0.026	11.340	0.000	
Equation 10	(Constant)		60.703	1.696	35.789	0.000	0.976
(SN)	PREMIUM	22	12.959	0.533	24.315	0.000	
	V (km/h)	22	-0.233	0.021	-11.05	0.000	
Equation 11	(Constant)		57.032	6.638	8.592	0.000	0.558
(SN)	MTD (mm)	26	11.435	2.741	4.171	0.000	
	V (km/h)	26	-0.258	0.075	-3.426	0.002	

Sand Patch Texture versus Laser Based Texture

An excellent correlation was found between the Mean Profile Depth (MPD) and Root Mean Square (RMS) of macrotexture from ARAN laser profiler with a r value of 0.994. However, no correlation was found in this study between the RMS and MTD as well as between the MPD and MTD probably due to limited data. Further close study is recommended to correlate the MTD and laser based macrotexture.

CONCLUSIONS

This study was directed at evaluating the true effect of AC mix properties and PCC surface texture on macrotexture and surface friction. Among the studied surfaces, deep transverse broom texture on PCC surface has shown to exhibit the highest skid resistance with recorded *BPN* of 83 for a *MTD* of 1.84 mm. Exposed aggregate on PCC pavements may not be a preferred texture because of loss of sand microtexture with washed mortar. A *MTD* of about 1.8 mm was shown to provide the maximum friction on textured PCC surface. Good correlation was found between the PCC surface friction (*BPN*) and *MTD* (R^2 of 0.906 and 0.958).

Among the studied AC surfaces, the SP mix with premium aggregates has shown to exhibit the highest *SN* with a relatively low *MTD*. SMA surface with the greatest *MTD* has exhibited relatively lower *SN* as compared to *SP* surface despite both mixes containing premium aggregates. ROGC with a good *MTD* has shown to exhibit the lowest *SN*. All these indicated that surface friction is a complex function of many factors including surface texture, aggregate qualities and actual tire-pavement contact. The relationship between the *MTD* and aggregate gradation in AC mixes was very good with R^2 values of 0.843. The model correlating the S*N* with *MTD* and vehicle's speed has shown a fair correlation with R^2 of 0.558. Alternatively, the *SN* model incorporating the aggregate quality code has shown an excellent R^2 of 0.976 indicating that aggregate quality is of outmost importance than anything else in achieving the desired friction.

BPN was shown to correlate well with the *SN* where a *BPN* of 55 would mean a SN_{64} of 42 agreeing well with previous recommendations and rejecting the assumption that *BPN* is an indicator of surface microtexture and low speed friction. Being the interior properties of the AC mixture, none of the AV and VMA showed correlation with the *SN*. The skid number-speed gradient was shown be different for individual mix (not something universal) indicating that separate relationship should be developed for best utilization of the available friction from each surface mix.

ACKNOWLEDGEMENTS

The authors gratefully acknowledge the financial support of Cement Association of Canada (CAC) and the Natural Science and Engineering Research Council (NSERC) of Canada for financial support to this research. The contributions of Dufferin ready mix, Dynatest and ROADWARE are also acknowledged.

REFERENCES

ACI Committee 325 (1988). "Texturing Concrete Pavements." *ACI Materials Journal,* American Concrete Institute, Detroit, Michigan, No. ACI 325.6R-88.

Ahammed, M. A., and Tighe, S. L. (2008). "Statistical Modeling in Pavement Management: Do Your Model(s) Make Sense?" Accepted for Publication in the *Journal of Transportation Research Board*, TRR (Available in CD-ROM Proceedings, 87th Annual TRB Meeting), Washington D.C.

Balmer, G. G., and Hegmon, R. R. (1980). "Recent Developments in Pavement Texture Research." *Journal of the Transportation Research Board,* Washington D.C., Transportation Research Record, No. 788, 28-33.

Boscaino, G., Pratico, F. G., and Vaina, R. (2004). "Texture Indicators and Surface Performance in Flexible Pavements." *CD-ROM Proceedings, 5th Symposium on Pavement Surface Characteristics- SURF 2004*, Toronto, Ontario.

Corley-Lay, J. B. (1998). "Friction and Surface Texture Characterization of 14 Pavement Test Sections in Greenville, North Carolina." *Journal of the Transportation Research Board,* Washington D.C., Transportation Research Record, No. 1639, 155- 161.

Dahir, S. H. M., and Gramling, W. L. (1990). "Wet Pavement Safety Programs." *NCHRP Synthesis of Highway Practice 158*, Transportation Research Board, Washington D.C.

Ergun, M., Iyinam, S., and Iyinam, A. F. (2005). "Prediction of Road Surface Friction Coefficient Using Only Macro- and Micro-texture Measurements." *Journal of Transportation Engineering,* American Society of Civil Engineers, Volume 131, No. 4, 311-319.

Forster, S. W. (1989). "Pavement Microtexture and Its Relation to Skid Resistance." *Journal of the Transportation Research Board,* Washington D.C., Transportation Research Record, No. 1215, 151-164.

Fwa, T. F, Choo, Y. S., and Liu, Y. (2003). "Effect of Aggregate Spacing on Skid Resistance of Asphalt Pavement." *Journal of Transportation Engineering,* American Society of Civil Engineers, Volume 129, No. 4, 420-426.

Goodman, S. N., Hassan, Y., and El Halim, A. O. A. (2006). "Preliminary Estimation of Asphalt Pavement Frictional Properties from Superpave Gyratory Specimens and Mix Parameters." *CD-ROM Proceedings, 85th Annual Meeting of Transportation Research Board,* Washington D.C.

Grady, J. E., and Chamberlin, W. P. (1981). "Groove-Depth Requirements for Tine-Textured Pavements." *Journal of the Transportation Research Board,* Washington D.C., Transportation Research Record, No. 836, 67-76.

Henry, J. J., and Saito, K. (1983). "Skid-Resistance Measurements with Blank and Ribbed Test and Their Relationship to Pavement Texture." *Journal of the Transportation Research Board,* Washington D.C., Transportation Research Record, No. 946, 38-43.

Hoerner, T. E., and Smith, K. D. (2002). "High Performance Concrete Pavement: Pavement Texturing and Tire-Pavement Noise." *Federal Highway Administration,* USA, Report No. FHWA-DTFH61-01-P-00290.

Hoerner, T. E., Smith, K. D., Larson, R. M., and Swanlund, M. E. (2003). "Current Practice of Portland Cement Concrete Pavement Texturing." *Journal of the Transportation Research Board,* Washington D.C., Transportation Research Record, No. 1860, 178-186.

Hosking, J. R. (1987). "Relationship between Skidding Resistance and Accident Frequency: Estimates Based on Seasonal Variation." *Transport and Road Research Laboratory (TRRL),* Department of Transport, Crowthorne, U. K., Report No. RR 76.

Kennedy, C. K, Young, A. E., and Butler I. C. (1990). "Measurement of Skidding Resistance and Surface Texture and the Use of Results in the United Kingdom." *Surface Characteristics of Roadways: International Research and*

Technologies, American Society for Testing and Materials, Philadelphia, ASTM STP 1031, 87-102.

Mahone, D. C., McGhee, K. H., McGee, J. G. G., and Galloway, J. E. (1977) "Texturing New Concrete Pavements." *Journal of the Transportation Research Board,* Washington D.C., Transportation Research Record, No. 652, 1-9.

Noyce, D. A., Bahia, H. U., Yambo, J. M., and Kim, G. (2005). "Incorporating Road Safety into Pavement Management: Maximizing Asphalt Pavement Surface Friction for Road Safety Improvements." *Draft Literature Review and State Surveys,* Midwest Regional University Transportation Center Traffic Operations and Safety (TOPS) Laboratory.

Olek, J., Weiss, W. J., and Garcia-Villarreal, R. (2004). "Relating Surface Texture of Rigid Pavement with Noise and Skid Resistance." Purdue University, West Lafayette, Indiana, Report No. SQDH-2004-1 (Final Report HL 2004-1).

Rizenbergs, R. L, Burchet, J. L, and Warren, L. A. (1976). "Relation of Accidents and Pavement Friction on Rural, Two-Lane Roads." *Journal of the Transportation Research Board,* Washington D.C., Transportation Research Record, No. 633, 21-27.

Sandberg, U., and Ejsmont, J. A. (2002). "Tyre/Road Noise Reference Book." *INFORMEX Ejsmont & Sandberg Handelsbolag,* Harg, Kisa, Sweden.

Transport Canada (2004). "Canadian Motor Vehicle Traffic Collision Statistics: 2004." *Road Safety in Canada: An Overview,* <http://www.tc.gc.ca> (November 7, 2006).

Wambold, J. C, Henry, J. J., and Yager, T. (2004). "NASA Wallops Tire/Runway Friction Workshops 1994-2003." *Proceedings, 5th Symposium on Pavement Surface Characteristics- SURF 2004,* Toronto, Ontario.

Wambold, J. C. (1988). "Road Characteristics and Skid Resistance." *Journal of the Transportation Research Board,* Washington D.C., Transportation Research Record, No. 1196, 294-299.

Yager, T. J., and Buhlmann, F. (1982). "Macrotexture and Drainage Measurements on a Variety of Concrete and Asphalt Surfaces." *Pavement Surface Characteristics and Materials,* American Society of Testing and Materials, Philadelphia, ASTM STP 763, 16-30.

Evaluating International Roughness Index Data Quality at Project and Network Levels

G. P. Ong[1], Aff. M. ASCE and K. C. Sinha[2], Hon. M. ASCE

Abstract

Pavement management can be considered to consist of network and project levels. Regardless of pavement management level, pavement condition data is an integral component and data quality can profoundly affect decision-making at all levels of pavement management. This paper therefore attempts to evaluate the quality of pavement roughness data (i.e. international roughness index or IRI) that is collected at project and network levels. IRI data quality collected at the project-level was found to be affected by the individual run, the wheel path, and the lane at which the profile is measured. IRI data quality collected during routine survey at the network level can be affected by the wheel path profile used to evaluate IRI. The variability of the IRI data was further quantified. Given the IRI collected from routine network-level surveys, a proposed relationship is developed to determine an IRI that is suitable for use in project-level pavement management applications.

Introduction

Pavement condition surveys provide valuable information to allow pavement management at network and project levels. This information is used at the network level in budget allocation between different funding sources and between different sub-agencies, priority setting and programming of maintenance, rehabilitation and reconstruction activities, and predicting future level of service of a given network for

---

[1] Postdoctoral Research Associate, School of Civil Engineering, Purdue University, 550 Stadium Mall Drive, West Lafayette, IN 47907-2051; PH (765) 494-2255; FAX (765) 496-7996; email: ongr@purdue.edu
[2] Olson Distinguished Professor, School of Civil Engineering, Purdue University, 550 Stadium Mall Drive, West Lafayette, IN 47907-2051; PH (765) 494-2211; FAX (765) 496-7996; email: ksinha@purdue.edu

a given budget; and are used at the project level in the selection of appropriate maintenance, rehabilitation and reconstruction strategies for a project site.

The international roughness index (IRI) is a critical performance measure used in pavement management. In recent decades, routine network-level IRI data collection is advocated by state highway agencies. Detailed IRI data collection on specific project sites are performed in cases of warranty projects, development of deterioration models, estimation of work quantities for specific projects or other projects of interest. Table 1 shows how project-level and routine network-level IRI data collection methods differ and how the collected IRI data could be used in project and network-level pavement management.

Table 1. Applications and Measurement of Roughness Data at Project and Network Levels

Description	Network Level	Project Level
Applications	• Describe present status • Predict future status (deterioration curves of roughness vs. time or loads) • Basis for priority analysis and programming	• Quality assurance (as-built quality of new surface) • Create deterioration curves • Estimate overlay quantities
Measurement	• IRI measurement using laser profilers for one lane and one run • IRI measurement using inertial profilers for one lane and one run	• IRI measurement using laser profilers for multiple lanes and runs • IRI measurement using inertial profilers for multiple lanes and runs • Rod and level survey measurements

There is an emerging interest to assess IRI data quality. Recent research done to date focuses on evaluating IRI data quality collected at specific project sites. For example, Evans and Eltahan (2000) attempted to identify possible causes affecting IRI data quality in the Long-Term Pavement Performance (LTPP) experimental pavement sections. Similarly, Perera et al. (2006) found that project-level IRI values computed from the profile data obtained from inertial profilers were accurate and were in good agreement between the reference device and profiler. Yin et al. (2006) made recommendations to control the variability of longitudinal profile data. While these works address the issue of IRI accuracy within a project site, few relate how the difference in data collection techniques during routine network-level data collection and detailed project-level data collection can affect IRI.

This paper therefore attempts to evaluate the IRI data quality collected during routine network-level data collection and that collected at the project level. First the paper describes the data collection procedures and experiments performed to evaluate the data quality at the network and project levels. The IRI data quality collected at project and network levels are next evaluated. Finally, the paper shall study how IRI collected during routine network-level data collection can be related to that collected during project-level data collection.

Methodology

A total of 80 asphalt pavement sections in the state of Indiana were selected. Each pavement section was one mile long. From Table 1, the collection of IRI at network and project levels differs in the sampling rate (in terms of lanes) and the number of runs. In this paper, the network-level and project-level IRI data collection procedures are based on that practiced in the state of Indiana (INDOT, 1997). Details of network and project level data collection procedures are as follow:

- *Routine network level data collection*: For each one-mile pavement section, a routine run is performed by the data collection vehicle on the driving lane of the highway. The left and right wheel path profiles are collected and the IRI evaluated for each pavement section.
- *Detailed project level data collection*: For the same pavement section tested during the routine network level data collection, three consecutive IRI runs are performed for every lane (i.e. driving and passing lanes) on the highway. The left and right wheel path profiles are collected during each run and the IRI evaluated for each pavement section.

Data was collected from 1998 to 2006 as part of the State's annual network-level pavement condition survey. The annual network-level IRI data collection takes place between June and August each year. Data is collected on non-wet days and under relatively good weather conditions. For the pavement sections used in this paper, project-level IRI measurements were performed during the same time frame used for network level data collection.

IRI is measured on each section using a data collection vehicle (Class I laser profiler). Lasers mounted on the front bumper of the vehicle measure the pavement's longitudinal profile, which is used to evaluate the IRI. Measurements of the longitudinal profile are taken approximately every 0.125 inch, according to the AASHTO and ASTM standards (AASHTO, 2000; ASTM, 2005).

IRI Data Quality using Project Level Data Collection Procedures

The quality of IRI data collected using project-level data collection procedures can be affected by: (a) variation in IRI between consecutive runs on the same pavement section; (b) variation in IRI between wheel paths within a single run; and (c) variation in IRI collected on different lanes within the pavement section.

Run-to-Run Variation

Run-to-run IRI data quality can be determined by evaluating the coefficient of variation of the three runs performed on the same pavement section. For each pavement section, the coefficient of variation can be determined for each wheel path using Equation (1).

$$(CV)_{ij} = \frac{\sigma_{ij}}{\mu_{ij}} \qquad (1)$$

where $(CV)_{ij}$ is the coefficient of variation for the ith pavement section and jth wheel path, σ_{ij} is the standard deviation obtained from the three runs performed on the ith pavement section and jth wheel path, and μ_{ij} is the mean IRI obtained from the three runs performed on the ith pavement section and jth wheel path. This would give a representation on the precision of the IRI collected at project level.

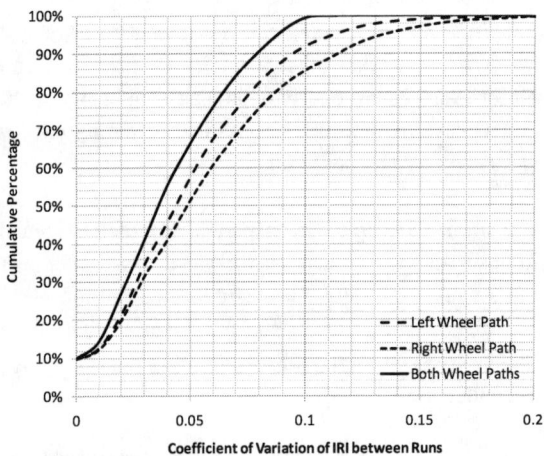

Figure 1. Coefficient of Variation of IRI between Different Runs

Figure 1 shows the cumulative frequency plot comparing coefficients of variation for IRIs obtained from the left wheel path profile, right wheel path profile, and the average of the left and right wheel path profiles. From this figure, it is observed that:

- More than 85% of IRI obtained from the left wheel path profile and more than 90% of IRI obtained from the right wheel path profile have a coefficient of variation of less than 10%.
- The left wheel path tends to produce IRI of better data quality (i.e. lower coefficient of variation) compared to the right wheel path. This is due to a higher deterioration of the pavement near the shoulders in the driving lane.
- The median coefficients of variation of the IRI obtained from the left and right wheel path profiles are found to be 4.5% and 4.8% respectively.
- IRI obtained from the average profile of both wheel paths has a coefficient of variation of less than 10% and a median coefficient of variation of 3.6%. This measure for IRI provides a greater level of precision compared to IRI evaluated from profiles of individual wheel path.

Path-to-Path Variation

Recognizing that there is a difference in data quality for IRI evaluated from the left and right wheel path profiles, the effect of wheel path on data quality is examined. Figure 2 relates the IRI calculated from the left and right wheel path profiles. It is noticed that IRI evaluated from the left wheel path profile tends to be lower than IRI evaluated from the right wheel path profile. A paired t-test was performed to test for differences in the IRIs evaluated from the left and right wheel path profiles. It was found that the IRIs values obtained from the left and right wheel path profiles were different at a 95% significance level (see Case I in Table 2).

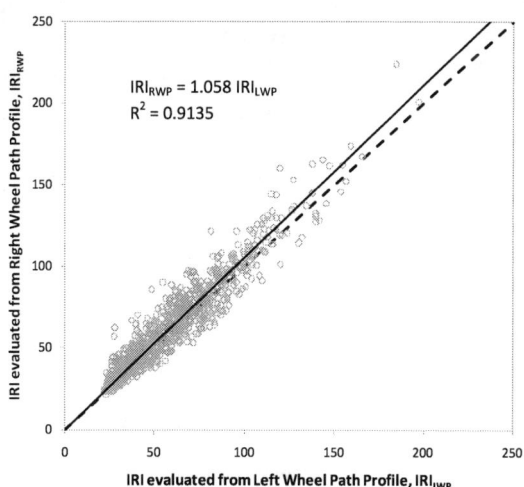

Figure 2. Comparison between IRI evaluated from left and right wheel paths

Table 2. Test for Differences in Mean IRI under Different Scenarios

Cases	Case I		Case II		Case III		Case IV	
Scenario	LWP	RWP	LWP	RWP	LWP	RWP	Drive	Pass
Mean IRI	54.70	56.03	55.46	58.37	54.04	53.78	59.73	56.46
Variance of IRI	728.8	806.9	795.0	791.4	666.4	829.4	747.5	735.5
Number of Observations	1274	1274	646	646	629	629	629	629
Pearson Correlation	0.9484		0.9521		0.9497		0.9002	
Hypothesized Mean Difference	0		0		0		0	
Degrees of freedom	1273		645		628		628	
t-stat	-5.23		-8.50		0.71		6.72	
Critical t	1.962		1.963		1.963		1.963	

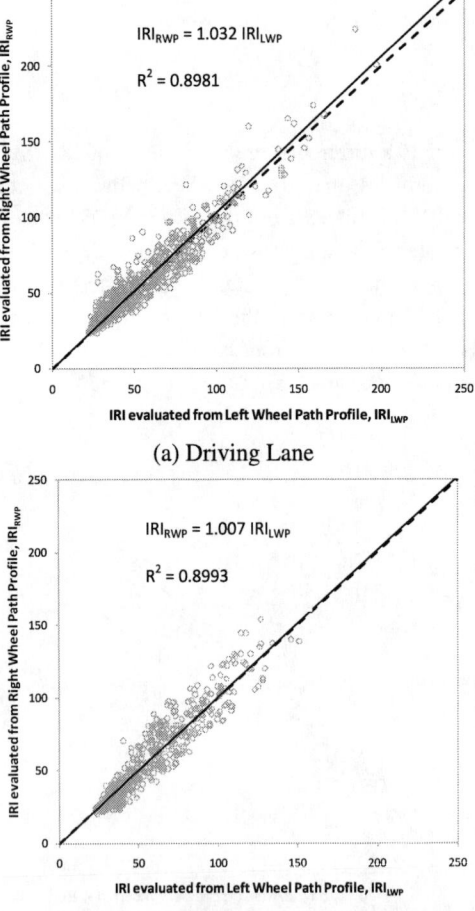

(a) Driving Lane

(b) Passing Lane

Figure 3. IRI evaluated from left and right wheel paths for different lanes

Figures 3(a) and 3(b) show the relationship between the IRI obtained from the left and right wheel path profiles, for the driving and passing lanes respectively. It is observed from these figures that there is a clear difference in IRI evaluated from the left and the right wheel paths of the driving lane, but a small difference in those evaluated from the passing lane. Paired t-tests performed on these data showed that there is a difference in IRI obtained from the left and right wheel path profiles for the driving lane at a 95% significance level, but no significant difference between the IRIs evaluated from the left and right wheel path profiles for the passing lane at a 95% significance level (shown in Cases II and III in Table 2 respectively).

Lane-to-Lane Variation of IRI

Recognizing that the lane which the profiles are tested on could have an effect on the IRI data quality, the effect of profile-testing on the driving and passing lane on the average IRI (i.e. IRI evaluated from the average profiles of the two wheel paths) is studied. The average IRI is a better parameter for evaluating pavement sections at project level due to a smaller coefficient of variation (and hence more precise representation of IRI).

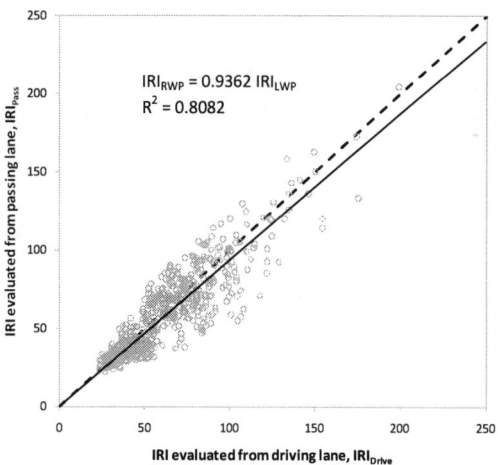

Figure 4. Average IRI evaluated from driving and passing lanes

It was found that there is a significant difference in average IRIs obtained from the driving lane and the passing lane, at a 95% significance level (shown in Case IV in Table 2). Figure 4 illustrates the average IRI evaluated from the driving and passing lanes. It was found that IRI evaluated from profile of the driving lane tends to be higher than that obtained from the profile of the passing lane. This is expected since the driving lane is subjected to heavier traffic loadings (truck and slow moving traffic) and faster deterioration (hence a higher IRI).

IRI Data Quality using Network Level Data Collection Procedures

As mentioned in the previous section, one run is performed on the driving lane of the mile-long pavement section to determine the longitudinal profile during routine network-level pavement condition surveys. Under such a data collection scenario, the possible cause of variation within a particular run would be the variation in IRI values obtained from different wheel paths. This section thus evaluates the wheel-path to wheel-path variation in the IRI collected at network level. Furthermore, it was recommended in the previous section that the average IRI be used. Hence, the coefficient of variation of the average is assessed in this section.

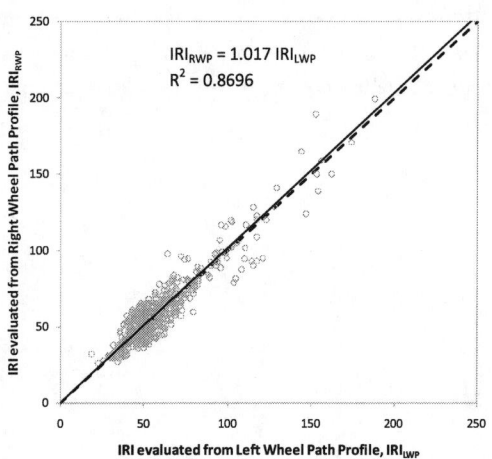

Figure 5. Comparison between IRI evaluated from left and right wheel paths

Table 3. Test for Differences in Mean IRI from Left and Right Wheel Paths

Description	LWP	RWP
Mean IRI	59.71	61.23
Variance of IRI	545.3	580.6
Number of Observations	646	646
Pearson Correlation	0.9341	
Hypothesized Mean Difference	0	
Degrees of freedom	645	
t-stat	-4.055	
Critical t (95% significance level)	1.963	

Figure 5 relates the IRI calculated from the left and right wheel path profiles. It is noticed that the left wheel path tends to produce a lower IRI than the right wheel path. A paired t-test was performed and it was found that IRI evaluated from the left and right wheel path profiles were significantly different at a 95% significance level (see Table 3). This further reinforces results shown in the earlier section, which indicates that there is a difference in IRI evaluated from the left and right wheel path profiles.

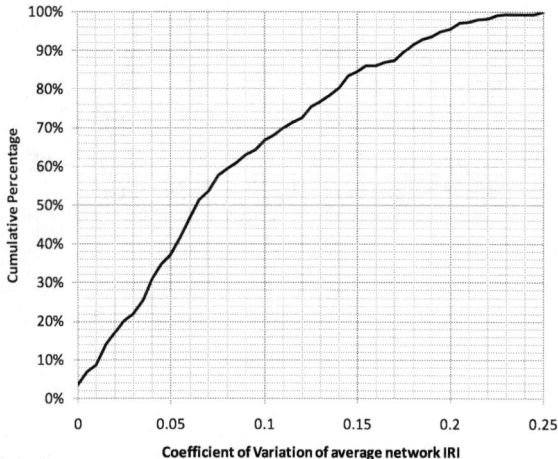

Figure 6. Coefficient of variation of average IRI collected at network-level

Figure 6 shows the cumulative frequency plot depicting the coefficient of variation of the average IRI collected at the network level. It can be observed that more than 95% of the average IRI has a coefficient of variation of less than 20% and more than 65% of the data has a coefficient of variation of less than 10%. Compared to results shown in the previous section, it can be observed that IRI collected during routine network level surveys tends to be more variable than the IRI collected during specific project-level evaluation.

Relating Network and Project Level IRI

Realizing the differences in network-level and project-level IRI data collection methods and the possible sources of variability, it is next sought to relate these IRIs. This is motivated by the need to make use of the data collected during routine network surveys for site-specific applications such as determining IRI of individual pavement sections for project selection, deterioration modeling and calibration for the mechanistic-empirical pavement design guide.

The average IRI obtained from the average profile of the left and right wheel paths is used because of a higher data quality, as shown in the earlier parts of this paper. Recognizing that spatial differences in profiles collected in different runs could affect the quality of the data, the spatial differences were corrected to ensure that the profiles collected at the project and network levels are "tied" prior to analyses.

Figure 7 illustrates the comparison of IRI data collected during routine network-level survey and detailed project-level IRI evaluation. It is found that the IRI can be related by the following relationship:

$$IRI_{project} = 0.971(IRI_{network}) \tag{2}$$

where $IRI_{project}$ is the IRI of a specified project site and $IRI_{network}$ is IRI of the same site evaluated from routine network-level survey. Figure 8 indicates a relatively good agreement between the IRI collected at the network level and the IRI collected at the project level.

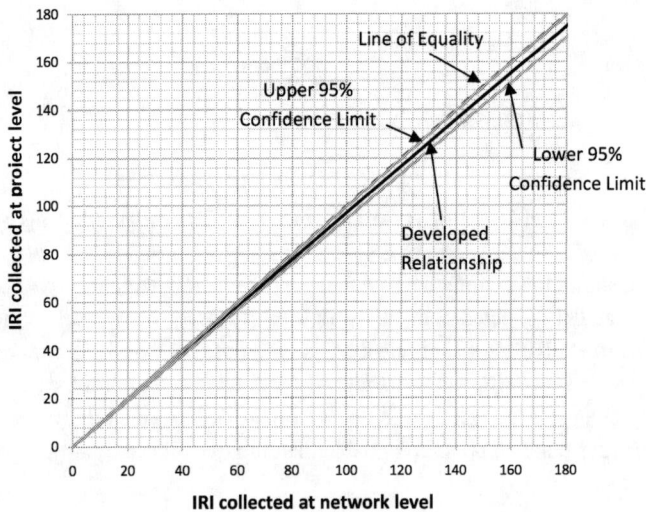

Figure 7: Variability in IRI data

The 95% confidence limits shown in Figure 8 show that a rougher pavement tends to exhibit a larger variation, compared to the smaller variation exhibited by a smooth pavement surface, albeit the overall accurate IRI data. The upper and lower 95% confidence limits shown in Figure 8 are shown in Equations (3) and (4) respectively:

$$IRI_{project} = 0.998(IRI_{network}) \tag{3}$$

$$IRI_{project} = 0.944(IRI_{network}) \tag{4}$$

This equation could be used to determine the range of IRI (i.e. variation) for a project site given the IRI collected from routine network-level surveys.

The relationships described in Equations (2) to (4) allow pavement managers to make decisions on maintenance, rehabilitation and reconstruction at planning and network levels with consideration to the IRI data quality collected at the network level. In addition, there are potentials in using such models to relate the mechanistic-empirical IRI models developed at the project level to the IRI deterioration models used in network-level pavement management systems. However, the relationships shown in this paper are only applicable to Indiana's highways and are specific to the IRI data collection methodology used in Indiana. Further improvements can be made to the models to include regional differences, seasonal variations and other environmental conditions.

Conclusions

The quality of pavement condition data can affect decision-making at project and network levels. With the need for a fast and cost-effective pavement condition data collection, routine network-level pavement condition assessment has become the norm for many state highway agencies. This paper has evaluated the quality of IRI collected from routine network-level surveys and detailed project-level evaluation. The data quality collected during project-level evaluation was shown to be affected by the individual run, the wheel path and the lane at which the IRI is evaluated. The quality of the IRI data collected from routine network-level surveys was also found to be affected by the wheel path used to evaluate IRI. The use of IRI evaluated from the average profile of wheel paths is recommended as it provides better data quality. A relationship is also proposed to determine the average IRI of a specific project site and its variation, given the average IRI evaluated from the network-level survey from the same site.

Acknowledgements

This study was conducted under the Joint Transportation Research Program. The authors acknowledge the support of the Federal Highway Administration and the Indiana Department of Transportation.

Disclaimer

The contents of this paper reflect the views of the authors, who are responsible for the facts and the accuracy of the data presented. The contents do not necessarily reflect the official views or policies of the Indiana Department of Transportation or the Federal Highway Administration. This paper does not constitute a standard, specification, or regulation.

References

AASHTO. (2000). *Standard Practice for Determination of International Roughness Index (IRI) to Quantify Roughness of Pavements*, Designation PP37-00.

ASTM. (2005). *Standard Practices for Simulating Vehicular response to Longitudinal Profiles of a Vehicular Traveled Surface*. ASTM Standard E1170-97. Annual Book of ASTM Standards.

Evans, L.D. and Eltahan, A. (2000). *LTPP Profile Variability*. FHWA-RD-00-113 Final Report. FHWA, Washington D.C.

INDOT. (1997). *Pavement Condition Data Collection Manual*. Indiana Department of Transportation, Indianapolis, IN.

Perera, R.W., Kohn, S.D., and Wiser, L.J. (2006). "Factors Contributing to Differences Between Profiler and the International Roughness Index," *Transportation Research Record*, 1974, 81–88.

Yin, H., Stoeffels, S.M. and Antle, C.A. (2006). "Profile Data Variability in Pavement Management: Findings and Tools from LTPP." *Proceedings of the 2006 Airfield and Highway Pavement Specialty Conference*, I. L. Al-Qadi, eds., ASCE, Reston, Virginia, 984-995.

Managing Utility Installation/Maintenance Activities in Advance to Reduce Pavement Utility Cuts Using Spatiotemporal Objects Database

Chien-Cheng Chou[1], Yi-Ping Chen[2], and Chien-Ming Chiu[3]

[1]Assistant Professor, Department of Civil Engineering, National Central University, No.300 Jhongda Rd., Jhongli, Taoyuan 32001, Taiwan, Phone: 886-3-422-7151 ext.34132, Fax: 886-3-425-2960, E-mail: ccchou@ncu.edu.tw
[2]Graduate Research Assistant, Department of Civil Engineering, National Central University, Taiwan, E-mail: vicky74722@hotmail.com
[3]Graduate Research Assistant, Department of Civil Engineering, National Central University, Taiwan, E-mail: lion5123@gmail.com

Abstract

As more and more utility installation and/or maintenance activities are located in highly congested urban roadways, frequent pavement utility cuts in such areas may cause more traffic disruption as well as deteriorate pavement life and quality. Utility owners normally need to obtain permits from public road authorities before commencing utility activities; however, public road authorities in Taiwan currently just issue permits without trying to coordinate and communicate with utility owners involved to schedule their utility-related activities in a more consecutive way. An information model based on the spatiotemporal objects database technique was proposed to help public road authorities identify the utility activities that might be combined together to avoid unnecessary pavement utility cuts. In the proposed model, constraints pertaining to pavement moratorium, utility clearance distance and traffic conditions were considered. The software architecture is discussed, followed by research conclusions.

Introduction

As more and more people dwell in urban areas, there is an increasing number of utility installation and/or maintenance activities located in such areas that make a great impact on paved roads. The steady escalation of the internet penetration rate demonstrates the need that more communication equipment such as fiber broadband lines or wireless access points will be deployed along major transportation systems in the near future. In order to provide new services or maintain deteriorating utility networks, utility owners have to cut pavements open, install new utility facilities or fix problems identified, backfill proper materials, and restore road surfaces. Reports showed that in the District of Columbia, there were over 5,000 utility cuts in 1996 and over 6000 cuts in 2000 (Wilde et al. 2003); in New York City, more than 250,000 cuts a year were made in 2000, and the number increased by 8% each year (Khogali and Mohamed 1999). Researchers indicated that utility activities in the U.K. rank as the second major cause of traffic disruptions with estimated delay costs of $13 billion dollars. Additionally, uneven pavement surfaces due to frequent utility cuts may further result in driver annoyance and other safety issues (Jesen et al. 2005). Others pointed out pavement utility cuts as a major problem in the transportation infrastructure of the U.S., creating serious financial stress on public road authorities (Wilde et al. 2003). In the M-Taiwan project, the government in Taiwan planned to build 6,000 km of optical fiber transmission lines during 2004-2007 to provide most residents with high-speed internet services (Lin 2005). All of the above studies reveal that pavement utility cuts are inevitable and may bring about numerous challenges associated with costs, safety, pavement maintenance, etc.

Since utility owners normally need to obtain permits from public road authorities before commencing utility activities, encouraging utility owners to work together is generally recognized by public road authorities as a potential solution to reduce pavement utility cuts. However, in Taiwan, most public road authorities currently just issue permits to utility owners without trying to coordinate and communicate with utility owners involved to schedule their utility-related activities in a more consecutive way. Hence, a managerial tool that can keep track of the schedule and geometric boundary of every planned utility activity is highly desired. Any potential cooperation between utility owners that might reduce pavement utility cuts would be detected by such tool, and public road authorities can use the suggestions generated to persuade the utility owners into working jointly.

To this end, this research aims at investigating an information model that can help public road authorities manage utility activities to reduce pavement utility cuts. Literature review regarding problems associated with utility activities and their overall impact is described first. The proposed information model that is designed to best describe spatial and temporal properties of planned utility activities is presented next, followed by model exploitation and evaluation of the prototype's software architecture. The tasks required to validate the model are described, and research conclusions are made finally.

Literature Review: Reducing Pavement Utility Cuts

Several approaches to reducing pavement utility cuts have been proposed and investigated. All were designed to cause minimum disturbance to traffic and to have a low impact on the environment. Briefly, these approaches can be categorized into two types: technology-based and policy-based approaches (Wilde et al. 2003). The technology type of approaches such as the trenchless technique focuses on construction methods, practices, tools, etc. that can be employed to perform the utility work. The policy type of approaches focuses on how to allocate construction resources, how to manage organizations and teams, and how to communicate and coordinate with project stakeholders in order to control frequent pavement utility cuts. The technology-based approaches usually accompany high initial cost and short history of proven success, whereas the policy-based approaches often involve incentives, disincentive, and changes of permit procedures that utility owners must follow to complete their works (Wilde et al. 2003). In fact, more researchers investigate the technology-based approaches. The policy-based approaches have less attention in the literature, and few researchers recognize the trend that since the number of pavement utility cuts is escalating, public road authorities might need a managerial tool that can assist project managers in condensing or rescheduling the work schedules of planned pavement utility cuts performed by different utility owners in order to minimize the impact on the traveling public. For example, assume that utility company A will install new facilities in a street during certain days, and utility company B will perform maintenance activities in the same area but one month later after completion of A's work. If both A's and B's work schedules are flexible, the public road authority might be able to persuade A or B to reschedule their work to perform consecutively. Identifying the project circumstances where two or more different utility activities can be combined together is very important to public road authorities because the influential area of pavements due to utility cuts can be carefully calculated so that both the number of utility cuts and the affected area can be minimized.

In addition to reducing pavement utility cuts, public road authorities also face a challenge regarding the increasing number of interferences among the planned but not-yet-deployed utility facilities. Public road authorities or other public agencies may prescribe the clearance distances between certain types of utility lines. For instance, the Taipei City government regulates any gas pipeline to have at least 5-6m of horizontal separation. To calculate the influential area of a pavement utility cut requires consideration of these planned utility facilities. Research showed that utility permit procedures may take significant time because public road authorities need to consider myriad factors determining whether to issue the permit and coordinate with other organizations to address concerns such as environmental and archaeological issues (Chou et al. 2007). Assume that a utility company is in the progress of acquiring its permit for placing new pipelines, and another utility company would like to submit its permit application to install new facilities. Without proper coordination and communication within the two utility owners, the public road authority might not be able to detect any possible clearance violation of the new utility facilities and still issue the permits. Hence, additional pavement utility cuts may be needed when one of the utility owners performs adjustments to fix the problem. The clearance violation might be resolved if the road segment involving the two utility owners will be rehabilitated jointly because there will be a coordination meeting hosted by the public road authority to address each party's concerns. This is due to the fact that sometimes utility owners are willing to discuss with each other if the public road authority is involved. Intermediate utility facilities to keep up the service during utility work are another possible source of utility interferences. For instance, if water main lines are underneath temporal power distribution poles, serious problems such as voltage shortage or overloading may happen. Therefore, managing the interferences of planned utility activities is becoming a total nightmare from public road authorities' perspective. To reduce pavement utility cuts, public road authorities require a systematical approach to effectively and efficiently manage any future utility installation and/or maintenance activities.

Overall, if public road authorities would like to better manage the utility activities, they might need to deal more with the utility work schedules and use a computerized tool to precisely depict the spatial information of each planned utility activity. The following section elaborates more temporal and spatial requirements of each utility activity so that an information model to capture such requirements can be proposed.

Temporal Properties of a Utility Activity

To schedule a set of activities, traditional techniques such as Program Evaluation and Review Technique (PERT) emphasize the importance of the constraints on the activities. Constraints involve time or resources-related rules that govern the execution sequences to complete the project. From public road authorities' perspective, since utility activities are performed by different utility owners, sharing resources among these companies is rare, and oftentimes utility owners do not want to overlap their work schedules. The primary constraints in this type of work thus become spatiotemporal-related requirements. For instance, if a utility company will install a new service line in the area, is there any active construction plan located in the same place? When will the plan start and when will it end? The location and the duration are the major spatiotemporal resources that cannot be easily shared in most of the utility activities. If one activity needs a particular spatiotemporal resource, the other activities cannot occupy without proper coordination.

One advanced database technique that has emerged as a main focus of many spatiotemporal information systems such as the digital battlefield in the military is to keep track of object locations over time and to support temporal queries about future locations of the objects (Wolfson 2002; Wolfson et al. 1998). Called moving objects database (MOD) or spatiotemporal objects database (SOD), this technique aims to deal with geometries changing over time and to simplify the data update process through use of dynamic attributes (Guting and Schneider 2005), thereby having the potential for eliminating or reducing some of the associated challenges and complications. The way SOD employs to process the time dimension for each moving object may serve as a starting point for utility activity modeling.

In SOD, there are two time dimensions associated with each time-sensitive attribute: valid time and transaction time (Tansel et al. 1993; Guting and Schneider 2005). The valid time refers to the time in the real world when an event occurs or a fact is valid. The transaction time refers to the time when a change is recorded in the database. Formal definitions for D_v (valid time) and D_t (transaction time) are listed as follows:

$$D_t = \{t_0, t_1, ..., t_i, ..., now\}$$
$$\forall t', t'' \in D_t \setminus \{now\} : t' < t'' < now \vee t'' < t' < now \vee t' = t'' < now$$

$$D_v = \{t_0, t_1, \ldots, t_i, \ldots, now, \ldots\} \cup \{\infty\}$$
$$\forall t', t'' \in D_v \setminus \{\infty\}: t' < t'' < \infty \vee t'' < t' < \infty \vee t' = t'' < \infty$$

For example, the duration of a utility activity is defined by a manager as from January 1, 2009 to March 1, 2009, which is in the D_v domain. The manager enters this activity information, including the geometric boundary and the schedule, to a computer on November 1, 2008, which is in the D_t domain, but the schedule is changed on January 5, 2009. The new schedule is from February 1, 2009 to April 1, 2009. There is another utility activity scheduled to be performed from March 1, 2009 to May 1, 2009. The manager enters the second utility activity to the computer on December 1, 2008 (see Figure 1). In fact, the three database records regarding the two utility activities must be persisted individually since each data entry may be associated with a permit application and fees. A public road authority may ask the utility owner of the first activity to pay additional fees for the time period between January 1, 2009 and January 5, 2009, although no one indeed performs any work at that time. Recording the two time dimensions of each utility activity helps project stakeholders retrieve not only latest date but historical one for future auditing purposes.

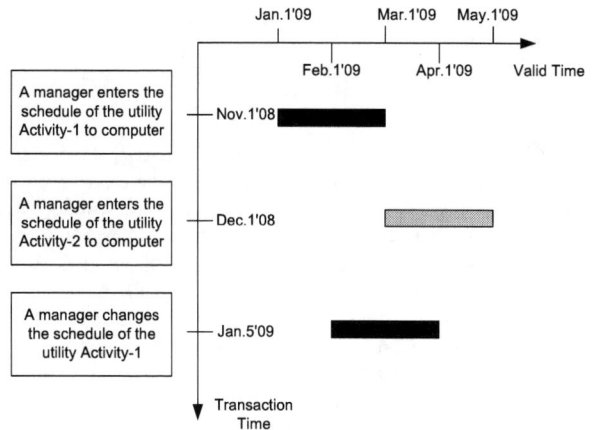

Figure 1. Valid time and transaction time for two utility activities.

Further, the duration of a utility activity may be associated with a different time scale. For example, if an accident destroys a power switch facility, the problem due to switching over voltages needs to be fixed within several hours. However, leaks of a water main lines may take months to fix the problem. The need of a multi-scale time hierarchy associated with each utility activity is evident; however, current relational database techniques cannot provide an appropriate solution to satisfy this need.

Software Architecture Analysis

In addition to the temporal requirements of a utility activity usually not implemented in current GIS commercial tools, traditional GIS techniques do not fit the internet architecture due to lack of support of the multi-user environment. Since utility permits are often created by respective utility owners at their offices, the modern three-tiered architecture aiming at dividing information presentation, processing and database operations into three layers of software components might be more adequate for our research problem. Based on the three-tiered architecture, a spatial database that can provide both common relational database and GIS functions is needed. An open-source spatial database, i.e., PostGIS, is employed for this research so that the research team can customize some default functions. Briefly, PostGIS is an add-on component designed to conform to the Open Geospatial Consortium (OGC) specification. Its relational database functions are provided by PostgreSQL, a prestigious open-source relational database. Using PostgreSQL with PostGIS, one can easily reuse the robust relational and spatial database engines. As noted before, traditional GIS techniques concentrate on providing an integrated solution to users, include map data presentation, processing and persistence; however, prevalent internet applications change this traditional approach of GIS. Indeed, the new OGC specification covers almost all GIS functions pertaining to data processing. Commercial database products such as Oracle also apply the same architecture as PostgreSQL and PostGIS.

The Model

The proposed information model is shown in Figure 2. Elaboration of each class in the model is described in this section. Basically, the model consists of three main parts: (1) roadway classes, including RdNetwork, Road, RdSegment, MaintenancePlan, and Resurface; (2) utility classes, including UtilPart, UtilLine, UtilFacility, Abandonment, UtilSection, UtilNetwork, UtilEnterprise, and UtilDivision; and (3) permit classes, including Permit and UtilActivity. The roadway classes contain information regarding a road network itself and its maintenance plans. The utility classes are designed to capture existing utility lines and facilities, including abandoned ones that are still buried in the field. The permit classes are the reflection of the future utility plan that may have conflicts with other permits, existing utilities, and road maintenance plans. The elements of the model are described as follows:

RdNetwork. This class represents the concept of a road network. A road network object is designed to group road objects. For instance, in a city, the public road authority has four divisions. Each division is responsible for managing one quarter of the city's roads. Hence, four road network objects exist in the model.

Road. This class represents the concept of a road, e.g., Riverside Drive. A road object contains a set of road segment objects. The road class has six attributes. The

"Type" attribute is designed to record the road's type, e.g., way, line, drive, road, street, avenue, and boulevard. The "Line" attribute is a "linestring" type in PostGIS, which describes a road as a set of lines. The last three attributes are used to record the traffic flow information of a road.

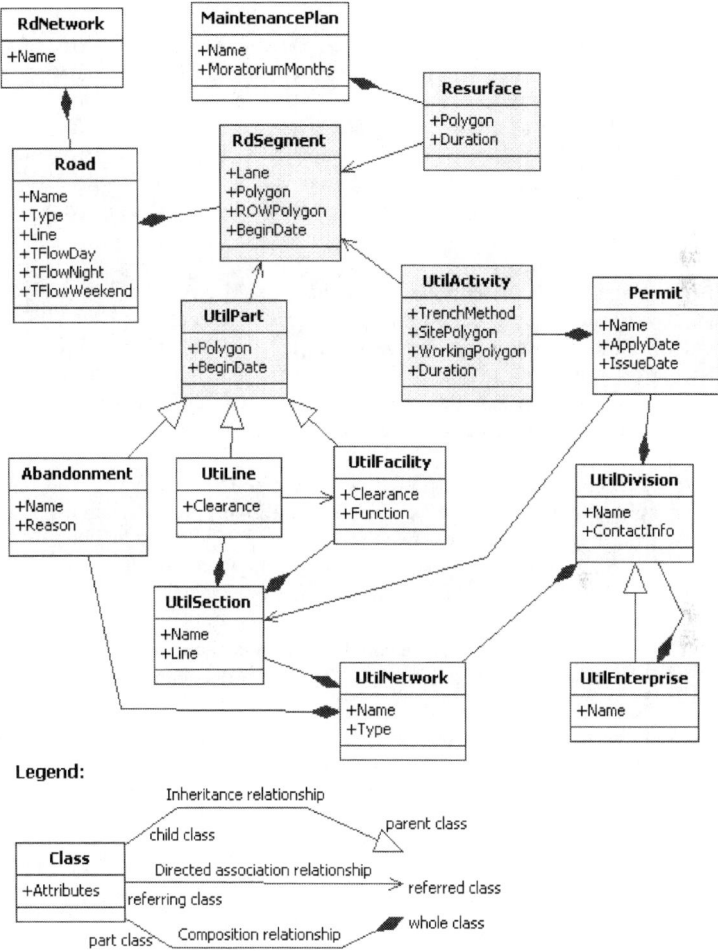

Figure 2. Model for spatiotemporal road and utility information.

RdSegment. This class represents the concept of a road segment. A road segment object is the building block to construct a road object. The road segment class has four attributes. The "Lane" attribute is designed to record how many lanes the given road segment has. The "Polygon" attribute can be used to depict the geometry of the segment. The "ROWPolygon" attribute can be used to depict the right-of-way (ROW) geometry of the segment. The "BeginDate" attribute is used to record when the road segment is open to serving the traveling public. Note that this attribute pertains to the valid time dimension so the "EndDate" attribute is set to forever.

MaintenancePlan. This class represents the concept of a road maintenance plan. Most public road authorities will not allow utility owners to cut their newly paved road segment, which forms a constraint the manager scheduling the utility activities needs to consider. A road maintenance plan object can have many resurface objects, which represent the resurface activities and will be described next. The "MoratoriumMonths" attribute is used to store how long the road segment cannot be cut. The value of this attribute should be applicable to all resurface activities of the plan.

Resurface. This class represents the concept of a resurface activity. In our proposed model, the resurface activity pertains to a certain road segment; hence the influential area of a resurface activity must reside within the boundary of the corresponding road segment. The "Polygon" attribute denotes the influential area of a resurface activity. The "Duration" attribute records the start and end dates of a resurface activity. Note that a resurface activity can be a past, present, or future event.

UtilPart. This class represents the abstract concept of a utility facility. This class is a base class, and three classes, i.e., UtilFacility, UtilLine, and Abandonment, inherit from it. Basically, a road segment may have zero to many utility facilities, and the "Polygon" attribute depicts the boundary of a utility facility. The "BeginDate" attribute records the date when this facility becomes operational.

UtilFacility. This class represents the concept of an actual utility facility. Because it is derived from the "UtilPart" class, only the "Function" attribute and the "Clearance" attribute are defined. A utility facility object can be used to depict any facility or appurtenance with a clearance distance and must be contained in a utility section object.

UtilLine. This class represents the concept of an actual utility pipeline. Because it is derived from the "UtilPart" class, only the "Clearance" attribute is defined. A utility line object can be used to depict any pipeline with a clearance distance and must be contained in a utility section object. Note that a utility facility object, like a pole, can have two or more utility line objects, like a power line and a communication line suspended in the same pole.

Abandonment. This class represents the concept of an actual abandoned utility facility. Because it is derived from the "UtilPart" class, only the "Name" and "Reason" attributes are defined. An abandonment object can be used to depict any utility facility or pipeline that still exists in the field but no longer provides any service. An abandonment object must be contained in a utility network object

UtilSection. This class represents the concept of a utility section, which contains a set of utility part objects. The "Line" attribute is a "linestring" type in PostGIS, which describes a utility section as a set of lines. A utility section object must be contained in a utility network object and may contain zero or more utility permits.

UtilNetwork. This class represents the concept of a utility network. A utility network object is designed to group utility section objects. For instance, in a city, the water utility network can be divided into many utility sections. Each section can be represented by a utility section object. The "Type" attribute is used to record the type of the utility network, e.g., water, wastewater, power distribution, gas, telecommunication, etc. The "Name" attribute records the name of the network.

UtilDivision. This class represents the owner of a utility network. It includes a person's contact information so that public road authorities can get in touch with him or her when any emergent event on the utility network occurs. A utility division object manages one or more utility networks. The "Name" attribute records the name of the organization.

UtilEnterprise. This class represents the concept of the utility division's business entity and inherits from the "UtilDivision" class. For example, a city government may be responsible for the water supply system, the wastewater treatment system, and the power distribution system. There must be three engineering divisions in the city government, so "UtilEnterprise" represents the city government whereas "UtilDivision" represents each engineering division. The Enterprise-Division structure can be easily extended to model any infrastructure owners' organizations.

Permit. This class represents the concept of a utility permit. A utility division may apply for several utility permits for each of their utility work. Similarly, a utility permit may pertain to one or more utility divisions who are all responsible for the utility work. A utility permit object can contain many utility activity objects. The two date attributes record when to submit the permit application and when to receive the approval.

UtilActivity. This class represents the concept of a utility activity. Each utility activity resides in one road segment. The "Duration" attribute records the duration of a given utility activity. The "TrenchMethod" attribute can be "trench," "trenchless" or "pole." The "SitePolygon" attribute contains the geometry of the trench or activity, whereas the "WorkingPolygon" attribute contains the geometry of the area needed for the work.

It should be noted that the "TranDate" attribute is added to each class in the proposed model because it represents the time when a given object is updated. The transaction time dimension is applicable to each class, whereas the valid time dimension is applicable to some of the classes in the proposed model with two formats. The first format of the valid time dimension is represented as the "BeginDate" attribute, which exists in the "RdSegment" and "UtilPart" classes. Since the road segments and the utility lines and facilities will not be demolished, the "EndDate" attribute of the valid time dimension is meaningless for the two classes. However, the "Duration" attribute of the "Resurface" and "UtilActivity" classes actually uses the start and end date fields to record the time period of a given activity.

Exploiting the Model

In this section, several hypothetical examples to exploit the capabilities of the model are discussed. The SQL language with system-defined or our customized spatiotemporal functions is used to explain how a spatiotemporal query can be done.

Finding Any Spatiotemporal Activity Violating the Constraints. As noted before, generally two or more utility activities performed by different utility owners can not happen in the same place during the same period of time, unless they have been well coordinated in advance. Assume our database has accumulated enough road and utility infrastructure baseline data and recorded many utility permits data. Among these permits, it is difficult and time-consuming for a project manager to manually check all utility activities that violate the above constraint. Additionally, the pavement moratorium constraint of a road segment, the clearance distance of a utility facility or pipeline, and the traffic condition of a road all need careful examination before a utility permit can be issued. In SOD, the problem can be solved by the following steps:

1. Let Set1 = select * from UtilActivity where Duration.EndDate >= NOW
   // find the utility activity data set that the work items have not been finished

2. Let Set2 = select * from Set1 as a, Set1 as b where st_intersect(a.Polygon, b.Polygon) is not null and a <> b
   // list any pair of two objects that constitute a shared polygon, which means these two utility activities share the same place. Note that st_intersect is an OGC-defined function

3. Let Set3 = select * from Set2 where t_overlap(a.Duration, b.Duration) and a <> b
   // because Set2 means the two activities occupy the same place, Set3 finds out whether their time durations are overlapped. Note that t_overlap is a customized function

4. Let Set4 = select * from Set1 join RdSegment join Resurface where st_intersect(Set1.Polygon, Resurface.Polygon) is not null and

t_overlap(Set1.Duration, (NOW, Resurface.Duration.EndDate + Resurface.MaintenancePlan.MoratoriumMonths))
// if the space needed by Resurface and Set1 is the same, and if the work duration is overlapped, Set4 stores these records

5. Let Set5 = select * from Set1 join RdSegment join UtilPart where st_intersect(Set1.Polygon, expand(UtilPart.Polygon, Clearance)) is not null
// if the space needed by expanding the clearance distance of a utility facility and Set1 is the same, Set5 stores these records

6. Let Set6 = select * from Set1 join RdSegment join Road where TFlow > certain_traffic_flow_valuee
// if the space needed by Set1 is the road with heavy traffic flows, Set6 stores these records

The result set is (Set3 union Set4 union Set5 union Set6), which means these planned utility activities violate one of the constraints.

Listing the Utility Activities That Can Be Performed Consecutively. In order to minimize the pavement utility cuts, the project manager would like to list the utility activities that are planned to be performed in the approximate same place but at different times. These activities can be rescheduled to be performed consecutively to reduce pavement utility cuts. Assume the minimum distance between the two activities that can be combined to be performed jointly is d. Assume "consecutively" means the difference between the end date of one activity and the start date of another consecutive activity is n days or less. In SOD, the problem can be solved by the following steps:

1. Let Set1 = select * from UtilActivity where Duration.EndDate >= NOW
// find the trench data set that the work items have not been finished

2. Let Set2 = select * from Set1 as a, Set1 as b where st_dwithin(a.Polygon, b.Polygon, d) and a <> b
// list any pair of two objects located nearby, which means these two utility activities will be performed in the approximate same place. Note that st_dwithin is a OGC-defined function

3. Let Set3 = select * from Set2 where (t_overlap(a.Duration, b.Duration) or a.Duration.EndDate + n >= b.Duration.StartDate) and a <> b
// Set3 finds out whether their durations are overlapped or very close

4. Let Set4 = Set2 – Set3
// the remaining data items are those activities that can be rescheduled. Set 4 is the result set

A simulation tool is currently under development. The tool will use the proposed model to simulate events of each utility activity in a small city in Taiwan. The water supply system, power distribution lines, communication lines, and natural gas pipelines are depicted in the tool. Their baseline and interdependency data are recorded in the database. With interdependency data in the database, users of the tool can find out the useful information to rescheduling.

Conclusion

With the ever-increasing demand for a streamlined analysis of utility activities to reduce pavement utility cuts, public road authorities are making substantial efforts to improve the decision-making process regarding how to optimize the work plan. Because an information model is fundamental to be used to calculate and reschedule the utility activities, and because modern database technologies have provided powerful spatial query capabilities, the time dimension for the utility activities scheduling such as the one in this research can help project managers retrieve relevant information on demand. This research has designed a UML-based model that can describe spatiotemporal information with constraints. Further implementation and evaluation of the proposed model proposed is needed in order to demonstrate how such information technology can help reduce pavement utility cuts.

Acknowledgements

The research was supported in part by National Science Council (NSC) of Taiwan under Grant NSC-97-2218-E-008-004.

References

Blair, J.S. (2003). "Utility Relocations on Construction Projects: A Contractor's Perspective." *89th Annual Purdue Road School Conference*, Purdue University, Lafayette, IN.

Chou, C.C., Caldas, C.H., and O'Connor, J.T. (2007) "Decision Support System for Combined Transportation and Utility Construction Strategy." *Transportation Research Record: Journal of the Transportation Research Board*, 1994, 9-16.

GAO. (1999). "Transportation and Infrastructure: Impacts of Utility Relocations on Highway and Bridge Projects." *GAO/RCED-99-131*, United States General Accounting Office, Washington, DC.

GAO. (2002). "Highway Infrastructure: Preliminary Information on the Timely Completion of Highway Construction Projects." *GAO-02-1067T*, United States General Accounting Office, Washington, DC.

Guting R.H., and Schneider, M. (2005) *Moving Objects Databases*, Morgan Kaufmann, San Francisco, CA.

Jensen, K.A., Schaefer, V.R., Suleiman, M.T., and White D.J. (2005) "Characterization of Utility Cut Pavement Settlement and Repair Techniques" *Proceedings of the 2005 Mid-Continent Transportation Research Symposium*, Iowa State University, Ames, Iowa, August, 2005.

Khogali, W.E.I., and Mohamed, E.H. (1999) "Managing Utility Cuts: Issues and Considerations." *APWA International Public Works Congress*, NRCC/CPWA.
Lin, J.D. (2005) *The M-Taiwan Project*, Department of Transportation, Taiwan.
Ney, J. (2001). "Utility Project Development Issues." *Proceedings of the 2001 AASHTO / FHWA Right of Way and Utilities Conference*, AASHTO / FHWA, Portland, OR.
Tansel, A., Clifford, J., and Gadia, S. (1993) *Temporal Databases: Theory, Design, and Implementation*, Benjamin-Cummings, San Francisco, CA.
Wilde, W.J., Grant, C., and White G.T. (2003) "Controlling and Reducing The Frequency of Pavement Utility Cuts." *Transportation Research Board Annual Meeting*, Washington, D.C., January, 2003.
Wolfson, O., Xu, B., Chamberlain, S., and Jiang, L. (1998) "Moving Objects Databases: Issues and Solutions." *Proceedings of the Tenth International Conference on Scientific and Statistical Database Management*, IEEE, Capri, Italy, 111-122.
Wolfson, O. (2002) "Moving Objects Information Management: The Database Challenge." *Lecture Notes in Computer Science*, Springer Berlin / Heidelberg, 2382, 15-26.

Dynamic Modulus and Fatigue Testing of Lightly Cementitiously Stabilized Granular Pavement Materials

Piratheepan Jegatheesan[1] and C.T. Gnanendran[2]
School of Aerospace, Civil and Mechanical Engineering (ACME)
University of New south Wales at Australian Defence Force Academy
Canberra, ACT 2600, AUSTRALIA.
[1]j.piratheepan@adfa.edu.au and [2]r.gnanendran@adfa.edu.au

Abstract: This paper examines the characterization of a granular material lightly stabilized with slag-lime cementitious binder particularly utilizing the monotonic and cyclic load IDT testing method. An extensive laboratory investigation was carried out to determine the mechanical properties of the lightly stabilized granular material and to establish fatigue life relationships by IDT testing and typical results obtained from this ongoing research are presented in this paper. This study shows that the static stiffness modulus and dynamic stiffness modulus of the stabilized material increases with binder content and showed little change with moisture content variation around the optimum moisture content. Fatigue life relationships were established using two methods, namely the approach of determining the number of cycles for 50% reduction in the stiffness compared to the initial stiffness and the energy ratio method, and they both showed similar linear relationships. It is therefore concluded that cyclic load IDT testing can be used reliably to characterize lightly stabilized granular materials with slag-lime in terms of their strength, stiffness modulus and fatigue life relationships.

Key words: IDT testing, fatigue life, stabilized granular material, slag-lime stabilization, characterization, lightly stabilized material

1. Introduction

Cementitiously stabilized granular materials have been used satisfactorily and economically for the construction and rehabilitation of roads all over the world for the past few decades. Pavement modification by stabilisation using cementitious binders is a low cost rehabilitation method that is practically useful for expansive, weak and/or wet subgrades and base materials. A binder should be selected such that it doesn't create shrinkage cracking in the stabilized material which could otherwise affect its strength and stiffness. The slow hydration of slag-lime and fly ash - lime satisfy these requirements and are therefore well suited to be used as binders for granular stabilization as they reduce the heat generated and have low tendency to shrink and crack upon cooling (Foley and Group 2001a).

The major distress modes involving cementitiously stabilized granular materials in pavements are fatigue cracking, permanent deformation and thermal cracking. Amongst the major distress modes, fatigue is a phenomenon in which pavement layers are subjected to repeated stress levels due to the traffic loading leading to cracking. Fatigue properties are generally derived through laboratory and field testing. The determination of elastic properties such as stiffness modulus and fatigue life relationships of cementitiously stabilized pavement materials is an important consideration in the mechanistic design of pavements which contain base and sub-base layers consisting of cementitiously stabilized granular materials. The design of such pavements with cementitiously stabilized granular materials is typically based on their in-situ flexural stiffness (AUSTROADS 2004). Because of the difficulties in preparing and handling a beam of cementitiously stabilized pavement material at low levels of binder content, Indirect Diametric Tensile (IDT) test method has recently been suggested as a possible alternative for economically obtaining repeatable and reliable stiffness characteristics for these materials (Foley and Group 2001b; Gnanendran and Piratheepan 2008).

This paper presents typical results of an ongoing detailed laboratory investigation on the characterization of a lightly cementitiously stabilized granular material with slag-lime in terms of the dynamic modulus and fatigue life. In particular, this paper presents the results of a laboratory investigation involving monotonic and cyclic load IDT testing with internal displacement measurements for slag-lime stabilized granular materials. The objective is to asses whether the stiffness and fatigue life of a lightly stabilized material could be determined from cyclic load IDT testing. Details of the testing methods adopted and typical results are also presented in the paper.

2. Fatigue Criteria for Cemented Materials

According to the current AUSTROADS (2004), fatigue relationships have been derived for cemented materials having various moduli and these relationships are used to give an indication of fatigue life. These relationships were derived from overseas research work and only limited information is available on the in-service fatigue behaviour of cemented materials used in Australia. The relationship between the maximum tensile strain in cemented materials produced by a specific load and the allowable number of repetitions of that load is given by the following expression:

$$N = RF \left[\frac{(113000/E^{0.804} + 191)}{\mu\varepsilon} \right]^{12} \tag{1}$$

where,

N = allowable number of load repetitions to fatigue

$\mu\varepsilon$ = the tensile strain (in microstrain) produced by the load

E = Modulus of cemented materials (MPa)

RF = Reliability factor for cemented materials fatigue typically 1.0 for 95% project reliability.

This general fatigue criterion has some serious restrictions and is valid for cemented materials with moduli ranging from 2000 to 10000 MPa. Due to the restrictions and the unavailability of sufficient data, AUSTROADS recommends that laboratory fatigue testing of cementitiously stabilised materials be carried out using flexure test, direct tension testing and indirect tension testing. The flexural bending test is most preferred as it resembles the field conditions. However, due to the difficulties in this test method as discussed earlier, IDT testing method could be carried out preferably in conjunction with field trials or by adopting relationships contained in the literature.

3. Laboratory investigation undertaken

The experimental investigation involved the study of the IDT strength, stiffness and fatigue life characteristics of a typical granular base material stabilized lightly with slag-lime (i.e. less than 6% by weight). The moisture - dry density relationships of the parent material as well as the stabilized material were determined initially to select a suitable range of moisture content (MC) for this investigation as shown in Figure 1. This was followed by the study of the stiffness and strength characteristics of the granular base material stabilized lightly with small percentages of slag-lime binder at the selected moisture conditions under monotonic and cyclic load IDT testing regimes in this investigation. Details of the materials used and the different tests that were performed are discussed below.

The stabilizer chosen for this experiment was slag-lime which is a blend of Ground Granulated Blast Furnace Slag and hydrated lime in the ratio of 85 to 15 by dry weight. The parent material selected for this research was a freshly quarried crushed rock granular base material - maximum size 19mm, obtained locally in Canberra, Australia from Mugga Quarry supplied by Boral Resources (Country) Pty Ltd. The crushed rock was identified as rhyodacite porphyry (an acid sub-volcanic igneous rock – abundantly porphyritic and otherwise quite finely crystalline, unweathered, moderately altered), which consisted hard, strong, angular fragments of unweathered, subtly reddish medium dark grey acid porphyry, The Atterberg limits determined on the material passing Number 325 sieve in according to the Australian Standard were: liquid limit = 18 and plastic limit = 15. The wet and dry strengths of the parent material were 157 and 217 kN respectively.

Getting representative samples of such a granular material is very difficult and it often leads to inconsistency of particle proportions in samples. This leads to considerable variability of properties determined from different samples. To minimize such inconsistency, the reconstituted material with an unchanged (or consistent) material grading shown in Figure 2, was adopted for this research. This was achieved by sieving a large batch of granular material obtained from the quarry through standard sieves, separating the material into different particle size ranges in different containers and then remixing them at suitable weight proportions to get the specified unchanged grading. The adopted grading for the reconstituted sample, hereafter referred as parent material, shown in Figure 2 was essentially the same for all the samples tested in this research program.

The IDT tests were conducted on 28 days cured samples prepared by gyratory compaction with 500 kPa vertical pressure and 250 gyrations that achieved approximately 95% Standard Proctor maximum dry density at optimum moisture content. In the IDT testing setup, an internal measurement setup with a pair of LVDT and Perspex strips, as shown in Figure 3, onto the test sample was used to measure more accurate reading. This setup avoids the disturbance by the external vibration and rocking action (Gnanendran and Piratheepan, 2008). The IDT strength and static stiffness modulus (SSM) were determined from a series of monotonic loading tests with a vertical deformation rate of 1 mm/min and the dynamic stiffness modulus (DSM) was determined from a series of cyclic loading tests with a frequency of 3 Hz sinusoidal loading. The sinusoidal type of cyclic load IDT tests were performed at different levels of maximum loads for each sample. The maximum cyclic load for each load level was taken as a percentage of the average ultimate failure load from the monotonic loading test. The procedure for determining the DSM and the SSM was described by Gnanendran and Piratheepan (2008). The fatigue testing were performed on IDT samples with 3, 4 and 5% binder contents (BC) and 9% molding moisture content. This test was a stress controlled with a frequency of 3 Hz sinusoidal loading.

A series of unconfined compression tests was carried out on 28 – days cured stabilized samples with the same binder – moisture combinations in accordance with Australian Standard 1141.51 (AS 1996) (Piratheepan et al. 2008). The results showed that the unconfined compressive strength (UCS) ranged from 2.776 MPa to 7.370 MPa for the addition of 3 – 5% slag-lime.

4. IDT strength and Stiffness modulus

The IDT strength and static stiffness modulus (SSM) for a particular stabilized granular base mix were determined from monotonic loading tests performed on a set of 3 samples at a vertical deformation rate of 1 mm/min. The IDT strength of the cementitiously stabilized granular material was calculated using the following elastic theory solution:

$$IDT\ strength\ (MPa) = \frac{2000 \times P}{\pi \times D \times t} \quad (2)$$

where P is the applied failure load in kN; D is the diameter of the specimen in mm and t is the thickness of the specimen in mm. Figure 4 shows the variation of IDT strength with binder content. It is clear from the curve that IDT strength significantly increased with binder content and always high at 9% moisture content for all binder content cases.

The static stiffness modulus (E_{static}) of the stabilized material was determined by applying the following equation (based on BSI 1993) for the linear part of the load versus deformation curve obtained from the monotonic load IDT test (see Gnanendran and Piratheepan 2008 for further details):

$$E_{static\ (MPa)} = 1000 \times \frac{P \times (\upsilon + 0.27)}{t \times \delta} \quad (3)$$

where P is the load (in kN) and δ is the corresponding horizontal tensile deformation (in mm), υ is the Poisson's ratio, and t is the thickness of the specimen (in mm). The variation of SSM with binder content shown in
Figure 5 reveals that the SSM increased with BC for all MC cases.

The dynamic stiffness modulus (DSM) was determined from cyclic loading tests performed at a frequency of 3 Hz adopting sinusoidal loading. The cyclic load IDT tests were performed at different levels of maximum loads, referred hereafter as the sub-maximal cyclic load, for each sample. Thus, each sample was tested at several sub-maximal cyclic load levels, ranging from 30 to 70% of the ultimate failure load in steps of 10%, and for a certain number of cycles at each sub-maximal load level (e.g. 200 cycles per sub-maximal cyclic load). The DSM (S_m) of the cementitiously stabilized granular material was determined from the following equation proposed by Gnanendran and Piratheepan (2008):

$$S_m = \frac{1000 \times (P - k_1 \times b) \times (v + 0.27)}{t \times \delta} + k_2 \times b \qquad (4)$$

where, b is the BC in %, k_1 and k_2 are material constants and all other terms were as used in the previous equation. For the particular parent material – slag – lime binder combination, $k_1 = 0.235$ and $k_2 = 270$. The DSM increased with the BC for all MC cases as shown in Figure 6. From the IDT test investigation undertaken on a particular lightly cementitiously stabilized granular material, the IDT strength is sensitive to the MC and increased with BC and SSM and DSM are not sensitive to MC but significantly increased with BC.

5. Fatigue testing

Fatigue testing was performed on IDT samples as a stress controlled cyclic load test at a frequency of 3Hz of sinusoidal loading pattern carried out at different maximum stress levels. Controlled stress tests are more widely used for asphaltic concrete and bituminous materials because it reproduces site conditions in which load is applied to the pavement structure and is the mode from which fatigue design equations are deduced (Khalid 2000). Therefore, in this particular investigation of slightly cementitiously stabilized granular materials, a repeated stress of constant amplitude was applied on IDT samples which were prepared at 3, 4 and 5% of binder content and 9% moisture content. The test was continued until the stiffness reduced to about 40% of the initial stiffness.

The fatigue failure is typically defined as the number of (load) cycles at which the pavement structure has failed due to fatigue. Defining the fatigue failure is a very important consideration in establishing fatigue life and it is usually defined as the number of load cycles when its stiffness reduces to a certain percentage of the initial stiffness or breaking of the sample. In this particular investigation, the fatigue failure of lightly cementitiously stabilized granular material under IDT testing was defined by two methods; i.e. the stiffness reduced by 50% of its initial stiffness value and an approximate energy ratio method. In the approximate energy ratio method, the energy released by the material during the fatigue test was used to define the fatigue failure. The energy ratio is defined by the following equation:

$$\text{Energy ratio} = \frac{W_0}{\left(\dfrac{W_n}{n}\right)} \qquad (5)$$

Where, n is the number of load cycles to failure; W_0 is the dissipated energy/cycle at the start of the test; and W_n is the dissipated energy at the nth cycle. The dissipated energy, W, is expressed by the equation (Khalid 2000):

$$W = \pi\sigma\varepsilon \sin\theta \tag{6}$$

where σ, ε and θ are the stress, strain and phase lag respectively. Hence, the equation (5) is written as:

$$\text{Energy ratio} = \frac{n \times (\pi\sigma_0 \varepsilon_0 \sin\theta_0)}{\pi\sigma_n \varepsilon_n \sin\theta_n} \tag{7}$$

By substituting ε ($\varepsilon = \sigma/E$, where E is dynamic stiffness modulus) into the equation (6):

$$\text{Energy ratio} = \frac{n(\sigma_0^2 E_n \sin\theta_0)}{\sigma_n^2 E_0 \sin\theta_n} \tag{8}$$

For a controlled stress test, stress is constant (i.e. $\sigma_0 = \sigma_n$), hence:

$$\text{Energy ratio} = \frac{nE_n \sin\theta_0}{E_0 \sin\theta_n} \tag{9}$$

E_0 is constant which can be removed to simplify the equation (9) without causing too much change in the shape of the function. Also E_n is the loss modulus at cycle n from IDT testing. It is evident from the experimental data that the change in $\sin\theta$ is zero (i.e. $\sin\theta_0 = \sin\theta_n$) for cementitiously stabilized materials and can be removed from the equation (9). Thus, the equation for the energy ratio:

$$\text{Energy ratio (ER}_\sigma) = n \times E_n \tag{10}$$

Figure 7 shows two typical plots of dynamic stiffness modulus calculated by the equation (4) and the corresponding energy ratio against the number of load cycles applied to the IDT samples. The number of cycles to failure by energy ratio method was found from Figure 7 by the corresponding number of cycles at the maximum energy ratio.

For each specimen, the initial maximum horizontal tensile strain, ε_{max}, at the centre of the specimen was calculated using the following equation:

$$\varepsilon_{max} = \frac{\sigma_{max}(1+3\upsilon)}{S_m} \tag{11}$$

Figure 8 shows the linear variations of the failure numbers of cycles defined by the two methods against the initial maximum horizontal tensile stains calculated using equation (11). From the figure, two important observations were made. Firstly, it can be seen that the fatigue lives have similar linear patterns for both methods defined and are higher for materials with higher DSM (i.e. higher with higher binder content) and secondly, the fatigue lives are almost equal according to both the methods defined. Thus, the two different failure criteria produced almost the same fatigue lives for a lightly cementitiously stabilized granular material with slag-lime.

6. Conclusions

This paper examines the characterization of granular materials lightly stabilized with slag-lime cementitious binder and the use of monotonic and cyclic load IDT testing to determine the stiffness characteristics of these materials and the IDT fatigue testing for characterizing them. An extensive laboratory investigation was carried out to determine the mechanical properties of the stabilized material and the fatigue life relationships by IDT testing. The main conclusions can be summarized as follows:

1) The addition of slag-lime is effective in enhancing the IDT strength. The strength value increased by around 50% when the binder was increased from 3 to 4%. The target MC also influenced the IDT strength and it was found to be beneficial to work on the high side of the optimum MC of the stabilized material rather than on the lower side.

2) The static stiffness modulus and dynamic stiffness modulus estimated were found to increase with BC and showed little variations to MC for the narrow range of MC variation (close to OMC) investigated in this study.

3) The fatigue life relationships derived from the two methods (i.e. failure defined as the number of cycles for 50% reduction of stiffness compared to the initial stiffness and the energy ratio method) showed similar linear relationships.

4) From the results obtained in this ongoing laboratory investigation, it can be concluded that the IDT testing can be used reliably to characterize lightly cementitiously stabilized granular materials with slag-lime in terms of strength, stiffness modulus and fatigue life relationships.

Acknowledgements

The authors would like to thank Mr. David Sharp and Mr. Jim Baxter for their technical assistance during the experimental work reported in this paper.

Reference:

1. ASTM C 295, (2003). Standard Guide for Petrographic Examination of Aggregates for Concrete, 2003.
2. ASTM D 698, (1991). Standard Test Methods for Laboratory Compaction Characteristics of Soil Using Standard Effort (12 400 ft-lbf/ft3 (600 kN-m/m3)).
3. AUSTROADS. (2004). "Guide to the Structural Design of Road Pavements."
4. Foley, G., and Group, A. S. E. (2001a). "Contract report -Effect of Design, Construction and Environmental Factors for Long-term Performance of Stabilised Materials."
5. Foley, G., and Group, A. S. E. (2001b). "Contract report-Mechanistic Design Issues for Stabilised Pavement Materials."

6. Gnanendran, C. T., and Piratheepan, J. (2008). "Characterization of a Lightly Stabilized Granular Material by Indirect Diametrical Tensile Testing." *International journal of pavement Engineering (Under review)*.
7. Khalid, H. (2000). "A comparison between bending and diametral fatigue tests for bituminous materials." *Materials and Structures*, 33(7), 457-465.
8. Piratheepan, J., Gnanendran, C. T., and LO, S.-C. R. (2008). "Characterization of Cementitiously Stabilised Granular Materials for Pavement Design Using Unconfined Compression and IDT Testings with Internal Displacement Measurements." *International journal of pavement Engineering (Under review)*.
9. White, G. W. and Gnanendran, C. T. (2005). 'The influence of compaction method and density on the strength and modulus of cementitiously stabilised pavement materials'. *The international Journal of Pavement Engineering*, Vol. 6, No 2, June 2005, 97-110
10. Yeo, R., Vuong, B. and Alderson, A. *Contract Report-Towards National Test Methods for Stiffness and Fatigue Characterisation of Stabilised Materials*, 2002, ARRB Transport Research, RC2028-002, 23 July.

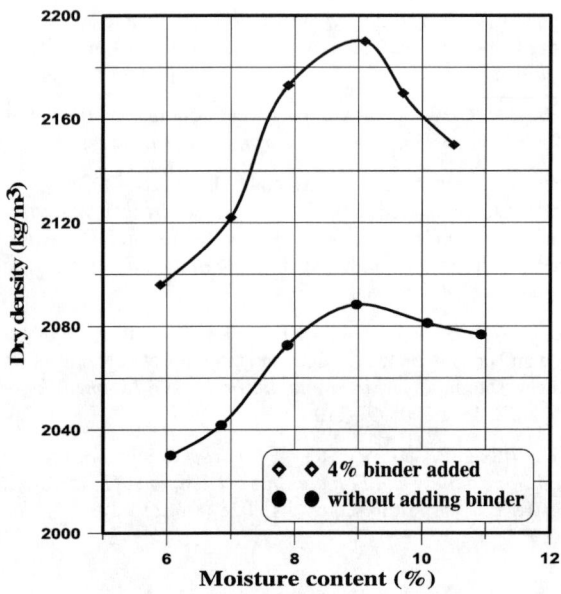

Figure 1 Variation of dry density versus moisture content (from Gnanendran and Piratheepan 2008)

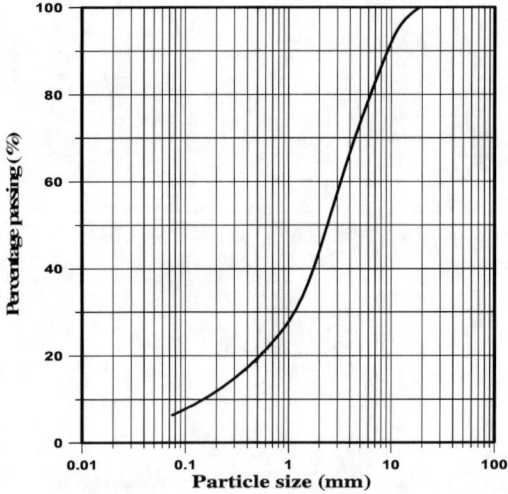

Figure 2 Particle size distribution of the parent material (from Gnanendran and Piratheepan 2008)

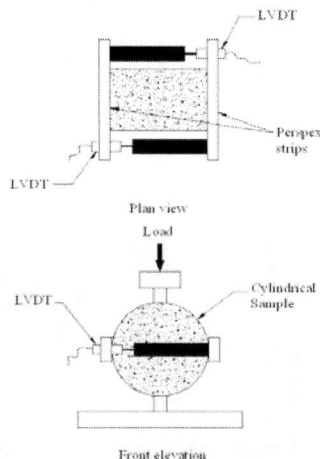

Figure 3 Schematic diagram of LVDT setup

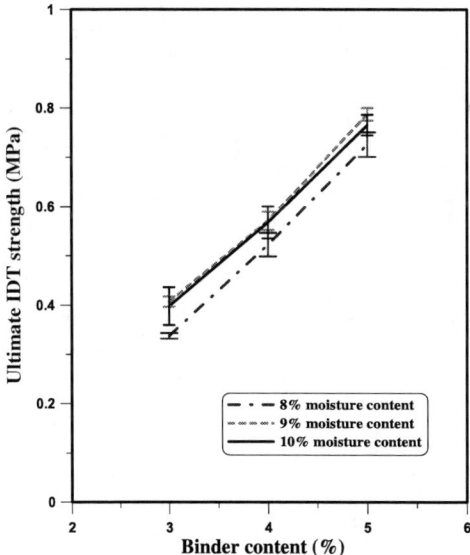

Figure 4 Variation of ultimate IDT strength versus binder content (from Gnanendran and Piratheepan 2008)

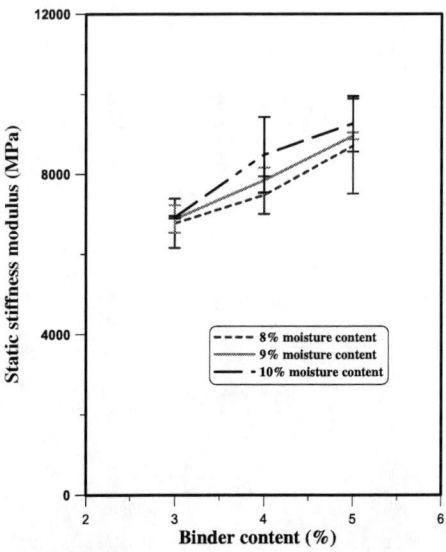

Figure 5 Variation of static stiffness modulus versus binder content (from Gnanendran and Piratheepan 2008)

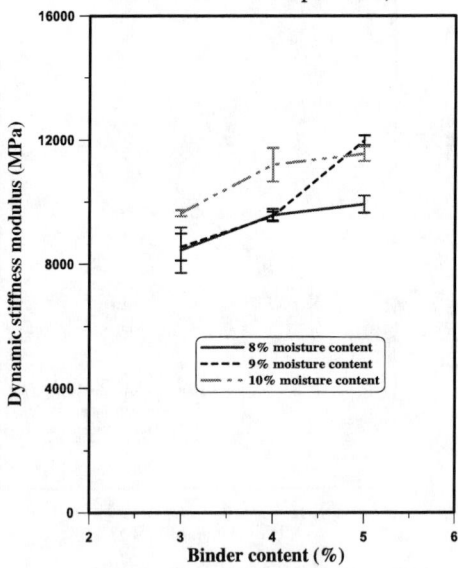

Figure 6 Variation of dynamic stiffness modulus versus binder content (from Gnanendran and Piratheepan 2008)

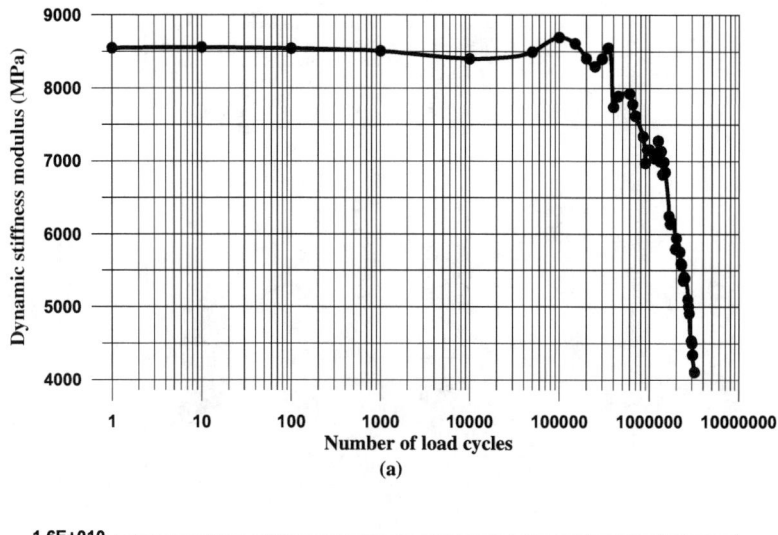

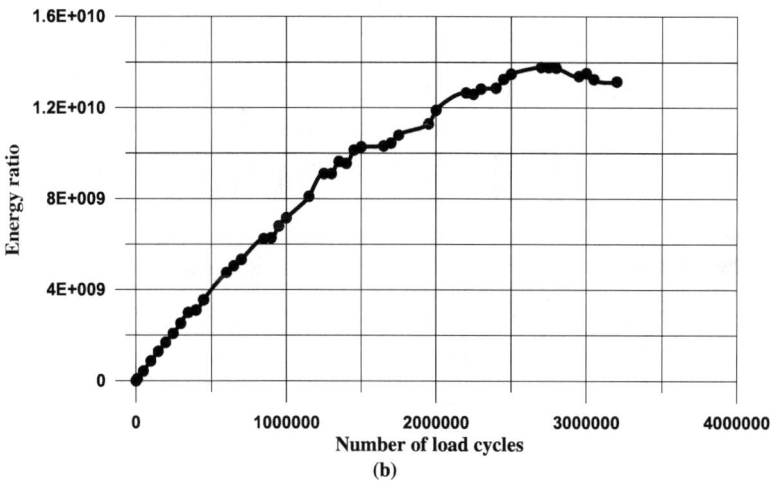

Figure 7 (a) Variation of dynamic stiffness modulus with increasing number of load cycles; (b) Variation of energy ratio with number of load cycles

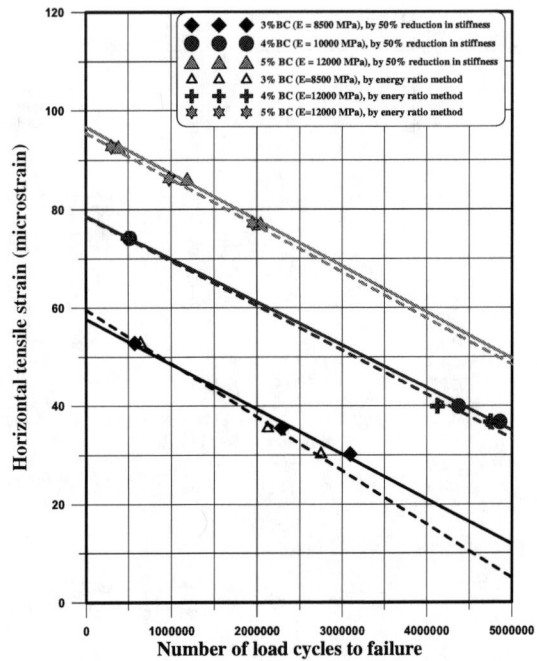

Figure 8 Fatigue life relationships between Initial horizontal tensile strain and number of load cycles to failure by IDT testing

Author Biographies

1. Mr. Jegatheesan Piratheepan is a PhD Candidate in the School of Aerospace, Civil and Mechanical Engineering, University of New South Wales at Australian Defence Force Academy, Canberra, Australia.

2. Dr. Carthigesu Thiagarajah Gnanendran is a Senior Lecturer in geomechanics and pavements in the School of Aerospace, Civil and Mechanical Engineering, University of New South Wales at Australian Defence Force Academy, Canberra, Australia.

Performance Evaluation of Asphalt Pavement with Fly Ash Stabilized FDR Base: A Case Study

Haifang Wen[1] and Bruce Ramme[2]

[1] Department of Civil & Environmental Engineering, University of Wisconsin at Madison
1415 Engineering Drive, Madison, WI 53706, Tel: 608-262-1777, Email: hwen2@wisc.edu (Corresponding Author)
[2] We Energies, 333 W. Everett Street, Milwaukee, WI 53290-0001, Tel: 414-221-2442, Email: bruce.ramme@we-energies.com

ABSTRACT

With the demands for environmental-benign technology for highway construction, recycling in-place pavement materials for reconstruction or rehabilitation has become an important re-construction alternative. For asphalt pavement reconstruction/rehabilitation, full-depth reclaimed (FDR) asphalt pavement materials is a viable option, which pulverizes the existing asphalt layer and/or base materials as a new base. In some occasions, to increase the support of base consisting of recycled pavement materials, additives, such as cement, emulsion, or self-cementing Class C fly ash, are added as a binder. Self cementing Class C fly ash is both economical and an environmentally-friendly approach which makes beneficial use of a coal combustion by-product.

This paper addresses the performance evaluation of an asphalt pavement with Class C fly ash stabilized FDR asphalt pavement materials as the base. The performance was evaluated using distress survey, falling weight deflectometer (FWD), and mechanistic-empirical (M-E) design program, as a case study. It was found that within the three year after the construction, the modulus of Class C fly ash stabilized FDR base has continuously increased, as a result of continuing pozzolanic reactions. However, it was also found that in the sixth year after the construction, the modulus of base began to decline, when compared to the third year results. It is believed that freeze-thaw effects may have reduced the strength of the stabilized base course. A visual distress survey found that top-down cracking from the asphalt surface began to occur three years after the construction as the predominant distress, followed by thermal cracking and rutting. The verification of performance using M-E design program found that the M-E design program underpredicted the amount of top-down cracking and thermal cracking and did reasonably well for prediction of the rutting. It is believed that the stiff base could be a reason for the top-down cracking and that friction between the HMA layer and base contributed to the thermal cracking.

INTRODUCTION

With the demands for environmentally friendly technologies for highway construction, recycling in-place pavement materials for reconstruction or rehabilitation has become an important construction alternative. For asphalt pavement reconstruction/rehabilitation, full-depth reclamation (FDR) of asphalt pavement materials is a viable option, which pulverizes the existing asphalt layer and/or base materials as a new base [1,2]. In some occasions, to increase the support of the base consisting of recycled pavement materials, binders, such as cement, emulsion, or self cementing Class C fly ash, are added as a stabilizer [3,4,5]. It is both economical and environmentally-friendly to beneficially use self cementing Class C as a byproduct of coal combustion.

Self cementing Class C fly ash, a coal combustion product from lignite or sub-bituminous coal obtained as a result of the power generation process, has been used extensively over a wide range of construction applications. Its self-cementing property is valuable in developing strength of concrete or other mineral mixtures. Each year, approximately 68 million tons of fly ash are produced in the U.S.A. About 46 million tons were placed in landfills resulting in significant costs and use of land [6].

The design of an asphalt pavement with Class C fly ash stabilized FDR base course has been empirical, referring to the design methods used for asphalt pavement with other types of stabilized base courses, such as cement or lime being the additives. There has been a lack of evaluation of field performance of asphalt pavements with fly ash stabilized FDR base course. In addition, with the development of the mechanistic-empirical pavement design guide, it is imperative to evaluate the effectiveness of the M-E design of asphalt pavement with fly ash stabilized FDR base course. This paper addresses the six-year performance evaluation of an asphalt pavement with Class C fly ash stabilized FDR base course, using falling weight deflectometer (FWD) and M-E design program, as a case study.

PROJECT DESCRIPTION

County Trunk Highway (CTH) JK in Waukesha, Wisconsin was selected as a test section. It was reconstructed in October of 2001, using FDR materials as pavement base course. Fly ash was used to stabilize the FDR pavement base course in place for CTH JK. CTH JK is located in Waukesha County, Wisconsin and the project segment runs between CTH KF and CTH K, with a project length of 1,009 m. It is a two-lane road with an average daily traffic (ADT) count of 5,050 vehicles in year 2000 and a projected ADT of 8,080 in design year 2021. The existing pavement structure consisted of approximately a 127 mm asphalt concrete surface layer and a 178 mm granular base course. The new pavement structure consists of a 127 mm asphalt concrete layer and a 305 mm Class C fly ash stabilized FDR base course. The truck percentage on CTH JK in 2000 was 5%. Laboratory tests indicated that unconfined compressive strengths of 1.72 MPa and 2.62 MPa at optimum moisture contents of 5% were obtained after seven day curing, respectively.

The pulverized mixes were compacted and graded to form the base for a 127 mm thick new asphalt overlay. Construction of CTH JK consisted of pulverization, application of 8% fly ash and 5% water, compaction and grading, and placement of the new asphalt overlay. For detailed construction procedure, the readers are referred to reference 7.

FIELD PERFORMANCE EVALUATION

The nondestructive deflection testing is one of the primary techniques for determining the in situ structural capacities of pavement. The 1993 *AASHTO Guide for Design of Pavement Structures* [8] describes Falling Weight Deflectometer (FWD) testing as a means of evaluating the conditions of existing pavement. The M-E design guide also proposes the use of FWD for existing pavement evaluation [9]. In this study the FWD was used to evaluate the performance of fly ash stabilized FDR.

KUAB 2M Falling Weight Deflectometer (FWD) tests were conducted to evaluate the field performance of CTH JK. The impact load used in this study was approximately 40KN. The pavement surface deflections were recorded by seven sensors located at 0, 0.3, 0.46, 0.61, 0.91, 1.22, and 1.52m from the center of loading plate. The FWD tests were performed at an interval of 30.5m.

Deflection

The pavement deflections under the impact load indicate the structural capacities of existing pavement, including subgrade, base course and surface layer. Deflections may be either correlated directly to pavement performance or used to determinate the in situ material characteristics of pavement layers.

Twenty three FWD tests were conducted on CTH JK in 2001, 2002, 2003, and 2006. The average deflections measured by the sensors are shown in Table 1.

TABLE 1. Average Deflection Measurements

		D_0, mm	$D_{0.3}$, mm	$D_{0.46}$, mm	$D_{0.61}$, mm	$D_{0.91}$, mm	$D_{1.22}$, mm	$D_{1.52}$, mm
Year 2001	Mean	0.234	0.177	0.152	0.129	0.099	0.074	0.056
	Std. Dev.	0.045	0.037	0.031	0.027	0.023	0.019	0.016
	C.V. (%)	19.3	20.6	20.7	21.1	23.2	25.5	27.9
Year 2002	Mean	0.138	0.115	0.106	0.094	0.075	0.062	0.049
	Std. Dev.	0.029	0.024	0.022	0.020	0.017	0.015	0.013
	C.V. (%)	19.6	19.9	19.5	19.4	20.6	22.3	24.2
Year 2003	Mean	0.151	0.125	0.110	0.098	0.080	0.065	0.051
	Std. Dev.	0.028	0.025	0.022	0.020	0.017	0.015	0.012
	C.V. (%)	18.5	19.8	20.0	20.2	21.7	23.1	24.3
Year 2006	Mean	0.19	0.16	0.14	0.12	0.09	0.07	0.06
	Std. Dev.	0.041	0.034	0.030	0.026	0.021	0.018	0.014
	C.V. (%)	21.19	22.04	22.18	21.98	23.06	23.79	25.05

Backcalculation of Layer Modulus

The measured deflection data was used to backcalculate the properties of each pavement layer. Modulus 5.1 and Michback programs were used in the preliminary analysis. It was found that both programs yielded close results. It was decided to continue the backcalculation using only Michback program, as Michback has been used by the authors. Based on the backcalculation of pavement deflection from FWD tests, the elastic modulus of fly ash treated CIR recycled asphalt base course increased from 1.23 Mpa in 2001 to 1.84 Mpa in 2002, 2.26 Mpa in 2003, and 1.48Mpa in 2006. The results indicated that the structural capacity of the fly ash stabilized CIR recycled asphalt base course developed significantly within the first three year after construction. This is due to the pozzolanic reaction in the mixes containing Class C fly ash. However, the FWD test results in 2006 indicated that the strength of the fly ash stabilized base course was reduced. This could be due to the freeze-thaw attacks, as well as repeated traffic loads.

Structural Number and Layer Coefficient

The structural number of the pavement was backcalculated from surface deflection, as follows [10]:

$$SN = \left[1.49 \times (ET)^3\right]^{1/3} \tag{1}$$

$$Log_{10}(ET)^3 = 5.03 - 1.309 Log_{10}(AUPP) \tag{2}$$

$$AUPP = \frac{1}{2}(5D_0 - 2D_{0.3} - 2D_{0.61} - D_{0.91}) \tag{3}$$

where: SN = structural number of pavement, mm,
ET^3 = flexural rigidity of pavement, mm,
AUPP = area under the pavement profile, mm, and
D_i = surface deflection, mm.

The structural coefficient for fly ash stabilized CIR material, α_2, was calculated as follows:

$$\alpha_2 = \frac{SN - \alpha_1 h_{HMA}}{h_{base}} \tag{4}$$

Where: h_{HMA} = thickness of HMA layer, mm, and
h_{base} = thickness of base course, mm.

The structural coefficient of asphalt concrete layer was calculated based on the 1993 *AASHTO Guide for Design of Pavement Structures*, as follows:

$$\alpha_1 = 0.40 \times \log(\frac{E}{3000 Mpa}) + 0.44 \tag{5}$$

where E is the laboratory resilient modulus of asphalt concrete, MPa.

According to 1993 *AASHTO Guide for Design of Pavement Structure*, the backcalculated layer modulus of asphalt concrete could be up to three times higher than the resilient modulus obtained in the laboratory. Therefore, the backcalculated modulus value of the asphalt layer was converted into resilient modulus and was input in Equation 6. The structural coefficient obtained form Equation 6 was input in Equation 5.

The structural coefficient of asphalt concrete obtained from Equation 4 was input in Equation 5. A structural coefficient of 0.19 was found for the fly ash stabilized base course in CTH JK at the time of testing in 2006. The structural coefficients were 0.16 in 2001, 0.23 in 2002, 0.248 in 2003, and 0.19 in 2006. Figure 1 shows the backcalculated modulus and layer coefficient for the stabilized FDR base course. Since the layer coefficient is a measure of the relative ability of the material to function as a structural component of the pavement, the moduli and layer coefficients indicated a strength gain within the three year and a reduction of structural capacity of the fly ash stabilized base course in CTH JK after the third year. As mentioned earlier, this deterioration could be due to either freeze-thaw attacks or repeated loads.

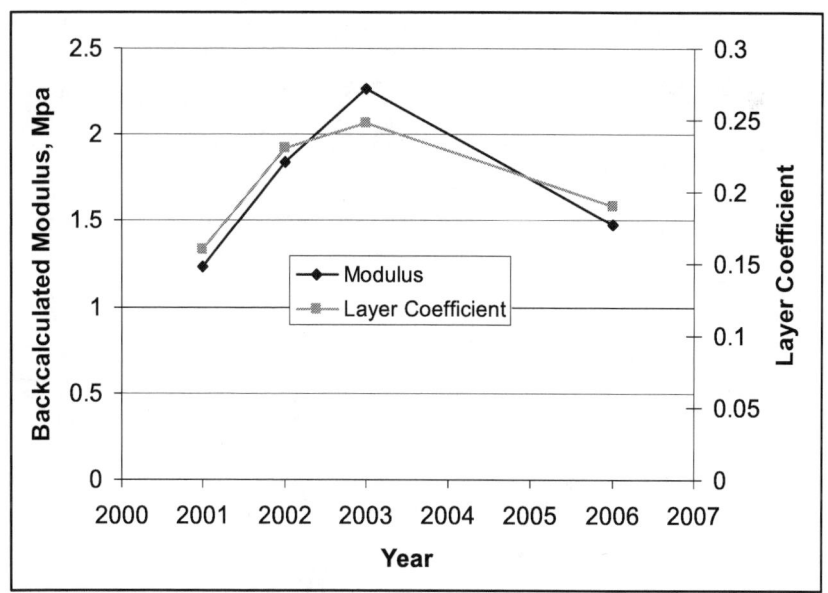

Figure 1. Backcalculated Modulus and Layer Coefficient of Fly Ash Stabilized Base

DISTRESS SURVEY

Condition surveys were conducted to evaluate the physical condition and distresses of the pavement in CTH JK in 2002, 2003, 2004, and 2007. Figure 2 shows the mapping of distresses in the pavement. Three types of distresses were observed: transverse cracking, wheel-path longitudinal cracking, and rutting. Also, minimal raveling was noticed. There are no alligator cracks in the CTH JK pavement. The results of distress survey is shown in Table 2.

The transverse cracking is believed to be the thermal cracking, as a result of cyclic thermal stresses and/or rapid temperature drop. The thermal cracking started in 2003 from a single crack and continued to grow. By May of 2007, a total of 51m of thermal cracks were observed. This corresponds to 56m/km (or 296ft/mile) of thermal cracking in the pavement. It is noted that the asphalt binder PG grade used in this pavement was the grade specified by Wisconsin Department of Transportation which meets the climate conditions in Wisconsin, in accordance with Superpave PG grade specifications.

Top-down cracking found in this pavement is the longitudinal cracking at the edge of the wheelpath. The stresses and strains induced by the wheel loads initiates the cracks by tension and propagates downwards. According to the M-E design guide, the pavement is prone to top-down cracking if the HMA layer is too thick (e.g. perpetual pavement) or the base support is too strong (stabilized base course). By May of 2007, a total of 347m of top-down cracking was identified. This corresponded to 379m/km (or 2005ft/mile) to top-down cracking in the pavement.

The rutting of pavement was surveyed in the inner wheel path of pavement in CTH JK. It was found that the rut depth in pavement of CTH JK had an average of 1.78mm. By May of 2007, the rut depth was found to be 6.9mm (0.27 inch).

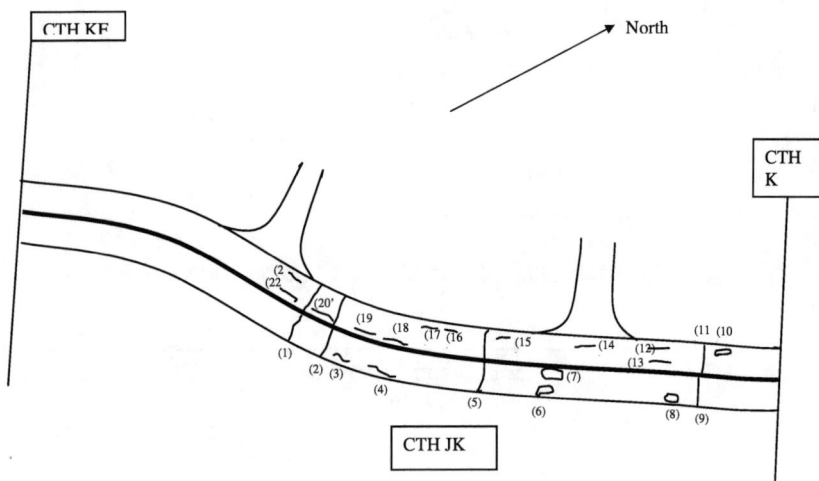

Figure 2. Mapping of Distresses in CTH JK Pavement in 2004

VERIFICATION USING M-E DESIGN

The M-E design guide has been developed to accurately design and predict the performance of pavements. It would be plausible to evaluate the effectiveness of M-E design for this type of pavement. The performance of this pavement was predicted using the M-E design guide program. The design life used in the M-E design was 20 years.

Table 2. Distress Survey Results of CTH JK Pavements

Months	Thermal Cracks m	Rut Depth mm	Top-Down Cracks m
12	0.00	1.02	0
24	4.16	1.52	0
36	29.27	1.78	16.16
48	40.24	N/A	241.30
69	51.22	6.95	347.32

Traffic

For the traffic information, the annual average daily traffic (AADT), traffic growth rate, and percentage of truck were input to the program. Default values were used for other parameters, such as vehicle class distribution, hourly adjustment, axle load distribution factors.

Materials

For the hot mix asphalt, an E-3 mix was used in this pavement. Level 3 input was used by providing the aggregate gradation and binder grade (PG-58-28). For the Class C fly ash stabilized RPM base course, the initial backcalculated modulus (one month after construction of base course) from FWD tests was divided by three to obtain the resilient modulus value, in accordance with the recommendation of the 1993 *AASHTO Guide for Design of Pavement Structures*. The modulus of rupture was estimated to be 20% of the unconfined compressive strength, in accordance with the recommendation of the M-E design guide. The subgrade of this pavement consists of clay and the modulus was based on the backcalculated results. Other input used the default value.

Climate

The climate from the closest national weather service station (Milwaukee, Wisconsin) was used for the M-E design program. The water table was estimated to be 14 feet from the ground.

Performance

The predicted performance by M-E design program was investigated. The performance of this pavement predicted by the M-E design program was compared to that of the field data.

Fatigue Cracking

As mentioned in the distress survey, top-down cracking was found in CTH JK while no bottom-up cracking was identified. According to M-E design guide [9], the pavement is prone to top-down cracking if the stabilized base course is used, or there is a thick HMA thickness. The thickness of HMA in CTH JK is 127mm and does not belong to a thick HMA layer. The predicted top-down cracking by M-E design guide with reliability and measured top-down cracking is shown in Figure 3. It is found that the M-E design program underestimated the susceptibility of asphalt with stabilized base course to top-down cracking. There are two possible reasons for the top-down cracking (1) the fatigue cracking model of asphalt pavement with stabilized base was not calibrated in M-E design guide, or (2) the deterioration of base course caused the top-down cracking, as indicated by the backcalculated base course modulus from FWD test results. It is believed that the former one in true, as the weakened base generally results in bottom-up cracking which was not found in this pavement, instead of top-down cracking. It is noted that once the top-down cracking was initiated, the progression rate was rapid, when compared to that predicted by the M-E design guide.

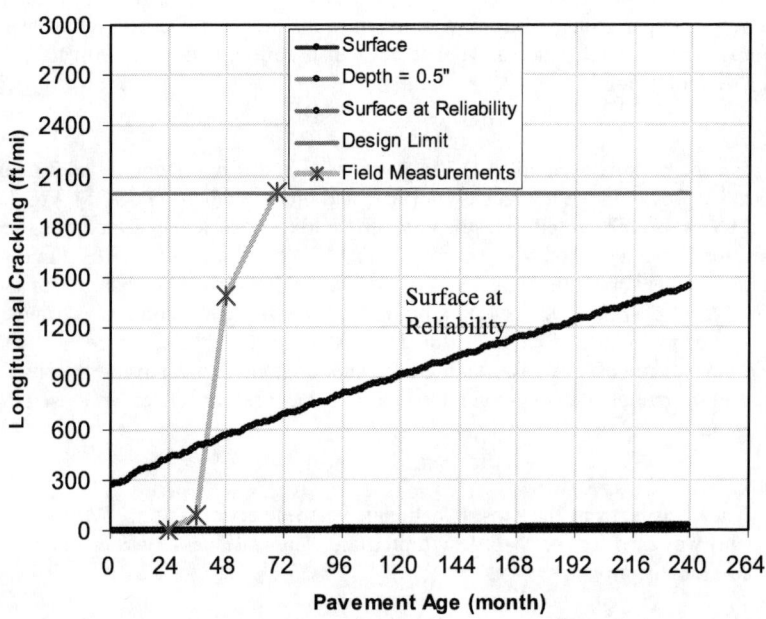

Figure 3. Predicted and Measured Top-Down Cracking

Thermal Cracking

Thermal cracking was found two years after the construction and continue to grow. Figure 4 shows the thermal cracking measured in the field and predicted by the M-E design guide. It seems that the M-E design guide underestimated the amount of thermal cracking of this pavement. This may be due to the excess friction between HMA layer and stabilized base course. The continuous hydration of fly ash after the construction may increase the bond between base course and HMA, when compared to crushed aggregate base course. The constraint from base course could contribute to thermal cracking [11].

Rutting

The rutting in this pavement has been minimal within three years after which the rutting was accelerated. It is noted that the FWD tests results after the third indicated a decrease in strength of the base course. After six years of service, the rutting depth was about 6.95mm which correlated well with the rutting depth predicted by the M-E design guide. The predicted and measured rut depth was shown in Figure 5.

Overall, the M-E design reasonably predicted the rutting, but underestimated the top-down cracking and thermal cracking. It should be noted that the performance models of asphalt pavement with stabilized base course was not subject to calibration, which could introduce error.

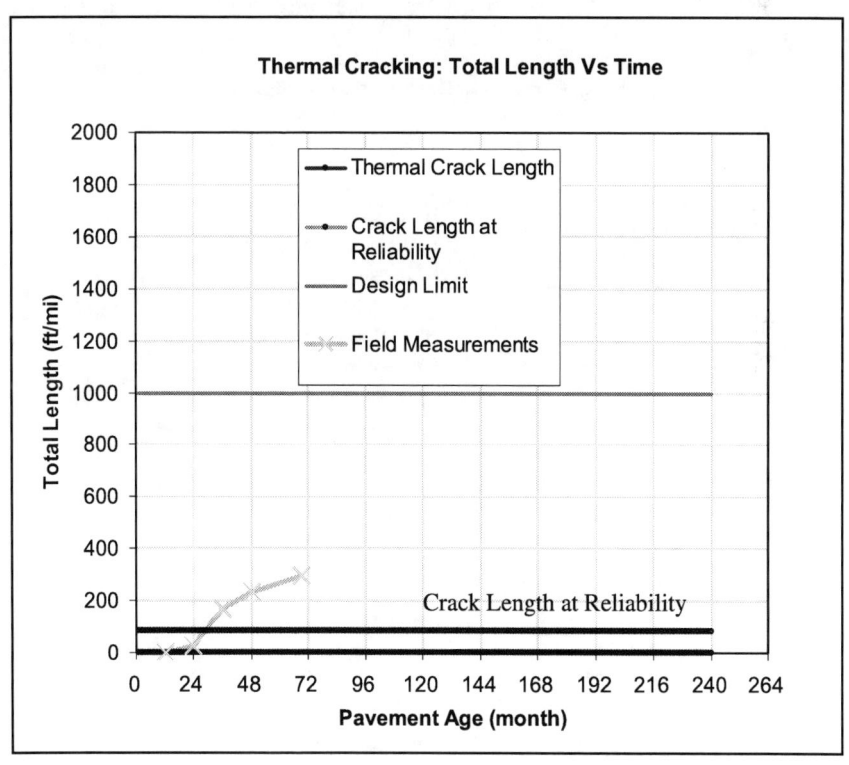

Figure 4. Predicted and Measured Thermal Cracking

CONCLUSION

FDR is becoming a widely used rehabilitation technique by highway agencies. It is also environmentally friendly and cost-effective to utilize Class C fly ash to stabilize FDR mixtures. County Trunk Highway (CTH) JK in Waukesha, Wisconsin was selected as a test section. The performance was evaluated using distress survey, falling weight deflectometer (FWD), and the mechanistic-empirical (M-E) design program, as a case study. It was found that within the three year after the construction, the modulus of Class C fly ash stabilized FDR base has continuously increased, as a result of ongoing pozzolanic reactions. However, it was also found that in the sixth year after the construction, the modulus of base decreased, when compared to the third year results. It is believed that freeze-thaw action weakened the strength of the stabilized base course. Distress survey found that top-down cracking occurred three years after the construction as the predominant distress, followed by thermal cracking, and rutting. The verification of performance using M-E design program found that the M-E design program underpredicted the amount of top-down cracking and thermal cracking and reasonably well for the rutting. It is believed that the stiff base could also be a reason for the top-down cracking and the excess friction between HMA layer and base which may have contributed to the thermal cracking.

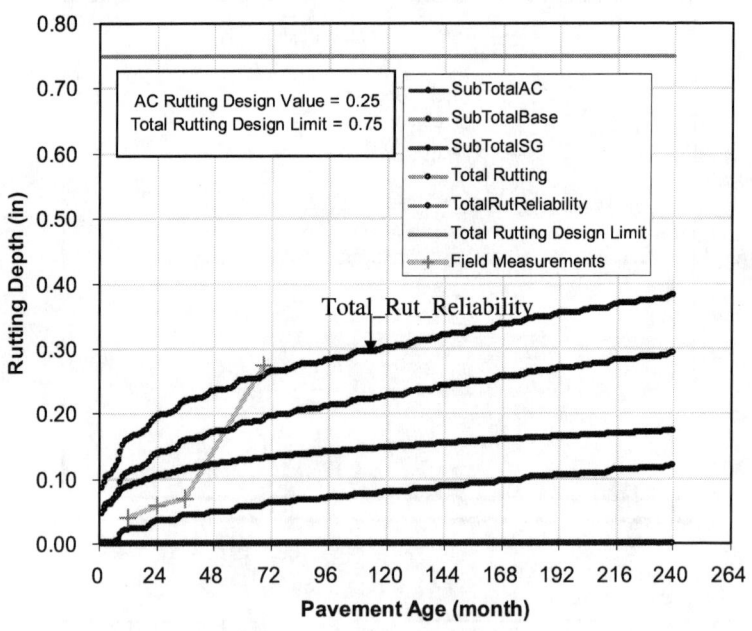

Figure 5. Predicted and Measured Top-Down Cracking

ACKNOWLEDGEMENT

The authors wish to thank Wisconsin Department of Transportation for providing the FWD testing for this project.

REFERENCES

1. Kearney E. J. and Huffman J. (1999), "Full-depth Reclamation Process." *Transportation Research Record*, No. 1684, Transportation Research Board, National Research Council, Washington D.C., pp203-209.
2. Wilson J., Fischer D., and Martens K. (1998), "Pulverize, Mill & Relay Asphaltic Pavement & Base Course." Construction Report, No. WI-05-98, Wisconsin Department of Transportation.
3. Cross S.A. and Fager G.A. (1995), "Fly Ash in Cold Recycled Bituminous Pavement." *Transportation Research Record*, No 1486, Transportation Research Board, National Research Council, Washington D.C., pp.49-56.
4. Cross S.A. and Young D. (1997), "Evaluation of Type C Fly Ash in Cold In-Place Recycling." *Transportation Research Record*, No 1583, Transportation Research Board, National Research Council, Washington D.C., pp.83-90.
5. Mallick R.B., Bonner D.S., Bradbury R.L. (2002), Andrews J.O., Kandhal P.S., and Kearney E.J., "Evaluation of Performance of Full-Depth Reclamation Mixes." *Transportation Research Record*, No 1809, Transportation Research Board, National Research Council, Washington D.C., pp.199-208.
6. American Coal Ash Association (2002), "2001 Coal Combustion Product (CCP) Production and Use."
7. Wen H., Tharaniyil M., and Ramme B. (2003), "Investigation of Performance of Asphalt Pavement with Fly Ash Stabilized Cold In-Place Recycled Base Course." *Transportation Research Record*, No 1819, Vol. 2, Transportation Research Board, National Research Council, Washington D.C., pp.27-31.
8. American Association of State Highway and Transportation Officials (1993),AASHTO Guide for Design of Pavement Structures. Washington, D.C..
9. Applied Research Associates, Inc. (2004), "Guide for Mechanistic-Empirical Design of New and Rehabilitated Pavement Structures,"
10. Crovetti, J. (1998), "Design, Construction, and Performance of Fly Ash Stabilized CIR Asphalt Pavements in Wisconsin", Wisconsin Electric-Wisconsin Gas, Milwaukee, Wisconsin.
11. Asphalt Institute. "Superpave Mix Design Method SP-2," Lexington, KY.

Investigation into the use of cement stabilized sand in road pavement construction in Bangladesh.

Waliur Rahman[1], Richard Freer-Hewish[1], Gurmel S. Ghataora[1].

[1]Department of Civil Engineering, the University of Birmingham, B15 2TT, UK.
mwr133@bham.ac.uk, g.s.ghataora@bham.ac.uk, PH: (+) 44-121-4145153.

ABSTRACT

In recent times as good quality materials have become rarer, use of out-of-specification materials in structural parts of road pavement construction have been investigated. This is particularly attractive in developing countries where good quality materials may be prohibitively expensive. This paper describes an investigation into the use of cement stabilized fine to medium sand in the construction of roads undertaken entirely in Bangladesh. The laboratory results indicated that the unconfined compressive strength and constructibility of sand-cement mixes were acceptable for road pavement layers. The sand-cement material with 8 to 10% cement proved to be adequate for subbase layers for heavily trafficked roads. A pilot field trial showed the constructability of a pavement layer using fine to medium sand and 10% cement mix. The un-surfaced trial road only experienced a few minor cracks after six months of construction.

1. INTRODUCTION

Engineering properties of some soils are such that those soils cannot fulfill the requirements of particular construction work alone, for example; some expansive clayey soils are not suitable to be used in road subgrades. Improving the properties of these soils sometimes become essential to meet the pavement requirements. The improvement of the engineering properties of a soil involves a stabilization process usually through mixing with an additive such as, cement, lime, bitumen, or calcium chloride. These processes are normally named after the additives. Cement stabilization is widely accepted at present time. The purpose is to improve the strength and durability of the structures constructed with natural soils by the addition of cement to the soils as an additive. It is evident that earth structures such as, embankments, roads, highways, airport runways, dams require soils with sufficiently good engineering properties such as, low plasticity, high bearing capacity, low settlements and high modulus. In the developing countries conventional construction materials are being adopted extensively. However, there appears to be ample scopes of minimizing the cost of construction and making the structures more durable by using locally available materials through chemical stabilization.

2. BACKGROUND

In regions of high rainfall the performance of unbound pavement layers and also the behavior of subgrade soils on soaking have long been points of concern for the engineers. During the monsoon period, moisture enters into the pavements from the flooded roadside drainage system. Subsurface layers weaken due to inundation and can result in shear failure manifested by rutting in the upper layers. This also results in the punching of granular upper layers into the weakened subgrades. The hot and humid climate in the countries in Asia particularly in the Indian Sub-Continent can result in weathering of black top in the case of sealed roads. The cracks in the weathered asphalt permit water to enter from surface. Under wheel load hydrodynamic stresses develop and consequently lead to creation of potholes, raveling, alligator cracking and rutting. One way to counteract these problems are to bind near surface pavement layers. This paper reports on the work carried out to investigate the potential of using cement to stabilize fine to medium sand for roadbase or subbase layers of a pavement. Labor intensive construction is very common in developing countries due to cheap labor and high cost of construction equipments. To compete with the conventional pavement materials, stabilized soils have to be constructible using similar techniques.

3. FINE TO MEDIUM SAND

River sand from the central part of Bangladesh (near Dhaka) was used in this investigation. The fine to medium grained sand used was uniformly graded with fineness modulus of 0.9 to 1.1. Particle size distributions of the sands are shown in Figures 1.

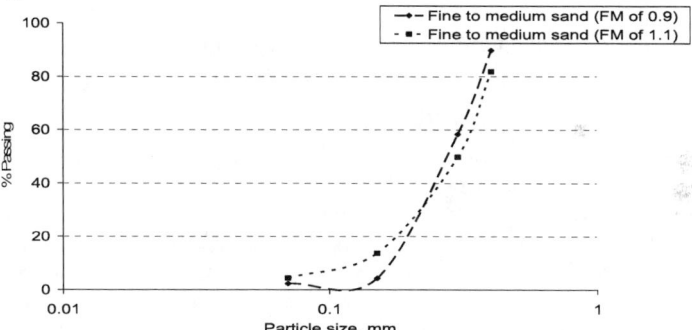

Figure 1: Particle size distribution of fine to medium sands.

Fine to medium silt from the western part of Dhaka city was also investigated during this study, however, this investigation has not been included in this paper.

4. STRENGTH OF SAND-CEMENT MIXES

Fine to medium sand was mixed with a range of cement and water content and compacted for preparing unconfined compressive strength (UCS) specimens. The UCS of the mixes and the comparisons with the specifications are shown in Figure 2. The mixes with sand and 8 and 10% cement (7 days UCS, 1.64 and 2.35 MPa) complied with the Transport Research Laboratory specifications (TRL Overseas Road Note 31, 1993) for roadbases.

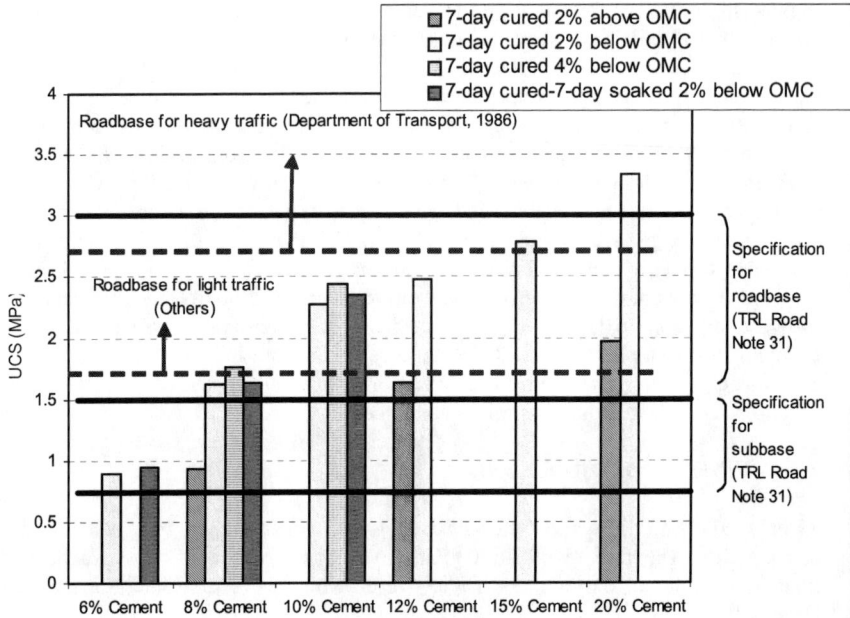

Figure 2: UCS values of sand-cement specimens with a range of moisture and cement content.

Other specifications provided by Davidson (1960), Maclean and Lewis (1963) require 7-day UCS of 1.72 MPa for roadbases of light traffic. Department of Transport (1986), used in the United Kingdom, specifies a minimum compressive strength of 2.76 MPa for soil-cement roadbases of heavily trafficked roads. The results show that sand-cement mixes with 8 and 10% cement are acceptable for roadbases of light traffic considering the UCS results. However, due to adverse field conditions, possibility of improper mixing, lack of quality control and lack of legal enforcement of maximum axle load, it would be safer to use the sand-cement mixes in the subbase courses of heavily trafficked roads instead of roadbases of light traffic. More than 6 million Equivalent Standard Axle Load (ESAL) was considered here as heavy traffic (TRL Overseas Road Note 31, 1993).

Amount of water present in any mix has considerable effect on the strength and density obtained with the soil-cement mix. Similar to the natural soils, cement-treated soils exhibit the same type of moisture-density relationship. Maximum dry density is achieved for a given compacting effort with a particular amount of moisture in the mix, which is the OMC. In case of compressive strength, the specimen of maximum dry density does not necessarily provide maximum strength. In general, the moisture content for maximum strength tends to be on the dryer side for sandy soils (Felt 1955 and Niazi 1996). The sand-cement mixes follow the water cement ratio theory of concrete because concretes are normally mixed and compacted with the water contents below their OMC values.

Fine to medium sand mixed with 20% cement at ±2% of optimum moisture content (OMC) to prepare UCS samples (OMC=13.5%). The 7-day cured strength of the specimen at -2% of OMC showed much higher strength (3.34 MPa) in compared to the specimen at +2% of OMC (1.97 MPa). The same tests with 8 to 12% cement showed similar results. Therefore, specimens for the sand-cement mix with 8% cement (OMC=12.25%) and 10% cement (OMC=12.5%) were tested for UCS at moisture contents -2% and -4% of OMC. The specimens at -4% of OMC gave slightly higher strength than those at -2% of OMC. Therefore, in case of sand-cement mixes the compaction should probably be carried out at moisture content -2% to -4% of OMC to obtain the maximum strength

The soaked tests were performed according to Test 12 BS 1924 (British Standard, 2002). The compressive strengths of 14-day cured specimens were compared with 7-day cured and 7-day soaked specimens. The strength reduction on soaking for specimens with 8 and 10% cement were 9.4 and 8.95% respectively. According to the British Standards the strength reduction should not be more than 20% (British Cement Association, 1998). These soaked tests also showed negligible changes in the volumes as well as the weights of the specimens.

5. POSSIBLE APPLICABILITY

Many researchers have suggested due to the possibility of reflection cracking that the soil-cement materials could perform better in the subbase layers rather than in the roadbases. Arnold (2000) reported that the stabilized materials should not be accepted as roadbase materials unless the tensile strength and shrinkage are within the allowable values. This requirement will probably limit the amount of cement added to a maximum of 2%. However, higher cement contents could be considered if there are provisions for additional maintenance requirements to seal cracks when comparing alternatives. Therefore, an "upside-down" pavement could be acceptable. The use of granular roadbases above the stabilized layer could arrest the shrinkage and traffic associated cracks to propagate in the surface courses. Illustrating the problems of reflection cracking Williams (1978) stated,

"In the case of cement bound material and soil-cement, restrictions on their use in roadbases for heavily trafficked roads have effectively designated them as subbase materials in either flexible or rigid pavements".

Figure 3, 4, 5: bed preparation And in-situ mixing.

Djakfar (1999) tested several pavements using different combinations of crushed stone and soil-

Figure 6, 7, 8: in-situ mixing and compaction.

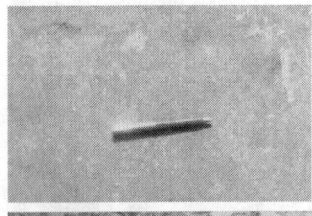
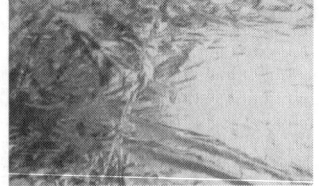

Figure 9, 10: Hardened sand-cement subbase and curing.

cement bases and subbases with the same wearing course. As many as 9 test sections were constructed. The author discovered that the combination of crushed stone base over a soil-cement subbase, known as an inverted pavement, provided the best performance in resisting rutting and retarding the occurrence of reflection cracking. Furthermore, the stabilized subbase layer stiffens the overall pavement system, thus reducing the likelihood of physical degradation occurring in the unbound upper layer (Dunlop 1980).

According to Technical Recommendations for Highways of South Africa (TRH 13) (1986), cracks in bituminous surface have been observed mostly in pavements where the cement stabilized layer has been used in the base courses. However, the post-cracked phase of a cement-stabilized subbase adds substantially to the useful life of a pavement. The suitable material for the subbase layers have been obtained by performing the standard tests as discussed in Section 4 above.

6. PILOT CONSTRUCTION

Fine to medium sand mixed with 10% cement showed satisfactory UCS values for subbase layers. However, the diversified field conditions in terms of material variability and construction quality required to be considered to assess whether the materials could perform well in the field. A pilot field trial was therefore, considered to prove the validity. A small section was constructed to observe the primary surface erosion and cracking of the surface and durability of the material. A small section of an existing road was chosen and the existing conventional granular subbase was removed manually (see Figure 3). The field trial was constructed following the in-situ labor intensive process, which is very common in the developing countries.

6.1 Pilot Section

A 3.6m by 3m section was constructed with the selected material. The fine to medium sand was spread into 0.13m deep excavation and all the materials were mixed manually (see Figure 4, 5, 6)

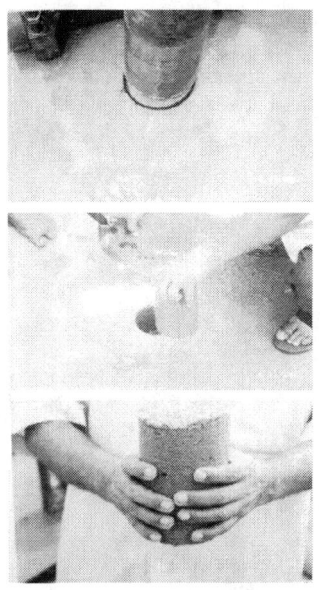

Figure 11, 12, 13: Core cutting from the subbase.

Figure 14: Core crashing in a concrete crusher.

as would be the case in practice, particularly in the Sub-Continent. Prior to starting the construction the moisture in the sand was adjusted in the field to provide moisture content to -1% of optimum moisture content of the mix. Laboratory results showed better performance for mixes at lower moisture contents (-2 to -4%); however, it was predicted that some water would be lost due to evaporation during mixing in the field conditions.

When final mixing was completed on a half of the strip (1.8 meter by 1.5 meter), an eight-ton smooth wheeled roller was passed twice (see Figure 7, 8) and then another half was prepared and compacted in the same way. Finally, total section was compacted by passing the roller 4 times over the surface until the surface looked hard and dense (see Figure 9). Curing of the trail section was carried out by covering it with a polythene paper (see Figure 10).

The dry density of the compacted material was measured from the 7-day cured core specimens. According to Williams (1978), the field dry density measurement should be made at least 4 hours after the completion of the compaction and preferably within a period of 24 hours. However, it was not possible to measure field dry density using the sand replacement method after 4 hours of compaction, as the material became stiff very quickly. The dry density achieved was approximately 96% of the laboratory density according to the measured average dry density of 7-day cured cores.

6.2 Core strength and laboratory results

Cores were cut in 3 different location of the pilot section and tested for compressive strength (see Figure 11, 12, 13 and 14). The core strengths for 2:1 cylindrical samples were 2.52 MPa, 2.77 MPa and 3.02 MPa for 7, 14 and 28 days curing period respectively. Core strengths were slightly higher than the strengths found in the laboratory (see Figure 15). This may have happened because the fine to medium sand used in the trial had a better grading with an FM of 1.1 whereas in the laboratory mixes sand of FM 0.9 was used. The other reasons might be due to warmer curing condition in the field and also mixing in the field may not have been as thorough as in the laboratory and the cores taken from the places where cement content might have been a little higher due to variable mixing. In the trial construction the weight of the sands was determined by loosely pouring the materials in a container and after that the weight of further sands were assumed by pouring the same container loosely. As the weight was assumed during the construction the cement content might have been a little higher. However, the results showed good repeatability (see Figure 15). The soil-cement material introduced in this paper has shown acceptable labor intensive constructability even without using a mixer. The only equipments used were an 8-ton smooth wheeled roller for compaction and shovels for mixing. However, for a large scale use of this material a mixing plant required to be used.

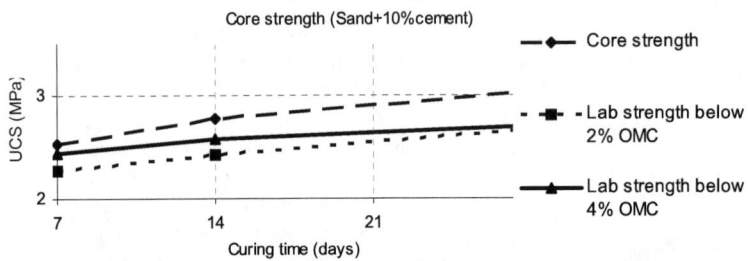

Figure 15: Comparison among core strength, with sand FM 1.1, of the trial and laboratory strength, with sand FM 0.9, and 10% cement.

The compaction of the soil-cement was adequate with the smooth wheeled roller; however, there were some overlapping of materials which was expected due to the presence of sand. This overlapping will help to keep the surface of the soil-cement subbase rough for good bonding with upper layers. The compaction of the soil-cement should be carried out within 2 hours of mixing to avoid hardening of the mix during compaction.

For labor intensive construction of the stabilized layer it would always be safer to use 10% cement with the sand though 6 and 8% cement gave acceptable results for subbase layers. For equipment based construction with close quality control 8% cement mixed with fine to medium sand should be adequate for subbase layers.

7. CONCLUSIONS

The following conclusion are drawn from this investigation
- Mixes with both 8% and 10% cement showed sufficient strengths for using in roadbase layers. However, these mixes should be used in subbase layers if the possibility of variable mixing and compaction in field conditions are anticipated.
- The in-situ labor intensive construction process has proved to be applicable for soil-cement mixes.
- The trial section showed good core strengths and also in-situ labor intensive mixing was good enough to show very little variation among the individual cores. After more than six months of construction the trial did not show any cracks or rutting.

REFERENCES

Arnold, G. (2000). "Performance Based Specifications for Road Construction and Materials". 'Unbound Aggregates in Road Construction' *Dawson, A. R. and Balkan, A. A., Eds.*, Rotterdam, Netherlands. pp. 183-191

British Standard (2002). "Stabilized Materials for Civil Engineering Purposes, Part 2: Methods of Tests for Cement-Stabilised and Lime-Stabilised Materials". BS 1924-2:1990.

Davidson, D. T. (1960). "Comparison of Type I and Type III Portland Cement for Soil Stabilisation". Highway Research Board Bull 267. pp 28-45.

Department of Transport (1986). Specifications for Highway Works, 6th Edition. H. M. S. O. London.

Djakfar, L. (1999). "Implementation of ALF Results to Designing Flexible Pavements in Louisiana". *UMI ProQuest Digital Dissertations. Index of Thesis.* Website Address: http;//wwwlib.umi.com/dissertation/fullcit/9926822 date 06/11/02.

Dunlop, R. J. (1980). "A Review of the Design and Performance of Roads Incorporating Lime and Cement Stabilised Pavement Layers". Australian Research Record, Vol. 10, No. 3. pp 12-26.

Felt, E. J. (1955). "Soil and Soil Aggregate Stabilisation" Highway Research Board Bulletin 108, National Academy of Science-National Research Council Publication 359. pp. 138-162.

Maclean, D. J. and Lewis, W. A. (1963). "British Practice in the Design and Specificaiton of Cement-Stabilised Bases and Sub-bases of Roads". Highway Research Record 36. pp 56-76.

National Institute for Transport and Road Research (1986). "Cementitious Stabilisation in Road Construction". Technical Recommendations for Highways 13. *National Institute for Transport and Road Research of the Council for Scientific and Industrial Research, South Africa.*

Niazi, Y. (1996). "Stabilisation of Desert Sand with Desert Clay Plus Lime, and Cement Kiln Dust in Desert Road Construction". A PhD. Thesis of School of Civil Engineering, Faculty of Engineering, University of Birmingham.

Transport Research Laboratory (1993). "A Guide to the Structural Design of Bituminous-Surfaced Roads in Tropical and Sub-Tropical Countries". *Overseas Road Note 31. 4th Edition*, TRL, Crowthorne, Berkshire, UK.

Williams, R. I. T. (1978). "Cement Stabilised Materials". Development in Highway Pavement Engineering-1, *Edited by Pell, P. S., Applied Science Publishers. UK.*

Sustainable Base Course

Casimir J. Bognacki, P.E[1] and Marco Pirozzi[2]

[1] Chief of Materials, The Port Authority of NY & NJ, PA Technical Center, Jersey City, NJ 07310, USA, Phone: (201) 216-2984; Fax: (201) 216-2949, cbognack@panynj.gov

[2] Materials Engineer, The Port Authority of NY & NJ, PA Technical Center, Jersey City, NJ 07310, USA, Phone: (201) 216-2337; Fax: (201) 216-2092 mpirozzi@panynj.gov

Abstract

At existing airports, re-aligning and replacing taxiways is a common practice due to reconfigurations of the operational area or just normal wear and tear. For example, with the arrival of the new "Super-Jumbo Jet" aircrafts, such as the Airbus A380, our nation's taxiways will need widening and increased load capacity to accommodate the large aircrafts. It is common practice when removing and re-aligning taxiways to dispose of the material and replace it with new material, either asphalt or other base courses. The Port Authority wanted to try and reuse some of this material to develop sustainable design, and reduce material costs.

Preliminary research in the Port Authority of NY & NJ Materials Laboratory investigated the possibilities of reusing existing asphalt pavement (RAP), lime-cement-fly ash base course (LCF) and blending it with Portland cement to create a sustainable base course. This research was successful and lead to the option of using either a new asphalt base or a blend of existing pavement and Portland cement in a contract to upgrade the taxiways at John F. Kennedy International Airport. The low bidder chose to use recycled pavement materials and blend it with cement on-site.

This sustainable option not only proved to be economical but efficient as well. The on-site blending of Portland cement with the removed pavement material saved in travel time and reduced truck traffic on the surrounding infrastructure. This paper will discuss in detail the following topics related to sustainable base courses:
- Preliminary laboratory results
- Sustainable base course mix design
- Batching and placement methods
- QC/QA procedures
- Test results

Introduction

Each year the popularity of air transit increases. With cheaper commercial airline rates and the increased usage of "just-in-time" shipping, airport congestion and delays are becoming the operational norm not the exception. According to the International Air Transport Association, between 2005 and 2009 international passenger routes will grow 5.5% while international freight will grow 6.3%. To accommodate the increase in traffic, US airports are undergoing various tactics that will ultimately allow this growth while minimizing delays and problems with its costumers. Airports are trying to accommodate the future "super-jumbo" jet class of airplanes such as the Airbus A-380 and the Boeing 747-800. Other airports are re-configuring their operational area to facilitate more flight slots in a given period. Either decision potentially requires large reconfigurations or rehabilitations of taxiways and runways. The Port Authority of NY & NJ analyzed the different options available to them and decided to widen and strengthen their existing taxiways. This decision led to the investigation of cement-treated base courses as a substitute for the conventional asphalt base course to increase the load capacity of the pavement. After numerous lab studies were conducted, a final, sustainable mix design consisting of recycled asphalt pavement (RAP), lime-cement fly-ash (LCF), sand and cement proved to be a very successful venture.

History and Preliminary Laboratory Results

The history of a sustainable base course can be traced back to the use of roller-compacted concrete and lean concrete. Roller compacted concrete (RCC) is identical to conventional concrete except that it has a much lower water content and therefore essentially zero slump. This type of concrete is usually placed using dump trucks, a pavement spreader and compacted by vibratory rollers. Lean concrete is categorized by a low cement factor. Extensive research into different mix designs that would combine the characteristics of these two types of concrete was performed in the materials laboratory of the Port Authority of NY & NJ. The original mix designs, listed in Table 1, were used as a starting point from which to improve upon. As can be seen from this table, a low cement factor and a large aggregate content was targeted to achieve the desired targets.

Eventually it was proposed to substitute the coarse aggregate with RAP. The concrete specimens cast using the RAP aggregate replacement was shown to produce similar results with respect to the RCC and lean concrete specimens. It was a benefit to use the RAP because the recycling process was more economical. Instead of hauling the milled asphalt to an off-site landfill, the removed product was now carted to a section of the airport where the contractor built an on-site plant to process the blend, add cement, sand, LCF and water. (LCF is an antiquated base course that consisted of excavated brown sand that was stabilized with hydrated lime, cement, fly-ash and water. The material was then placed and compacted prior to the installation of the top course pavement.) The end product, a sustainable base course, was actually born from a fiscal motivation.

Table 1.

July 9, 2003	Roller Compacted	Lean Concrete
Cement	170	250
Fly Ash	170	250
Sand	1475	1100
Stone, #467	1900	2000
Water	187	275
W/C	0.55	0.55
Air Content	N/A	N/A
Slump	N/A	4 "
Compressive strength		
7 days	320 psi	1250 psi

Preliminary Research

Mix designs targeted aggregate replacements with LCF and RAP at different percentages by weight. It was unknown at the time which replacement would perform better and which would provide the most economical solution. Ultimately, after various testing in the laboratory it was determined that a variation on Mix # 4 would be accepted. It has been referred to colloquially as "econocrete".

Table 2.
Refined Mix Designs

March 11, 2004	Mix1	Mix 2	Mix 3	Mix 4	Mix 5
Cement	500	500	500	500	500
"Brown" sand	3100	1600	N/A	N/A	2300
"Grey" sand (LCF)	N/A	N/A	3100	1600	N/A
RAP (3/8")	N/A	1600	N/A	1600	800
Water	370	422	418	390	400
W/C	0.74	0.84	0.7	0.78	0.8
Air Content	3.8	4.8	4.4	6.1	4.2
Slump	3.75	3.75	4.25	4.5	3.5
Compressive strength					
7 days	430	750	585	715	870
14 days	565	1000	740	855	990
28 days	775	1200	1080	1005	1390

Batching and Placement Methods

The Port Authority contract documents do not specify the means and methods that the contractor must use when performing the agreed upon work. Therefore, the following methods were developed by the contractor with some suggestions and input from Port Authority engineers. The process consists of batching the material in an onsite plant and then placing the material in dump trucks. The material is then hauled to the placement site.

The batch plant is a mobile plant that is constructed piecemeal in the contractor's construction yard. It consists of two aggregate hoppers, a cement silo, a conveyer belt and an auger system to blend all the material together. At the end of the process, another conveyer belt carries the finished product into the back of a waiting dump truck.

The process began by removing the existing pavement and base course through a traditional mill and removal process. The reclaimed pavement material, both asphalt product and LCF, was screened to the proper gradation sizes (Fig. 1) and placed into stockpiles in the contractor's yard. Stockpiles for RAP, LCF and brown sand were kept separate in order to achieve the desired blend. Before each day's production, the stockpiles were examined for proper gradation and moisture by both the contractor's quality control personnel and the owner's quality assurance representative. The cement used for production was stored in a silo connected to the batch plant along the assembly line.

Figure 1. Screen & Gradation Process   **Figure 2. Cement Addition Process**

Figure 3. Final Product

During production, loaders gathered the appropriate material from the stockpiles and place them into specific hoppers at the beginning of the assembly line. The plant is completely automated and the exact weights of RAP/LCF and sand drop onto the conveyor belt at the appropriate times. Then the designated amount of cement was dropped on top of the mix (Fig. 2) and the combination was carried to the next stage. A large auger mixer was calibrated periodically in order to add the correct amount of water as the RAP, LCF and cement mixture was carried through the plant by the conveyer. The amount of mix water could be varied but it usually ranged between 9-12% of the total weight. This entire process was computer-controlled and the amount of water added could be changed to reflect changes in aggregate moisture or placement consistency. As a rule of thumb, the initial batch was done at the low end of the specified water content. It was then examined for consistency and if need be additional water was added. This was done so that if there was moist ground or a chance of rain, the day's production was not lost. Once the mixing portion of the process was complete, the final product, econocrete, was taken by another conveyor belt that to large tandem dump trucks (Fig. 3). The tandem trucks were large 40-ton capacity trucks that were able to handle 22 cubic yards of material per trip.

At each location, forecasted work for that day was formed by either the surrounding land at grade level or traditional plywood forms. The sub-grade was for the most part sand that was graded and compacted, ready to receive the econocrete. Once the dump truck arrived at the location, the econocrete was dumped into small, manageable piles. A dozer then spreads the econocrete in order to achieve a lift thickness of no more than eight inches. The design drawings specified four different depths depending on the service loading of the overlying pavement: taxiway, heavy shoulder, light shoulder and erosion areas (Fig. 4a, 4b). A survey crew working for the contractor made sure that all elevations and thickness were per the contract documents. Pavement rollers compacted the econocrete until it reached optimum density, as tested by both the contractor's quality control manager and the Port Authority field inspector. Once the material had achieved acceptable density results, it was sawcut at one hundred foot intervals perpendicular to the taxiway centerline to allow for cracking. Afterwards, the top lift was sealed with a low-grade asphalt membrane. This was done to aid in the curing process of the econocrete. At this point the base course was ready to receive the designated aeronautical pavement.

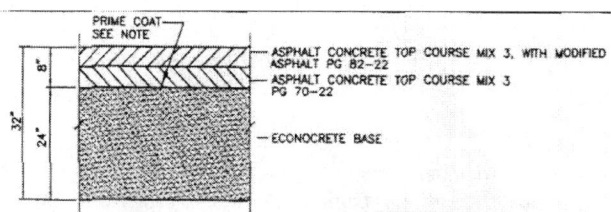

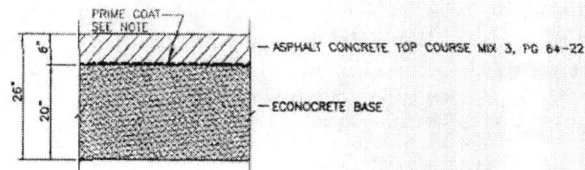

Figure 4a. Design Details

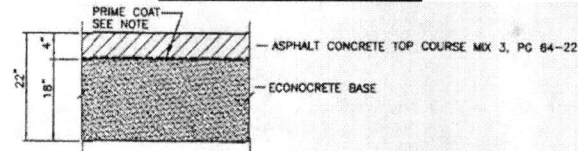

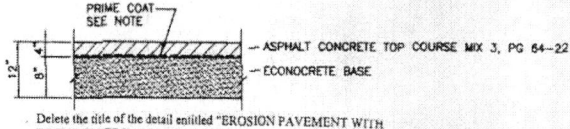

NOTES:

SAWCUT CEMENT TREATED BASE PRIOR TO PLACING ASPHALT CONCRETE TOP COURSE AT 100' MAXIMUM INTERVALS PERPENDICULAR TO THE TAXIWAY CENTERLINE.

PRIME COAT SHALL MEET ALL REQUIREMENTS OF SPECIFICATION SECTION 02561-3.010 EXCEPT THAT IT SHALL BE AT AN APPLICATION RATE OF 0.25 TO 0.50 GALLONS PER SQUARE YARD.

Figure 4b. Design Details

Quality Control and Quality Assurance Procedures

The Quality Control and Quality Assurance procedures are more extensive and complex with the sustainable base course when compared to traditional asphalt or concrete. As part of the contractor's quality control plan, the aggregate gradation control chart, which is specific to the econocrete mix used, should be kept posted. Should there be any field problems; a gradation can be run and cross-checked against the required gradation for the econocrete mix design. It is recommended that the chart contain the running average of the last five gradations tests performed. If gradation is indeed a problem, this will identify whether or not the screening process is the cause of the problem.

The contract should contain action limits for falling out of gradation and moisture content specification. This will allow the contractor to take corrective action prior to a large loss in the quality of the material. A suspension limit should also be included in the contract, if the moisture content is below optimum or if the econocrete is not compactable. Corrective action must also be taken whenever two in-place density measurements are below 98% of the target or a single measurement is below 95% of the target. Strict requirements should also be specified for the materials to be used for the sustainable base course. All reclaimed/recycled aggregate should be from pavement that is located within the project limits and should be required to meet the gradation requirements listed below (Table 3a, 3b).

Table 3a.
JFK 134.105 Gradation

Sieve Sizes	RAP	Grey Sand/LCF	Brown Sand
2-1/2"	100	100	100
2"	100	100	100
1-1/2"	100	100	100
1"	97	98	98
3/4'	95	96	97
1/2"	85	91	93
3/8"	80	87	91
#4	60	77	83
#8	42	71	79
#16	29	64	74
#30	17	52	63
#50	9	28	28
#100	4	9	7

Table 3b.
JFK 134.105 Combined Gradation

	Combined Gradation
1-1/2"	100
1"	90-100
1/2"	80-100
#8	50-85
#30	30-70
#100	0-10

The Port Authority's conventional procedure of inspecting the placement of an asphalt base course requires only one plant inspector verifying that the asphalt is batched correctly. Field placement of base courses is typically not inspected by either the contractor's Quality Control or the owner's Quality Assurance personnel. Sustainable base course testing is a three-step process. First, the optimum density and moisture for that particular lot are determined, which is done at a mobile lab adjacent to the plant. The inspector executes a one-point proctor and calculates the moisture content for each lot of material that is produced. The inspector then observes the placement of the econocrete and records on contract drawings the locations that were placed that day. A day's production is considered a lot of material and each lot is then divided into at least 3 sublots, depending on the quantity of material produced that day. A random system of placing cores is used and 4 cores are cut from each sublot, 2 for compressive testing at 14 days and 2 for testing at 28 days. Using this method, the cores are then transported to the lab where they are prepared for testing. This preparation involves trimming the cores to eight inches, where possible. If for some reason eight inches could not be obtained, the greatest length to diameter ratio is tried for larger than one to one. The cores are then sulfur capped and tested according to ASTM C-617 and ASTM C-39. The material quality is a payment item for most Port Authority contracts and in the case of econocrete, the two categories that must pass specifications are field density and compressive strength.

The following is a summary of the Port Authority of NY & NJ specification number 02242 which details the process for a sustainable base course. An influence on this specification was the FAA specification P-306 which deals with cement treated base. Initially, the contractor must submit a mix design to be approved by the Materials Engineering Unit. This design must show results of at least 1,000 psi in 14 days and 1,500 psi in 28 days. In addition, the weight loss of a sample after 12 cycles in a freeze-thaw cabinet must not exceed 14%. The average percent density of the compacted econocrete should be greater than 98% of the maximum laboratory density. Finally, the minimum average 28-day strength results should be 1,250 psi.

Present Example

A current contract that utilizes the sustainable base course is JFK 134.105 "The Relocation of Taxiway A and Rehabilitation of Taxiway B." The contractor submitted a mix design that was approved which can be seen below.

Table 4.
JFK 134.105 Approved Mix Design

Cement	Initially 600, lowered to 500
LCF	1500
RAP	1500
Gradation 50/50 blend	Percent Passing
1.5"	100
1"	91.8
.5"	78.2
#8	51.9

#30	35.6
#100	6.4

Current pavement rehabilitation contracts at John F. Kennedy International Airport have allowed the contractor to bid with either an asphalt base course or sustainable base course option. The winning bidder, (lowest, responsive responsible bidder) included the sustainable base course in his package. When asked about his bid items, it was made very clear that economy and efficiency were the main motivations for choosing to include the sustainable rather than the traditional base course. The construction methods were more streamlined under this system as opposed to the conventional way of pavement rehabilitation. The contractor was given a sufficiently large plot of land on the airport to construct the on-site batch plant. During the removal phase of the project, the RAP and LCF were hauled to this site. The recycled material was then processed, graded and used to create the sustainable base course. This was then transported back to the jobsite during the placement phases of the contract. Many steps that are part of the normal removal and rehabilitation of pavement were excessive and unnecessary with the usage of econocrete. The project schedule was greatly compressed because transit time for hauling and material delivery was significantly reduced.

Test Results

The initial laboratory test results were bordering on the strong side, between 2,000 and 3,000 psi. Unlike typical cement based structures, a base course that is too strong is undesirable. Large flexural loads on the base course with cause extensive cracking if it is too rigid to deflect with those loads. These cracks will then propagate to the overlay and cause reflective cracks and rapid deterioration of the pavement. As a result, the cement content was lowered to produce laboratory design strength of 1,250 psi. Once the contractor began to mix and place the base course, the compressive strength tests were consistently 400 psi above the required 1,250 psi at 28 days. In a value engineering effort, the contractor recommended reducing the cement content from 600 to 500. After it was shown to provide passing test results, this change was approved.

The test results have been satisfactory with failures occurring few and far between. The average 28-day compressive strength results (based on ASTM C-39) were 1650 psi, with a standard deviation of 260 psi and a coefficient of variation of 0.16. The field tests have proven that batching the material has been very consistent and that the contractor has continually supplied the job with a quality product.

Green Construction

The idea of the sustainable base course also provides a green product that benefits both the owner and the community. Currently the environmental movement is a very popular and strong force throughout the construction industry. Our industry is one of the most notorious polluters of the environment and therefore anything that can be done to minimize this negative effect is looked at very favorably. The concept of a sustainable base course is one that is simplistic and it can be easily substituted for conventional asphalt base courses.

Four aspects of the econocrete base course process help categorize itself as a green construction process. These aspects are: the management of natural resources, reducing energy consumption, reducing emissions, and lowering community impact.

A large percentage of construction materials delivered to a jobsite are wasted or inefficiently used. This results from poor design and planning but also from specifications that are too lenient when it comes to the selection of materials. One way to remedy this is to initiate the idea of re-using and recycling materials during the planning stages of the job. This will facilitate the acceptance of a sustainable project and aid in specifying certain items that the contractor may otherwise not use. Restricting the percentage of new material usage also does away with the obvious pollution caused by delivering this material to the jobsite.

The construction industry is a large contributor of pollution to the environment. Much of this is attributed to emissions arising from the various heavy machinery and vehicles used throughout the construction process. However, aside from the emissions, noise and dust caused by deliveries and on-site activities are a nuisance to the local community. Part of the responsibility of the owner is to minimize this negative effect by steering the contractor into better practices through specifications.

The construction industry consumes energy through:
1. The extraction, processing, manufacturing and transport of construction materials and products
2. The transport of product manufacturing wastes.
3. Construction and demolition activities.
4. The transport of construction and demolition wastes.
5. Operations on construction sites such as contractors' offices.

The transport of materials for construction and maintenance is a major energy issue. Through the contract specifications, the Port Authority is mandating the use of local materials and recycled products. This has greatly reduced the number of and length of vehicle journeys required in construction and maintenance activities.

The Authority takes great pride in its dealing with the local community. The respect for people, especially those directly affected by a construction project, is one of the top priorities during construction. One way this is done is by incorporating the opinions of those affected and making decisions to minimize the impact it will have on the community. The local population does not only include those within a close proximity of the construction site but also those that live and work along the major transportation routes to and from the jobsite. Minimizing the traffic and transportation demands on the infrastructure is also considered a major concern of the Authority.

The use of the sustainable base course addresses in a positive way each of the above issues. It manages natural resources efficiently and effectively because more than eighty percent by weight is recycled material. Only the cement used is new production that has to be transported to the site. Using econocrete also lowers the total emissions for the production of the base course. Instead of producing each individual portion of the mix and transporting it to the site to be placed, the recycled material is taken from the jobsite and batched at an on-site plant. A truck now travels in one day what it would have in a single delivery trip. The energy consumption is also less because again econocrete is made from recycled material and batched on-site. When compared with an asphalt base course, the energy needed for the production of econocrete is much less. Finally, econocrete is great when the impact to the community is considered. Since almost the entire production process happens on the jobsite, the surrounding community is impacted less and for the most part do not even know that construction is occurring. This is a product that succeeds on every environmental level.

Summary

Econocrete is a cheap and effective alternative for airfield base courses. The extensive lab testing and subsequent usage in the field has given us more than enough faith in the performance and economy of the product. It was very telling that a low bid contractor informed us that the econocrete allowed him to be the low bidder on the contract. What should not be lost however, is the fact that econocrete is a product that is also beneficial to the environment. The savings on energy consumption and vehicle emissions because of the nature of the recycled constituents allow it to be sustainable and profitable. It is not a stretch of the imagination that more airports throughout the United States will move towards a sustainable base course in the near future.

References

1. Shilstone, James M., "Concrete Mixture Optimization," Concrete International, Vol. 12 No. 6 , pp. 33-39, June 1990.
2. The Port Authority of NY & NJ. Contract Specification Division 2, Section 02242, "Cement Treated Base".
3. "Building Better Roads: Towards Sustainable Construction". Complied by Centre for Sustainability for The Highways Agency.

Minimum Standards for Using Recycled Materials in Unbound Highway Pavement Layers

Dr. Athar Saeed, P.E., A. M. ASCE [1]
Dr. Michael I. Hammons, P.E., M. ASCE [2]

[1] Research Engineer, Air Force Research Laboratory (AFRL/RXQD)
(Principal Engineer, Applied Research Associates, Inc.)
AFRL/RXQD – Deployed Systems Branch
104 Research Road, Bldg 9738, Tyndall AFB, Florida 32403
Phone: 850 283 3718, Fax: 850 283 2035, athar.saeed@tyndall.af.mil
(**corresponding author**)

[2] Research Engineer, Air Force Research Laboratory (AFRL/RXQD)
(Principal Engineer, Applied Research Associates, Inc.)
AFRL/RXQD – Deployed Systems Branch
104 Research Road, Bldg 9738, Tyndall AFB, Florida 32403
Phone: 850 283 3718, Fax: 850 283 2035, michael.hammons@tyndall.af.mil

ABSTRACT

Recycled asphalt pavements (RAP) and recycled concrete pavements (RCP) were evaluated for use as an unbound base/subbase material. The objective was to recommend procedures for performance-related testing and selection of recycled hot-mix asphalt (HMA) and Portland cement concrete (PCC) materials for use as aggregates in unbound pavement layers, singularly or in combination with other materials. Performance-related tests were conducted on RAP and RCP samples containing three constituent aggregate types (gravel, crushed limestone and granite) providing a range of performance from poor to excellent. The recycled materials were blended with a virgin aggregate known to provide good performance in unbound pavement layers. Requirements for test parameters for recycled materials were established to evaluate recycled materials' suitability for use in particular traffic and climatic conditions. The research results are implemented as a decision chart incorporating aggregate shear strength, stiffness, toughness, and frost susceptibility to provide a measure of the performance potential of a particular recycled aggregate.

INTRODUCTION

Demand pressures on the construction industry have made aggregates produced by recycling asphalt and concrete pavements highly attractive and economically viable (*Saeed 2006, Ferragut 2001, Dumitru 2000*). The United States Geological Survey (USGS) estimates that about 2.241 billion metric tons of crushed stones were used in pavement construction in 2006 in the United States (*USGS 2007*), of which about 108 million metric tons were used in graded base/subbase construction. The transportation infrastructure uses about 95 percent of all aggregates (virgin and recycled) manufactured in the United States (*TRB 2007*).

Unbound pavement layers in flexible and rigid pavements generally serve to provide 1) a working platform, 2) structural layers for the pavement system, 3) drainage layers, 4) frost-free layers and 5) select fill material (sometimes as part of the working platform). The properties of recycled aggregates greatly influence their performance as unbound granular pavement layers. Failure of an unbound pavement layer results in pavement distresses. Fatigue cracking, rutting/corrugations, depressions and frost heave of flexible pavements are distresses (performance parameters) that can result from poor performance of aggregate in unbound base and subbase layers. Similarly, cracking, pumping/faulting/loss of support, frost heave and erosion in rigid pavements can result from poor performance of subbase layers (*Saeed 2008*).

Factors contributing to distresses in both rigid and flexible pavements due to the poor performance of unbound layers include 1) shear strength, 2) density, 3) gradation, 4) fines content, 5) moisture level, 6) degradation during construction, under repeated load and freeze-thaw cycling, 7) particle angularity and surface texture and 8) drainability. Recycled aggregate properties that were determined to affect performance of unbound pavement layers are shear strength, frost susceptibility, durability, stiffness and toughness (*Saeed 2008, 2001*).

For this study, tests were conducted on RAP and RCP containing three different constituent aggregates (crushed limestone, granite and gravel) to provide a range of materials with poor to excellent performance. The recycled materials were blended with a virgin aggregate known to provide good performance in unbound pavement layers. Laboratory test data were analyzed, and the following tests were found to produce statistically significant performance indicators of recycled aggregates in unbound pavement layers:

- Screening tests (sieve analysis, moisture-density relationship)
- Toughness (Micro-Deval)
- Stiffness (resilient modulus)
- Shear strength (static triaxial and repeated load at optimum moisture content [OMC] and saturated)
- Frost susceptibility (tube suction)

Requirements for test parameters for recycled materials were established to evaluate recycled materials' suitability for use in particular traffic and climatic conditions. The research results are implemented as a decision chart incorporating aggregate shear strength, stiffness, toughness, and frost susceptibility to provide a measure of the performance potential of a particular aggregate. Environmental concerns arising from the use of RAP and RCP as unbound base material were outside the scope of the research effort; a number of studies have investigated these concerns and the results are summarized by Saeed et al (2006).

This paper reports laboratory test results and the procedure used to develop the minimum standards for RAP and RCP materials for use as unbound base.

RESEARCH OBJECTIVE

The research objective was to recommend procedures for performance-related testing and selection of recycled HMA and PCC materials for use as aggregates in unbound pavement layers, singularly or in combination with other materials. The research included evaluating existing aggregate tests that have been known to predict pavement performance for their applicability to RAP and RCP and to develop new tests or modify existing tests. This paper reports laboratory test results and the procedure used to develop the minimum standards for RAP and RCP materials for use as unbound base.

RESEARCH APPROACH

The research approach included a literature search and phone interviews with individuals representing state highway agencies and relevant industry groups. NCHRP Report 453 (*Saeed 2001*) served as the initial guide for the literature search. The telephone interviews provided information on agencies' practices regarding recycling of RAP and RCP as unbound aggregate in base/subbase layers. The approach also included the selection of pavement performance parameters that may be influenced by the properties of recycled aggregate in unbound pavement layers, the identification and evaluation of recycled aggregate properties that affect pavement performance parameters, and identification and evaluation of current aggregate test procedures and potential techniques that can be used to measure relevant recycled aggregate properties. Complete research results are available elsewhere (*Saeed 2008*); this paper summarizes laboratory testing and the minimum material standards developed for RAP and RCP unbound base/subbase layers.

SELECTION OF TEST METHODS AND MATERIALS

Earlier work by Saeed (2008) and Saeed et al (2006, 2001) summarizes that unbound pavement layers in flexible and rigid pavements are used to provide a 1) working platform for construction, 2) frost blanket (frost-free layers), 3) drainage layer and 4) structural layer for the pavement system. Fatigue cracking, rutting/corrugations, depressions and frost heave of flexible pavements can be attributed, at least in part, to poor performance of granular base and subbase layers. Cracking, pumping, faulting, loss of support, frost heave and erosion of rigid pavements can also be attributed to poor performance of granular base and subbase layers.

Saeed (2008) summarized that the performance of pavements built with unbound base and subbase layers incorporating RAP and RCP can be affected by physical, chemical and mechanical properties of the recycled aggregate particles (particle properties) and the proportion in which they are mixed with virgin aggregate (zero to 100 percent). Recycled aggregate mass properties that are considered relevant to their use in unbound pavement layers are listed in Table 1; Table 2 shows the linkage between these properties (performance parameters) and laboratory test measures.

TABLE 1. Recycled material mass properties versus applicability (*after NCHRP 598*).

Mass Property of Material	Relevance of Mass Property to the Use of Recycled Material as					
	Structural Layer	Construction Platform	Drainage Layer	Frost Blanket	Control Pumping	Select Fill
Shear Strength	Y	Y	N	N	N	N
California Bearing Ratio, CBR	Y	Y	N	N	N	Y
Cohesion & Angle of Internal Friction	Y	N	N	N	N	N
Resilient or Compressive Modulus	Y	Y	Y	Y	Y	Y
Density	Y	Y	N	Y	Y	Y
Permeability	N	N	Y	Y	Y	N
Frost Resistance	Y	N	Y	Y	N	Y
Durability Index	Y	N	Y	Y	Y	N
Resistance to moisture damage	Y	N	N	N	N	N

The project team considered performance predictability, precision, accuracy, practicality and cost to select test methods to determine the performance-related properties of RAP and RCP; complete details are provided by Saeed (2008) elsewhere. The following tests were selected:

- Screening tests (sieve analysis, moisture-density relationship)
- Toughness (Micro-Deval)
- Stiffness (resilient modulus)
- Shear strength (static triaxial and repeated load at OMC and saturated)
- Frost susceptibility (tube suction)

Tests, following appropriate ASTM (1995) or AASHTO (1995) protocols, were conducted on RAP and RCP containing three different constituent aggregates (crushed limestone, granite and gravel) to provide a range of performance from excellent to poor; test materials with expected performance potential included:

- RCP with granite from South Carolina (RCP-GR-SC), excellent
- RCP with limestone from Illinois (RCP-LS-IL), very good
- RCP with gravel from Louisiana (RCP-GV-LA), good
- RAP with limestone from Mississippi (RAP-LS-MS), good
- RAP with granite from Colorado (RAP-GR-CO), fair
- RAP with gravel from Louisiana (RAP-GV-LA), poor

The test gradations included a typical dense graded base layer (DGBL) and an open graded drainage layer (OGDL). Tests were also conducted on 50 percent blends of recycled materials with virgin aggregate. A typical limestone aggregate from Georgia with excellent performance history was used for blending.

Table 2. Linkage between aggregate properties and performance (*after NCHRP 598*).

Pavement type	Performance parameter	Related aggregate property	Test measures
Flexible	Fatigue Cracking	Stiffness	Resilient modulus, Poisson's ratio, gradation, fines content, particle angularity and surface texture, frost susceptibility degradation of particles, density
	Rutting, Corrugations	Shear Strength	Failure stress, angle of internal friction, cohesion, gradation, fines content, particle geometrics (texture, shape, angularity), density, moisture effects
	Fatigue Cracking, Rutting, Corrugations	Toughness	Particle strength, particle degradation, particle size, gradation, high fines
		Durability	Particle deterioration, strength loss
		Frost Susceptibility	Permeability, gradation, percent minus 0.02 mm size, density, nature of fines
		Permeability	Gradation, fines content, density
Rigid	Cracking, Pumping, Faulting	Shear Strength	Failure stress, angle of internal friction, cohesion, gradation, fines content, particle geometrics (texture, shape, angularity), density, moisture effects
		Stiffness	Resilient modulus, Poisson's ratio
		Toughness	Particle strength, particle degradation, gradation
		Durability	Particle deterioration, strength loss
		Permeability	Gradation, fines content, density
	Cracking, Pumping, Faulting, Roughness	Frost Susceptibility	Permeability, gradation, percent minus 0.02 mm size, density, nature of fines

SUMMARY OF TEST RESULTS

Because RAP and RCP materials are generally used as an unbound structural layer, most laboratory tests were conducted on samples meeting a target gradation similar to a DGBL. A few tests were conducted on samples prepared to a gradation similar to an OGDL. The target gradations are based on typical gradations provided in the *Aggregate Handbook (NSA 1991)*, which were adjusted based on results of telephone interviews and literature search.

Test results and observations are summarized in the following sections; laboratory test results and analysis of data are provided by Saeed (*2008*) elsewhere.

Toughness and Abrasion Resistance

Material toughness and abrasion resistance characteristics were determined using the Micro-Deval test (AASHTO TP58-00). RCP and RAP in as-received condition exhibited more material loss than virgin aggregate material or 50 percent blends of RAP and RCP with virgin aggregate. For some materials, it was apparent that the test results were affected by the amount of fines that were produced during testing.

The test method was able to differentiate (statistical significance $\alpha=0.05$) between RAP and virgin aggregate and recycled and virgin aggregate. The test did not differentiate between other tested material combinations. For example, the test method could not differentiate between recycled materials.

Durability

Recycled materials and virgin aggregate durability characteristics were determined using the Canadian Freeze-Thaw test (MTO LS-614). These data show that RCP samples and 50 percent blends of RCP with virgin aggregate had the largest material loss. It appears that the results are affected by the production of excess fines from recycled materials during testing resulting from the disintegration of the cement paste on the aggregate particles. The test method differentiated between recycled and virgin aggregates and between RCP and virgin aggregate but was insensitive to other material combinations.

Frost Susceptibility

Frost susceptibility of the recycled and virgin aggregates was determined using the tube suction test (*Saarenketo 1995*). Aggregates were considered acceptable, marginal, or unacceptable if the dielectric constant was less than 10, between 10 and 16 and greater than 16, respectively. Statistical test results on Tube-Suction test data at 5 percent test significance indicated that the test method differentiated between recycled (RAP, RCP, combined) and virgin aggregates.

Static Triaxial Test

The static triaxial test was conducted in accordance with AASHTO T 234 on each sample at confining stresses of 0, 34.5, 103.4 kPa to determine the shear strengths at OMC. The test yielded different results for different conditions, although relatively small in some cases. For example, similar results were obtained for materials containing limestone and gravel. Statistical analysis of test data also showed similar trends. Test results were different for RAP and RCP samples with a p-value of 0.027 (statistically different). However, other comparisons were statistically inconclusive.

Repeated Load Triaxial Test

The repeated load triaxial tests were conducted to obtain a relative measure of the resistance of tested materials to permanent deformation. The test procedure, described in NCHRP Report 453 (*Saeed 2001*), was modified to ensure proper testing of recycled materials; sample preparation and seating load requirements were modified.

The number of load repetitions required to cause failure (defined as 10 percent permanent strain) was also used to evaluate resistance to permanent deformation (*Saeed 2008*). Virgin aggregate DGBL exhibited the highest resistance to permanent deformation. The test was terminated after 10,000 cycles at which the average permanent deformation was only 3.67 percent. Figure 1 shows the repeated load triaxial test results for wet (saturated) and dry (at OMC) tests on DGBL gradations.

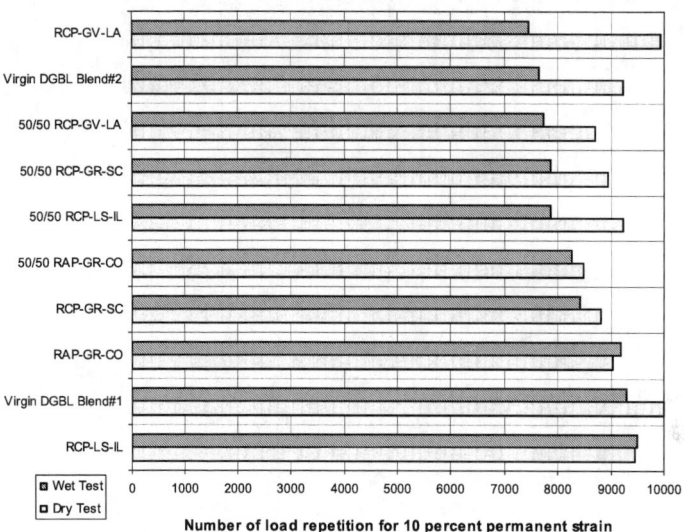

Figure 1. Failure load repetitions for wet and dry repeated load triaxial tests.

Figure 2 shows the "mean" maximum deviator stress at 15 psi (103.4 kPa) confining pressure sorted by aggregate type, tested gradation and material for tests conducted on unsaturated and saturated samples (OGDL samples were tested under unsaturated conditions only). Test results indicate differences between different conditions; although, the differences between saturated and unsaturated tests were relatively small. Statistical significance test results indicated similar trends; the test method correctly differentiated between different materials.

The materials selected for laboratory tests were expected to provide a range of expected performance as indicated by shear strength. At 1 percent strain, 100 percent RAP material had the lowest strength, followed by 50 percent blends of RAP with virgin aggregate. The 100 percent RCP and virgin aggregate samples had the highest strengths; the 50 percent RCP blends with virgin aggregate had the second highest shear strengths. Shear strengths estimated at 3 percent strain provided somewhat different order.

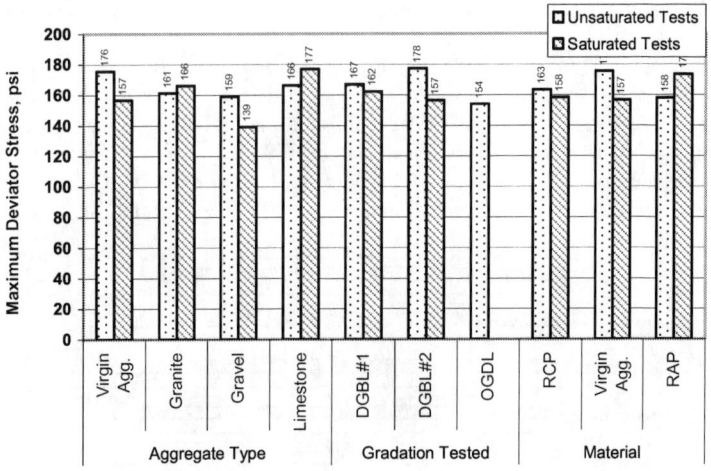

Figure 2. Repeated load triaxial test results at 15 psi confining pressure.

Resilient Modulus Test

The resilient modulus (or stiffness) was estimated at different bulk stresses from data obtained during repeated load triaxial tests; RCP-GR-SC was the least stiff material. Generally, virgin aggregate and 50 percent blends of recycled materials with virgin aggregate exhibited higher stiffness than 100 percent recycled materials. Statistical analysis of resilient modulus data indicate that test data reveal differences between different materials.

Test Method Selection Summary

The performance potential of an unbound pavement layer depends on its dry and wet shear strength, resistance to freeze-thaw (durability), toughness and frost susceptibility. These properties were evaluated using selected tests in a laboratory investigation. Also, screening tests were conducted to characterize recycled materials. Based on results of the laboratory investigation, Saeed (2008) found the following tests to relate to performance:

- Screening tests (sieve analysis, moisture-density relationship)
- Toughness (Micro-Deval test)
- Stiffness (Resilient modulus)

- Shear strength (static triaxial and repeated load at OMC and saturated)
- Frost susceptibility (Tube suction test)

SELECTION OF RECYCLED MATERIALS FOR INTENDED USE

Recycled materials can be selected for use in a particular traffic and climatic condition. NCHRP Report 453 (*Saeed 2001*) summarizes an approach for evaluating aggregates using selected test parameters, performance ratings and traffic and climatic categories. In this approach, tests are conducted in sequence and results are compared to suggested performance levels for specific traffic and climatic ranges. Three traffic levels are proposed:

- Low traffic (<100,000 ESALs/years),
- Medium traffic (100,000 – 1,000,000 ESALs/year) and
- High traffic (>1,000,000 ESALs/year).

The climatic conditions of moisture (high/low) and temperature (freezing/not freezing) are based on the AASHTO (*1993*) definitions. Table 3 shows the significance levels of traffic, moisture and climate combinations on a scale of 1 to 4, where 4 is most significant and 1 least significant on aggregate performance potential (*Saeed 2008, 2001*). Proposed ranges for selected test parameters that relate to performance for various levels of climatic and traffic condition are shown in Table 4; these ranges determine the traffic and climatic conditions where these recycled materials and their blends can be used.

Table 3. Significance level of intended use on aggregate performance potential.

Temperature Condition[1]	Moisture Condition[2]	Traffic[3]		
		High	Medium	Low
Freezing	High	4	4	3
	Low	4	3	2
Non Freezing	High	3	2	2
	Low	3	2	1

Table 4. Recommended tests and test parameters for levels of intended use.

Tests (Test Parameters)		Traffic	H		M		H		L	M	L
		Moisture	H	L	H	L	H		L	H	L
		Temperature	F		NF		F			NF	
Micro-Deval Test (percent loss)			< 5 percent		< 15 percent		< 30 percent			< 45 percent	
Tube Suction Test (dielectric constant)			≤ 7		≤ 10		≤ 15			≤ 20	
Static Triaxial Test (Max. deviator stress)	OMC, σ_c = 5psi		≥ 100 psi		≥ 60 psi		≥ 25 psi			Not required	
	Sat, σ_c = 15psi		≥ 180 psi		≥ 135 psi		≥ 60 psi			Not required	
Repeated Load Test (Failure deviator stress)	OMC, σ_c = 15psi		≥ 180 psi		≥ 160 psi		≥ 90 psi			Not required	
	Sat, σ_c = 15psi		≥ 180 psi		≥ 160 psi		≥ 60 psi			Not required	
Stiffness Test (Resilient modulus)			≥ 60 ksi		≥ 40 ksi		≥ 25 ksi			Not required	

Based on laboratory test results summarized in NCHRP Report 598 (*Saeed 2008*) and Tables 3 and 4, the following intended use of tested recycled materials is highlight.

- Toughness Test
  o Recycled materials are generally appropriate for use in medium to low traffic conditions in non-freezing climates with low and high moisture contents.
  o RCP-GR and RAP-GV seems appropriate for use in high traffic areas with non-freezing temperatures or in low and medium traffic areas in freezing climates with low moisture conditions.
- Frost Susceptibility Test
  o RCP materials can be used in medium traffic with no freezing temperatures (significance level 2) and low traffic with no freezing temperatures and low moisture (significance level 1).
  o RAP and 50 percent blends with virgin aggregate are appropriate for use in high traffic conditions.
- Static Triaxial Test
  o Generally RCP materials and their blends with virgin aggregate are appropriate for use in extreme traffic and climatic conditions.
  o RAP is appropriate for use in conditions representing significance level of 3 (high traffic in non-freezing temperatures, medium traffic level in freezing temperature in the presence of low moisture and low traffic level in freezing temperatures).
- Repeated Load Triaxial Test
  o RCP and 50 percent RCP blend with limestone and granite are appropriate for use in conditions representing significance level 3 (i.e., high traffic level in non-freezing temperatures, medium traffic level in freezing temperature in the presence of low moisture and low traffic level in freezing temperatures).
  o RAP and 50 percent RAP blends are generally appropriate for use in conditions representing significance level 2.

- Material Stiffness
  - Most recycled materials and 50 percent blends with virgin aggregate were deemed appropriate for use in conditions representing significance level 3.

No attempt was made to further separate the results based on the constituent aggregate type in the recycled materials due to limited data.

SUMMARY AND RECOMMENDATIONS

This research highlighted that recycled materials (RAP and RCP) are a viable alternative to virgin aggregate unbound base/subbase construction. Properties of recycled aggregates that affect their performance as unbound base/subbase layers include shear strength, stiffness, toughness, durability, frost susceptibility and permeability. Shear strength and stiffness (resilient modulus) have a much greater influence on the performance of an unbound aggregate layer than the other properties.

Tests that relate to the field performance of recycled materials used in unbound base/subbase layers include 1) screening tests (sieve analysis, moisture-density relationship), 2) toughness (Micro-Deval test), 3) stiffness (resilient modulus), 4) shear strength (static triaxial and repeated load at OMC and saturated) and 5) frost susceptibility (tube suction test). Results from these tests could be used to determine the appropriate use location of most RAP and RCP materials available.

The proposed ranges of traffic and climatic conditions where recycled materials (RAP and RCP) and their blends can be used are based on work reported in NCHRP Reports 453 (*Saeed 2001*) and 598 (*Saeed 2008*). These ranges need to be confirmed and validated using accelerated pavement tests and/or in-service test pavement evaluations. Results from accelerated pavement tests and/or in-service test pavement evaluations will also help refine these ranges.

ACKNOWLEDGEMENTS

This research was conducted under the auspices of the National Cooperative Highway research Program (NCHRP) Project 4-31; complete research results are documented in NCHRP Report 598, *"Performance Related Tests of Recycled Aggregates for Use in Unbound Pavement Layers."* The authors would like to acknowledge NCHRP, the technical panel and other project team members for their help and support throughout this research effort.

REFERENCES

AASHTO (1993). American Association of State Highway and Transportation Officials, Guide for Design of Pavement Structures, Washington, DC.

AASHTO (1995). American Association of State Highway and Transportation Officials (AASHTO), *Standard Specifications for Transportation Materials and*

Methods of Sampling and Testing, Part I: Specifications. 17th Edition, Washington, D.C.

ASTM (1995). American Society for Testing and Materials, "Road and Paving Materials; Paving Management Technologies," *Annual Book of ASTM Standards*, Vol. 4.03, Philadelphia, PA.

Dumitru, I., Munn, R. and Smorchevsky, G. (2000). "Progress Towards Achieving Ecologically Sustainable Concrete and Road Pavements in Australia." WASCON 2000, The Fourth International Conference on the Environmental and Technical Implications of Construction with Alternative Materials, Leeds/Harrogate, United Kingdom, Proceedings, pp 107-120.

Ferragut, T. (2001). "Partnership for Sustainability – A New Approach for Highway Materials." *Federal Highway Administration Report on the Houston Workshop, October 9-11, 2000*, Washington, D.C., 36 pp.

Holtz, K. and Eighmy, T. T., "Scanning European Advances in the Use of Recycled Materials in Highway Construction." Public Roads, Vol. 64, No. 1 (July-August 2000), Federal Highway Administration, Washington, DC (2000).

National Stone Association, *The Aggregate Handbook*. National Stone Association, Washington, DC (1991).

Saarenketo, T., and Scullion, T. Using Electrical Properties to Classify the Strength Properties of Base Course Aggregates, Research Report 1341-2. Project No. 1341. Texas Transportation Institute, College Station, TX. July 1995.

Saeed, A., Hall, J.W. and Barker, W., 2001. "Performance-Related Tests of Aggregates for Use in Unbounded Pavement Layers," *NCHRP Report 453*, Project 4-23, National Cooperative Highway Research Program, Washington D.C.

Saeed, A., Hammons, M, Rufina, D., and Poole, T., 2004. "Evaluation, Design, and Construction Techniques for Airfield Concrete Pavement Used as Recycled Material for Base," *Final Report IPRF-01-G-002-03-5*, Innovative Pavement Research Foundation, Airport Concrete Pavement Technology Program, Skokie, Illinois.

Saeed, A, 2008. "Performance Related Tests of Recycled Aggregates for Use in Unbound Pavement Layers," *NCHRP Report 598*, Project 4-31, National Cooperative Highway Research Program, Washington D.C.

Transportation Research Board (2007). "TR News, Highway Design and Construction – A 2020 Vision," *Transportation Research News*, Nov. - Dec. 2007, Number 253, The National Academies, Washington, D.C.

USGS 2007   United States Geological Survey, 2007. "2006 Minerals Yearbook – Stone, Crushed," USGS, United States Department of the Interior, Washington, D.C.

Effect of Aircraft Load Wander on Unbound Aggregate Pavement Layer Stiffness and Deformation Behavior

Phillip Donovan[1] and Erol Tutumluer[2], M. ASCE

Abstract

Load wander in airport pavements is much wider than highway wander due to the varying paths of different aircraft gear configurations and applied wheel loading locations. Full scale pavement testing of aircraft loads at the FAA's National Airport Pavement Test Facility (NAPTF) indicates that wander can negate the stiffening in unbound granular layers, known as the shakedown effect, and make them prone to developing increased deformations on subsequent aircraft passes when compared to channelized highway traffic. As part of the research activities at the FAA's Center of Excellence for Airport Technology (CEAT) established at the University of Illinois, dynamic response data from airport pavement test sections were collected due to passing of each of the 6-wheel B777 type and the 4-wheel B747 type gears for various combinations of applied load magnitudes and loading sequences (application order and stress history effects), traffic directions, gear spacings, and wander positions and sequences. The field data analyzed showed that the permanent deformation during a complete wander cycle was negated due to aircraft wander, indicating recurring particle movements and rearrangements in the unbound pavement layers. Using the multi-depth deflectometer data and the concepts of base damage index and base curvature index from heavy weight deflectometer testing, the load wander was determined to cause high percentages of residual responses per wheel pass in the granular layers that were indicative of reduced strength and modulus properties for potentially higher damage accumulations during the initial stages of NAPTF test section trafficking to failure.

Keywords: Airport pavements, aircraft wander, aggregate base/subbase, deformation

Introduction

In airport pavements, individual aircraft load wander patterns are wide ranged and variable depending on the different aircraft gear configurations, and are normally distributed in nature to make runways and taxiways experience non-channelized traffic loading. Early data collected in the 1970's indicate wander widths of 70 in. (1778 mm) for taxiways and 140 in. (3556 mm) for runways (Ho Sang, 1975). The standard deviations were reported by Ho Sang (1975) as 30.5 in. (775 mm) for a taxiway and 60 in. (1524 mm) for a runway. The wander width was defined by the zone containing 75% of the aircraft centerlines (1.15 standard deviations on either

---

[1] Graduate Research Assistant, PH (217) 333-6975; FAX (217) 333-1924; email: pdonova3@uiuc.edu
[2] Associate Professor, PH (217) 333-8637; FAX (217) 333-1924; email: tutumlue@uiuc.edu
   Department of Civil and Environmental Engineering, University of Illinois, Urbana, IL 61801.

side of the mean value). Note that it is not only individual aircraft wander that affects pavement performance. Each aircraft has a unique gear configuration and accordingly, is expected to induce additional load wander that is not associated with lateral deviation of the individual aircraft position. To illustrate this fact and also to further elaborate the difficulty of studying wander effects, Figure 1 shows the wide variation of the transverse gear wheel locations of some common large aircraft.

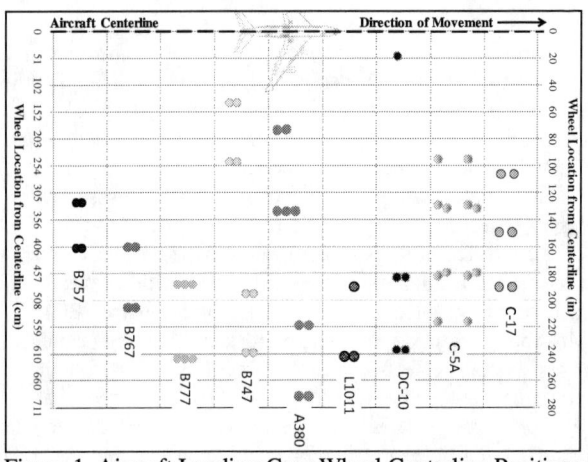

Figure 1. Aircraft Landing Gear Wheel Centerline Positions

Pavement unbound aggregate layers are said to shakedown when the application of additional loads cause the granular layer to consolidate, gain strength with time, and stabilize with little additional residual deformation. This process is seen in the field as well as with repeated load triaxial testing. The recent successful research efforts on the aggregate shakedown concept by Werkmeister et al. (2002) identified three zones of shakedown as: A – plastic shakedown, B – plastic creep, and C – incremental collapse. In range A (plastic shakedown), the residual strain rate decreases quickly and eventually, the layer shows no further residual deformation with additional load repetitions. Range B (plastic creep) initially shows a decreasing residual strain rate but as the number of load cycles increase, the residual strain rate resumes an upward climb, eventually leading to incremental collapse. This behavior has been attributed to grain abrasion caused by the large resilient deformations seen in this stress range. The grain abrasion is thought to decrease the angle of internal friction by polishing the grain contact points thus lowering the coefficient of friction between grains. This causes more residual deformation with additional load cycles without increasing the applied stress. In range C (incremental collapse), it is probable that due to the high stress range both grain abrasion and particle crushing combine to quickly destroy, i.e., permanently deform in an excessive manner, an unbound aggregate layer. This region is characterized by a slower reduction in the residual strain rate than range A or B and a quick resurgence of the strain rate after a very limited number of load cycles. It is also likely that for all shakedown ranges, any particle rearrangement that occurs due to stress will relieve some small amount of the residual compressive stress in an unbound layer that was induced by compaction and preloading of the layer; which in turn will cause additional rutting.

Data from testing at the Federal Aviation Administration's (FAA's) National Airport Pavement Test Facility (NAPTF) using new generation aircraft loads on asphalt pavement indicate that a sequential wander pattern causes residual deformation to be recovered, potentially reducing or even negating the shakedown effect. What has been seen is that the downward residual deformation (rutting) caused by a pass of heavily loaded landing gear is canceled by the upward residual deformation (heave) resulting from the pass of the same gear offset by wander (Hayhoe and Garg, 2002). This interaction indicates a rearrangement of the particles in the unbound layers of the pavement system. The particle rearrangement in turn reduces the strength of the unbound layer causing future load applications to cause more residual deformation, referred to hereafter as the "anti-shakedown" effect. It is thought that the strength reduction is due to two factors: (1) a less dense particle matrix and (2) grain abrasion which reduces the coefficient of friction between particle contact points, as seen in range B shakedown behavior.

This paper presents NAPTF full scale pavement test section findings from the analyses of heavy weight deflectometer and multi-depth deflectometer data and identifies deformation trends in the unbound aggregate layers due to applied aircraft gear loading with wander. Important conclusions will be drawn on the effects of load wander on pavement damage accumulation to eventually help improve our understanding of airport pavement damage mechanisms and performance prediction.

NAPTF Testing

The FAA's NAPTF located at the William J. Hughes Technical Center close to the Atlantic City International Airport was built to analyze the effects of New Generation Aircraft (NGA) on pavements. The NAPTF was constructed to generate full-scale tests in support of the investigation of airport pavements subjected to complex NGA gear loading configurations. Full-scale pavement tests were conducted using a specially designed 1.2-million-pound (5.3 MN) test vehicle which can apply loads of up to 75 kips (333.6 kN) per wheel on two landing gears with up to six wheels per gear (total of 12 wheels for a load capacity of 900 kips [4 MN]).

The test vehicle at NAPTF is supported by rails on either side which allow the load to be varied according to the testing protocols. The vehicle can be configured to handle single, dual, dual-tandem, and dual-tridem loading configurations. The wheel and gear spacing can be varied. The maximum tire diameter is 56 in. (1420 mm) and maximum tire width is 24 in. (610 mm). Vehicle control can be automatic or manual. Traffic tests can be run in a fully automatic control mode at a travel speed of 5 mph (8 km/h). This speed represents aircraft taxiing from the gate to the takeoff position. It is during this maneuver that maximum damage occurs to the pavement because the aircraft is fully loaded and speed is low. Wheel loads are programmable along the travel lanes and the lateral positions of the landing gears are variable up to plus or minus 60 in. (1524 mm) from the nominal travel lanes to simulate aircraft wander.

The first series of tests conducted are referred to as Construction Cycle 1 (CC1) tests. The Boeing 777 (B777) type landing gear tested in the north testing lane was a six wheel dual-tridem configuration with dual wheel spacing of 54 in. (1372 mm) and tridem axle spacing of 57 in. (1448 mm). The wheel loads were set to 45 kips (200.2 kN) and the tire pressure was set to 189 psi (1.3 MPa). The complete six wheel strut load was 270 kips (1.2 MN). Traffic was applied at 5 mph (8 km/h).

The south wheel track was loaded with a four-wheel dual-tandem type representing a Boeing 747 (B747) gear configuration. The dual wheel spacing was 44 in. (1,118 mm) and the tandem axle spacing was 58 in. (1,473 mm). Wheel loads of 45 kips (200.2 kN) per wheel similar in magnitude to the B777 loading case were applied to give a strut load of 180 kips (800.8 kN). The load carriage containing both struts is a continuous system, therefore traffic speed for the B747 and B777 match.

NAPTF Multi-depth Deflectometer Data

The CC1 NAPTF nominal pavement test section details are shown in Figure 2. Multi-depth deflectometers (MDDs) were used to record the deflections of the pavement system under applied traffic loading. The MDD sensors were located at critical locations within the asphalt, unbound aggregate, and subgrade layers (see Figure 2). There were seven MDD sensors in each MDD sensor stack with two stacks used per test section per gear path and with an additional stack recording the response at the center of the test section where no traffic traveled. The MDDs work by recording the deflection of the individual sensors in relation to an "anchor" sensor that is buried below the zone of influence of the anticipated loads. The surface sensor is actually the only sensor to be directly connected to the anchor; the other sensors measure deflections in relation to the surface sensor. The absolute movement of an individual sensor is then calculated by subtracting the sensor reading from the surface sensor reading. Accordingly, individual layer response is calculated by subtracting the lower sensor reading from the higher sensor reading.

NAPTF Load Wander Patterns

To account for aircraft wander, the test passes or load applications were divided into nine wander positions spaced at intervals of 9.84 in. (250 mm). One complete wander pattern consists of 66 vehicle passes (33 traveling East and 33 traveling West) with five wander sequences per wander pattern. Each pass at a wander position traveled from the West end to the East end first and then traveled back along the same path East to West. The gear was then moved to the next wander position. Figure 3 shows the complete wander pattern, the wander position trafficking order, and the wander sequences. A wander sequence is based on a block of trafficking which simulates the normal distribution. Each position was traveled a different number of times based on a normal distribution with a standard deviation that is typical of multiple gear passes on airport taxiways, i.e., 30.5 in. (775 mm). The nine wander rows covered 87% of all traffic (approximately 1.5 standard deviations). Figure 4 shows the wander positions applied at the NAPTF full-scale pavement tests to scale for both the B777 type dual tridem and B747 type dual tandem (hereafter referred to as the B777 and B747 gear/wheel) gear trafficking paths.

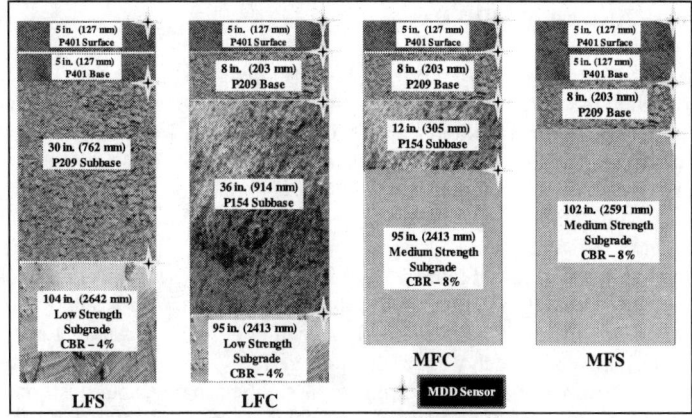

Figure 2. Cross Section Details of NAPTF CC1 Test Sections

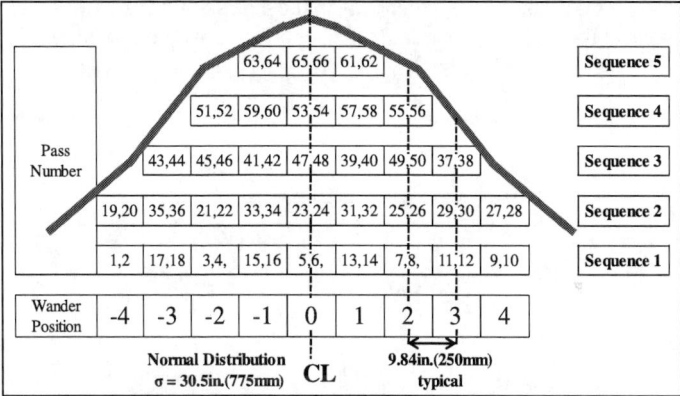

Figure 3. 66 Pass Wander Pattern and Wander Sequence

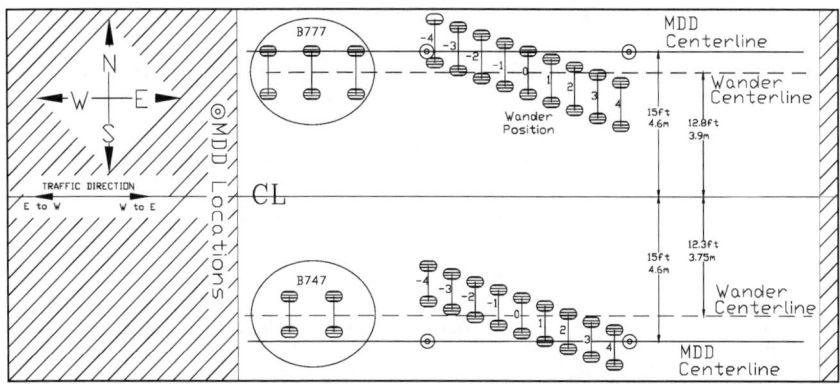

Figure 4. B777 (top) and B747 (bottom) Gear Locations for Each Wander Position

Preliminary Analysis of MDD Data

The initial assessment of the collected MDD readings seemed erratic and random; however, when the data were separated by wander position, travel direction, and wander sequence distinct patterns emerged. Figure 5a, for example, shows the separation of the residual deflection data for wander position 0 for the P209 base layer and Figure 5b shows the separation of the same data for the P154 subbase layer. It was only through the separation of the MDD data that analyses of the layer responses to individual wander positions and sequences could be clearly accomplished as indicated in Figure 5. The P154 subbase layer was often thicker and certainly of lower quality than the P209 base layer and thus one would expect more residual deformation in this layer as clearly shown in Figure 5b. Another observation was that the first pass on each wander position in the West to East direction typically caused the most response and the return pass along the same wander position showed significantly less residual deflection. This finding clearly indicates that shakedown was occurring in the unbound aggregate layers; however, because the wander position shifted every other pass, the granular layer did not stabilize and thus the residual deformations continued to increase. This phenomenon would not be visible without data separation.

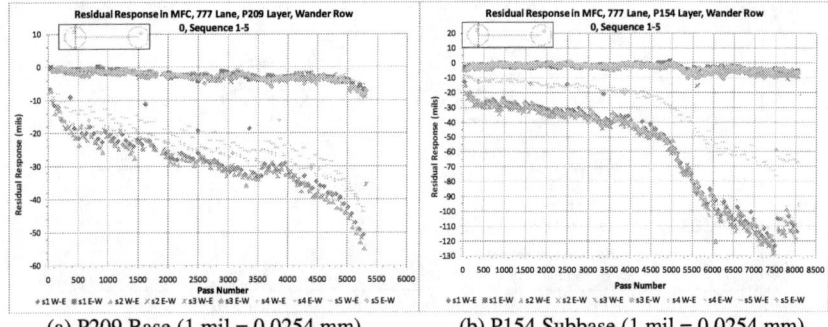

(a) P209 Base (1 mil = 0.0254 mm)   (b) P154 Subbase (1 mil = 0.0254 mm)
Figure 5. Residual Response Accumulated in the MFC Section Granular Layers

Figure 5b clearly shows that shakedown did occur when the wander pattern was kept narrow. As the 66 pass wander pattern goes from sequence 4 to sequence 5 (s4 to s5), the residual deflection caused by the West to East pass on wander position 0 decreased 50% yet the other sequences all had similar responses under wander position 0. Note that the gear loading of sequence 5 (s5) placed one gear wheel closest to the MDD line along wander positions -1, 0, and 1 (see Figure 3).

NAPTF Heavy Weight Deflectometer Testing

The heavy weight deflectometer (HWD) is a nondestructive pavement evaluation device similar to a falling weight deflectometer (FWD) with the only difference being a larger loading capacity for the HWD machine. The principle behind the FWD tests is applicable as for any weight/deflection measurement test; deflection can be used as an indicator of the condition and strength of a pavement system. Higher deflections point to a degraded or weaker pavement system. The FWD device uses instantaneous loads (20 to 60 milliseconds) to induce deflections. Deflection can be induced by static loads, but the falling weight is indicative of the transient nature of actual traffic and the impact load can be changed by varying the drop height. As part of the pavement nondestructive evaluation efforts, HWD tests can be conducted regularly during testing to help determine the degradation of the pavement system.

The FAA used a KUAB Model 240 HWD in their CC1 testing and has a standard procedure for routine HWD tests. According to Garg and Marsey (2004), they use a 12 in. (305 mm) segmented loading plate, a pulse width of 27-30 msec., and four drop heights consisting of a 36 kip (160 kN) seating drop followed by impact loads of 12 kips, 24 kips, and 36 kips (53.4, 106.8, and 160 kN). The deflection basin is recorded for each drop with sensors placed radially from the center of the load at 0 in. (D_0), 12 in. (D_1), 24 in. (D_2), 36 in. (D_3), 48 in. (D_4), and 60 in. (D_5) (0, 30, 60, 90, 120, and 150 cm) offsets. As the distance of the sensor increases from the center of the load, the depth to the effective layer properties increases. For example, the deflection of sensor D_2 may indicate the combined properties of the layers deeper than 6 in. (15 cm) while deflection of sensor D_4 could correlate to layer properties deeper than 12 in. (30 cm). The deflection at the center of the loading plate, D_0, is a function of the loading plate diameter, the applied load, and the pavement structure as a whole while deflection at the outermost sensor, D_5, is predominately controlled by the subgrade properties (Garg and Marsey, 2002). Depending on the thickness of the pavement system layers, sensors D_1 to D_4 may provide insight into base and subbase layer properties.

Analyses of the MDD and HWD Data

The analyses of the MDD data from the MFC test sections indicate that the P209 base and P154 subbase layers did not consolidate and shakedown during testing (Donovan and Tutumluer, 2008). Figure 6 shows the residual deflection response in each layer of the MFC test section on the B777 lane, wander position 0, and West to East loading direction for wander sequences 1-4. Figure 6 shows the percentage of the total residual response computed in each layer. Note that wander position 0 has one gear wheel directly over the MDD location. If shakedown did occur in the unbound aggregate layer, one would expect that the percent of residual response by layer would decrease for the P209 and P154 layers. However, as shown in Figure 7, this did not happen with increasing pass number. The percent of the residual response remained relatively constant for each layer as the testing progressed, i.e., 8% for P401, 36% for P209, 46% for P154, and 10% for the subgrade. An exception can be seen in the P401 and P209 layers (see Figure 7). An apparent shakedown or compaction occurred in the P401 layer as the initial 1000 passes shows a decrease in the percent of residual response from the P401 layer. Interestingly, this compaction in the P401 layer was counteracted by an increase in the percent of residual response by the P209 layer for the same 1000 passes. In general, there is a slight decrease in the percent of residual response by the P154 while the percent of residual response by the subgrade is increasing, and this correlates with the subgrade failure of the MFC section at around 12,000 passes.

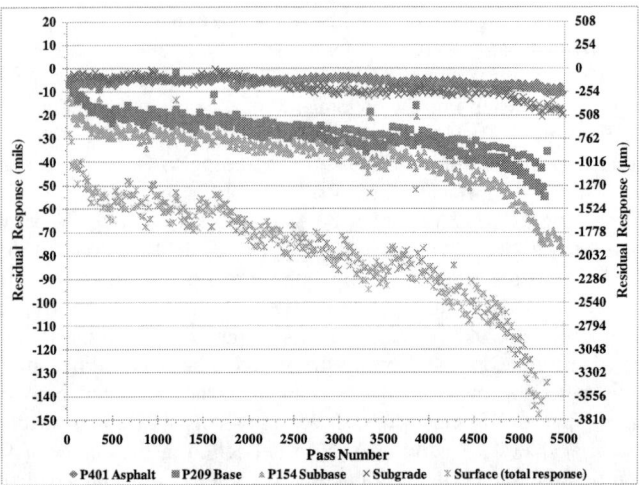

Figure 6. Residual Response of Each Layer on the MFC B777 Lane (Wander Position 0, West to East Loading Direction, and Wander Sequences 1-4)

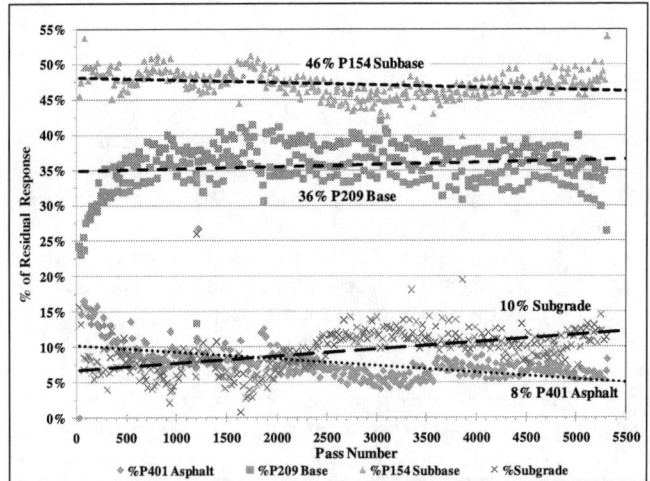

Figure 7. Layer Percentage of the Total Residual Response on the MFC B777 Lane (Wander Position 0, West to East Loading Direction, and Wander Sequences 1-4)

The MFC section B777 lane was declared failed after 12,000 passes due to 4 to 6 in. (10 to15 cm) of rut development with upheaval outside the traffic lane and asphalt surface cracking within the traffic lane (Hayhoe and Garg, 2003). Due to the excessive movements of the layers, the MDD data are only reliable up to 5500 passes, however, Figures 6 and 7 show that shakedown did not occur and that failure was likely the result of shakedown range B behavior. Again, range B shakedown is characterized by an initial shakedown phase where residual deflections taper off, but then a dramatic increase in residual deflections after a limited number of cycles caused by the gradual reduction in the bearing capacity of the unbound materials from constant particle rearrangement. Post traffic trench studies found that the subgrade had penetrated into the P154 subbase layer due to lateral movement of the P154 unbound aggregate particles after the tensile stress caused by loading exceeded the residual compressive stress in the layer (Hayhoe and Garg, 2003). From the shakedown range comparison, this is analogous to the final range B behavior. Though this failure may seem to be range C, incremental collapse, the number of passes was too high to correlate with the sudden dramatic failure seen in range C.

Figures 8 and 9 show data similar to those given in Figures 6 and 7 but this time for the MFC B747 lane. Only wander sequences 1-3 are shown for clarity in Figure 6 as wander sequences 4 and 5 showed reduced residual deformation due to shakedown during the complete 66 pass wander pattern (which was negated once the pattern restarted on pass 67). Wander position 1 had one gear wheel directly over the MDD. Once again, the percent of residual response in each layer remains relatively constant. This time, however, the P154 layer dominates the residual response with 53% of the residual response coming from the P154 layer, 20% from the subgrade, 18% from the P209 layer, and 13% from the P401 layer. It is interesting to note the reduction in percent of residual response from the P401 layer and the gradual increase in the percent of residual response by the P209 layer. This is likely due to range B behavior where the constant particle rearrangement slowly degraded the P209 layer while at the same time the P401 layer consolidated due to traffic.

HWD data and the associated deflection basin parameters can be used to determine the degradation of layers in a pavement system (Gopalakrishnan and Thompson, 2004). Using HWD test results, Gopalakrishnan and Thompson (2004) investigated the conventional Base Damage Index (BDI, D_1-D_2) and Base Curvature Index (BCI, D_2-D_3) values for the NAPTF test sections and found that the BDI was 10-20% higher than the BCI. However, the most common FWD algorithms may not account for the thicker pavement layers tested in the NAPTF since D_3 may no longer represent subgrade behavior. Therefore, in our analyses of the HWD data, a modified BDI was used to identify the degradation in the base/subbase and a modified BCI for the subgrade to more clearly see the relative contributions in damage accumulation.

The BDI is related to the base layer modulus and is normally calculated as D_1-D_2 (deflection at D_1, 12 in. or 305 mm, minus D_2, the deflection recorded at 24 in. or 610 mm); however due to the thickness of the layers in the NAPTF tests, the BDI has been redefined as D_1-D_3 (D_3 is the deflection at 36 in. or 915 mm). Because non-traffic lane HWD tests were conducted at the same time and pavement temperature as the traffic lane tests, the BDI value can be normalized for temperature by dividing the traffic lane BDI by the non-traffic lane BDI. The BDI is then compared to the BCI which is also modified and redefined as D_3-D_5 (instead of D_2-D_3) and is also divided by the non-traffic lane BCI to normalize for temperature. The BDI and BCI are also normalized with respect to load by dividing the respective traffic lane or non-traffic lane value by their D_0 value. The BDI should be larger than the BCI if the base is sustaining the most damage. If the damage in both layers is equivalent, then the ratio between the BDI and BCI should be 100%, but if the damage in the unbound layers is higher, then the ratio should be greater than 100%

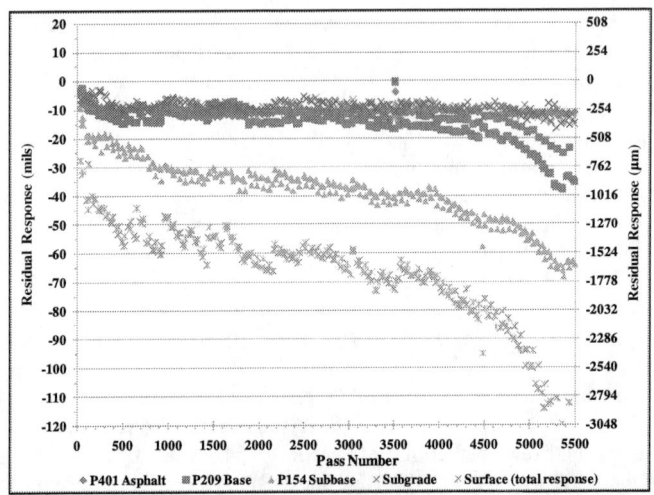

Figure 8. Residual Deflection Response of Each Layer on the MFC B747 Lane (Wander Position 1, West to East Loading Direction, and Wander Sequences 1-3)

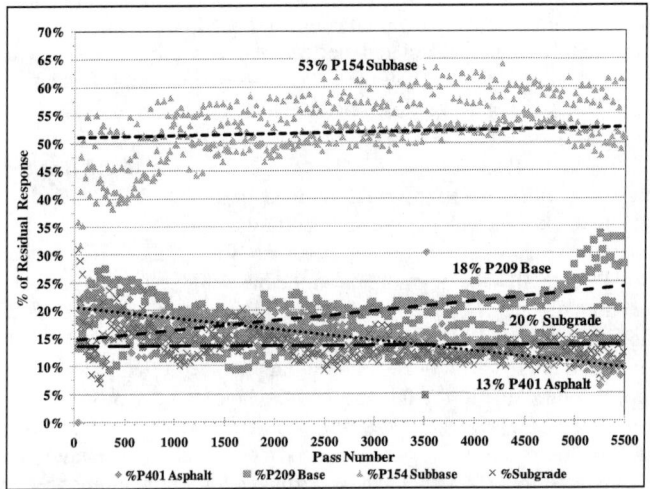

Figure 9. Layer Percentage of the Total Residual Response on the MFC B747 Lane (Wander Position 1, West to East Loading Direction, and Wander Sequences 1-5)

Figure 10 shows that BDI to BCI ratio is in general higher than 100% in the MFC test section, which clearly indicates more damage or residual deflection accumulation was observed in the unbound aggregate layers than in the subgrade throughout trafficking. The MFC 747 lane sustained up to 75% more damage in the unbound aggregate than in the subgrade while the 777 lane sustained only 20% more damage. From these findings, one could deduce that the 6-wheel B777 loading was relatively not as damaging to the granular layers when compared to the B747 dual tandem gear configuration. Further, Figure 10 also demonstrates the range B behavior of the pavement system. The unbound aggregates initially sustained the most damage, but after 5000 passes, the stresses increased in the subgrade due to the decreased stiffness and load bearing capacity of the unbound aggregates as part of the particle movements and rearrangements in the granular layers due to aircraft wander. For the declared failure of the MFC section B777 lane after 12,000 passes, the BDI/BCI ratio yet decreases below 100%.

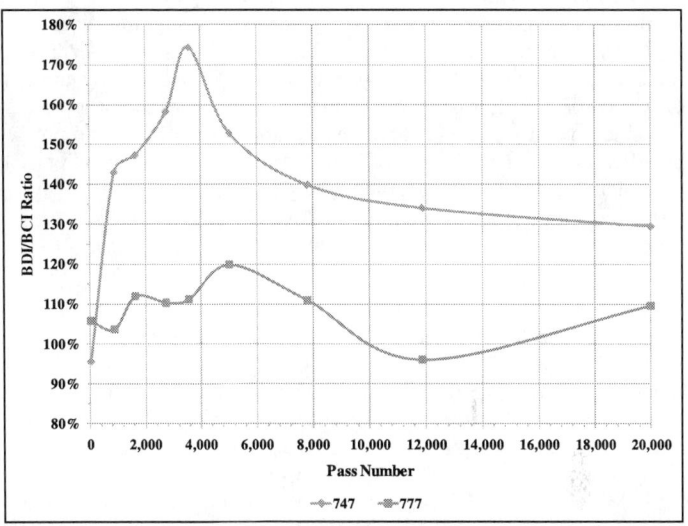

Figure 10. Ratio of Base Damage to Subgrade Damage in MFC Section (BDI/BCI)

Conclusions

Load wander has a dramatic effect on the stability and strength of the unbound aggregate layers because it causes constant particle rearrangement. Stresses that would normally result in stable range A shakedown behavior actually cause range B behavior with eventual collapse of the system at a greatly reduced number of load cycles. Multi-depth deflectometer (MDD) and Heavy Weight Deflectometer (HWD) data of the Federal Aviation Administration's (FAA's) National Airport Pavement Test Facility (NAPTF) medium strength conventional flexible pavement (MFC) construction cycle 1 test sections indicate that shakedown did not occur to densify the unbound aggregate layers under initial trafficking to failure. The MDD data showed that the percent of residual deflection responses in the granular P209 base and P154 subbase layers were relatively constant which is in contradiction to the shakedown theory where the unbound aggregate layers should consolidate and thus reduce their contribution to permanent deformation. HWD deflection basin interpretations proved that under a 9-track, 66-pass wander sequence of dual-tridem and dual-tandem type aircraft gear loadings, the unbound aggregate layers degraded much more quickly than the subgrade. This phenomenon resulted in a cascading degradation of the pavement system with the unbound aggregate layers degrading first followed in quick succession by the subgrade. These findings contribute greatly to our understanding of the field damage mechanisms to eventually help improve the development of realistic pavement performance prediction models.

Acknowledgements

This paper was prepared from a study conducted in the Center of Excellence for Airport Technology (CEAT). Funding for CEAT is provided in part by the Federal Aviation Administration. The CEAT is maintained at the University of Illinois at Urbana-Champaign. Ms. Patricia Watts is the FAA Program Manager for Air Transportation Centers of Excellence and Dr. Satish Agrawal is the FAA Airport Technology Branch Manager. The contents of this paper reflect the views of the authors who are responsible for the facts and accuracy of the data presented within. The contents do not necessarily reflect the official views and policies of the Federal Aviation Administration. This paper does not constitute a standard, specification, or regulation.

References

Donovan, P. and Tutumluer, E. (2008). "The Anti-shakedown Effect," In Proceedings of the 1st International Conference on Transportation Geotechnics, Nottingham, UK. August 25-27, 2008.

Garg, N. and Marsey, W.H. (2002). "Comparison between Falling Weight Deflectometer and Static Deflection Measurements on Flexible Pavements at the National Airport Pavement Test Facility NAPTF," In Proceedings of the Federal Aviation Administration Airport Technology Transfer Conference, Chicago, IL.

Garg, N. and Marsey, W.H. (2004). "Use of HWD to Study Traffic Effects on Flexible Airport Pavement Structure," Road Materials and Pavement Design, Vol. 5/3, pp. 385-396.

Gopalakrishnan, K. and Thompson, M.R. (2004). *Performance Analysis of Airport Flexible Pavement Subjected to New Generation Aircraft.* FAA COE Report No. 27, Department of Civil Engineering, University of Illinois, Urbana, IL.

Hayhoe, G.F. and Garg, N. (2002). "Subgrade Strains Measured in Full-Scale Traffic Tests with Four and Six-wheel Landing Gears," In Proceedings of the FAA Airport Technology Transfer Conference, Atlantic City, NJ, USA, May 5-8, 2002.

Hayhoe, G.F., and Garg, N. (2003). "Post Traffic Testing on Medium-Strength Subgrade Flexible Pavements at The National Airport Pavement Test Facility," In Proceedings of the ASCE Airfield Pavement Specialty Conference, Las Vegas, NV.

Ho Sang, V.A. (1975). *Field Survey and Analysis of Aircraft Distribution on Airport Pavements.* Report No. FAA-RD-74-36. U.S. Federal Aviation Administration.

Werkmeister, S., Numrich, R., and Wellner, F. (2002). "The Development of a Permanent Deformation Design Model for Unbound Granular Materials with the Shakedown-Concept," In Proceedings of the 6th International Symposium on the Bearing Capacity of Roads and Airfields (BCRA), Lisbon, Portugal, 24-26 June 2002, Balkema (Rotterdam), pp. 1081-1098.

Pavement Subgrade Evaluation and Value Engineering Solution for H-JAIA End-Around Taxiway (Taxiway Victor)

Raghuram N. Tadimalla[1], Richard L. Boudreau[2], Subash Kuchikulla[3], Viswanath Dokka[4] and Robert C. Briggs[5]

Abstract

The Hartsfield-Jackson Atlanta International Airport opened a critical end-around taxiway in April 2007 to reduce operational delays for air and ground traffic. As part of the construction specifications, the Contractor was required to prepare three feet of soil subgrade by undercutting and recompacting in order to achieve a minimum design modulus of subgrade reaction (k-value) of 90 pci. A value engineering alternative was utilized to verify the capacity of in-situ soils using a portable, light weight deflectometer (LWD) and laboratory resilient modulus tests to model the pavement subgrade modulus.

A validation of direct k-value (from LWD testing) was performed by comparing to the k-value obtained from resilient modulus testing. The final phase of the test program involved the computation of in-situ subgrade modulus from LWD for the remaining part of the Taxiway Victor to demonstrate the suitability of in-situ soils thus eliminating majority of the traditional subgrade preparation, as specified originally in project documents. The LWD offered an added advantage of fast turn-around results after the collection of field data.

The paper provides a background of existing subgrade issues, the test program methodology and conclusions from this value-engineering approach.

Introduction

The Hartsfield-Jackson Atlanta International Airport (H-JAIA) constructed a critical end-around taxiway (Taxiway Victor) to reduce operational delays. Considered the second most important capacity enhancement project at H-JAIA, the 3,000-foot taxiway was built around the 8R (west) end of Runway 8R-26L to accommodate

---

[1] Project Manager, Accura Engineering and Consulting Services, Inc., 3342 International Park Drive, Atlanta, GA 30316, PH (678) 522-0934, EIT, raghurambo@gmail.com
[2] President, Boudreau Engineering, Inc., 5392 Blue Iris Court, Norcross, GA 30092, PH (404) 388-1137, PE, M. ASCE, rlboudreau@comcast.net
[3] Executive Vice-President, Accura Engineering and Consulting Services, Inc., 3342 International Park Drive, Atlanta, GA 30316, PH (404) 867-2183, subashatl@gmail.com
[4] Project Engineer, Metals & Materials Engineers, LLC, 1039 Industrial Court, Suwanee, GA 30024, PH (678) 382-1316, viswanath.dokka@gmail.com
[5] Senior Engineer, Dynatest Consulting, Inc., 13953 US Highway 301 South, Starke, FL 32091, PH (904) 964-3777, PE, rbriggs@dynatest.com

taxiing of Group IV aircraft (wingspan from 118' to < 171') (FAA 1989) around the runway while simultaneously executing departures on the runway. Taxiway Victor was commissioned in April of 2007. This pavement feature alleviated the need to halt departures in order to allow taxiing aircraft to cross.

The project involved excavation of 1.1 million cubic yards of existing soils to build the proposed pavement system on suitable soil subgrade that would support the anticipated aircraft loads. The boring logs from the geotechnical investigations (R&D/LAW 2002, ACCURA/MACTEC 2003) of the area revealed that either partially weathered rock or bed rock was encountered at a depth of 2 to 15 feet below the design subgrade elevation. Additionally, groundwater was encountered at elevations higher than the design elevations of the subgrade underlying the pavement system. The alignment of the taxiway also consisted of a variety of soils with a wide range of natural moisture contents, typically in the high range (20% to 25%) with upto 46% in some locations.

To achieve a minimum design modulus of subgrade reaction (k-value) of 90 pci, the project specifications required the Contractor to undercut and re-compact three feet of soil subgrade. To follow the specifications, the Contractor needed to undertake two tasks: (1) undercut three feet of existing soils along the entire 3,000 feet of the taxiway and (2) adopt procedures to bring down the high natural moisture contents of different types of in-situ soils or import suitable select materials in order to prepare the subgrade to the required degree of compaction. The second task was particularly daunting and time-consuming part of the project that was on a critical path to meet the project completion deadline.

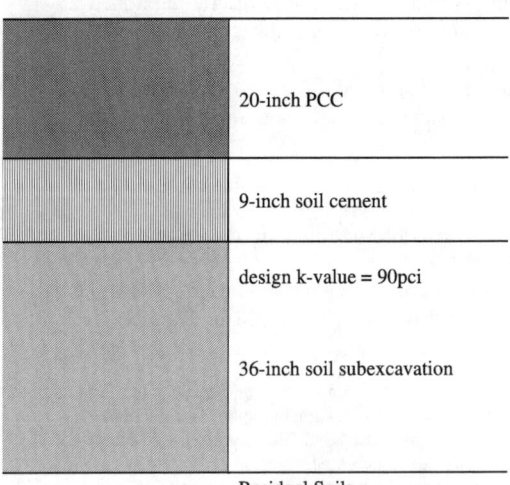

Figure 1. Design Pavement Cross-Section

A value engineering alternative was proposed to the Contractor to exhibit the suitability of in-situ soils to support the design pavement system (9 inches of soil-cement layer underneath 20 inches of PCC pavement, shown in Figure 1) with minimal subgrade preparation. This alternative would save time and money and enable the Contractor to meet the project schedule while still meeting the design requirements.

Field Sampling and Testing

The methodology involved a pilot phase of the test program to include approximately 600 feet of the taxiway from Taxiway Hotel tie-in. The pilot phase implemented field sampling and testing by performing two tasks: (1) Subgrade evaluation to determine the in-situ k-value in the soil foundation by performing resilient modulus testing on UD (undisturbed) samples and (2) Light Weight Deflectometer (LWD) testing to derive a direct k-value.

Task 1 – K-value from UD Samples: Once the Contractor completed excavation to grades representing elevations immediately below the PCC layer, five locations, B1 to B5 (Figure 2), were selected for Shelby tube sampling of subgrade soils. The top 9 inches of soil was removed from each location using a backhoe. This top 9 inches of soil was the zone of soil that was to be scarified and treated with cement (later it was decided to be replaced with M10 sand, treated with cement and compacted). Immediately beneath the top 9 inches of soil, a 10-inch Shelby tube was hand-driven using a sliding weight hammer normally used for dynamic cone penetrometer (DCP) testing. This tube sample was identified at the top (T) sample. In addition to the top one foot of subgrade soil sampled at each of the five locations, tube samples at two of the locations (B2 and B3) were obtained representing a 2-foot zone (middle, or M) and 3-foot zone (bottom, or B) below the anticipated soil-cement layer. Each tube was hand excavated following the drive, capped and taped, and returned to the laboratory for immediate processing and testing.

Location B1 was selected at a distance of 185 feet west of the existing edge of pavement tie-in to Taxiway Hotel as a starting point. The area between this location and Taxiway Hotel was excavated to a depth of three feet and backfilled with an on-site recycled concrete materials mixture prior to the field sampling and testing. The material was too coarse to drive Shelby tubes through and this area was not included in the evaluation.

Task 2 - Non-Destructive Testing: A Dynatest portable LWD was used to collect load-deflection data at numerous locations along the plan dimensions of the new taxiway structure. Testing was conducted at 85 discrete locations along the initial 400-foot longitudinal alignment - beginning at the west edge of the exposed edge joint of Taxiway Hotel. Tests were conducted at 5 transverse locations (centerline, 15-feet left and right of centerline, and 35-feet left and right of centerline) at longitudinal locations every 25 feet. These locations are shown in Figure 2.

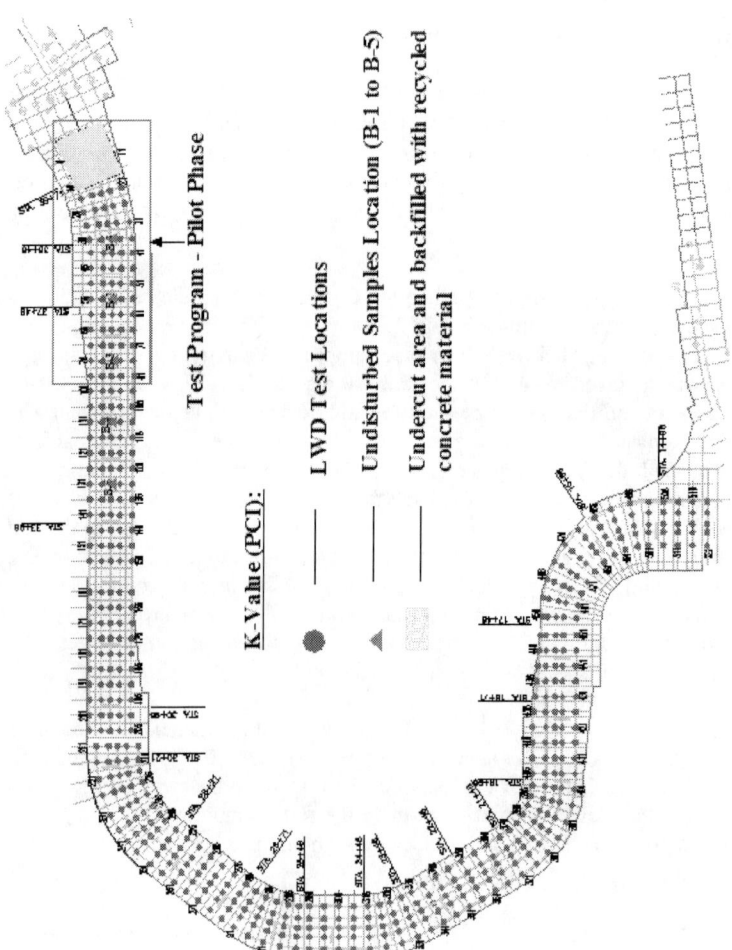

Figure 2. LWD Locations and Borings (B1 to B5) of the Test Program

The LWD is a testing device used by civil engineers to evaluate the physical properties of unbound pavement layers prior to or during pavement construction. This could include (but is not limited to) highways, local roads, and airport runways. The device is about the size of an upright vacuum cleaner and is transportable by hand, making it a convenient tool to evaluate in-situ elastic moduli of the different pavement layers quickly and efficiently (Nazzal 2007).

The LWD is designed to impart a load pulse to an unbound pavement surface layer. The load pulse is regulated to simulate the load produced by a rolling vehicle wheel. The load is produced by dropping a weight onto a set of rubber buffers, which transmit the load to the pavement structure through a circular load plate. A load cell mounted on top of the load plate measures the load imparted to the pavement surface. Deflection sensors (up to 3 geophones) measure the pavement deflection at the center of the load plate, and optionally, at two additional distances from the plate center.

LWD data is most often used to calculate stiffness-related parameters of a pavement structure. These parameters include modulus as well as K-values of unbound materials.

A diagram of the LWD is shown in Figure 3. The load plate, in this case a 300 mm diameter plate, rests on the testing surface. Above that is the LWD body, containing the load cell, geophone, and electronics. Above the LWD body are the rubber buffers which soften and extend the load pulse which is generated when the stainless steel weight is released. Finally, near the top, is the LWD handle which is equipped with a trigger release (see Figure 4 and Figure 5) for dropping the weight.

Figure 3. The LWD with hand-held computer for data collection

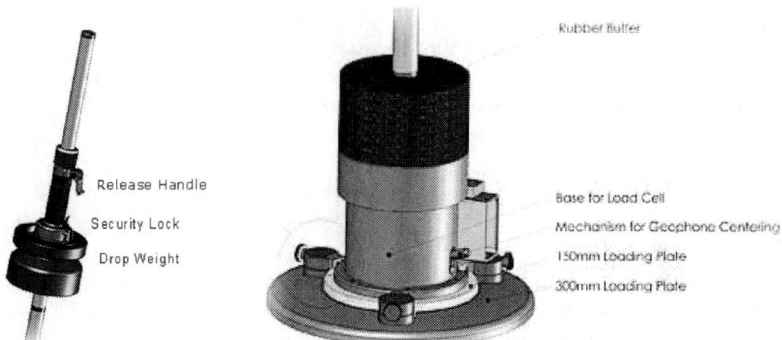

Figure 4. LWD handle, weight and release mechanism

Figure 5. LWD loading plates, load cell and buffer pads arrangement

The Dynatest LWD requires no reference measurements and provides a simple, cost effective alternative to time-consuming and expensive static plate bearing testing. The LWD is ideal for Quality Assurance / Quality Control on subgrade, subbase and thin flexible pavement constructions to verify that specifications are met. It can also be used to identify weaknesses, leading to further tests using FWDs and other material analysis techniques.

Results and Analyses

Unit Weight and Resilient Modulus Testing: Each UD tube sample was immediately brought to the laboratory, extruded, trimmed and measured for physical dimensions (mass, height and diameter) prior to testing for resilient modulus properties in accordance with AASHTO T307-99. With the exception of the tube driven at location 4 (485 feet west of Taxiway Hotel exposed pavement edge, 5 feet south of centerline), each tube yielded a good specimen for testing. The specimen extruded from the tube driven at location 4 contained numerous diagonal cracks through blocky and seamed, hard soil, thus could not be measured for unit weight nor tested for resilient modulus properties. The results of the testing program are contained in Table 1.

The modulus properties obtained from the resilient modulus testing program were used in a plate load simulation using the elastic layer computer program WESLEA. For the simulations, modulus values ranging from 2,000 to 7,000 psi were used to model the subgrade soil. A rigid 30-inch diameter plate was loaded to 10 psi pressure and surface deflections were computed from the WESLEA program. From these simulations, the k-value was computed (see Table A in the Appendix). Results indicated that in-situ k-values of 90-125 pci appeared reasonable and expected. The resilient modulus value of 2,100 was the value at which (at semi-infinite depth) a k-value of 90 pci was derived. Thus, the analysis showed that values below 2,100 psi provided less that 90 pci k-value support and values in excess of 2,100 psi provided greater than 90 pci k-value support. From the geotechnical investigations, the project site did not contain deep soil deposits, but rather had bedrock at relatively shallow depths (less than 20 feet). The depth to bedrock positively affected the k-value results.

Table 1 – Unit Weight and Resilient Modulus Properties of UD Samples

Sample No.	Depth*	Soil Properties			Modulus Regression Coefficients**			Modulus Value at S_c=10psi, S_3=3psi (psi)
		Wet Density (pcf)	Moisture (%)	Dry Density (pcf)	K1	K2	K5	
B1 (T)	0-12"	109.6	26.6	86.5	4,582	-0.10446	0.34663	5,272
B2 (T)	0-12"	111.3	22.3	91.0	3,507	-0.14112	0.36371	3,779
B2 (M)	12-24"	110.9	22.1	90.9	2,935	-0.12346	0.34869	3,240
B2 (B)	24-36"	116.8	16.7	100.1	3,078	-0.10139	0.38580	3,724
B3 (T)	0-12"	101.9	23.9	82.2	1,474	-0.04754	0.43142	2,122
B3 (M)	12-24"	104.8	24.5	84.1	1,495	-0.06225	0.41933	2,053
B3 (B)	24-36"	101.8	26.7	80.3	2,212	-0.08370	0.34337	2,660
B4 (T)	0-12"	Sample could not be tested for Unit Weight or Resilient Modulus Properties						
B5 (T)	0-12"	106.0	29.4	81.9	1,352	-0.06952	0.42792	1,843

* The depth of the sample is measured from the bottom of the proposed soil-cement layer
** From constitutive model: $Mr = K1(S_C)^{K2}(S_3)^{K5}$

Light Weight Deflectometer (LWD): Raw data from the test program was used to normalize to 1095 lbs (the load applied to the 300mm plate to achieve 10 psi pressure). The results were post processed to determine a dynamic modulus of elasticity (E_0) of the composite structure (with a depth of stress influence of 3 to 5 feet). The dynamic modulus of elasticity measured ranged from 700 to 11,400 psi and averaged 2,420 psi. These results were in close agreement with the data obtained from the resilient modulus testing of UD samples.

The dynamic modulus of elasticity was then transformed to a static modulus of elasticity using the conversion factor E_0 dynamic x 0.75 = E_0 static (Poulsen 1978). The static modulus of elasticity was then used in the layered elastic computer program to simulate the plate load test using a 30-inch diameter plate again loaded to 10 psi of vertical pressure. The deflection resulting from this loaded plate was computed and used to determine the k-value. The results indicated that the existing in-situ soils provided average k-value of 102 pci in the area of evaluation. The LWD data also showed that the subexcavated and backfilled area (rectangle shaded yellow in Figure 2) provided less k-value support than the undisturbed zone.

The pilot phase locations were tested with LWD at various stages of the project to assess the improvement in the k-value to support the pavement system: (1) Initial testing during field sampling and testing phase, (2) after the underdrain installation to monitor the change due to lower moisture contents in subgrade soils, (3) at soil-cement bottom elevation and (4) at soil-cement top elevation after cement treatment and curing. The k-values for locations 1 to 85 for the above stages are shown in Figure 6.

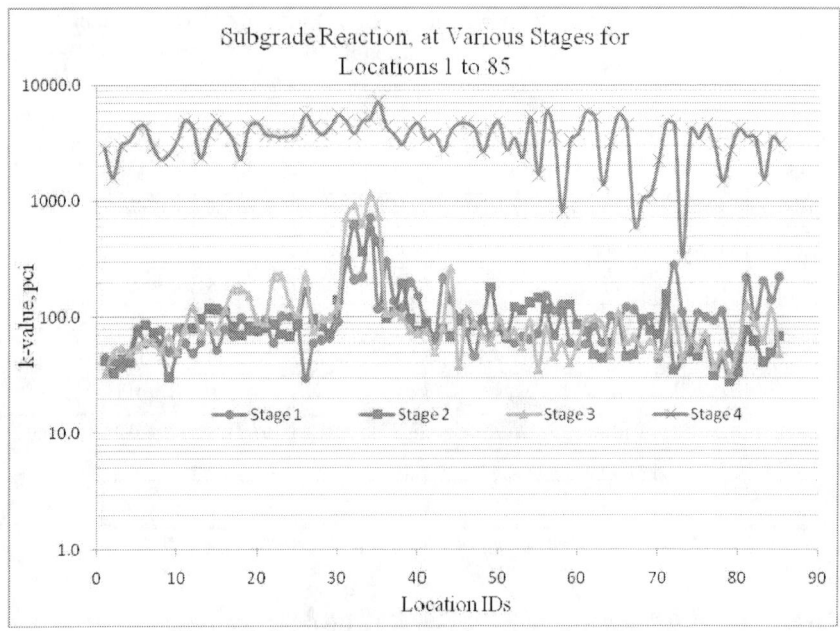

Figure 6. Subgrade support values at various stages

The LWD testing was performed at one or more of the above stages for the remaining portion of the taxiway. See Table 2 for summary of LWD testing.

Table 2. LWD Location and Testing Stage Summary

LWD Locations	Station		LWD Testing Stages
	From	To	
1 – 85	40+37.5	36+37.5	1, 2, 3, 4
101 – 235	36+12.5	29+37.5	2, 3, 4
236 – 280	29+12.5	27+12.5	3
86 – 100	26+87.5	26+37.5	3
281 – 295	26+12.5	25+62.5	3
296 – 525	25+37.5	14+12.5	In-situ soils at soil-cement top elevation

Conclusions and Recommendations

A significant portion of the taxiway subgrade evaluated in the pilot phase of the testing program possessed the design value of at least 90 pci of pavement foundation support. The LWD testing identified locations that were weak in the k-value support. Based on the data collected, 8-foot deep underdrains were installed for 500 feet of the taxiway in place of shallow depth underdrain system specified in the project drawings. The design change resulted in improved soil support for pavement system. The stage 4 LWD testing revealed that the project-specific M10 sand cement-treated base offered k-value support substantially higher than the design value of 300 pci. Over $1 million was saved due to implementation of this value engineering alternative that eliminated most of the traditional subgrade preparation.

The methodology utilized in this value engineering alternative to traditional subgrade preparation procedures should be considered for evaluating subgrades for suitability to the design support k-values. Implementation of similar value engineering alternatives could result in substantial savings in time, materials and labor for critical projects.

Acknowledgments

We would like to thank contractor's representatives Ron Hamilton, Martin Porges, Albert Rice and Steve Hausler for their coordination in implementing this value engineering approach. We also extend our sincere thanks to Frank Hayes and Kathy Masters from Department of Aviation, City of Atlanta for their support in the implementation of this alternative.

References

Federal Aviation Administration (FAA) (1989), *Advisory Circular: 150/5300-13 – Airport Design*, Washington, D.C.

R&D Testing and Drilling, Inc. (R&D) / LAW Engineering, Inc. (LAW) (2002), *Geotechnical Report For Planning*, Atlanta, Georgia.

Accura Engineering and Consulting, Inc. (ACCURA) / MACTEC Engineering and Consulting, Inc. (MACTEC) (2003), *Geotechnical Report for 08R End-Around Taxiway Design Study*, Atlanta, Georgia.

Nazzal, M.D., Abu-Farsakh, M.Y., Alshibli, K., Mohammad, L. (2007), *Evaluating the LFWD Device for In Situ Measurement of Elastic Modulus or Pavement Layers*, Transportation Research Board, Annual Meeting CD-ROM, Washington, D.C.

Poulsen, J., Stubstad, R. N. (1978), *Laboratory Testing of Cohesive Subgrades: Results and Implications Relative to Structural Pavement Design and Distress Models*, Transportation Research Board, 671, 84-91.

APPENDIX

Table A. Plate Load Simulation, Taxiway Victor Subgrade Prep Test Program
Hartsfield-Jackson Atlanta International Airport

Layer 1 Depth = D_1 Modulus = E_1 Poisson's ratio = v_1

Layer 2 Depth = D_2 Modulus = E_2 Poisson's ratio = v_2

$\sigma = 10$ psi, plate diam = 30 inches (radius = 15 in.)

WESLEA Results:

Trial	Comments	D_1	E_1	v_1	D_2	E_2	v_2	defl. (in.)	est. k-value	AASHTO k
Run1		48	2,000	0.45	semi-infinite	3,000	0.45	0.11067	90	103
Run2		0	0	0	semi-infinite	3,000	0.45	0.07975	125	155
Run3		0	0	0	semi-infinite	4,000	0.45	0.05881	170	206
Run4		0	0	0	semi-infinite	5,000	0.45	0.04785	209	258
Run5		0	0	0	semi-infinite	6,000	0.45	0.03987	251	309
Run6		0	0	0	semi-infinite	7,000	0.45	0.03418	293	361
Run7		48	3,000	0.45	semi-infinite	3,000	0.45	0.07975	125	
Run8		48	4,000	0.45	semi-infinite	3,000	0.45	0.06397	156	
Run9		48	5,000	0.45	semi-infinite	3,000	0.45	0.05431	184	
Run10		48	6,000	0.45	semi-infinite	3,000	0.45	0.04773	210	
Run11		48	7,000	0.45	semi-infinite	3,000	0.45	0.04294	233	
Run12		36	2,000	0.45	semi-infinite	3,000	0.45	0.10788	93	
Run13		36	1,500	0.45	semi-infinite	3,000	0.45	0.13553	74	
Run14		36	1,500	0.45	semi-infinite	8,000	0.45	0.11875	84	
Run15		36	1,500	0.45	semi-infinite	10,000	0.45	0.11662	86	
Run16		24	2,000	0.45	semi-infinite	3,000	0.45	0.10275	97	
Run17		24	1,500	0.45	semi-infinite	3,000	0.45	0.1251	80	
Run18		24	1,500	0.45	semi-infinite	8,000	0.45	0.10126	99	
Run19		24	1,500	0.45	semi-infinite	10,000	0.45	0.09824	102	
Run20		24	2,500	0.45	semi-infinite	3,000	0.45	0.08907	112	

AASHTO k predicted from Eqn.provided on page II-44 (1993 Guide): $K_{est} = M_r/19.4$

Performance of flexible pavements over two subgrades with similar CBR but different soil types (Silty Clay and Clay) at the FAA's National Airport Pavement Test Facility

Navneet Garg[1], Gordon F. Hayhoe[2]

[1] Airport Technology R&D Branch, FAA's William J. Hughes Technical Center, Atlantic City Intl. Airport, NJ 08405, U.S.A; Phone: (609)485-4483; Email: navneet.garg@faa.gov

[2] Airport Technology R&D Branch, FAA's William J. Hughes Technical Center, Atlantic City Intl. Airport, NJ 08405, U.S.A; Phone: (609)485-8555; Email: gordon.hayhoe@faa.gov

Abstract

The National Airport Pavement Test Facility (NAPTF) is located at the FAA William J. Hughes Technical Center, Atlantic City International Airport, New Jersey. It is used to generate full-scale pavement response and performance data for development and verification of airport pavement design criteria. During the Construction Cycle 5 – Test Strip, four flexible pavement test items were subjected to accelerated traffic tests using representative aircraft landing gear configurations. Pavements were constructed over two subgrades with similar CBR (approximately 3) but different soil types (silty clay and clay). Two test items (LFC1-N and LFC1-S) were designed to fail in approximately 100 passes (4-wheel gear and 55,000 lbs wheel load) and the pavement structure consisted of 2.5-inch P401 HMA surface, 8-inch P209 crushed stone base, and 16-inch P154 subbase. The other two test items (LFC2-N and LFC2-S) were designed to fail in approximately 2000 passes and the pavement structure consisted of 2.5-inch P401 HMA surface, 8-inch P209 crushed stone base, and 24-inch P154 subbase. The test items exhibited permanent deformations at the surface (evidenced by rut depths) of over 4 inches and upheaval at the sides of the ruts in excess of 1 inch. This paper presents the results from Heavy Weight Deflectometer (HWD) tests, trafficking tests (rut depth measurements) and posttraffic tests (trenching study). The trenching involved removal of the P-401 AC layer, the P-209 base, and the P-154 subbase layer to reveal the subgrade interface and subsequent subgrade layers below. Tests conducted on the pavement component layers included CBRs, in situ densities and moisture contents. Layer interface profile measurements from trench walls clearly show shear flow in the subgrade, with vertical movement of the subgrade material in the upheaval areas. This information will be used for developing thickness design procedures and load evaluation of airport pavements.

Introduction

The FAA's NAPTF is located at the FAA William J. Hughes Technical Center, Atlantic City International Airport, New Jersey. The primary purpose of the NAPTF is to generate full-scale pavement response and performance data for development and verification of airport pavement design criteria. It is a joint venture between the FAA and the Boeing Company and became operational on April 12, 1999. Additional information about the test facility is available elsewhere (http://www.airporttech.tc.faa.gov). A construction cycle at the NAPTF includes test pavement construction including instrumentation, traffic tests to failure, posttraffic testing (includes trenching activities and other tests), and pavement removal. A typical construction cycle (CC) at the NAPTF is shown in figure 1.

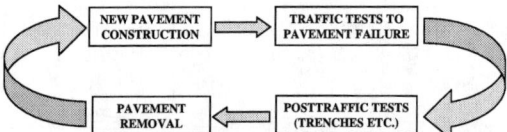

Figure 1. Construction cycle at the NAPTF

During the Construction Cycle 5 – Test Strip (CC5-TS), four flexible pavement test items were subjected to accelerated traffic tests using representative full-scale aircraft landing gear configurations. Pavements were constructed over two subgrades with similar CBR (approximately 3) but different soil types (silty clay and clay). Two test items were designed to fail in approximately 100 passes and the other two test items were designed to fail in approximately 2000 passes (4 wheel gear and 55,000 lbs wheel load). The test items were traffic tested to complete structural failure as evidenced by permanent deformations at the surface (evidenced by rut depths) of over 4 inches (102 mm) and upheaval at the sides of the ruts in excess of 1 inch (25 mm).

This paper presents the results from Heavy Weight Deflectometer (HWD) tests, trafficking tests (rut depth measurements) and posttraffic tests (trenching study). The trenching involved removal of the P-401 AC layer, the P-209 base, and the P-154 subbase layer to reveal the subgrade interface and subsequent subgrade layers below. Tests conducted on the pavement component layers included CBRs, and in situ densities and moisture contents. Layer interface profile measurements from trench walls clearly show shear flow in the subgrade, with vertical movement of the subgrade material in the upheaval areas. This information will be used for developing thickness design procedures and load evaluation of airport pavements.

Objectives for CC5-TS Testing

The objectives for CC5-TS were as follows:
1. Document Pavement Failure - Make a movie of the structural failure of a pavement (similar cross-section to CC3 LFC-1) designed to fail in approximately 100 passes.
2. Study Feasibility of Using Dupont Clay as Low-Strength Subgrade - Previous studies at NAPTF (1,2) have shown that the current low-strength subgrade, County Sand & Stone Clay (CSSC), loses moisture from the upper layers over

time and the CBR increases compared to the "as-constructed" CBR. Whereas, DuPont clay (currently used as medium-strength subgrade) is better in holding the moisture and maintaining the "as-constructed" CBR. An attempt was made to place DuPont clay at a target CBR of 3.
3. Pavement Performance Comparison Over Two Different Subgrade Types - Compare the performance of flexible pavements over two subgrades with similar CBR but different soil types (silty clay and clay). One set of test items was designed to fail in approximately 100 passes, and the other set of test items was designed to fail in approximately 2000 passes.
4. Moisture Control in P-154 Subbase - Previous studies at NAPTF have shown that the moisture from the P-154 subbase collects at the interface between the subbase and subgrade and in turn reduces the CBR of the subgrade. An attempt was made to control the moisture in the P-154 subbase by placing the subbase at a moisture content lower than optimum.

CC5-TS Pavement Test Items

Figure 2 shows the pavement cross sections constructed during CC5-TS.

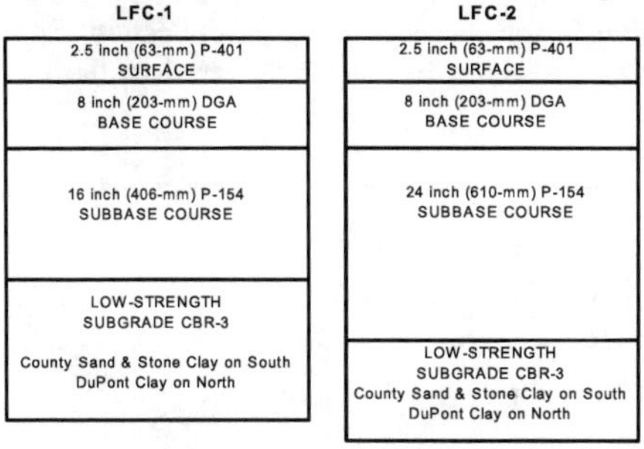

LFC-1	LFC-2
2.5 inch (63-mm) P-401 SURFACE	2.5 inch (63-mm) P-401 SURFACE
8 inch (203-mm) DGA BASE COURSE	8 inch (203-mm) DGA BASE COURSE
16 inch (406-mm) P-154 SUBBASE COURSE	24 inch (610-mm) P-154 SUBBASE COURSE
LOW-STRENGTH SUBGRADE CBR-3 County Sand & Stone Clay on South DuPont Clay on North	LOW-STRENGTH SUBGRADE CBR-3 County Sand & Stone Clay on South DuPont Clay on North

Figure 2. CC5-TS Pavement Test Items

The top three feet of existing County Sand and Stone Clay (CSSC) subgrade was removed between stations 0+00 and 150+00 and was replaced by DuPont clay on the north and CSSC on the south. The test items were named LFC-1 and LFC-2 (L-low strength subgrade, F-flexible pavement, C-conventional base). LFC-1N and LFC-2N had DuPont clay (CH soil, liquid limit 66%, plasticity index 33%) as the subgrade. In test items LFC-1S and LFC-2S, CSSC (MH-CH soil, liquid limit 53%, plasticity index 19%) was used as the subgrade. As per ASTM D1557 (Method C), the optimum moisture content for CSSC and DuPont clay was 16.6% and 19% respectively. The corresponding maximum dry densities were 109.7 pcf for CSSC and 102.1 pcf for DuPont clay. The subgrade material was tilled until the material was at the target moisture content (25% for CSSC and 36% for DuPont clay), was placed in 6 to 8 inch thick lifts, and compacted using a rubber tire roller. The subgrade lift acceptance criterion was based on CBR, and the CBR measurements on top of the completed subgrade are listed in Table 1.

Table 1. CBR at Top of the Subgrade (subgrade acceptance test results)

Test Item	DuPont Clay (North)	CSSC (South)
LFC-1	3.3	3.3
LFC-2	3.35	3.1

Heavy Weight Deflectometer Tests

Test item LFC-1 was 30-feet long (Station 0 to 30), and test item LFC-2 was 50-feet long (station 50 to 100). Heavy weight deflectometer (HWD) tests were performed to study uniformity of the pavement structure in LFC-2 using the FAA's Kuab HWD equipment. The summary of test results (at 36,000 lbs load) is shown in Table 2 and Figures 3 and 4. The results show that the pavement structure within the wheel track in LFC-2 was fairly uniform.

Table 2. Summary of Pavement Uniformity HWD Tests (at 36,000 lbs load)

Test Item	Wheel Track	Deflection Summary (mils)	D0 0 in Offset	D1 12 in Offset	D2 24 in Offset	D3 36 in Offset	D4 48 in Offset	D5 60 in Offset	D6 72 in Offset
LFC-2	North Wheel Track (DuPont Clay Subgrade)	Minimum	94.98	65.34	36.52	21.48	13.93	10.10	8.31
		Maximum	107.64	74.12	42.63	25.59	16.57	11.73	9.32
		Mean	99.92	69.52	39.34	23.09	14.89	10.79	8.80
		Std. Dev.	3.95	2.70	1.87	1.31	0.82	0.47	0.31
		COV, %	3.95	3.89	4.74	5.67	5.54	4.33	3.53
LFC-2	South Wheel Track (CSSC Subgrade)	Minimum	94.14	64.83	34.66	20.04	13.17	9.67	7.89
		Maximum	113.91	79.22	45.40	27.19	17.21	11.83	9.07
		Mean	106.07	74.28	42.07	24.90	15.89	11.12	8.63
		Std. Dev.	6.46	5.15	3.59	2.30	1.28	0.67	0.37
		COV, %	6.09	6.94	8.54	9.23	8.03	6.06	4.28

As shown in Figure 3, the mean peak center deflection (D0) in the south side pavement structure was about 6 percent higher than the north side pavement structure.

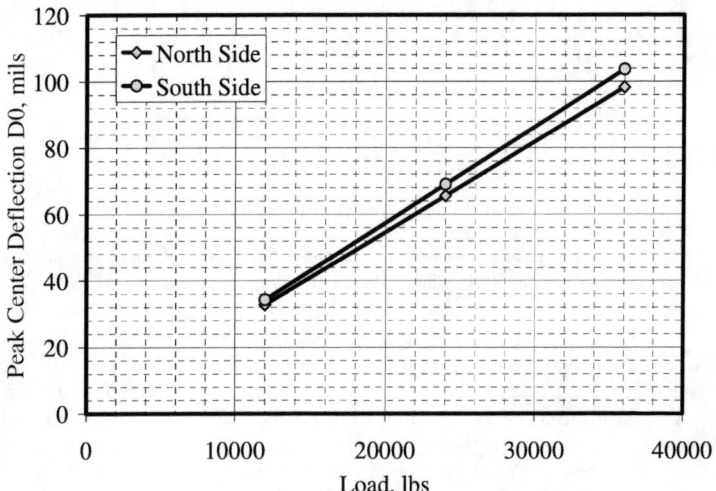

Figure 3. Peak Center Deflections D0 for Test Item LFC-2
(from pavement uniformity tests)

Figure 4 shows deflection D6 (at 72-inch offset from the center of the plate). Previous studies have shown that deflection D6 (at 72-inch offset) is a good indicator of subgrade strength since the contribution of overlying layers is small. The results in figure 4 show that the two subgrades exhibited similar stiffness. The average value of D6 deflection for the two subgrade types differed by less than 2 percent (Table 2).

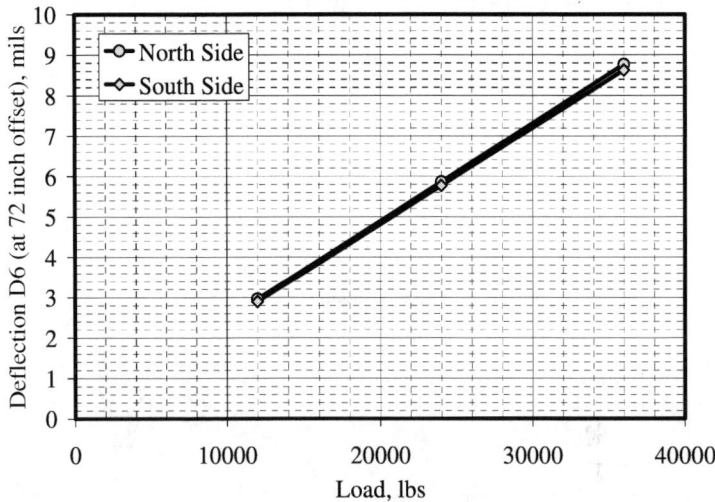

Figure 4. Deflections D6 (at 72 inch Offset from center of plate) for Test Item LFC-2 (from pavement uniformity tests)

Traffic Testing Parameters and Failure Criteria

The test items were trafficked by a four-wheel dual tandem (2D) configuration at 1372-mm (54-in) dual spacing and 1449-mm (57-in) tandem spacing. Wheel loads were set at 245 kN (55,000 lbs) each. This gives "strut" loads of 979 kN (220,000 lbs). Traffic speed was 4 km/h (2.5 mph). A fixed wander pattern was applied to the traffic during the tests. The wander pattern consisted of 66 repetitions, 33 traveling east and 33 traveling west. The transverse position of the gears was changed only at the start of the eastward repetitions. That is, westward repetitions always had the wheels following in the same paths as in the preceding eastward repetition. The wander pattern was designed to simulate a normal distribution with standard deviation of 775 mm (30.5 in) (equivalent to a taxiway distribution for design). The distribution of the transverse wheel positions is not random, but consists of nine equally spaced wheel paths at intervals of 260 mm (10.25 in). The pattern is similar to the patterns used in previous full-scale airport pavement traffic tests, except that one wander cycle includes a larger number of repetitions than were typically used before (*3*).

The failure criteria used for flexible pavements are: 1) 25.4-mm (1-in) upheaval outside the traffic lane, signifying structural shear failure in the subgrade or other supporting layers; and 2) surface cracking to the point that the pavement is no longer waterproof, signifying complete structural failure of the surface layer. These are the same criteria as defined in (*3*) for the multiple wheel heavy gear load (MWHGL) test series run by the U.S. Army Corps of Engineers Waterways Experiment Station. A detailed explanation for selection of this failure criterion is given in reference (*4*).

Pavement Performance During Traffic Testing

Test Item – LFC-1:
One of the objectives for the tests run on test item LFC-1 was to document pavement failure. The deformations in such a weak pavement structure under large wheel loads are of such magnitude that the movement of the pavement structure can be seen by the naked eye. A movie was made of the structural failure of the pavement. The straightedge rut depth measurements are shown in Figure 5. The pavement section on the CSSC subgrade was declared failed after 54 passes. The pavement section on the DuPont clay subgrade lasted 114 passes. As mentioned earlier in the paper, test item LFC-1 was designed to fail in 100 passes.

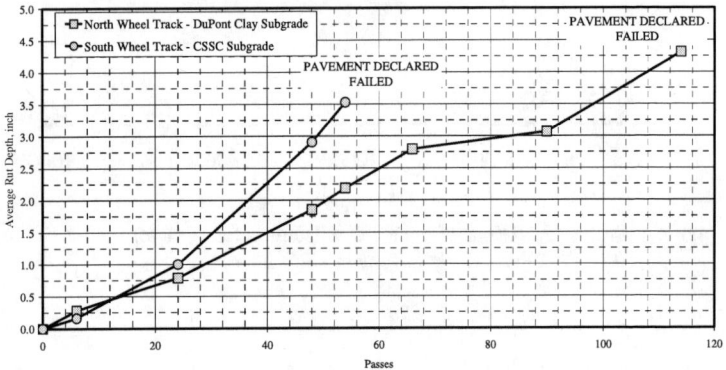

Figure 5. Straight Edge Rut Depth Measurements in Test Item LFC-1

Test Item – LFC-2:
The straightedge rut depth measurements in test item LFC-2 are shown in Figure 5. Test item LFC-2 was designed to fail in 2000 passes. The pavement section on DuPont clay subgrade (north wheel track) was declared failed after 1716 passes. The south wheel track (CSSC subgrade) pavement test item lasted for 2112 passes.

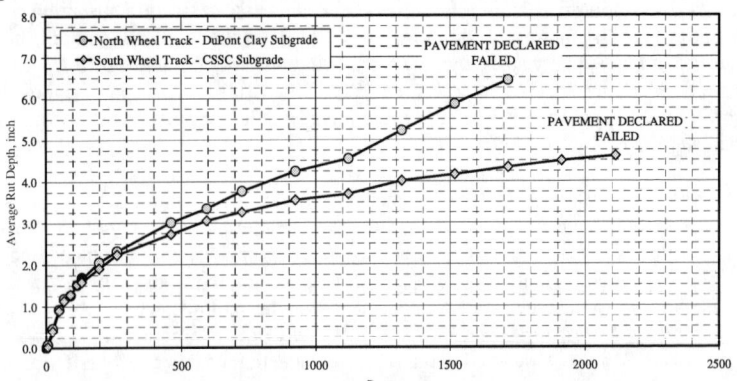

Figure 6. Straight Edge Rut Depth Measurements in Test Item LFC-2

Posttraffic Testing

Two trenches were excavated perpendicular to the centerline of test items LFC-1 and LFC-2 at the locations of rut depth measurements on the test pavements. The purpose of the trenches was to conduct posttraffic investigation into the failure mechanism of the pavement structure. The trenching involved removal of the P-401 AC layer, the P-209 crushed stone base, and the P-154 subbase layer to reveal the subgrade interface and subsequent subgrade layers below. The final trench dimensions were 60 feet (18.3 m) long (across the width of the test pavement) and four feet (1.22 m) wide. Tests and measurements were performed on the various layers of the pavement structure.

Figure 7 shows the layer profile measurements made in test item LFC-1. The figure shows that the rutting was contributed by the subgrade and subbase. The rutting in the subgrade was also accompanied by upheaval, which is indicative of shear failure in the subgrade. Test item LFC1 exhibited very large vertical pavement deformations right from the beginning of traffic testing and shear failure in the subgrade occurred before a significant amount of consolidation or shear failure of the subbase took place. This resulted in a "W" shaped transverse rut profile at the surface of the pavement in the south wheel track that is characteristic of failure at low pass levels.

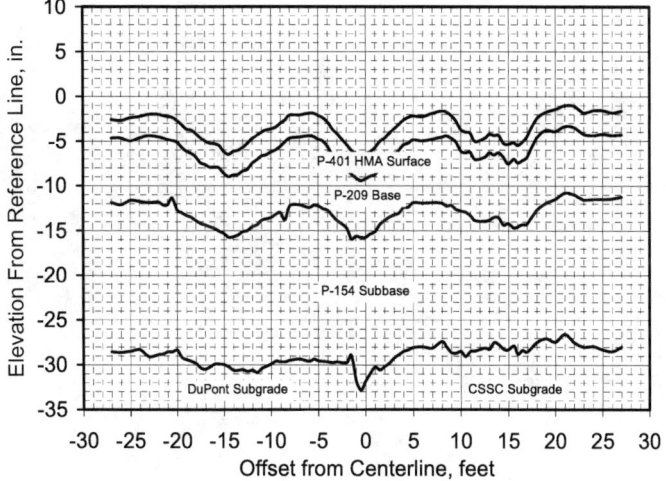

Figure 7. Profile Measurements from Posttraffic Trench Face in Test Item LFC-1

Layer profiles from the posttraffic trench in test item LFC-2 are shown in Figure 8. Minimal rutting was observed in the P-209 base layer. Most of the rutting was contributed by the subbase and subgrade. Previous studies at NAPTF *(2 and 5)* have shown that a major portion of the rutting in subbase layer may be caused by densification rather than shear failure. In test item LFC-2, very high levels of consolidation and strength increase of the subbase layer were measured. Subgrade intrusion into the subbase layer was observed in the north wheel track and is shown in Figure 9.

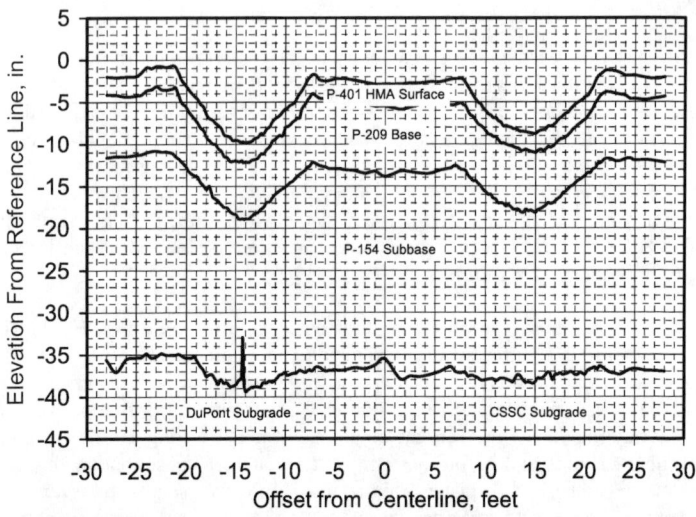

Figure 8. Profile Measurements from Posttraffic Trench Face in Test Item LFC-2

Figure 9. DuPont Subgrade Intrusion into P-154 Subbase in Posttraffic Trench Face in Test Item LFC-2

Tables 3 and 4 show a summary of the CBR tests performed in the posttraffic trenches at the subgrade surface in test items LFC-1 and LFC-2 respectively. In test item LFC-1 (Table 3), it was observed that the mean CBR values for the

CSSC subgrade were closer to the as-constructed CBR values (Table 1). The CBR values for the DuPont clay subgrade reduced and an increase in moisture content was observed. The increased moisture is believed to have been contributed by the subbase layer (non plastic fines).

Table 3. Summary of CBR Tests from Posttraffic Trenches in Test Item LFC-1

Test No.	Soil Type	Depth from Pavement Surface, inch	Offset from CL feet	Moisture Content percent	CBR Penetration No.			Mean
					1	2	3	
1	DuPont Clay	0.0	-26	36.61	2.40	2.70	3.00	2.7
2		0.0	-18	37.48	1.73	2.70	2.00	2.1
3		0.0	-13	37.23	3.10	3.00	2.00	2.7
4		0.0	-3	37.08	2.40	2.40	2.40	2.4
5	CSSC	0.0	3	25.20	3.33	3.00	2.87	3.1
6		0.0	12	25.36	3.00	3.07	3.20	3.1
7		0.0	19	25.57	2.67	2.73	2.53	2.6
8		0.0	26	25.59	3.40	3.00	3.20	3.2

In test item LFC-2, similar behavior was observed in the case of the DuPont clay subgrade, with increased moisture content and reduced CBR. In the case of the CSSC subgrade, an increase in CBR values was observed. This phenomenon has been observed in CSSC subgrades at the NAPTF in previous construction cycles. This is one of the reasons to replace the CSSC subgrade with DuPont clay subgrade.

Table 4. Summary of CBR Tests from Posttraffic Trenches in Test Item LFC-2

Test No.	Soil Type	Depth from Pavement Surface, inch	Offset from CL feet	Moisture Content percent	CBR Penetration No.			Mean
					1	2	3	
1	DuPont Clay	0.0	-25	37.91	2.13	2.60	2.40	2.4
2		0.0	-14	39.18	2.10	2.20	2.40	2.2
3		0.0	-12	38.57	2.13	2.27	2.70	2.4
4		0.0	-3	38.39	2.60	2.20	2.20	2.3
5	CSSC	0.0	4	25.60	3.90	4.13	3.80	3.9
6		0.0	13	24.86	3.20	3.50	4.00	3.6
7		0.0	17	24.25	3.60	4.00	3.60	3.7
8		0.0	26	25.36	4.13	4.13	4.00	4.1

Conclusion/Summary

The main objectives for CC5-TS were to document pavement failure, study the feasibility of using Dupont Clay as low-strength subgrade, and compare pavement performance over two different subgrade types. The results from the full-scale testing on two flexible pavement test items have been presented in this paper. The study showed that it is feasible to use DuPont clay as the low strength subgrade and that it maintains the as-constructed CBR if the moisture in the subbase used at NAPTF is limited to about four percent (which is the equilibrium moisture content for this material in its compacted condition). The pavement life observed for the two test items was close to their design lives. The results obtained from CC5-TS will be added to the existing database of full-scale test results and will be used to improve the calibration of the subgrade failure model in the FAA's airport pavement thickness design procedure.

Acknowledgement

The work described in this paper was supported by the FAA Airport Technology Research and Development Branch, AAR-410, United States of America, Dr.

Satish K. Agrawal, Manager. The contents of the paper reflect the views of the authors, who are responsible for the facts and accuracy of the data presented within. The contents do not necessarily reflect the official views and policies of the FAA, USA. The paper does not constitute a standard, specification, or regulation.

References
1. Hayhoe, G.F., 2004. *Traffic Testing Results from the FAA's National Airport Pavement Test Facility.* Proceedings of 2nd International Conference on Accelerated Pavement Testing, University of Minnesota, Minneapolis, U.S.A.
2. Hayhoe, G.F., Garg, N., 2003. *Posttraffic Testing on Medium-Strength Subgrade Flexible Pavements at the National Airport Pavement Test Facility.* Proceedings of Specialty Conference: Airfield Pavements – Challenges and New Technologies, Las Vegas, U.S.A.
3. Ahlvin R.G., H.H. Ulery, R.L. Hutchinson, and J.L. Rice. *Multiple-Wheel Heavy Gear Load Pavement Tests Volume I Basic Report.* Technical Report S-71-17, U.S. Army Engineer Waterways Experiment Station, November, 1971.
4. Hayhoe, G.F., Garg, N., Dong, M., 2003. *Permanent Deformations During Traffic Tests on Flexible Pavements at the National Airport Pavement Test Facility.* Proceedings of Specialty Conference: Airfield Pavements – Challenges and New Technologies, Las Vegas, U.S.A.
5. Hayhoe, G.F., Garg, N., 2006. *Traffic Testing Results From CC3 Flexible Pavements at the FAA'S NAPTF.* Proceedings of 10th International Conference on Asphalt Pavements 2006, Quebec City, Canada.

USE OF RECYCLED CONCRETE AS UNBOUND BASE AGGREGATE IN AIRFIELD AND HIGHWAY PAVEMENTS TO ENHANCE SUSTAINABILITY

Dr. Athar Saeed, P.E., A. M. ASCE [1], and
Dr. Michael I. Hammons, P.E., M. ASCE [2]

[1] Research Engineer, Air Force Research Laboratory (AFRL/RXQD)
(Principal Engineer, Applied Research Associates, Inc.)
AFRL/RXQD – Deployed Systems Branch
104 Research Road, Bldg 9738, Tyndall AFB, Florida 32403
Phone: 850 283 3718, Fax: 850 283 2035, athar.saeed@tyndall.af.mil
(**corresponding author**)

[2] Research Engineer, Air Force Research Laboratory (AFRL/RXQD)
(Principal Engineer, Applied Research Associates, Inc.)
AFRL/RXQD – Deployed Systems Branch
104 Research Road, Bldg 9738, Tyndall AFB, Florida 32403
Phone: 850 283 3718, Fax: 850 283 2035, michael.hammons@tyndall.af.mil

ABSTRACT

This research evaluated the use of recycled concrete aggregate (RCA) from airfield and highway pavements to enhance sustainability based on engineering, economic, and environmental criteria. The objective was to develop evaluation and construction guidelines by defining the minimum standards for RCA as unbound base. The research approach included contacts with industry representatives, technical assessments, site visits, and performance review of airfield pavements with unbound RCA layers. RCA can be used as unbound base material if produced from uncontaminated PCC. All virgin aggregate tests and their limits are applicable to RCA except the sulfate soundness test, which is waived for RCA due to the incompatibility of PCC components with the chemical reactants used in the test. RCA should not be used where there is a potential for sulfate exposure from subgrade soils, ground water, or other external sources. RCA from ASR-distressed PCC can be used considering site conditions. RCA is not a hazard to the environment. An economic analysis can be conducted by considering initial material and construction costs for both RCA and virgin aggregate.

INTRODUCTION

The United States Geological Survey (USGS) estimates that about 2.241 billion metric tons of crushed stones were used in pavement construction in 2006 in the United States. About 108 million metric tons were used in graded base/subbase construction. Large economic and environmental benefits will result from a small increase in the amount of recycled materials (recycled concrete aggregate [RCA] and recycled asphalt pavement [RAP]) used to replace the virgin aggregate in pavement construction while extending the limited supply of natural aggregates. The USGS reports that development of aggregate resources is "being constrained by urbanization, zoning regulations, increased costs, and environmental concerns." RCA is produced from the demolition of existing portland cement concrete (PCC) pavements and normally consists of high-quality aggregate particles. Recent data indicate that approximately 2.9 million metric tons of RCA was produced and used in the United States (*USGS 2007*). The Federal Highway Administration (FHWA) has led the research in using RCA is pavement construction with the assistance of the recycled materials research centers, the Transportation Research Board (TRB), State departments of transportation (DOTs), and industry associations. The FHWA has conducted numerous feasibility studies and demonstration projects including *User Guidelines for Waste and Byproduct Materials* (*Chesner 1998*). More recently, National Cooperative Highway Research Program (NCHRP) completed Projects 4-21 and 4-31 on the use of RCA and RAP in transportation, while NCHRP Project 25-9 evaluated the environmental impact of (highway) construction and repair materials on surface and groundwater (*Saeed 2008, 2006*).

RESEARCH OBJECTIVE

The research objective was to develop evaluation and construction guidelines by defining the minimum standards for RCA as unbound base. The developed guidelines serve to assure owners that pavements constructed with RCA unbound base will perform successfully. The guidelines further assist designers by providing the engineering rationale for using unbound RCA base. Economic benefits result from reduced construction costs associated with RCA base. Replacing a portion of virgin aggregate with RCA lessens demand for aggregates thereby reducing the environmental impacts associated with aggregate mining and production.

The research objectives were met by conducting the following tasks:

1) State-of-the-knowledge survey (literature search/review and industry interviews)
2) Case study reviews (RCA production and usage, and specifications review)
3) Performance-related physical and mechanical properties tests on RCA
4) Establish minimum material standards for evaluation, design, and construction

STATE-OF-THE-KNOWLEDGE SURVEY

Literature Search and Review

RCA predominantly replaces virgin aggregate in construction of granular, cement-treated, or econocrete subbase layers and, to a lesser extent, in hot mix asphalt (HMA) and PCC surface layers. Approximately 68 percent of RCA is used as subbase (*Wilburn 1998*). Saeed et al. (*Saeed 2008, 2006, 1997, 1996, 1995*) investigated the use of RCA in granular bases and developed a specification based on laboratory tests.

Upon removal PCC slabs are usually hauled to a central processing facility and broken into smaller pieces. Next reinforcing steel and dowel bars are removed by magnetic separation. The broken PCC is then crushed and screened to produce the specified gradation using conventional equipment. In addition to aggregates, processed RCA has hardened cement paste that holds together smaller aggregate particles. The amount of cement paste attached to aggregate in RCA depends on the process used to produce RCA and the properties of the original concrete (*Saeed 2006, 2008*). Each aggregate size is stored separately.

The RCA production process affects the particle size and shape properties. Using a jaw crusher as the primary crusher and a rotating crusher as the secondary crusher produces the best particle grading and shape. Laboratory tests performed on different sources of PCC show consistent results (*Saeed 2006, 2008*). There is little or no RCA particle breakdown during material handling and construction.

Aggregates account for 10 to 14 percent of the total construction cost (excluding right-of-way and engineering costs). Aggregate recycling is especially economical where the hauling distance for virgin aggregate exceeds 80 km. Crushing costs are generally the only cost associated with recycling concrete pavements. Costs of hauling aggregate and disposing of the old PCC are eliminated and the costs of breaking, removing, separating steel, and transporting are considered incidental (*Saeed 2006*).

The grading of RCA is similar to the grading of crushed stone aggregate. RCA fines are non-plastic, and hardened cement paste attached to the aggregate results in changes in the general aggregate characteristics relative to virgin aggregate that include 1) lower specific gravity, 2) more surface texture, 3) greater water absorption, 4) higher optimum moisture content, 5) higher sulfate soundless loss, and 6) less abrasion resistance. Typical processed RCA properties are:

- Specific Gravity
  - Coarse (plus No. 4 sieve): 2.2 to 2.5
  - Fine (minus No. 4 sieve): 2.0 to 2.3
- LA Abrasion Loss (%): 20 – 45 (coarse)
- California Bearing Ratio (%): 94 to 184
- Absorption (%): Coarse: 2 to 6
  Fine: 4 to 8
- $MgSO_4$ Soundness Loss (%)
  - Coarse: 4 or less
  - Fine: less than 9

Mechanical properties of RCA and virgin aggregate should be compared only under similar testing conditions relevant for RCA use using samples prepared with comparable techniques. Under these conditions, RCA has the several advantages over conventional aggregates such as 1) higher California bearing ratio, 2) higher shear strength, 3) higher rutting resistance (lower permanent strains), and 4) higher resilient modulus, M_R (stiffness).

One study observed that there is an increase in stiffness with time due to hydration properties (*Arm 2000*). Previous research has demonstrated that the residual potential for cement hydration was too small to be of practical use (especially for conventional PCC), and RCA is more likely to be carbonated to the point that hydration potential would not exist (*Poole 1994*). Evaluation of RCA use in pavement base courses indicated that RCA materials often tend to form crusts that could give a false impression of stiffening. These crusts usually form when calcium hydroxide reacts with atmospheric carbon dioxide to form relatively insoluble calcium carbonate. If these crusts form, they have the potential to inhibit drainage through poorly designed filter fabric drainage systems (*Pomeroy 1981, Hansen 1992, Kibert 1994*).

A number of studies have investigated environmental concerns arising from the use of RCA as unbound base material. The effluent has a relatively high pH value at the source, which mitigates rapidly away from the source. No heavy metals are released under alkaline conditions. However, trace amounts of heavy metals well below Environmental Protection Agency (EPA) guidelines are sometimes released under acidic conditions (*Kuo 2002*), which generally do not exist in field conditions. There are no EPA restrictions on using RCA on the site from which it is obtained.

RCA has a proven history of use as base, subbase, fill, and drainage layers within the pavement structure; construction and performance have been excellent (*Saeed 2006, 2008*). The only documented failure of a RCA base was at Holloman AFB in New Mexico, which heaved and expanded due to sulfate attack (*Rollings 2003, 1996*).

Chemical durability of RCA is a concern. Even sulfate-resistant PCC when recycled into base and fill has proven vulnerable to sulfate attack. Until more is understood or effective countermeasures are found, RCA should not be used where exposure to sulfates is likely (*Saeed 2006*).

Results of Interviews with Industry Representatives

Industry representatives provided information on their policies, practices, experiences (including past studies), and perspectives on RCA base material. Project owners, designers, and construction contractors provided RCA relative information on 1) use policy, 2) pavement design considerations, 3) material specifications, 4) construction and constructability issues, and 5) performance observations (*Saeed 2006, 2008*).

There are no restrictions on the use of RCA as unbound base, so all respondents allow and/or use up to 100 percent RCA base. RCA gradation dictates the proportion of virgin aggregate to be added. Economic savings provide the impetus for using RCA, and the choice of whether to use RCA was left to the contractor. The lack of quality virgin aggregate in certain areas and the availability of PCC from an existing or airport pavement also encourage the use of RCA.

RCA is generally considered equivalent to virgin aggregate material for pavement design purposes. RCA is generally treated as virgin aggregate for structural design purposes with 100 percent replacement of virgin aggregate with RCA. RCA has to meet the same specifications requirements (laboratory tests) as virgin aggregate. If RCA meets the grading and Atterberg limits requirements, which it usually does, then it is simply plugged into the system as another aggregate source. Grading and maximum particle size are mostly used as criteria for accepting and control of RCA base. Density tests are typically used for quality control during construction.

Generally, environmental testing of virgin aggregate or RCA is not required when the intended use is as unbound base aggregate. Usually, current virgin aggregate specifications are modified to allow RCA use.

No constructability related concerns were expressed for RCA use relative to virgin aggregates. A few contractors indicated a preference for using RCA over virgin aggregate due to ease of compaction. More importantly, material degradation was not a problem when vibratory rollers are used for compaction. The performance of RCA unbound base is usually reported to be excellent.

ENVIRONMENTAL AND ECONOMIC ASSESSMENTS

RCA must be assessed for value based on technical, environmental, societal, and economic considerations (*Saeed 2006, 2004, 1997, 1996, 1995*). The technical, economic, and environmental aspects of recycling are quantifiable, and Saeed (*Saeed 1996*) proposed an objective methodology for societal assessment.

Environmental Assessment

RCA is being used successfully as base course without evidence of environmental problems. Environmental tests of RCA are usually not required provided the PCC used to produce RCA is obtained from the same location where RCA will be used. The basis for this practice is that PCC is not considered hazardous (because of the slow rate that some chemical constituents could be leached from the material). Even though RCA has an increased surface area compared to its parent concrete, the rate of release of chemical constituents from RCA in contact with groundwater and/or infiltrated with surface moisture remains extremely slow (*Saeed 2006*).

Published information indicates that using RCA as unbound base material is safe, and the potential for environmental hazards is very low or nonexistent. Leaching rates under realistic environmental base conditions are nonexistent to very low at best due to several properties of concrete: low diffusion rates of most substances in PCC, chemical immobilization by interaction with hydrated cement paste, and the relative durability of PCC forms (even relatively small pieces of RCA, as represented in the TCLP [Toxicity Characteristic Leaching Procedure] test) to chemical degradation at all but very low sustained pH's, which are never found in paving base applications. Effects of dilution, environmental degradation of toxic organic compounds and adsorption onto soil particles appears to be a major buffering mechanism (*Saeed 2006*).

Economic Assessment

Economic factors are an important consideration in selecting RCA as a base material. While a life cycle cost analysis (LCCA) would be preferable, the experience data base required to defend such a computation is not currently available. Therefore, the economic assessment is based on initial materials and construction costs. Evaluating the economic viability of RCA involves calculating the tangible costs of using RCA as aggregate and comparing costs with the costs of using virgin aggregate materials. One of the major expenses of using virgin aggregate is the disposal of existing PCC pavement as waste in compliance with current environmental regulations.

The economic assessment should be limited to factors that can be determined rationally and defended as having a direct impact at the project level. The societal benefits, though significant, are none-the-less subjective and thus can be difficult to quantify and defend for a project-level economic assessment. Therefore, societal benefits are more appropriately addressed at the policy level.

In the case of virgin aggregates, several costs must be quantified including 1) cost of virgin aggregate meeting the project specifications delivered to site, 2) cost to demolish and remove the existing PCC, 3) cost of disposal of existing PCC in accordance with environmental regulations, 4) cost of placement of virgin aggregate, 5) cost of compaction, and 6) cost of quality control/quality assurance (*Saeed 2006*).

For RCA, the costs quantified in the economic analysis should include 1) cost of removal of existing PCC, 2) cost for land for crusher, 3) cost of crushing, screening, and stockpiling, 4) cost to meeting environmental regulations at crusher/stockpile site, 5) cost of hauling to/from crushing/stockpiling site, 6) cost of placement of RCA, 7) cost of compaction of RCA, and 8) cost of quality control/quality assurance of RCA

TECHNICAL ASSESSMENT

Comprehensive case studies were conducted on eight sites representing dense graded and open graded base courses with benign, sulfate attack, alkali-silica reactivity (ASR) and D-cracking conditions (see TABLE 1). Site details are provided and discussed elsewhere (*Saeed 2006*).

TABLE 1. Candidate sites for comprehensive case studies.

Base Type	Site Conditions			
	Benign	Sulfate Attack	ASR	D-Cracking
Dense Graded	• Shaw AFB • North Auxiliary Field at Charleston AFB	• Holloman AFB	• Mountain Home AFB • Atlanta-Hartsfield Jackson International	• Grand Forks AFB
Open Graded	• Offutt AFB • US 167, Dubach, LA			

Detailed review of case study sites indicated that RCA is used successfully at commercial and DOD airports as well as highway projects as base and compacted fill material. In most cases, RCA exceeded the virgin aggregate requirements. RCA has been produced successfully to meet open and dense graded aggregate requirements.

Construction using RCA is the same as construction using virgin aggregate materials. Contractors that have experience with RCA construction generally prefer RCA to virgin aggregate for reasons that include 1) easier compaction, 2) perceived better than virgin aggregate, 3) easier to handle during construction, and 4) stable working platform allowing work to continue even when wet. Special construction equipment is not required. Contractors used different means to spread RCA. Some contractors preferred spreaders to spread plant-mixed RCA base material to control segregation, while others felt that a motor grader or a bulldozer did an adequate job of spreading RCA dumped in windrows. The method used was usually based on the contractor's experience. Static and vibratory rollers have been used successfully to compact RCA with no degradation. The contractors noted that RCA has a relatively high water demand, and RCA should not be compacted to a moisture level below optimum.

Experience is the key to using RCA as base course aggregate, and some potential problems can be avoided with careful planning. Some of these include (*Saeed 2006*):

- High water demand: RCA typically compacts at higher water content relative to virgin aggregate material.
- Segregation: RCA segregation problems can be traced to poor stockpiling when RCA is produced or when it is spread into lifts before compaction.
- Grading control: RCA grading before and after compaction should be specified to alleviate concerns that RCA tends to degrade during construction.
- Plant operations: Proper crushing plant operation is the key to producing quality RCA that is crushed and not just separates the aggregate from the cement mortar.

The literature search reported several instances where RCA has been tested for conformance with general virgin aggregate material requirements. Questions still remained about the adequacy of RCA produced from distressed PCC, especially PCC that has had ASR or D-cracking. Limited laboratory tests on distressed RCA from formerly Pease AFB (ASR distressed) and Grand Forks AFB (D-cracked) provided insight into this aspect of producing and using RCA. Detailed test results are reported by Saeed et al. (*Saeed 2006*).

Information gathered during laboratory testing pointed to the fact that RCA is comparable to virgin aggregate material for base course application. Tests on RCA produced from ASR-distressed and D-cracked PCC, supplemented with experience from similar research efforts and other project tasks, indicated the following:

- Sieve Analysis
  - RCA could be produced and blended to provide the desired gradation
- Moisture-Density
  - RCA has a higher optimum moisture content (OMC) and lower maximum dry density (MDD) than virgin aggregate
- Static Triaxial Shear
  - RCA from distressed PCC is comparable to typical virgin aggregate
- Repeated Load Triaxial Shear
  - RCA permanent deformations are comparable to virgin aggregate material
  - RCA compared well with a typical virgin aggregate at a failure permanent deformation strain of 10 percent.

No chemical reaction tests were conducted. Deleterious chemical reactions in PCC (such as ASR) are expected to continue when RCA is used as unbound base layer aggregate. However, this should not be a significant source of damage in unbound base course applications because the porosity in the structure of unbound base should allow expansion of the reaction product without deleterious stress within the structure. This expectation is supported by the absence of known cases of pavement failures due to residual ASR in unbound base. Given the current level of concern about ASR in pavements and the absence of quantitative knowledge, Saeed et al (*Saeed 2006*) have proposed a method to assess the use of RCA from ASR-distressed PCC in unbound base courses according to the importance of use.

MATERIAL STANDARDS

Material requirements for RCA base course are more stringent than those of an aggregate base course. A base course constructed with crushed material will usually have higher strength and stability. Published literature, limited laboratory tests, and specifications from current projects were studied to develop minimum material standards for RCA (*Saeed 2006*). These minimum material standards allow designers to use RCA with confidence. RCA is treated as virgin aggregate material and has to meet all virgin aggregate material requirements. Some of the requirements are relaxed where these are not applicable for obvious reasons.

Aggregate

RCA is to be derived from PCC pavement. RCA should consist of clean, sound, and durable crushed particles and be free of silt, clay, organic matter, HMA, steel reinforcement, or other objectionable material. An incidental amount of recycled asphalt concrete pavement could be present in the RCA, and to control the amount asphalt concrete, overlays should be removed from the surface prior to pavement removal and crushing. Also, full-slab asphalt concrete panels (used as a replacement for a removed PCC slab) should be removed. A reasonable number of small asphalt concrete patches (less than a full slab panel) may be incorporated with the RCA.

Aggregate Grading

Aggregate mass grading is determined in accordance with ASTM C136. The specified aggregate grading refers to grading of the stockpiled material. The maximum aggregate size should be limited to 50 mm.

Percent Abrasion Loss

The Los Angeles Abrasion test has long been used as an index for aggregate toughness. The test, as described in AASHTO T96 or ASTM C131, is included in aggregate material specifications. The LA Abrasion test loss is generally limited to 40 percent. However, this limit can be increased to 45 percent for RCA. This practice is allowed at RCA construction sites with no discernable impact on performance.

Flat and Elongated Particles

The shape of the aggregate particle has long been used to judge the potential of an aggregate to resist permanent deformation. Flat and elongated (F&E) particles can break during mixing, hauling, and placing, and especially under compaction, and ultimately change aggregate grading. F&E particles can also influence ease of compaction. An excess of such particles can be detrimental to good performance.

The shape of the aggregate particles in terms of the percentage of F&E particles is assessed in accordance with test procedure ASTM D4791. A flat particle is defined as one having a ratio of width to thickness greater than 3, and an elongated particle is defined as one having a length to width ratio greater than 3. The amount of F&E particles is limited to 30 percent. Some of the more restrictive specifications have limited the amount of F&E particles to 20 percent on material retained and passing the 12.5-mm sieve.

Percent of Fractured Particles

Fractured particles consist of crushed RCA particles containing fine and coarse aggregates within the concrete matrix and are not limited to single aggregate particles without attached matrix. A higher percentage of fractured particles contribute to an increase in shear strength. The percentage of fractured particles is determined in accordance with ASTM D5821. A fractured particle is defined as having two or more fractured faces with the area of each face being at least 75 percent of the smallest midsectional area of the piece. For contiguous fractures, the angle between the fractures planes should at least be 30 degrees to be considered as two fractured faces. The fractured particles should at least be 50 percent of the by weight of the material retained on each specified sieve.

Soundness Test

Aggregate durability or resistance to weathering is determined using the $MgSO_4$ or $NaSO_4$ soundness test conducted in accordance with ASTM C88 (AASHTO T104). This test simulates the weathering action using crystallization of soluble salts in aggregate pores. Test results are represented as percent loss. This requirement is usually waived for RCA materials because this test is chemically unsuited. The sulfate component of sodium or magnesium sulfate salts reacts with the concrete mortar, leading to erroneous results.

Liquid and Plasticity Limits

The liquid and plasticity limits of the aggregate fraction finer than the 0.425-mm (No. 40) sieve are determined in accordance with AASHTO T89 and T90, respectively. Atterberg Limits are indexes defined as the moisture contents at which the fine content (passing No. 200 [75 μm] sieve) changes from one state into another (i.e., from solid to semi-solid as moisture increases beyond the plastic limit). The Atterberg Limits are specified for the completed course. The portion passing the No. 40 (425 μm) sieve is specified to be nonplastic or have a liquid limit of less than 25 and plasticity of less than 5.

CONCLUSIONS

- RCA can be used as unbound base material if produced from uncontaminated PCC. RCA should not be used where there is a potential for sulfate exposure from subgrade soils, ground water, or other external sources.
- RCA is not a hazard to the environment. Localized environmental effects from raised pH in leachate are insignificant. Discharge of heavy metals or organics from common sources of RCA, if any, is insignificant.
- Hydration has not been shown as a chemical phenomenon in RCA. The perception that unbound RCA base can gain strength through hydration is not supported.
- An economic analysis can be conducted by considering initial material and construction costs for both RCA and virgin aggregate. RCA typically is a better economic option, considering transportation costs of virgin aggregate and disposal costs of PCC.
- All the virgin aggregate tests and their limits are applicable to RCA except the sulfate soundness test. The sulfate soundness test is waived for RCA due to the incompatibility of PCC components with the chemical reactants used in the test.
- RCA from ASR-distressed PCC can be used considering site conditions.

RECOMMENDATIONS

- The adoption of these guidelines will facilitate the use of RCA in pavement construction providing serviceable pavements while accruing economic and environmental benefits.
- The use of ASR-distressed RCA should be based on an evaluation of the site conditions, severity of ASR, and other factors. RCA manufactured from PCC with mild severity ASR can be used as an unbound base in most airfield pavement applications. However, in case of RCA manufactured from PCC with aggressive severity ASR, recommendations by Saeed et al. (*Saeed 2006*) should be followed.

ACKNOWLEDGEMENTS

This research was conducted under the auspices of Innovative Pavement Research Foundation (IPRF) funding by Federal Aviation Administration (FAA). The authors would like to acknowledge IPRF, FAA, the technical panel and other project team members for their help and support throughout this research effort.

REFERENCES

Arm, M. (2000). "Self-Cementing Properties of Crushed Demolishing Concrete in Unbound Layers: Results from Triaxial Tests and Field Tests." *WASCON 2000 Proceedings,* pp 579-587.

Chesner, W., Collins, R., MacKay, M. and Emery, J. (1998). *User Guidelines for Waste and Byproduct Materials in Pavement Construction,* FHWA Report FHWA-RD-97-148, Federal Highway Administration, McLean, Virginia.

Chini, A. and Kuo, S. S., Duxbury, J. P., Monteiro, F.M.B., Mbwambo, W.J. (1998). "Guidelines and Specifications for the Use of Reclaimed Aggregates in Pavement," Florida Department of Transportation Final Report for Contract BA 509, State Materials Office, Gainesville, FL.

Federal Aviation Administration (FAA) (2004). "AC 150/5370-10B – Standards for Specifying Construction of Airports – Part V," Draft Advisory Circular, US Department of Transportation, FAA.

Federal Aviation Administration (FAA) (2005). "Engineering Brief No. 150 – Accelerated Alkali-Silica Reactivity in Portland cement concrete pavements exposed to runway deicing chemicals." Engineering Briefs, US Department of Transportation, FAA.

Hansen, T. C. (1992). *Recycling Demolished Concrete and Masonry.* Report of Technical Committee 37-DRC Demolition and Reuse of Concrete, RILEM. E&FN Spon, 316pp.

Kibert, C. J. (1994). "Concrete/Masonry Recycling Progress in the USA." in Lauritzen, 1994, pp 83-91.

Kuo, S.-S., Mahgoub, and H.S., Nazef, A. (2002). "Investigation of Recycled Concrete Made with Limestone Aggregate for a Base Course in Flexible Pavement," *Transportation Research Record 1787,* Transportation Research Board, pp. 99-108.

Mack, J.W., Solberg, C. E., and Voigt, G.F. (1993). "Recycling Concrete Pavements," *Aberdeen's Concrete Construction*, vol. 38, no. 7, July 1993, pp. 470-473.

Pomeroy, C. D. (1981). "Report on Workshop 3 - 'Reuse of Concrete' (Other Than as Aggregate for Concrete)." in Kreijger, pp. 343-345.

Poole, T. (1994). "Recycled Concrete: Investigation of Properties of Fine Material from Crushed, Hydrated Mortars." Waterways Experiment Station, Vicksburg, MS, USA.

Rollings, R., Burkes, J.P., and Rollings, M.P. (1999). "Sulfate Attack of a Cement Stabilized Sand," *Journal of Geotechnical and Geoenvironmental Engineering*, vol. 125, no. 5, pp. 364-372.

Rollings, M.P. and Rollings, R.S. (2003). "Sulfate Attack on Bound Bases," Transportation Research Board 2003 Annual Meeting CD-ROM, Washington, D.C.

Saeed, A., Hudson, W. R., and Anaejionu, P. (1995). "Location and Availability of Waste and Recycled Materials in Texas and Evaluation of their Utilization Potential in Roadbase," Research Report 1348-1, Center for Transportation Research, The University of Texas at Austin, Austin, Texas, 40 pp.

Saeed, A. and Hudson, W. R. (1996). "Evaluation and the Use of Waste and Reclaimed Materials in Roadbase Construction," Research Report 1348-2F, Center for Transportation Research, The University of Texas at Austin, Texas, Austin, Texas.

Saeed, A. and Hudson, W. R. (1997). "Recycled Materials in Roadbase," *Proceedings*, 5th Annual ICAR Symposium, International Center for Aggregates Research, The University of Texas at Austin, Austin, Texas, pp. B1-2-1–B1-2-11.

Saeed, A., Hall, J. W. and Barker, W. (2001). "Performance-Related Tests of Aggregates for Use in Unbounded Pavement Layers," *NCHRP Report 453*, Project 4-23, National Cooperative Highway Research Program, Washington D.C.

Saeed, A., Hammons, M., Rufino, D., and Poole, T. (2006). "Evaluation, Design, and Construction Techniques for Airfield Concrete Pavement Used as Recycled Material for Base," *Final Report IPRF-01-G-002-03-5*, Innovative Pavement Research Foundation, Airport Concrete Pavement Technology Program, Skokie, Illinois.

Saeed, A. (2008). "Performance Related Tests of Recycled Aggregates for Use in Unbound Pavement Layers," *NCHRP Report 598*, Project 4-31, National Cooperative Highway Research Program, Washington D.C.

United States Geological Survey (2007). "2006 Minerals Yearbook – Stone, Crushed," USGS, United States Department of the Interior, Washington, D.C.

Wilburn, D. R. and Goonan, T. G. (1998). "Aggregate from Natural and Recycled Sources, Economic Assessment of Construction Applications – A Materials Flow Analysis," *USGS Circular 1176*, United States Department of Interior, Washington, D.C.

Location and Timing of Fatigue Cracks on Jointed Plain Concrete Pavements

Jacob E. Hiller[1], M.ASCE and Jeffery R. Roesler[2], A.ASCE

ABSTRACT

The prediction of fatigue cracking in jointed concrete pavements has traditionally focused on transverse cracking initiating from the bottom of the slab and propagating both up and across the slab width. However, field surveys of several locations, particularly in the Western United States indicate that alternative fatigue cracking mechanisms, such as top-down and bottom-up longitudinal cracking, top-down transverse cracking, and corner cracking, exist in substantial quantity. To better understand these alternative cracking mechanisms and to account for such mechanisms in both the analysis and design of jointed concrete pavements, a mechanistic analysis software program named RadiCAL was developed. Mechanistic parameters such as built-in and cyclical curling, axle spacing effects, load transfer, etc. was utilized to predict the critical fatigue crack location using the method of linear fatigue damage accumulation. Several sites in California that have exhibited both traditional and alternative fatigue cracking mechanisms were examined using RadiCAL to validate the proposed mechanistic analysis principles for stress and fatigue damage development.

1 INTRODUCTION

1.1 Background

With the development of mechanistic-empirical (M-E) design methods to predict fatigue cracking under a variety of climatic, material, pavement geometries, and loading conditions, advancements in pavement technology now allow for better understanding of alternative fatigue failure modes in rigid pavements. The vast

---

[1] Department of Civil and Environmental Engineering, Michigan Technological University, 1400 Townsend Drive, 201F Dillman Hall, Houghton, Michigan, USA; PH (906) 487-3053; FAX (906) 487-1920; e-mail: jhiller@mtu.edu

[2] Department of Civil and Environmental Engineering, University of Illinois at Urbana-Champaign, 205 N. Mathews Ave., 1211 Newmark Civil Engineering Laboratory, MC-250, Urbana, Illinois, USA; PH (217) 265-0218; FAX (217) 333-1924; e-mail: jroesler@uiuc.edu

majority of M-E design has focused on transverse fatigue cracking criterion initiated by loads placed at the mid-slab edge, as this would accommodate the major distress seen on many rigid pavement sections. However, condition surveys of pavements without dowels or tied shoulders in California and Washington have shown that longitudinal and corner cracking occur as frequently as transverse cracking (Mahoney et al. 1991, Harvey et al. 2000b, Roesler et al. 2000). However, no current M-E design methodology directly accounts for these alternative failure modes.

1.2 Built-in Curling of Concrete Slabs

Studies (Hatt 1923, Hveem 1951, Armaghani et al. 1987, Yu et al. 1998) have shown that many factors can cause an upward curling of the slab, which suggests that transverse joints loading deserves greater consideration in concrete pavements. If large positive temperature gradients (temperature of slab is hotter on top than bottom) exist as the concrete hardens shortly after construction, the slab will curl upward as the pavement cools and reach a zero-gradient condition. This built-in curling can reach magnitudes of 0.55 °C/cm or more (Eisenmann and Leykauf 1990). Differential drying shrinkage can also add to this built-in curling effect and can be more pronounced in drier, less humid climates as those found in the western United States. Permanent upward curling of concrete slabs has been noted for many years in the state of California (Hveem 1951). This phenomenon is affected by factors such as construction temperature conditions, curing methods, slab geometry, base type, and restraint conditions. This permanent curl can fundamentally change the baseline for stress development in rigid pavements (Hiller and Roesler 2005a).

To quantify this factor, Rao and Roesler (2005) developed a term called the Equivalent Built-In Temperature Difference (EBITD). This value can be backcalculated in a variety of ways including falling weight deflectometer testing, surface profiling, joint deflection or multi-depth deflectometers measurements with mechanical and/or environmental loading. This value, as defined by Rao and Roesler, takes into consideration the built-in temperature difference from construction, permanent differential shrinkage, and reversible moisture gradients as well as creep of the concrete which could negate some of this permanent curl. This term is expressed as the negative value of the slab's temperature differential required to achieve full-support of the slab with the underlying layers. Without restraint from a tied shoulder or doweled transverse joint, EBITD levels have been found to be highly negative and can approach -30ºC in extreme cases (Rao and Roesler 2005). The addition of the EBITD to the actual temperature difference (ΔT) in the slab can be expressed as an equivalent ΔT (ΔT_{eq}) which would characterize all of the curling components present in the slab.

An important input needed for evaluating rigid pavements is the inclusion of this EBITD value into the analysis. The addition of the EBITD level effectively shifts the frequency distribution of the daily cyclical slab ΔT to ΔT_{eq} as demonstrated in Figure 1 for the California Central Valley climatic region (Sacramento). In this case, an EBITD level of -20ºC can shift the ΔT_{eq} distribution almost entirely into the negative temperature difference region, as seen in Figure 1. This EBITD would limit the occurrences of a downward curled slabs even under the largest positive ΔT. This condition also limits the extreme stress cases that typically drive fatigue damage development at the bottom of the slab, mid-way between the transverse joints.

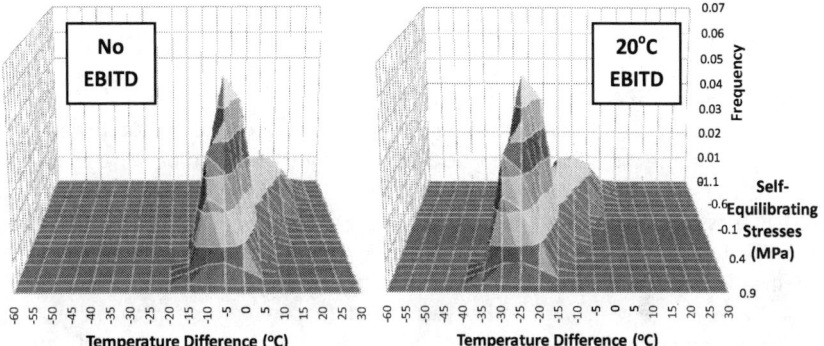

Figure 1. Frequency Distribution Shift of Temperature Differential for Sacramento, California due to EBITD including Non-Linear Self-Equilibrating Stresses at the Top and Bottom of the Slab.

1.3 Fatigue Transfer Functions

In rigid pavement fatigue analysis, traditional models for fatigue utilize a semi-logarithmic relationship between the stress ratio (maximum stress to the strength of the material) and the number of repetitions to failure. The newly completed Mechanistic-Empirical Pavement Design Guide (MEPDG) (ARA 2007) utilizes such a transfer function in assessing jointed concrete pavement fatigue.

$$Log(N) = 2.0 \left(\frac{MOR}{\sigma_{max}} \right)^{1.22} \qquad (1)$$

where MOR = concrete modulus of rupture;
σ_{max} = max tensile stress applied during cyclic loading from load and curling
N = number of loading cycles to failure (at 50% reliability)

When considering both the maximum stress developed from mechanical loading in conjunction with residual bending stresses from temperature and built-in curling effects, a stress range approach to concrete pavement fatigue can be applied. Many researchers (Murdock and Kesler 1958, Awad 1974, Tepfers 1979, Rao 2005) have documented the influence of stress range on the fatigue life of concrete. By incorporating the stress range concept into fatigue-based design, residual stresses from EBITD and cyclic curling can be considered. Tepfers (1979) proposed a stress range model for concrete fatigue as seen in equation (2), which can be utilized to determine the allowable number of loads for a given load/climate combination.

$$\frac{\sigma_{max}}{MOR} = 1 - \beta(1-R)\log_{10}N \quad (2)$$

where $R = \sigma_{min}/\sigma_{max}$ residual flexural stress in slab before load
β = calibration coefficient (0.0685 for concrete by Tepfers)

1.4 Development of RadiCAL

With the importance of built-in curling and its potential effects on rigid pavement cracking phenomenon, a deterministic analysis program for rigid pavements named RadiCAL (Rigid Pavement Analysis for Design in CALifornia) by Hiller and Roesler (2005b) was developed to reproduce the observed corner and longitudinal cracking in California. RadiCAL specifically uses statistical distributions of inputs such as traffic classifications, load spectra, axle spacing distributions, and climatic influences in conjunction with design parameters such as traffic counts, built-in curl level, slab geometry, load transfer level, etc. RadiCAL calculates both the level of fatigue damage as well as the locations of damage using Miner's Hypothesis (Miner 1945) for designing against premature fatigue cracking in the transverse and longitudinal directions on the top and bottom of the slab. This program can utilize many fatigue transfer functions in the analysis of concrete pavement sections, including those in equations (1) and (2). More information on the development, inputs, and features of RadiCAL can be found in Hiller and Roesler (2005a, 2005b) and Hiller (2007).

The development of RadiCAL and the subsequent addition of self-equilibrating stresses from non-linear temperature profiles (Hiller 2007) provide an excellent analysis tool for evaluating the impact of individual input parameters on both the level and location of critical fatigue damage in JPCPs. However, one major drawback is that the stress prediction algorithm, damage calculation procedure, and fatigue models used in RadiCAL are not calibrated with in-service field sites. This study aims to assess the predictive power of RadiCAL in terms of matching fatigue damage mechanisms of several California sites through a "design verification" process. This process utilizes RadiCAL as a forensic tool to match the predominant fatigue cracking modes, while ignoring the timing of crack initiation due to the lack of information.

2 DESIGN VERIFICATION OF IN-SERVICE SECTIONS

2.1 Background

Using the RPPR (Rigid Pavement Performance/Rehabilitation) database (Smith et al. 1998), several projects in California with the occurrence of fatigue cracking at multiple locations on the slab were examined as a preliminary design confirmation of the predicted damage locations. The University of California Pavement Research Center (UCPRC) has also overseen a vast data acquisition process conducted by

Stantec for numerous flexible, rigid, and composite pavement sections within the state of California. Correspondingly, the UCPRC has shared much of this data to conduct some trial design verification studies of RadiCAL. Using the two data sources, three of these sections of jointed plain concrete pavements (Table 1) with a variety of fatigue cracking mechanisms are shown using site specific climatic data, EBITD, geometry, and load spectra in RadiCAL and compared with the actual cracking patterns noted during recent condition surveys.

Table 1. Data for Three In-Service Jointed Concrete Pavements Sections in California.

Section #	CA1-3	04-N253	04-N284
Route / Direction	I-5 Northbound	I-80 Eastbound	US-101 Northbound
County	San Joaquin	Solano	Sonoma
Climatic Region	Central Valley	Central Valley	North Coast
PCC Thickness (cm)	21.3	25.4	25.4
Base Thickness (cm) / Type	13.7 / CTB	19.1 / CTB	11.4 / AC
Subbase Thickness (cm)	61	25	15
Joint Spacing (m)	3.7, 4.0, 5.8, 5.5	4.6	3.7, 4.3, 4.0, 4.6
Transverse (% Slabs Cracked)	18	82	6
Longitudinal (% Slabs Cracked)	15	88	11
Corner (% Slabs Cracked)	n/a	9	3
Average EBITD (°C)	-13	-5	-6
Average LTE (%)	20	50	50
Average k-value (kPa/mm)	70	40	55

2.2 RRPR Site CA 1-3

Section CA1-3 is a small section of a larger jointed plain concrete pavement test section on northbound I-5 in Tracy, California (Central Valley climatic region). It was built in 1971 with joints designed to be skewed with no load transfer devices. The EBITD value of -13°C was estimated from the FWD corner deflection and ΔT in the slab was compensated for using the method from Rao and Roesler (2005). However, if erosion of the base exists, this process would erroneously equate the additional deflections to a more extreme EBITD value.

Using the backcalculated EBITD value, measured load transfer, climatic zone, slab thickness, site specific load spectra from northbound I-5 in Tracy (Lu et al. 2001), and variable joint spacing, predicted relative damage profiles using stress range (Figures 2a and 3a) and maximum stress fatigue approaches (Figures 2b and 3b) were generated for 3.7m and 4.6m joint spacings as these are the only joint spacings available for analysis using RadiCAL. The relative damage profiles give an indication of the location of damage along the edges of the slab where fatigue cracking would initiate from without noting the timing of the crack initiation. Relative damage above the slab indicates top-down fatigue damage and vice versa in these profiles.

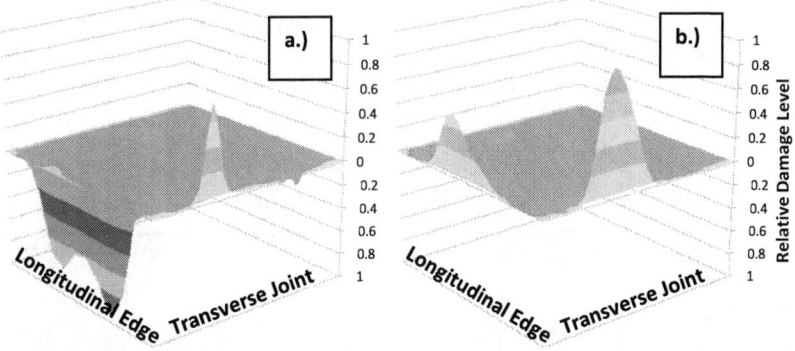

Figure 2. RadiCAL Relative Damage Profiles of Section CA 1-3 at 3.7m Joint Spacing with a.) Stress Range and b.) MEPDG Maximum Stress Damage Analysis.

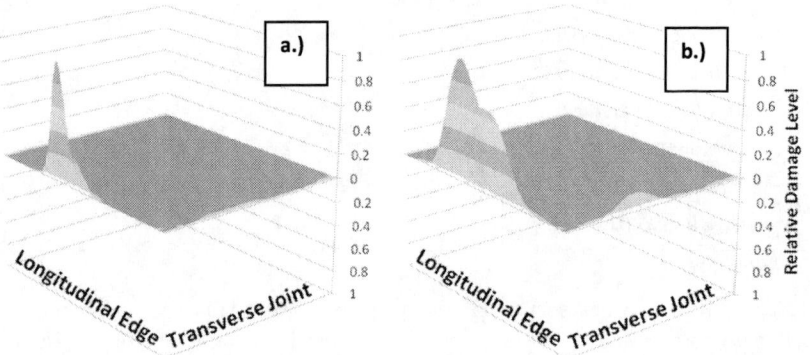

Figure 3. RadiCAL Relative Damage Profiles of Section CA 1-3 at 4.6m Joint Spacing with a.) Stress Range and b.) MEPDG Maximum Stress Damage Analysis.

Section CA1-3 exhibited 18% of slabs cracked transversely and roughly 15% of the slabs with longitudinal cracks. The damage profile in Figure 3a using a stress range approach would project a higher percentage of top-down transverse cracking with only a slight probability of longitudinal cracking (traffic would be traveling parallel to the longitudinal edge). For the shorter joint spacing in Figure 2a using stress range, both bottom-up transverse cracking and longitudinal cracking between the wheelpaths is predicted. While this damage profile does not exactly match the distresses found on CA 1-3 for all joint spacings, the probability for both transverse and longitudinal cracking does exist. With the assumption that no temperature gradient existed in the slab during the time of the FWD testing, this may have lead to an improper calculation of EBITD. Daytime testing of this section could add -3ºC to -6ºC to the EBITD level, thereby increasing the longitudinal fatigue cracking potential due to more unsupported corners of the slab (Hiller and Roesler 2005a)

Figure 2b shows the damage profile using a maximum stress fatigue approach on the shorter 3.7m slabs. Just as with the stress range approach, both top-down transverse and longitudinal cracking potential is shown in this case. When modeling 4.6m joint spacing slabs (Figure 3b), the predicted damage profile predicts top-down transverse cracking near midslab, but also a smaller potential for longitudinal cracking to form at the transverse joint. This matches the cracking found on section CA 1-3 slightly better than using the stress range approach for both joint spacings available for analysis in RadiCAL.

2.3 UCPRC Site 04-N253

Section 04-N253 is located on eastbound Interstate-80 in Solano County, between San Francisco and Sacramento (Central Valley climate). This segment was originally constructed in 1946 with a widening of the outside lane in 1963. During FWD testing, only one slab was found in the 150m site that was fully intact for backcalculation, exhibiting an EBITD value of -5ºC. Since no replicates were available to assess the level of permanent built-in curling site 04-N253, this value is fairly unreliable. Unlike the previous section 04-N249, this section is severely damage with respect to fatigue cracking with 82% of slabs with transverse cracking, 88% of slabs with longitudinal cracking, and 9% of the slabs with corner cracking. One difference with this site in comparison with CA1-3 is that the longitudinal cracks appear to be the original fatigue failure mechanism as most of the transverse cracks are arrested when encountering longitudinal cracks.

Figure 4 shows the fatigue damage profiles for (a.) stress range fatigue equation and (b.) MEPDG fatigue equation. In both of these cases, the fatigue damage predicts primarily transverse cracking. Since this site exhibits large amount of longitudinal cracking neither fatigue algorithm matches well.

As the calculated EBITD value is quite moderate at -5ºC, this predicted critical damage mechanism is not unexpected. To produce longitudinal cracking for this geometry, load spectra, climate, and shoulder type, an EBITD value around -8ºC to -11ºC would need to exist as shown in Figure 5. With the exception of the alternative fatigue cracking patterns, no quantitative evidence in terms of surface profile or backcalculation of EBITD exists to supports a more negative level of EBITD due to the fact that only one slab was fully intact for FWD testing. Therefore, the other conclusion is that RadiCAL does not properly characterize the damage profile for this section for the original input assumptions.

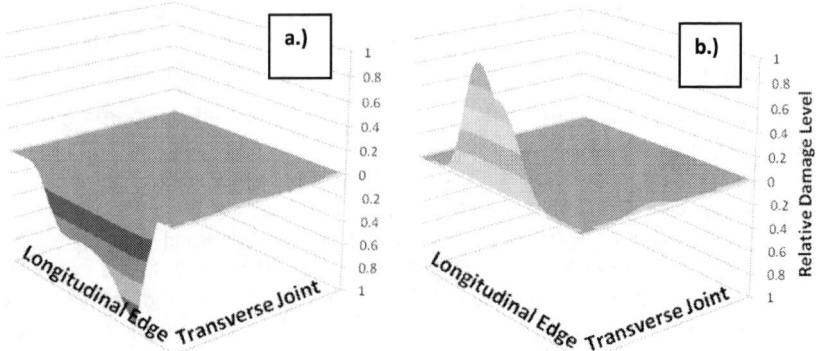

Figure 4. RadiCAL Relative Damage Profiles of Section 04-N253 with a.) Stress Range and b.) MEPDG Maximum Stress Damage Analysis.

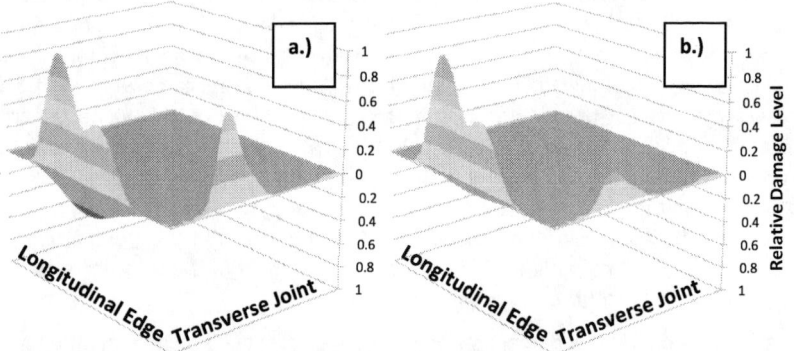

Figure 5. Relative Damage Profile of Section 04-N253 with MEPDG Maximum Stress Damage Analysis for an EBITD of a.) -8ºC and b.) -11ºC.

2.4 UCPRC Site 04-N284

Section 04-N284 resides along the North Coast of California, which possesses a distinctly more temperate climate than the previous two sections. Constructed in 1993, this section exhibits little fatigue cracking with approximately 6% of slabs cracked transversely, 11% of slabs cracked longitudinally, and 3% exhibiting corner cracking. Backcalculation of FWD data reveals a low EBITD level of -6ºC. Due to the variable joint spacing on site 04-N284, fatigue damage analyses were conducted for a shorter 3.7m joint spacing (Figure 6) and longer 4.6m joint spacing (Figure 7).

For both the shorter (Figure 6a) and longer (Figure 7a) joint spacing using the stress range approach, the critical damage location is transverse cracking offset from the mid-slab edge. Bottom-up longitudinal cracking is probable for both joint spacings but its relative damage level is only 0.1. The locations of these damages in relation to the damage along the longitudinal edge could potentially lead to corner cracking initiating roughly 0.6m from the corner of the slab if the crack propagates in that direction. The predicted longitudinal cracking using this method is small, which is not consistent with the primary cracking mechanism observed for this section.

For 3.7m joint spacings, the MEPDG transfer function predicts a high level of damage (Figure 6b) at both the transverse joint (top-down between the wheelpaths) and the longitudinal edge of the slab (top-down near mid-slab). In this section, the longitudinal cracking is more prevalent on the slabs with shorter joint spacing as stresses from combined curl and load at the mid-slab edge are reduced. This trend using the MEPDG fatigue function in RadiCAL appears to be a better predictor of longitudinal cracking for this site's conditions. On this site, the observed corner cracks are quite large in nature, initiating 1.2 to 1.8m from the corner of the slab. This also matches the trend predicted by RadiCAL using the MEPDG fatigue function.

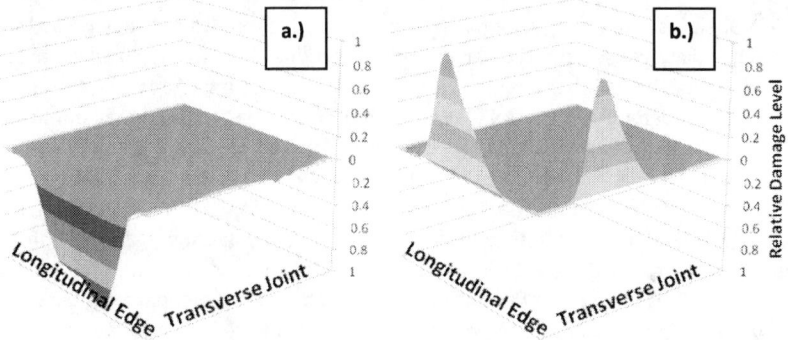

Figure 6. RadiCAL Relative Damage Profiles of Section 04-N284 at 3.7m Joint Spacing with a.) Stress Range and b.) MEPDG Maximum Stress Damage Analysis.

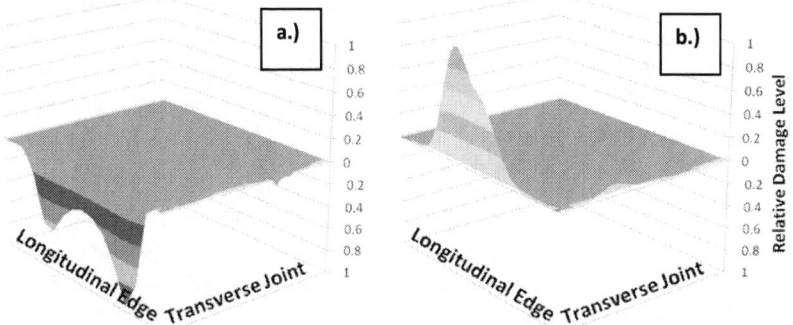

Figure 7. RadiCAL Relative Damage Profiles of Section 04-N284 at 4.6m Joint Spacing with a.) Stress Range and b.) MEPDG Maximum Stress Damage Analysis.

3 DISCUSSION

Analysis of these three jointed concrete pavement sections in California show that proper characterization of the EBITD to account for built-in curling is imperative for prediction of both traditional and alternative fatigue cracking mechanisms. Sites such as 04-N253 on I-80 do not provide enough forensic data to properly conduct a design verification or calibration of such method for use in future design and should therefore not be included in such activities. Further design verification analyses of numerous sections in California can be found in (Hiller 2007).

In addition, proper characterization of the temperature curling mechanisms in terms of an actual temperature profile in the slab during FWD testing is critical. The EBITD backcalculation process cannot separate curling or warping from several sources. As the largest source of curl in most cases, cyclic temperature curling needs to be accounted for in backcalculation of more permanent, non-reversible sources of curling in concrete slabs.

The use of stress range fatigue algorithms in RadiCAL tended to produce cases that could potentially cause longitudinal or corner cracking, but not as significantly as with the use of the MEPDG fatigue function for these sections. As the stress range function developed by Tepfers (1979) was calibrated for beam data, it may not be an accurate representation of slab fatigue behavior. Calibration of the β coefficient to field slab data should improve its predictive capability. Domenichini and Marchionna (1981) calibrated this β coefficient for field slabs simply using the Westergaard stress solution (1927) at the mid-slab edge location and found a change from Tepfers' default β value of 0.0685 to 0.0954 to account for factors not present in laboratory beam testing. This increase in the β coefficient would create a steeper fatigue curve, thereby reducing the load repetitions required to initiate significant fatigue damage in a linear damage accumulation process like that used in RadiCAL.

4 CONCLUSIONS

Three JPCP sections in northern California covering two distinct climatic zones (North Coast and Central Valley) were analyzed with site specific geometry, shoulder considerations, and load spectra to attempt to validate the predictive power of RadiCAL in terms of fatigue cracking locations. Limited FWD backcalculation on each of these sites permitted site-specific load transfer and EBITD values to use in RadiCAL as well. Each of these sites was analyzed with non-linear temperature considerations for both the Tepfers' stress range and MEPDG maximum stress fatigue transfer functions to determine the best predictor of crack location. Unfortunately, a wealth of data did not exist for either the RPPR or UCPRC sites in terms of the timing of crack initiation or changes in traffic levels and thus engineering judgment was used on the timing of initial cracking mechanisms.

When numerous replications of EBITD where available, the maximum stress approach using the MEDPG transfer function in RadiCAL tended to produce fairly realistic damage profiles in comparison with observed cracking patterns. For the section 04-N253 with little intact slabs for EBITD backcalculation, neither fatigue transfer function predicted the observed field cracking patterns well, thereby indicating that a higher level of EBITD could have previously existed on that section before massive fatigue crack propagation occurred.

While a clear picture was not found in terms of the best fatigue transfer function to predict locations of cracking, the results show that the use of several functions can be used in conjunction with calibration to design against both traditional and alternative jointed concrete pavement fatigue cracking mechanisms.

ACKNOWLEDGEMENTS

The research presented herein was conducted under a contract from the University of California Pavement Research Center and the support of Caltrans. Financial assistance was also provided through the FHWA Eisenhower Transportation Fellowship program. Additional work in this area has been funded by the Illinois Department of Transportation through the Illinois Center of Transportation and the Illinois Chapter of the American Concrete Pavement Association. The financial assistance received from all sources is greatly appreciated.

REFERENCES

ARA, Inc. 2007. *Interim Mechanistic-Empirical Pavement Design Guide Manual of Practice*. Final Draft. NCHRP Project 1-37A.

Armaghani, J.M., Larsen T.J., & Smith, L.L. 1987. Temperature Response of Concrete Pavements, *Trans. Research Rec. 1121*, TRB, National Research Council, Washington, D.C., 23-33.

Awad, M.E. & Hilsdorf, H.K. 1974. Strength and Deformation Characteristics of Plain Concrete Subjected to High Repeated and Sustained Loads, *Abeles Symp., Fatigue of Concrete*, ACI Publication SP-41, 1-13.

Domenichini, L. & Marchionna, A. 1981. Influence of Stress Range on Plane Concrete Pavement Fatigue Design, *Proc., 2nd Int. Conf. on Concrete Pavement Design*, Purdue University. West Lafayette, IN, 55-65.

Eisenmann, J., & Leykauf, G. 1990a. Effects of Paving Temperature on Pavement Performance, *Proc., 2nd Int. Workshop on the Theoretical Design of Concrete Pavements*, Siquenza, Spain.

Harvey, J.T., Roesler, J.R., Farver, J. & Liang, L. 2000b. *Preliminary Evaluation of Proposed LLPRS Rigid Pavement Structures and Design Inputs*, Final Report, FHWA/CA/OR-2000/02, UCPRC, Richmond, California.

Hatt, W. K. 1923. The Effect of Moisture on Concrete, *ASCE Transactions*, 89, 271-315.

Hiller, J.E. 2007. *Development of Mechanistic-Empirical Principles for Jointed Plain*

Concrete Pavement Fatigue Design. Ph.D. Dissertation. University of Illinois at Urbana-Champaign, Urbana, Illinois.

Hiller, J.E., & Roesler, J.R. 2005a Determination of Critical Concrete Pavement Fatigue Damage Locations Using Influence Lines, *J. of Trans. Eng.*, ASCE, 131(8), 599-607.

Hiller, J.E., & Roesler, J.R. 2005b *User's Guide for Rigid Pavement Analysis for Design in California RadiCAL Software*. Version 1.2, UCPRC, Richmond, CA.

Hveem, F. N. 1951. Slab Warping Affects Pavement Joint Performance, *J. of American Concrete Inst.*, 47, 797-808.

Lu, Q., Le, T., Harvey, J.T., Lea, J., Quinley, R., Redo, D. & Avis, J. 2001 *Truck Traffic Analysis using WIM Data in California,* Draft Report, UCPRC, Richmond, CA.

Mahoney, J., Lary, J.A., Pierce, L.M., Jackson, N.C., & Barenberg, E.J. 1991. *Urban Interstate Portland Cement Concrete Pavement Rehabilitation Alternatives for Washington State*, Washington State DOT, Report No. WA-RD 202.1, Seattle, WA.

Miner, M.A. 1945. Cumulative Damage in Fatigue, *Transactions of the ASME*, Vol. 67, A159-A164.

Murdock, J.W. & Kesler, C.E. 1958. Effect of Range of Stress on Fatigue Strength of Plain Concrete Beams, *J. of the American Concrete Inst.*, 30(2), 221-231.

Rao, S. 2005. *Characterization of Built-In Curling, Damage, and Cracking in Accelerated-Pavement-Tested Restrained Concrete Slabs at Palmdale, CA*. Ph.D. Thesis. University of Illinois at Urbana-Champaign, Urbana, Illinois.

Rao, S., & Roesler, J.R. 2005. Nondestructive Testing of Concrete Pavements for Characterization of Effective Built-In Curling, *ASTM J. of Testing and Evaluation*, 33(5), 356-363.

Roesler, J.R., Harvey, J.T., Farver, J. & Long, F. 2000. *Investigation of Design and Construction Issues for Long Life Concrete Pavement Strategies*, Final Report, FHWA/CA/OR-2000/04, UCPRC, Richmond, California.

Smith, K.D., Wade, M.J., Peshkin, D.G., Khazanovich, L. Yu, H.T. & Darter, M.I. 1998. *Performance of Concrete Pavements; Volume II—Evaluation of In-service Concrete Pavements*. FHWA-RD-95-110, Washington, DC.

Tepfers, R. 1979. Tensile Fatigue Strength of Plain Concrete, *J. of the American Concrete Inst.*, 76, 919-933.

Yu, H.T., Khazanovich, L., Darter, M.I. & Ardani, A. 1998, Analysis of Concrete Pavement Responses to Temperature and Wheel Loads Measured from Instrumented Slabs, *Trans. Research Rec. 1639*, TRB, National Research Council, Washington, D.C., 94-101.

Maximizing Pavement Design for Highway Design Build Projects
Lessons Learned from I-5 Everett HOV

Authors: Kurt Pedersen, PE, CH2M HILL, Los Angeles, CA; Dan Peterson, PE, CH2M HILL, Corvallis, OR

About the Authors:

Kurt Pedersen, PE, has over 11 years of experience in the design and construction of roads and highways in Washington, California, and Michigan. He served as the Roadway Task Lead on the I-5 Everett HOV Design Build Project. Mr. Pedersen has provided construction support throughout his career, giving him the understanding of the challenges that a contractor faces during construction.

Dan Peterson, PE, has over 22 years of experience is the design of pavements. He has provided pavement design services and consultation for numerous roadway improvement projects for the cities of Portland, Tualatin, and Eugene, Oregon; the counties of Washington, Clackamas, Marion, and Multnomah, Oregon; the Oregon Department of Transportation (ODOT), Nevada Department of Transportation (NDOT), Washington State Department of Transportation (WSDOT); Port of Seattle; Reno Transportation Commission; Utah Transit Authority; and various other private and public agencies.

I   Abstract:

In traditional Design-Bid-Build projects, the engineering consultant generally has no financial incentive to minimize the construction cost as long as it stays within the client's budget. As a result, the design may be conservative, resulting in needless dollars being spent during construction.

Design-Build projects require all participants involved to rethink their traditional roles on a project. The Owner no longer dictates the final design, but instead sets the design parameters, requiring the Design-Builder to determine the final design. Teamed with the contractor, the designer now has to look at the financial impacts of the design, considering material costs, procurement, subcontractor involvement, traffic and schedule. Therefore, the designer and contractor together can develop alternatives that meet the design requirements, while reducing costs. The contractor is being paid on a lump sum basis, so minimizing impacts to cost and schedule, while still meeting the design parameters, means more money in the Design Builder's pocket.

The focus of this presentation is on a case study of the pavement design and construction for the I-5 Everett HOV Project located about 30 miles north of Seattle, Washington. The Design-Builder will present on the challenges of working as a Joint Venture between designers and contractors and the impact it had on the final product.

The goal is to use real world experience to increase awareness and improve the effectiveness of future Design-Build projects.

II Introduction

A Project Definition

The I-5 Everett HOV is a Design-Build project constructed by a Joint Venture of Atkinson Construction, LLC and CH2M HILL Constructors, Inc. The project extends from SR-526 (Boeing Freeway) to north of US-2 in Everett, Washington. A high occupancy vehicle (HOV) lane was added in each direction with auxiliary lanes at numerous locations. The mainline widening consisted of both hot-mix asphalt (HMA) and portland cement concrete pavements (PCCP), and the ramps consisted of new HMA pavement and HMA overlays. Twenty two bridges were widened and 20 retaining walls constructed. One of the key features of the project was eliminating the left northbound exit to Broadway Avenue and constructing a new single point urban interchange (SPUI) at 41st Street.

B Project History

I-5 through Everett had become one of the most congested sections of freeway in the State of Washington. HOV lanes existed south of SR-526, but ended south of the City, narrowing from 4 lanes to 3 lanes in each direction. Once funding was obtained with the recent gas tax increases, it became possible to eliminate the bottleneck in Everett. WSDOT chose a Design Build delivery method as a way to expedite the completion of the project. The design of the $220,000,000 project began in April 2005, with construction starting shortly after. Completion was in June of 2008.

III Design-Build vs. Traditional Design-Bid-Build

A Risk

The biggest risk for the Design-Builder is bidding on a project with many unknowns. At the time of bid, design is only 15% complete. The extremely compressed schedule requires the designer to develop construction level drawings within 2-3 months of award. The contract is a lump sum award, with limited change orders only due to a change of scope. The contractor is not paid on a unit price, so any overruns due to overexcavation, or extra materials will not be covered.

WSDOT also assumes risk. No longer does the Department have 2 months to review plans at each level of design. Review periods are generally 1-2 weeks, which results in more responsibility given to the designer for the accuracy and completeness of the design. Day to day inspection was turned over to the Design Builder. WSDOT still provided inspection, but the frequency was less.

B Reward

There are potential benefits for everyone. Going from a 15% design to a completed project in just over three years shaved two years off of a traditional design-bid-build project. This benefits the community by providing traffic relief sooner. By going to construction two years sooner, construction cost escalation rates were reduced, saving WSDOT money. The Design-Build process also allows for more flexibility in developing innovative solutions. The contractor and the designer can work together, developing solutions that can result in reduced construction costs, faster construction methods, and better results. Any cost savings developed and awarded after the bid were split between the Department and the Design-Builder.

IV   Design Tasks

A   Geotechnical Investigation

Falling Weight Deflectometer (FWD) testing was performed on both shoulders in both directions at 500-foot intervals by Pavement Consultants, Inc. of Seattle, WA using a Dynatest 8081 Heavy Falling Weight Deflectometer. The FWD was equipped with an 11.8-inch diameter loading plate and deflection sensors were located 0, 11.8, 24, 36, 48, 60 and 72 inches from the center of the plate. At each test location, testing was conducted at three load levels ranging from approximately 7,000 pounds to approximately 14,000 pounds. Upon completion of FWD testing, the collected deflection and location data were reduced. The subgrade composite resilient modulus, M_R, was calculated for each deflection sensor at each test point using recorded loading information and deflection with the following relationship:

$$M_R = \frac{P(1-u)}{pDd}$$

where:

M_R = dynamic resilient modulus in pounds per square inch
P = applied load in pounds
u = Poisson's ratio, assumed as 0.25
D = distance from center of applied load in inches
d = deflection at D in inches

The minimum calculated value at each test point was taken as the appropriate M_R value at that location. This resulted in subgrade modulus values ranging from 12,000 psi to 106,000 psi with most values between 20,000 and 30,000 psi. Modulus values over 50,000 psi were not included in the data analysis and were considered outliers in the data set.

Per discussions with WSDOT during the task force meetings, WSDOT practices are to apply a correlation factor of 0.75 to 1.0 to raw FWD data to determine subgrade resilient modulus values for design purposes, based on past experience. A correlation factor of 0.75 was used for the I-5 Everett project. Average values were determined for each shoulder, inside and outside, in each direction. The average values ranged from 17,520 psi to 20,821 psi for the four areas. For design purposes, a subgrade resilient modulus value of 18,000 psi was chosen for all areas. This value is indicative of sandy soils with fines (SM/SP/SC) that have a California Bearing Ratio (CBR) value greater than 10 and an R Value greater than 40, which is the predominate soil type in the area of the project. A graph of milepost vs resilient modulus values is shown in Figure 1.

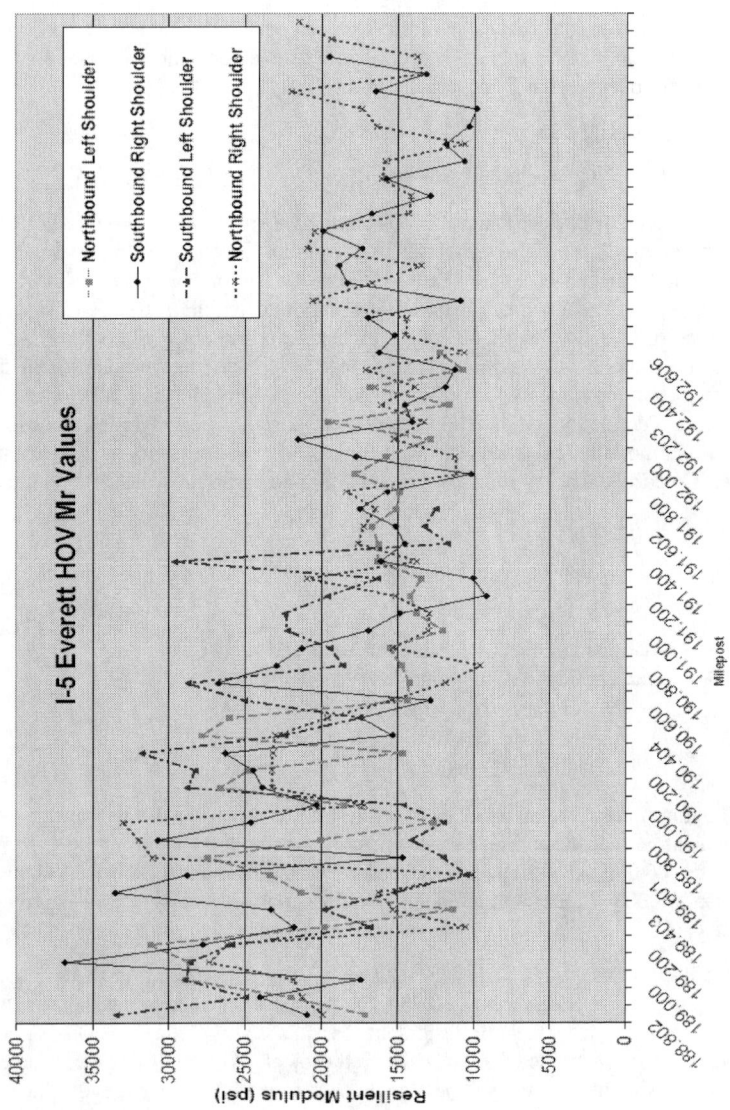

Figure 1 – Resilient Modulus Values

B   Pavement Design

1.   Design Methodology and Procedure

A pavement design task force was set up to meet periodically during the course of design to discuss pavement design parameters and construction procedures. The task force consisted of members from CH2M HILL, Atkinson Construction and the Washington Department of Transportation. The WSDOT Pavement Guide, Volume 1, and the 1993 AASHTO Guide for the Design of Pavement Structures were mandatory references required by WSDOT to use for the project. The WSDOT Pavement Guide Interactive along with the DarWin software program were also used to supplement the design and analysis of the pavement structures. Per the WSDOT Pavement Guide, the values shown in Table 1 were used for input into the DarWin program to determine the required thickness for each pavement layer using the layered analysis approach.

Table 1 – Design Inputs

Variable	Flexible (MMA) Pavement	Rigid (PCCP) Pavement
Initial Serviceability	4.5	4.5
Terminal Serviceability	3.0	3.0
Reliability	95%	95%
Overall Standard Deviation	0.50	0.40
HMA Layer Coefficient	0.44	-----
HMA Drainage Coefficient	1.0	-----
HMA Elastic Modulus	400,000	-----
CSBC Layer Coefficient	0.13	-----
CSBC Elastic Modulus	25,000 psi	-----
CSBC Drainage Coefficient	1.0	-----
28-day Mean PCC Modulus of Rupture	-----	650 psi
28-day Mean Elastic Modulus of Slab	-----	4,000,000 psi
Mean Effective K-value (HMA over CSBC)	-----	400 pci

Load Transfer Coefficient, J (JPCP)	-----	2.7
Overall Drainage Coefficient, Cd	-----	1.0

Existing Shoulder Evaluation

Existing shoulders had to be analyzed for adequacy to handle temporary traffic during construction. The existing shoulders were examined for current structural capacity vs. anticipated traffic. Existing shoulders were analyzed using the Condition Survey Method as described in Chapter III, Section 5.4.5 of the AASHTO Guide. Core data provided by WSDOT was used to determine average asphalt depths along the shoulders. Overly large (0.80 feet and higher) asphalt depths were not included to determine the average value. Core depths are provided in Table 2.

Table 2 – Existing Shoulder Pavement Depths

	Existing Shoulder Pavement Depths			
	NORTHBOUND		SOUTHBOND	
	Inside	Outside	Inside	Outside
MP187.5	0.83'	0.52'	0.89'	0.72'
MP188.0	0.82'	0.42'	0.80'	0.64'
MP188.5	0.82'	0.95'	0.80'	0.52'
MP189.0	0.49'	0.59'	0.74'	0.60'
MP189.5	0.42'	0.45'	0.48'	0.54'
MP190.0	0.62'	0.60'	0.48'	1.20'+
MP190.5	0.51'	0.44'	1.20'+	1.05'
MP191.0	0.56'	0.40'	0.96'	1.05'
MP191.5	0.61'	0.50'	1.08'	0.88'
MP192.0	1.00'	0.80'	0.63'	0.36'
MP192.5	0.89'	0.50'	Skip	0.56'
MP193.0	0.78'	0.90'	0.19'	1.20'+
MP193.5	0.62'	1.20'+	Skip	0.53'
MP194.0	0.51'	0.50'	0.50'	Skip
MP194.5	0.60'	0.85'	0.52'	0.60'
MP195.0	Skip	Skip	0.45'	1.20'+

The pavement design task force concluded that the overall depth of the shoulder structure was assumed to be equivalent to the overall depth of the adjacent travel lane with the crushed surfacing base course (CSBC) course making up the difference between the overall depth and the asphalt concrete core thickness. Also, the CSBC could have full value (layer coefficient of 0.13) while the HMA layer would have a reduced value. Based on visual observations, a layer coefficient of 0.30 was given to the existing HMA layer.

An effective Structural Number (SNeff) was determined from the layer coefficients and layer thicknesses to compare with the required Structural Number (SNreq) determined for the duration of temporary traffic. The SNreq was determined by using the traffic information provided WSDOT and the WSDOT ESAL calculator provided in the interactive design guide, and applying to the length of time for traffic. The analysis showed that the shoulders will be adequate for traffic up to the maximum of 20 months.

Additional coring was performed at 100-foot intervals at locations along the inside and outside shoulders for verification and to provide additional information on existing depths of asphalt concrete. The information showed that existing asphalt depths along the shoulder ranged from 1 inch to 12 inches with most areas between 4 inches and 8 inches. The WSDOT Pavement Design Guide recommends a minimum asphalt thickness of 0.35 feet for shoulders. The AASHTO Pavement Design Guide recommends a minimum of 3 inches of asphalt concrete for ESAL loads of 500,000 to 2,000,000.

Therefore, in areas where the existing shoulder will be used for temporary traffic that has less than 4 inches of asphalt concrete, it was recommended to mill and inlay the existing pavement with HMAC to provide a total of at least 4 inches of HMA. This may be accomplished by milling one or two inches and replacing with two or three inches of new HMA for existing sections with 4 inches of asphalt, depending on grade requirements, or by total removal of the asphalt layer and replacement with at least 3 inches of new HMA. In areas where the existing HMA is 3 inches or less, it is recommended to totally remove and replace the existing asphalt concrete with at least 3 inches of new HMA.

Traffic Evaluation

Traffic numbers provided in the RFP were used for the design of the mainline and ramp paving. Design 18-kip equivalent single axle loads (ESALs) of 200,000,000 for the mainline (all lanes-one direction) and 40,000,000 ESALs for the ramp were used for analysis. A lane distribution factor of 0.60 for the outside lane was used based on the values provided in the WSDOT Pavement Guide Interactive for 3 lanes (does not include HOV lane). The Interactive Guide provides a range of 60 percent to 80 percent for the lane distribution factor. It is assumed that in an urban environment with a high occupancy level, that the actual lane distribution factor will be on the lower end of the range and that the truck traffic will be more evenly distributed across all lanes. Therefore, a lane distribution factor of 0.60 was selected, resulting in a design ESAL load of 120,000,000 for mainline pavements.

Likewise, a lane distribution factor of 0.60 was selected for analysis of the temporary shoulders, using the lower end of the range for 3 lanes. As previously discussed, the WSDOT ESAL calculator, provided in the WSDOT Interactive Guide, and Pavement Management System Historical Data provided by WSDOT gave truck numbers that were used to determine design ESAL loads for temporary shoulders.

2. Pavement Types and Selection

Paving sections were determined for mainline, ramps, ramp shoulders, and city streets. Both flexible and rigid sections were determined for both mainline and ramp paving. Shoulders on the mainline have the same paving section as the adjacent travel lane. Shoulders on the ramp sections were constructed to meet the minimum requirements as stated in the WSDOT Pavement Design Guide, Volume 1, which require 0.35' of HMA over 0.35' of CSBC and sufficient depth of gravel base to extend to the bottom of the mainline base course. City Streets were be constructed per City of Everett Standard Detail 301 as shown in Figure 3.

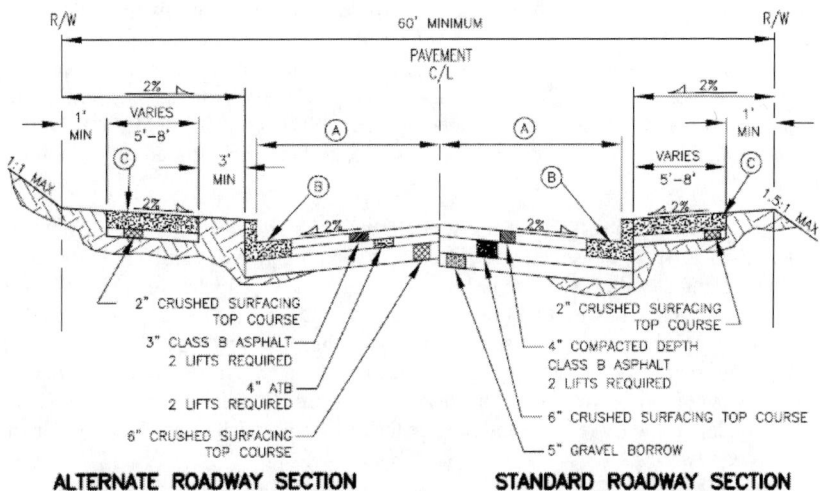

Figure 3 – City of Everett Standard Sections

Materials used for the pavement types met WSDOT specifications. Mix design requirements and material specifications are summarized in Table 3.

Table 3 – Material Specifications

Material	Project Specifications
PCC Mix Design	Section 5-05.3(1) of the WSDOT Standard Specifications (650 psi flexural strength at 14 days)
Dowel Bars	Stainless Steel Clad
Tie Bars	WSDOT Section 9-07.6

Joint Material	WSDOT Section 9-04.2
Asphalt Concrete Mix Design	– Superpave Mix Design – Volumetric testing required for mix design preparation and test section (test section required each year) – 100 gyration mix – ½" Nominal Aggregate Size -- PG 64-22 asphalt cement
Crushed Surfacing Base Course	WSDOT Section 9-04.2(1)

Pavements for the 41st Avenue Interchange were evaluated separately as part of the proposal for the 41st Avenue Single Point Urban Interchange Design. The pavement sections for the project, including the 41st Avenue Interchange are included in Table 4.

Table 4 – Pavement Thicknesses

Pavement Area	Layer Thickness, Inches		
	PCCP	HMA	CSBC
Mainline PCCP – HOV lanes, general purpose lanes, auxiliary lanes, shoulders	12-1/2	4	4
Mainline HMA	—	12	6 to 9-1/2 (varies to match exst)
Ramps -PCCP	9	4	4
Ramps - HMA	—	11	6
Ramp Shoulders	—	4	12-1/2
41st Avenue Ramps	9	—	4
	—	11	5
41st Avenue Auxiliary Lane	12-1/2	4	4
	—	11	6-1/2
41st Avenue Ramp Shoulder (HMA Section)	—	4	12-1/2
41st Avenue Ramp Shoulder (PCCP Section)	—	4	13

V Constructability

A Pavement Types

In the south segment, the mainline widening and shoulders were constructed with HMA. In the central and north segments, the mainline widening and shoulders were constructed with PCCP. All ramps were constructed with HMA.

The HMA widening in the south segment did not create any unexpected challenges. The work zone was wide and continuous. This was by far the easiest to construct.

The PCCP widening was more challenging. The 13 inches of PCCP were constructed over 4 inches of HMA and 4 inches of CSBC. Separate subcontractors constructed each of the layers, requiring difficult scheduling. With numerous ramps and bridges, the work zones were very short, resulting in frequent mobilization and demobilization. The work zone width was generally very tight, making it very difficult for the equipment to maneuver. Some of the areas had to be poured by hand, because there was not enough room.

Trying to maintain traffic made the construction of the ramps very challenging. Using HMA made the process easier, because of the ability to partially construct a ramp. Even with the HMA, the ramps were the greatest challenge to fully reconstruct. They became the focal point of trying to develop alternative designs where the existing pavement did not need to be removed.

B Maintenance of Traffic (MOT)

Three mainline lanes were required to be open in each direction during the day, but could be reduced to one lane at night. The ramps had to be open during the day, Monday through Friday, but could be closed at night. On a very limited basis, ramps could be closed for a weekend.

In general, the ramps were being widened on the outside to allow vehicle storage for ramp metering. A portion of the existing ramp shoulder would become the second lane for the ramp, and 12 feet of widening for the rest of the second lane and the new shoulder would be constructed further out. The original design by the Design-Builder called out for a full reconstruction of the ramps. The consequences of such a design were not fully thought out.

First, the existing shoulder would be closed, providing a work zone to construct the outside. About 6 feet of the shoulder had to remain off limits to the contractor to provide a 2' shoulder, barrier, and a 2-foot slide on the barrier. Effectively, only 12 to 14 feet could be constructed on the outside. This did not leave enough room to shift traffic for the following stage. A weekend closure to reconstruct the existing shoulder to become thick enough for ramp traffic would be required.

Second, the existing and proposed profiles did not match very well. The proposed profiles were often lower than the existing pavement. As traffic shifted from the existing onto the new up to 6 inches of temporary asphalt had to be constructed and eventually discarded.

By considering alternative pavement designs other that a full reconstruction and profiles that match or are higher than the existing pavement, maintenance of traffic can be simplified. HMA grinds and structural overlays can be constructed over many night closures and can be fully opened to traffic during the day.

C   Subcontractor Scheduling

Due to having 22 bridges and 20 retaining walls on this project, the prime contractor, Atkinson, was a structural contractor. Therefore, almost all roadway construction was done by subcontractors. Being able to schedule separate subcontractors for PCC removal, HMA grinding, earthwork, HMA pavement, PCCP pavement, concrete barrier, guard rail, and striping created enormous challenges in keeping the project on schedule. All of the subcontractors were very busy on other projects and, at times, were unable to return until the following week. Any delay caused by either design or delay by a previous subcontractor created a domino effect of rescheduling and delays of multiple subcontractors.

With closely spaced ramps and bridges, only small sections could be constructed at any one time. Segments that only required two weeks of actual work would often take two months to complete due to the scheduling conflicts. At times, the subgrade would be exposed to rain for weeks until the following subcontractors were available, resulting in additional overexcavation of the saturated soils.

By modifying the design of a pavement section, the number of subcontractors can be reduced. Raising profiles to eliminate the need for a full depth reconstruction is one option. With profile and superelevation corrections, the overall quantity of asphalt may increase significantly, but this is often offset by the earthwork savings and schedule savings.

VI   Examples of Changes During Construction and Alternative Section

An opportunity was missed when developing the proposal. WSDOT only required that the ramps to be constructed with HMA. They did not require that they be reconstructed. The proposal submitted by the Design-Builder showed that all of the ramps were to be reconstructed. Once construction time approached on the ramps, it was realized how difficult and expensive reconstructing the ramps would be. Changing the design at this point to anything less than full reconstruction meant giving half of the savings back to WSDOT. If the design were changed at part of the bid, all of the savings could have been kept by the Design-Builder, or the bid price could have been reduced.

Per the RFP, the mainline was to be designed for 200 million ESALs over all lanes in one direction, and all ramps were to be designed for 40 million ESALs. Any simplification in the design of the pavement had the ability to reduce the duration and construction cost of any particular segment. In particular, additional pavement analysis and modifying profiles to allow for HMA grind and structural overlays significantly simplified the process. Instead of multiple traffic shifts and long weekend closures, a single nighttime closure for HMA grinding followed by multiple nights of structural overlays accomplished the same final product. The following are two examples of alternative designs that saved a combination of money and time for the Design Builder, while still meeting the design requirements.

A   JR1/JR2 Ramp (WB US-2 to NB I-5 and Everett Avenue onramps to NB I-5)

The existing and proposed geometry for this ramp is very tight, and unusual. These two ramps merge together and then merge with mainline in a very short distance. The exit to Marine View Drive is less than ¼ mile away. With so many ramps entering and exiting, maintaining traffic was very difficult. The existing ramps were 6 to 12 inches below the projected mainline grade. The design profiles eliminated these grade differences, raising the final grades of the pavements.

The existing ramps were 8" HMA over 4" of CTB over 4" ATB. The shoulders were 8" HMA over 8" CSBC. With the changes in alignment, ramp traffic would be driving on the existing shoulders. Two alternatives were considered:

1. Grind the existing HMA ramp and shoulder pavement and provide enough structural overlay to save the shoulders. This required more asphalt due to the thinner existing shoulders, but reduced the total square footage of reconstruction. A 2-inch HMA grind and 7-inch HMA overlay would be required.

2. Grind the existing HMA ramp pavement and reconstruct the existing shoulders. A 2-inch HMA grind and 5-inch HMA overlay would be required. Once MOT was considered, this option was eliminated. Both the gore between the ramp and mainline and the gore between the ramps would have required reconstruction. The contractor did not have enough room to reconstruct the ramps without closing the ramps.

Although the square footage saved was rather low (14,000 square feet), the reconstruction of this area would have required three consecutive dry weekends in October, which is highly unlikely, and hundreds of tons of temporary paving to tie in between the stages. Instead, only one weekend closure was required. Final paving was not finished until March, when the weather was better, and the paving subcontractor was available.

B   Existing HMA Shoulders in the South Segment

WSDOT had identified the existing shoulders in the south segment as an area that may need reconstruction. Only one side was widened in each direction, saving the opposite shoulder. The existing shoulders varied from 4 to 8 inches of HMA over 6 to 16 inches of CSBC. Mainline was thicker, only requiring a 2-inch HMA grind and 2-inch inlay. WSDOT pavement manual requires full depth of mainline shoulders for new construction or reconstruction, but does not address how to analyze remaining service life for shoulders. The question arose, how many ESALs will a shoulder be exposed to over its design life?

Design traffic numbers in ESALs were provided in the RFP. For the northbound shoulder, the adjacent lane is an auxiliary lane, connecting the on-ramp from SR-526 and the off-ramp to Broadway/41st Street. This is considered as an extended ramp. Therefore, the ramp design ESAL of 40 million was selected. It is assumed that the shoulder will receive 5 percent of the traffic from the adjacent lane. Therefore, the design traffic for the northbound shoulder is 2 million ESALs.

For the southbound shoulder, the adjacent lane is an HOV lane. Currently the bus traffic is low in this segment, but is expected to increase as the population density increases over the next 50 years. For traffic estimating purposes, it is assumed that 15% of all mainline ESALs (bus only), or 30 million ESALs, would be in the HOV lane. It is also assumed that 5 percent of the adjacent lane will drive on the shoulder. Therefore, the design traffic for the southbound shoulder is 1.5 million ESALs.

After analysis, only short segments of the shoulders required reconstruction due to inadequate strength. Fortunately, most of those locations fell in areas where the shoulder had to be reconstructed for the installation of new storm drains. Instead of a full reconstruction, 130,000 square feet of shoulder were able to be preserved with only a 2-inch HMA grind and 2-inch inlay.

VII   Lessons Learned

1. Read the Request for Proposal (RFP) twice. The RFPs for Design-Build projects are performance based. Pavement designs, typical sections, profiles, and superelevations are generally preliminary. The freedom to modify any of these in order to create a better, less costly project is great. The engineers that have created the preliminary designs have not thought out all scenarios and have not designed beyond 15%-30%. Modifications can be made, but need to be clearly identified in the proposal developed by the Design-Builder.

2. Spend extra money on analyzing the existing pavement conditions. Bores and cores only provide pavement thicknesses. Without additional analysis using better tools like FWDs, a more conservative pavement design will be required. Many of the ramp overlays on this project probably required thinner HMA overlays than what were actually constructed. Every dollar spent for extra HMA constructed was money thrown away by the Design-Builder.

3. Work extremely closely with the contractor. Contractors know very well how to build things the cheapest way possible. If there is a less expensive way to construct pavement, consider it. Engineering analysis costs much less than construction dollars. Always keep in mind that quality cannot be compromised just to save money, but constructing a pavement that far

exceeds the requirements of the RFP is wasting money for the Design Builder.

4. Pay attention to profiles and cross sections. It is easier to build up than to lower pavement. Designing profiles that are 2" to 4" higher than the existing often makes room for the additional pavement required to bring the pavement design up to the required ESALs. Design that lower profiles for no reason requires full reconstruction and cost. Thoroughly analyze the working cross sections. Due to changes in horizontal alignment, design speeds, and superelevation, what appears to be to be a good profile may not work in cross section.

5. Don't remove existing pavement unless the earthwork and paving are coming shortly. That dry subgrade can turn to mud very quickly, requiring additional overexcavation. There is no such thing as a dry season in Seattle. It just rains less in the summer.

6. Review the designs about a month prior to the construction of a particular segment with the contractor. Everyone is in a rush at the beginning of a DB project. Reviews are cursory until the actual construction begins. Problems will arise due to MOT or scheduling that may affect the preferred design. Alternatives that were not originally discussed will arise once everyone gets focused on a segment.

7. Keep an open mind. There is no perfect answer when it comes to pavement design. There are many alternatives, and empirical designs that are not practical will be costly to construct.

VIII References

1. American Association of State Highway and Transportation Officials, Guide for Design of Highway Pavement Structures, 1993.

2. Washington Department of Transportation, Pavement Guide, Volumes 1, September 2004.

3. Washington Department of Transportation, Pavement Design Guide Interactive CD.

4. Washington Department of Transportation, Standard Specifications for Road, Bridge and Municipal Construction, 2004.

5. Washington Department of Transportation, Request for Proposals, I-5 Everett/HOV, 2004

6. City of Everett, Washington, Design and Construction Standards and Specifications

SYNTHESIS ON COMPOSITE PAVEMENT SYSTEMS: BENEFITS, PERFORMANCE, DESIGN, AND MECHANISTIC ANALYSIS

Orlando Núñez[1], Gerardo W. Flintsch[2], and Brian K. Diefenderfer[3]

[1]Graduate Student, Center for Sustainable Transportation Infrastructure, Virginia Tech Transportation Institute, 3500 Transportation Research Plaza, Virginia Tech, Blacksburg, VA, 24061; PH (540) 231-1659; FAX (540) 231-1555; e-mail: onunez@vt.edu

[2]Associate Professor, The Via Department of Civil and Environmental Engineering Director, Center for Sustainable Transportation Infrastructure, Virginia Tech Transportation Institute, 3500 Transportation Research Plaza, Virginia Tech, Blacksburg, VA 24061; PH (540) 231-9748; FAX (540) 231-1555; e-mail: flintsch@vt.edu

[3]Research Scientist, Virginia Transportation Research Council, 530 Edgemont Road, Charlottesville, VA 22903; PH (434) 293-1944; FAX (434) 293-1990; e-mail: brian.diefenderfer@vdot.virginia.gov

ABSTRACT

Composite pavement systems have shown good potential for becoming a cost-effective pavement alternative for new construction of high volume roads. This paper discusses the potential benefits that composite pavement structures (flexible layer over a rigid layer) have to offer during their long service lives, investigates the structural design approaches used nationally and internationally, and compares the theoretical performance of typical composite pavement systems. Typical structural packages obtained using various design methodologies are evaluated, and the mechanistic responses (stresses and strains) of typical composite pavements are contrasted with that of a comparable flexible pavement. Fatigue and rutting models are used to further study the behavior of the composite pavements present in terms of these distresses. The results suggest that composite pavement structures can provide a long-lasting and effective alternative for roadways carrying very heavy traffic.

INTRODUCTION

There are several types of composite pavement structures; however, in this study, a composite structure is defined as a multi-layer structure where there is a flexible layer (topmost layer) over a rigid layer. The flexible (asphalt concrete) layer (e.g., dense-graded hot-mix asphalt [HMA], stone matrix asphalt [SMA], open-graded friction course [OGFC], etc.) provides a smooth, safe, and quiet driving surface; whereas, the rigid layer (e.g., cement-treated base [CTB], roller-compacted concrete [RCC], continuously-reinforced concrete pavement [CRCP], etc.) provides a stiff and strong base. This high-modulus rigid base tends to change the traditional pavement concept in which the layers' moduli decrease as depth increases. In composite structures, the stiffness of the base (rigid layer) is greater than that of the surface layer (HMA).

Composite structures are also known as semi-rigid or flexible composite structures in other countries. These pavements have been widely used in roads where there is a high traffic volume (50+ million equivalent single axle loads [ESALs]), heavy trucks (which translates to high ESALs), and the designer seeks long-life pavements with minimum rehabilitation (replacement of the wearing surface).

OBJECTIVE

The objective of this paper is to provide a synthesis of composite pavement systems which include an overview of the potential benefits, their past performance in the United States and the world, and methods used for their design to date. In addition, the typical composite structures were further investigated using the multi-layer analysis software MICH-PAVE *(1)*.

COMPOSITE PAVEMENTS OVERVIEW

Composite pavements may present various structures due to the different combination of layers. Table 1 presents the typical materials properties for each layer.

TABLE 1 Typical Composite Pavement Materials and Layer Characterization

Layer No.	Material	Elastic Modulus (psi)[1]	Poisson's Ratio[2]	Modulus of Rupture (psi)[1]
1	HMA	500,000[†]	0.35	N/A
2	Portland Cement Concrete (PCC) *or*	4,000,000	0.15	650
	RCC *or*	3,500,000	0.15	600
	Lean mix concrete *or*	2,000,000	0.15	450
	CTB *or*	1,000,000	0.20	200
	Soil Cement	500,000	0.20	100
3	Base *and/or*	30,000	0.35	N/A
	Subbase	20,000	0.35	N/A
4	Subgrade (compacted)	7,500[*]	0.40	N/A

[†] Tested at 68 °F and 0.1 cycles/sec.   [1] Typical Values *(2)*; 1 MPa = 145.0377 psi
[*] CBR ≈ 5

Potential Benefits

According to the literature reviewed, a composite pavement system has the potential to provide a beneficial scenario for both the HMA and the rigid base, and thus may improve the overall pavement performance. Some of these benefits include:

- The stiff layer provides a strong base support for the HMA, which causes tensile stress/strain conditions at the bottom of the HMA layer to significantly decrease. This reduces the likelihood of direct, bottom-up, alligator cracking to occur *(2)*.
- The high-quality HMA surface contributes to ride quality and driver comfort by providing a smooth and quiet driving surface.
- The HMA surface (which can be periodically replaced) helps preserve the structural integrity of the rigid layer and prevents the intrusion of deicing salts and surface water that may attack the rigid base, resulting in a long-life system.
- The HMA surface reduces the temperature gradient in the rigid layer. Since the HMA is the layer directly exposed to the weather and climate, it acts as an insulating layer, thus reducing some of the effects of drying shrinkage during the curing period, and temperature stresses during the service life.

Past Performance

Composite pavement has been constructed for decades. In Europe, countries such as the United Kingdom, the Netherlands, Hungary, Denmark, and Spain are known for their use of long-life semi-rigid structures in their road network.

In the United Kingdom (U.K.), as of 1999, there were 649 km of composite pavements which had been constructed between 1959 and 1987 and had carried between 8 and 97 million ESALs. A composite pavement performance study concluded that there was considerable variability in the performance of these composite structures, particularly in the thickness of the asphalt overlays during maintenance. The current U.K. Pavement Design Guide includes a section for the design of flexible composite pavements to carry heavy traffic levels of at least 100 million ESALs (MSA) *(3)*.

Parry et al. *(4)* reviewed the experiences with composite pavements in Europe. This study found that composite pavements in the United Kingdom, the Netherlands, and Hungary were performing satisfactorily in terms of rutting, cracking, and deflections. The expected life of a semi-rigid pavement structure was found to be statistically longer than that of a flexible one, and that semi-rigid (composite) structures with a total thickness of HMA layers of 250 mm (9.8 in.) could achieve a long-life even under heavy traffic. Moreover, observations confirmed that composite structures can have unexpectedly long lives (i.e., they may be classified as long-life pavements).

The use of composite pavements in Spain is very wide, as reported by Jofre and Fernandez *(5)*. Composite pavement structures in Spain are called semi-rigid pavements because they do not tend to use a concrete pavement (PCCP) as the base; instead, they use different types of rigid bases that differ in the cement content and aggregate type used for their construction. Typical rigid bases include soil-cement, gravel-cement, lean-mix concrete, and compacted hydraulic concrete. Some of the technical aspects that should be controlled when constructing semi-rigid structures include reflective cracking control or mitigation, assurance of full bonding (full-friction) between all layers —especially between the bituminous layer and the rigid one. The study concluded that semi-rigid structures have proven to perform satisfactorily with a high level of confidence throughout the years *(5)*.

In the United States, composite pavements have been usually the result of rehabilitated PCCP. This rehabilitation consists of HMA overlays of deteriorated rigid pavement, thus creating a composite structure. This treatment has been widely used to restore functional and structural performance of existing pavements and to increase the structural capacity to accommodate additional or heavier traffic. Although these pavements are not strictly composite pavements, their performance was also included to increase the number of available sections.

Composite Pavements Deterioration

A composite pavement structure, throughout its service life, may develop different type of distresses. The distresses that affect composite pavements, according to Von Quintus et al. *(6)*, are very similar to those of flexible pavements because of the exposure that the asphalt concrete layer has in the composite structure. The distresses may be grouped into three major categories: fracture (cracking), distortion, and disintegration. These distresses could potentially affect the performance and structural capacity of composite pavements. However, the majority could be mitigated with a high-quality HMA mix, adequate overall structural design, and appropriate constructive procedures.

Several research studies, *(2, 6, 7)*, have agreed that reflective cracking (also known as reflection cracking) is a major distress type in composite pavements. Reflective cracks are cracks that occur in the asphalt surface course of the composite pavement and that coincide with cracks with appreciable width or joints in the underlying layer. They are caused by the relative horizontal and vertical movements of these cracks or joints caused by temperature cycles and/or traffic loading.

Reflective cracks are undesirable in a composite pavement structure as they tend to undergo a progressive width increase, permitting the leakage of surface water to the layer beneath, and causing serious raveling and disintegration of the asphalt surfacing adjacent to the cracks *(8)*. When a crack has a considerable width, it acts as a joint and high stress intensity is generated at this location. The contraction and expansion of the rigid layer tends to open and close this "joint" causing a significant change in width; as a result, the tensile stresses induced at the bottom of the HMA surface layer exceed the strength of the asphalt overlay and a reflective crack is initiated.

When a Chemically Stabilized Material (CSM) is used as the rigid base (e.g., CTB), drying shrinkage during the curing period is a major cause for the cracking of the base. The reasons that contribute to shrinkage cracking occurrence, which then lead to reflective cracks, include material characteristics, construction procedures, traffic loading, and restrain imposed on the base by the subgrade (9). These cracks occur in the early life of the pavement and reduce the structural capacity of the base; thus, this reduction should be incorporated in the structural analysis.

A study conducted in the United Kingdom suggested that reflective cracks often start at the surface of the road and propagate downwards to meet the existing crack in the underlying concrete base. The researchers suggest that a crack will occur when the tensile stress and related strain conditions induced by traffic and/or temperature changes exceed the mixture's breaking strength. At high temperatures, stress relaxation will avert these stresses and strains reaching a level that may cause cracking. Conversely, at low temperatures, the tensile condition will persist, thus pavement cracking will be more likely. In addition, aging of the binder, which is more severe at the surface of the road, results in a progressive increase in the elastic modulus of the asphalt and a decrease of stress relaxation capabilities. This "stiffer" surface further increases the possibility of cracking (10).

To mitigate and control reflective cracks, various methods and techniques could be used. These include the use of crack relief layers, pre-cracking (microcracking) of the cemented base, and use of geotextiles (paving fabrics) (9). The identification of the most effective reflective cracking mitigation approaches is still under investigation. For example, the use of a continuously reinforced concrete base may minimize (or eliminate) the development of reflective cracking.

DESIGN APPROACHES

Although there is no current widely accepted standard in the United States for designing composite pavement structures, several agencies in the United States and overseas have developed guidelines, manuals, and procedures for designing composite pavements based on experience and/or mechanistic analysis.

Comparison of Design Alternatives

To compare the output (thicknesses mainly) and layer recommendations from the different design methodologies, it was important to design composite pavement systems for a fixed set of conditions (inputs) and compare the results. Therefore, the various design procedures were followed to design composite pavement structures for the same input parameters (e.g., traffic, subgrade, design life). Table 2 shows the basic design inputs; the typical values used for the material properties of each layer were presented in Table 1.

Composite pavement structures were designed according to all the design procedures and methods described in each agency publication. The agencies' designs used for the analysis were the American Association of State and Highway Transportation Officials (AASHTO) (11), U.K. flexible-composite document (12), U.S. Army and Air Force pavement guide (13), Danish Road Institute semi-rigid design (14), and Illinois Department of Transportation (IDOT) composite pavement design guide (15). Figure 1 presents the cross sections of all the designed composite structures. The

AASHTO options are based on two different approaches using the 1993 guide. The AASHTO 1 is the design of a flexible pavement with a CTB, thus making it a semi-rigid pavement. The AASHTO 2 is based on the design of a flexible overlay on top of a rigid pavement with no distresses.

TABLE 2 Parameters Used for the Design of Composite Structures

Parameter	Value
Design life	40 years
Traffic	50,000,000 ESALs[a] 58,230 ADT[b] 12% trucks
Reliability	95% (AASHTO design) 75% (Danish design)
PSI_o[c]	4.5
PSI_f	3.0

[a] ESAL = equivalent single-axle load
[b] ADT = annual daily traffic   [c] PSI = present serviceability index

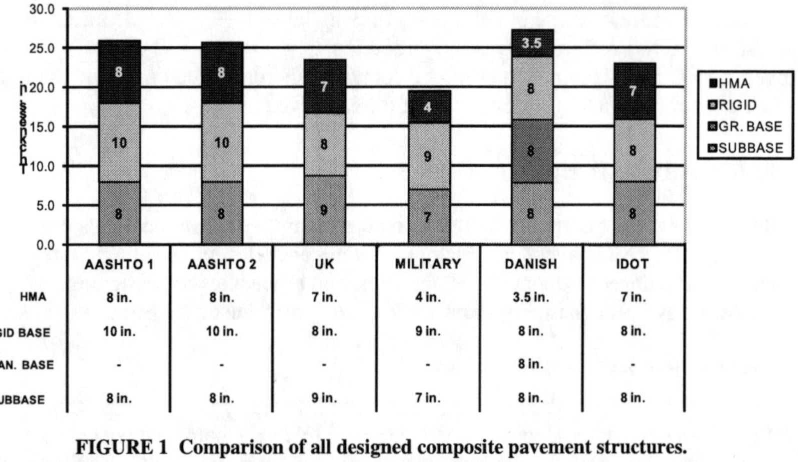

	AASHTO 1	AASHTO 2	UK	MILITARY	DANISH	IDOT
HMA	8 in.	8 in.	7 in.	4 in.	3.5 in.	7 in.
RIGID BASE	10 in.	10 in.	8 in.	9 in.	8 in.	8 in.
GRAN. BASE	-	-	-	-	8 in.	-
SUBBASE	8 in.	8 in.	9 in.	7 in.	8 in.	8 in.

FIGURE 1 Comparison of all designed composite pavement structures.

The composite structures shown in Figure 1 ranged from a total thickness of 20 to 28 in and may be grouped into three design groups with similar thicknesses in the HMA and rigid base layer. Group 1 designs (the two AASHTO alternatives) consist of an 8-in HMA surface course over a 10-in rigid. In the AASHTO 1 alternative (flexible pavement procedure with a CTB) the structural coefficient of the HMA, $a_1 = 0.47$, was greater than the CTB, $a_2 = 0.27$. This suggests that the structural capacity provided by the CTB is lower than that of the HMA; which is somehow contra-intuitive. In the AASHTO 2 alternative (rigid pavement procedure with HMA rehabilitation) the structural package was the same as the AASHTO 1 alternative. Group 2 (U.K. and IDOT alternatives) designs are very similar to one another; the package is composed of an HMA of 7 in and a rigid base of 8 in. The thickness of the HMA is required to reduce the temperature gradient in the rigid layer, reduce the tensile stress at the bottom of the rigid base, and provide a thicker medium that would help mitigate reflective cracking. The rigid bases consisted of a lean-mix concrete (2,000,000 psi) and a PCC for the U.K. and IDOT procedures, respectively. The thicknesses obtained in this group were chosen as the typical composite pavement to be analyzed using mechanistic modeling in the following section. The main reason is the vast experience that both agencies had with their design procedures for new composite pavements. This was especially true in the U.K., which according to the literature *(3, 4)* is one of the countries that has the most experience investigating, designing, and constructing composite pavement systems.

Group 3 (military and Danish alternatives) designs have the lowest thicknesses for the HMA surface layer. Although these thicknesses appear to be too low according to most of the literature, the Washington State Department of Transportation specifies, based on experience, that a 4-in. HMA thickness is thought to be thick enough to retard reflective cracking *(16)*. It is interesting that the Danish alternative is the only one that proposes the use of a granular base layer underneath the rigid base and above the subbase layer. The presence of this granular base layer could account for the lower modulus of the subbase (14,500 psi) and of the subgrade (5,800 psi) used as fixed values in their design table. In addition, this alternative had the lowest HMA surface thickness (3.5 in.).

MECHANISTIC ANALYSIS OF COMPOSITE PAVEMENT STRUCTURES

A simplified mechanistic-based analysis was performed to understand and model pavement behavior and responses (e.g., stress, strain, deflections). The MICH-PAVE software, available as freeware, was used to model the mechanistic responses of the composite structures. MICH-PAVE is a non-linear finite element software for the analysis of flexible pavements based on the multi-layer elastic theory. The program calculates displacements, stresses, and strains within the pavement structure due to a single circular wheel load. A user-defined mesh can be visualized using the software, and the nodes that compose the mesh are used to compute pavement responses at specific locations at both vertical and radial distances from the applied load *(1)*.

Mechanistic analyses were performed on various composite structures to understand their behavior as various rigid bases were used, including a granular base to exemplify a "conventional" pavement. The material properties for each layer are given in Table 1. The structural analysis was conducted assuming full bonding between all layers. The composite structure analyzed included the following layers:
- Surface course: HMA layer: 4 in of thickness
- Base course (granular, soil cement, CTB, lean mix, RCC, or PCC): 8 in
- Subbase: granular subbase: 6 in
- Subgrade: compacted subgrade; at least 12 in

A single 9,000-lb load with a tire pressure of 120 psi was used. Some of the results of the mechanistic analysis are presented in Figure 2. The following findings may be inferred by studying the plots:

- Horizontal Stresses: The tensile stress at the bottom of the HMA for the granular base becomes compressive stress when rigid layers are used as the base material. This increases the fatigue life of the HMA layer significantly. The maximum tensile stresses at the bottom of the rigid bases are considerably larger than those of the granular base. As a result, this location—bottom of the rigid layer— becomes the critical location that needs to be considered when designing the composite structure. If the stresses at the bottom of this layer are such that a crack is initiated, reflective cracking is likely to occur in the HMA layer.
- Horizontal Strains: The strain responses confirm the statements made from the Tensile Stresses plot. In this strains plot, a noticeable decrease of magnitude in both tensile (negative) and compressive (positive) strains can be observed for all composite pavement structures.
- Vertical Strains: The overall magnitude of the critical vertical strains is higher in the granular base than in the rigid bases. The compressive strain responses in the subbase and subgrade are also notably higher in the pavements with granular base than in those when a rigid base was used. By decreasing the compressive strains in the subbase and especially the subgrade, the likelihood of rutting due to subbase or subgrade deformation is diminished.

The critical responses determined in the mechanistic analysis were then used in the un-calibrated fatigue and HMA rutting models recommended by the proposed Mechanistic-Empirical Pavement Design Guide (MEPDG) *(2)* to evaluate these distresses. The results of the (bottom-up) fatigue analysis are shown in Figure 3. The number of load repetitions to HMA fatigue is much greater in a pavement with a cement-bound base (e.g., soil cement) than in pavement with a granular base. The table in Figure 3 shows an infinite number of load repetitions for the HMA on CTB, lean mix, RCC, and PCC base courses; this is because when any of these bases are used, the strain at the bottom of the HMA becomes very small (CTB case) or compressive in nature (lean mix, RCC, and PCC cases) and the flexible layer is highly unlikely to fail due to fatigue cracking. Moreover, it can be observed that for composite pavements, the base is the layer that controls the design, in terms of fatigue, as it would fail earlier than the HMA. In the case of RCC and PCC fatigue evaluation, the repetitions were determined to be infinite because the stress ratio (SR) terms after a load was applied for RCC and PCC were 0.17 and 0.16, respectively. The fatigue behavior of RCC was assumed to be similar to that of conventional PCC as recommended by the American Concrete Institute (ACI) *(17)*.

The results of the HMA rutting model is shown in Figure 4; they suggest that as the stiffness of the base increases, the rut depth in the HMA layer increases as well. The rut depth line at 0.5 in., as shown in the figure, represents the value typically allowed by various transportation agencies and the Asphalt Institute *(18)*. The HMA rutting results show that for 50,000,000 18-kip load repetitions, the typical flexible pavement was the only structure that met the 0.5-in. rut depth criterion. All the composite pavement structures presented greater (up to 0.83-in.) degrees of permanent deformation due to the high number of load repetitions.

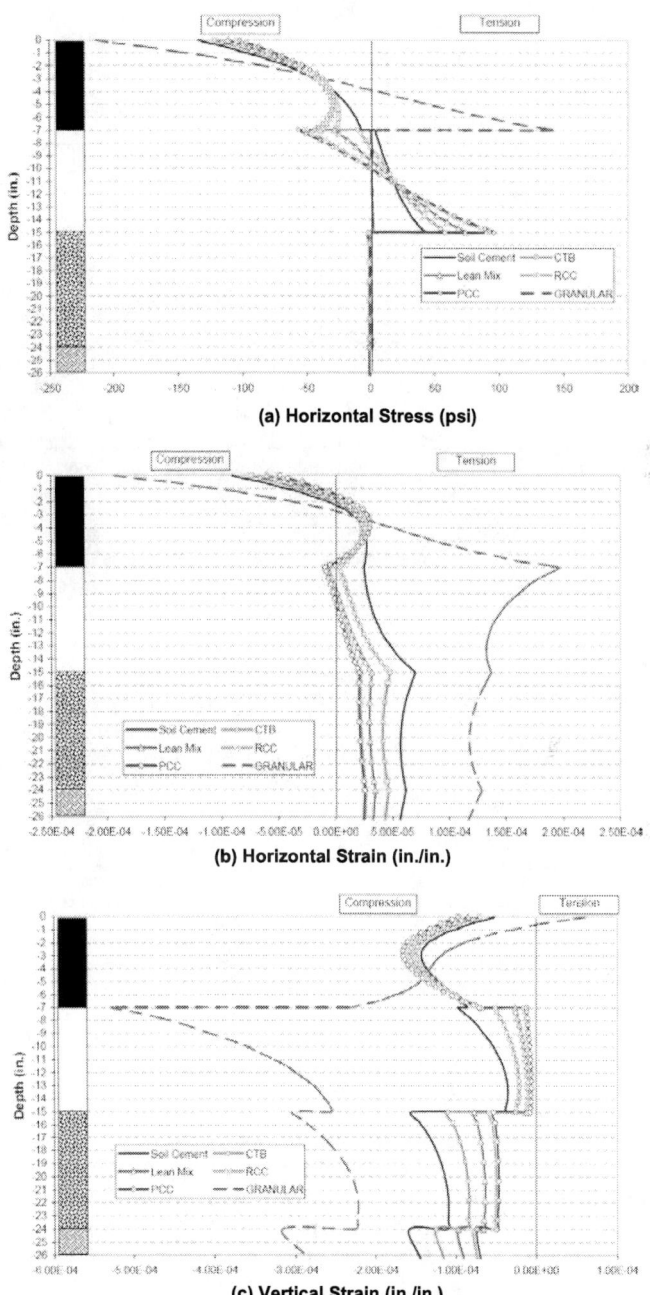

FIGURE 2 Summary of the pavement mechanistic analysis.

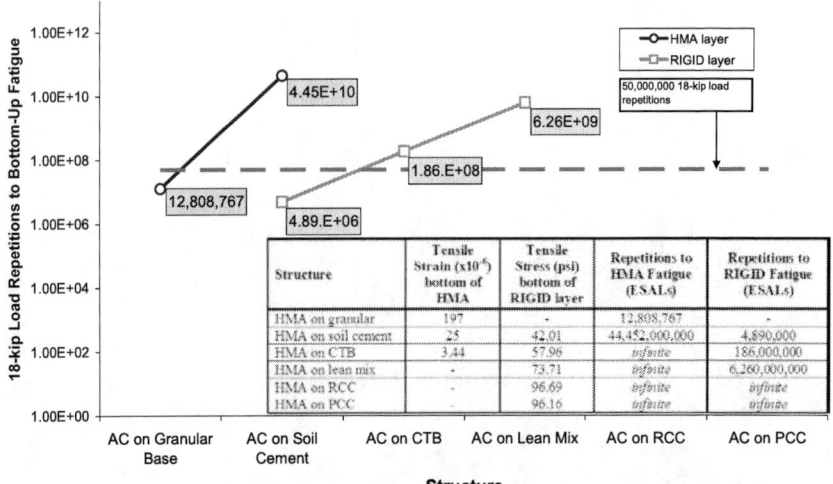

FIGURE 3 HMA and rigid base repetitions to fatigue failure.

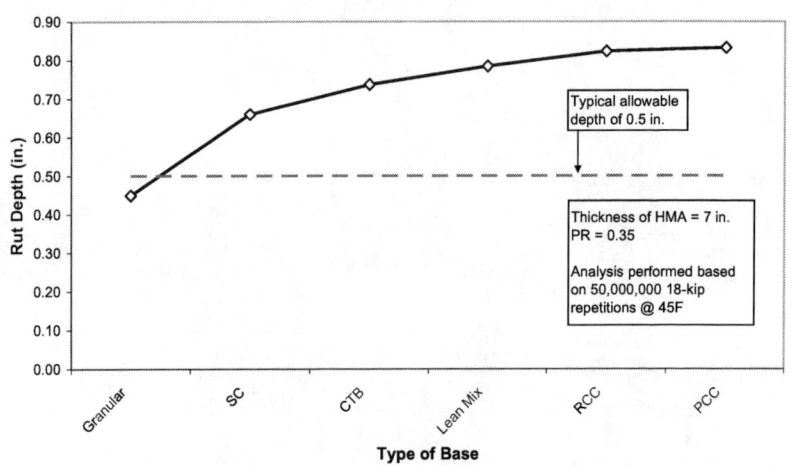

FIGURE 4 HMA rut depth versus base type.

It is noted, however, that the computed rut depth (Figure 4) for all the structures (both flexible and composite) assumed *no* rehabilitation operations at any time during the 50,000,000 load applications. Therefore, if a functional rehabilitation is applied at any time during the service life of the pavement, part of the permanently deformed HMA would be replaced. Since the upper region of this layer is the one that contributes the most to the rutting (i.e., it is the region that has the highest accumulated plastic strains as a result of the vertical strain responses shown in Figure 2) the structure may never reach the rut limit of 0.5-in.

SUMMARY AND CONCLUSION

Composite pavements are structures that have shown potential benefits structurally, functionally, and economically. According to the literature, many countries (e.g., the U.K., Spain) that have used composite pavement systems in their main road network have had a positive experience in terms of functional and structural performance. A composite structure, comprised of an HMA surface over a rigid base, is expected to provide high levels of serviceability during a long service life, thus making these types of pavements an attractive alternative for high traffic highways. Some of the main findings of the study are summarized following:

- Based on field experience, research studies, and the simplified mechanistic analysis presented in this paper, the use of a stiffer base such as CTB, Lean-Mix Concrete, RCC, or PCC, can eliminate bottom-up fatigue failure in the HMA. Horizontal strain conditions at the bottom of the HMA layer were found to decrease significantly in terms of tension as stiffer bases are used, and in some cases became compressive in nature. The rigid base becomes the critical layer and reflection cracking may need to be controlled.

- As the modulus of the base layer increases, horizontal strains are greatly reduced for all layers of the pavement structure (and specifically the stiff base layer). This suggests that the HMA layer should not fail due to fatigue. However, higher vertical strains that can be observed in the upper part of the HMA when stiffer bases are used suggest that HMA rutting will be prone to affect composite pavements more so than flexible pavements. However, no rutting due to subgrade or subbase permanent deformation will affect the HMA in composite pavement systems as opposed to flexible pavements.

- New construction composite pavements are expected to have a prolonged service life with good performance in terms of fatigue cracking and rutting caused by unbound layers' permanent deformations. Regarding rutting in the HMA itself, a mill-and-replace of the top 2 in. of the HMA layer could potentially mitigate this functional issue.

- Although reflective cracking may be a cause of concern, this distress could be retarded and potentially mitigated, if an appropriate reflective crack control method/technique is used. This distress could also be mitigated (or eliminated) with the use of continuously reinforced concrete bases.

ACKNOWLEDGEMENTS

This paper was produced under the joint sponsorship of the Virginia Tech Transportation Institute, the Virginia Transportation Research Council and the Virginia Department of Transportation. Drs. Linbing Wang and Edgar de León from VTTI contributed to the conception and development of the project.

REFERENCES

1. Harichandran, R. S., and G. Y. Baladi. *MICHPAVE User's Manual*, Department of Civil and Environmental Engineering, Michigan State University, East Lansing, MI., 2000
2. *Guide for Mechanistic-Empirical Design of New and Rehabilitated Pavement Structures.* www.trb.org/mpedg. National Cooperative Highway Research Program, Transportation Research Board, Washington, D.C., 2004.
3. Merrill, D., A. V. Dommelen, and L. Gaspar. A Review of Practical Experience Throughout Europe on Deterioration in Fully-Flexible and Semi-Rigid Long-Life Pavements. *International Journal of Pavement Engineering*, 7(2), 2006, pp. 101-109.
4. Parry, A. R., S. Phillips, J. F. Potter, and M. E. Nunn. UK Design for Flexible Composite Pavements. *Eighth International Conference on Asphalt Pavements*, Volume I. 1997.
5. Jofre, C., and R. Fernandez. "El Empleo de Pavimentos de Suelocemento en España." Madrid (Spain), 2004.
6. Von Quintus, H. L., F. N. Finn, W. R. Hudson, and F. L. Roberts. *Flexible and Composite Structures for Premium Pavements*, Report Nos. FHWA-RD-81-154 and -155, Vol. 1 and 2, FHWA, U.S. Department of Transportation, 1979.
7. Smith, R.E., R. P. Palmeri, M.I. Darter, and R.L. Lytton. Pavement Overlay Design Procedures and Assumptions, Vol. 3: Guide for Designing Overlay. Publication FHWA-RD-85-008. FHWA, U.S. Department of Transportation, 1984.
8. Breemen W. V. Discussion of Possible Designs of Composite Pavements. *Highway Research Record*, No. 37 "Composite Pavement Design", 9 Reports, January, 1963.
9. Adaska, W. S., and D. R. Luhr. Control of Reflective Cracking in Cement Stabilized Pavements. *5th International RILEM Conference*, Limoges, France, May 2004.
10. Nesnas, K., and M. E. Nunn. A Model for Top-Down Cracking in Composite Pavements. TRL Limited. 2004.
11. AASHTO, *Guide for Design of Pavement Structures*, American Association of State Highway and Transportation Officials, Washington, D.C., 1993.
12. Nunn, M. "Development of a more versatile approach to flexible and flexible composite pavement design." *TRL Report TRL615*, Highways Agency, Berkshire, U.K., 2004.
13. UFC. *Pavement Design for Roads, Streets, and Open Storage Areas, Elastic Layered Methods*, Unified Facilities Criteria (UFC), Departments of the Army and the Air Force, Washington, D.C., 2002.
14. Thogersen, F., C. Busch, A. and Henrichsen, A. "Mechanistic design of semi-rigid pavements—An incremental approach." *Report 138*, Danish Road Institute, Hedenhusene, Denmark, 2004.
15. IDOT. "Structural Design of Composite Pavements." *Pavement Design*, Chapter 54, Illinois Department of Transportation, Springfield, IL, 2002.
16. WSDOT. *Pavement Construction—Surface Preparation*, <http://training.ce.washington.edu/wsdot/modules/07_construction/07-2_body.htm> (Nov. 18, 2007), 2007.
17. Huang, Y. H. (2004). *Pavement Analysis and Design*, Pearson Prentice Hall, Upper Saddle River, NJ.

Recalibration of Airport Pavement Structural Design System

Greg White

Principal Pavement Engineer
GHD Pty Ltd, Brisbane, Australia
gregory.white@ghd.com.au

Abstract

Aircraft pavement thickness design remains tied to empirical relationships derived from full scale test data. With the introduction of the six wheel landing gears of the A380 and B777 the current failure criteria for the Australian developed pavement design program Aircraft Pavement Structural Design System (APSDS) were no longer valid.

APSDS was recently recalibrated using a greater number of aircraft at different masses and a greater range of aircraft coverages than were used initially. The S77-1 pavement thicknesses against which the calibration was performed were generated using the replacement Alpha Factor values endorsed by the International Civil Aviation Organization (ICAO) in October 2007. Replacement values of the calibration constants have now been developed for the following three situations:

- A single failure criterion over all subgrade California Bearing Ratios (CBRs) and wheel configurations.

- A criterion for each value of CBR but combining all wheel configurations.

- A criterion for each subgrade CBR and for each wheel configuration (1, 2, 4 and 6-wheel).

The second scenario produced significant improvement in the agreement between S77-1 and APSDS pavement thicknesses when compared to that provided by the current calibration constants. By considering each aircraft wheel configuration separately, a more significant improvement was achieved. Combining all wheel configurations and subgrade CBRs into a single failure criterion provided the least improvement to the current APSDS calibration.

The influence of subgrade CBR and aircraft wheel configuration were found to be significant and separate failure criteria should be adopted for each subgrade and each wheel configuration. This finding contrasts with the approach by the FAA in LEDFAA 1.3. LEDFAA 1.3 adopts a single failure criterion for all CBRs and all wheel configurations.

Introduction

Aircraft pavement thickness design, even when performed with mechanistic-empirical tools, remains tied to the results of full scale testing conducted between 1940 and the early 1970s by the US Army Corps of Engineers. This testing culminated in the publishing of the S77-1 design method in 1977 (Pereira, 1977) which remains the basis for calibration of most flexible aircraft pavement design methods today. For layered elastic and other mechanistic-empirical design tools, one or more failure criteria are required to relate the indicator of damage to an allowable number of repetitions. Calibration of such tools involves determining failure criteria that produce calculated pavement thicknesses which are, on average, as close as practical to the empirical relationship.

The recalibration of Aircraft Pavement Structural Design System (APSDS) to S77-1 is presented. This recalibration is based on the methodology developed by White (2007). The importance of wheel configuration on pavement thickness is demonstrated and justifies the need to treat one, two, four and six wheel aircraft gears separately during the calibration process.

The input parameters are detailed following discussion regarding the limitations of the original calibration effort. The calibration methodology and outcomes are also described. The pavement thicknesses generated using the recommended calibration constants are compared to those generated by S77-1 as well as the original APSDS calibration constants.

S77-1 Pavement Design

S77-1 relates subgrade CBR to the require pavement thickness for a given single wheel load. To allow the S77-1 relationship to be plotted on a single graph, pavement thickness was normalized to tyre contact area and CBR was normalized to tyre pressure. With the introduction of multiple wheel aircraft gear in the 1940s, the concept of equivalent single wheel loads was introduced to allow the S77-1 design method to continue to be used. Equivalent single wheel loads were calculated to be the single wheel load that produced the same maximum deflection in the subgrade as the multiple wheel gear of interest. Using deflection as the basis for determining equivalence overstated the damaging effect of multiple wheel gears. Consequently, pavement thickness correction factors, known as Alpha Factors, were introduced for different aircraft wheel configurations. Some researchers have hypothesised that the need for Alpha Factors could be avoided by the adoption of a strain-based method for comparing the damaging effect of various wheel configurations.

The S77-1 curve provides a pavement thickness of a predetermined composition. The standard S77-1 pavement structure is shown in **Figure 1**. P401, P209 and P154 are standard designations for the described materials, utilised by the US Federal Aviation Administration (FAA) for design and specification purposes (FAA, 1995). It is noted that the S77-1 curve is a best-fit curve to the full scale test data, with no built-in factors of safety.

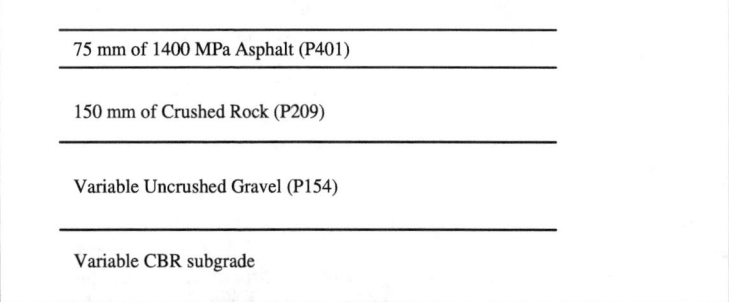

Figure 1   **Standard S77-1 Pavement Structure (after Pereira, 1977)**

Airport Pavement Structural Design System

Mechanistic-empirical or (layered elastic) design for aircraft pavement thickness determination was introduced into regular practice in the mid 1990s by the introduction of the FAA's LEDFAA software (FAA, 2004). At about the same time the Australian developed APSDS also became commercially available (MINCAD, 2000). Both LEDFAA and APSDS utilise strain as the indicator of damage.

APSDS is a specialised version of the road design tool Circly (MINCAD, 1999). APSDS incorporates the unique feature of modeling all the aircraft in all their wandering positions across the pavement surface. This approach negates the requirement for the traditional Pass-to-Coverage Ratio (PCR) and provides a Cumulative Damage Factor (CDF) for all aircraft at all locations across the pavement width. When an APSDS pavement thickness is compared to a S77-1 calculated thickness, a PCR is required to convert the modeled number of aircraft passes (in APSDS) to an equivalent number of coverages (for S77-1).

Chicago Criteria

The original calibration of APSDS was performed not against full scale test results of the US Army Corps of Engineers, but against S77-1 pavement thicknesses. This calibration exercise is reported by Wardle, *et al* (2001). Calibration involved determining APSDS failure criteria that optimized the average difference in total thickness between APSDS and S77-1 for a range of aircraft and a number of subgrade moduli. Calibration against S77-1 requires acceptance of the appropriateness of S77-1 for determining the thickness requirements for aircraft pavements. Whilst S77-1 was developed for relatively limited aircraft load and pavement scenarios, it has proven to be a reliable basis for aircraft pavement thickness determination over many years and remains the most appropriate anchor to empirical performance of aircraft pavements.

The original calibration constants are referred to as the Chicago Criteria. The Chicago Criteria were derived based on the assumption that pavement life was independent of aircraft wheel configuration. White (2007) found this to be an invalid assumption with greatly improved agreement S77-1 and APSDS possible when different aircraft wheel configurations were considered separately. A number of other limitations were also identified in the generation of the Chicago Criteria. These limitations included:

- **Aircraft.** Only aircraft with two and four wheel gear configurations were considered. With the introduction of the B777 and the A380, six wheel gears are now also important. Designing for single wheeled aircraft such as the military F111 and FA18 is also considered to be important as these can often govern pavement thickness requirements.
- **Aircraft Passes.** Only two levels of aircraft coverages were considered for each aircraft. These were 10,000 and 100,000 coverages. In practice pavements are designed to cater for a much greater range of aircraft numbers.
- **Aircraft Masses.** All aircraft were modeled at their maximum mass. Aircraft commonly operate at significantly below their maximum mass and Australian design practice is to consider the operating (landing and take off) masses expected over the design life.
- **Subgrade values.** The generation of the Chicago Criteria was performed at CBRs 3, 6, 10 and 15% only. Values of the Chicago Criteria for other subgrade CBRs were estimated by non-linear interpolation. The validity of this approach was not assessed during the original calibration.
- **Alpha Factors.** The Alpha Factors used for the calculation of S77-1 thicknesses were revised by ICAO in 2007 (ICAO, 2007) as a result of full scale test results at the National Airport Pavement Test Facility (NAPTF) (Hayhoe, 2006). Whilst not a short coming at the time that the Chicago Criteria were derived, their continued use following ICAO ratification of the revised Alpha Factors would be inappropriate.

The Chicago Criteria are all of the form shown in **Equation 1** and the values of the Chicago Criteria calibration constants are detailed in **Table 1** for select subgrade CBRs.

$$N = (K/\varepsilon)^B \quad \quad \quad \textit{Equation 1}$$

Where: N = the predicted life in terms of repetition of subgrade strain ε.
ε = the induced vertical strain at the top of the subgrade.
K = material constant, a calibration constant.
B = material exponent, a calibration constant.

Table 1   Chicago Criteria calibration constants

Subgrade CBR (%)	K	B
3	0.0032	9.5
6	0.0030	10.9
10	0.0024	15.0
15	0.0020	23.6

For the Chicago Criteria, the differences between APSDS and S77-1 thicknesses had a reported median of around 60 mm. Some scatter between APSDS and S77-1 thicknesses is always expected, primarily resulting from the difference in utilising strain (in APSDS) (MINCAD, 1999) and deflection (in COMFAA) (Ahlvin, et al 1971) as the basis for comparing the damaging effect of various aircraft wheel configurations. Deflections attenuate much more slowly than strains as one moves horizontally away from under an aircraft tyre. This results in greater wheel interaction being modeled by deflection-based design tools and results in some scatter when comparing results from deflection and strain based tools.

Calibration Methodology

White (2007) presented a methodology for the recalibration of APSDS which was based on iterative refinement of the calibration constants. This methodology:

- Expanded the aircraft to include six wheel aircraft gears.
- Considered a range of aircraft pass levels.
- Considered a range of aircraft masses.

White (2007) used CBR 6% as an example of the methodology developed. It was recommended that the following be adopted for the full recalibration of APSDS:

- The average of the differences in thicknesses for all design scenarios considered be used as the criterion for determining the optimum calibration constant values.
- Precision level of two decimal places for k and no decimal places for B.
- Aircraft of different wheel configurations be considered separately as this significantly decreases the average difference between S77-1 and APSDS thicknesses.

S77-1 Pavement thicknesses

S77-1 pavement thicknesses were determined using the computer program COMFAA. In October 2007, ICAO formally advised member states of changes to the Alpha Factors based on the full scale testing carried out at the NAPTF (ICAO, 2007). A new version of COMFAA that incorporated the new Alpha Factors was produced by FAA in 2006 (FAA, 2006). It is this version of COMFAA that has been used for this calibration work. For the 2007 calibration White (2007) used the traditional Alpha Factors in determining S77-1 thicknesses, as the acceptance of the new alphas had not been confirmed at that time.

S77-1 thicknesses were calculated for each design scenario, which included the aircraft, mass, number of passes and the subgrade CBR.

APSDS Pavement thicknesses

To allow the large number of APSDS thicknesses to be determined for each trial pair of values of K and B, an automated version of APSDS was developed. Verification of this software was performed by calculating a number of pavement thicknesses and comparing them to manually calculated thicknesses using a standard version of APSDS.

Calibration Process

Aircraft

In order to allow the calibration process to cater for the widest possible range of aircraft types, single, dual, dual-tandem and dual-tridem gear configurations were considered. Two aircraft of each wheel configuration were selected as detailed in **Table 2**.

Table 2   Aircraft details

Aircraft	Gear Configuration	Maximum Mass (t)	Tyre Pressure (MPa)
FA18	Single	24	1.72
F111	Single	51	1.48
BAe146	Dual	41	0.88
B737-800	Dual	79	1.36
B747-400	Dual-tandem	397	1.38
A340-600	Dual-tandem	366	1.38
B777-200	Dual-tridem	244	1.28
A380-800	Dual-tridem	562	1.34

Only the six wheel gear of the A380 was considered. Similarly, the belly gear of the A340 was omitted and only one of the two sets of B747 gears was included.

Traffic Variables and Constants

For each aircraft, 100%, 80% and 60% of the maximum mass was considered to allow the calibration to cover the range of typical operating masses of each aircraft. Tyre pressures were maintained at the standard pressure for each aircraft as is understood to be common operational practice. Aircraft passes spanned the range of typical designs from 100 to 100,000 passes of each aircraft.

Aircraft wander was set to a standard deviation of 773 mm. This was selected as being the aircraft wander statistic for taxiways in APSDS. For COMFAA, a PCR was required for each aircraft. PCRs were calculated using the method prescribed by the US Army Corps of Engineers (Barker and Brabston, 1975). For a constant tyre pressure, PCRs vary with aircraft mass, as a result of the changing width of the tyre contact area. However, designers commonly adopt a single PCR for each aircraft, reasoning that the effect on calculated pavement thickness is usually negligible and some design methods consider aircraft at their maximum anticipated take off mass only (ie reduced mass landings are ignored in the FAA advisory circular).

For any aircraft, at 100% and 60% of its maximum mass, a difference in PCR of 29% will result. For a B747 aircraft at 10,000 coverages on a subgrade CBR 6% pavement, this difference resulted in a 26 mm (or 2%) difference in pavement thickness. It was therefore considered that the additional effort in calculating and applying mass-specific PCRs for the conversion of passes to coverages was justified for calibration purposes. The PCRs utilised are shown in **Table 3**.

Table 3   Pass-to-Cover Ratios

Aircraft	PCRs		
	100% Mass	80% Mass	60% Mass
FA18	8.70	9.70	11.20
F111	5.50	6.16	7.12
BAe146	3.90	4.36	4.65
B737-800	3.63	4.06	4.69
B747-400	1.72	1.92	2.22
A340-600	1.85	2.09	2.39
B777-200	1.34	1.43	1.73
A380-800	1.30	1.46	1.68

Pavement Details

To allow direct comparison with the S77-1 curve, the standard S77-1 pavement was considered in all cases. This is shown in **Figure 1**. Crushed rock base and uncrushed gravel sub-base were sub-layered and assigned moduli values utilising the Barker and Brabston method (White, 2006). Subgrade values of CBR 3% to 15% were considered. Subgrade moduli values were calculated using **Equation 2**.

$$Modulus\ (MPa) = 10 \times CBR\ (\%) \quad\quad\quad\quad\quad Equation\ 2$$

Iterative refinement

As White (2007) found that the Chicago Criteria were not necessarily a good estimate for the new calibration constants, a grid of trial values were selected. The point of minimum difference within this grid of values was determined and the then the K and B values were iteratively refined until the minimum average difference between S77-1 and APSDS was determined for each design scenario considered. Initially CBR 3, 6, 10 and 15% were assessed in this manner and then interpolated values were used as seed values for other CBRs.

Analysis of Results

At the completion of the recalibration process, the recommended calibration constants were summarized. In some cases, non-optimum values of K and B were adopted in order to provide a smooth transition of values across the range of subgrade CBRs. Such changes only increased the average difference between S77-1 and APSDS thicknesses by less than 1 mm and were therefore not detrimental to the calibration of APSDS. Pavement thicknesses returned by the recommended calibration constants were compared to S77-1 thicknesses as well as the Chicago Criteria pavement thicknesses. The calibration criteria were also compared to those developed by the FAA for use in LEDFAA. The importance of subgrade and wheel configuration on the calibration constants was also analyzed.

Recommended Calibration Constants

Table 4 details the calibration constants developed for the all aircraft wheel configurations considered together, as per the Chicago Criteria. **Table 5** details the calibration constants returned when each wheel configuration is considered separately.

Table 4   Recommended Calibration Constants for combined wheels

CBR	K	B
3%	0.0031	10
4%	0.0033	10
5%	0.0034	10
6%	0.0035	10
7%	0.0036	10
8%	0.0037	10
9%	0.0038	10
10%	0.0039	10
11%	0.0040	10
12%	0.0040	10
13%	0.0041	10
14%	0.0041	10
15%	0.0041	10

Table 5   Recommended Calibration Constants for separate wheels

CBR	1 wheels/gear		2 wheels/gear		4 wheels/gear		6 wheels/gear	
	K	B	K	B	K	B	K	B
3%	0.0051	7	0.0041	8	0.0036	9	0.0030	10
4%	0.0054	7	0.0044	8	0.0038	9	0.0031	10
5%	0.0057	7	0.0045	8	0.0039	9	0.0032	10
6%	0.0059	7	0.0045	8	0.0040	9	0.0033	10
7%	0.0060	7	0.0046	8	0.0040	9	0.0034	10
8%	0.0060	7	0.0047	8	0.0041	9	0.0035	10
9%	0.0061	7	0.0047	8	0.0042	9	0.0036	10
10%	0.0061	7	0.0047	8	0.0042	9	0.0037	10
11%	0.0061	7	0.0048	8	0.0043	9	0.0038	10
12%	0.0061	7	0.0048	8	0.0044	9	0.0039	10
13%	0.0061	7	0.0049	8	0.0044	9	0.0040	10
14%	0.0062	7	0.0049	8	0.0044	9	0.0040	10
15%	0.0062	7	0.0050	8	0.0045	9	0.0040	10

As White (2007) utilized the 2004 version of COMFAA, which did not include the new Alpha Factors, the constants in **Tables 4 and 5** replace those presented by White (2007) for CBR 6%.

Comparison with S77-1

The average difference between S77-1 and APSDS thicknesses are shown in **Table 6**. The differences are also expressed as a percentage of the S77-1 total pavement thickness. Median differences were also calculated and were slightly lower, but similar, to the average differences in all cases. It can be seen from **Table 6** that the average difference between S77-1 and APSDS thicknesses ranges from 5% (for separate wheel configuration criteria at higher CBRs) to 10% (for combined wheel configurations at lower CBRs). The average difference between S77-1 and APSDS over all design scenarios was 30 mm (6%) for the separate wheel configuration criteria. For the combined wheel configuration criteria, the average difference increased to 42 mm (8%). These compare favorably to the Chicago Criteria, with an average difference of 85 mm (17%).

Table 6   Average differences between S77-1 and APSDS

CBR	Combined Wheels/Gear		Separate Wheels/Gear	
	Difference	Percentage	Difference	Percentage
3%	79 mm	10%	61 mm	7%
4%	68 mm	10%	50 mm	7%
5%	58 mm	10%	41 mm	6%
6%	50 mm	10%	34 mm	6%
7%	43 mm	9%	29 mm	6%
8%	39 mm	9%	26 mm	5%
9%	36 mm	9%	24 mm	5%
10%	33 mm	8%	22 mm	5%
11%	29 mm	8%	21 mm	5%
12%	26 mm	7%	21 mm	5%
13%	25 mm	7%	20 mm	5%
14%	23 mm	7%	20 mm	5%
15%	21 mm	6%	20 mm	5%

Comparison with Chicago Criteria

The Chicago Criteria pavement thicknesses had a reported median difference with S77-1 thicknesses of 60 mm for the design scenarios considered at the time of their development. The Chicago Criteria were based on a significantly narrower range of design scenarios compared to the recommended criteria and were generated against the 2004 version of COMFAA, prior to the adoption of the revised Alpha Factors.

When compared to the S77-1 thicknesses calculated using the revised Alpha Factors and across all the design scenarios considered herein, the Chicago Criteria average difference ranges from 10% (82 mm) for CBR 3 to 25% (113 mm) for CBR 13. Across all subgrade CBRs, the average difference was 85 mm (17%). The greatest average difference occurred at the CBRs whose K and B values were interpolated during the Chicago Criteria determination. This suggests that interpolation was not appropriate.

The recommended K and B values contained in **Tables 4 and 5** provide for significantly improved fit between S77-1 and APSDS than the Chicago Criteria did. This is considered to be a result of the broader design scenarios that the recommended values were derived for, the inclusion of the six wheel gears in the calibration process and the fact that the Chicago Criteria were determined from the pre-Alpha Factor revised version of COMFAA.

Importance of Subgrade

An additional analysis was performed which determined the calibration constants that would be recommended if a single K and B value were to be adopted for all aircraft wheel configurations and all subgrade CBRs. The values returning the lowest average difference to S77-1 pavement thicknesses were:

- K. 0.0037.
- B. 10.

The average difference between APSDS and S77-1 pavement thicknesses was 53 mm (10%). This compares to an average difference of 42 mm (8%) and 31 mm (6%) for the combined wheels/separate CBR and separate wheels/separate CBR criteria respectively. It is therefore concluded that the retention of separate values of K and B for each subgrade CBR value is critical to maximising the level of agreement between APSDS and S77-1 achieved by the recalibration process.

Importance of Wheel Configuration

Over all subgrade conditions, the difference between the average APSDS and S77-1 pavement thickness reduced from 42 mm (8%) to 31 mm (6%) when separate K and B values were adopted for each wheel configuration (1, 2, 4 and 6 wheels). The significant improvement in agreement between S77-1 and APSDS suggests that the adoption of separate failure criteria for each wheel configuration is justified.

Some researchers have hypothesized that moving from a deflection based design system (S77-1) to a strain based design system (APSDS) could negate the requirement for Alpha or other factors that distinguish between aircraft of differing numbers of wheels. The analysis undertaken shows that aircraft wheel configuration is an important determinant of pavement thickness required. Alpha or similar factors therefore remain important. Version 5 of APSDS, scheduled for release in 2008, will include the ability to maintain separate K and B values for each wheel configuration.

Comparison to LEDFAA

The failure criterion for deformation of the subgrade for flexible pavements in the FAA's LEDFAA is of the form of **Equation 1** (FAA, 2004). The relationship between strain and the repetitions of the strain to cause failure does not depend upon either the subgrade CBR or wheel configuration. The K and B values are detailed in **Table 7**.

Table 7   LEDFAA Calibration Constants

Coverages	K	B
≤ 12,100	0.0040	8.1
> 12,100	0.002428	14.21

By adopting single values for K and B for all subgrade CBRs and all wheel configurations, the FAA appear to assume that the induced vertical strain at the top of the subgrade and whether the number of coverages is above or below 12,100 are the only determinants of pavement thickness required. Previous versions of LEDFAA provided for K and B values that varied with subgrade CBR. The more complex the model used in the failure criteria, the closer the agreement with S77-1 will be. A more complex model should be adopted where the improvement in fit to the empirical data justifies the increased complexity. Based on the recalibration of APSDS, a similar modification could be made to LEDFAA, if the FAA believed that the expected increase in agreement with S77-1 justified the additional complexity.

Conclusions

APSDS has been recalibrated to S77-1. This recalibration process has provided for significant improvement of the 2001 effort of Wardle, Rodway and Rickards by including more aircraft at various masses and repetitions as well as including the new ICAO-endorsed Alpha factors and the six wheel gears of the A380 and B777. The recalibration process and analysis of the outcomes have allowed the following conclusions to be drawn:

- In contrast to the 2001 calibration effort, the values of B were found to be constant over all subgrade CBRs from 3% to 15%.

- The new combined-wheels failure criteria produced better agreement between S77-1 and APSDS than the Chicago criteria, with an average difference of 42 mm (8%) compared to 85 mm (17%) for the Chicago criteria.

- The new separate-wheel failure criteria produced even greater agreement with an average difference of just 30 mm (6%).

- The adoption of a common failure criterion over all subgrade CBRs and wheel configurations resulted in a reduced agreement between APSDS and S77-1 with an average difference of 53 mm (10%).

- Aircraft of different wheel configurations should be assigned different failure criteria, using the calibration constants detailed in **Table 5**.

References

Ahlvin, R. G., et al. (1971). Multiple-wheel heavy gear load pavement tests. Technical Report S71-17. Volume 1. US Army Corps of Engineers. Waterways Experiment Station. Vicksburg, USA.

Barker, W. R. and Brabston, W. N. (1975). Development of a Structural Design Procedure for Flexible Aircraft Pavements. Report No S77-17. US Army Corps of Engineers. Waterways Experiment Station, Vicksburg, USA.

FAA. (1995). Airport Pavement Design and Evaluation. Advisory Circular 150/5320-6D. Federal Aviation Administration of the United States. February. Washington.

FAA. (2004). *Computer Program LEDFAA Help file*. Version 1.3. Federal Aviation Administration. William J. Hughes Technical Centre. Atlantic City. June. Available from www.airporttech.tc.faa.gov/naptf.

FAA. (2006). *Computer Program COMFAA Help file.* Version 11/2006. Federal Aviation Administration. William J. Hughes Technical Centre. Atlantic City. Available from www.airporttech.tc.faa.gov/naptf.

Hayhoe, G. F. (2006). New Alpha Factor Determination as a Function of Number of Wheels and Number of Coverages. Letter report submitted to the Director of Airport Safety and Standards. Federal Aviation Administration. 19 November.

ICAO. (2007). Revised Alpha Factor Values for Calculation of Aircraft Classification Number (ACN) on Flexible Pavements. Letter to member states. International Civil Aviation Organization. Montreal, Canada. 16 October.

Mincad. (1999). CIRCLY 4 Users' Manual. Mincad Systems Pty Ltd. Richmond. Australia. February.

Mincad. (2000). APSDS 4 Users' Manual. Mincad Systems Pty Ltd. Richmond. Australia. September.

Pereira, A. T. (1977). Procedures for Development of CBR Design Curves. Instruction Report S77-1. US Army Corps of Engineers. Engineer Waterways Experiment Station. Washington.

Wardle, L., Rodway, B. and Rickards, I. (2001). 'Calibration of Advanced Flexible Aircraft Pavement Design Method to S77-1 Method'. In Proceedings Advancing Airfield Pavements. American Society of Civil Engineers' 2001 Airfield Pavement Specialty Conference. Illinois. 5-8 August. (Buttlar, W. G. and Naughton, J. E, eds.) pp 192-201.

White, G. W. (2006). 'Material Equivalence for Flexible aircraft Pavement Thickness Design'. Proceedings 2006 Airfield and Highway Pavements Specialty Conference. Refereed paper number 17121. American Society of Civil Engineers. Atlanta, United States of America. 1-3 May.

White, G. W. (2007). 'Towards Calibration of APSDS for Six Wheel Gear Loads'. Proceedings 2007 FAA Worldwide Airport Technology Conference and Exposition. Refereed paper number P07004. Federal Aviation Administration. Atlantic City, United States of America. 15-18 April.

Acknowledgement

Greg White would like to thank his employer, GHD Pty Ltd, for supporting the preparation of this paper and attendance at this conference.

Greg would also like to thank Bruce Rodway for his assistance with the preparation and review of this paper and Leigh Wardle for the provision of the automated APSDS software and advice regarding the original calibration work.

Pavement Design Issues and Embankment Construction for the Second Runway at Cancún International Airport, Mexico

George Nowak P.Eng.[1], Ing. Hector Saldivar Moguel[2] and Ing. David Martinez Salazar[3]

[1]Deputy Airports Practice Manager, Hatch Mott MacDonald, 2800 Speakman Drive, Mississauga, Ontario, Canada; Ph: 905-403-3981; email: george.nowak@hatchmott.com
[2]Manager of Construction, Aeropuertos del Sureste – Cancún, Carret. Cancún-Chetumal Km. 22, Cancún, Q. Roo, Mexico; Ph: +52 (998) 848 7289; email: hsaldivar@asur.com.mx
[3]Subdirector of Infrastructure and Planning, Aeropuertos del Sureste – Cancún, Carret. Cancún-Chetumal Km. 22, Cancùn, Q. Roo, Mexico; Ph: +52 (998) 848 7268; email: davidmtz@asur.com.mx

Abstract

Cancún International Airport (CUN) located on the tip of the Yucatán peninsula in southeast Mexico is the second largest airport in Mexico. It served more than 11.3 million passengers in 2007 - the most international passengers in Latin America and Cancún is the biggest tourist destination in Mexico and the Caribbean. Aeropuertos del Sureste (ASUR) group has a concession contract for 9 airports in the southeast of Mexico and under their Master Development Agreement are committed to the construction of a second (parallel) runway at Cancún International Airport by the end of 2009 in order to meet the ever increasing demand for air services at this key worldwide tourist destination.

This paper describes the key pavement geotechnical issues, design methods and construction materials and techniques utilized to engineer and build the embankment for the new 2800 metre (9200 foot) runway. The main design and construction issues covered include: discussion of the greenfield site featuring variable karst topography with the potential for extensive surface and hidden subterranean cavities; pavement design criteria including aircraft loading and minimum elevation for hurricane flooding; high water table and drainage by infiltration; design of runway and taxiway stabilized flexible pavements to FAA criteria including utilization of FAARFIELD methodologies; development of heavyweight proofrolling methods to confirm suitability of natural subgrade and identification of cavities; and, utilization of local materials and construction methods for construction of the subgrade embankment and pavement structure layers.

Introduction and Background

The proposed new runway for Cancun International Airport was originally planned as an independent parallel of 2400 m (7874 ft) in length spaced at 1500 m (4921 ft) from the existing runway in accordance with the 2003 Airport Master Development Plan. During the Runway and Airside Planning Review (Arup Canada, 2006) for this project, a runway length of 2800m (9186 ft) was determined to be suitable to handle all landing aircraft and the vast majority of aircraft taking off from Cancún. The 2800 m runway is unlikely to ever need extension, in particular so long as the 3500 m existing runway is also available. The runway location has been set inside the airport boundary so as to accommodate approach lighting and to avoid penetration of the obstacle limitation surfaces by surrounding structures and trees. The runway separation distance from the existing runway was reduced to 1420 m to avoid tree clearing outside the airport property but still provides the capability for simultaneous independent operations for the two runways. The electrical towers to the northwest have been dismantled (after Hurricane Wilma) and a new transmission line corridor was provided farther to the north, such that these towers will not be an operational constraint.

Since Mexico is a signatory to ICAO with no variations, the runway was design to ICAO aerodrome standards (Annex 14). The new runway is ICAO Code 4E (FAA Group V) and the design included allowances for future upgrade to Code 4F (Group VI). The runway width is 45 m (150 ft) and provisions for widening to 60m (200ft) have been made in the geometric configuration of the runway cross section and pavement structures. A short section of parallel taxiway spaced to suit long term upgrade to Code F operations has been designed for initial construction. This section of taxiway permits queuing to the 30R end for take-off, and accommodates a rapid exit at 1905m from the 12L threshold. The potential to extend this parallel taxiway for the full length of the runway has also been planned.

A single cross-over taxiway with a Code F taxiway bridge over the main airport access road is required to connect the existing terminal area with the new parallel runway. All taxiways are planned to ICAO Code F requirements (25 m wide). Fillets are designed to the most demanding aircraft which is currently the Airbus A340-600.

Site Topography and Subsurface Conditions

The airport is located approximately 18 km (11.3 miles) south of the city of Cancún and about 5 km (3.1 miles) inland from the coast of the Caribbean Sea. Ground elevations range from 1.2 m (4 ft) above sea level (ASL) to 9.1 m (29.8 ft) ASL along the centerline of the runway and its approach areas to the airport property limits. The area of the new runway was heavily wooded with trees and scrub generally ranging from 10 to 20 m above the ground surface. The area was extensively cleared for the runway/taxiway strip and approaches, and the relatively thin topsoil and cleared materials have been screened for organics and rock fill. The organic component of the topsoil and the cleared vegetation has been mulched and mixed, and are now stockpiled for future reinstatement on the runway and taxiway graded areas.

The Yucatán Peninsula is primarily comprised of sea-bed and coral reef deposits with the deposits in the Cancún area being some of the youngest soil and rock on the peninsula. The geology at the airport has three main elements: limestone rock, weathered limestone (known locally as "sascab") and topsoil/fill. The upper crust of the limestone deposits are where the karst structures exist. The geological nature of the area and large amount of precipitation has created the karst features through the dissolution of the calcite material. As the water flows to the Caribbean Sea, the rock has been dissolved and creates underwater caves. Ultimately, when these caves become large enough to generate surface instability, they collapse and form sinkholes that fill with water and are referred to locally as "cenotes".

Runway Location and Elevation

The new parallel runway 12L-30R, rapid-exit Taxiway H, threshold 30R Taxiway G and interconnector Taxiway F are shown in Figure 1.

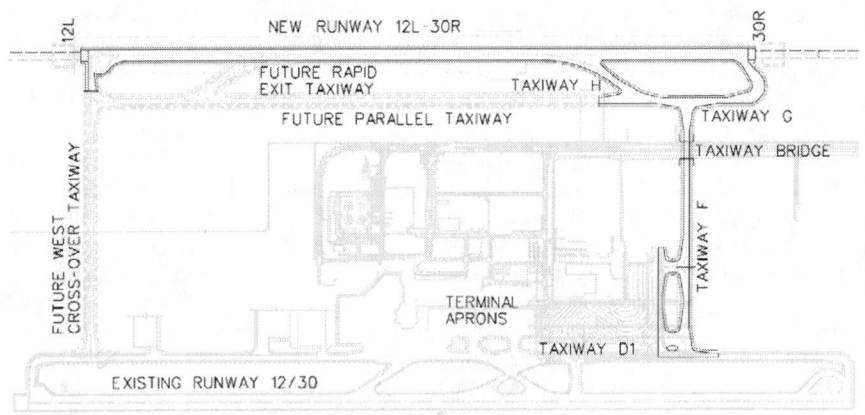

Figure 1 New Runway 12L-30R and Taxiways

The runway profile has been adjusted to minimize overall runway fill requirements as well as potential rock cut in the graded area. The minimum runway centerline elevation is the based on an assessment of the actual flooding level during Hurricane Wilma which hit the Yucatán in 2005 and stayed in the vicinity of the Cancún for three days depositing over 1200 mm (47 inches) of rain. The 4.26 m ASL elevation was selected as the minimum centerline elevation on the new runway to avoid flooding of the runway surface since the existing runway had this minimum elevation and its pavement surface was not inundated during Wilma. The other constraint was the obstacle limitation surface on rising ground to the highway. At the west end the runway was also raised in order to avoid deep cuts in the graded area. The two graded area profiles varied from the runway centerline profile so that the graded area transverse slope could be varied from 2.5% in deep fills to 1.0% in cut areas in order to minimize the earthworks quantities.

Cenote Mapping and Subgrade Rehabilitation

One of the key elements in the design of the new runway and taxiway pavements at Cancún was to ensure that: 1) for pavement areas that were cleared and cleaned of topsoil all the visible cenotes were rehabilitated; and, 2) that potential (invisible) cenotes that had not broken as yet would not affect the pavement performance under aircraft loading.

For the first case, all visible cenotes in the aircraft maneuvering areas (and future Code F widening area) were mapped and investigated in detail. This occurred during and following the clearing and topsoil removal operations. On the new runway and taxiways there were 134 cenotes mapped into four categories based on shape and surface visibility. The four types included: medium to large circular cenotes (largest was approximately 13m in diameter); deep cenotes with step vertical sides and wide openings; small surface holes that lead to larger cavities; and, medium surface holes only 300 mm deep that might lead to deeper holes. In general all cenotes were probed along the perimeter with drilling equipment to determine their subsurface extent. Those cavities with small surface holes were probed and mapped until the depth of rock cap over the cavity was approximately one metre thick at which point it was concluded that the cenote cavity limits would not affect surface pavement performance and the cavity was excavated and repaired only to the one metre limit.

The method used to rehabilitate and fill the cenotes generally involved the removal of all organic and unsuitable material, placement of rockfill to within 500 to 700mm of the existing stripped surface, placement of lean concrete mix on the rockfill and up the sides to ensure that fill materials would not escape into auxiliary cenote flow channels and finally backfilling with select granular fill to the surrounding stripped levels (See Figure 2). The technique varied widely since the cenotes were not perfectly round and in many cases extensive hand excavation was required to rehabilitate the snake-like channels that the cenote water flows formed beneath the surface. This method of rehabilitation was very successful and had been developed by the ASUR staff and used on the Cancun Terminal 3 aircraft parking apron area about 18 months earlier.

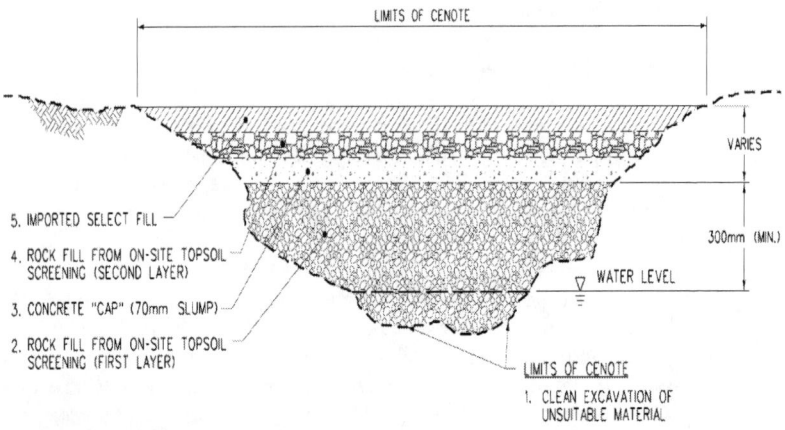

Figure 2 Typical (Circular) Cenote Repair Method

Photos of Cenote Excavation and Rehabilitation as per Typical Method

Subgrade Proofing

While all visible cenotes were mapped and treated, the design staff and ASUR recognized that the whole operation of tree clearing combined with topsoil removal would naturally cover up small cenote holes that could lead to larger cavities or there could also be cavities beneath the surface that could have a thin surface rock crust that could be broken under embankment loading or future aircraft pavement loading. (It should also be noted that the stripping operation broke through at various locations and thus identified cenote locations that were probed during mapping.) In order to ensure that all pavement areas were examined for cenotes and that the stripped subgrade was capable of supporting the design loading, it was proposed that all pavement areas be subjected to proofrolling using a 445 kN (100,000 lb) proofroller with four wheels on a single axle. This proofroller has been used on aircraft pavement in Canada for over 50 years and is used to check for weak and unsuitable subgrades by testing the pavement structure at a level in the subgrade or subbase that produced a overload ratio of 2.0 relative to the loading produced on the subgrade by the design aircraft at the pavement surface. In case of Cancún, the proofrolling was done where possible on the stripped subgrade regardless of the embankment fill thickness that was going to be placed. In many instances the stripped ground was too uneven for proofrolling so a thin layer of select fill (usually less than 300mm) was placed to smooth out the surface. The proofrolling did not find any additional cenotes by breaking through the rock surface, and based on the design pavement structure, the embankment construction (filling operation) was allowed to proceed.

Photo of 445 kN (100,000 lb) Subgrade Proofroller

Pavement Design Criteria and Design Methodologies

ASUR's terms of reference stipulated that only a flexible pavement design would be considered based on their own internal assessment and the fact that the existing runway had a flexible pavement that had performed well for over 20 years.

The pavement design involved the following elements:

1. Reviewing the available pavement history on the existing Runway 12-30

2. Utilizing field CBR and AASHTO plate load testing results for the T3 apron earthworks construction and selecting suitable design parameters based on using similar local fill (sascab) materials for select runway fill.
3. Developing a rational air traffic movement forecast for the new runway with the recognition that the runway would be used initially for arrivals traffic and in the long term as an independent arrival/departures runway as traffic growth continued at Cancún.
4. Utilize both FAA Layered Elastic Airport Pavement Design (LEDFAA v. 1.3 June 2004), FEDFAA (v. 2.0 Beta, 2006) and Public Works and Government Services Canada (PWGSC) ASG-19 Pavement Structural Design methodologies for comparative flexible pavement design.
5. Select a pavement structure using a limited number of processed local materials.

Using the LEDFAA and FEDFAA software, several options were analyzed using several CBR values. The CBR value selected was the estimate of 28.5% converted from Cancún Airport's recently completed Terminal 3 apron design "k" value of 320 pci. It should be noted that the FAA software methods do not work with design subgrade CBR values exceeding 30 since they are effectively better than standard subbase (CBR 20). The PWGSC method (Transport Canada standard) was also used for comparing with the LEDFAA and FEDFAA software output. The PWGSC method is more conservative since aircraft movement data is not a design input.

A number of CBR values were estimated from grain size classifications of potentially saturated local sascab material and from correlation curves from the AASHTO plate load testing that was carried out on the select granular fill used on the Terminal 3 apron. Ultimately a Design CBR value of 28.5% was selected, which was considered a conservative value based on the T3 compaction and plate load tests, but also allows for complete saturation of the subgrade. The select fill used on the T3 apron is consistent with the gradations and physical characteristics of the fill to be used on the new parallel runway and taxiways.

For the PWGSC method, a saturated CBR of 35 was selected for the sascab material in doing the flexible design calculations. This was solely based on soil classification.

The pavement design calculations were based on the aircraft mix taken from the Runway and Airside Planning Report (Arup Canada, 2006).

Aircraft	Number of Movements 2002-2005	Percentage of all Aircraft Movements (281,061 Total)	Number of Movements (2002)	Number of Movements (2004)
B757 All Series	38,532	13.71%	8,024	15,807
Airbus A320	29,955	10.66%	6,940	9,942
MD-80/81/82/83/87/88	21,512	7.65%	7,541	6,884
B737-100/200 Series	19,200	6.83%	5,189	6,882
DC9-20/30/40/50	18,701	6.65%	4,882	5,791
B727 All Series	17,520	6.23%	9,366	3,010
B737-800	16,251	5.78%	5,637	4,676
Airbus A319	15,175	5.40%	1,207	8,472
B737-300	12,531	4.46%	3,071	4,604
B767 All Series	13,123	4.67%	3,167	4,882
A330	6,361	2.26%	1,527	2,300
B737-700	5,945	2.12%	1,181	2,940
Fokker 100	5,344	1.90%	3,261	38
B737-900	4,188	1.49%	1,117	1,424
B737-400	1,704	0.61%	846	348
L1011	1,283	0.46%	640	217
B747-100/200/300	872	0.3%	50	472

Table 1 – Current Aircraft Movements at Cancun International Airport (2004)

Two options of aircraft mix were considered, one with a 62% increase in forecast movements (to 2018), and the other one without the 62% increase. The mix with the 62% increase allows for traffic growth to mid-term of an initial 20-year lifespan for the flexible pavement. However, the new runway will primarily handle arrivals traffic and not considering a growth in traffic allows for this reduced loading while still using the maximum takeoff weights for all aircraft as compensation for reduced arrival weights.

Runway Pavement Section

There were 26 different LEDFAA, FEDFAA and PWGSC pavement section alternatives considered with both unbound and cement/asphalt treated base courses and both with and without the 62% increase in traffic. The results of the owner - preferred layer combination using a conservative CBR value of 28.5% and the 62% traffic increase was as follows:

LEDFAA: 125mm (5") Hot Mix Asphalt (HMA)
 105mm (4") Cement Treated Base Course
 210mm (8") Crushed Aggregate Subbase

FEDFAA: 125mm (5") HMA
 105mm (4") Cement Treated Base Course
 195mm (7.5") Crushed Aggregate Subbase

PWGSC: 125mm (5")   HMA
                  250mm (10")   Cement Treated Base Course
                  195mm (7.5") Crushed Aggregate Subbase

Selected Pavement Structures:

OPTION 1   OPTION 2
125mm (5") HMA   125mm (5") HMA
150mm (6") Cement Treated Base   105mm (4") Cement Treated Base
150mm (6") Cr. Aggregate Subbase   105mm (4") Cement Treated Subbase

Option 1 was based on the minimum thickness value for stabilized base from FAA AC 150/5320-6D Paragraph 706 b. The subbase thickness is the remaining value to conform to the output thickness of 12 inches (300mm) for base and subbase from the software output. The actual design thickness from the LEDFAA software indicated 105 mm cement treated base thickness and a 210 mm crushed aggregate base thickness. By complying with the minimum 150 mm stabilized layer thickness we could reduce the crushed granular to 150 mm at a conservative equivalency (to cement treated base) of 1.33. Option 1 was selected for the final design. It is interesting to note that using reduced traffic forecasts showed that the pavement achieved a cumulative damage factor (CDF) of 1.0 after 31 years rather than the 20-year normal design period.

Embankment Section and Subgrade Requirements

The typical runway fill section is shown in Figure 3. In order to comply with the minimum "Wilma" runway centerline elevation as well as being able to create the approach slopes for the taxiway bridge over the main access road, the new runway and taxiway project is essentially a "fill" project with most of the cut being topsoil removal for reprocessing. The proportions are about 90% fill and 10% cut, with the deepest fill being about six metres (19 feet).

The embankment design makes maximum use of all available materials and also allows the use of both select granular fill – the local "sascab" and local rock fill which actually can be broken down into a granular material with special compaction equipment which is illustrated in subsequent photographs.

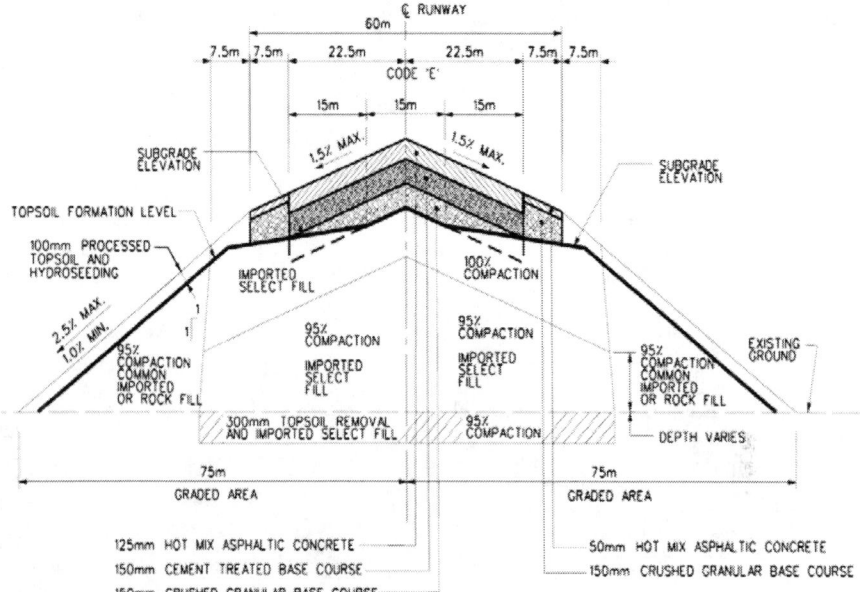

Figure 3 Typical Runway 12L-30R Embankment Fill Section

In order to allow for future Code F pavement upgrades, the 7.5m (24.6 ft) wide Code E shoulder pavement will have to be converted to accept aircraft loading. This widening to Code F would be based on using 70% of the pavement structure on the outside pavement edges rather than the full depth in the central keel of the runway. To accommodate this future requirement the bottom (subgrade level) of the 70% structure was extended to the Code E shoulder area. This will avoid excavating deeper in the future. The Code E surface of the thinner section of granular base and thinner section of hot mix asphalt could then be removed and replaced with the remaining aircraft bearing layers at a later date. Of course, the future Code F shoulders would not have to follow this pattern and would be a thinner pavement structure from the outset and this area would be located on the 2.5% transverse slope zone, which is acceptable for shoulders. In addition, the primary airfield lighting cables will have been placed outside the Code F pavement edge and both the aircraft pavement widening and shoulders could be constructed at some future date without any major shutdowns.

Embankment Material Characteristics

Three types of fill material were generally utilized for the embankment construction. In the lower section of the aircraft bearing pavement both the local rockfill and the screening rock from the topsoil removal were utilized. In the graded areas both the rockfill and screened rock were allowed and the rockfill at the surface was overlaid with the screened topsoil and clearing mulch, combined with a sprigged grass selected for Cancún. Under the aircraft bearing pavement areas, a minimum 300 mm (12 inches) of select granular fill (sascab) was specified and compacted to 100% Modified

Proctor field densities. The overall flexible pavement structure met the compaction requirements set out in FAA AC 150/5320-6D.

The typical select granular fill (sascab) gradation is shown in Figure 4 (right). Sascab can be described as decomposed limestone. It ranges in color from light beige to medium brown and is typically a light brown silty sand or sandy silt that is usually highly cementitious in its natural state. Sascab was extensively used by the Mayans as mortar, and more recently as a building and paving material owing to its cementitious properties and excellent strength once compacted. The specified minimum CBR value for the sascab is 35% but actual values typically are in the order of 135% to 170% on site.

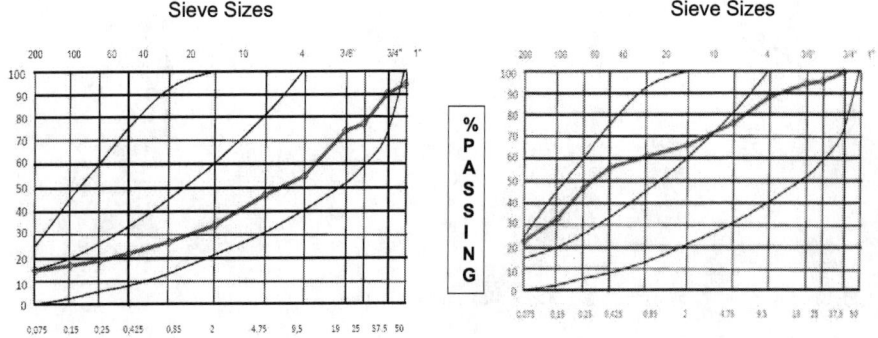

Figure 4 Gradation of Rockfill Broken Down on Site and Imported Sascab (right)

Embankment Construction and Equipment

Generally, the embankment construction equipment was conventional in nature, with the exception of the proofroller and the "inverted" sheepsfoot roller that was used to breakdown the granular rock fill. As noted previously the rock fill that was used in the lower layers of the embankment and in the runway and taxiway graded areas was not "rock" in the normal sense. This rock fill was essentially the weak limestone beach deposits, the weathered surface limestone or cemented sascab from quarry operations. It was relatively easy to break the rock down into a compactable gradation using the inverted sheepsfoot. In addition, there were some localized in-situ rock excavation areas on the runway strip. Again these were easily ripped using bulldozers with ripper bars. No blasting was necessary for on-site "rock" excavation.

Photos of Rockfill Reverse Sheepsfoot Breakdown Roller and Vibratory Compactor

Drainage

There is no system of storm drains or perimeter drains (perforated pipe) for the new runway and taxiway system. Due to the porous characteristics of the underlying limestone base, the runoff is allowed to percolate into the rock. In isolated areas or major catchment zones, drainage wells (typically 30 m (98 ft) deep with perforated 300 mm dia. piping) are drilled into the limestone rock. These drainage wells have an infiltration rate of from 10 to 17 liters/sec (158 to 269 USGPM) per well. Additional wells are added as required and are interconnected for infiltration load distribution into the limestone. The design storm was based on a 1 in 10 year return period used in Cancún with ponding allowed for a maximum of four hours in infield graded areas between runways and taxiways.

Schedule and Quality Control Testing

The earthworks for the new runway and taxiway system was divided into two projects: 1) the Runway 12L-30R proper, Runway 12L-30R graded areas and the portion of the rapid-exit Taxiway H and 30R threshold Taxiway G that was within the graded area; and, 2) the remaining fill up to the abutment for the Taxiway F bridge, all of Taxiway F and the connections to T3. The separation of the earthworks was tied to an early start on the runway coupled with the fact that ASUR was undertaking the clearing and topsoil removal under direct contracts in two stages.

The specifications laid out the materials testing frequency for each material type and the local supervision consultant is accurately documenting the subgrade repairs, proofrolling, aggregate testing and compaction testing on site.

Summary and Conclusions

The design and construction of the new parallel runway embankment at Cancún International Airport faced a number of engineering challenges with respect to the rehabilitation of the embankment base in the presence of the karst topography, porous limestone foundation rock, high water table and designing for potential hurricane flood levels in order to ensure a fully operation runway throughout its lifespan. Fortunately, the extensive mapping and probing of cenotes (sinkholes) on the runway and taxiway subgrades combined with the proven repair methods developed by ASUR during their construction of the Terminal 3 apron ensured the creation of a stable embankment platform. This mapping and repair of visible cenotes was validated by proofrolling all pavement subgrades with a 100,000 lb. Proofroller to ensure that any potential "invisible" cenotes were discovered and repaired. The flexible pavement design for the runway was primarily based on FAA methodologies and utilized the cementitious properties of the local select fill (sascab) and rockfill to achieve a high performance subgrade embankment structure that exceeded the minimum specified construction values and will ensure a superior performing pavement for years to come.

Acknowledgements

George Nowak would like to acknowledge the great assistance of Abraham Ahumada, EIT who prepared all the initial pavement design calculations and software runs and also developed the initial earthwork embankment profiles. George would also like to thank Paul Heslop and Tim Thompson who were responsible for the geotechnical investigations for the project on behalf of Arup. Finally, George would like to thank his co-authors, Hector and David from ASUR, for providing invaluable data on the cenote repair methods used on the T3 apron and the detailed cenote mapping on the runway.

References

1. Arup Canada Incorporated, Runway and Airside Planning Review Report, Toronto, Canada, April 2006.

2. FAA, Airport Pavement Design and Evaluation, Advisory Circular 150/5320-6D, Federal Aviation Administration of then United States, Washington, February 1995

3. FAA, Layered Elastic Airport Pavement Design (LEDFAA) (v. 1.3), June 2004.

4. FAA, Finite Element Design (FEDFAA) (v. 2.0 Beta), January 2006

5. International Civil Aviation Organization, Aerodromes Annex 14, Volume 1 Aerodrome Design and Operations, 4th Edition, July 2004.

6. Ove Arup & Partners Consulting Engineers PC, Geotechnical Design Basis Report, New York, October 2006.

7. Public Works Canada, Air Transportation, Architectural and Engineering Services, Manual of Pavement Structural Design, ASG-19, July 1992.

Quintin Watkins[1], Robert Rau[2] and Richard L. Boudreau[3]

Impact of New-Generation Aircraft at Hartsfield-Jackson Atlanta International Airport

Abstract

Rising and turbulent fuel costs have airlines challenged to meet the needs of future air travel. Recognizing this, aircraft designers/manufacturers have developed larger aircraft capable of flying greater distances, carrying more passengers and subsequently burning less fuel (per passenger). As the world's busiest airport, Hartsfield-Jackson Atlanta International Airport must be capable of evaluating each airlines needs to integrate new aircraft into their fleets, and the impact of doing so as it relates to operations and the structural performance of the airport's pavement network.

The airport consists of over 42 million square feet of airside concrete pavement, ranging from 16 to 22 inches in concrete thickness. As part of the Pavement Management Program, a large historical database has been developed and maintained since 1984. Part of this database contains structural response data from falling/heavy weight deflection testing, as well as layer thickness data. This information is crucial for evaluating the impact of new aircraft added to the fleet.

An evaluation of the structural impact on the existing pavement structures at the airport using the most recently collected (Summer-2007) structural response data is presented. Aircraft evaluated include the Boeing 787 and 777 as well as the Airbus A380, and results indicate areas which represent the highest risk of structural deterioration. Operational limitations are also examined, showing where pavement width, turning radius and gate capacity issues would limit the maneuverability of these aircraft.

Introduction

From its beginnings in the mid 1960s, the Atlanta airport has grown into an economic juggernaut recognized as the world's busiest airport in terms of annual aircraft operations and passenger volume. Aircraft operations have grown from roughly 400,000 operations in the late 1960s to 980,000 operations in 2007.

---

[1] Airside Area Manager, Department of Aviation, Hartsfield-Jackson Atlanta Intl. Airport, Atlanta, GA, 30337, PH (404) 530-5903, quintin.watkins@atlanta-airport.com
[2] Sr. Airfield Planning Manager, Department of Aviation, Hartsfield-Jackson Atlanta Intl. Airport, Atlanta, GA, 30337, PH (404) 530-5500, rob.rau@atlanta-airport.com
[3] President, Boudreau Engineering, Inc., 5392 Blue Iris Court, Norcross, Georgia 30092, PH (404) 388-1137, PE, M.ASCE, rlboudreau@comcast.net

Central to this growth is the infusion of larger aircraft, capable of flying longer distances and carrying more passengers. These new-generation aircraft include the Airbus A380 and the Boeing 777 and 787.

The impact these aircraft have on operations as well as potential problems with structural integrity of the pavement system are examined. Concerns with passenger capacity at gates, dimensions of the aircraft while parked at gate areas, maneuverability around the airside pavement network and the impact of load on the existing pavements throughout Hartsfield-Jackson Atlanta International Airport (ATL) are all critical in the decision to allow such aircraft operations.

Background

The pavement network at ATL is shown in Figure 1. The system includes 5 parallel runways; 2 located on the north side of the midfield concourses (Runways 8L-26R and 8R-26L), 2 located on the south side of the midfield concourses (Runways 9L-27R and 9R-27L, and the recently completed 5th runway located further south, spanning an 8-lane interstate freeway (Runway 10-28).

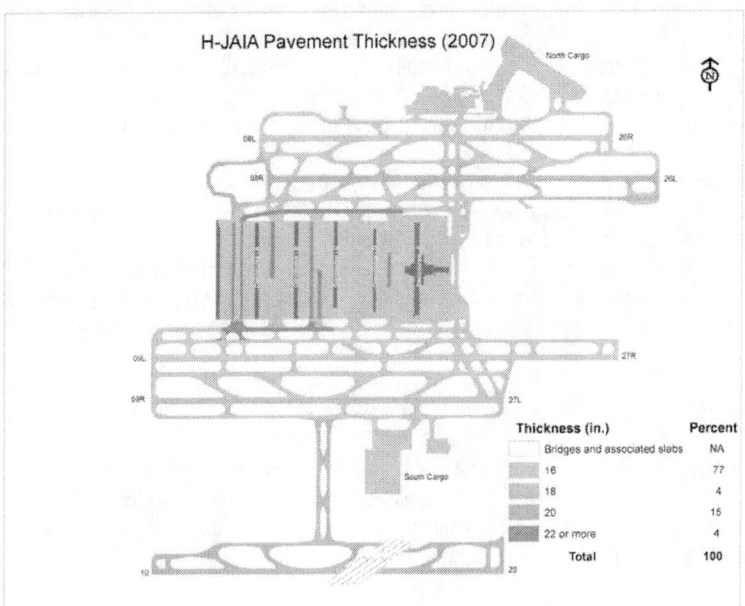

Figure 1 – GIS showing pavement network with PCC thickness attributed

Each of these runways is 150 feet wide and range in length from 9000 feet to 12000 feet. The oldest runways (9L-27R, constructed in 1972 and 8L-26R, constructed in 1984) were built with a 16-inch concrete section. The newest Runways (8R-26L, reconstructed in 2006 and 10-28, constructed in 2005) consist of a 20-inch concrete section. Exits, taxiways and apron pavements vary in age and construction, ranging from 16-22 inches of concrete at 1 to 40 years in age. The color scheme and legend in Figure 1 illustrates the concrete thickness distribution of the pavement contained within the network. In total, ATL contains approximately 42 million square feet of structural concrete pavement.

Fleet Mix at ATL. The City of Atlanta developed a traffic forecast model at Hartsfield-Jackson Atlanta International Airport in 2002. The data projections were based on actual data obtained in 2000 as well as anticipated influx of new-generation aircraft available at that time to make forecasts into the future, up to 2025. The data presented in Table 1 is a consolidated version of the data presented in the forecast plan. Although numerous other aircraft are included in the fleet (piston, turboprop, small jets, and numerous other narrow –body and wide-body jets), those shown in Table 1 represent two of the more common aircraft (B737-800 and B757-200) in addition to the new-generation aircraft central to this paper.

Table 1 – Aircraft Fleet Mix and Departure Forecasts at ATL

Aircraft Type	2010	% of Fleet	2015	% of Fleet	2020	% of Fleet	2025	% of Fleet
Boeing 737-800	91,186	15.1	111,260	16.1	121,281	15.9	131,302	15.7
Boeing 757-200	72,470	12.0	93,434	13.5	113,102	14.8	132,769	15.9
Boeing 777	6,845	1.1	6,943	1.0	7,780	1.0	8,616	1.0
Boeing 787								
Airbus A-380			555	0.1	735	0.1	914	0.1
All others	434,046	71.8	81,721	11.8	104,851	13.7	127,981	15.3
TOTAL	**604,547**		**690,936**		**763,716**		**836,496**	

Source: HAIA Aviation Forecast Update, Hartsfield Planning Collaborative, May 2002. Projections made from 2000 Departures.

In order to gain an appreciation for how these aircraft differ in size, select properties of each have been compiled and summarized in Table 2 (Burns and McDonnell, 2006). In addition, these aircraft are shown in Figure 2 superimposed in a simulated taxi to takeoff at the west end of Taxiway Mike leading to the 9L end of Runway 9L-27R. The wingspan and engine offset (84 feet) of the Airbus A-380 make this aircraft a challenge on taxiways with insufficient paved width (including paved shoulder) as the engine hangs above grass infields, thus making infield FOD a definite safety concern.

Table 2 – Aircraft Properties

Property	Boeing 737-800	Boeing 757-200	Boeing 777-200ER	Boeing 787-8	Airbus A-380
Operations (2015 projection)	222,520	186,868	13,886		1,110
Class	III	IV	V	V	VI
Passenger Capacity	184	186	440	250	555
Max. Gross Takeoff Weight (lbs)	172,500	240,000	506,000	484,000	1,235,000
Range (nautical miles)	3,000	4,000	4,500	8,000	9,400
Body Length (ft.)	129.5	155	209	186	239
Wing Span (ft.)	112.5	125	155	197	262
Tail Height (ft.)	41	45	61	56	79
Spacing of Nose to Nose Gear (ft.)	13.5	19	19		17
Spacing of Nose to Main Gear (ft.)	65	79	104		115
Engine Offset from Aircraft Centerline (ft.)	15.8	22	32		84
Engine Clearance Above Ground (ft.)	1.5	2.5	3.5		5.5

Figure 2 – Simulated Taxi to Takeoff for Various Fleet Aircraft at ATL

Operational Impacts

The present configuration of the runways, exits, taxiways and aprons at ATL were originally designed to the FAA Airplane Design Group V standard, which at the time represented the largest of the aircraft operated by the airline industry. Since then, the Group VI category has entered the airline industry and additional Group V aircraft with more complicated operational challenges for the airports today. Some of those challenges include:

- Runway and Taxiway Width
- Runway and Taxiway Shoulder Width
- Runway to Parallel Taxiway Separation
- Taxiway to Fixed or Moveable Object
- Taxiway to Taxiway Separation
- Fillet Size
- Terminal Gate Parking

FAA Design Standards - At present, ATL is evaluating the airfield to determine the necessary improvements required to accommodate these larger Group V and VI aircraft. FAA has published guidance to assist airports in achieving these standards. In some cases airports will need to make airfield modifications to meet the Group VI standard (widen runways/taxiways etc.) and/or in other cases the airports may prove to the FAA that some standards can be met through a Modification of Standards by proving an acceptable level of safety. This is strictly considered by a case by case basis.

Taxiway Fillets – Presently, the B777-300 and the A340-600 (Group V aircraft) have the longest wheel base which creates challenges for crews taxiing these aircraft. Many of the fillets at ATL cannot accommodate nose of centerline taxiing for these aircraft. Specific taxi routes were developed in concert with ATL Tower to taxi these aircraft in areas of the airfield where larger pavement is available. Also, the use of judgmental oversteer techniques and markings are being used to assist crews in taxiing these aircraft. As new pavements are constructed at ATL, these aircraft are now being considered when designing fillets.

Terminal Gates – The wingspan and the passenger capacity of Group VI size aircraft poses challenges for gating these aircraft at ATL (see Figure 3). At present, Group V aircraft are the largest aircraft accommodated on Concourse E. Group VI aircraft would limit the aircraft gauge of the adjacent gate in some cases. Additionally, the Group VI capacities necessitate the need for a minimum of two boarding bridges to provide an appropriate level of customer service in enplaning and deplaning passengers.

Figure 3 – Concourse E Gate Configuration (B737 vs. A380)

Structural Impacts

Stress determination is a critical factor in pavement modeling and performance evaluation. Factors which influence working stresses in concrete pavements include:

- Concrete layer thickness
- Concrete material properties (modulus, strength, Poisson's ratio)
- Load transfer mechanism and efficiency at joints
- Substrate support characteristics (k-value)
- External physical loads (# of tires, contact area or pressure, vertical load per tire)
- External environmental loads (curling due to temperature and moisture differentials)

Up until recently, pavements were designed to serve aircraft weighing up to 350,000 pounds with a DC-8-50 series landing gear configuration, a standard adopted by the Federal Aviation Administration in 1958 (FAA, 2008). Although aircraft gross weights have long exceeded 350,000 pounds, these larger aircraft were equipped with gear configurations (quantity and spacing of wheels) so as to not stress the pavements more than the referenced aircraft.

It was this design standard that lead to 16-inch concrete pavement sections at ATL in the 1970s and 1980s. Over time, aircraft manufacturers realized aircraft efficiency problems associated with adding extra wheels to limit pavement stresses. Because of this, aircraft manufacturers began to focus on designing aircraft weights and gear configurations that optimized efficiency. As these aircraft began to roll off the assembly line and air carriers began to add these to their fleets while retiring older aircraft, pavement stress analysis in the 1990s resulted in structural designs requiring 18 inches of concrete to support the heavier aircraft fleet of that generation.

Designs for new generation aircraft having multiple dual wheel (2D and 3D) landing gears, such as the Boeing B777 and Airbus A380 series, were not covered by the previous design procedures. The new advisory circular, AC 150/5320-6E, in conjunction with FAARFIELD (2008), provides the necessary information for thickness design when 2D/3D and complex aircraft gears are included in the aircraft mix. The trend for thicker pavement has continued, as now the design standard at ATL is 20 inches.

Stresses computed for this study were done so using the computer software FAARFIELD -- using the *Life* routine, as shown in Figure 4. The following fixed or variable inputs were used:

PCC thickness = variable (16.5, 18.5, and 20.5 inches)
Composite K-value = variable (200, 300, 400 and 500 pci)
  use 9-inches of soil cement at Esc = 250,000 psi in all cases
  use subgrade resilient modulus variables E_{sbgr} = 3900, 6000, 8500, 11000 psi
Aircraft Loading = variable (B737, B757, B777, B787 and A380)
PCC Modulus of Elasticity = 4.0×10^6 psi
PCC Modulus of Rupture = 700 psi (flex strength at 90 days)
Joint Transfer Efficiency (0% JTE or Free Edge condition, 30%, 50% and 70% JTE)

These variables were selected in order to develop a matrix of stresses that would represent those possible in the ATL pavement network. The results of the FAARFIELD stress analyses are summarized in Tables 3-5. The FAARFIELD output is represented for the free edge loading condition (JTE=0% column of these tables) in an output file named NikePCC, which can be viewed in any text editor to retrieve the computed maximum horizontal stress. Stress reductions were made for joint transfer efficiency of deflection (Sawan and Darter, 1979), which are represented by the last 3 columns of these tables. The computation resulting from the Figure 4 screen capture is represented in Table 3, row 1 of the A380 stresses (shown as bold font with double line cell border in the table). Shaded cells represent stresses exceeding 50 percent of the flexural strength, or stress ratio exceeding 0.50.

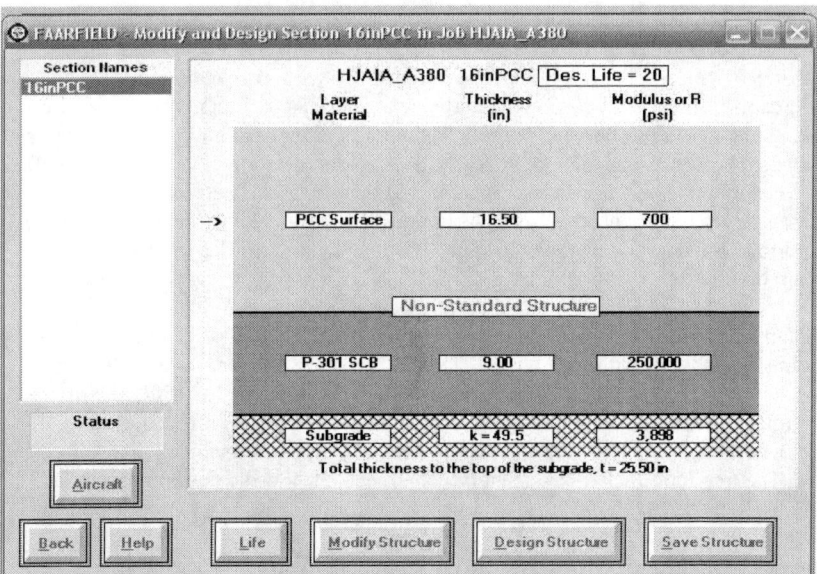

Figure 4 – FAARFIELD input screen (select *Life* button to obtain stress output)

Table 3 – FAARFIELD Edge Stresses for 16.5-inch PCC Pavement

Aircraft	Composite K-value	Maximum Stress (psi) PCC_{bottom}			
		JTE=0%	JTE=30%	JTE=50%	JTE=70%
B737-800	200	385	354	316	270
	300	371	341	304	260
	400	356	328	292	249
	500	344	316	282	241
B757-200	200	324	298	266	227
	300	308	283	253	216
	400	292	268	239	204
	500	278	256	228	195
B777-200ER	200	445	409	365	312
	300	420	386	344	294
	400	395	363	324	276
	500	374	344	307	262
B787-8	200	455	418	373	318
	300	430	396	353	301
	400	406	373	333	284
	500	386	355	317	270

Table 3 – FAARFIELD Edge Stresses for 16.5-inch PCC Pavement (Cont'd)

Aircraft	Composite K-value	Maximum Stress (psi) PCC_{bottom}			
		JTE=0%	JTE=30%	JTE=50%	JTE=70%
B787-9	200	442	407	363	309
	300	419	385	343	293
	400	397	366	326	278
	500	379	349	311	266
Airbus A380	200	455	419	373	319
	300	426	392	349	298
	400	401	369	329	281
	500	377	347	309	264

Table 4 – FAARFIELD Edge Stresses for 18.5-inch PCC Pavement

Aircraft	Composite K-value	Maximum Stress (psi) PCC_{bottom}			
		JTE=0%	JTE=30%	JTE=50%	JTE=70%
B737-800	200	313	288	257	219
	300	303	279	248	212
	400	294	270	241	206
	500	285	262	234	200
B757-200	200	265	244	218	186
	300	255	234	209	178
	400	245	225	201	171
	500	235	216	193	165
B777-200ER	200	363	334	297	254
	300	350	322	287	245
	400	333	307	273	233
	500	319	293	261	223
B787-8	200	373	343	306	261
	300	357	329	293	250
	400	340	313	279	238
	500	326	300	267	228
B787-9	200	362	333	297	253
	300	347	319	285	243
	400	332	305	272	232
	500	316	291	259	221
Airbus A380	200	376	346	308	263
	300	357	328	293	250
	400	340	313	279	238
	500	323	297	265	226

Table 5 – FAARFIELD Edge Stresses for 20.5-inch PCC Pavement

Aircraft	Composite K-value	Maximum Stress (psi) PCC$_{bottom}$			
		JTE=0%	JTE=30%	JTE=50%	JTE=70%
B737-800	200	257	236	211	180
	300	251	231	206	176
	400	245	225	201	172
	500	239	220	196	167
B757-200	200	220	202	180	154
	300	212	195	174	149
	400	205	189	168	144
	500	200	184	164	140
B777-200ER	200	301	277	247	211
	300	294	270	241	206
	400	283	260	232	198
	500	267	246	219	187
B787-8	200	310	285	254	217
	300	299	275	245	210
	400	286	263	234	200
	500	275	253	226	193
B787-9	200	300	276	246	210
	300	290	267	238	203
	400	278	256	228	195
	500	271	249	222	189
Airbus A380	200	314	289	257	220
	300	301	277	247	211
	400	289	266	237	202
	500	277	255	227	194

The results summarized in these tables indicate that the thinner (16.5-inch PCC shown in Table 3) pavements may develop stresses in excess of 50 percent of the tensile strength, thus may be more susceptible to fatigue damage. However, these thinner pavement sections should be structurally sufficient if joint transfer efficiencies of 70 percent or better are maintained throughout the year (seasonal fluctuations). The 18.5-inch PCC sections appear susceptible only if no load transfer is available, and a relatively low k-value support exists.

Based on years of structural response measurements with a heavy weight deflectometer, the pavement features at ATL have been characterized in a manner such that representative cold season and warm season pavement stiffness, k-value and joint efficiency responses can be quickly and efficiently used in database form to assess potential structural deficiencies for any aircraft at any location within the ATL network. Comparing the results in Tables 3 through 5 with parameters derived from the database, the pavement areas highlighted in Figure 5 are most susceptible to fatigue accumulation (i.e., stress ratios in excess of 0.5).

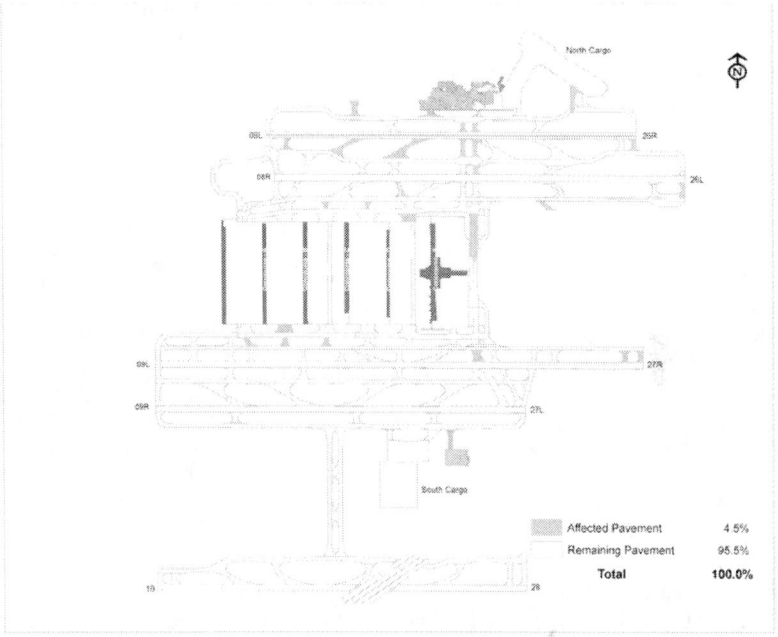

Figure 5 – GIS showing pavements susceptible to potential fatigue

Although the pavement areas highlighted in this figure are most susceptible to future fatigue-related damage, they are not expected to instantaneously fail when these new-generation aircraft traverse them. The ATL management can consider structural upgrades to these areas, pattern taxi-traffic to avoid these areas, or simply not allow these aircraft on the airport.

Conclusions

As discussed, there are many challenges that airports are facing regarding new generation larger aircraft.

- Maneuverability – ATL has many areas on the airfield which pose a challenge to aircraft with large wheelbases (A340-600 and B777-300). ATL has developed taxi routes in concert with ATL Tower giving flight crews larger pavement areas for maneuvering the aircraft. The addition of airfield markings and use of judgmental oversteer techniques has provided additional assistance for these aircraft.

- Capacity – Since ATL is designed for Group V aircraft, gating Group VI aircraft becomes a challenge. Adjacent gates to the Group VI aircraft would have to be limited in size because of the large wingspan. The Group VI aircraft also recommends servicing with two passenger boarding bridges which adds additional complexities to gating these aircraft at ATL.
- Structural – Pavements having less than 16.5 inches of concrete and less than 50 percent joint efficiency (deflection) are susceptible to potential future fatigue damage due to loads imposed by these new generation aircraft. Pavement areas with these characteristics total approximately 4.5% of the 42 million square feet of structural pavement at ATL.

Acknowledgements

The authors wish to thank Raghuram Tadimalla for his assistance in obtaining the GIS graphics for this paper.

References

Burns & McDonnell Aviation Services, *Aircraft Characteristics,* 9th Edition, 2006.

Federal Aviation Administration, *Airport Pavement Design and Evaluation,* Advisory Circular AC/150-5320-6E, United States Department of Transportation, Draft Copy for Review by the Airfield Pavement Committee Members of the American Society of Civil Engineers, 2008.

Federal Aviation Administration, *FAARFIELD Computer Software, Version 1.102, USER's Manual,* United States Department of Transportation, download from http://www.faa.gov/airports_airtraffic/airports/construction/design_software/, 2008.

Sawan, J.S, and Darter, M.I., *Structural Design of PCC Shoulders - Pavement Systems: Assessment of Load Effects, Design, and Bases,* TRB Record No. 725, pp. 80-88, 1979.

Landing Gear Factors for Pavements with FAA Ratings

Kenneth J. DeBord, P.E., M.ASCE[1]

[1]Civil Engineer, The Boeing Company, Airport Technology, P.O. Box 3707
MC 67-KR, Seattle, WA 98124, Ph: (425) 237-4294, Fax: (425) 237-8281,
email: kenneth.j.debord@boeing.com

Abstract

Many airports in the United States and around the world list their runway pavement ratings according to the Federal Aviation Administration (FAA) system of Dual (D), Dual-Tandem (DT), and Double Dual-Tandem (DDT). This system has been in place for many years, and it has enabled airport authorities to determine allowable gross weights by a method that is easy to understand and use. In recent times, however, the development of larger widebody aircraft by the manufacturers has resulted in pavement loadings that cannot be adequately expressed by this type of rating, and so most ratings are now represented by the Aircraft Classification Number/Pavement Classification Number (ACN/PCN) system.

Rather than develop new ratings for the 777 and other widebody aircraft, the FAA has gone ahead with a new design system, but left the current ratings in place. The ACN/PCN system will eventually take over as the rating of choice, but complete implementation appears to be far into the future, at least for U.S. airports. This has left the airport authority and many consultants with no guidance as how to evaluate an FAA-rated pavement which is traversed by aircraft such as the 777 or those with other unusual or non-standard gear configurations.

The result is that many pavement loading analyses may be both inadequate and misunderstood. For example, the Boeing 777 has a 6-wheel gear, and it is not defined by the FAA system. Also, an Airbus A330 has a similar, but much larger wheel geometry than the standard FAA gear, and even though its pavement loading characteristics are superior to the FAA Dual-Tandem (DT), proper pavement loading credit may not be given. In the military sector, the C-17 has a six-wheel gear that is unlike any other arrangement.

This paper proposes a method that adjusts the standard FAA rating to account for a variety of wheel sets such as those found in the larger widebody aircraft. Factors are developed that adjust the FAA rating for any widebody gear, including the 777. The method is easy to use, requiring only a use of a factor to arrive at an allowable gross weight. Pavement characteristics such as thickness, modulus, subgrade type and strength are not part of the standard FAA rating, and they are not required to utilize this method.

Keywords: ACN, PCN, FAA, Flexible Pavement, Rigid Pavement, Allowable Gross Weight

Introduction

Have you ever tried to determine the allowable gross weight of a 777 on a pavement that was rated something like D200, DT350, DDT800, ratings which don't reflect the gear configuration of this aircraft? What about the allowable gross weight of a 767 with a dual-tandem gear arrangement on a pavement having a rating of DT250? Is the 767 gear type close enough to the DT rating so that its allowable gross weight is well represented by the DT rating? What do you do for any large aircraft when only a dual rating, such as D175, is provided?

In review, the FAA system rates pavements in terms of aircraft having certain landing gear configurations. The Dual gear system is represented by D, followed by the allowable gross weight in thousands of pounds. The Dual-Tandem system is represented by DT, again accompanied by an allowable gross weight number. The third term, DDT, or Double Dual-Tandem, is specifically used for the 747 aircraft, which has four dual-tandem gear sets of wheels. DDT comes from the two sets of dual-tandem gear sets on each side of the aircraft. Other gear configurations, such as the 6-wheel gear on the 777, are not represented in this system. Many military gear configurations, such as T, ST, SBTT, TT, TDT, TRT, and TTT, are either similar to the FAA system or are for specialized aircraft. Although the military naming conventions are not discussed specifically in this paper, the methods developed herein apply to these types of landing gear also.

Recognizing that the three standard FAA rating types don't always match with current aircraft gear types, the FAA devised a new set of naming rules (1) that clearly defines the gear configurations of current aircraft. The gear sets that match the pavement ratings of D, DT, and DDT have been changed to D, 2D, and 2D/2D2, respectively. Although the gear naming convention has changed, the old pavement rating system is still in effect.

There are several aircraft with landing gear trucks having six or more wheels, and many aircraft have larger gear footprints or dimensions than in the past. The FAA has placed design charts in their advisory circular AC 150/5320-6D (2) for the new widebody aircraft, but they soon found it futile to try to publish a rating for each gear configuration. There are about 15 gear types that do not fit the standard D, DT, DDT system plus many more that may look similar, but are not the same as the gear configuration on which the ratings were based. The question becomes: how do these landing gear models relate to the FAA rating system? And how can this become reduced to a simple process of pavement loading comparison? These topics are the subject of this paper.

The impetus for developing landing gear factors really began with the 1994 introduction of the 6-wheel gear 777 aircraft. The question immediately arose as to how would this work with the standard FAA pavement rating terms of D, DT, and DDT? Although other widebodies, such as the 767 and A300 (both with DT gear) had been in operation for some time before the 777, there was not a great need to incorporate their superior pavement loading characteristics into the FAA system. The ACN/PCN system was in full use by then, and the FAA was beginning to rate its pavements with PCN's.

For those not familiar with these terms, ACN is the Aircraft Classification Number and PCN is the Pavement Classification Number. An airplane is allowed to operate on a pavement without restriction provided that the numerical value of the ACN is no greater than the PCN. The comparison of ACN to PCN is also subject to pavement type (F-flexible and R-rigid) and subgrade classifications of A, B, C, and D, which range from high to ultra low, respectively. Aircraft manufacturers are responsible for the ACN, subject to standard rules of calculation, while airports or airport authorities report the PCN. A complete description may be found in the International Civil Aviation Organization's (ICAO) Annex 14 publication (3).

However, in about the year 2000 the FAA stopped using PCN for major U.S. airports, and it removed PCN's already published for all of their airports in the Airport Information Publication (AIP). This action has left many airport engineers and consultants with a lack of direction and guidance on how to accommodate various aircraft rated with the FAA system.

It has become clear that the evaluation of pavement strength requirements of existing runways needs to be updated with some additional engineering tools, at least as long as the FAA ratings system is being used.

This method was developed as an interim procedure, meant only to be used until the FAA ratings are replaced with PCN's as established by airport authorities.

International units (SI) are not used in this paper because the FAA ratings are always shown in thousands of pounds. Adding SI conversions would not be appropriate.

Development of the Method

Originally, most jet transport aircraft had single aisle (narrowbody) configurations, and FAA ratings were based on this standard aircraft loading. For example, the Dual rating was based on gear similar to the 727, the Dual-Tandem rating was based on gear similar to the DC8-63, and the Double-Dual Tandem rating was based on the twin aisle (widebody) 747. These standard aircraft, with their gear configuration and gear dimensions are:

AIRFIELD AND HIGHWAY PAVEMENTS 2008

Standard Aircraft	Gear Type	Gear Dimension (inches)
727	D	34
DC8-63	DT (2D)	32x55
747	DDT (2D/2D2)	44x58

For comparison, other narrowbody aircraft gear configurations are shown below. Although the gear spacings on dual-gear aircraft vary, they are rated with the Dual allowable gross weight according to the FAA system. Likewise, the narrowbody aircraft with dual-tandem gear have spacings that vary from the DC8-63, but they also have allowable gross weights with the DT ratings. This arrangement is shown in the order of increasing dual gear dimension. The 727 and DC8-63 aircraft are repeated for comparison:

Aircraft	Gear Type	Gear Dimension (inches)
DC8-43	DT (2D)	30x55
A320 bogie	DT (2D)	30.7x39.5
DC8-63	DT (2D)	32x55
757	DT (2D)	34x45
707	DT (2D)	34x56
DC9	D	24 to 26
MD80	D	28.125
737 Classic	D	30.5
737 Next Generation	D	34.0
727	D	34.0
A318	D	36.5
A319	D	36.5
A320	D	36.5
A321	D	36.5

When the landing gear configurations of many widebody aircraft are compared with the standard DC8-63, the dimensions are significantly larger. However, except for the 777 and the body gears of the A380, all have the same DT configuration. The DC8-63 is repeated for comparison:

Aircraft	Gear Type	Gear Dimension (inches)
DC8-63	DT (2D)	32x55
A310	DT (2D)	36.5x55
A300	DT (2D)	38.5x60
767	DT (2D)	45x56
787-8	DT (2D)	51x57.5
A380 wing	DT (2D)	53.1x66.9
MD-11	DT (2D/D1)	54x64
A330	DT (2D)	55x78
A340	DT (2D)	55x78
777	3D	55x57x57
A380 body	3D	61x66.9x66.9

Knowing that these gear dimensions significantly exceed that of the DC8-63, it is reasonable that they also have better pavement loading characteristics. To further examine this thought, a comparison was made of the ACN of each aircraft at the same gross weight as the DC8-63. Similar trends are evident when comparing required thickness for individual aircraft at common traffic intensities:

- Gross weight   = 358,000 lb
- CG   = 95% weight on the main gear
- Tire pressure   = 196 psi
- Traffic intensity   = 5000 annual passes for 20 years (flexible pavement)
- Stress ratio   = 0.500 (rigid pavement)

Aircraft	ACN FB	ACN RB	required t-flex (in.)	required t-rigid (in.)
DC8-63	54.0	58.5	27.7	14.9
A310	50.3	52.8	26.4	14.3
A300	47.4	48.7	25.6	13.9
767	45.6	46.4	25.0	13.6
787-8	43.2	42.8	24.3	13.1
A380 wing (DT)	41.2	40.9	23.5	12.5
MD-11	31.7	30.6	20.6	11.2
A330	38.8	36.7	22.9	12.1
A340	31.0	29.7	20.5	11.0
777 (3D)	25.0	26.4	18.0	10.3
A380 body (3D)	23.2	25.1	17.3	9.8

For pavement loading in the ACN world, smaller is better, and the same applies when thickness requirements are evaluated. It is apparent that all of these widebody aircraft have better pavement loading characteristics for the same conditions than does the standard DC8-63.

FAA ratings are based on allowable gross weight, while ACN's are based on relative comparisons with standard parameters of traffic intensity, aircraft characteristics, and pavement features. Each method represents an allowable gross weight, and by equating the two at the same calculated weight, a valid comparison results which can be used regardless of the number of wheels or gear configuration.

For example, at ACN 50 FB, the DC8-63 allowable gross weight is 335,800 lb, and the 767-300ER allowable gross weight is 391,400 lb. The ratio is 391,400/335,800 = 1.17, indicating that there is a 17% improvement in pavement loading capability for the 767-300ER as compared to the DC8-63.

Applying this to an actual pavement, for a flexible pavement with a code B subgrade and a rating of DT300, the 767 rating would be 300 x 1.17 = 351. At other ACN's, pavement types, and subgrade codes, the ratios are:

Aircraft	40 FA	50 FB	60 FC	60 RB	80 FC	80 RC	100 RD
DC8-63 (lb)	311,900	335,800	336,000	360,000	412,200	396,500	430,600
767-300ER (lb)	355,300	391,400	391,700	430,400	470,600	467,400	497,700
Ratio (factor)	1.14	1.17	1.17	1.20	1.14	1.18	1.16

The gross weights of these two aircraft sometimes are at times shown greater than their respective maximum taxi weights (MTW), but these are only as examples of how the ratios can be calculated.

It can now seen that for a given DT pavement rating (which is based on a gear similar to the DC8-63), the allowable gross weight for other aircraft with DT gear arrangements can be determined by use of ratios or factors. The calculation of factors for pavement ratings other than the DT gear arrangement will be examined later.

When a FAA rating is recorded in 5010 forms or other publications, it is immediately observed as to what is not included with the rating. The following attributes are included with the PCN, and the first two appearing as part of the ACN. The codes are from the ICAO Annex 14 publication:

- pavement type or classification (F or R)
- subgrade code or classification (A, B, C, or D)
- tire pressure limitations (W, X, Y, or Z)
- method of calculation (T or U)

Similar parameters are also used in the FAA rating derivation for a particular pavement, but are not shown as a part of the rating. They are, along with traffic intensity, shown in the FAA advisory circular 150/5320-6D as requirements to determine the rating.

Since none of the above parameters are known with the FAA rating, with the possible exception of pavement type, it was decided to use average factors in this procedure. Average factors are found from the ratio of the aircraft in question and the standard FAA model at each of the four standard ACN subgrades and two pavement categories, resulting in a total of eight ratios. The average of the eight becomes the factor for that aircraft. Although the ratio varies somewhat as seen previously, depending on the reference ACN number, the pavement type, and the subgrade classification, the variations can be overlooked because the FAA rating itself is not precise, and these other attributes are not readily available with the FAA rating.

Application of the Method – Standard Aircraft

Looking at the standard aircraft in Table 1, and using a reference ACN 50, the allowable gross weight and ratios of each can be calculated. For example, the 727-200 allowable gross weight for ACN 50 FB is 190,000 lb. At ACN 50 FD, the DC8-63 allowable gross weight is 248,000 lb. Likewise, for ACN 50 RC, the allowable gross weight for the 747-400 is 661,000 lb.

It is immediately noticeable in the table that many of the ratios are 1.00. This results from the aircraft allowable gross weight being divided by itself and brings up a set of principles to be followed in application of the factors:

1. If the FAA rating is Dual (D), then all aircraft with dual gears will have a factor of 1.00. The FAA Dual rating does not distinguish between aircraft with varying dual landing gear and neither should the gross weight factor.
2. If the FAA rating is Dual-Tandem (DT), then all narrowbody aircraft with dual-tandem gear will have a factor of 1.00. The same reasoning exists for this case as for the Dual rating.
3. If the FAA rating is Double-Dual-Tandem (DDT), then all 747 aircraft will have a factor of 1.00 because this rating is meant specifically for the 747.

As application of these principles, consider the following ratings. For the purposes of this example, ignore the fact that some of the gross weights may exceed that of the aircraft MTW:

D200 DT350 DDT800

1. Aircraft such as the DC-9, MD-87, 737, 727, and A320 Dual will have allowable gross weights of 200,000 lb. Each has a narrowbody configuration.
2. Aircraft such as the DC8, 707, 757, and A320 bogie will have allowable gross weights of 350,000 lb, because they are all narrowbody configurations.
3. The allowable gross weight for any 747 is 800,000 lb.
4. Other aircraft, such as the 777 or A380, cannot be determined from these ratings because its gear pattern does not fit any of them. This is where the proposed gear factor method will have the greatest use.

In cases where only two ratings exist, such as D200 and DT350, the 747 must be factored according to its average DT ratio or factor, as seen below, and other aircraft are factored according to the above rules. Maximum values of allowable gross weight according to the aircraft MTW are not considered in this method, but that can be acknowledged when the values are published in a report. The 747-400 factor is the average ratio of the 747-400 to the DC8-63 in Table 1:

Aircraft	Gear Type	FAA Rating	Average Factor	Average Weight, lb
737-800	D	D200	1.00	200,000
757-300	DT (2D)	DT350	1.00	350,000
747-400	DDT (2D/2D2)	DT350	2.35	823,000

For a pavement with only a single rating of D200, it is required that both the 747 and the 757-300 be factored according to their Dual ratios from Table 1:

Aircraft	Gear Type	FAA Rating	Allowable Factor	Average Weight, lb
737-800	D	D200	1.00	200,000
757-300	DT (2D)	D200	1.73	346,000
747-400	DDT (2D/2D2)	D200	4.02	804,000

Table 1. Standard Aircraft ACN Ratios

727-200 (D)

Subgrade Code:	A	ACN-Rigid B	C	D	A	ACN-Flexible B	C	D	Average Ratio
GW at reference ACN* (1000 lb):	187	178	169	162	198	190	172	157	
RATIO TO 727-200	1.00	1.00	1.00	1.00	1.00	1.00	1.00	1.00	1.00 D
RATIO TO DC8-63	0.53	0.56	0.60	0.63	0.54	0.57	0.58	0.63	0.58 DT
RATIO TO 747-400	0.28	0.27	0.25	0.24	0.30	0.28	0.26	0.23	0.26 DDT

DC8-63 (DT)

Subgrade Code:	A	ACN-Rigid B	C	D	A	ACN-Flexible B	C	D	Average Ratio
GW at reference ACN* (1000 lb):	354	318	283	257	368	336	296	248	
RATIO TO 727-200	1.89	1.78	1.67	1.58	1.85	1.77	1.72	1.58	1.73 D
RATIO TO DC8-63	1.00	1.00	1.00	1.00	1.00	1.00	1.00	1.00	1.00 DT
RATIO TO 747-400	0.53	0.47	0.42	0.38	0.55	0.50	0.44	0.37	0.46 DDT

747-400 (DDT)

Subgrade Code:	A	ACN-Rigid B	C	D	A	ACN-Flexible B	C	D	Average Ratio
GW at reference ACN* (1000 lb):	842	750	661	593	836	777	689	555	
RATIO TO 727-200	4.50	4.22	3.91	3.65	4.22	4.08	4.01	3.53	4.02 D
RATIO TO DC8-63	2.38	2.36	2.34	2.31	2.36	2.45	2.45	2.16	2.35 DT
RATIO TO 747-400	1.00	1.00	1.00	1.00	1.00	1.00	1.00	1.00	1.00 DDT

* Reference ACN = 50

In the rare case of a single rating of DT350:

Aircraft	Gear Type	FAA Rating	Allowable Factor	Average Weight, lb
737-800	D	DT350	0.58	203,000
757-300	DT (2D)	DT350	1.00	350,000
747-400	DDT (2D/2D2)	DT350	2.35	823,000

Application of the Method – Widebody Aircraft

It has been shown that widebody aircraft have superior loading compared to the standard models, and the calculation of these factors can now be illustrated. In each case the reference ACN must be included along with the FAA rating in order to have comparable allowable gross weights.

Table 1 shows that at ACN 50 FB the DC8-63 allowable gross weight is 335,800 lb (rounded to 336,000), and from Table 2 the 767-300ER weight is 392,000 lb, for a ratio of 392,000/336,000 = 1.17. Coincidentally, the average ratio is also 1.17. For a pavement rating of DT350, the 767-300ER allowable gross weight is 350,000 x 1.17 = 410,000 lb. This use of the factor provides a 60,000 lb gross weight increase over the DT350 rating. The D and DDT ratings do not apply since the 767 has a DT gear.

Aircraft	Gear Type	FAA Rating	Average Factor	Allowable Weight, lb
767-300ER	DT (2D)	D200	2.03	----
767-300ER	DT (2D)	DT350	1.17	410,000
767-300ER	DT (2D)	DDT800	0.54	----

The same procedure applies for the A330-300 of Table 2:

Aircraft	Gear Type	FAA Rating	Average Factor	Allowable Weight, lb
A330-300	DT (2D)	D200	2.28	----
A330-300	DT (2D)	DT350	1.32	462,000
A330-300	DT (2D)	DDT800	0.60	----

If the only rating available was D200, then the A330-300 allowable weight would be:

200,000 x 2.28 = 456,000 lb

The third aircraft in Table 2 is the 6-wheel 777-300ER. In this case the gear pattern does not fit any FAA rating, so both the DT and DDT allowable weights are calculated. Normally, the greater allowable weight would be used from this table, but that depends on other judgmental factors such as pavement condition:

Aircraft	Gear Type	FAA Rating	Average Factor	Allowable Weight, lb
777-300ER	3D	D200	3.03	----
777-300ER	3D	DT350	1.75	613,000
777-300ER	3D	DDT800	0.80	640,000

For a different set of FAA ratings, the allowable gross weight would be:

Aircraft	Gear Type	FAA Rating	Average Factor	Allowable Weight, lb
777-300ER	3D	D150	3.03	----
777-300ER	3D	DT400	1.75	700,000

Table 2. Widebody ACN Ratios

767-300ER (2D)

Subgrade Code:	ACN-Rigid				ACN-Flexible				Average Ratio	
	A	B	C	D	A	B	C	D		
GW at reference ACN* (1000 lb):	424	378	334	299	423	392	348	279		
RATIO TO 727-200	2.27	2.13	1.98	1.84	2.13	2.06	2.02	1.78	2.03	D
RATIO TO DC8-63	1.20	1.19	1.18	1.16	1.15	1.17	1.17	1.13	1.17	DT
RATIO TO 747-400	0.63	0.56	0.50	0.45	0.63	0.59	0.52	0.42	0.54	DDT

A330-300 (2D)

Subgrade Code:	ACN-Rigid				ACN-Flexible				Average Ratio	
	A	B	C	D	A	B	C	D		
GW at reference ACN* (1000 lb):	478	432	383	342	452	428	391	324		
RATIO TO 727-200	2.56	2.43	2.27	2.10	2.28	2.25	2.27	2.06	2.28	D
RATIO TO DC8-63	1.35	1.36	1.36	1.33	1.23	1.27	1.32	1.30	1.32	DT
RATIO TO 747-400	0.71	0.64	0.57	0.51	0.68	0.64	0.58	0.48	0.60	DDT

777-300ER (3D)

Subgrade Code:	ACN-Rigid				ACN-Flexible				Average Ratio	
	A	B	C	D	A	B	C	D		
GW at reference ACN* (1000 lb):	646	558	473	409	655	609	537	428		
RATIO TO 727-200 ADV	3.46	3.13	2.80	2.52	3.30	3.20	3.12	2.73	3.03	D
RATIO TO DC8-63	1.82	1.76	1.67	1.59	1.78	1.81	1.81	1.72	1.75	DT
RATIO TO 747-400	0.96	0.83	0.71	0.61	0.98	0.91	0.80	0.64	0.80	DDT

C-17A (2T)

Subgrade Code:	ACN-Rigid				ACN-Flexible				Average Ratio	
	A	B	C	D	A	B	C	D		
GW at reference ACN* (1000 lb):	551	588	540	467	665	610	539	447		
RATIO TO 727-200 ADV	2.95	3.30	3.19	2.88	3.35	3.21	3.13	2.85	3.11	D
RATIO TO DC8-63	1.56	1.85	1.91	1.82	1.81	1.82	1.82	1.80	1.80	DT
RATIO TO 747-400	0.82	0.88	0.81	0.70	0.99	0.91	0.80	0.67	0.82	DDT

* Reference ACN = 50

Finally, consider the C-17A military aircraft of Table 2. It has a 6-wheel gear of unusual size and geometric characteristics, and it likewise does not fit the standard FAA rating system. However, by applying the same method as above, the allowable gross weight is calculated as 656,000 lb:

Aircraft	Gear Type	FAA Rating	Average Factor	Allowable Weight, lb
C-17A	2T	D200	3.11	----
C-17A	2T	DT350	1.80	630,000
C-17A	2T	DDT800	0.82	656,000

Conclusions and Recommendations

1. FAA pavement ratings do not adequately account for the superior pavement loading characteristics of widebody aircraft, nor do they represent other non-standard gear configurations such as some military and those with six wheels and greater.
2. Application of the gear factors to pavements with FAA ratings will recognize the superior loading characteristics of aircraft with widebody landing gear trucks.
3. Regardless of the FAA rating of the pavement, the allowable gross weight of any aircraft can be determined.
4. Pavements that are rated according to the FAA system should progress towards the PCN classification system in order to alleviate the necessity for these factors.
5. The methods presented herein only represent an interim solution until the ACN/PCN system is fully implemented.

References

1. FAA, "Standard Naming Conventions for Aircraft Landing Gear Configurations", Order 5300.7, October 2005
2. FAA, "Airport Pavement Design and Evaluation", Advisory Circular 150-5320-6D, July 1995
3. ICAO (International Civil Aviation Organization) International Standards and Recommended Practices, "Aerodromes"; Annex 14 to the Convention on International Civil Aviation, Volume 1, 4th Edition, Appendix A, July 2004.

Appendix

Table 3 contains a listing of gear factors for many common jet transport aircraft. FAA gear type nomenclature per FAA Order 5300.7 and other notations are also included.

- NB – Narrowbody (single aisle) aircraft
- WB – Widebody (twin aisle) aircraft
- MIL – Military aircraft
- LP – Low pressure tire

Table 3. Gear load factors for many common jet transport aircraft.

Aircraft	Gear Type	Body Type	D	DT	DDT
707-320C	DT (2D)	NB	1.85	1.00	0.46
717-200	D	NB	1.00	0.53	0.23
720B	DT (2D)	NB	1.74	1.00	0.43
727-100	D	NB	1.00	0.58	0.25
727-200	D	NB	1.00	0.59	0.25
727-200 LP	D	NB	1.00	0.58	0.25
737-200	D	NB	1.00	0.57	0.25
737-200 LP	D	NB	1.00	0.60	0.26
737-300	D	NB	1.00	0.57	0.25

Aircraft	Gear	Body			
737-300 LP	D	NB	1.00	0.58	0.25
737-400	D	NB	1.00	0.55	0.24
737-500	D	NB	1.00	0.56	0.24
737-600	D	NB	1.00	0.58	0.25
737-700	D	NB	1.00	0.58	0.25
737-700ER	D	NB	1.00	0.59	0.29
737-800	D	NB	1.00	0.57	0.25
737-900	D	NB	1.00	0.55	0.24
737-900ER	D	NB	1.00	0.56	0.24
737BBJ	D	NB	1.00	0.58	0.25
737BBJ-2	D	NB	1.00	0.57	0.25
747-200	DDT (2D/2D2)	WB	4.14	2.39	1.00
747-400	DDT (2D/2D2)	WB	4.02	2.35	1.00
747-400ER	DDT (2D/2D2)	WB	3.93	2.27	1.00
747-8F	DDT (2D/2D2)	WB	3.93	2.27	1.00
747-8P	DDT (2D/2D2)	WB	3.93	2.27	1.00
747SP	DDT (2D/2D2)	WB	4.12	2.38	1.00
757-200	DT (2D)	NB	1.73	1.00	0.43
757-300	DT (2D)	NB	1.70	1.00	0.42
767-200	DT (2D)	WB	2.02	1.17	0.54
767-200ER	DT (2D)	WB	2.02	1.17	0.54
767-300	DT (2D)	WB	2.02	1.17	0.54
767-300ER	DT (2D)	WB	2.03	1.17	0.54
767-400ER	DT (2D)	WB	1.95	1.13	0.52
777-200	3D	WB	3.07	1.77	0.82
777-200ER	3D	WB	3.07	1.77	0.82
777-200LR	3D	WB	3.06	1.76	0.81
777-300	3D	WB	2.95	1.70	0.78
777-300ER	3D	WB	3.03	1.75	0.80
777F	3D	WB	3.06	1.76	0.81
787-8	DT (2D)	WB	2.11	1.22	0.56
A300-600	DT (2D)	WB	1.84	1.06	0.49
A300B4	DT (2D)	WB	1.97	1.13	0.52
A310-200	DT (2D)	WB	1.84	1.06	0.49
A310-300	DT (2D)	WB	1.85	1.07	0.49
A318-100	D	NB	1.00	0.63	0.27
A319-100	D	NB	1.00	0.60	0.26
A320-200D	D	NB	1.00	0.59	0.25
A320-200DT	DT (2D)	NB	1.53	1.00	0.38
A321-200	D	NB	1.00	0.57	0.25
A330-200	DT (2D)	WB	2.31	1.33	0.61
A330-300	DT (2D)	WB	2.28	1.32	0.60
A340-200	DT (2D)	WB	2.75	1.59	0.73
A340-300	DT (2D)	WB	2.73	1.58	0.72
A340-500	DT (2D)	WB	3.38	1.95	0.89
A340-600	DT (2D/2D1)	WB	3.37	1.95	0.89
A380-800	2D/3D2	WB	5.44	3.14	1.44
C130	2S	MIL	1.28	0.74	0.32
C17A	2T	MIL	3.11	1.80	0.82

C5A	C5	MIL	5.84	3.36	1.55
CONCORDE	DT (2D)	NB	1.64	1.00	0.43
DC10-10	DT (2D)	WB	2.22	1.28	0.59
DC10-30/40	DT (2D/D1)	WB	2.80	1.62	0.74
DC8-63/73	DT (2D)	NB	1.73	1.00	0.43
DC9-32	D	NB	1.00	0.54	0.23
DC9-51	D	NB	1.00	0.53	0.23
KC-10A	DT (2D/D1)	MIL	2.72	1.57	0.72
KC-135E	DT (2D)	MIL	1.81	1.00	0.45
KC-135R	DT (2D)	MIL	1.83	1.00	0.46
L1011-1	DT (2D)	WB	2.24	1.29	0.59
L1011-100/200	DT (2D)	WB	2.28	1.32	0.60
L1011-500	DT (2D)	WB	2.27	1.31	0.60
MD-11ER	DT (2D/D1)	WB	2.59	1.50	0.70
MD-83	D	NB	1.00	0.53	0.23
MD-87	D	NB	1.00	0.53	0.23
MD-90-30	D	NB	1.00	0.52	0.23

George Washington Bridge
Asphalt-Wearing Course and Bond Coat Analysis

Jami M. Bjornstad[1], EIT, Casimir J. Bognacki[2], P.E., and Joseph Marsano[3]

[1] Assistant Materials Engineer, The Port Authority of NY & NJ Technical Center 241 Erie Street, Jersey City, NJ 07310; PH (201)216-2940; FAX (201)216-2949
[2] Chief of Materials Engineering, The Port Authority of NY & NJ Technical Center 241 Erie Street, Jersey City, NJ 07310; PH (201)216-2940; FAX (201)216-2949
[3] Construction Materials Inspection & Testing Supervisor, The Port Authority of NY & NJ Technical Center 241 Erie Street, Jersey City, NJ 07310; PH (201)216-2940; FAX (201)216-2949

ABSTRACT

The upper level of the George Washington Bridge is a flexible, orthotropic steel deck. In the past, it has been difficult to obtain a significant service life from asphalt-wearing courses due to this deck design coupled with a high traffic volume. Lately, the asphalt-wearing course on the upper level has seen accelerated fatigue because all truck traffic is redirected to the upper level for security reasons. The primary modes of failure have been fatigue of the wearing course and loss of bond to the steel deck.

In an attempt to increase the service life, The Port Authority of NY and NJ Materials Engineering Laboratory developed several asphalt mixture proportions with various polymer modifications. These asphalt mixtures were tested at the Rutgers Asphalt/Pavement Laboratory for fatigue and plastic deformation. Research and testing was also performed to improve the bond of the wearing course to the steel deck.

This paper will present the mixture proportions tested, the results of the laboratory testing, the mixture that was chosen, and the experiences encountered during the installation of the asphalt-wearing course.

INTRODUCTION

Connecting New Jersey and New York City, The George Washington Bridge (GWB) is the busiest bridge in the world. Thousands of people everyday rely on the GWB to commute to work, visit family and friends, or connect to the more entertaining areas of the region. This high traffic volume, coupled with a flexible orthotropic steel upper deck, creates a unique environment for the GWB asphalt pavement. In the past, there have been numerous repairs and repaving projects that attempted to achieve an extended service life of the wearing courses. With each new mixture design and bond coat selection, more opportunities are presented minimize the dominant failure modes of fatigue cracking and de-bonding. This paper will look at the common problems faced in past pavement systems on the upper level of the GWB, and how the current pavement system was selected and installed through the Upper Level Eastbound Pavement Replacement project (GWB-244.011).

HISTORY OF CONSTRUCTION

The GWB was designed in 1923 by the Swiss-born Architect and Engineer, Othmar H. Ammann. At the time it was built, the GWB was double the length of the longest span ever built, reaching 4,760 feet between anchorages. Today, the GWB stands as the only 14-lane suspension bridge in the world, and named a National Historic Civil Engineering Landmark by the American Society of Civil Engineers. Originally a 6-lane, single-level roadway, the GWB was opened to traffic on October 25, 1931, and serviced more than 5.5 million vehicles within the first year. Since then, various construction projects developed to increase the capacity of the bridge and alleviate the growing congestion in the area. The original roadway capacity was increased to 8-lanes in 1946 by opening the two originally unpaved center lanes. Ammann designed the bridge to hold an additional level for either transit service or another 6-lane roadway. Due to the high traffic demand, this addition became a necessity, and the current 6-lane lower level roadway was completed in 1962. Presently, The Port Authority of NY and NJ (PANYNJ) preserves the historic structure by focusing on projects that ensure safety and security for all commuters, improve traffic flow, and maintain the structural integrity of the bridge components.

TRAFFIC VOLUME AND LOAD

The GWB spans the Hudson River, connecting Fort Lee, New Jersey to Washington Heights in Manhattan, New York. The bridge is accessible to major highways of both New York and New Jersey, making the GWB an important passageway between the two states. Because of this accessibility, and the amount of businesses and attractions within the New York City area, the GWB sees heavy daily traffic. The yearly traffic count administered by the PANYNJ for 2006 showed 108,530,000 vehicles crossing the GWB. This converts to an Annual Average Daily Traffic (AADT) of 297,342, which is relatively high compared to the 2006 AADT of other popular bridges such as the Golden Gate Bridge at 100,000, or the Brooklyn Bridge at 145,000 (US Department of Transportation, Federal Highway Administration 2007).

The GWB facility also includes the George Washington Bridge Bus Station (GWBBS). This station is located on the upper level, east end of the GWB, and contains a multitude of bus services as well as subway connections. Many buses traveling to and from the GWBBS add to the heavy-loaded traffic on the upper level of the GWB. The main source of heavy-loaded traffic, however, is from trucks using the bridge as a regular route to New York City and New England. For an average weekday, buses make up about 1 percent of the total traffic volume, while trucks make up about 10 percent. After the events on September 11, 2001, all trucks were restricted from using the lower level of the GWB. Because of this, the upper level roadway has experienced a much higher load impact, which directly resulted in accelerated pavement fatigue and multiple repair and repaving projects. More specifically, eastbound lanes 2 and 4 (closest to the median barrier) have seen the most failures, which can be attributed to their direct continuity to Highway I-95.

This excessively higher load and constant stress must be taken into account when designing the asphalt-wearing course. Pavement material for the GWB should be dense and strong enough to withstand the applied loads. The amount of density achieved is mostly dependant on the compaction process on site; however the desired strength can be designed for by including larger aggregates and more stone-to-stone contact. The high traffic load, caused by trucks and buses, is a further concern when coupled with the slow traffic flow most common during morning and afternoon rush hours, with frequent acceleration and deceleration. Rutting and shoving are common results in this situation. To prevent this instability, designs often call for less asphalt binder, or a stiffer binder that performs well in higher temperatures. The use of angular shaped, rough textured aggregate can also reduce rutting and shoving by restricting movement within the mixture.

ORTHOTROPIC STEEL DECK

A notable characteristic of the upper level roadway is that it rests upon an orthotropic steel deck. An orthotropic deck is identified as being a solid steel plate that is used as the top flange of both the main girder and transverse floor beam systems. The original 1927 bridge design called for a modified 6-inch steel grid deck with a 2-inch concrete overfill. This deck system, however, began to deteriorate after 40 years of traffic exposure. The present-day orthotropic steel deck was installed in 1978. This updated system includes a 5/8-inch deck plate welded to inverted WT sections functioning as ribs, with deep tee sections and secondary floorbeams for stiffening. This deck provides a consistent and even subgrade for the overlay pavement; however there are issues of de-bonding due to the smoothness of the surface. The bond coats used in the past have not been able to maintain a strong bond, causing shoving of the asphalt overlay when exposed to the frequent accelerating and decelerating traffic. For projects on the upper level of the GWB, there must be a high quality bond coat that ensures the asphalt overlay can properly adhere to the steel deck.

The deck system also contains expansion joints throughout the main span spaced 60 feet on center. These joints break up the roadway into 60-foot panels, each having a width of 11 feet. For the GWB-244.011 project, 310 panels were repaved on the main span. The thickness of a panel is also determined by the steel deck components. Shown in Figure 1, a typical expansion joint on the upper level of the GWB contains an end dam on either side, reaching 1.5 inches above the deck to prevent material from entering the joint. The compacted pavement must therefore not exceed 1.5 inches. This thickness restriction is a major concern in the mixture design. The allowable aggregate stone size is dependant on the lift thickness in addition to the load types. According to Industry standards, the lift thickness should be about triple the size of the nominal maximum aggregate stone size. If the stone is too large for the lift, there is a danger of graveling or roughness of the roadway, and the degree of compacted density may also be reduced. For the GWB-244.011 project, a 3/8-inch nominal maximum size aggregate was used. Through a study conducted by Roman Wolchuck Consulting Engineers (RWCE), it was discovered the stress on the wearing course for the upper level of the GWB would be reduced by about 30-percent if the allowable thickness was increased to 2 inches (RWCE 2003).

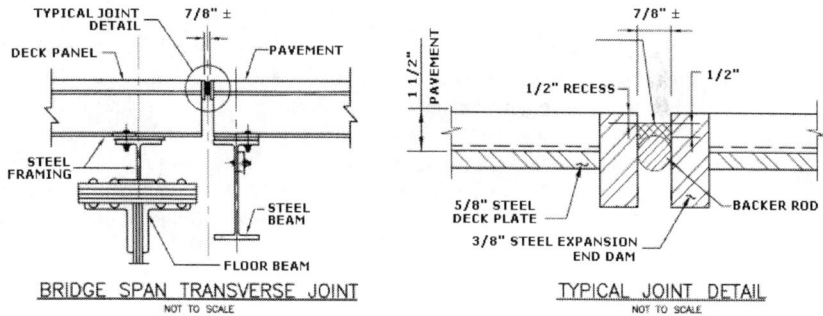

Figure 1: Expansion Joint Detail

More importantly, the expansion joints are necessary to accommodate the expanding and contracting of the steel deck in reaction to the ambient temperature. Without these joints, the structure would be in constant stress, resulting in early failure of the bridge deck system. This expanding and contracting of the steel, however, applies a cyclic, tensile, shear load on the overlay. If the wearing course is too stiff, reflective cracking from the bottom can occur, leading to exposure of the underlying steel. Exposing the steel deck leaves it susceptible to possible corrosion, added stress from freeze-thaw cycles, or a decrease in bond strength to the overlay. Therefore, the pavement should be a tight, dense-graded course that resists excessive cracking and prevents water seepage to the steel deck. Furthermore, the deck system constantly deflects and vibrates in response to the live loads acting on the structure. This deflection creates a cyclic flexural stress applied to the bottom of the overlay. These horizontal and vertical cyclic stresses increases the potential of fatigue failure, which is the dominant failure mode seen on the GWB upper level pavements. The overlay must be able to move with the deck system to increase its fatigue life. Generally, adding more asphalt binder improves flexibility by coating the aggregates and allowing more movement (RWCE 2003).

UPPER LEVEL EASTBOUND PAVEMENT REPLACEMENT PROJECT

After looking at the special conditions of the GWB and the most common failure modes experienced, one can better identify the most appropriate design to minimize fatigue, rutting, and de-bonding. The main issues specific to the GWB upper level roadway are the excessive traffic load and the orthotropic steel deck properties. One issue requires a dense, strong wearing course, while the other requires a more flexible mixture. According to RWCE, the use of conventional asphalt pavements on orthotropic decks has been unsuccessful in the past (2003). The GWB needs an innovative material that performs well under the two extreme conditions.

The installation phase of the GWB-244.011 project started in September 2007 to replace a 9-year-old wearing course that showed common failure modes, particularly in lanes 2 and 4, including cracking that exposed the steel deck. After only three years, 38 panels in lane 4 were replaced and about 40 more panels were replaced since. With much research and promising test results, the PANYNJ chose a Rosphalt 50 ® (Rosphalt) material pavement for lanes 2 and 4 combined with asphalt PG82-22 as a bond coat. The Rosphalt material was added to a 3/8-inch nominal max aggregate (I-5A) with asphalt binder PG64-22 mixture. Since lanes 2 and 4 required the most repairs in the past, they were in greater need of a special design. Other than the 8 test panels installed in 2006, this is the first time a Rosphalt pavement has been used as a PANYNJ permanent wearing course. The performance of these lanes will be closely monitored before planning to use the material on other lanes.

Rosphalt is a pulverized polymer modified powder additive that was developed by Royston Laboratories of Chase Corporation. This polymeric material was added during the plant mix process at a rate of 2.25-percent by the total weight of aggregate plus binder. This process requires no special equipment and can be used in standard batch or counter-flow drum plants. Once the Rosphalt is added, the material must be kept at very high temperatures to maintain workability. The out-of-plant temperature range is 410° to 450° F, which may pose a fire hazard at the plant concerning the baghouse, however the batch plant used for GWB-244.011 showed no issues. Both the field-placement and the compaction temperatures should be a minimum of 350° F. If the pavement is to be kept in this temperature range, the underlying bond coat must be able to properly cure and maintain adhesiveness at higher temperatures. PANYNJ tests show asphalt PG82-22 is able to perform well as a bond coat for this job.

Characteristics of both the Rosphalt material and the GWB itself require special considerations during installation. To minimize the risk of damaging the steel deck, milling was limited to the top inch of the overlay, while the remaining material was removed by scrapping or chipping and then sand blasted clean to SP-6. Once properly cleaned, the bond coat was applied at a rate of 0.25 gallons/ square yard. For compaction, the use of vibratory rollers and pneumatic rollers were restricted for the GWB-244.011 project. The concentrated vibrations of a vibratory roller may cause added stresses and deformations in the deck system. Meanwhile, the excessive heat of the Rosphalt pavement combined with a pneumatic roller causes the material to stick to the roller and pull up. For the GWB-244.011 project, a Cat 534B and an Ingersol-Rand DD-24 were used on static settings to achieve density requirements. The new mixture is highly homogeneous with a heavy concentration of Rosphalt polymers, creating a strong yet flexible wearing course that acts as a waterproofing barrier. This high concentration of polymers generates a high-density pavement, requiring fewer passes for compaction, and thus minimizing the danger of fracturing aggregates of the 1.5-inch lift. In addition, tests administered by the developers of Rosphalt show that when the material is added to a PG64-22, it meets Superpave binder standards of a PG94-34. This shows that Rosphalt creates a material that better withstands higher and lower temperatures without failure under an equivalent stress (Royston Laboratories 2008).

LABORATORY TESTING

RUTGERS ASPHALT / PAVEMENT LABORATORY

The Rutgers Asphalt/ Pavement Laboratory (RAPL), part of Rutgers University New Brunswick, often assists the PANYNJ by providing cutting-edge research and a larger variety of laboratory experiments. The RAPL recently conducted a study on PANYNJ Rosphalt mixture designs, which helped the PANYNJ predict the behavior of a Rosphalt pavement on the upper level of the GWB, and confirm that it is an appropriate material for the special bridge conditions. The PANYNJ provided three main mixtures with different proportions to compare in the laboratory tests: the Rosphalt additive mixture, the then used I-5A, PG76-22 with Polyester Fibers mixture, and a new Epoxy material mixture. The main tests conducted in the RAPL evaluated rutting potential and the occurrence of flexural fatigue, using methods that best correlate to measured field data.

Repeated Load Permanent Deformation Test

The Repeated Load Permanent Deformation Test subjects a sample to a repeated dynamic load while recording the amount of total and permanent deformation or strain. For this test, the RAPL followed the NCHRP 465 procedure, using a deviatoric cyclic stress of 25 psi on an unconfined specimen heated to 130° F. The actual test specimen was cored out from the center of a gyratory-compacted sample and trimmed to a height of 150 mm and diameter of 100 mm. As the specimen is subjected to the cyclic stress, the strain is measured though three linear variable differential transformers glued to the sides of the specimen at 120-degrees apart. The percent of permanent strain is directly proportional to the amount of loading repetitions; however, the rate of this permanent strain is dependant on the "zone" of the specimen. The Primary Zone has a faster rate of deformation due to the initial compaction of air voids. For samples with higher percent air voids, the permanent deformation would accumulate at a higher rate in the Primary Zone. Once the air voids are compacted, the rate of permanent strain decreases into the Secondary Zone. As the aggregates become overloaded and start to shift or crack under the pressure, the rate of permanent deformation increases to the final Tertiary Zone. This transition point is referred to as the Flow Number (Fn) and defines the rutting potential of a sample. A high Fn shows the mixture can withstand a higher load with minimal rutting. Table 1 shows the Rosphalt samples outperformed the PG76-22 mixtures in Fn and average permanent strain, demonstrating a high resistance to rutting (Bennert 2006).

Table 1: Deformation Test Results Summary

Mixture Type	Ave Fn (cycles)	Perm Strain 1,000 Cycles (%)	Perm Strain 10,000 Cycles (%)
Rosphalt 1% AVD, 1% AVC	>20,000	0.251	0.349
Rosphalt 1% AVD, 3% AVC	>20,000	0.260	0.373
Epoxy HMA	>20,000	0.018	0.020
I-5A, PG76-22, 5.9% AC	1,538	0.695	>2.0
I-5A, PG76-22, 6.3% AC	1,526	0.791	>2.0

AVD = Air Void Design, AVC = Air Void Compacted, AC = Asphalt Content
Note: The testing apparatus is limited to a maximum of 20,000 cycles

(Reproduced from Bennert 2006)

Flexural Beam Fatigue Test

The Flexural Beam Fatigue Test simulates the traffic-induced tensile and shear stresses in the bound layers that initiate fatigue cracking, a dominant GWB failure mode. According to Bennert, asphalt layers thinner than 5-inches are assumed to undergo constant strain, while layers greater than 8-inches undergo constant stress (2006). Since the thickness of the asphalt-wearing course on the upper level of the GWB is limited to 1.5-inches, the RAPL used the constant-strain test condition, following AASHTO T321 standards. The samples are first compacted with a vibratory compactor and trimmed to the dimension 380 mm long, by 65 mm wide, by 50 mm high. The RAPL tested each specimen in an environmental chamber at 15° C, with an applied tensile strain of 900 micro-strains, and a frequency of 2 loads per second. The samples were tested until approximately 3,000,000 cycles or until the flexural strength reached a chosen minimum value. From the test readings, the flexural stiffness of a sample can be calculated and measured against the corresponding load cycles. For this test, a failure condition is defined when the sample is weakened to 50-percent of its initial stiffness ($N_{f,50\%}$). A sample that is able to withstand more strain before reaching this failure condition represents a mixture that will have a longer fatigue life. Performance is based on both initial stiffness and air voids. As one can see from Table 2, the sample that had the longest fatigue life had a balance of stiffness and air voids. With percent air voids that are too high, the excess air would weaken the sample regardless of its stiffness; whereas if the sample were too stiff, there would be minimal elastic deformation before sudden brittle failure. The Rosphalt samples had $N_{f,50\%}$ values several orders of magnitude higher even at the worst air void and stiffness combinations (Bennert 2006).

Table 2: Fatigue Life Results Summary

Sample Type	Air Voids (%)	Initial Stiffness, So (Mpa)	Fatigue Life, $N_{f,50\%}$ (cycles)
Rosphalt 1% AVD, 1% AVC	1.4	1,347.7	2,832,294
Rosphalt 1% AVD, 3% AVC	2.7	782.9	3,191,433
Rosphalt 3% AVD, 3% AVC	3.1	891.4	2,939,057
Rosphalt 3% AVD, 5% AVC	4.5	766.6	259,538
I-5A, PG76-22, 5.9% AC	5.8	3,568.5	11,558
I-5A, PG76-22, 6.3% AC	5.1	2,803.7	17,712
AVD = Air Void Design, AVC = Air Void Compacted, AC = Asphalt Content			

(Reproduced from Bennert 2006)

The two tests show Rosphalt material can provide excellent fatigue and rutting resistance especially when paired with lower percent air voids. Considering the high quality of low temperature fatigue resistance, it is unique to see the same material outperform others in high temperature rutting resistance. An epoxy hot mix asphalt system also performed exceptionally well in the rutting and fatigue tests; however this material was not considered appropriate for pavement on the GWB due to its strict short delivery timetable necessary for a chemical reaction.

THE PORT AUTHORITY OF NEW YORK AND NEW JERSEY

In addition to the tests administered by the RAPL, the PANYNJ conducted in-house tests that further defined Rosphalt mixture properties. Garden State Parkway (GSP) samples and different bond coat samples were tested to help decide what pavement system works best under the GWB conditions. Testing of the new pavement system installed under the GWB-244.011 project was also conducted to determine the quality of the new pavement and whether it was acceptable to remain according to the conditions outlined by the project contract.

Permeability Tests

One favorable characteristic of a Rosphalt mixture is the waterproof barrier it creates for the orthotropic steel deck. To test this parameter, the PANYNJ took I-5, PG64-22 with Rosphalt cores from the GSP, and administered an in-house permeability test on the samples using a modified variable head test for soils. The specimen is placed in a permeameter, vacuum sealed, and then saturated with water at 77° F. A 0.30-square-cm standpipe above the sample reads the head with a marked upper limit and lower limit. The standpipe is filled with water to the upper limit and the seal below the sample is released. The time to decrease the head to the lower limit is recorded. The table below summarizes the results of this test including the calculated Coefficient of Permeability (K). The larger this K value, the more porous and less waterproof the material will be. Therefore, an ideal mixture for the GWB would be one that has a relatively low K value. For these tests, the PANYNJ set an upper limit of 1E-8 cm/sec for desired waterproof properties. Through testing, the average K value was significantly less at 6.72E-10 cm/sec and remained low regardless of the air voids.

Royston Laboratories also conducted permeability tests on the GWB test panel material using the ASTM D5084 test procedure. The resulting average K value was 3.80E-9 cm/sec. Based on these results, one can conclude the Rosphalt additive can create a consistently waterproof overlay material that will prevent seepage of liquids to the orthotropic steel deck.

Table 3: Permeability Test Results Summary

Sample Cross-Sectional Area, A (sq cm)	81.10	81.10	81.10
Sample Thickness, L (cm)	7.90	9.09	10.39
Sample In-Place Air Voids (%)	4.1	4.4	4.0
Initial Head, H_0 (cm)	91.44	91.44	91.44
Final Head, H_1 (cm)	85.34	79.38	77.60
Time (sec)	4010400	7120800	7120800
Permeability, K (cm/sec)	4.86E-10	6.45E-10	8.85E-10

Bond Coat Tests

To help determine the best bond coat for the upper level of the GWB, the PANYNJ conducted an in-house test for the tensile strength of several materials. I-5A, PG64-22 with Rosphalt plugs were created in the lab and bonded to a steel plate with different tack materials. The testing apparatus applied a tensile stress to the system and recorded the ultimate tensile stress, where the sample de-bonded to the plate. Ideally, the tensile strength should remain high under various temperatures; however the major focus for this study is the performance under higher temperatures due to the Rosphalt installation temperature requirements. The PANYNJ conducted the bond coat test at three temperatures for each sample (70, 100, and 125° F). The two best performing tack materials are compared in Figure 2. As one can see, the PG82-22 by Citgo outperformed the Royston material by an average of 10 psi. Although the creators of Rosphalt recommend the use of their own tack material, the PANYNJ laboratory tests showed the PG82-22 asphalt to be a more appropriate bond coat.

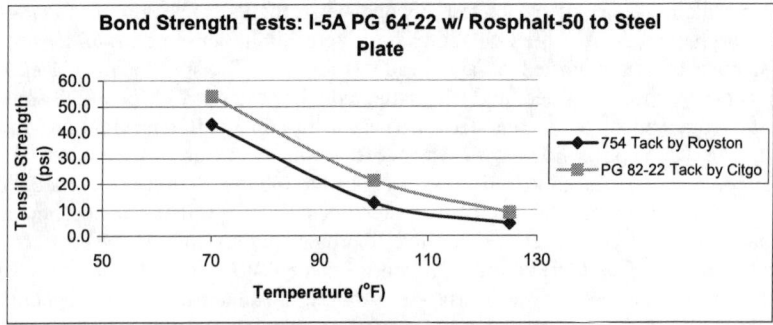

Figure 2: Bond Strength Comparison

Since de-bonding and shoving have been common and significant failure modes in the past, the PANYNJ further tested bond coats by installing eight test panels on the upper level of the GWB. A total of four different materials (Bridge Deck Membrane (BDM), Safetrack HW, Degadeck, and PG82-22) were tested as a bond coat for an I-5A, PG64-22 with Rosphalt overlay. The fast-curing BDM was unable to hold up to the high temperature of the Rosphalt material, and the overlay began to slide within weeks of installation. The Safetrack HW and the Degadeck bond coats lasted, but caused issues during compaction. The heat from the Rosphalt softened the bond coat materials, causing the pavement to slide under the roller. By allowing the system to rest before compaction, the sliding was reduced, however this resting period could not be extended too long without losing workability of the Rosphalt overlay. The four resulting panels did appear to perform well and were not replaced under the GWB-244.011 project. The best performing bond coat was the PG82-22, showing no sliding during installation or after compaction. These two PG82-22 panels also remained under the GWB-244.011 project.

Furthermore, the area between the New York Tower and the New York Anchorage on lane 4 has experienced the most severe de-bonding and shoving failures in the past. This failure mode is attributed to the downward-sloped grade and the sudden deceleration of vehicles ready to exit the bridge. In September 1998, 10 panels in that area were paved using Eliminator, a Methyl Methacrylate (MMA) membrane from Stirling Lloyd Group, with broadcasted aggregate for a mechanical interlock between the steel deck and the wearing course. This new system improved the performance of the asphalt-wearing course. Under the GWB-244.011 project, the overlay of those particular panels was milled down to ¼-inch of pavement, then tacked with the PG82-22 bond coat, and paved with the I-5A, PG64-22 with Rosphalt system. Keeping ¼-inch of the existing system creates a rougher subgrade for the asphalt overlay, increasing the friction and decreasing the ability to slide.

Field Tests

As production continued under the GWB-244.011 project, the PANYNJ took core samples from each night of production to test quality and determine payment. This data was compared to data from plant testing during corresponding production nights. If a failure as defined by the project contract was discovered from either sample source, the PANYNJ called for the failed section to be repaved. One notable property of the Rosphalt additive is the allowance of a tighter mixture, which contributes to its low permeability. Using the ASTM D2726 test, the resulting average Mat In-Place Voids (MIPV) of the field cores was 3.3-percent, which is low compared to the average value of around 6-percent for previous upper level, eastbound mixtures. It is important to note the MIPV values for the GWB upper level are usually 1-percent higher than for other sites due to the compaction restrictions. Since the mixture will be used as both a wearing course and a waterproof barrier, the MIPV upper limit is lowered to 4.3-percent. Keeping a conventional mixture at this range presents a high risk of rutting and shoving. With the Rosphalt additive, the high concentration of polymers increases the softening point, keeping the material flexible while maintaining stability. Using the ASTM D6927 test, the resulting average Marshall Stability value for the GWB-244.011 cores was 3217-pounds, with each sample exceeding the production lower limit of 1800 pounds as shown in the Figure 3 below. The field test data shows the overall installation of the new system was successful, warranting bonuses for a majority of the production nights.

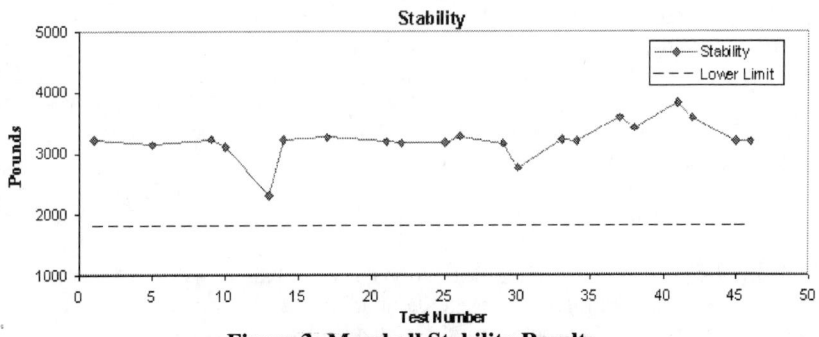

Figure 3: Marshall Stability Results

DISCUSSION

The upper level of GWB presents an environment that places a high stress on the asphalt pavement system. A combination of high-loaded, heavy traffic and a constantly moving, flexible orthotropic steel deck requires an innovative mixture design. Through laboratory research and preliminary in-field observations, the Rosphalt additive showed to create a pavement system that is stable enough to resist the effects from typical traffic, flexible enough to move with the steel deck, and waterproof to protect the bridge members below. Through the GWB-244.011 project, the PANYNJ hopes to achieve an extended service life of 10 to 15 years. Minimizing maintenance and repaving will lead to less spending and fewer interruptions in traffic flow, positively impacting commuters and visitors of the area. If the new Rosphalt pavement system continues to perform well, the PANYNJ may include this material in future pavement projects.

REFERENCES

Bennert, Thomas (2006). "Rutgers University Test Report Rosphalt 50." *Laboratory Research for the Port Authority of NY & NJ*. Rutgers Asphalt Pavement Laboratory, Rutgers University Center for Advanced Infrastructure Technology, New Brunswick, NJ.

Roman Wolchuck Consulting Engineers (2003). "George Washington Bridge Orthotropic Deck Studies Summary Report." *Analysis for the Port Authority of NY & NJ*. Roman Wolchuck Consulting Engineers, Jersey City, NJ.

Royston Laboratories (2008). "Rosphalt Asphalt Additive." *Royston Chase Specialty Coatings, Highway Products*. Retrieved February 15, 2008, from http://www.roystonlab.com/highway_pages/rosphaltadditive.html.

U.S. Department of Transportation Federal Highway Administration (2008). "Bridge Traffic Counts." Retrieved February 12, 2008, from www.fhwa.dot.gov.

Comparison of Butt and Notched Wedge Longitudinal Joints Constructed in Connecticut

A. Zofka[1], J. Mahoney[2], S. Zinke[3] and G. Shaffer[4]

[1] Assistant Professor, Department of Civil and Environmental Engineering, University of Connecticut, 261 Glenbrook Road, Unit 2037, Storrs, CT 06269-2037; Phone: 860.486.2733; Fax: 860.486.2298; e-mail: azofka@engr.uconn.edu

[2] Program Director, Connecticut Advanced Pavement Laboratory, Connecticut Transportation Institute, University of Connecticut, 179 Middle Turnpike, U5202, Storrs, CT 06269-5202; Phone: 860.486.9299; Fax: 860.486.2294; e-mail: james.mahoney@uconn.edu

[3] Research Engineer, Connecticut Advanced Pavement Laboratory, Connecticut Transportation Institute, University of Connecticut, e-mail: scott.zinke@uconn.edu

[4] Transportation Engineer II, Connecticut Department of Transportation, 2800 Berlin Turnpike, Rm 4209, Newington, CT 06131; Phone: 860.594.3477; e-mail: gregg.shaffer@po.state.ct.us

Abstract

Proper construction of longitudinal joints in asphalt pavements significantly contributes to their longevity and reduces joint-related pavement distresses, such as surface raveling, joint cracking and opening. Despite the apparent importance of the joint, there is no single construction practice that is widely accepted in the US. Most states have their own approaches and requirements regarding the construction techniques and quality assurance (QA) activities associated with longitudinal joints. Recently, the Connecticut Department of Transportation and Federal Highway Administration sponsored a research study that investigated two types of longitudinal joints: traditional vertical butt joints and notched wedge joints. The data was collected from two re-surfacing projects in Connecticut that utilized both joint types. Each project was divided randomly into multiple sections and the density data was collected from each section. The density was measured at the joint location as well as on both sides of the joint immediately after compaction. Two methods were employed to determine the density. First, the in-place density was measured using the nuclear density gauge. Next, the cores were taken from the same locations to measure the density in the laboratory using the vacuum seal method. This paper presents the experimental setup and the construction details on both joint types. The collected data is statistically analyzed and factors influencing the final density for each project are discussed. Both density measurement methods are compared and the correction factor was established between the two methods using additional cores taken from the mat away from the joint.

Introduction

The most common construction method in the US for the pavement longitudinal joint is the vertical joint, traditionally called a 'butt' joint. There are several different methods and tools for use in constructing the longitudinal joint. Some alternatives to the vertical joint are wedge joints (wedge indicates a tapered edge), notched wedge joints, different alternative rolling patterns, edge material restraining equipment, joint compacters, joint adhesives and joint re-heaters (*Akpinar 2004, Kandhal et al. 1997, Kandhal et al. 2002, Toepel 2003, TRC E-C-105*). All of these alternatives are intended to increase the quality and durability of the joint through either increasing adhesion between adjacent paved lanes and/or by changing the geometry of the connection. There have been several attempts to create machinery, attachments and alternative products in efforts to gain higher compaction levels at the joint and thus increase the longevity and overall performance of the pavement structure (*Denehy 2005, Estakhri 2001, Fleckenstien 2002, Marquis 2001*).

Traditional butt joints have been the customary method used in constructing longitudinal joints in hot mix asphalt (HMA) pavements in Connecticut in past years. The longitudinal joints on many Connecticut roadways have cracked or pulled apart thus expediting premature failure of the roadway and causing safety hazards to bicyclists, motorcyclists and pedestrians. The anticipated cause for this joint failure is a lack of material at the joint during the compaction phase of construction. Over the course of the expansion and contraction of the pavements due to thermal cycling, the area of the longitudinal joint generally does not contain enough material to fully recover from the thermal cycle. This results in a void area at the interface of the two paver passes. As time progresses and further thermal cycling takes place, this void space increases in size to the point where it may be comparable to the thickness of the wearing surface. This, in addition to safety hazards, allows water and incompressible materials to penetrate between pavements layers and cause further deterioration of the pavement.

To slow the rate at which longitudinal joints fail, proper construction techniques that ensure a high density and the proper amount of material along the longitudinal joint and compaction effort are essential. Increased longitudinal joint densities ensure there is enough material present to allow for the vertical thickness increases without requiring the material at the longitudinal joint to split in order to conform to the dimensional changes of the pavement.

Standard Joint Types in Connecticut

The traditional and foremost method for constructing longitudinal joints in Connecticut is a butt joint which joins together the warm material from the second pass to the cold material from the first pass creating an essentially vertical interface. Achieving adequate density on the cold edge of the longitudinal joint is difficult because at the time of its compaction, there is no lateral confinement to compact it against. Therefore, the unconfined edge is able to move laterally when the downward compaction force is applied. Theoretically, the ideal compaction method would provide some sort of lateral confinement on both edges of the pass such that the density at the longitudinal joint would approach the same density found at the center of the mat where it is expected and generally observed to be higher. This type of compaction is not practical for typical construction situations. Thus it would be beneficial to develop a joint construction method to minimize all of these problems.

Construction of Traditional Butt Joint in Connecticut

The traditional butt joint is constructed by butting the edge of the second paver pass on the hot side with the edge of the first paver pass on the cold side. To obtain proper density at the joint location, compaction of the joint commences immediately after the placement of the material. As stated in New England Transportation Technician Certification Program (NETTCP) Paving Inspector Manual (*NETTCP, 2006*), the hot material from the second paver pass is placed against the edge of the first pass and an overlap of 1 to 1.5 inches should be used in order to ensure an adequate amount of material for compaction. The thickness of the overlap should be in a range of 25% (1/4 in/in) of the compacted thickness of the wearing layer. Figure 1 shows this concept and similar figure can be found in the NETTCP manual. This method was used on both projects investigated in this study.

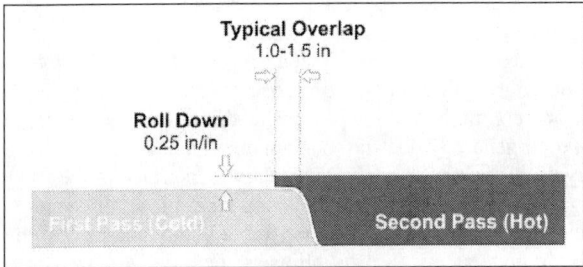

Figure 1. Typical Butt Joint Construction

Construction of Notched Wedge Joint in Connecticut

The notched wedge joint was formed in the study using a Contractor supplied device attached within the wing of the paver (Figure 2).

Figure 2. Notched Wedge Forming Device

The device was designed to create a notched wedge joint to meet the State's trial specifications. This tool allowed for adjustment in the formation of the wedge in its length and slope. The depth of the notch is also adjustable. To compact the wedge, a vibrating plate compactor was used. The plate is connected to the paver and is set just behind the wing directly over the wedge (Figure 3). The resulting notched-wedge joint is shown in Figure 4.

Figure 3. Wedge Compaction Device and Setup

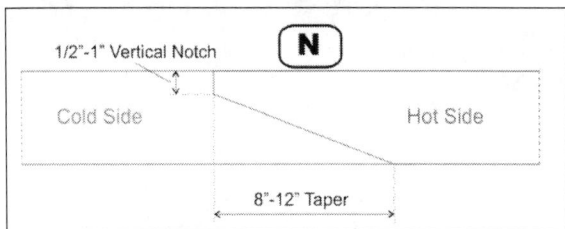

Figure 4. Notched Wedge Joint

Study Descriptions

The purpose of this study was to evaluate the constructability and durability of an alternate Hot Mix Asphalt (HMA) longitudinal joint method, the notched wedge joint, and compare the measurable properties of this construction method with those of the traditional joint construction method used in Connecticut. The notched wedge joint was investigated to potentially improve the State's standard longitudinal joint method known as a butt joint. Constructability includes the time, effort, equipment to form and compact the material at the joint and the resulting in-place density upon completion. Durability of both methods which includes the long term performance of the joints will be evaluated in the near future according to their ability to delay the formation of cracks at the joint as well as minimizing the width of the crack that forms.

The entire joint study consisted of several sites however results from only two project sites are presented in this article. Both projects were randomly divided into several study sections as presented in Table 1. In one of the projects (Windsor, I-91), five out of eight Notched Wedge sections were finalized after a several days of traffic crossing the open joint, i.e. hot side was paved at later time than the cold side. To account for that in this study, both section types were treated separately. In all study sections for all joint types, nuclear density measurements were taken 1 foot on the cold side of the joint, 6 inches on the cold side of the joint, on the joint itself, 6 inches on the warm side of the joint and 1 foot on the warm side of the joint. Each measurement consisted of the average of 2 one-minute readings at each point. This created a density profile across the joint as presented in Figure 5. For ground-truth values, five cores were taken from each study section, i.e. 1 foot on the cold side, 6 inches on the cold side, on the joint, 6 inches on the warm side and 1 foot on the warm side (see Figure 5). Each core was placed five longitudinal feet from the previous core. In all, there were 50 nuclear density measurements performed in each study section as well as five cores which were taken into the laboratory and measured volumetrically using Corelok^{TM}.

Table 1. Number of Study Sections within Each Project

Joint	Project	
	Berlin (Rt. 15)	Windsor (I-91)
Butt	2	5
Notched-Wedge*	3	3
Notched-Wedge**	0	5

* hot side finalized the same time (as cold side)
** hot side finalized after couple of days

In addition to cores taken from the vicinity of the joints, a limited number of cores were also collected from the mid-sections of the paving lane away from the joint. At each core location, the density was measured using a nuclear gauge before coring and the two sets of density values were later used to establish some correlation between both density measurement methods.

Berlin, Route 15

Rt. 15 in Berlin Connecticut was one of the project sites used in this study. The project was paved on the nights of September 6th and 7th, 2006. The original pavement consisted of Portland Cement Concrete (PCC) base overlaid with HMA layer. The existing HMA surface was first milled off at a depth of 75 mm (3 inches). A 25 mm (1 inch) leveling course of Superpave 9.5 mm (0.375 inch) traffic level 3 was placed over the milled surface prior to the wearing surface consisting of a (50 mm) 2 inch course of Superpave 12.5 mm (0.5 inch) traffic level 3. The notched wedge joint method was applied to the top course between the right and left travel lanes in the northbound direction only. Longitudinal joints for the right shoulder and left turn lanes consisted of the standard butt joint. The southbound lanes consisted of the standard butt joint method for all longitudinal joints.

Windsor, I-91

Interstate 91 in Hartford was paved at night from July through October 2007. The bituminous concrete surface was first milled at a depth of 75mm (3 inches). A 25mm (1 inch) leveling course of Superpave 4.75 mm traffic level 2 mix was placed over the milled surface prior to the wearing surface which consisted of a 50mm (2 inch) course of Superpave 12.5mm (0.5 inch) traffic level 4. The notched wedge joint was used on the surface course in the mainline travel lanes only. It was not used on the right or left shoulders, High Occupancy Vehicle (HOV) lanes, separator lane, or most ramps. These other areas utilized the standard butt joint.

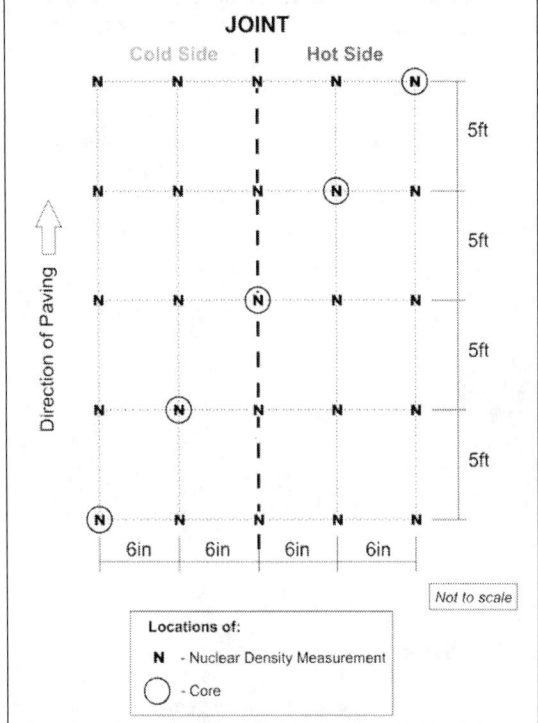

Figure 5. Typical Study Section

Data Comparison

The data collected in the two projects presented above was analyzed and compared in several ways. First, the potential application of correction factors between nuclear measurements and core densities measured in the lab was evaluated. Next, nuclear density measurements were used to plot spatial distributions of the densities along and across the joints. Furthermore, Analysis of Variance (ANOVA) was used to determine what factors are statistically significant and Bonferroni pairwise comparison was employed to distinguish between different levels of a given factor. Similar analysis was performed next on the core densities using however only average density values within each project-joint combination. Finally a simple ratio between nuclear and core densities was statistically analyzed to compared both joints types and at the same time their accuracy with the core densities.

It should be mentioned that before data was analyzed, the cores with densities higher than 2% from the nuclear readings were discarded and excluded from further analysis. It was assumed that such large error between both methods generally indicates a broken and damaged core resulting in lower density values.

Correction Factors from the Mat and Joint Readings

A recent report (*Padlo et al, 2005*) published by Connecticut Advanced Pavement Lab presented a method to determine a correlation factor which can be applied to the nuclear density gauge values to match core density values. It was found that this factor depends on the mix formula used in a given project. For the sake of this article, a simpler approach was incorporated: it was originally assumed that the correction factors depend solely on the core density values and the data collected in both projects along the joint as well in the mat was used to develop the correction factor as a function of the core density. The correction factor was calculated as a percentage ratio of the difference in densities to the nuclear gage density and it is presented in Figure 6.

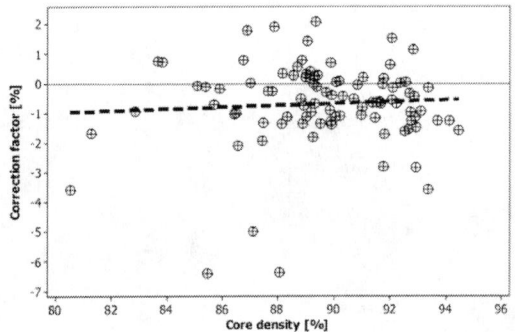

Figure 6. Correction factor as a function of core density

The least-square linear fit line is relatively flat suggesting that the correction factor in fact does not vary significantly with the core densities for asphalt mixtures used in this study. The potential application (or lack) of such correction factor should not influence the main conclusions of the analysis and hence, in the rest of this article, nuclear gage densities without any corrections were used. The full analysis with corrected nuclear gage densities is available through ConnDOT research report (*Zinke et al. 2007*).

Analysis of Nuclear Density Measurements

Nuclear density measurements were used to plot spatial distributions of the densities along and across the joints separately for each joint type. The distributions which are presented in Figure 7 suggest that for the sections with butt joints there is a clear decrease in density at the joint location that also spreads towards the cold side of that joint. For the notched wedge sections, there is some decrease in density at 6 inches into the cold side which corresponds to the actual location of the joint at the top of the pavement. In order to statistically evaluate these visual observations, Analysis of Variance (ANOVA) method was employed. *Joint type* together with *longitudinal* and *transverse locations* were set as factors, each with their corresponding levels. For the nuclear density measurements, ANOVA was performed separately for each project site. In both cases it was found that longitudinal direction is not a significant factor that provides explanation for the vertical trend of the density contours in Figure 7.

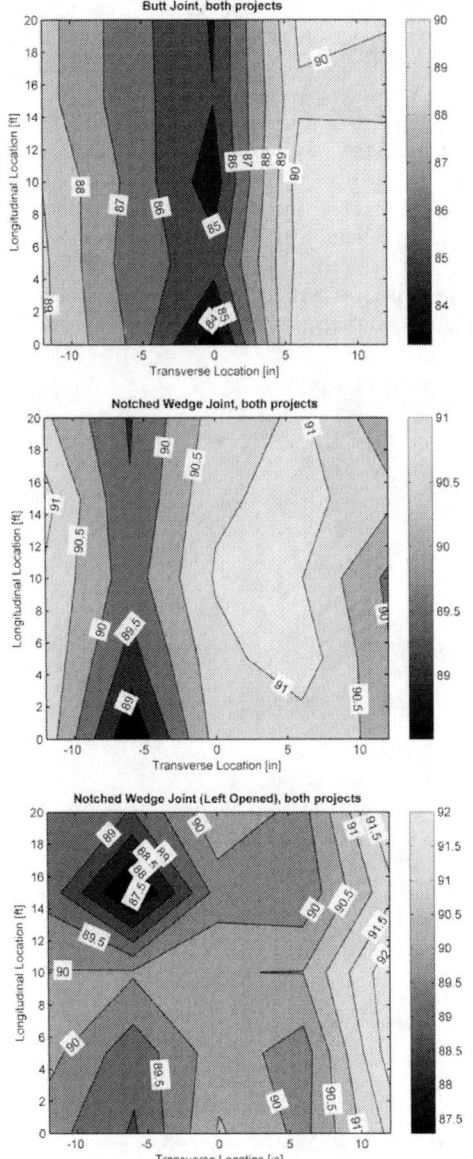

Figure 7. Average Nuclear Density measurements for both projects in, a) Butt joint, b) Notched Wedge, c) Notched Wedge joint sections (ones that were closed after couple of days)

ANOVA for each project site confirmed that joint type, transverse location and their mutual interaction are significant as might be assumed from Figure 7. The interaction plots for these factors separately for each project site are presented in Figure 8. The vertical bars represent 95% confidence intervals for the means at each transverse location. For both site it can be observed that the butt joint gives the lowest density at the joint location. It was also confirmed by Bonferroni pairwise comparison between different levels of the transverse location. In fact, for the Berlin site, the entire cold side with butt joint has a statistically different (lower) density than the notched wedge sections. In the case of the notched wedge sections, the Berlin site has all transverse locations except for 12 inches at the cold side statistically the same whereas Windsor site shows lowest density at 6 inches on the cold side that is consistent with the observation made for Figure 7. It is very likely that the lower density on the cold side and at the joint is due to the lack of lateral confinement during the first paver pass.

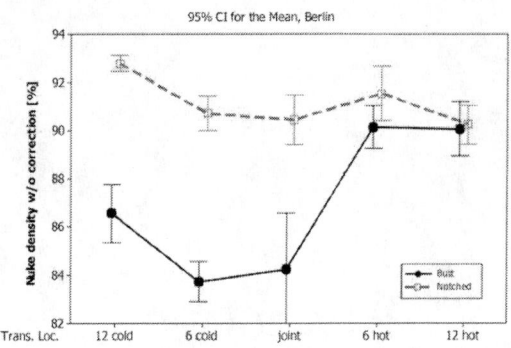

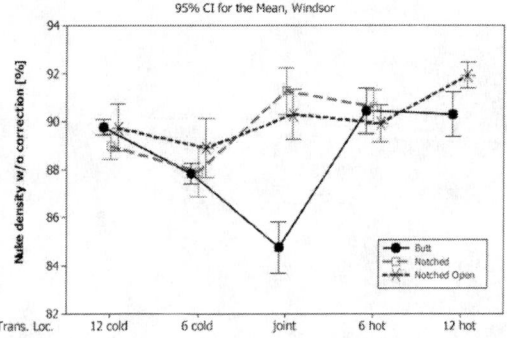

Figure 8. Interaction plots between joint type and transverse direction for nuclear density measurements for, a) Berlin, b) Windsor project site

Analysis of Core Densities

For core density comparison, both notched wedge joint datasets were merged which was possible since the ANOVA comparison showed no statistical difference between the means of both groups. Also, due to a small total number of cores, the cores from both projects sites were analyzed together and a potential *Project Site* factor was not considered which in result gave at least five (5) replicates for each factor-level combination. The interaction plot between the *joint type* and *transverse location* factors is shown in Figure 9 where again vertical bars represent 95% confidence intervals for the means.

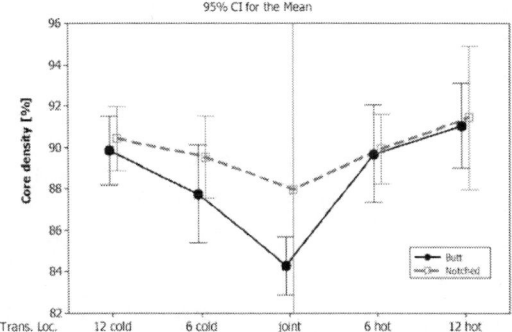

Figure 9. **Interaction plot between joint type and transverse direction for core density measurements**

Figure 9 shows that the average density for the butt joint is almost 4% lower than for the notched joint at the joint location. But due to the large variance of the cores at the notched wedge (where there are only two cores available for the analysis), both joint types have statistically the same density at the joint location. Bonferroni pairwise comparison also indicates that in the case of the butt joint, all locations are the same except for the joint location which is significantly lower than the rest. For the notched wedge sections, all locations are statistically the same which translates to a constant density profile across the joint. This observation differs from the observation made for the nuclear density where the notched wedge joint had the smallest density values on the cold side of the joint.

Comparison of Notched Wedge Joint and Butt Joint versus Core Density

One of the possible methods to simultaneously compare the densities from both construction methods using core densities is to calculate a ratio between nuclear measurements and core measurements and to perform ANOVA keeping all pertinent information in the factors. Similar to core density analysis, both notched wedge datasets were combined. Moreover, the *transverse location* factor had only 4 levels due to insufficient data at the location 12 inches on the hot side. The comparison of the ratio between nuclear and core densities is shown in Figure 10. The first observation which is true for both joint types, is that the ratio mean stays below 1 at the cold side of the joint and becomes greater than 1 at the joint location and on the hot side. This could be improved if the nuclear density measurement used in this analysis were adjusted using some correction factor procedure. Another interesting observation is that both lines – each for different joint type, behave in a very similar manner running almost parallel to each other. These visual observations were confirmed by ANOVA where *joint type* factor was not statistically significant and *transverse location* factor was indeed significant. It was expected that the considered ratio depends on the density and since the density varies with the transverse location then this ratio should change with this location as well. The fact that this ratio is not statistically different between considered joint types may be caused by limited sample sizes and large variances. From the interaction plot presented in Figure 10 it can be also concluded that on average considered ratio is closer to equality (ratio=1) for the notched wedge sections.

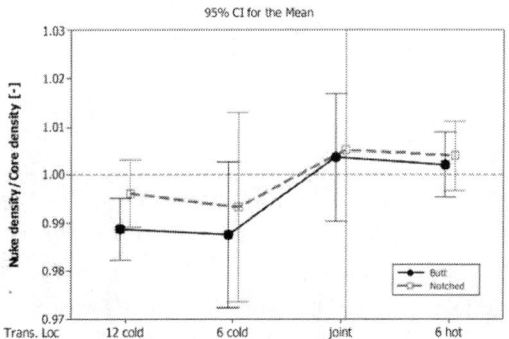

Figure 10. Interaction plot for the ratio between nuclear gage and core densities

Conclusions

ANOVA method indicates that nuclear density measurements along longitudinal directions are statistically the same for a given combination of project site, joint type and transverse location. Also ANOVA for each project site confirmed that joint type, transverse location and their mutual interaction are significant.

Results from the core density measurements indicate that the density profile across the joint is statistically constant for the notched wedge joint. In the case of the butt joint, the density is significantly lower at the joint location than the rest of locations. These conclusions are slightly different for the nuclear density measurements where the butt joint gives the lowest density at the joint location and notched wedge joints produce the lowest density either at the joint or 6 inches into cold side depending on the project site. It should be noted that these observations might change when more core density data is obtained from different project sites since a significant number of cores were not considered in this study due to their damage.

It was also found that the ratio between nuclear gage and core densities does not depend on the joint type but it is statistically different for different transverse locations. The ratio stays below one at the cold side of the joint and becomes greater than one at the joint location and the hot side.

It is recommended that additional projects be identified for which the notched wedge joint would be utilized in order to collect a sufficient amount of nuclear density and core data. This would provide better insight as to the overall performance, density and constructability of this joint method.

Acknowledgments

This study was sponsored by Federal Highway Administration (FHWA) and Connecticut Department of Transportation (ConnDOT). Their support is gratefully acknowledged. The results and opinions presented are those of the authors and do not necessarily reflect those of the sponsoring agencies.

References

Akpinar, Muhammet Vefa., Mustaque Hossain. *Longitudinal Joint Construction for Hot Mix Asphalt Pavements.* Report No. K-TRAN: KSU-98-4, Final Report. Kansas State University. Manhattan, Kansas. March, 2004.

Denehy, Edward J. *Constructability of Longitudinal Construction Joints in Hot-Mix Asphalt Pavements with Sealers to Retard Future Deterioration.* New York State Department of Transportation. 2005.

Estakhri, Cindy K., Thomas J. Freeman, Clifford H. Spiegelman. *Density Evaluation of the Longitudinal Construction Joint of Hot-Mix Asphalt Pavements.* Report No. FHWA/TX-01/1757-1. Texas Transportation Institute. April, 2001.

Fleckenstien, John L, David L. Allen, David B. Schultz Jr. *Compaction at the Longitudinal Construction Joint in Asphalt Pavements.* (Research Report No. KTC-02-10/SPR208-00-1F. Kentucky Transportation Center. College of Engineering. University of Kentucky. Lexington, Kentucky. May, 2002.

Kandhal, Prithvi S., Rajib B. Mallick. *Longitudinal Joint Construction Techniques For Asphalt Pavements.* NCAT Report No. 97-4. National Center for Asphalt Technology. Auburn University, Alabama. August, 1997

Kandhal, Prithvi S., Timothy L. Ramirez, Paul M. Ingram. *Evaluation of Eight Longitudinal Joint Construction Techniques for Asphalt Pavements in Pennsylvania.* NCAT Report No. 02-03. National Center for Asphalt Technology. Auburn University, Alabama. February, 2002.

Marquis, Brian. *Longitudinal Joint Study. Final Report.* Federal Experimental Report 96-2. State of Maine Department of Transportation. September, 2001.

New England Transportation Technician Certification Program (NETTCP), *Hot Mix Asphalt Paving Inspector Certification Manual.* Version 2.0, January, 2006.

Padlo, P., Mahoney, J., Aultman-Hall, L,. Zinke, S., *Correlation of Nuclear Density Readings with Cores Cut From Compacted Roadways.* Report # CT-2242-F-05-5. Connecticut Transportation Institute. University of Connecticut. August, 2005

Toepel, Amanda. *Evaluation of techniques for Asphaltic Pavement Longitudinal Joint Construction* Final Report. Report No. WI-08-03. WisDOT Highway Research Study No. 93-08. Division of Transportation Infrastructure Development, Bureau of Highway Construction Pavements Section, Wisconsin Department of Transportation. November, 2003.

Transportation Research Circular, TRC E-C105, *Factors Affecting Compaction of Asphalt Pavements,* Transportation Research Board, General Issues in Asphalt Technology Committee, September 2006

Zinke, S., Mahoney, J., Shaffer, G., *Comparison of the Use of a Notched Wedge Joint vs. Traditional Butt Joints in Connecticut,* CAP Lab report # 2249-1-07-3

A Quest for Successful Pavements in Texas

Carlos M. Chang Albitres[1]
Paul E. Krugler[2]
and
Ahmed Eltahan[3]

ABSTRACT

Building successful pavements requires understanding the factors that contribute to making a pavement succeed. The first question to ask is what a successful pavement is. To answer this question and gather information from successful pavement sections, a two-year project was conducted by the Texas Transportation Institute (TTI) for the Texas Department of Transportation (TxDOT).

Information about flexible pavements identified as superior performers has been collected and stored in a database with a user-friendly web-based interface. This database will be used to effectively learn from prior successes. Lessons learned in the past through experience will contribute to better understanding the factors that make a pavement succeed. The aim is to help practitioners improve technical specifications and testing techniques, which will in turn result in more consistent construction of premium pavements. This paper describes the methodology to identify a pavement as successful and briefly describes the Texas Successful Flexible Pavements Web Site.

WHAT IS A SUCCESFULL FLEXIBLE PAVEMENT?

TxDOT currently uses pavement condition scores, distress scores, and ride scores to characterize performance. Performance levels to qualify as a successful pavement can be expressed through these key parameters. However, pavement success includes consideration of functional performance as well as structural performance. Pavements subjected to different levels of traffic will certainly have different performance and maintenance costs. The performance-maintenance cost relationship varies with the pavement type. Undoubtedly, pavement performance and maintenance costs are key factors to identify a successful pavement.

---

[1] Associate Transportation Researcher, Texas Transportation Institute, CE/TTI Building 601H, 3136 TAMU, College Station, TX 77843-3136, (979) 862-2981, Fax (979) 845-1701, Email: C-Chang-Albitres@TTIMAIL.tamu.edu

[2] Research Engineer, Texas Transportation Institute, 1106 Clayton Lane, Suite 300E, Austin, TX 78723, (512) 4670952, Email: p-krugler@ttimail.tamu.edu

[3] Transportation Engineer II, Project Director, Materials and Pavements Section, Construction Division, Texas Department of Transportation, 4203 Bull Creek Rd. # 39, Austin, TX 78731, USA, (512) 467-3993, Fax: (512) 465-3681, Email: aeltaha@dot.state.tx.us

A nominated list of 75 flexible pavement sections which were considered as successful was assembled based on feedback from TxDOT's engineers. For these sections, records from TxDOT's Pavement Management Information System (PMIS) were reviewed. A rather wide range in the performance criteria was observed despite the fact that each of these pavements had been nominated as a particularly successful pavement. This wide range in performance criteria evidenced the difficulty in establishing a single set of criteria that reasonably identifies successful performance under the myriad of climatic, geographic, traffic, and local material factors involved across the state. The insight gained from the MIS records led to the creation of a definition that requires compliance with a number of the selected performance criteria, but not all of them, for a pavement's performance to be considered successful. Also, engineering judgment is recognized in the definition as an important element in pavement performance evaluations *(Krugler, P., et al., 2007).*

Discussion of Definition Criteria

A series of criteria arose as primary candidates for use in the definition of successful flexible pavement performance. These criteria included:
- age of the pavement section,
- drainage conditions,
- design service life,
- environmental factors (geographic location),
- maintenance history and treatment costs,
- material properties,
- pavement distresses,
- safety,
- serviceability (ride quality),
- structural adequacy of the pavement structure,
- subgrade, and
- traffic level (ADTs, ESALS).

Many of these factors are interrelated. For example, a section without distresses over its service life will more than likely be structurally adequate for the level of traffic and environmental factors acting upon the pavement structure. On the other hand, it is unlikely that a section with poor drainage conditions will have served without manifesting significant distresses over time. An important factor that does need to be included is the level of maintenance expenditures which have been required to adequately maintain the pavement performance being obtained. These discussions and preliminary analysis lead to focusing on only key performance criteria that, together, capture virtually all the factors affecting pavement performance, either directly or indirectly.

Another important consideration in selecting criteria was the ease with which TxDOT might be able to apply them. Not all data and information may be readily available even though they are potentially very valuable in defining successful performance.

From this perspective, it was determined that the criteria should bear upon measurable and objective parameters for identifying successful pavement performance. Analyzing the factors mentioned as potential parameters, it was concluded that the following factors should bear upon the determination of successful flexible pavement performance:

> Age of the Pavement,
> Cumulative Design Loading,
> Pavement Condition Score,
> Pavement Distress Score,
> Pavement Ride Score,
> Traffic Level, and
> Maintenance Expenditures (pavement-related).
> All of these factors are readily available to TxDOT personnel in PMIS records.

In order to characterize each of these factors, and to establish the criteria for identifying successful flexible pavement sections in Texas, it was found that the rating methods used in TxDOT's PMIS offer the best and the most practical solutions.

Age of the Pavement: The age of the pavement section is an obvious factor to include in the criteria. While age is only an indirect indicator of the amount of traffic that has been carried, it's a direct indicator of the length of time that the pavement has been exposed to environmental conditions. On less traveled rural roadways, age can become at least as definitive an indicator of superior performance as cumulative traffic loading. Another age-related aspect is that determination of "successful" pavement performance is time-dependent. A pavement section may meet "successful" criteria in its early stage of life but later on rapidly deteriorate and no longer be described as successful. The age categories included in the definition of successful performance are 0-7 years, 8-14 years, and above 14 years.

Cumulative Design Loading: The degree to which a pavement withstands traffic loading in comparison to its design loading is a most important indicator of successful performance. The definition criteria, therefore, must be flexible enough to correctly evaluate a pavement which has already exceeded its service life, regardless of pavement age. A pavement in reasonable condition after surpassing design traffic loadings should be considered successful.

Pavement Condition Score: The condition score provides a single descriptor of the overall pavement condition. This parameter combines ride quality and pavement distress characteristics of the pavement. The condition score scale ranges from 0 to 100. Pavement sections with condition scores from 90 to 100 are considered in very good condition, while above 70 in good condition. It is expected for a successful pavement to be in either very good or good condition, depending upon its stage of service life. A different minimum value for the condition score is established in the criteria for each of the pavement age categories. The criteria for successful performance also include the requirement for relatively low variability of condition scores within a successful pavement section. The maximum amount of variability allowed increases with increasing pavement age. Uniformity in performance is desirable and believed to be a strong indicator of quality in construction. Although the condition score is a good overall indicator of performance, as it combines pavement distress and ride quality characteristics, it was concluded that this parameter alone would be inadequate to identify successfully performing pavements. It was decided that the criteria in the definition of successful performance should also include independent distress score and ride quality factors, thereby stressing a specific minimum level of quality for each parameter.

Pavement Distress Score: The distress score reflects the degree of visible surface deterioration observed by pavement raters on an annual basis. Table 7 shows distress score classes defined by TxDOT. Like the condition score, it is expected that a successful pavement be in either very good (90-100) or good condition (70-89) from a distress rating standpoint, depending on the current age and stage of its service life. Relatively low variability is also a requirement.

Pavement Ride Score: The ride score expresses the ride quality on a scale from 0.1 (roughest) to 5.0 (smoothest). As with other rating criteria, it is expected that a successful pavement be in either very good (4.0-5.0) or good condition (3.0 – 3.9) for ride quality, depending on the current age and stage of its service life. Relatively low variability is again a requirement.

Traffic Level: The traffic level is currently expressed in terms of average daily traffic for establishing traffic categories within the definition of successful performance. Table 1 shows the traffic categories included in the definition.

Table 1. Traffic Classes (*TxDOT 1992*).

ADT	Class	Description
0 – 500	1	Low
501 – 10,000	2	Médium
Over 10,000	3	High

Maintenance Expenditures: A pavement section in very good condition with a high condition score, high distress score, and high ride score may not actually perform in a successful manner if maintenance treatment costs over its service life are above the average maintenance costs in the area. It is possible that the high pavement scores are the result of excessive maintenance work that has been required. For this reason, pavement-related maintenance costs over a period of years are considered a crucial factor in the criteria for identifying successful flexible pavement sections. The maximum average annual pavement maintenance costs included in the definition vary by traffic level, with $600 per lane-mile allowed for low traffic pavements, $900 per lane-mile allowed for medium traffic pavements, and $800 per lane-mile allowed for high traffic pavements. The selection of a lower allowable dollar rate for high traffic pavements compared to medium traffic pavements was based on demonstrated differences seen in PMIS maintenance cost records for nominated pavements.

Definition for Successful Flexible Pavement Performance

All of the criteria elements discussed in the previous section were combined in the definition of successful flexible pavement performance. A table containing criteria involved with the determination follows the verbal definition. This definition, as with other tools of pavement engineers, should be applied carefully and engineering judgment is a necessary element.

"A successful flexible pavement is defined as a structure that has met performance expectations over its service life with only normally expected levels of maintenance for its age, materials utilized, traffic loads, and local conditions." There are seven criteria recommended to assist in identifying a successful pavement section, as follows:

annual maintenance expenditure average,
minimum condition score average,
standard deviation of condition scores,
minimum distress score average,
standard deviation of distress scores,
minimum ride score average, and
standard deviation of ride scores.

To be identified as a successful pavement, it is recommended that the pavement section be at least six years old and meet the maintenance expenditure criteria plus at least four of the other individual criteria listed above and in Table 2. If the section does not meet these requirements, but extenuating circumstances exist, engineering judgment should be used in determining if performance is considered successful.

The Texas Successful Flexible Pavements Web Site and Database

The pavements described in this web site should be considered representative of the many successfully performing flexible pavements in Texas. The process of selecting the initial flexible pavements to include in this web site began with a solicitation of nominations from TxDOT's 25 geographically located district offices. The districts were asked to nominate pavements that their staffs believe have performed in a superior fashion considering all factors involved. Preliminary information was gathered about each nominated pavement section, and each nominating district was visited to view and discuss the pavement section's history with district personnel. The final selection of pavements considered the need to represent a wide variety of flexible pavement structure types, asphalt mixture types, and the broad geographical area of the state.

Table 2. Criteria for Identifying Successful Pavement Sections in Texas *(Krugler, et al., 2007)*

Parameter	ADT	Age of the Pavement Section						Beyond Design Life	
		From 0 to 7 years		From 8 to 14 years		Above 14 years			
		Minimum	Std. Dev.	Minimum	Std. Dev.	Minimum	Std. Dev.	Minimum	Std. Dev.
Condition Score	0 to 500	90	6	85	6	80	6	70	6
	501 to 10,000	90	8	85	8	80	8	70	8
	Above 10,000	90	10	85	10	80	10	70	10
Distress Score	0 to 500	92	6	88	6	84	6	75	6
	501 to 10,000	92	8	88	8	84	8	75	8
	Above 10,000	92	10	88	10	84	10	75	10
Ride Score	0 to 500	3.2	0.6	3.0	0.6	2.8	0.6	2.5	0.6
	501 to 10,000	3.6	0.7	3.4	0.7	3.2	0.7	2.8	0.7
	Above 10,000	3.8	0.8	3.6	0.8	3.4	0.8	3.0	0.8
3-Year Average Pavement Maintenance Expenditure	0 to 500	Below $ 600 / lane-mile		Below $ 600 / lane-mile		Below $ 600 / lane-mile		Below $ 600 / lane-mile	
	501 to 10,000	Below $ 900 / lane-mile		Below $ 900 / lane-mile		Below $ 900 / lane-mile		Below $ 900 / lane-mile	
	Above 10,000	Below $ 800 / lane-mile		Below $ 800 / lane-mile		Below $ 800 / lane-mile		Below $ 800 / lane-mile	

Notes:
- The average and standard deviation of all score values is to be determined using all scores from within the pavement section being considered.
- Pavement maintenance-related annual expenditures do not include construction programmed seal coats.
- Pavement maintenance costs in the table are based on 2005 cost information.

Registered web site users and guests have access to all pavement information included in the database. Three security access levels are provided for registered users, allowing TxDOT's web site administrator broad flexibility during implementation. The web site features online functionality for new pavement nominations by registered users. Figure 1 shows the login screen.

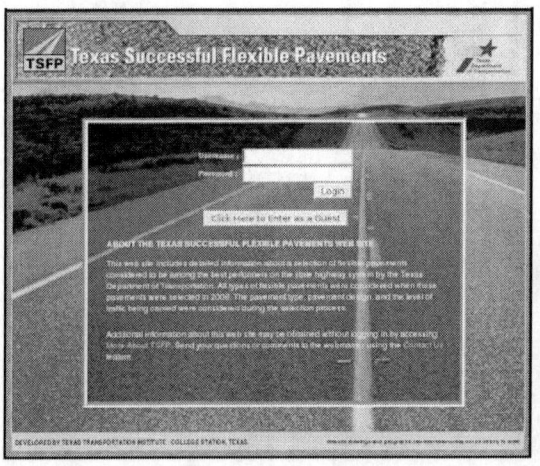

Figure 1. Texas Successful Flexible Pavements Web Site Login Screen.

Figure 2 shows the home page for registered users of the web site. Guests observe the same home page, but without the capability to nominate pavement sections. Blue and red stars on the map are links to detailed information about approved pavements and nominated pavements, respectively, at those locations. Separate drop down lists of approved and nominated pavements provide alternative methods of pursuing information about specific pavements.

The database provides TxDOT area engineers and district pavement engineers with quick access to flexible pavement designs of various types which have been particularly successful. It also provides valuable information for materials engineers to evaluate adequacy of specification criteria on an ongoing bases. When accessing a pavement's files in the database, the web site first provides the user an overview of information about the pavement section. The overview includes a description of the pavement and lists the factors believed to be instrumental in its particularly successful performance. As shown in Figure 3, a pavement cross-section is also provided to show the type of pavement structure and thicknesses of the various pavement layers.

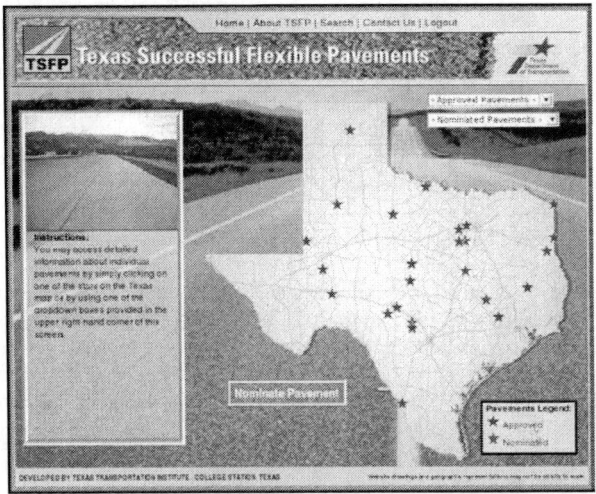

Figure 2. Texas Successful Flexible Pavements Web Site Home Page.

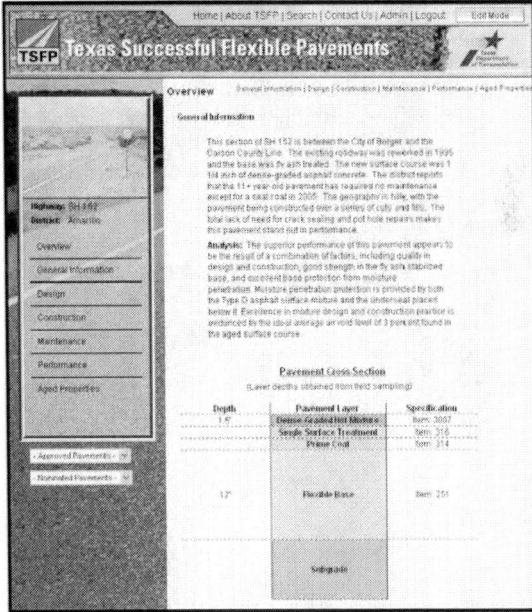

Figure 3. Upper Portion of Overview Screen – SH 152, Amarillo District.

Below the cross-section the overview screen continues and provides the user some of the more frequently desired information items about the pavement. If the user desires to access more detailed information, the navigation options shown under the photograph of the highway in the side bar box are the portals to all additional pavement information contained in the database. The web site divides information into the following major navigation categories: General Information, Design, Construction, Maintenance, Performance, and Aged Properties.

Premium Flexible Pavement Sections in Texas

The initial core set of premium flexible pavements is composed of 25 sections. The 25 pavement sections were selected from the initial list of 75 nominated sections. After receiving all nominations, each pavement section was visited. Pavement conditions were visually noted, photographed, and additional information was obtained. Unique subgrade conditions, unusual traffic considerations, and any unique aspects of construction and maintenance were discussed during these visits. PMIS pavement information was also analyzed along with maintenance expenditures over a three-year period.

A goal of the selection process was that the initially selected pavements include all commonly used types of flexible pavement structures and that these pavements would be distributed throughout the varied geographic and climatic regions of the state. Other considerations in the selection process were that the list should include a variety of material types, pavement designs, and levels of traffic. Because of these non-performance related selection factors, and because the database could only include a limited number of pavements during the two-year research project, the selected pavements do not constitute an exclusive list of the best performing Texas flexible pavements. A number of the other nominated pavements provide equally impressive performances. Figure 4 shows the geographic location of the 25 premium pavement sections and Table 3 includes de list of sections.

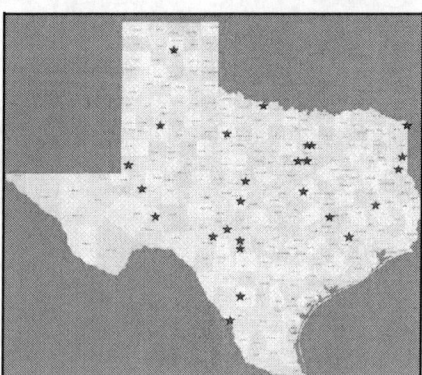

Figure 4. Geographic Locations of Selected Pavements.

Table 3. Initial Successful Flexible Pavements in the Database.

TxDOT District	Highway Designation	Pavement Structure Type	Pavement Age, Years	2006 Average Daily Traffic	2006 Percent Trucks in ADT
Abilene	US 380	Surface Treatment(s) over Flexible Base	40	1,693	13
Amarillo	SH 152	Thin Asphalt Concrete	11	1,250	9
Atlanta	IH 20 (WB)	Composite (Asphalt Surfaced Concrete)	10	15,017	39
Atlanta	IH 30	Composite (Asphalt Surfaced Concrete)	10	21,289	28
Atlanta	US 59	Thick Asphalt Concrete	17	5,940	23
Brownwood	US 67	Thick Asphalt Concrete	26	7,140	18
Brownwood	US 190	Surface Treatment(s) over Flexible Base	29	1,450	20
Bryan	SH 21(EB)	Medium Asphalt Concrete	14	5,206	15
Fort Worth	BIH 35	Composite (Asphalt Surfaced Concrete)	14	4,800	5
Fort Worth	SH 183	Composite (Asphalt Surfaced Concrete)	21	14,660	10
Fort Worth	SH 121	Composite (Asphalt Surfaced Concrete)	21	73,845	7
Fort Worth	SH 171	Medium Asphalt Concrete	14	6,738	14
Houston	SH 6	Thick Asphalt Concrete	11	8,576	17
Laredo	FM 1472	Thick Asphalt Concrete	5	20,167	17
Lubbock	FM 1585	Surface Treatment(s) over Flexible Base	20	2,442	6
Lufkin	US 287	Thick Asphalt Concrete	37	1,850	35
Odessa	IH 10	Surface Treatment(s) over Flexible Base	28	2,155	55
Odessa	US 385	Medium Asphalt Concrete	8	2,110	16
Odessa	SH 176	Medium Asphalt Concrete	9	2,250	32
San Angelo	IH 10	Medium Asphalt Concrete	37	3,963	29
San Angelo	US 377	Surface Treatment(s) over Flexible Base	57	324	32
San Antonio	FM 2771	Surface Treatment(s) over Flexible Base	42	865	16
San Antonio	RM 2828	Surface Treatment(s) over Flexible Base	37	896	29
Waco	FM 3223	Medium Asphalt Concrete	14	14,820	6
Wichita Falls	FM 3492	Surface Treatment(s) over Flexible Base	10	1,015	11

Table 4 breaks down the distributions of pavement structure categories and traffic levels among the selected pavements. Approximately two-thirds of the 25 pavements are in the medium traffic category, one was selected from the low traffic category, and the rest were in the high traffic category. The types of pavement structures represented are well distributed except that several additional thin asphalt concrete pavements would have been desirable.

Table 4. Numbers of Pavements in Structural Categories and Traffic Levels.

Traffic Level (ADT)	Thick Asphalt Concrete Pavement (ACP)	Medium Asphalt Concrete Pavement (ACP)	Thin Asphalt Concrete Pavement (ACP)	ACP over Concrete Pavement	Surface Treatment over Flexible Base	Total
High > 10,000	1	1	0	4	0	6
Medium 501– 10,000	4	5	1	1	7	18
Low < 500	0	0	0	0	1	1
Totals	5	6	1	5	8	25

The task of sampling and testing the 25 flexible pavements began soon after the pavements were jointly selected by the research team and TxDOT. The objective of the sampling and testing of the aged pavements was to learn as much as possible about why the pavements had performed as well as they had. In some respects, the field work was similar to—but the inverse of—the more typical forensic pavement evaluations for determining why a pavement had failed to adequately perform. A one-mile length of a single lane of each pavement was selected for sampling and field testing. The lane was selected to well represent the entire pavement. Both in situ pavement testing and laboratory testing were performed. A number of nondestructive pavement tests were conducted on the selected pavements. The pavement field tests included ground penetrating radar (GPR), falling weight deflectometer (FWD), and dynamic cone penetrometer (DCP). In most cases, the GPR test run was performed prior to the site visit for obtaining field samples. The FWD and DCP tests were generally conducted at the time of field sampling *(Krugler, P., et al., 2007)*.

CONCLUDING REMARKS

The development of a methodology for identifying and documenting successful flexible pavements in a database that uses a web site friendly interface is the key for learning about best practices that should be encouraged in the future. Some of the findings after conducting a two-year research study are:

a. While the success of each pavement appears to have somewhat differing factors involved, in general, superior performance may be attributed to the combined result of good construction practice, high-quality materials, conservative inputs made in design stage, less traffic than anticipated, and timely maintenance.

b. Pavement performance data from Pavement Management Systems can be used effectively as an initial screen in identifying premium pavements. The analysis of TxDOT's PMIS data from the 25 pavement sections reflects superior pavement performance over time. Average condition scores, distress scores, and ride scores from data available from 1998 through 2006 indicate very good performance and substantiate the selections of these pavements by the districts.

c. GPR colormaps display uniform pavement profiles for the selected pavements, an indication of uniformity in materials production, materials quality, and use of good pavement construction practices. Colormaps were consistent with observations of layer depths and occasional anomalies made in the field during pavement sampling.

d. Most pavement layer back-calculated modulus values were within or above the range of expected typical values for the types of materials involved.

e. TxDOT now possesses a populated database with 25 premium flexible pavement sections and an on-line tool to access this database. The on-line tool can be used for collecting additional information about particularly successful flexible pavements. The information already available on the web site is a valuable resource to area engineers and district pavement engineers interested in comparing their design practices and typical job control test results to pavements that have performed notably well.

f. The greatest ongoing need for data in the Texas Successful Flexible Pavements database is the nomination and inclusion of additional pavements. This need soon became the focus of strategizing for updating capabilities. It is envisioned that the Successful Flexible Pavements Database will continue growing as TxDOT districts nominate additional pavement sections for inclusion. The quest for identifying and documenting successful flexible pavements in Texas will continue in the following years.

ACKNOWLEDGMENTS

This research study was conducted for the Texas Department of Transportation (TxDOT), and the authors thank TxDOT for their support in funding this research study.

REFERENCES

Krugler, P., Chang Albitres, C., Scullion, T. and Chowdhury, A., (2007), "Analysis of Successful Flexible Pavement Sections in Texas – including Development of a Web Site and Database", Research Report No FHWA/TX-08/0-5472-1, College Station, Texas.

Texas Department of Transportation, (1992), Administrative Circular 5-92, February 13.

TALES FROM THE PACIFIC
AIRPORT ACTIVITIES AND EXPERIENCES IN MICRONESIA
Frank V. Hermann P.E., M ASCE
Senior Aviation Engineer, Leo A Daly, Inc., Honolulu, Hawaii, frankh@lasvegas.net

Abstract:

This paper will discuss some interesting experiences working on airfields in Micronesia and Hawaii. None are unique but issues that are readily resolved in the continental United States become more difficult at airfields on remote islands.

First Case study. The Palau International Airport runway, which had been overlain over 10 years earlier, was shedding aggregate at an alarming rate. To reduce the risk of FOD a heavy application of emulsion was applied. It performed satisfactorily for one year. The subsequent project proceeded well until the contractor unilaterally changed the surface lift aggregate gradation. After study, the surface was removed and replaced.

Second Case study. At the Yap International Airport the PCC apron placed in 1999 had expanded by 2003 due to ASR. Forensic studies resulted in saving the cement treated base. The Pavement is being removed and replaced at this time.

Third Case study. Grass grows fast all year in Micronesia. Constructing 3' wide concrete pads around signs and other equipment results in faster mowing without damage to the items.

Fourth Case Study; Miscellaneous Conformance Issues. Some issues apply to several airfields but are often resolved on a case-by-case basis by the local FAA office. It would be desirable to have formal guidance from FAA Headquarters so that uniform standards are applied whenever possible.
   a) Fences on, or essentially on, the safety area and a penetration of FAR surfaces.
   b) Public roads outside the Runway safety area where the 5 meter (15 foot) requirements cannot be achieved.
   c) Applicable pavement design, geometric and use guides for service and security roads that must be on the runway safety area.
   d) Guidance for turnarounds at runway ends where no parallel taxiway can be provided.

Fifth Case Study. Benefits of a Permanent Pavement Design. The reef runway at Honolulu International Airport is cited as an example.

Introduction. The discussions in this paper are based on experiences in the Pacific islands generally identified as Micronesia. These islands, except for the Territory of Guam, had come under Japanese control after the First World War and were central in the events of World War Two. The Islands were placed under United Nations control after World War II and the United States was designated Trustee. Since then, many of the islands have been granted various forms of independence but are still semi-tied to the United States through various treaties. Guam is a territory of the United States. The areas discussed are shown on Figure 1.

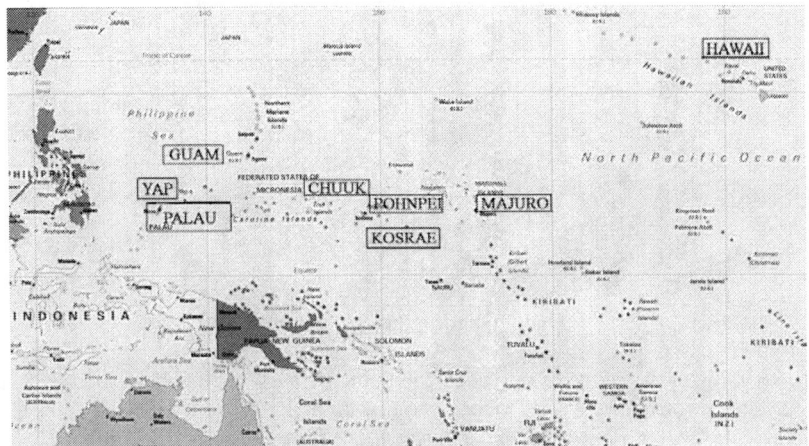

Figure 1, Major Islands in Micronesia

First Case Study. Palau Runway Disintegration Temporary Repair. At the Palau International Airport a skid resistant overlay over 10 years old was disintegrating at an accelerating rate. Over 32 kilograms (70 pounds) of loose aggregate was swept up daily presenting a serious risk of FOD damage. The original runway was over 20 years old. Aircraft traffic had increased both in volume and weight, so reconstruction was needed. Design to reconstruct the runway was starting but construction start was over a year away.

The skid resistance overlay apparently had consisted of aggregate ranging from 4.77mm to 19mm (number 4 to 3/4 inch) aggregate with a minimum of finer material. Over time the finer material had eroded leaving a surface consisting of larger size aggregate. The asphalt bond was breaking down with time leaving larger particles free to break out of the surface. The runway surface is shown in Figure 2.

Figure 2, Disintegrating Surface of Palau Runway

The resulting open graded surface allowed sealing with a heavy application of emulsion. Two engineers and the senior contractor manager were personally on site full time during the application that took about one week. Adjustments of the emulsion application rate were made on a very localized basis. The final application averaged over 1.35 liters per square meter (0.3 gallons per square yard). It was estimated that some areas received less than half this amount while other areas received over three times that amount. Because the application was placed under careful supervision no flushing or excess asphalt occurred. The resultant sealing performed satisfactorily for over one year before removal for new pavement.

This type of sealing is not generally applied to an airport runway due to a high risk of the surface becoming slippery. The lesson is that it can be a valid solution to a specific problem, but a high level of senior personnel on site during the work is critical to its success.

Palau Runway Reconstruction. Reconstruction involved removal of the porous friction course, leveling of the transverse and longitudinal profiles and strengthening with bituminous concrete. The overall runway profile was modified slightly to reduce asphalt quantities. The airport worked with the airlines to provide minimum 8-hour workdays five days each week. Longitudinal and/or transverse tapers were provided at the end of each work period and removed at start of work the next workday.

The Contractor did very good work up to the surface lift where he unilaterally changed the aggregate gradation to obtain a "Good Looking" surface. This sanded surface would not hold groves and shoved easily under aircraft wheels. Our initial desire was to leave the material in place and overlay it. However we were concerned that the material might be subject to striping when water penetrated the new surface or might be so far out of specification that it would shove and rut.

Cores of the material were obtained, but testing of the cores is difficult and expensive. The cores can only be imported into the United States by a laboratory certified to receive soil type materials that may contain plants, seeds, diseases, etc. Finding labs so certified that could also perform the tests was more difficult than expected. A California laboratory was found and they made the tests.

The primary concern was the gradation and striping characteristics of the material. There are no precisely applicable tests for the finished product cored out of the runway. We ran tests normally used to design a mixture or check materials during construction. The extraction test performed on the cores verified that the design asphalt percent had been used, but the aggregate contained too much fine material. Striping tests indicated that the material was marginal. Recognizing that although these tests were not fully appropriate the professional evaluation was that there would be a high risk of future surface disintegration if the material were overlain. In addition, the Contractor was unable to provide satisfactory test reports indicating the material conformed to the specifications. The surface was removed and replaced with a new two-inch thick bituminous concrete surface conforming to FAA specifications.

The primary reason this event occurred was insufficient on site construction management. No full time construction manager was provided. Local inspectors, who were very qualified to perform their tasks, were the only onsite staff representing the owner. A minimum one-week site visit costs over six thousand dollars. Funding was provided for only a few site visits by the construction surveillance team. Most of these visits were used up early in the project. Local inspectors watched and did a very good job of controlling and documenting the fieldwork. However no review of the day-to-day laboratory testing was performed. When the gradation changed the field inspectors did not observe it. The lesson learned is that sufficient funds are critical for adequate construction surveillance staff and independent testing.

Second Case Study. Alkali Silica Reaction (ASR) Damage at Yap International Airport. The Portland Cement Concrete (PCC) apron placed in 1999 had expanded by 2003 due to ASR. The rapid reaction was partly due to ideal climate conditions. Temperatures year round average 29.5 degrees C (85 degrees F) day and night and rainfall can be expected every two days or more even in the dry season.

Perhaps the most interesting observation was that several engineers had visited and studied the site and none recognized the problems as ASR. There were several theories including subgrade (schist) movement, aircraft breaking or turning forces, typhoon related movements and other speculative causes. No engineer or geotechnical person working in this region of the world had ever heard of or seen ASR. To the best of our knowledge, it had never happened before in this part of the world. This experience illustrates the need to keep all professionals aware of issues that may not have previously affected them.

The apron consisted of 40.6cm (16 inch) thick Portland cement concrete over 30.5cm (12 inch) thick cement treated base resting on the native schist subgrade. It was obvious that the concrete pavement had undergone ASR, but it was unknown whether the underlying CTB had also reacted. It was desirable to determine if the CTB was unaffected by reaction and could remain. This would be a considerable time and cost savings.

Forensic tests were made on site. Cores of the concrete were taken and then 0.9m by 0.9m (3 foot by 3 foot) holes cut through the PCC. As the demolition approached the bottom of the PCC layer, work was performed carefully so the undisturbed interface of the PCC and CTB could be observed. It was found that the surface of the CTB was polished smooth. The surface of the Cement treated base is shown in Figure 3.

Figure 3. Cement Treated Base Surface

Cores were then taken of the CTB. The CTB was then removed again using care to observe the surface of the schist. The schist surface was unscarred and textured similar to what would be expected of the original CTB surface immediately prior to placing the CTB. This surface is shown in figure 4.

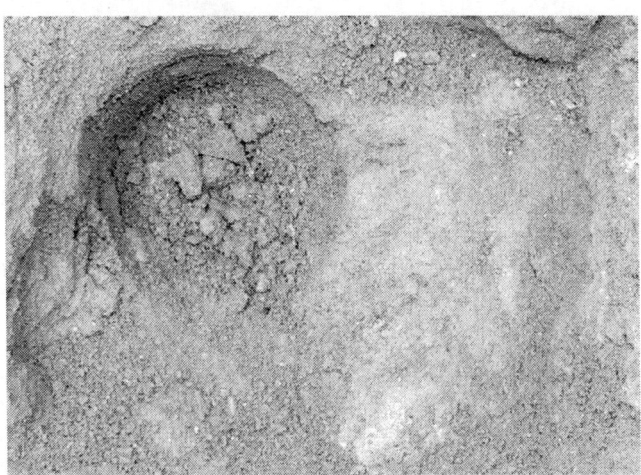

Figure 4. Native Schist Subgrade Surface

Additional excavations were made at the edges of the apron. It was observed that the PCC edge extended several inches outward past the underlying CTB edge. An FAA handhole in the apron was also observed. The top 40.6cm (16 inches) of it had been sheared off and shifted 17.8cm (7 inches) eastward.

The cores were visually examined and then subject to petrographic analysis. The visual examination revealed that the aggregate was different in each layer. The PCC showed typical range of aggregate while the CTB appeared to be manufactured using local dredged coral. The Petrographic analysis confirmed that the PCC had undergone ASR while the CTB was not affected. The design reused the cement treated base saving considerable time and money.

There is no good aggregate on Yap. Coral dredging had been a source of excellent subbase material, but is no longer available. All aggregate, including the subbase, has to be imported. The concrete aggregate was tested for potential alkali reaction prior to importing it. ASTM C-1260 was specified with 16-day expansion specified but the tests run to 28 days to obtain long range reaction. The material will again be tested after arrival on island, but this time using the actual cement and other additives on-site for the project. Class F fly ash is required in the mix. The percentage will be determined during the design phase and may be adjusted based on the second set of ASTM C-1260 tests.

The apron had to be kept open for airline and aircraft operations. Due to the risk of FOD, the airline used tugs, pulling the aircraft from the taxiway to the terminal and back. They agreed to continue this procedure during construction making the planning much easier.

Third Case Study. Protecting Equipment From Mower Damage. Grass grows fast in the Pacific islands. Constant year round mowing is essential. The signs, REILS, PAPIS, etc. were all built with bases similar to the standard FAA design. Mowers try to avoid hitting things by keeping away, but then the remaining wall of grass must be hand-mowed. There is a constant challenge for the mower operator to get as close as he can to the items since all too often he will do the hand mowing. We also observed that the hand mowing was not performed as often as the machine mowing. It was not unusual to see a thin wall of grass around some of the items, although rarely on the front side. We also observed broken signs and mounts caused by collusions.

The solution is obvious and easy. At the Palau airport we placed a 0.91m (3 foot) wide concrete sidewalk around each of the items. On our final visit the maintenance crew expressed sincere thanks to us for making their work so much easier. At Palau most of the items were new and the larger pads were included in the cost of the new items. At the Yap project this work is a bid item in the contract. All the fixtures had been installed in a prior contract or were FAA items such as the PAPI and REILS. The Yap bid price for the concrete apron around the existing items is $1,936 for each of the 15 items, or a total of $29,040. While this price seems somewhat high, it reflects the realistic cost of imported materials and the necessity to work between flights and not leave open pits when the airport is open to aircraft.

Fourth Case Study; Miscellaneous Conformance Issues. During many years of work we have encountered many situations where not every FAA requirement can be met. Often there is a workaround, or after negation an exemption is made to the requirement. Paragraph 6 in FAA Advisory Circular 150/5300-13 clearly implies that exceptions can be made. Reference is made to appendices that provide guidance. However some issues have arisen more frequently in recent years. Safety and security are enforced more strictly. Negotiating problems on a case-by-case basis is time consuming and unpredictable. It has led to situations where some waiver exists at one airport but is denied at another. Our recommendation is that FAA gather information on proposals and the decisions and make this information available. As a trend develops, provide nationwide guidance to all regions so that as issues are resolved the experience can be shared.

a) Fence Violations of Criteria. Security fencing has become a difficult item at several Pacific island airports. The airfields were built to the then current criteria of 150m (500 foot) wide safety areas with only a 60m (200 foot) extended safety area. Where these airports are located adjacent to water, the runway centerline was generally placed as low as possible due to costs of embankment. The transverse slopes are generally as flat as possible for the same reason. The controlling item for the transverse slopes was the minimum height for the backside of the shore protection system. Generally the shore protection consists of rubble rocks that begin immediately at the edge of the safety area. A typical condition is shown in Figure 5.

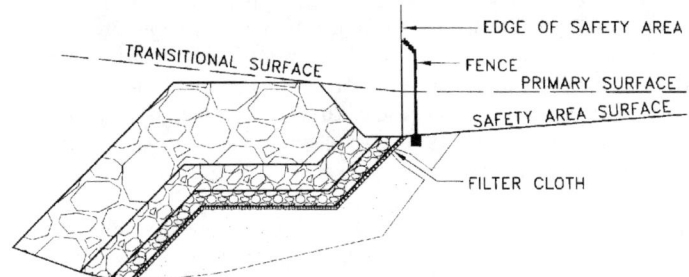

Figure 5. Fence at Edge of Runway Safety Area

A fence cannot be placed on the rubble armor due to its open rubble structure and potential for overtopping in severe storms. At best there may be one or two feet of width outside the safety area where the fence might be placed. Some of these structures use a fabric membrane to keep the fine soils of the embankment from washing out through the rubble armor. Penetrating that fabric will result in loss of fines and eventually drainage channels forming thought the safety area.

Given these conditions, the ideal place for the fence would be on the outside edge of the safety area. If the fence is located somewhere within the first three feet of the outside edge, the construction problems are avoided. Such a fence is a violation of the safety area. Or is it? FAA criteria for the safety area for these airports are 500 feet (150 meters) width. But 150 meters converts back to 492.13 feet. Thus if we could dimension the existing safety area in meters we would have about 2.4 meters (7.87 feet) available at both edges for the fence. Our recommendation is a formal concurrence from FAA headquarters that in cases where the fence cannot be placed outside the safety area it may be placed beyond the 75-meter (246.06 foot) distance from the centerline.

The next issue is the height of the fence in these areas where no alternative location is available. Regardless of which side of the safety area it is placed, the fence will penetrate an FAR Part 77 surface. However it will not penetrate the Runway Protection Zone. Thus it appears that it is not an obstacle as viewed by flight standards, but is still a FAR 77 violation. Generally this penetration will be less than three feet, and the major penetration is the barbed wire. However during discussions at one site the suggestion was made to provide red lights on the fence at undetermined intervals. This was not implemented, but at this site the project has been deferred and the final decision has not been made. Again a national guidance is desired.

b) Roads On The Runway Safety Area. Service and security roads at these airports must be on the runway safety area for reasons discussed above. They are located close to the edge of the safety area where the security staff can observe the fence and violations of the airfield security from their vehicles. There is no clear guidance for pavement design or road geometrics. Highway and road design standards are used, but there is no guidance as to whether the design vehicle should be a pickup, small truck or large truck. If road geometry is to permit travel by large rescue trucks, at what speed should the curves be designed? It would be desirable to have a set of standards to guide all airfield designers. Pavement strength should be designed to permit appropriate passes of heavy rescue trucks. (Or should it?) A relevant question is should service roads be designed for ten passes of the most critical aircraft since the roads are on the aircraft load bearing safety area.

The remaining question is use of these service roads during aircraft operations. Generally no vehicles should be on them during aircraft operations since the roads are on the runway safety area. However at some of the airports it had been considered desirable to locate one of the rescue vehicles along the runway. Our observations are that with the newer trucks and trained staffs the trucks are now parking beyond the safety area. If the airport rescue staff obtains permission to park alongside the runway should a road be provided from this location to the runway? At this point we become involved as designers, and no clear guidance is available.

c) Public Roads Adjacent to the Runway Safety Area. On some of these islands the available width from the ocean to the lagoon drop-off is only slightly more than 150m (500 feet). The public road is close to the edge of the runway safety area and the paved surface is below the FAR Part 77 transitional surfaces. Most passenger vehicles are also below the surface. However the Part 77 requirement for a 4.57m (15-foot) clearance over the pavement cannot be met. Again every airport is evaluated independently and the result is a mixture of waivers and conditions. Our request again is for a set of clear guidance. For example, private vehicles might be allowed to use the road during aircraft operations, but vehicles over a specified height would be prohibited. However since the road edge is less than 90m (300 feet) from the runway centerline is this also a security problem and should all vehicles be prohibited during aircraft operations?

d) Turnarounds at Runway Ends. When these airports were designed 30 years ago, the anticipated aircraft were of the B-727 type. These aircraft could turn easily on the 45m (150 foot) wide runways. As the airlines move into larger aircraft the width is somewhat confining. Turnarounds have been added at some airports and will be added at the others. Taxiway edge lights are provided on the turnarounds as well as edge stripes and shoulder marking where appropriate.

Observations indicate that pilots are reluctant to use the turnarounds. No clear reason has been found. Night visits confirm that at night and in marginal weather the turnarounds may appear as black holes. The relatively bright, white, runway edge lights seem to visually reinforce this conception. Discussions have revealed that there are at least four issues for which there is no design guidance.

1) One issue is the design of the turnaround centerline stripe. These stripes have been designed similar to the lead in lines at the terminal buildings to assist the pilot in positioning the turned aircraft as close to the departure threshold as practical.

However there are no ground support personnel to guide the aircraft. Pilots have a genuine concern for the precise position of their wheels and during the turns there is limited visibility from the cockpit even on the best of days. It would be desirable for FAA to use their relations with pilot groups to discuss this problem and if the present design is unsatisfactory determine a suitable striping plan. As an alternative some supplemental signage or guidance may assist the pilot in making a safe turn with confidence.

2) The second issue is the lack of a standard sign for turnarounds. Using a letter to indicate a "taxiway" exit could be misleading and result in a pilot turning too steep and running off the edge of the narrow turnaround. The word "TURNAROUND" seems too long while a simple word like "TURN" might be misinterpreted. An international symbol, word or letter/numeral combination might also be considered.

3) The third issue is visibility of the centerline at night. Site visits at night confirm that the yellow stripe is not as readily visible as one would expect. Visibility becomes even more limited in inclement and rainy weather. We suggest FAA consider providing centerline taxiway lights on these turnarounds. The location and alignment of these lights should be based on the stripe layout resulting from the first item above.

4) The fourth issue is whether the turnarounds should be on one side or both sides of the runway. To put this in perspective, at most of these airfields there is no land past the safety area. When an aircraft is on the turnaround centerline it is visually very close to the drop-off. By placing the centerline on both sides the aircraft would not have to go as far from the runway alignment and would still have adequate turning width. These two conditions are shown in Figure 6. Again this would seem to be a desirable item to discuss with pilot groups to provide recommended design guides.

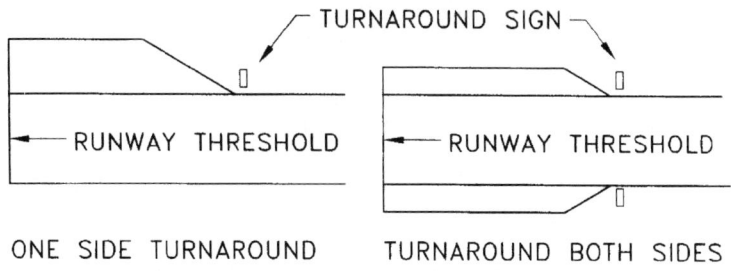

Figure 6. Turnarounds at Runway Ends

Fifth Case Study. Benefits of a Permanent Pavement Design. The Reef Runway at Honolulu International Airport was designed in the early 1970's when the Boeing 747 aircraft was just coming on line. Discussions indicated that the ultimate aircraft might weigh 680 metric tons (1.5 million pounds). The State of Hawaii directed that the runway and taxiway complex be designed to accommodate this large aircraft.

The pavement design entered uncharted areas at the time. Working with the FAA, the Corps of Engineers and the Asphalt Institute two designs were prepared. The first was for the 747 and other heavy aircraft of the time at maximum gross weights. The second was for a future aircraft with landing gear somewhat between the 747 and C-5A with a load of 680 metric tons (1.5 million pounds). The final design was for the future heavy aircraft, but left off the top 12.7cm (5 inches) of asphalt surface.

The key difference between the two designs was in the subgrade, subbase, base and stabilized base layers. These were essentially over designed for the aircraft of the time. The design required the contractor to excavate to minus 0.3m (one-foot) sea level, place coral subgrade to elevation plus 0.9m (three feet) and proof roll at this level. The coral subbase was then placed at 95 percent compaction. An aggregate base course was placed above that to protect the coral subbase. An asphalt-stabilized base was the next lift followed by the initial 12.7cm (five inch) asphalt surface pavement. The contractor, for construction purposes, placed the dredged coral fill directly on the future pavement alignments, thereby surcharging these areas. This simplified the placement of the initial fill and made it easier to achieve the proof rolling requirements. The typical pavement section is shown in Figure 7.

REEF RUNWAY PAVEMENT SECTION
HONOLULU INTERNATIONAL AIRPORT

The runway opened to air traffic in October 1977 and has now been in service for over 30 years. Generally the pavement is still in good condition and maintenance has been relatively inexpensive. The Asphalt Institute has promoted a concept of permanent pavement. This runway and the associated taxiways appear to be such a project.

A question might be whether the state spent too much money at the time. The total cost of the entire reef runway project was about 83 million dollars. This included several miles of taxiways, relocation of military recreation facilities, a new fire station, considerable environmental work and other incidental requirements to enable the facility to operate. The extra costs for the overbuilt pavement were probably less than six million dollars. Considering that the facility has served so long and will continue to serve in the future it appears to have been a good investment.

Recommendations:

1. Adequate funding for full time inspection is critical to a successful project.
2. FAA should create a knowledge base system where experiences could be uploaded and made available for others to read. For example, I wrote a comprehensive paper on fence issues at the airports and offered alternative suggestions. Others have done the same. The FAA and TSA reaction to these suggestions should be added to the knowledge base for use by others.
3. Information must be disseminated broadly by means that will lead to its being learned and used throughout the world.
4. Permanent pavements do not cost much more and are worth a lot more in savings.

Subject Index

Page number refers to first page of paper

Absorption, 111
Aggregates, 40, 111, 134, 465, 497
Aging, 195
Aircraft, 465, 573
Airport runways, 122, 279, 289, 301, 322, 334, 346, 358, 442, 465, 477, 487, 497, 547, 560, 573, 632
Asphalt pavements, 16, 74, 229, 301, 370, 423, 453, 596, 608
Asphalts, 40, 65, 86, 134, 147, 171, 182, 195, 205, 213, 222, 241, 253

Bangladesh, 434
Base course, 442
Bridges, 596

Case reports, 346, 423
Cements, 434
Comparative studies, 147
Composite materials, 535
Computer software, 86, 397
Concrete, 40, 111, 122, 497
Concrete pavements, 301, 358, 370, 453, 509
Connecticut, 608
Costs, 28, 535
Cracking, 74, 99, 222, 509, 608
Creep, 159
Curing, 213

Damage, 28, 134
Databases, 385, 397
Deformation, 16, 86, 465
Design/build, 521
Deterioration, 53
Discrete elements, 65

Elasticity, 159

Embankments, 560
Experimentation, 222, 229

Fast track construction, 289
Fatigue, 182, 509
Fatigue life, 410
Finite element method, 1, 16, 74
Flexible pavements, 1, 53, 171, 487, 585, 620
Florida, 147, 289
Fly ash, 423

Georgia, 573
Geosynthetics, 16
Granular materials, 410
Gravity, 111

Hawaii, 632
Highways and roads, 521
History, 322

In situ tests, 1

Laboratory tests, 182, 213
Life cycles, 334
Loads, 74, 99, 267, 465

Material properties, 182, 301
Measurement, 111, 147
Mechanical properties, 253
Mexico, 560
Mixtures, 65, 85, 147, 171, 182, 195, 205, 213, 229, 253, 267, 370
Moisture, 134, 159

Neural networks, 40
New Jersey, 487, 596
New York, 596

Optimization, 122, 241

Parameters, 53
Pavement design, 171, 521, 547
Pavements, 28, 99, 267, 385, 397, 410, 434, 465, 477, 535, 585
Poisson ratio, 159
Predictions, 346
Probability, 53

Ratings, 585
Recycling, 213, 453, 497
Regression models, 370
Rehabilitation, 279, 289
Research, 241
Roughness, 385

Service life, 358
Simulation, 65, 195

Soft soils, 122
Stabilization, 410, 423, 434
Stiffness, 53, 465
Structural design, 569
Subgrades, 477, 487
Sustainable development, 442, 497

Temperature, 53, 205, 222, 267
Tensile strength, 213
Texas, 620

Value engineering, 477
Vehicles, 74
Virginia, 279, 322, 334, 346, 358

Weight, 585

X rays, 40

Author Index

Page number refers to first page of paper

Ahammed, M. Alauddin, 370
Akisetty, Chandra K., 205
Albitres, Carlos M. Chang, 620
Al-Qadi, Imad L., 1, 74
Amirkhanian, Serji N., 195, 205
Arambula, Edith, 40

Baek, Jongeun, 74
Bardt, D. R., 289
Bhattacharjee, Sudip, 267
Bjornstad, Jami M., 596
Bognacki, Casimir J., 442, 596
Boudreau, Richard L., 477, 573
Briggs, Robert C., 477

Carvalho, Regis L., 86
Chen, Yi-Ping, 397
Chiu, Chien-Ming, 397
Chou, Chien-Cheng, 397
Collop, A. C., 159

Daniel, Jo Sias, 267
Darter, Michael J., 322
DeBord, Kenneth J., 585
Dedmon, C., 301
Diefenderfer, Brian K., 535
Diefenderfer, Stacey, 182
Ding, Qingjun, 229, 241
Dokka, Viswanath, 477
Donovan, Phillip, 465

Eltahan, Ahmed, 620

Flintsch, Gerardo W., 535
Freer-Hewish, Richard, 434
Fuselier, Gary K., 279, 322, 334, 346

Gandhi, Tejash, 195

Garg, Navneet, 487
Ghataora, Gurmel S., 434
Gibson, Nelson, 40
Gnanendran, C. T., 410
Goh, Shu Wei, 171
Grubbs, Joseph S., 322, 334

Hale, W. M., 122
Halsey, T. L., 122
Hammons, Michael I., 453, 497
Hayhoe, Gordon F., 487
Hearon, Amy, 182
Herman, Frank V., 632
Herrin, Stanley M., 346, 322
Heymsfield, E., 122
Hiller, Jacob E., 509
Holland, T. Joseph, 53
Hu, Shuguang, 229, 241

Im, Soohyok, 213

Jegatheesan, Piratheepan, 410

Khattak, Mohammad J., 253, 134
Kim, Minkwan, 99
Kim, Yongjoo, 213
Krugler, Paul E., 620
Kuchikulla, Subash, 477
Kutay, M. Emin, 40
Kyatham, Vikram, 134, 253

Leahy, R. B., 301
Lee, Hosin "David", 213
Lee, Soon-Jae, 205
Li, Peng, 222
Liu, Juanyu, 222
Liu, Xinquan, 229, 241
Liu, Yu, 65

Mahoney, J., 608
Mallick, Rajib B., 267
Marsano, Joseph, 596
McNerney, Michael T., 358
McQueen, Roy D., 322, 334
Miao, Yinghao, 16
Mills-Beale, Julian, 111
Moguel, Hector Saldivar, 560
Monismith, C. L., 301

Nowak, George, 560
Núñez, Orlando, 535

Ong, G. P., 385

Pedersen, Kurt, 521
Peters, K. D., 28
Peterson, Dan, 521
Petros, Katherine, 40
Ping, W. Virgil, 147
Pirozzi, Marco, 442
Popescu, L., 301

Rahman, Waliur, 434
Ramme, Bruce, 423
Rau, Robert, 573
Roesler, Jeffery R., 509

Saeed, Athar, 452, 497
Salazar, David Martinez, 560
Schwartz, Charles W., 86

Shaffer, G., 608
Shen, Fan, 241
Sinha, K. C., 385

Tadimalla, Raghuram N., 477
Taherkhani, H., 159
Thuma, Richard G., 279
Tighe, Susan L., 370
Timm, D. H., 28
Turochy, R. E., 28
Tutumluer, Erol, 99, 465

Vélez-Vega, E. M., 289
Vitillo, Nicholas, 53

Wang, Hao, 1
Watkins, Quintin, 573
Wen, Haifang, 423
White, Greg, 547
Wu, Xuewei, 229

Xiao, Yuan, 147

Yip, Peter K., 279
You, Zhanping, 65, 111, 171, 253,
Youtcheff, Jack, 40

Zaghloul, Sameh, 53
Zhang, Jinxi, 17
Zinke, S., 608
Zofka, A., 608